T0349042

ADVANCED POWER GENERATION SYSTEMS

ADVANCED POWER GENERATION SYSTEMS

By

IBRAHIM DINCER AND CALIN ZAMFIRESCU
University of Ontario Institute of Technology
2000 Simcoe St N
Oshawa, Ontario L1H 7K4
Canada

ELSEVIER AMSTERDAM • BOSTON • HEIDELBERG • LONDON • NEW YORK • OXFORD
PARIS • SAN DIEGO • SAN FRANCISCO • SINGAPORE • SYDNEY • TOKYO

Elsevier
Radarweg 29, PO Box 211, 1000 AE Amsterdam, Netherlands
The Boulevard, Langford Lane, Kidlington, Oxford OX5 1GB, UK
225 Wyman Street, Waltham, MA 02451, USA

First edition **2014**

Library of Congress Cataloging-in-Publication Data
A catalog record for this book is available from the Library of Congress

British Library Cataloguing in Publication Data
A catalogue record for this book is available from the British Library

For information on all **Elsevier** publications
visit our web site at store.elsevier.com

ISBN: 978-0-12-383860-5

Working together
to grow libraries in
developing countries

www.elsevier.com • www.bookaid.org

Contents

Acknowledgments

Bringing a book to fruition is not an easy job since it requires lots of effort and time tirelessly put upon. This particular book on *Advanced Power Generation* has caused many sleepless nights, without tears, but with smiles. We have extensively benefited from our previously published works, especially with the graduate students of Dr. Ibrahim Dincer, namely, Pouria Ahmadi, Eda Cetinkaya, Rami El-Emam, Ahmet Ozbilen, Hasan Ozcan and Tahir Ratlamwala for various illustrations and case studies. We sincerely acknowledge their assistance. Dr. Dincer also acknowledges the support provided by the Turkish Academy of Sciences.

Last but not least, we warmly thank our wives, Gulsen Dincer and Iuliana Zamfirescu, and our children Meliha, Miray, Ibrahim Eren, Zeynep and Ibrahim Emir Dincer, and Ioana Zamfirescu. They have been a great source of support and motivation, and their patience and understanding throughout this book have been most appreciated.

Ibrahim Dincer and Calin Zamfirescu
Oshawa, June 2014

Preface

Energy has been a critical issue for humankind over the centuries, and it has shaped the past, is shaping the present, and will definitely shape the future. Our existence depends essentially on it and the way we generate, convert, transform, transport, and utilize it. The picture has become even more complicated with environmental consequences and sustainability issues. When we look at the dimensions of energy, power generation stands out as the most crucial commodity and is the main drive behind the economies.

Due to increasing energy, environmental, and sustainability issues, we need to go beyond conventional practices and hence conventional power production technologies, systems, and applications. It has become a kind of ultimate target to make power generation systems more efficient, cost effective, and more environmentally friendly and to develop potential options to fulfill these requirements. This book offers a unique coverage of the conventional and novel power generating systems in a harmonized manner with renewables and includes our recent developments on system integration and multigeneration. It is expected to serve the power generation community by providing comprehensive tools for system design, analysis, assessment, and improvement.

Being a relatively new, the emerging technology of power generation using advanced thermodynamic cycles is found as subject matter scattered in specialized research journals and a few books, written at an advanced level and devoted exclusively to particular multiple output thermodynamic cycles. There is a scarcity of publications that introduce all these cycles for power generation in a single volume to a beginner. Therefore, the proposed book will fill the void of a much needed and relatively inexpensive textbook at the post-graduate level in the area of power generation. While teaching a particular course on power generation cycles to post-graduate engineering students, the authors themselves felt this need, which inspired them to write the proposed book.

This book is essentially intended to serve a research-oriented textbook with a comprehensive coverage of fundamentals and main concepts, which can be used for system design, analysis, assessment, and improvement, and it therefore includes practical features in a usable format often not included in other solely academic textbooks. There is also room for this book to be used by senior undergraduate and graduate students in mainstream engineering fields (such as mechanical and electrical engineering) and new engineering programs on energy and/or power. Due to its extensive coverage it can also serve as a primary reference book for practicing engineers and researchers.

The book consists of ten chapters which nicely amalgamate all power-related aspects, ranging from basic fundamentals to novel power generating systems, by considering efficiency, environment, and sustainability issues. The introductory chapter addresses general concepts, fundamental principles, and basic aspects of thermodynamics, energy, entropy, and exergy. These topics are covered in a broad manner, so as to furnish the reader with the background information necessary for subsequent chapters. Chapter 2 provides detailed information on energy, environment, and sustainable development, while Chapter 3 focuses on traditional fossil fuels and potential alternative fuels or fuel sources. Chapter 4 gives a distinct coverage on hydrogen and fuel cells and their systems for power generation. Chapter 5 dwells on conventional power generating systems and their design, analysis, assessment, and improvement. Chapter 6 is about nuclear power for power production with a comparative analysis and assessment, while Chapter 7 takes a comprehensive focus on renewable energy-based systems and applications for power production. Chapter 8 offers advanced power generating systems with illustrative examples and case studies. Chapter 9 delves into integrated multigeneration systems for power production, and Chapter 10 makes a good close with descriptions of novel power generating systems from our own works and some literature studies. Incorporated throughout are many illustrative examples and case studies, which provide the reader with a substantial learning experience, especially in areas of practical application. Complete references are included to point the truly curious reader in the right direction. Information on topics not covered fully in the text can, therefore, be easily found.

We hope this book brings a new dimension to power generating practices and helps the community implement better solutions for a better future.

Ibrahim Dincer and Calin Zamfirescu
Oshawa, June 2014

1

Fundamentals of Thermodynamics

1.1 INTRODUCTION

In the modern world electric power is a widespread and intensely used commodity. Electric power is in demand almost everywhere: it powers appliances, provides lighting, moves vehicles, runs machines, powers computers, and the internet, and it is widely used in industrial processes to produce all sorts of goods and commodities.

Electricity appears to be the most versatile form of energy transport because it can be converted into many other forms without major loss and used in many ways at a competitive cost. For example, electric power can be converted completely into heat using electric heaters. Also, electric power can be converted into mechanical (motive) power with a very high efficiency using electric motors. Moreover, electric power can be converted into light using devices such as light emitting diodes and incandescent lamps. Electric power can also be used to convert water into hydrogen and oxygen by electrolysis producing a fuel, a substance which stores energy in chemical form. Other chemicals can be produced electrochemically using energy generated by electric power.

The high efficiency of electric power conversion is explained by the flow of electric charges. In physics, "power" is defined as the rate of energy transfer. Electric power represents the rate of transfer of electrical energy due to movement of electric charges (generally electrons). This flow of electrons cannot be observed by the naked eye. Flow of electricity through a wire can be compared in general terms to the flow of water through a duct. Similarly, "flow of heat" is a thermal energy transfer due to a gradient of temperature.

Thermodynamics differentiates two kinds of energy: (i) organized (such as mechanical, electrical or electromagnetic, photonic, gravitational etc.), and (ii) disorganized (thermal energy or "heat"). According to the second law of thermodynamics (SLT), which will be detailed further in this chapter, organized forms of energy can be completely converted into any other form of energy. However, thermal energy cannot be fully converted into organized forms of energy due to intrinsic irreversibilities.

"Thermodynamics" comes from "therme" which means heat in Greek and "dynamis," the Greek word for power. Therefore, a common example is given as the conversion of thermal energy into mechanical energy. Thermodynamics is commonly defined as the science of energy and entropy. However, we find this partly incomplete and inconsistent and modify it as "thermodynamics is the science of energy and exergy". By this definition we can compare both energy and exergy since they possess the same units and can correspondingly be used for performance assessment and comparison of thermodynamic systems.

Any power generation system must be designed in accordance to the principles of thermodynamics. Most traditional power generation systems use a combustion process to generate high-temperature heat from a conventional fuel (coal, petroleum, natural gas, or uranium). Then, using a power cycle, thermal energy is converted into the motive power of a rotating shaft. The generated mechanical energy is then converted into electrical energy by an electric generator, producing electric power.

Novel power generation devices such as fuel cells are able to directly convert the chemical energy of a fuel into electricity. Typically, a fuel cell consumes a synthetic fuel such as hydrogen or methanol, or it may consume a fossil fuel such as natural gas. Electrochemical reduction and oxidation reactions occur in a fuel cell generating electric power. In addition, multi-output-generation systems are currently in development throughout the world. These systems can generate multiple outputs such as electric power, heating, cooling, hydrogen, drinking water, and synthetic fuels.

In this chapter, concepts and methods of thermodynamics are introduced for use in the design, analysis and assessment of power generation systems. The first and second laws of thermodynamics are introduced and discussed, and the use of mass, energy, entropy and exergy balance equations are written and explained in detail. The efficiency of thermodynamic systems, devices, and processes is defined in relation to energy and exergy, and various relevant examples are given.

1.2 THERMODYNAMIC PROPERTIES AND BASIC CONCEPTS

In this section the most important thermodynamic concepts and thermodynamic properties are introduced. The fundamental concept of thermodynamics is the *thermodynamic system* proposed by Carnot (1824). By definition, a thermodynamic system is separated by a real or imaginary boundary from the rest of the universe, denoted as the *surroundings*. For the purpose of a thermodynamic analysis, the universe is divided into two parts, the thermodynamic system and the surroundings. A thermodynamic system that can exchange energy but does not exchange matter (mass) with its surroundings is said to be a *closed thermodynamic system* or *control mass*. If a closed thermodynamic system does not exchange energy in any form with its surroundings then the system is an *isolated system*. If a thermodynamic system can interact with its surroundings by mass transfer and energy transfer then the system is denoted as an *open system* or *control volume*. Two main forms of energy transfer exist; these are work and heat and will be discussed in detail later. Figure 1.1 illustrates the two types of systems: (a) a closed system and (b) an open system. Figure 1.2 depicts an insulated thermodynamic system.

The *universe* represents the cosmic system, consisting of matter and energy. In addition, *matter* is defined as particles with rest mass, including *quarks* and *leptons*, which are able to combine and form *electrons*, *protons* and *neutrons*. *Chemical elements* consist of matters formed by combinations of protons, neutrons, and electrons, whereas electrons are a class of leptons.

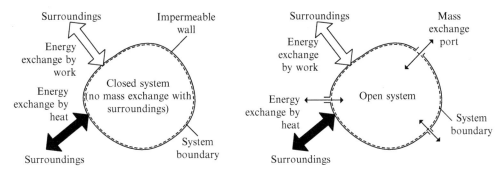

FIGURE 1.1 Thermodynamic systems: (a) a closed system, (b) an open system.

FIGURE 1.2 Isolated system—a particular type of closed system.

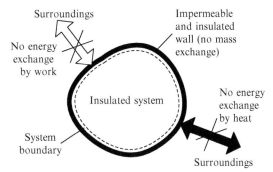

The interaction between a thermodynamic system and its surroundings is affected by an intermediary of *forces* or *force fields*. A *field* represents a region of the universe affected by forces of a given kind. Consequently, it is possible for a field to have no matter. However, the contrary is not true. A field may exist anywhere in the universe even where matter (such as particles with rest mass) is nonexistent.

A particular thermodynamic system is the ideal *vacuum* characterized by the lack of any form of matter but the presence of fields. Some examples of fields are gravitational, magnetic, and electric. Bulk matter or *substance* is formed by groups of atoms, molecules, and clusters of atoms and molecules. There are four *forms of aggregation* of substances, *solid*, *liquid*, *gas*, and *plasma*.

In physics there are seven fundamental quantities, and their units are defined according to the "International System of Units" (SI). All other quantities and units—denoted as "derived"—can be determined from these fundamental quantities based on physical laws. Not all fundamental quantities of the SI system are directly relevant to thermodynamics. For example, *luminous intensity* may not be very important in thermodynamics when compared to temperature. However, derived quantities such as specific volume and pressure are very important. Table 1.1 presents the physical quantities used in thermodynamics which

TABLE 1.1 Main Measurable Quantities Used in Thermodynamics

Quantity	Definition	Measurement Unit
Length (l)	Geometric distance between two points in space	m—meter; the length of the path traveled by light in a vacuum during a time interval of 1/299,792,458 of a second
Mass (m)	Quantitative measurement of inertia that is the resistance to acceleration	kg—kilogram; the weight of the International Prototype of the kilogram made in platinum-iridium alloy
Specific volume (v)	Represents the volume of unit mass	m^3/kg—cubic meter per kilogram; the volume of a mass of 1 kg of a substance or assembly of substances
Time (t)	A measurable period that measures the progress of observable or non-observable events	s – second; duration of 9,192,631,770 periods of the radiation corresponding to the transition between the two hyperfine levels of the ground state of the cesium 133 atom
Velocity ($\boldsymbol{v}$)	Rate of position vector change with respect to a reference system. $\boldsymbol{v} = \mathrm{d}l/\mathrm{d}t$	m/s—meter per second; a change of position of 1 m occurring during the time interval of 1 s
Acceleration (a)	Rate of velocity vector change. $a = \mathrm{d}\boldsymbol{v}/\mathrm{d}t$	m/s^2—meter per squared second; a change of velocity of 1 m/s occurring during a time interval of 1 s
Temperature (T)	A measure of kinetic energy stored in a substance which has a minimum of zero (no kinetic "internal" energy)	K—Kelvin; the fraction 1/273.16 of the thermodynamic temperature of the triple point of water (having the isotopic composition defined exactly by the following amount of substance ratios: 0.00015576 mol of ^{2}H per mole of ^{1}H, 0.0003799 mol of ^{17}O per mole of ^{16}O, and 0.0020052 mol of ^{18}O per mole of ^{16}O)
Force (F)	A measure of action capable of accelerating a mass. $F = m\,a$	N—Newton; the force which accelerates a body of 1 kg with 1 m/s^2
Pressure (P)	Represents the force exerted by unit of area. $P = F/A$	Pa—Pascal; represents the force of 1 N (Newton) exerted on 1 m^2 surface
Electric current (I)	Rate of flow of electric charge	A—Ampere; constant current which, if maintained in two straight parallel conductors of infinite length, of negligible circular cross section, and placed 1 m apart in a vacuum, would produce between these conductors a force equal to 2×10^{-7} N/m of length

Sources: ISU (2006).

are directly measurable. Measurable quantities are important because they can offer quantitative information regarding the thermodynamic state of a system. Other quantities can be determined based on measurable quantities and relevant equations. Table 1.2 presents a number of physical constants relevant in thermodynamics. The table shows the constant name, its symbol, its value and measurable unit, and a brief definition. Some constants are fundamental constants of physics (e.g., Planck's constant) or standard parameters (e.g., standard atmospheric pressure).

TABLE 1.2 Some Important Constants in Thermodynamics

Constant	Value	Definition
Number of Avogadro	$N_A = 6.023 \times 10^{26}$ molecules/kmol	Ratio of constituent entities of a bulk substance to the amount of substance. $N_A = N/n$
Gravitational acceleration	$g = 9.80665$ m/s^2	Standard gravitational acceleration represents the gravitational force (G) per unit of mass. $g = G/m$
Standard atmospheric pressure	$P_0 = 101.325$ kPa	Pressure of the terrestrial atmosphere at sea level in standard conditions
Standard ideal gas volume	$V_0 = 24.466$ m^3/kmol	Volume of 1 mole of ideal gas at $P_0 = 101.325$ kPa, $T_0 = 298.15$ K
Planck's constant	$h = 6.626 \times 10^{-37}$ kJ s	Indicates the magnitude of energy of a quanta (particle) which expresses the proportionality between frequency of a photon and its energy according to $E = h \cdot v$
Speed of light in a vacuum	$c = 299,792,458$ m/s	Maximum speed at which matter and information can be transported in the known cosmos
Stefan–Boltzmann constant	$\sigma = 5.670373 \times 10^{-8}$ W/m^{2}K^4	Constant in Stefan-Boltzmann law expressing the proportionality between fourth power of temperature and black body's emissive power
Universal gas constant	$\mathcal{R} = 8.314$ J/mol K $\mathcal{R} = Pv/T$	Represents a measure of kinetic energy of 1 mol of an ideal gas at molecular level
Boltzmann constant	$k_B = 1.3806 \times 10^{-23}$ J/K. $k_B = \mathcal{R}/N_A$	Represents a measure of kinetic energy of one molecule of ideal gas
Elementary charge	$e = 1.60218 \times 10^{-19}$ C	Electrical charge carried by a single proton
Faraday's constant	$F = 96,485$ C/mol	Electric charge of 1 mol of electrons

Some important physical properties derived from the measurable quantities listed in Table 1.1 are introduced next. The *amount of* matter (or the *number of* moles) represents the number of unambiguously specified entities of a substance such as electrons, atoms, or molecules. The unit of the amount of matter is a *mol*. According to ISU (2006), this represents the amount of substance in a system which contains as many elementary entities as there are atoms in 0.012 kg of ^{12}C. Another important property is the *molecular mass* of a substance, which represents the ratio between the mass and the amount of matter, namely

$$M = m/n \tag{1.1}$$

where n represents the number of moles (or the amount of matter).

The action of forces produces changes in the magnitude and/or direction of velocity of a particular piece of matter. The magnitude of force is proportional to the mass of the matter on which it acts and to *acceleration*, which is defined as the variation of *velocity* in time. Velocity is the rate of change of position of a piece of matter with respect to a reference system (a fixed landmark in space). If one denotes l the position vector of a piece of matter (material point) then velocity is expressed by the derivative

$$\vec{v} = \mathrm{d}\vec{l}/\mathrm{d}t \tag{1.2}$$

and the acceleration is defined by

$$\vec{a} = \mathrm{d}\vec{v}/\mathrm{d}t \tag{1.3}$$

therefore, the vector force is expressed by the following equation

$$\vec{F} = m\,\vec{a} \tag{1.4}$$

The total force exerted on a surface is denoted as *pressure*. By definition

$$P = F/A$$

where F is the magnitude of force in the direction perpendicular to the surface. The unit for pressure is Pa (Pascals), where $1\,\mathrm{Pa} = 1\,\mathrm{N/m}^2$.

Often in thermodynamics one deals with the concept of *flow*, which represents continuous movement of substance along a stream. In addition to carrying matter, a flow can transport energy and do useful work. In Table 1.3 a number of term related to the concepts of flow and work are defined. First, the average velocity in a duct is defined. Here a duct is defined as any tube transporting a fluid or any kind of port at the inlet or an outlet of a control volume. The average velocity is defined by an integral duct cross section. The average velocity (v) is the volumetric flow rate ($\dot{V}$) which represents the rate of volume transport at a cross section. Mass flow rate ($\dot{m}$) is defined as the rate of mass transport at a cross section. Specific volume (v, in m^3/kg) is the parameter that connects the volumetric and mass flow rates.

One of the ways in which a flow can transfer energy is by work. Work is a measure of energy change due to matter displacement by a force. In solid mechanics, work is defined as the scalar product of the vectors of force ($\vec{F}$) and displacement ($\vec{l}$)

$$W = \vec{F} \cdot \vec{l}$$

Consequently, work is proportional to the magnitude of force and the magnitude of displacement, but also depends on the relative direction of the force and the displacement. For example, if the displacement direction is perpendicular to the direction of force it can be inferred that this displacement is not produced by the respective force but has other causes; in this case, the work is nil. On the contrary, if force and displacement have the same direction, then the work of the force is the maximum possible and equal to the force magnitude and the displacement. Table 1.3 (item #5) illustrates the work of gravitational force which is developed when a system is displaced in the gravitational field. When the displacement is in a vertical direction then the force and displacement vectors act in the same direction, therefore the work is

$$W = mgz \tag{1.5}$$

where z represents the displacement with respect to a reference system.

TABLE 1.3 Definitions Table of Some Important Notions in Thermodynamics

Notion and Definition Relationship	Explanatory Sketch

1. Average velocity cross sectional velocity, ν

$$\nu \equiv \frac{1}{A}\int_A \nu\, dA$$

2. Volumetric flow rate, $\dot{V}$

$$\dot{V} \equiv \nu A$$

3. Mass flow rate, $\dot{m}$

$$\dot{m} \equiv \frac{\dot{V}}{v}$$

where v is specific volume

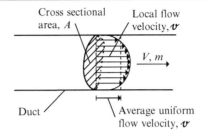

4. Mechanical work

$$W \equiv \vec{F}\cdot\vec{l} = Fl\cos(\alpha)$$

5. Gravity work

$$W = Gz = m\,g\,z$$

where $g = 9.81$ m/s^2 is the gravitational acceleration

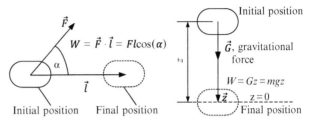

6. Boundary work

$$dW_b = F\,dl = PA\,dl \equiv P\,dV$$

$$W_b \equiv \int_1^2 P\,dV$$

where

$dl =$ infinitesimal boundary displacement

$A =$ moving boundary area

Note: work is the area below the path of process; if the process is cyclic then the net work is the area enclosed by the cycle path

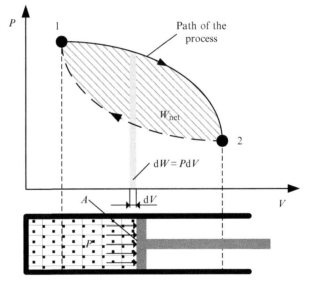

7. Flow work

$$dW_f = d(F\,l) = d(P\,A\,l) \equiv d(P\,V)$$

$$W_f \equiv \int d(P\,V) = PV$$

8. Specific flow work

$$dw_f = d(P\,v)$$

where $v =$ specific volume

9. Specific flow work rate

$$dw_f = \dot{m}\,d(Pv)$$

where $\dot{m} =$ mass flow rate.

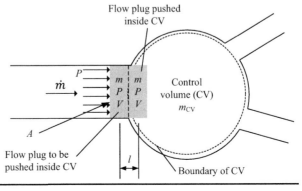

If the boundary of a thermodynamic system or control volume moves, then it produces or consumes work because it displaces matter. The work done by boundary movement is referred to as *boundary work*. As illustrated in the Table 1.3 (item #6), boundary work is expressed by

$$W_b = \int_1^2 P \, dV \qquad (1.6)$$

It is interesting to remark that the boundary work for the process 1–2 equals the area of the process path represented by the *P–v* diagram (see Table 1.3, #6). A *process path* is a graphical representation of all thermodynamic states through which a system passes during a transformation. In a cyclical process for which the path 2–1 is different from the path 1–2 as illustrated in the *P–v* diagram, the system produces or consumes work continuously. The net work exchanged with the surroundings (W_{net}) is represented by the area enclosed by the cycle. Note that this area is positive if work is delivered to surroundings and negative if work is supplied to the system from the surroundings. This observation is the basis of sign convention in thermodynamics which will be explained later.

Furthermore, in order for a system to exchange work with its surroundings it must be either a closed system with a movable boundary or an open system (control volume). Isolated systems cannot exchange work or any kind of energy with their surroundings.

Another form of work defined in Table 1.3 is the *flow work* which is characteristic of control volumes only. Flow work is a measure of the action of pushing (or pulling) a volume of fluid inside a control volume. This represents a flow-displacement work. The flow work is equal to the product of pressure and volume. Specifically, it becomes

$$w_f = Pv \qquad (1.7)$$

or, in differential form, $dw_f = d(Pv)$.

Many devices such as turbines, compressors, and pumps are open systems that perform or consume work. Both flow work and boundary work are exchanged when operating these devices. The work generated or consumed by an open system (control volume) is denoted as *useful work* and represents the difference between boundary work and flow work: $W_u = W_b - W_f$. In differential form, it is written as

$$dw_u = P \, dv - d(Pv) = -v \, dP \qquad (1.8)$$

therefore, the useful work becomes

$$W_u = -\int v \, dP$$

The ability to do work quantifies a change in *energy* which is a scalar quantity evidenced by indirect measurement of energy changes caused by various interactions of a thermodynamic system with its surroundings. There are several forms of energy that fall into two categories: microscopic and macroscopic. There are two basic energy exchange (transfer) mechanisms, namely work and heat. As described above, work refers to that energy transfer which takes place by displacement of matter, either microscale or macroscale. Heat transfer is a different mechanism of energy exchange related to kinetic energy of the molecules or atoms. Heat amount, Q, is measured with the same unit as work and energy, the Joule. The energy exchange by heat will be detailed later.

Energy can manifest in two macroscopic forms, kinetic and potential. Kinetic energy is the amount of energy due to movement. Kinetic energy of a body of matter is defined with respect to the reference system for velocity. Variation of kinetic energy due to a change in velocity between ν_1 and ν_2 is

$$\Delta KE = \frac{1}{2} m \left(\nu_2^2 - \nu_1^2 \right)$$

When a body moves in a force field, its total macroscopic energy is a sum of kinetic and potential energy. When the change in associated potential energy of displaced matter does not depend on the path taken but only on the initial and final positions of the displacement, the force field is denoted as *conservative*. Any conservative force field has an associated potential energy. There are many types of potential energies, each with an associated type of force field:

- gravitational,
- elastic, produced by elastic forces,
- magnetic, produced by magnetic forces,
- electrostatic, produced by Coulomb forces,
- chemical, produced by Coulomb forces while chemical reactions occur (during chemical reactions atoms and electrons are rearranged, releasing or absorbing energy; this energy is electrical and manifests among nuclei, electrons, and molecules),
- thermal, due to the kinetic energy of molecules and their relative position,
- nuclear, caused by various nuclear forces (weak, strong).

The potential energy variation due to displacement of a thermodynamic system in a gravitational field between elevations z_1 and z_2 is equal to the mechanical work of gravitational force. From Table 1.3, the expression of potential energy variation is written as

$$\Delta PE = mg(z_2 - z_1)$$

Internal energy represents a summation of many microscopic forms of energy including vibrational, chemical, electrical, magnetic, surface, and thermal. The internal energy is proportional to the average force that molecules exert on the system boundary. Internal energy can be introduced based on the concept of *ideal gas*. According to the definition, ideal gas represents a special state of matter which can be delimited by a system boundary. Ideal gas is formed by a number (N) of freely moving molecules or atoms with perfect elastic behavior at collisions with each other and with the system boundary. It is assumed that:

- all particles have rest mass ($m > 0$; the particles are not photons),
- the number of particles with respect to the system volume is small,
- the total volume of particles is negligible with respect to system volume,
- collisions of particles with each other is much less probable than the collisions with the system boundary.

Internal energy is an extensive quantity which plays the role of a state function. According to the thermodynamic definition a *state function* is the property of a system that depends only on current state parameters. When a change occurs, the state function is not influenced by the process in which the transformation is performed. Variation of internal energy can be expressed based on the *specific internal energy* which is defined by the relationship

$$u = \frac{U}{m}$$

where m is the system mass; as inferred from the definition, the specific internal energy is an intensive property. Using the specific internal energy, the following expression for a system's internal energy is obtained for a closed system

$$\Delta U = m(u_2 - u_1)$$

The shape of the thermodynamic system can be arbitrary, but for simplicity we assume here a cubic volume as indicated in Figure 1.3. The average particle velocity is denoted with ν and the cube edge is l; it follows that the time between two collisions, approximated by the time needed for the particle to travel between two opposite walls, is

$$\Delta t = \frac{2l}{\nu}$$

Based on the momentum conservation law, the force exerted by one particle on the wall, during collision, is

$$F = \frac{m\nu - m(-\nu)}{\Delta t} = \frac{m\nu^2}{l} \tag{1.9}$$

One also assumes that there is a uniform distribution of particle collisions for the three Cartesian directions; thus only one-third of particles exert force on a wall. Therefore, the pressure expression is determined by dividing the force (given by Equation (1.9)) with wall area A

FIGURE 1.3   A thermodynamic system filled with ideal gas.

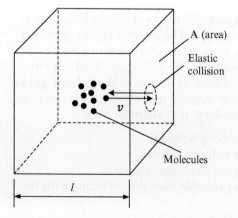

$$P = \frac{1}{3} N m \frac{v^2}{lA} = \frac{Nm}{3V} = \frac{1}{3} \rho v^2 \tag{1.10}$$

where $V = lA$ is the volume of the thermodynamic system and A is the surface area of the face, and

$$\rho = \frac{Nm}{V}$$

is the gas density. The kinetic energy of a single gas particle can be expressed based on the average particle velocity; in this respect, Equation (1.10) is solved for v^2 and it results in

$$KE = \frac{1}{2} m v^2 = \frac{3PV}{2N} \tag{1.11}$$

The *degree of freedom* of monoatomic gas molecules is DOF$=3$, because there are only three possible translation movements along Cartesian axes. According to its thermodynamic definition, *temperature* (T) is a measure of the average kinetic energy of molecules per degree of freedom. The quantitative relation between temperature and kinetic energy of one single molecule is according to

$$\frac{KE}{DOF} \equiv \frac{1}{2} k_B T \tag{1.12}$$

where k_B is the Boltzmann constant defined in Table 1.2. Solving Equation (1.12) for T results in the thermodynamic expression for temperature, as follows:

$$T \equiv \frac{2 KE}{k_B DOF} \tag{1.13}$$

Kinetic energy expression from Equation (1.11) can be introduced in Equation (1.13) and is obtained by

$$T = \frac{PV}{k_B N} = \frac{PV}{k_B (N/N_A) N_A} = \frac{PV}{k_B n N_A} = \frac{Pv}{k_B N_A}$$

where v is the *molar specific volume*

$$v \equiv \frac{V}{n}$$

and n is the amount of substance (number of moles). Therefore, the following thermodynamic definition of temperature is obtained, which is equivalent with that given in Equation (1.13)

$$T \equiv \frac{Pv}{k_B N_A} = \frac{Pv}{\mathcal{R}} \tag{1.14}$$

where $\mathcal{R}$ (the *universal gas constant*) is given as follows:

$$\mathcal{R} = k_B N_A$$

The value of the universal gas constant is given in Table 1.2. Observe that Equation (1.14) relates pressure, temperature, and specific volume of ideal gas. This relationship can be equivalently written as $Pv = \mathcal{R}T$ which represents the *ideal gas law*, where v is the molar specific volume. It is possible to express the ideal gas equation on a mass basis if one introduces the *gas constant*

$$R = \frac{\mathcal{R}}{M} \tag{1.15}$$

The ideal gas equation can be expressed using the gas constant R (in kJ/kmol K) and mass specific volume v (in m^3/kg) as follows:

$$Pv = RT \tag{1.16}$$

Here pressure, temperature, and specific volume represent a set of thermodynamic properties that completely define the *thermodynamic state* and the state of aggregation of a substance (e.g., solid, liquid, gas). The thermodynamic state of a system can be modified via various interactions, among which heat transfer is one. Heat can be added or removed from a system. When heat is added the internal energy of the system can increase, while the opposite happens when heat is removed. If heat is added the state of aggregation of the system passes from solid to liquid to gas.

In Figure 1.4 the state diagram of water is exemplified. On the abscissa the heat addition to a thermodynamic system enclosing pure water is represented. On the ordinate the average temperature of the system is shown. For this diagram we assume that the internal energy of the system is zero at 250 K when the state of aggregation is that of subcooled ice. Ice reaches its melting point at 273.15 K. During the melting process, an ice-water mixture is formed. Due to phase change, the temperature remains constant (see figure), although heat is continuously added. At the end of the melting process all water is in a liquid state of aggregation. Water is further heated, and its temperature increases until it reaches the boiling point at 373.15 K. Additional heating produces boiling which evolves at constant temperature, while a water and steam mixture is formed. The boiling process is completed when all liquid water is transformed in steam. Further heating leads to temperature increase and generation of superheated steam.

The pressure versus specific volume diagram of water is presented in Figure 1.5. This plot indicates the saturation lines where liquid and vapor reach the saturation temperature at a

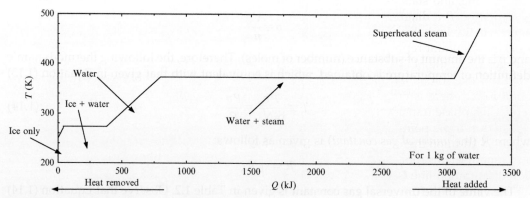

FIGURE 1.4 State diagram of water for 101.325 kPa.

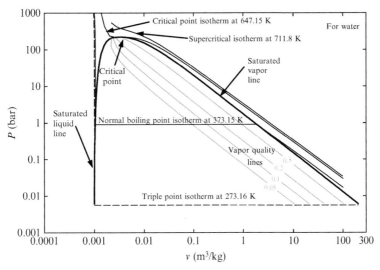

FIGURE 1.5   Pressure versus specific volume diagram for water.

given pressure. Observe that the specific volume of saturated vapors is 1,000 times higher for normal boiling point isotherm. The normal boiling point isotherm corresponds to a temperature of 373.15 K. The specific volume along the boiling pallier can be expressed based on *vapor quality*, which is defined based v, v', v'', the specific volumes of mixture, saturated liquid, and saturated vapor, respectively, according to

$$x = \frac{(v - v')}{(v'' - v')}$$

where specific volumes are evaluated at the same pressure (or temperature). It can be observed that vapor quality varies between 0 and 1, with value 0 for saturated liquid and 1 for saturated vapor.

Another useful diagram that shows the relationship between pressure, temperature, specific volume, and state of aggregation plots temperature versus specific volume. Figure 1.6 presents the $T–v$ diagram for water, including the saturation lines, isobars, and constant vapor quality lines. Similar diagrams can be constructed for other substances.

The triple point represents that thermodynamic state where solid, liquid, and vapor can coexist. The triple point of water occurs at 273.16 K, 6.117 mbar, and the specific volume is 1.091 dm^3/kg for ice, 1 dm^3/kg for liquid water, and 206 m^3/kg for vapor. Below the triple point isobar there is no liquid phase. A sublimation/desublimation process occurs which represents phase transition between solid and vapor.

Between triple point isobar and critical point isobar three phases exist: solid, liquid, and vapor. In addition there are defined thermodynamic regions of subcooled liquid, two-phase, and superheated vapor. The subcooled liquid region exists between the critical isobar and the liquid saturation line (see Figure 1.6). The two-phase region is delimited by the liquid saturation line at the left, vapor saturation line at the right, and triple point isobar at the bottom. Superheated vapor exists above the vapor saturation line and below the critical isobar. At

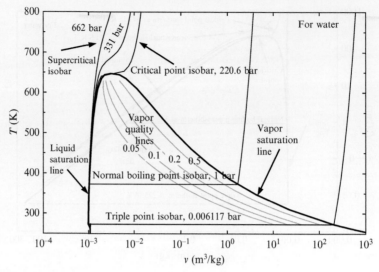

FIGURE 1.6   Temperature-specific volume diagram for water.

temperatures higher than the temperature of critical point and above the critical isobar there is a thermodynamic region denoted as *"supercritical fluid region"* where the substance is neither liquid nor gas, but has some common properties with gas and with liquid; supercritical fluids will be discussed in detail in other chapters of the book.

The *pressure versus temperature diagram* is also an important tool that shows phase transitions of any substance. Figure 1.7 presents the *P–T* diagram of water. There are four regions delimited in the diagrams: solid, vapor, liquid, and supercritical fluid. The phase transition

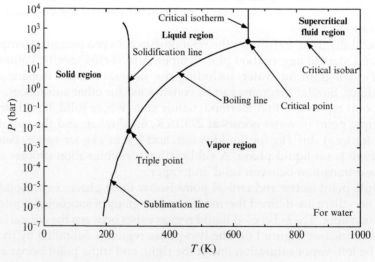

FIGURE 1.7   Pressure versus temperature diagram of water. *Data from Haynes and Lide (2012).*

lines are sublimation, solidification, boiling, critical isotherm, and critical isobar; the last two lines are represented only for the supercritical region (at pressure and temperature higher than critical).

1.3 EQUATIONS OF STATE AND IDEAL GAS BEHAVIOR

An equation of state of a substance is a functional relationship which relates thermodynamic parameters and has the purpose of describing in a complete form the interrelation among state variables. For example, if one specifies the pressure and temperature one must be able to calculate with the help of an equation of state the specific volume and the specific internal energy. If these quantities can be determined, any other thermodynamic parameter can be calculated using standard thermodynamic relations.

According to the so-called *state postulate* a simple compressible system is completely specified by two independent, intensive properties. Usually the following thermodynamic properties are taken as independent variables: pressure, specific volume, and temperature. Once the independent variables are specified all other thermodynamic parameters can be derived for any particular substance and any given thermodynamic state. There are two types of equations of state:

- *thermal equation of state*—as a function of P, v, T
- *caloric equation of state*—as a function of u, v, T, where u represents the specific internal energy

The ideal gas law—expressed by Equation (1.16)—is a thermal equation of state because it relates pressure, specific volume, and temperature. The caloric equation of the state of ideal gas can be derived from Equation (1.11) which represents the kinetic energy of a single molecule. Accordingly, for 1 mol the internal energy is N_A times higher, and, because $\mathcal{R} = N_A k_B$, the caloric equation of state for ideal gas is

$$u = 0.5 \text{ DOF } \mathcal{R} T$$

Recall that for monoatomic gases DOF = 3, therefore, the internal energy, according to caloric equation of state, is $u = 1.5$ DOF $\mathcal{R} T$. The ideal gas model is valid only for low pressure and large specific volume. The model becomes inexact at high pressure and temperature in the vicinity of critical point, when gas becomes much dense and is influenced by the vapor–liquid transition. In addition, the gas compressibility decreases greatly close to saturation line and critical point. In order to account for compressibility of a substance, the *compressibility factor Z* is introduced, which is a measure of alleviation of actual specific volume using the prediction of ideal gas law for the same pressure and temperature conditions. Compressibility factor thus expresses the deviation of a substance from ideal gas behavior as defined below:

$$Z \equiv \frac{Pv}{RT}$$

where specific volume is expressed on mass basis.

The order of magnitude is about 0.2 for many fluids. For accurate thermodynamic calculations compressibility charts can be used, which express compressibility factor as a function of pressure and temperature. In this way, an equation of state obtained based on compressibility factor is written as follows:

$$Pv = ZRT$$

where the compressibility factor is a function of pressure and temperature.

According to the *principle of corresponding states*, the compressibility factor has a quantitative similarity for all gases when it is plotted against reduced pressure and reduced temperature. The reduced pressure is defined by the actual pressure divided by the pressure of the critical point

$$P_r = \frac{P}{P_c}$$

where subscript c refers to critical properties and subscript r to reduced properties. Analogously, the reduced temperature is defined by

$$T_r = \frac{T}{T_c}$$

Compressibility charts showing the dependence of compressibility factor on reduced pressure and temperature can be obtained from accurate P, v, T data for fluids. These data are obtained primarily based on measurements. Accurate equations of state exist for many fluids; these equations are normally fitted to the experimental data to maximize the prediction accuracy. A generalized compressibility chart

$$Z = f(P_r, T_r)$$

is presented in Figure 1.8; the chart was obtained with Engineering Equation Solver software (EES, 2013), with which compressibility factor has been calculated for thirteen fluids and averaged.

There are many equations of state published in the literature. One of the most basic equations of state is that of *Van der Waals*, which is capable of predicting the vapor and liquid saturation line and a qualitatively correct fluid behavior in the vicinity of the critical point. This equation is described by the following expressions

$$\text{Thermal equation of state}: P_r = 8T_r/(3v_r - 1) - 3/v_r^2$$

$$\text{Caloric equation of state}: u_r = 4\mathcal{R}T_r$$

FIGURE 1.8 Generalized compressibility chart obtained for 13 fluids [Engineering Equation Solver software is used for Z prediction with real gas equations of state].

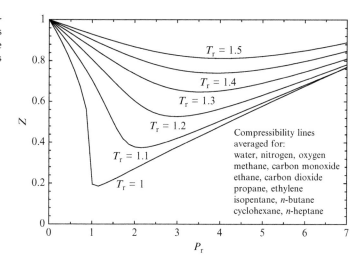

TABLE 1.4 Simple Thermodynamic Processes and Corresponding Equations for Ideal Gas Model

Process	Definition	Equation	Work Expression
Isothermal	$T=$const.	$P_1 v_1 = P_2 v_2$	$w_{1-2} = P_1 v_2 \ln(v_2/v_1)$
Isochoric	$v=$const.	$\dfrac{P_1}{T_1} = \dfrac{P_2}{T_2}$	$w_{1-2} = 0$
Isobaric	$P=$const.	$\dfrac{v_1}{T_1} = \dfrac{v_2}{T_2}$	$w_{1-2} = P(v_2 - v_1)$
Polytropic	$Pv^n=$const.	$\dfrac{P_2}{P_1} = \left(\dfrac{V_1}{V_2}\right)^n = \left(\dfrac{T_2}{T_1}\right)^{n/(n-1)}$	$w_{1-2} = \dfrac{1}{n-1}(P_2 v_2 - P_1 v_1)$
General	P, v, T vary constant mass	$\dfrac{P_1 v_1}{T_1} = \dfrac{P_2 v_2}{T_2}$	$w_{1-2} = \displaystyle\int_1^2 P dv$

where subscript r indicates reduced quantities. The reduced volume and the reduced internal energy are defined by

$$v_r = \frac{v}{v_c}; \quad u_r = \frac{u}{(P_c v_c)}$$

In thermodynamics, there are some special processes—of practical importance—during which P, or v, or T remains constant. If there is no heat exchange with the exterior ($dq=0$) then the process is called adiabatic. If a process is neither adiabatic nor isothermal it can be modeled as polytropic. Table 1.4 summarizes the principal features of simple processes for ideal gas. Figure 1.9 presents a $P-v$ diagram for four simple processes with ideal air, modeled as ideal gas.

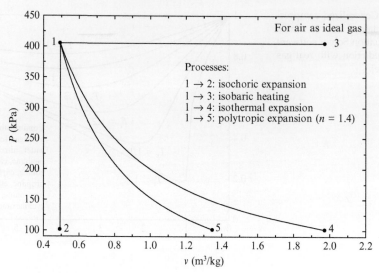

FIGURE 1.9 Simple processes with ideal gas represented in $P-v$ diagram.

1.4 LAWS OF THERMODYNAMICS

In thermodynamics there are two main laws (first and second law) which describe the behavior of thermodynamic systems regardless of their type. There is also a third law of thermodynamics—commonly called the zeroth law which refers to the state of thermodynamic equilibrium. In general, thermodynamic analysis assumes that thermodynamic systems are at equilibrium, although for some cases non-equilibrium thermodynamics methods may be the better choice for analysis. Two thermodynamic systems are said to be in *thermal equilibrium* if they cannot exchange heat, or, in other words, they have the same temperature. Two thermodynamic systems are in *mechanical equilibrium* if they cannot exchange energy in the form of work. Two thermodynamic systems are in *chemical equilibrium* if they do not change their chemical composition. An insulated thermodynamic system is said to be in *thermodynamic equilibrium* when no mass, heat, work, chemical energy, etc. is exchanged between any parts within the system.

The *zeroth law of thermodynamics* is a statement about thermodynamic equilibrium expressed as follows: "if two thermodynamic systems are in thermal equilibrium with a third, they are also in thermal equilibrium with each other." A system at internal equilibrium has a uniform pressure, temperature, and chemical potential throughout its volume.

"Energy is neither created nor destroyed." This is one of the statements of the *First Law of Thermodynamics* (FLT) which is known as the energy conservation principle. The FLT can be phrased as "you can't get something from nothing." If one denotes E the energy (in kJ) and ΔE_{sys} the change of energy of the system, then the FLT for a closed system undergoing any kind of process is written in the manner illustrated in Figure 1.10. There are two mathematical forms for FLT, namely on an amount basis or on a rate basis. These mathematical formulations are indicated by

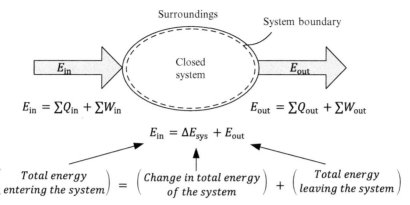

$E_{in} = \sum Q_{in} + \sum W_{in}$ $E_{out} = \sum Q_{out} + \sum W_{out}$

$$E_{in} = \Delta E_{sys} + E_{out}$$

$$\left(\begin{array}{c} Total\ energy \\ entering\ the\ system \end{array} \right) = \left(\begin{array}{c} Change\ in\ total\ energy \\ of\ the\ system \end{array} \right) + \left(\begin{array}{c} Total\ energy \\ leaving\ the\ system \end{array} \right)$$

FIGURE 1.10   Explanatory sketch for the first law of thermodynamics.

$E_{in} - E_{out} = \Delta E_{sys}$, on amount basis, and $\dot{E}_{in} - \dot{E}_{out} = dE_{sys}/dt$, on rate basis

Energy can be transferred to or from a thermodynamic system in three basic forms, namely as work, heat, or through energy associated with a mass crossing the system boundary. Therefore one has

$$E_{in} = \sum Q_{in} + \sum W_{in} \quad and \quad E_{out} = \sum Q_{out} + \sum W_{out}$$

and the variation of system energy—assuming that the mass of the closed system is m—is

$$\Delta E_{sys} = m(e_2 - e_1)$$

where index 1 represents the initial state and index 2 the final state and e is the specific total energy of the system, comprising internal energy, kinetic energy, and potential energy, expressed as follows:

$$e = u + \frac{1}{2}v^2 + gz$$

In classical thermodynamics there is, however, a sign convention for work and heat transfer which is the following:

- the heat is positive when given to the system, that is $Q = \sum Q_{in} - \sum Q_{out}$ is positive when there is net heat provided to the system.
- the net work, $W = \sum W_{out} - \sum W_{in}$ is positive when work is generated by the system.

Using the sign convention, the FLT for closed systems becomes

$$Q - W = \Delta E_{sys}$$

The above equation can be expressed on mass specific basis as follows:

$$q - w = \Delta e_{sys}$$

where $Q = mq$ and $W = mw$ and $\Delta E_{sys} = m(e_2 - e_1)$. Furthermore, FLT can be written in differential form as follows:

$$de = dq - dw$$

Taking into account that for a closed system $dw = Pdv$, it results that

$$de = dq - Pdv$$

With the assumption that the changes in kinetic and potential energies are negligible, the FLT for a closed system becomes

$$du = dq - Pdv \tag{1.17}$$

If the system is a control volume, then the energy term will comprises the additional term of flow work. In this case the total specific energy of a flowing matter is

$$\theta = u + Pv + 0.5\,v^2 + gz$$

The term $u + Pv$ from above equation represents a state function named *enthalpy* (or flow energy), denoted with h

$$h = u + Pv$$

Using enthalpy formulation, the FLT for a control volume that has neither velocity nor elevation is $d(h - Pv) = dq - Pdv$ which can be expanded to the following equation: $dh - Pdv - vdP = dq - Pdv$. Thus, the FLT for control volumes (open systems) becomes

$$dh = dq + v\,dP \tag{1.18}$$

In addition, one remarks that the total specific energy of a flowing matter can be written with the help of enthalpy

$$\theta = h + 0.5\,v^2 + gz \tag{1.19}$$

The SLT for control volume, using the sign convention for heat and work, is formulated mathematically, in rate form, in the following way:

$$\dot{Q} + \sum_{in} \dot{m}\theta = \dot{W} + \sum_{out} \dot{m}\theta + \frac{d(me)}{dt} \tag{1.20}$$

An important consequence of the FLT is that the internal energy change resulting from any process is independent of the thermodynamic path followed by the system during thermodynamic transformations. In turn, the rate at which the internal energy content of the system changes is dependent only on the rates at which heat is added and work is done. Observe from Equation (1.17) that internal energy is a function of specific volume; recall that internal energy is also a function of temperature. Therefore one has $u = u(T,v)$ and, accounting for Equation (1.17), the total derivative of internal energy is

$$du = \left(\frac{\partial u}{\partial T}\right)_v dT + \left(\frac{\partial u}{\partial T}\right)_T dv = C_v\,dT - P\,dv$$

The above equation defines the *specific heat at constant volume* C_v, which represents the amount of heat required to increase the temperature of a constant volume system by $1°K$; as is observed from above equation, $du = dq = C_v dT$. The specific heat is equal to the change in internal energy with temperature at constant volume

$$C_v \equiv \left(\frac{\partial u}{\partial T}\right)_v \tag{1.21}$$

In addition it results in the following thermodynamic identity

$$P \equiv -\left(\frac{\partial u}{\partial u}\right)_T$$

Similarly, from Equation (1.18), enthalpy is a total derivative of temperature and pressure, $h = h(T, P)$. The total differential form of enthalpy becomes

$$dh = \left(\frac{\partial h}{\partial T}\right)_P dT + \left(\frac{\partial h}{\partial P}\right)_T dP = C_p \, dT + v dP$$

an expression that defines the *specific heat at constant pressure* C_p, which represents the amount of heat required to increase the temperature of a system evolving at constant pressure at $1°K$. The specific heat is the change of enthalpy with temperature at a constant pressure defined according to

$$C_p \equiv \left(\frac{\partial h}{\partial T}\right)_P \tag{1.22}$$

In addition it is observed from above equation that for an isobaric process the heat exchange is governed by the equation $dh = dq = C_p dT$, and the following thermodynamic identity is noted:

$$v \equiv \left(\frac{\partial h}{\partial P}\right)_T$$

For an ideal gas it is remarked that $dh = d(u + Pv) = du + R \, dT$, or by replacing dh and du with $dh = C_p \, dT$ and $du = C_v \, dT$ one obtains the well-known Robert Meyer equation

$$C_p = C_v + R$$

For ideal gas it was demonstrated above that kinetic energy associated with molecules is the same as its internal energy, $u = 1.5RT$. It can be remarked that for ideal gas the internal energy is a function of temperature only. Therefore, specific heat for ideal gas is $C_v = 1.5R$ and $C_p = 2.5R$. For ideal gases, internal energy is a function of temperature only, so $du = c_v \, dT$. From the definition $h \equiv u + Pv$ it can be inferred that the enthalpy of ideal gas ($Pv = RT$) is dependent enthalpy only because $h(T) = u(T) + RT$. In conclusion one has $dh = C_p dT$. In addition, from enthalpy definition it results that $dh = du + R \, dT$.

The ratio of specific heat at constant pressure and constant volume is known as the *adiabatic exponent*

$$\gamma = \frac{C_p}{C_v}$$

which for monoatomic ideal gas has the value 1.4, and a value of $5/3 = 1.67$ for diatomic gas.

Observe that the specific heat of ideal gas is constant. In some literature sources an ideal gas with constant specific heat is denoted as "*perfect* gas," whereas gases that behave according to ideal gas law $(Pv = RT)$ but show specific heat variation with temperature are denoted as "*ideal* gases."

Another thermodynamic law is the *second law of thermodynamics*, which provides a way to predict the direction of any process in time, to establish conditions of equilibrium, to determine the maximum attainable performance of machines and processes, and to assess quantitatively the irreversibilities and determine their magnitude for the purpose of identifying ways of improvement of processes. Also SLT is the fundamental law based on which the absolute scale of temperature is defined. Furthermore, SLT is useful in evaluating various relevant thermodynamic properties based on experimental data.

In order to introduce SLT the concepts of *reversibility* and *irreversibility* must be discussed. It is known that a thermodynamic process is reversible if during a transformation both the thermodynamic system and its surroundings can be returned to their initial states. All infinitesimal changes through which the process evolves must be reversible, such that the overall process is reversible. A process performed by a thermodynamic system is irreversible if it cannot return to the initial state because of dissipation that occurs (irreversibly) by friction, heat rejection, electrical, chemical, or mechanical losses etc. In any real application there is dissipation; therefore, reversibility is only an idealization that is important for theoretical point of view. Reversible processes can be characterized as follows:

- *Externally reversible*—a process with no associated irreversibilities outside the system boundary.
- *Internally reversible*—a process with no irreversibilites within the boundary of the system during the process.
- *Totally reversible*—a process with no irreversibilities within the system or surroundings.

The classical statements of SLT which say in essence that heat cannot be completely converted into work, although the opposite is possible (a form of work can be completely converted into any other form of work) are as follows.

- **The Kelvin-Plank statement**. It is impossible to construct a device, operating in a cycle (e.g., a heat engine), that accomplishes only the extraction of heat energy from some source and its complete conversion to work. This simply shows the impossibility of having a heat engine with a thermal efficiency of 100%.
- **The Clausius statement**. It is impossible to construct a device, operating in a cycle (e.g., a refrigerator and heat pump), that transfers heat from the low-temperature side (cooler) to the high-temperature side (hotter).

The Clausius inequality provides a mathematical statement of the SLT, namely

$$\oint \frac{dQ}{T} \leq 0 \tag{1.23}$$

where the circular integral indicates that the process must be cyclical.

At the limit when the inequality becomes zero then the process is reversible (ideal situation). A useful mathematical artifice is to attribute to the integral from Equation (1.23) a new physical quantity. This will define entropy generated by the system S_{gen}

$$S_{gen} = -\oint \frac{dQ}{T} \tag{1.24}$$

Any real process must have positive generated entropy; the following cases may thus occur: (i) $S_{gen} > 0$, real, irreversible process; (ii) $S_{gen} = 0$, ideal, reversible process; (iii) $S_{gen} < 0$, impossible process. Therefore, the mathematical formulation of the SLT is

$$S_{gen} \geq 0$$

Entropy is an extensive property. Generated entropy of a system during a process is a superimposition of entropy change of the thermodynamic system and the entropy change of the surroundings

$$S_{gen} = \Delta S_{sys} + \Delta S_{surr} \tag{1.25}$$

For a reversible process $S_{gen} = 0$, Equation (1.25) shows that entropy change of the system is the opposite of the entropy change of the surroundings:

$$\Delta S_{rev} = -\Delta S_{surr} = \left(\frac{Q}{T}\right)_{rev}$$

For an irreversible process the following inequality must be true

$$\Delta S_{sys} > \Delta S_{surr} = \left(\frac{Q}{T}\right)_{surr}$$

which is explained by the existence of generated entropy that is always positive.

Although the change in entropy of the system and its surroundings may individually increase, decrease, or remain constant, the total entropy change (the sum of entropy change of the system and the surroundings or the total entropy generation) cannot be less than zero for any process. Note that entropy change along a process 1–2 is defined based on the integral

$$S_{1-2} = \int_1^2 \frac{dQ}{T}$$

therefore by differentiation one obtains

$$dQ = T\,dS \tag{1.26}$$

Entropy quantifies the molecular random motion within a thermodynamic system and is related to the *thermodynamic probability* (p) of possible microscopic states as indicated by the Boltzmann equation

$$S = k_B \ln p$$

The engineering usefulness of SLT consists of the ability to define and predict the performance limits of systems. As is known, a heat engine is a device operating based on a specific thermodynamic cycle, which eventually generates useful work. In the reverse process, a heat pump consumes work in order to increase the temperature of a working fluid and generate heating. A refrigerator is similar to a heat engine, except that it generates cooling. The environment plays the role of a large *thermal energy reservoir* at a specified temperature. A thermal energy reservoir is a thermodynamic system with relatively large thermal energy capacity that can absorb or deliver large quantities of heat without observable changes in its temperature.

A *heat source* represents a thermal reservoir capable of providing thermal energy to other systems. A *heat sink* represents a thermal reservoir capable of absorbing heat from other systems. The environment can play roles of both heat source or heat sink, depending on the application. A heat engine operates cyclically by transferring heat from a heat source to a heat sink. While receiving more heat from the source (Q_H) and rejecting less to the sink (Q_C), a heat engine can generate work (W). As stated by the FLT, energy is conserved, thus $Q_H = Q_C + W$.

A typical "black box" representation of a heat engine is presented in Figure 1.11a. According to the SLT the work generated must be strictly smaller than the heat input, $W < Q_H$. The thermal efficiency of a heat engine—also known as energy efficiency—is defined as the

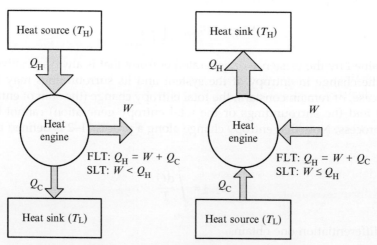

FIGURE 1.11   Conceptual representation of a heat engine (a) and heat pump (b).

net work generated by the total heat input. Using notations from Figure 1.11a, energy efficiency of a heat engine is expressed (by definition) with

$$\eta = \frac{W}{Q_H} = 1 - \frac{Q_C}{Q_H} \tag{1.27}$$

If a thermodynamic cycle operates as a refrigerator or heat pump, then its performance can be assessed by the *Coefficient of Performance (COP)*, defined as useful heat generated per work consumed. As observed in Figure 1.11b, the energy balance equation (EBE) for a heat pump is written as $Q_C + W = Q_H$, according to SLT $Q_H \geq W$ (this means that work can be integrally converted in heat). Based on its definition, the coefficient of performance is

$$COP = \frac{Q_H}{W} = \frac{Q_H}{Q_C + W} \tag{1.28}$$

The Carnot cycle is a fundamental model in thermodynamics representing a heat engine (or heat pump) that operates between a heat source and a heat sink, both of them being at constant temperature. This cycle is a conceptual (theoretical) cycle and was proposed by Sadi Carnot in 1824. The cycle comprises fully reversible processes, namely two adiabatic and two isothermal processes. Figure 1.12 illustrates two Carnot cycles for air, modeled as ideal gas: (a) is a power cycle, and (b) is a reversed (refrigeration or heat pump) cycle. The efficiency of the Carnot cycle is independent of the working fluid which performs the cyclical process. Based on the definition of the Carnot cycle it results that $s_2 = s_3$ and $s_4 = s_1$ (for a heat engine). The heat transferred at source and sink is $Q_H = T_H(s_3 - s_4) = T_H(s_2 - s_1)$ and $Q_L = T_L(s_2 - s_1)$. Therefore, energy efficiency of the Carnot cycle is

$$\eta = 1 - \frac{T_L}{T_H} \tag{1.29}$$

and COP of the reversed Carnot cycle

$$COP = \frac{T_H}{T_H - T_L} \tag{1.30}$$

Carnot efficiency expressed by Equations (1.29) and (1.30) is a useful criterion to assess practical heat engines, refrigerators, heat pumps, and other energy conversion systems with respect to the idealized reversible heat engine. Accordingly, energy efficiency (η) and coefficient of performance (COP) of a reversible thermodynamic cycle (Carnot) is the highest possible, and any actual (irreversible) cycle has smaller efficiency ($\eta_{rev} > \eta_{irrev}$) and ($COP_{rev} > COP_{irrev}$).

The Carnot cycle is useful for development of the absolute temperature scale, which is a scale independent of the properties of the substance used for measurement (e.g., mercury, alcohol, water, etc.). As it results from Equations (1.27) to (1.30) for reversible heat engines, one can obtain

$$\left(\frac{Q_H}{Q_L}\right)_{rev} = \frac{T_H}{T_L} \tag{1.31}$$

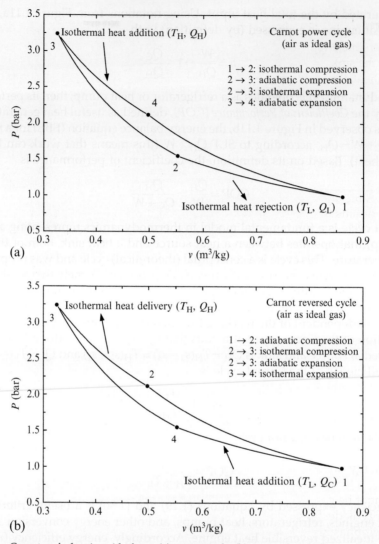

FIGURE 1.12   Carnot cycle for air as ideal gas: (a) power cycle, (b) reversed cycle (heat pump).

where index rev signifies "reversible processes." The absolute temperature scale is developed by assigning a temperature value to the thermodynamic state of the triple point of pure water; this value is $T_0 = 273.16$ K. Thus, if a heat engine operates between T and T_0 then the temperature T can be expressed as a function of reference temperature T_0, as follows:

$$T = T_0 \frac{Q}{Q_0} \tag{1.32}$$

Therefore, absolute temperature can be determined by the measurement of heat transfer at source and sink of a reversible heat engine. At the lowest limit, the temperature is 0 K, namely, when the reversible engine operates between reference temperature T_0 at the source and absolute zero at the sink. In this case, according to Equation (1.32) the heat delivered to the sink is nil and the heat absorbed from the source is completely transformed into work because in this idealized case the Carnot efficiency must be 1 (a limit never attainable). At 0 K the entropy of non-vibrating systems such as crystals is 0.

1.5 EXERGY

Exergy represents the maximum work which can be produced by a thermodynamic system when it comes into equilibrium with its surrounding environment. This statement assumes that at an initial state there is a thermodynamic system which is not in equilibrium with the environment. In addition it is assumed that—at least potentially—mechanisms of energy (and mass) transfer between the system and the environment exist, such that eventually the system can evolve so that equilibrium condition will eventually occur. The system must at least exchange work with the environment. Another remark is that exergy definition assumes the existence of a reference environment. The system under the analysis will interact only with that environment.

In many practical problems, the reference environment is assumed to be the earth atmosphere, characterized by its average temperature and pressure: $P_0 = 101.325$ kPa, $T_0 = 298.15$ K. In some problems, when reacting systems are present, the chemical potential of the reference environment must be specified. In such cases, thermodynamic equilibrium will refer to all possible interactions; one can say that a system is in thermodynamic equilibrium with the environment if it has the same temperature with it (thermal equilibrium), the same pressure with it (mechanical equilibrium), and the same chemical potential with it. Therefore, exergy includes at least two components, one thermomechanical and one chemical.

Exergy cannot be conserved. Any real process destroys exergy as, similarly, it generates entropy. Exergy is destroyed and entropy is generated due to irreversibilities. Dincer and Rosen (2013) emphasize that exergy is not only a quantity expressing the maximum potential of a system to do work in a defined environment, but also an effective measure of the potential of a substance to impact the environment.

The exergy of a closed (non-flow) thermodynamic system comprises four terms, namely physical (or thermomechanical), chemical, kinetic, and potential. In brief, total exergy of a non-flow system is written as

$$Ex_{nf} = Ex_{ph} + Ex_{ch} + Ex_{ke} + Ex_{pe} \tag{1.33}$$

As detailed by Dincer and Rosen (2013) the exergy of a flowing stream of matter Ex_f represents the sum of the non-flow exergy and the exergy associated with the flow work of the stream $Ex_{fw} = (P - P_0)V$, therefore

$$Ex_f = Ex_{nf} + (P - P_0)V \tag{1.34}$$

The physical exergy for a non-flow stream is defined by

$$Ex_{ph} = (U - U_0) + P_0(V - V_0) - T_0(S - S_0) \tag{1.35}$$

where U is internal energy, V volume, and S entropy of closed system which is in non-equilibrium with the environment, T_0 is the reference temperature of the surroundings environment, and index 0 refers to the values of the parameters when the system is in thermomechanical equilibrium with the environment.

The kinetic and potential exergies of the system equal the kinetic and potential energy respectively, which are given by known formulas such as

$$Ex_{ke} = \frac{1}{2}m\nu^2$$

where m is the system mass and ν is its (macroscopic) velocity. Moreover, the potential energy is

$$Ex_{pe} = mg(z - z_0)$$

where z is the system elevation and z_0 is a reference elevation of the environment (e.g., ground level).

Consider a system which is in thermomechanical equilibrium with the reference environment (it has the same temperature and pressure as the environment—T_0, P_0), but it is not in chemical equilibrium with the reference environment because it has another chemical composition. Chemical exergy represents the maximum work that can be extracted during a process where the system composition changes to that of the environment. The chemical exergy depends on the difference between the chemical potential of system components $\left(\sum n_i \mu_i^0\right)$ (being in thermomechanical equilibrium but not in chemical equilibrium) and the chemical potential of system components when they are brought into chemical equilibrium with the environment, $\left(\sum n_i \mu_i^{00}\right)$. Therefore, the chemical exergy of the system is

$$Ex_{ch} = \sum n_i \left(\mu_i^0 - \mu_i^{00}\right) \tag{1.36}$$

Chemical exergy is important in power generation systems based on fuel combustion or power generation systems based on electrochemical processes (e.g., fuel cells) and whenever chemical reactions or other processes such as mixing, absorption, desorption, separation processes, and concentration processes occur.

There are two main components of chemical exergy: exergy due to chemical reaction and exergy due to concentration difference. When a chemical compound is allowed to interact with the environment, chemical reactions may occur, involving unstable species. Eventually, more stable species are formed and further reaction is not possible. Moreover, if system compounds have a different concentration of phase than that corresponding to the environment, then various processes such as dilution or concentration may occur until there is no difference in concentration between system components and the environment. When there is no potential of chemical reaction with the environment and all system components are at the same concentration as the environment, then the system is in complete chemical equilibrium with the environment.

TABLE 1.5 Components Molar Fraction and Standard Chemical Exergy for Terrestrial Atmosphere

Component	N_2	O_2	H_2O	Ar	CO_2	Ne	He	Kr
C (molar %)	75.67	20.34	3.03	0.92	0.03	0.0018	0.00052	0.000076
Ex^{ch} (kJ/mol)	0.69	3.95	8.67	11.62	20.11	27.10	30.16	34.93

Chemical exergy results from chemical (by reactive processes) and physical (by diffusion processes, etc.) interaction between the system under consideration and the environment. Thence, in order to be able to estimate it chemical exergy, a reference environment must be defined in terms of temperature T_0, pressure P_0, and chemical composition. Let us analyze first the chemical exergy due to concentration difference. For this purpose we assume the earth atmosphere as the reference environment, formed of eight chemical compounds with the molar concentrations listed in Table 1.5. Let us assume that one of the chemical compounds listed in the table is released to the atmosphere at T_0 and P_0 and 100% molar concentration; denote this state with 1. A dilution process occurs at the end of which the component will have the same concentration as that listed in the table, y_i; this state is denoted with 2. The maximum work extractable from processes 1–2 represents the exergy due to concentration difference and is given by $\Delta Ex^{ch}_{conc} = Ex_1 - Ex_2$.

According to Equations (1.34) and (1.35) one has

$$\Delta Ex^{ch}_{conc} = (U_1 - U_2) + (P_1 V_1 - P_2 V_2) - T_0 (S_1 - S_2) = T_0 (S_2 - S_1) \tag{1.37}$$

where in Equation (1.37) one accounts for the fact that, for an isothermal process, $U_2 = U_1$ and $P_2 V_2 = P_1 V_1$. Furthermore, according to the FLT, $T dS = dU + P dV$. Therefore, for an isothermal process of ideal gas for which $dU = 0$ and $d(PV) = 0$, one has $T dS = d(PV) - v dP$, or $T dS = -v dP$. Further, by accounting for ideal gas law

$$\frac{v}{T} = \frac{\mathcal{R}}{P}$$

for the isothermal process it is obtained that

$$dS = -\mathcal{R} \frac{dP}{P}$$

which by integration from state 1 to 2 results in

$$S_2 - S_1 = -\mathcal{R} \ln\left(\frac{P_2}{P_1}\right) \tag{1.38}$$

In Equation (1.38) one notes that $P_1 = P_0$ and $P_2 = P_i = y_i P_0$, where index i refers to one of the components of the atmosphere. Therefore, from Equations (1.37) and (1.38) one obtains

$$Ex^{ch}_{conc, i} = -\mathcal{R} T_0 \ln(y_i) \tag{1.39}$$

Data reported in Table 1.5 is calculated with Equation (1.39). In order to derive an expression for the reactive component of chemical energy, the notion of *Gibbs free energy* must be introduced.

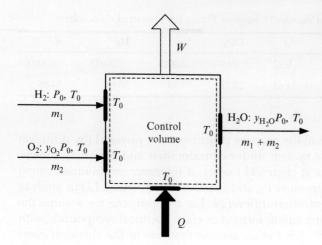

FIGURE 1.13 Control volume model for hydrogen oxidation reaction.

Let us determine the maximum work extraction from hydrogen when it is oxidized with oxygen from the terrestrial atmosphere. In this respect one assumes a reversible fuel cell system as depicted in Figure 1.13. Basically, the fuel cell is viewed as a control volume. Hydrogen is provided at T_0, P_0 to the control volume, and therefore it is in thermomechanical equilibrium with the environment. Oxygen is provided to the control volume at its concentration in the environment, denoted y_{O_2}.

According to the FLT for the system in Figure 1.13 one has

$$Q = (h_{H_2O} - h_{H_2} - 0.5h_{O_2}) + W$$

and according to the SLT one has

$$s_{gen} = s_{H_2O} - s_{H_2} - 0.5s_{O_2} - \frac{Q}{T_0}$$

Therefore, eliminating Q from the above two equations, the work extraction is

$$W = (h_{H_2} - T_0 s_{H_2}) + 0.5(h_{O_2} - T_0 s_{O_2}) - (h_{H_2O} - T_0 s_{H_2O}) - T_0 s_{gen}$$

Observe in the above relation the term $h - Ts$. This term is identified as a *state function* and it defines the *Gibbs free energy* denoted with g

$$g = h - Ts$$

Gibbs free energy has a major importance in determining equilibrium conditions for chemical reaction systems that normally operate at constant pressure. The chemical equilibrium of a closed system is obtained when

$$\Delta G = m \, \Delta g = 0$$

In addition, Gibbs free energy can be used to determine the maximum work related to chemical processes. Returning to the analysis of "isothermal" hydrogen oxidation, the following equation for maximum work can be deduced

$$w_{max} = ex_{H_2}^{ch} = g_{H_2}(T_0, P_0) + 0.5 g_{O_2}(T_0, y_{O_2} P_0) - g_{H_2O}(T_0, y_{H_2O} P_0) \tag{1.40}$$

where it is taken into account that for work to be maximum the process must be reversible. Therefore, $s_{gen} = 0$. From Equations (1.40) and (1.37) it results that

$$Ex_{H_2}^{ch} = g_{H_2}(T_0, P_0) + 0.5 g_{O_2}(T_0, P_0) - g_{H_2O}(T_0, P_0) - 0.5 Ex_{O_2}^{ch} + Ex_{H_2O}^{ch}$$

In the above equation one recognizes the standard Gibbs free energy of formation reaction of water ($H_2 + 0.5 O_2 \rightarrow H_2O$), which is $\Delta^f g_{H_2O}^0 = g_{H_2O}(T_0, P_0) - g_{H_2}(T_0, P_0) - 0.5 g_{O_2}(T_0, P_0)$, therefore the chemical exergy of hydrogen can be expressed as

$$ex_{H_2}^{ch} = -\Delta^f g_{H_2O}^0 - 0.5 ex_{O_2}^{ch} + ex_{H_2O}^{ch}$$

From thermodynamic tables (see Appendix B), it is found that the formation Gibbs energy of water is -237.12 kJ/mol. Thus, chemical exergy of hydrogen can be calculated as follows: $ex_{H_2}^{ch} = -(-237.12) - 0.5 \times 3.95 + 8.67 = 243.8$ kJ/mol. In a similar manner one can calculate the chemical exergy of any chemical element. If a substance is not present in the atmosphere then the reference for zero chemical exergy is the most stable state of that substance in seawater. There are tables of chemical exergy of elements in past literature data. A recent source for tabulated data of the chemical exergy of elements is presented by Rivero and Grafias (2006). Table 1.6 gives the chemical exergy of some of the most frequently encountered chemical elements in power generation. Observe that for hydrogen, the standard value for exergy differs from that calculated above; the standard value is 236.12 kJ/mol.

Standard chemical exergy of elements is useful for calculation of the standard chemical exergy of chemical compounds provided that their Gibbs energy of formation is known. For example, Gibbs free energy of water can be calculated based on an approach similar to that illustrated above for hydrogen. Instead of assuming a fuel cell as previously, the control volume will function as an electrolysis cell supplied with liquid water. It is easy to demonstrate that the chemical exergy of water can be calculated with

$$ex_{H_2O}^{ch} = \Delta^f g_{H_2O}^0 + 0.5 ex_{O_2}^{ch} + ex_{H_2}^{ch} = -237.12 + 0.5 \times 3.92 + 236.12 = 0.96 \text{ kJ/mol}$$

A general equation for the chemical exegy calculation of any chemical compound can be derived from the example above for water chemical exergy. In order to determine the chemical exergy of a compound, one must know its standard Gibbs free energy of formation, $\Delta^f g^0$. Then, using $\Delta^f g^0$ and the standard exergy of the elements, the following formula is used to determine the chemical exergy of the compound

TABLE 1.6 Standard Chemical Exergy of Some Elements

Element	B	C	Ca	Cl$_2$	Cu	F$_2$	Fe	H$_2$	I$_2$	K	Mg
ex^{ch}, kJ/mol	628.1	410.27	729.1	123.7	132.6	505.8	374.3	236.12	175.7	336.7	626.9
Element	Mo	N$_2$	Na	Ni	O$_2$	P	Pb	Pt	Pu	Si	Ti
ex^{ch}, kJ/mol	731.3	0.67	336.7	242.6	3.92	861.3	249.2	141.2	1100	855.0	907.2

Source: Rivero and Grafias (2006).

$$\text{ex}^{\text{ch}} = \Delta^{\text{f}} g^0 + \sum_{\text{element}} \left(v\, \text{ex}^{\text{ch}} \right)_{\text{element}} \qquad (1.41)$$

where v is the stoichiometric factor representing the number of moles of element per mole of chemical compound.

Let us give an example of the application of Equation (1.41) for calculation of the standard chemical exergy of carbon dioxide. Appendix B presents tabulated data for the formation enthalpy, entropy, and Gibbs free energy of chemical compounds. For carbon dioxide it is found that $\Delta^{\text{f}} g^0 = -394.35$ kJ/mol. The formation reaction of carbon dioxide is $C + 0.5\,O_2 \rightarrow CO_2$. From Table 1.6, the standard chemical exergy of carbon (graphite) is 410.27 kJ/mol, while that of oxygen is 3.92 kJ/mol. Thus, from Equation (1.41),

$$\text{ex}^{\text{ch}}_{\text{CO}_2} = -394.35 + 410.27 + 0.5 \times 3.92 = 17.9 \text{ kJ/mol}$$

The standard chemical exergy of methane can be calculated similarly based on the formation reaction $C + 2\,H_2 \rightarrow CH_4$. The standard formation Gibbs energy for methane is $\Delta^{\text{f}} g^0 = -50.53$ kJ/mol and the standard chemical exergy of hydrogen is 236.12 kJ/mol. Therefore, for methane one has

$$\text{ex}^{\text{ch}}_{\text{CH}_4} = -50.53 + 410.27 + 2 \times 236.12 = 832 \text{ kJ/mol}$$

1.6 BALANCE EQUATIONS FOR THERMODYNAMIC ANALYSIS

Thermodynamic analysis is required for assessing any kind of energy conversion system. In any thermodynamic analysis, both the first and second laws must be invoked in order to obtain a complete picture. Assessment criteria are generally based on energy and exergy efficiency which are also useful for comparing various deign options among each other and for identifying the ideal—maximum—margin of system performance under imposed working conditions. Thermodynamic analysis by first and second law can pinpoint the energy losses but also losses of work potential due to the irreversibilities.

Thermodynamic analysis is generally based on four types of balance equations which will be presented here in detail. These are mass balance equation (MBE), energy balance equation (EBE), entropy balance equation (EnBE) and exergy balance equation (ExBE).

1.6.1 Mass Balance Equation

The effect of mass addition or extraction on the energy balance of control volume is proportional with the *"mass flow rate,"* defined as the amount of mass flowing through the cross section of a flow stream per unit of time. For a control volume—according to the *"conservation of mass principle"*—the net mass transferred to the system is equal to the net change in mass within the system plus the net mass leaving the system.

The mass balance equation for a general control volume can be written as follows (see Figure 1.14):

$$\text{MBE}: \sum \dot{m}_{\text{in}} = \sum \dot{m}_{\text{out}} + \frac{\text{d}m_{\text{CV}}}{\text{d}t} \qquad (1.42)$$

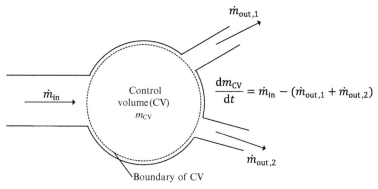

FIGURE 1.14 Illustrative sketch for mass balance equation.

Note that if flow is steady ($dm_{CV}/dt = 0$) then, from Equation (1.42),

$$\text{MBE steady flow}: \sum \dot{m}_{in} = \sum \dot{m}_{out}$$

1.6.2 Energy Balance Equation

The EBE is an expression of the FLT with a sign convention relaxed. Therefore the variation of system energy between states 1 and 2 is

$$\Delta E_{sys} = m\Delta e_{sys} = m\left[\left(u_2 + \frac{1}{2}v_2^2 + gz_2\right) - \left(u_1 + \frac{1}{2}v_1^2 + gz_1\right)\right] \qquad (1.43)$$

In Equation (1.43) the total specific energy of a non-flowing thermodynamic system (or fluid) is defined as the sum of the specific internal energy and kinetic and potential energies of the system, namely

$$e = u + \frac{1}{2}v^2 + gz \qquad (1.44)$$

Given that the system is closed and its mass does not vary, the EBE can be written on a per mass basis and in rate form; this is as follows:

$$\text{EBE}_{\text{Closed System}}: \sum \dot{q}_{in} + \sum \dot{w}_{in} = \sum \dot{q}_{out} + \sum \dot{w}_{out} + \frac{d\dot{e}}{dt} \qquad (1.45)$$

In amount formulation the EBE for a closed system that changes its state from 1 to 2 is

$$\text{EBE}_{\text{Closed System}}: e_1 + \sum q_{in} + \sum w_{in} = e_2 + \sum q_{out} + \sum w_{out}$$

The following EBE for open systems (or control volumes) which account for the existence of flow work and boundary work can be formulated

$$\text{EBE}_{\text{Open System}}: \sum_{in} \dot{m}\theta + \sum \dot{Q}_{in} + \sum \dot{W}_{in} = \sum_{out} \dot{m}\theta + \sum \dot{Q}_{out} + \sum \dot{W}_{out} + \left[\frac{d(me)}{dt}\right]_{sys} \qquad (1.46)$$

where $\left[\dfrac{d(me)}{dt}\right]_{sys}$ represents the rate of change of total energy of the system.

In Equation (1.46), m is the system mass and e is the total specific energy defined by Equation (1.44). The total energy of a flowing matter (θ) which represents the sum of internal energy, flow work, kinetic energy, and potential energy is introduced by Equation (1.46). As can be remarked, the total energy of a flowing matter is defined by

$$\theta = u + Pv + \frac{1}{2}\mathit{v}^2 + gz = h + \frac{1}{2}\mathit{v}^2 + gz \tag{1.47}$$

If the control volume is stationary (its displacement velocity is nil), $\mathit{v} = 0$; thus there is no kinetic energy associated with the system. In addition, if there is negligible change of elevation in the system one may neglect the potential energy term. For a stationary open system with no potential energy variation, the EBE becomes

$$\text{EBE}_{\text{Open System}} : \dot{Q}_{\text{in}} + \dot{W}_{\text{in}} + \sum_{\text{in}} (\dot{m}h) = \dot{Q}_{\text{out}} + \dot{W}_{\text{out}} + \sum_{\text{out}} (\dot{m}h) + \left[\frac{\mathrm{d}(me)}{\mathrm{d}t}\right]_{\text{sys}} \tag{1.48}$$

Another particular form of EBE is that for steady-flow systems. In a steady-flow system, flows of mass and energy that cross the system boundary are steady. A steady fluid flow infers a flow of which parameters such as mass flow rate, pressure, temperature, etc. do not change in time. Given that mass flow rate, heat transfer rates, and work transfer rates are constant, the integration of the following equations $\mathrm{d}(mh) = (\dot{m}h)\mathrm{d}t$, $\mathrm{d}Q = \dot{Q}\mathrm{d}t$, and $\mathrm{d}W = \dot{W}\mathrm{d}t$ between initial state 1 and latter state 2 of the open system is straightforward. Thus, the EBE can be written in amount form as

$$\text{EBE}_{\text{Steady Flow}} : \dot{m}_1 e_1 + \dot{Q}_{\text{in}} + \dot{W}_{\text{in}} + \sum_{\text{in}} (\dot{m}h) = \dot{m}_2 e_2 + \dot{Q}_{\text{out}} + \dot{W}_{\text{out}} + \sum_{\text{out}} (\dot{m}h) \tag{1.49}$$

If in addition of to steady flow the system has a steady state, then there is no net energy change associated with the open system. This type of operation is called a *steady-state steady-flow* (SSSF). In this case the EBE becomes

$$\text{EBE}_{\text{SSSF}} : \dot{Q}_{\text{in}} + \dot{W}_{\text{in}} + \sum_{\text{in}} (\dot{m}h) = \dot{Q}_{\text{out}} + \dot{W}_{\text{out}} + \sum_{\text{out}} (\dot{m}h) \tag{1.50}$$

Almost all thermal power generation systems operate at steady-state or quasi steady-state except for periods of startup and shutdown. Because non-steady-state periods are very small relative to the time of steady operation, the most relevant balance equations for power generation systems are those corresponding to SSSF.

1.6.3 Entropy Balance Equation

The SLT can be expressed in the form of an EnBE, which states that for a thermodynamic system entropy input plus generated entropy is equal to entropy output plus change of entropy within the system. The EnBE is schematically illustrated in Figure 1.15.

The EnBE is a mathematical statement derived from SLT, namely

$$\text{EnBE} : \sum S_{\text{in}} + S_{\text{gen}} = \sum S_{\text{out}} + \Delta S_{\text{sys}} \tag{1.51}$$

The EnBE postulates that the entropy change of a thermodynamic system is equal to entropy generated within the system plus the net entropy transferred to the system across its boundary (that is, the entropy entering minus the entropy leaving). Here the entropy change

FIGURE 1.15 Explanatory sketch for the entropy balance equation—a statement of SLT.

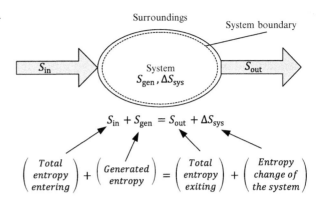

of a system is equal to the difference between final and initial entropy $\Delta S_{sys} = S_2 - S_1$. The EnBE can be also expressed on a rate basis:

$$\text{EnBE}: \sum \dot{S}_{in} + \dot{S}_{gen} = \sum \dot{S}_{out} + \frac{\mathrm{d}S_{sys}}{\mathrm{d}t} \tag{1.52}$$

Entropy can be transferred across a system boundary by substance flow and by heat, but it cannot be transferred by work. When a heat transfer occurs across a boundary then the associated entropy transfer results from the differential equation

$$\mathrm{d}Q = T\,\mathrm{d}S \tag{1.53}$$

Therefore, entropy transferred across the system boundary or along a process 1–2 due to heat becomes

$$S_{1-2} = \int_{1}^{2} \frac{\mathrm{d}Q}{T} \tag{1.54}$$

The general EnBE takes a special form for closed systems. For a closed system there is no mass transfer at the system boundary. Therefore, entropy can be transferred only by heat. On a rate basis the EnBE is

$$\text{EnBE}: \sum_{in} \left(\int \frac{\mathrm{d}\dot{Q}}{T} \right) + \dot{S}_{gen} = \frac{\mathrm{d}S_{sys}}{\mathrm{d}t} + \sum_{out} \left(\int \frac{\mathrm{d}\dot{Q}}{T} \right) \tag{1.55}$$

If the closed system is also adiabatic, then there is neither entropy transfer due to mass nor due to heat. In this case the EnBE simplifies to

$$\dot{S}_{gen} = \frac{\mathrm{d}S_{sys}}{\mathrm{d}t} \tag{1.56}$$

The entropy change of a system can be expressed based on specific entropy according to $\mathrm{d}S_{sys} = \mathrm{d}(ms)$, which after integration gives $\Delta S_{sys} = m_2 s_2 - m_1 s_1$; if the mass of the system remains constant, then $\Delta S_{sys} = m(s_2 - s_1)$. EnBE for an open system (control volume) has the following expression in rate form

$$\text{EnBE}_{\text{CV}} : \sum_{\text{in}} \left(\int \frac{d\dot{Q}}{T} \right) + \sum_{\text{in}} \dot{m}s + \dot{S}_{\text{gen}} = \frac{dS_{\text{CV}}}{dt} + \sum_{\text{out}} \left(\int \frac{d\dot{Q}}{T} \right) + \sum_{\text{out}} \dot{m}s \qquad (1.57)$$

where the index CV stands for "control volume." If the flow is steady (that is, there is no temporal variation of parameters) it implies that the mass enclosed in the control volume and specific entropy of the control volume remain constant in time; thus $S_{\text{CV}} = m_{\text{CV}} s_{\text{CV}}$ is constant.

Consequently, the EnBE for a steady flow through a control volume is

$$\text{EnBE} : \sum_{\text{in}} \left(\int \frac{d\dot{Q}}{T} \right) + \sum_{\text{in}} \dot{m}s + \dot{S}_{\text{gen}} = \sum_{\text{out}} \left(\int \frac{d\dot{Q}}{T} \right) + \sum_{\text{out}} \dot{m}s \qquad (1.58)$$

In addition for Equation (1.58) if one assumes a control volume with a single flow entrance and a single flow exit, the mass balance equation becomes $\dot{m}_{\text{out}} = \dot{m}_{\text{in}} = \dot{m}$, thus the EnBE becomes

$$\text{EnBE} : \sum_{\text{in}} \left(\int \frac{d\dot{Q}}{T} \right) + \dot{m}(s_{\text{in}} - s_{\text{out}}) + \dot{S}_{\text{gen}} = \sum_{\text{out}} \left(\int \frac{d\dot{Q}}{T} \right) \qquad (1.59)$$

In the case that the process is adiabatic, there is no heat transfer across the system boundary, therefore, the EnBE simplifies to

$$\text{EnBE} : \dot{m}s_{\text{in}} + \dot{S}_{\text{gen}} = \dot{m}s_{\text{out}} \qquad (1.60)$$

The generated entropy is the sum of entropy change of the system and of its surroundings. There are three relevant cases that can be assumed at heat transfer across a system boundary for determination entropy generation. Consider a thermodynamic system which has a diabatic boundary. As illustrated in Table 1.7, the EnBE for this system is given by the difference between Q/T_0 and Q/T_{sys}. It is assumed in this case that there is no wall with a finite thickness at the system boundary. Therefore, the temperature profile has a sharp change. A more accurate assumption is to assume the existence of a wall at the boundary. In this case there will be a variation of temperature across the wall.

In the second case represented in Table 1.7, the entropy generation has to be calculated by integration accounting of the temperature profile. In the third case, in addition to a wall, one considers the existence of boundary layers at the inner and outer sides of the wall. Therefore, the entropy generation will be the highest in the third case.

1.6.4 Exergy Balance Equation

"*Exergy balance equation*" is a mathematical expression of the SLT statement. The ExBE introduces the term *exergy destroyed*, which represents the maximum work potential that is impossible to recover for useful purpose due to irreversibilities. For a reversible system, there is no exergy destruction because all work generated by the system can be made useful. There is a direct connection between exergy destruction, entropy generation, and the reference temperature of environment, namely

$$Ex_{\text{d}} = T_0 \Delta S_{\text{gen}} \qquad (1.61)$$

where for the condition $S_{\text{gen}} \geq 0$ there are three cases:

TABLE 1.7 Effect of Wall Assumptions Considered at Entropy Associated with Heat Transfer

Case 1: No Wall Effect Considered	Case 2: Wall ΔT Considered	Case 3: Wall and Boundary Layer

$$S_{gen,1} = \frac{Q}{T_0} - \frac{Q}{T_{sys}}$$

$$S_{gen,2} > \frac{Q}{T_0} - \frac{Q}{T_{sys}}$$

$$S_{gen,3} > S_{gen,2} > \frac{Q}{T_0} - \frac{Q}{T_{sys}}$$

Modified from Cengel and Boles (2010).

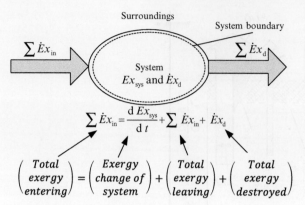

FIGURE 1.16 Explanatory sketch for the exergy balance equation.

- $Ex_d > 0$: irreversible process
- $Ex_d = 0$: reversible process
- $Ex_d < 0$: impossible process.

In Figure 1.16 there is an explanatory sketch to illustrate the incoming and outgoing exergies and exergy destructions and losses. In a thermodynamic system, the total exergy entering the system must be balanced by the total exergy leaving the system plus the change of exergy content of the system plus the exergy destruction. Exergy can be transferred to or from a system by three means: work, heat, and mass. Therefore, the ExBE can be expressed generally in rate form as

$$\text{ExBE}: \sum_{\text{in}} \left[\dot{W} + \dot{m}\varphi + \left(1 - \frac{T_0}{T}\right)\dot{Q} \right] = \frac{dEx}{dt} + \sum_{\text{out}} \left[\dot{W} + \dot{m}\varphi + \left(1 - \frac{T_0}{T}\right)\dot{Q} \right]$$
$$- P_0 \frac{dV_{\text{CV}}}{dt} + \dot{E}x_d \tag{1.62}$$

where $\varphi = (h - h_0) + T_0(s - s_0) + \frac{1}{2}\nu^2 + g(z - z_0) + \text{ex}^{\text{ch}}$ is the total specific exergy. If the flow is steady Equation (1.62) simplifies to

$$\text{ExBE}: \sum_{\text{in}} \left[\dot{W} + \dot{m}\varphi + \left(1 - \frac{T_0}{T}\right)\dot{Q} \right] = \sum_{\text{out}} \left[\dot{W} + \dot{m}\varphi + \left(1 - \frac{T_0}{T}\right)\dot{Q} \right] + \dot{E}x_d \tag{1.63}$$

If there is only a single stream then

$$\text{ExBE}: \sum_{\text{in}} \left[\dot{W} + \left(1 - \frac{T_0}{T}\right)\dot{Q} \right] = \sum_{\text{out}} \left[\dot{W} + \left(1 - \frac{T_0}{T}\right)\dot{Q} \right] + \dot{m}(\varphi_2 - \varphi_1) + \dot{E}x_d \tag{1.63}$$

If one denotes in Equation (1.63) that $\dot{W} = \sum \dot{W}_{\text{out}} - \sum \dot{W}_{\text{in}}$ and $\dot{Q} = \sum \dot{Q}_{\text{in}} - \sum \dot{Q}_{\text{out}}$ and assumes that there is no exergy destroyed, then the reversible work can be obtained as

$$\dot{W}_{\text{rev}} = \dot{m}(\varphi_1 - \varphi_2) + \sum \left(1 - \frac{T_0}{T}\right)\dot{Q} \tag{1.64}$$

Exergy due to work transfer (Ex^W) must be equal with the work, because by definition, exergy represents the maximum potential of a system to perform work

$$Ex^W = W$$

However, if the system has a moving boundary, then the exergy must include work done for boundary displacement; in such a case, the exergy is

$$Ex^W = W - P_0(V - V_0)$$

The exergy associated with mass transfer (Ex^m) is a characteristic only of a flowing system and can be expressed based on mass flow rate and the total specific flow exergy, φ

$$\dot{Ex}^m = \dot{m}\varphi$$

and the exergy due to heat transfer can be expressed based on the Carnot factor according to

$$\dot{Ex}^Q = \int \left(1 - \frac{T_0}{T}\right) d\dot{Q} \tag{1.65}$$

1.7 EFFICIENCY DEFINITIONS

Efficiency is a major criterion for assessment of systems, applications, and processes of any kind. The term efficiency originates mainly from thermodynamics. The attempt of assessing heat conversion into work led to its initial formulation as the "net work generated per total heat energy input." This efficiency criterion is based on the FLT, also called *energy efficiency*. Equation (1.27) expresses the energy efficiency for a heat engine.

The general efficiency expression of a system—as a measure of its performance and effectiveness—is represented by the ratio of useful output per required input

$$\eta = \frac{\text{Useful output}}{\text{Total input}} \tag{1.66}$$

If the system is an energy system then its input and output must be forms of energy. Therefore, for an energy system, energy efficiency is written as

$$\eta = \frac{E_{\text{useful}}}{E_{\text{input}}} \tag{1.67}$$

where E_{useful} is energy delivered in the desired (useful) form (sometimes denoted as energy output or as energy recovered), E_{input} is the energy input (sometimes called expended energy).

Any source of energy is characterized by a maximum potential of doing work. This maximum is the exergy, and, as discussed above, it represents the work generated by reversible processes. This is a measure of perfection because it assumes that there are no irreversibilities. In exergetic view, the efficiency must be the ratio between exergy associated to the useful output and the exergy associated to the input. One writes

$$\psi = \frac{Ex_{\text{useful}}}{Ex_{\text{input}}} = 1 - \frac{Ex_{\text{d,t}}}{Ex_{\text{input}}} \tag{1.68}$$

where Ex_{useful} is the exergy delivered as useful output (or exergy in product outputs or delivered exergy; in some cases this exergy can be also a recovered exergy), Ex_{input} is the exergy in inputs (sometimes this exergy is denoted as the exergy consumed by the system). In Eq. (1.68) the term $Ex_{d,t}$ refers to the so-called "total exergy destruction". The total exergy destruction represents the sum of the exergy destruction within the system ($Ex_{d,sys}$) and the exergy destruction at the interaction between the system and the surroundings ($Ex_{d,surr}$). Here ($Ex_{d,surr}$) represents an exergy lost ($Ex_{d,surr} = Ex_{d,loss}$). In this view, the exergy balance for the overall system is $Ex_{input} = Ex_{products} + Ex_{d,sys} + Ex_{d,loss} = Ex_{products} + Ex_{d,t}$.

Some books (e.g., Cengel and Boles, 2010) define the second law efficiency as the ratio of the actual work delivered as useful to the reversible work which would ideally be delivered under the same operational conditions. Because the reversible work is the highest possible it remains that the second law efficiency must be <1. Second law efficiency of heat engines—or power generation systems—is

$$\psi_{HE} = \frac{\dot{W}_{actual}}{\dot{W}_{rev}} \tag{1.69}$$

Example

Figure 1.17 represents a power generation system. In order to operate the system consumes exergy—denoted on the diagram with Ex_{cons}. Assume for Ex_{cons} to be the exergy of a fuel or high-temperature heat (e.g., concentrated solar radiation, geothermal heat, recovered waste heat). It is assumed further that the power generation system is equipped with a heat exchanger that operates reversibly and has the role of directing the exergy to the heat engine. In many practical power generation systems the role of this heat exchanger is played by the furnace or steam generator. The system is not necessarily capable of using all exergy provided and therefore, as shown in the figure, a part of it must be lost, Ex_{lost} Because the heat exchanger is assumed to be reversible, the ExBE is $Ex_{cons} = Ex_{used} + Ex_{lost}$ (see the figure). Note that here $Ex_{lost} = Ex_{d,surr}$.

There are two possible models for the power generation efficiency. As observed from the manner in which the system boundary is shown (see the figure), the model A assumes that the actual exergy input to the heat engine system is Ex_{used}. Therefore, in this case reversible work which can be potentially delivered by n ideal heat engine is $W_{rev,deliv} = Ex_{used} = Ex_{input}$ or $\dot{W}_{rev,deliv} = \dot{E}x_{cons} - \dot{E}x_{lost}$; thus the exergy efficiency is

$$\psi_{Model\,A} = \frac{\dot{W}_{deliv}}{\dot{E}x_{cons} - \dot{E}x_{lost}}$$

Model B considers a wider system boundary. The exergy balance for all system is in this case $Ex_{cons} = Ex_{deliv} + \sum Ex_d$, where $\sum Ex_d$ is the total exergy destruction. Recall that total exergy destruction is equal to the sum of exergy destruction due to internal irreversibilities ($Ex_{d,sys}$) and the exergy destruction due to the interaction with the environment ($Ex_{d,surr}$). The reversible work associated with Model B is equal to the exergy consumed, $\dot{W}_{rev,deliv} = \dot{E}x_{cons}$. Consequently, for Model B the exergy efficiency is

$$\psi_{Model\,B} = \frac{\dot{W}_{deliv}}{\dot{E}x_{cons}}$$

One notes that exergy efficiency based on Model B assesses the ability of a power generation system to convert the consumed exergy into work, but at the same time it quantifies the

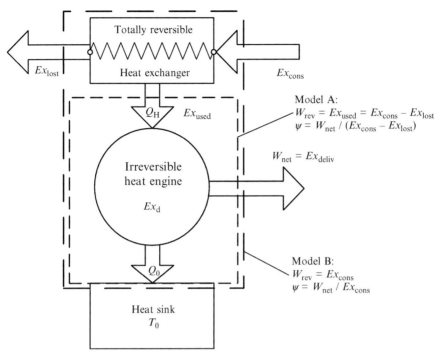

FIGURE 1.17 Thermodynamic model for exergy efficiency of power generation systems.

utilization fraction of the provided exergy. Keeping in mind that $Ex_{d,surr} = Ex_{lost}$ the *exergy utilization factor* can be defined with the following equation:

$$f = 1 - \frac{Ex_{lost}}{Ex_{cons}}$$

The exergy utilization factor takes values between 0 and 1. When $f = 1$ the source exergy is fully utilized by the power generation system; in this case the exergy destruction at system interaction with the surroundings is not existent: $Ex_{d,surr} = Ex_{lost} = 0$. Otherwise there are losses associated due to the interaction of the system with the surroundings. These losses reflect the incapacity of the system to utilize the source exergy integrally. With the help of the exergy utilization factor one obtains the following expression for the exergy efficiency of Model A:

$$\psi_{Model\,A} = \frac{\dot{W}_{deliv}}{f\dot{E}x_{cons}}$$

It is remarked that when $f = 1$ one has $\psi_{Model\,A} = \psi_{Model\,B}$. Exergy efficiency of a heat pump or a refrigerator according to the second law can be expressed as the ratio of COP to reversible work as follows:

$$\psi_{HP} = \frac{COP}{COP_{rev}} = \frac{\dot{W}_{rev,in}}{\dot{W}_{in}} \qquad (1.70)$$

where $\dot{W}_{rev,in}$ represents the work consumption by the reversible heat pump or refrigerator and $\dot{W}_{in}$ is the actual work consumption.

TABLE 1.8 Energy and Exergy Efficiency of Some Important Devices for Power Generation

Device	Equations
1. Turbine 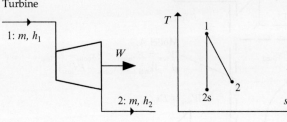	Balance equations: MBE: $\dot{m}_1 = \dot{m}_2 = \dot{m}$ EBE: $\dot{m}_1 h_1 = \dot{W} + \dot{m}_2 h_2$ EnBE: $\dot{m}_1 s_1 + \dot{S}_{gen} = \dot{m}_2 s_2$ ExBE: $\dot{m}[(h_1 - h_2) - T_0(s_1 - s_2)] = \dot{W} + \dot{E}x_d$ Efficiency equations: $\eta = \dfrac{\dot{W}}{\dot{W}_s} = \dfrac{\dot{m}(h_1 - h_2)}{\dot{m}(h_1 - h_{2s})}$ $\psi = \dfrac{\dot{W}}{\dot{E}x_{cons}} = \dfrac{\dot{W}}{Ex_1 - Ex_2} = \dfrac{\dot{m}(h_1 - h_2)}{\dot{m}[h_1 - h_2 - T_0(s_1 - s_2)]}$
2. Compressor 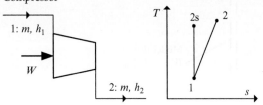	Balance equations: MBE: $\dot{m}_1 = \dot{m}_2 = \dot{m}$ EBE: $\dot{m}_1 h_1 + \dot{W} = \dot{m}_2 h_2$ EnBE: $\dot{m}_1 s_1 + \dot{S}_{gen} = \dot{m}_2 s_2$ ExBE: $\dot{W} = \dot{m}[h_2 - h_1 - T_0(s_2 - s_1)] + \dot{E}x_d$ Efficiency equations: $\eta = \dfrac{\dot{W}_s}{\dot{W}} = \dfrac{\dot{m}(h_{2s} - h1)}{\dot{m}(h_2 - h_1)}$ $\psi = \dfrac{\dot{W}_{rev}}{\dot{E}x_{cons}} = \dfrac{\dot{m}(h_{2s} - h_1)}{\dot{m}[h_1 - h_2 - T_0(s_1 - s_2)]}$
3. Pump 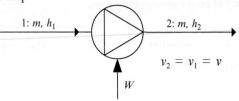	Balance equations: MBE: $\dot{m}_1 = \dot{m}_2 = \dot{m}$ EBE: $\dot{m}_1 h_1 + \dot{W} = \dot{m}_2 h_2$ EnBE: $\dot{m}_1 s_1 + \dot{S}_{gen} = \dot{m}_2 s_2$ ExBE: $\dot{W} = \dot{m}[h_2 - h_1 - T_0(s_2 - s_1)] + \dot{E}x_d$ Efficiency equations: $\eta = \dfrac{\dot{W}_s}{\dot{W}_{cons}} = \dfrac{\dot{m}v(P_2 - P_1)}{\dot{m}(h_2 - h_1)}$ $\psi = \dfrac{\dot{W}_{rev}}{\dot{E}x_{cons}} = \dfrac{\dot{m}[h_2 - h_1 - T_0(s_2 - s_1)]}{\dot{m}(h_2 - h_1)}$
4. Hydraulic turbine 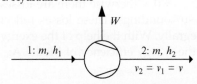	Balance equations: MBE: $\dot{m}_1 = \dot{m}_2 = \dot{m}$ EBE: $\dot{m}_1 h_1 = \dot{W} + \dot{m}_2 h_2$ EnBE: $\dot{m}_1 s_1 + \dot{S}_{gen} = \dot{m}_2 s_2$ ExBE: $\dot{m}[(h_1 - h_2) - T_0(s_1 - s_2)] = \dot{W} + \dot{E}x_d$ Efficiency equations: $\eta = \dfrac{\dot{W}}{\dot{W}_s} = \dfrac{\dot{m}(h_1 - h_2)}{\dot{m}v(P_1 - P_2)}$ $\psi = \dfrac{\dot{W}}{\dot{E}x_{cons}} = \dfrac{\dot{m}(h_1 - h_2)}{\dot{m}[h_1 - h_2 - T_0(s_1 - s_2)]}$
5. Diffuser and Nozzle 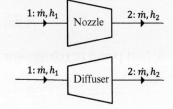	Balance equation: MBE: $\dot{m}_1 = \dot{m}_2 = \dot{m}$ EBE: $\dot{m}_1 h_1 = \dot{m}_2 h_2$ EnBE: $\dot{m}_1 s_1 + \dot{S}_{gen} = \dot{m}_2 s_2$ ExBE: $\dot{m}[(h_1 - h_2) - T_0(s_1 - s_2)] = \dot{E}x_d$ Efficiency equations: $\eta_{diffuser} = \dfrac{h_1 - h_{2s}}{h_1 - h_2} \quad \eta_{nozzle} - \dfrac{(\eta_1 - \eta_2)}{(\eta_1 - \eta_{2s})}$ and $\psi = 0$

In Table 1.8 efficiency formulations for some important devices for power generation are presented. The first device listed in the table is the turbine (#1), which is a work-producing apparatus operating on the principle of expansion of a fluid that eventually generates useful work. Although in practice there are heat losses from turbine shell, these are minor with respect to work generation. A high enthalpy flow enters the turbine, work is produced, and a lower enthalpy flow exits the turbine.

Isentropic efficiency is one of the most frequently used assessment parameters for turbines. Isentropic efficiency (η_s) is defined by the ratio of actual power generated to the power generated during an isentropic expansion. For an isentropic expansion there is no entropy generation; the turbine operation is reversible. Therefore, isentropic efficiency is a relative measure of alleviation from thermodynamic ideality. Exergy efficiency of a turbine is defined as the ratio of generated power to rate of exergy consumed. The rate of exergy consumed is given by

$$\dot{Ex}_{cons} = \dot{m}[h_1 - h_2 - T_0(s_1 - s_2)]$$

In the Table 1.8, the expansion process in the $T - s$ diagram is also represented. Process 1–2 is the actual process, while process 1–2s is the reversible process (isentropic).

A compressor is a device used to increase the pressure of a fluid at the expense of work consumption. Compressors are typically assessed by the isentropic efficiency which, for

TABLE 1.8   Energy and Exergy Efficiency of Some Important Devices for Power Generation—Cont'd

Device	Equations
6. Heat exchanger	Balance equations: MBE : $\dot{m}_{1h} = \dot{m}_{2h} = \dot{m}_h$ and $\dot{m}_{1c} = \dot{m}_{2c} = \dot{m}_c$ EBE : $\dot{m}_h(h_{1h} - h_{2h}) = \dot{m}_c(h_{2c} - h_{1c})$ EnBE : $\dot{S}_{gen} = \dot{m}_c(h_{2c} - h_{1c}) + \dot{m}_h(h_{2h} - h_{1h})$ ExBE : $\dot{m}_h[(h_{1h} - h_{2h}) - T_0(s_{1h} - s_{2h})]$ $\quad\quad + \dot{m}_c[(h_{1c} - h_{2c}) - T_0(s_{1c} - s_{2c})] = \dot{Ex}_d$ Efficiency equations: $\varepsilon = \dfrac{\dot{Q}_{cold}}{\dot{Q}_{max}} = \dfrac{(\dot{m}C_p\Delta T)_{cold}}{(\dot{m}C_p)_{min}(T_{1h} - T_{2h})}$ $\eta = \dfrac{\dot{Q}_{cold}}{\dot{Q}_{hot}}; \psi = \dfrac{(\dot{Ex}_{2c} - \dot{Ex}_{1c})_{cold}}{(\dot{Ex}_{1h} - \dot{Ex}_{2h})_{hot}}$ $\psi = \dfrac{\dot{m}_{cold}[h_{2c} - h_{1c} - T_0(s_{2c} - s_{1c})]}{\dot{m}_{hot}[h_{2h} - h_{1h} - T_0(s_{2h} - s_{1h})]}$
7. Mixing chamber or chemical reactor	Balance equations: MBE : $\dot{m}_1 + \dot{m}_2 = \dot{m}_3$ EBE : $\dot{m}_1 h_1 + \dot{m}_2 h_2 = \dot{m}_3 h_3$ EnBE : $\dot{m}_1 s_1 + \dot{m}_2 s_2 + \dot{S}_{gen} = \dot{m}_3 s_3$ ExBE : $\dot{m}_1 ex_1 + \dot{m}_2 ex_2 = \dot{m}_3 ex_3 + \dot{Ex}_d$ Efficiency equations: $\eta = \dfrac{(\dot{m}_1 + \dot{m}_2)h_3}{\dot{m}_1 h_1 + \dot{m}_2 h_2}$ $\psi = \dfrac{\dot{m}_2[h_3 - h_2 - T_0(s_3 - s_2)]}{\dot{m}_1[h_1 - h_3 - T_0(s_1 - s_3)]}$

the case of compressors, is the ratio of isentropic work to actual work. Balance and efficiency equations are listed in Table 1.8 (#2).

Pumps are organs used to increase the pressure of liquids at the expense of work input. The liquid is incompressible, therefore the power required to pump the liquid for a reversible process is

$$\dot{W}_s = \dot{m}v(P_2 - P_1)$$

Hydraulic turbines are devices that generate work from the potential energy of a liquid. The efficiency equations of hydraulic turbines are presented in Table 1.8 (#3). Denote with v the specific volume of the liquid. The reversible work generated by a hydraulic turbine is

$$\dot{W}_s = \dot{m}v(P_1 - P_2)$$

Nozzles and diffusers are adiabatic devices used to accelerate or decelerate a fluid, respectively. The balance equations for nozzles are indicated in Table 1.8 (#5). The exergy efficiency of nozzles is nil because they produce no work, although the expanded flow has work potential.

A heat exchanger is a device which facilitates heat transfer between two fluids without mixing. It is known that heat exchangers are assessed for effectiveness by the ratio of the actual amount of heat transfer to the maximum amount of heat able to be transferred. In Table 1.8 the heat exchanger effectiveness is denoted with ε. The maximum heat transfer rate is given by

$$\dot{Q}_{max} = \left(\dot{m}C_p\right)_{min}(T_{1h} - T_{2h})$$

For a heat exchanger the exergy source is derived from the hot fluid which during the process reduces its exergy. The exergy of the cold fluid represents the delivered exergy as the useful product of a heat exchanger. Regarding the energy exchange, ideally if there are no energy losses, all energy from the hot fluid is transferred to the cold fluid. However, some loses are unavoidable in practical systems; therefore one can define an energy efficiency of heat exchangers as the ratio of energy delivered to energy consumed. Regarding the exergy efficiency of a heat exchanger, this is given by exergy retrieved from the cold fluid divided by exergy provided by the hot fluid.

Many practical devices are used to mix streams. Mixing chambers accept multiple-stream inputs and have one single output. Mixers can be isothermal, or one can mix a hot fluid with a cold fluid, etc. A combustion chamber or a reaction chamber can be modeled from a thermodynamic point of view as a mixing device, and mixing is accompanied by chemical reaction. Let us assume a mixer with two inlet ports (see Table 1.8, 1 and 2) and one outlet port (called 3). Furthermore, one assumes that 1 is a hot stream and 2 a cold stream. Thus, the rate of exergy consumed by the system is

$$\dot{Ex}_{cons} = \dot{m}_1[h_1 - h_3 - T_0(s_1 - s_3)]$$

while the exergy delivered is given by

$$\dot{Ex}_{deliv} = \dot{m}_2[h_3 - h_2 - T_0(s_3 - s_2)]$$

There are many other devices which can be studied in a similar manner using balance equations and formulating process efficiency based on the first and second laws. One interesting case is that of electric heaters. Ideally, the energy efficiency of an electric heater is 100% because electric power can be completely converted in useful heat. However, the exergy efficiency of the electric heater must account for heat transfer processes due to the temperature difference between the hot wire and the surroundings. This introduces an exergy destruction or entropy generation.

Systems that use fuel as energy input must be assessed based on heating value and chemical exergy. For example the efficiency of a fuel-based heating device is

$$\eta = \frac{\dot{Q}_{\text{deliv}}}{\dot{m}_{\text{fuel}}\text{HHV}} \quad \text{and} \quad \psi = \frac{\dot{E}x^{\text{Q}}_{\text{deliv}}}{\dot{m}_{\text{fuel}}\text{ex}^{\text{ch}}}$$

where $\dot{Q}_{\text{deliv}}$ is the rate of heat delivered as useful product, HHV is the higher heating value, ex^{ch} is the chemical exergy of the fuel, $\dot{m}_{\text{fuel}}$ is the mass flow rate of the fuel, and $\dot{E}x^{\text{Q}}_{\text{deliv}}$ is the exergy associated with the delivered heat.

An important class of processes is found in psychrometrics, the science that studies the properties of moist air and is of great significance in many industrial sectors. Many processes of heating, cooling, and humidification, building insulation and roofing, and the stability, deformation, and fire resistance of building materials require psychrometrics analysis. In Table 1.9 the most important notions and properties of psychrometrics are described.

TABLE 1.9   Important Notions and Properties of Psychrometrics

Notion	Definition
Dry air	Purified atmospheric air without moisture (water vapors are removed)
Moist air	A binary mixture of dry air and water vapor; the mass fraction of water vapor can vary from nearly zero to 20 g H_2O per kg dry air depending on temperature and pressure
Saturated air	Moist air with water vapor at saturation condition at specified temperature and pressure
Dew point temperature, T_d	Temperature at which water vapor condenses when moist air is cooled at constant pressure
Dry bulb temperature, T_b	Actual temperature of moist air
Wet bulb temperature, T_w	Temperature corresponding to saturation at the actual partial pressure of water in moist air
Adiabatic saturation	An adiabatic process of humidification of moist air by adding water vapor until relative humidity becomes 100%
Adiabatic saturation temperature, T_s	Temperature obtained during an adiabatic saturation process when relative humidity reaches 100% at a given pressure
Vapor pressure, P_v	Partial pressure of water vapor in moist air: $P_v = P - P_a$, where P_a is the partial pressure of dry air
Relative humidity, φ	Mole fraction of water vapor in moist air divided by mole fraction in saturated air at the same temperature and pressure. Because air is ideal gas, mole fractions equal to partial pressures divided by total pressure. Therefore: $\varphi = P_v/P_s$, where P_s is saturation pressure of water at same temperature
Humidity ratio, ω	Mass ratio of water vapor to the amount of dry air contained in a given volume of moist air. By definition, $\omega = m_v/m_a = 0.622 P_v/P_a$. Note also that $\omega P_a = 0.622\varphi P_s$
Degree of saturation, x_s	Actual humidity ratio divided by humidity ratio of saturated air at same temperature and pressure

The total flow exergy of humid air may be expressed as a function of specific heat, as indicated in the past work of Wepfer et al. (1979)

$$\frac{ex}{T_0\left(C_{p,a} + \omega C_{p,v}\right)} = \left(\frac{T}{T_0} - 1 - \ln\frac{T}{T_0}\right) + \frac{(1+\widetilde{\omega})R_a \ln\frac{P}{P_0}}{C_{p,a} + \omega C_{p,v}} + \frac{R_a\left[(1+\widetilde{\omega}) \ln\frac{1+\widetilde{\omega}_0}{1+\omega} + \widetilde{\omega} \ln\frac{\widetilde{\omega}}{\widetilde{\omega}_0}\right]}{C_{p,a} + \omega C_{p,v}}$$

where $C_{p,a}$, $C_{p,a}$ are the specific heat of dry air and water vapor, respectively, R_a is the gas constant for air, and $\widetilde{\omega} = 1.608\omega$.

Balance equations are useful for evaluating efficiency of psychrometric processes. Five elementary processes will be analyzed next: sensible heating (or cooling), heating with humidification, cooling with dehumidification, evaporative cooling and adiabatic mixing. These processes are represented in the Molier diagram in Figure 1.18. The balance equations for each process in the process schematic are given in Table 1.10.

The first and simplest psychrometric process is 1–2, represented on the Molier diagram in Figure 1.18, which is a process of sensible heating. The reverse process 2–1 represents sensible cooling and has a similar thermodynamic treatment. In sensible heating process, heat is added to moist air and produces an increase of air temperature. During this process the mass of air and mass of water vapor do not change. Therefore the humidity ratio remains constant. The dry bulb temperature of air changes; consequently, the relative humidity must change. The energy input into the system is represented by the heat rate $\dot{Q}_{in}$. The useful gain of the process can be considered the amount of heat added to the air amounting to $\dot{m}_2 h_2 - \dot{m}_1 h_1$,

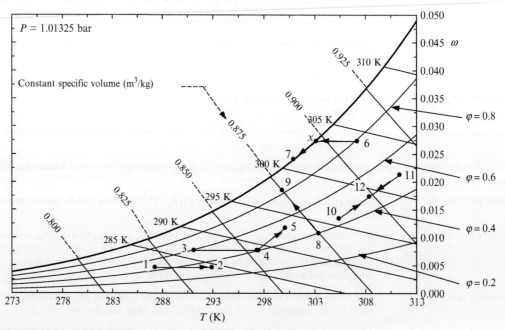

FIGURE 1.18 Molier diagram of moist air [generated with EES software, EES (2013)].

TABLE 1.10 Energy and Exergy Analysis of Fundamental Psychrometric Processes

Process and Diagram	Balance Equations
Process 1–2: sensible heating	MBE dry air: $\dot{m}_1 = \dot{m}_2$ MBE water vapor: $\dot{m}_{w1} = \dot{m}_{w2}$ EBE: $\dot{Q}_{in} + \dot{m}_1 h_1 = \dot{m}_2 h_2$ EnBE: $\dfrac{\dot{Q}_{in}}{T} + \dot{m}_1 s_1 + \dot{S}_{gen} = \dot{m}_2 s_2$ ExBE: $\dot{Q}_{in}\left(1 - \dfrac{T_0}{T}\right) + \dot{m}_1 ex_1 = \dot{m}_2 ex_2 + \dot{E}x_d$ where $ex = (h - h_0) - T_0(s - s_0)$
Process 3–4–5: heating with humidification	MBE dry air: $\dot{m}_3 = \dot{m}_4 = \dot{m}_5$ MBE for water vapor: $\dot{m}_{w3} + \dot{m}_{w4} = \dot{m}_{w5}$ where $\dot{m}_{w4} = \dot{m}_w$ (see diagram) EBE: $\dot{Q}_{in} + \dot{m}_3 h_3 + \dot{m}_w h_w = \dot{m}_5 h_5$ EnBE: $\dfrac{\dot{Q}_{in}}{T} + \dot{m}_3 s_3 + \dot{m}_w s_w + \dot{S}_{gen} = \dot{m}_5 s_5$ ExBE: $\dot{Q}_{in}\left(1 - \dfrac{T_0}{T}\right) + \dot{m}_3 ex_3 + \dot{m}_w ex_w = \dot{m}_5 ex_5 + \dot{E}x_d$ where $ex = (h - h_0) - T_0(s - s_0)$
Process 6–x–7: cooling with dehumidification	MBE dry air: $\dot{m}_6 = \dot{m}_7$ MBE water vapor: $\dot{m}_6 \omega_6 = \dot{m}_7 \omega_7 + \dot{m}_w$ EBE: $\dot{m}_6 h_6 = \dot{Q}_{out} + \dot{m}_w h_w + \dot{m}_7 h_7$ EnBE: $\dot{m}_6 s_6 + \dot{S}_{gen} = \dfrac{\dot{Q}_{out}}{T} + \dot{m}_w s_w + \dot{m}_7 s_7$ ExBE: $\dot{m}_6 ex_6 = \dot{Q}_{in}\left(1 - \dfrac{T_0}{T}\right) + \dot{m}_w ex_w + \dot{m}_7 ex_7 + \dot{E}x_d$ where $ex = (h - h_0) - T_0(s - s_0)$
Process 8–9: evaporative cooling	MBE dry air: $\dot{m}_8 = \dot{m}_9$ MBE water vapor: $\dot{m}_8 \omega_8 + \dot{m}_w = \dot{m}_9 \omega_9$ EBE: $\dot{m}_8 h_8 + \dot{m}_w h_w = \dot{m}_9 h_9$ EnBE: $\dot{m}_8 s_8 + \dot{m}_w h_w + \dot{S}_{gen} = \dot{m}_9 s_9$ ExBE: $\dot{m}_8 ex_8 + \dot{m}_w ex_w = \dot{m}_9 s_9 + \dot{E}x_d$ where $ex = (h - h_0) - T_0(s - s_0)$
Process 10–11–12: adiabatic mixing	MBE dry air: $\dot{m}_{10} + \dot{m}_{11} = \dot{m}_{12}$ MBE water: $\dot{m}_{10} \omega_{10} + \dot{m}_{11} \omega_{11} = \dot{m}_{12} \omega_{12}$ EBE: $\dot{m}_{10} h_{10} + \dot{m}_{11} h_{11} = \dot{m}_{12} h_{12}$ EnBE: $\dot{m}_{10} s_{10} + \dot{m}_{11} s_{11} + \dot{S}_{gen} = \dot{m}_{12} s_{12}$ ExBE: $\dot{m}_{10} ex_{10} + \dot{m}_{11} ex_{11} = \dot{m}_{12} ex_{12} + \dot{E}x_d$ where $ex = (h - h_0) - T_0(s - s_0)$

Note: state numbers for processes are in correspondence with process representation in Molier diagram shown in Figure 1.18.

which, according to EBE, is the same as $\dot{Q}_{in}$. Therefore, the energy efficiency of sensible heating of air

$$\eta_{1-2} = \frac{\dot{m}_2 h_2 - \dot{m}_1 h_1}{\dot{Q}_{in}}$$

is 100% provided that the system is well insulated and there are no heat losses to the environment. Another view on efficiency is the useful output represented by the total enthalpy of the heated stream $\dot{m}_2 h_2$ while the consumed input is the sum of total enthalpy of the input stream and the heat added to the system, $\dot{Q}_{in} + \dot{m}_1 h_1$. Based on this thinking, energy efficiency for process 1–2 is

$$\eta_{1-2} = \frac{\dot{m}_2 h_2}{\dot{Q}_{in} + \dot{m}_1 h_1}$$

The exergy efficiency formulations for the sensible heating of moist air process correspond to the above two definition variants for energy efficiency, namely

$$\psi_{1-2} = \frac{\dot{m}_2 ex_2 - \dot{m}_1 ex_1}{\dot{Q}_{in}(1 - T_0/T)}$$

corresponding to the first variant of energy efficiency, and

$$\psi_{1-2} = \frac{\dot{m}_2 ex_2}{\dot{Q}_{in}(1 - T_0/T) + \dot{m}_1 ex_1}$$

corresponding to the second variant of energy efficiency definition.

The process 3–4–5 schematically given in Table 1.10 represents heating with humidification of moist air. In this process, humidity in form of saturated water vapor is added to the air stream simultaneously with a process of heat addition. From a thermodynamic point of view, this process can be subdivided in two sub-processes 3–4 and 4–5. The process 3–4 is a process of sensible heating evolving at constant humidity ratio. In the subsequent process, 4–5, water vapor is added and the humidity ratio increases. Based on the first type of efficiency formulation presented above, the energy and exergy efficiency of the process becomes

$$\eta_{3-4-5} = \frac{\dot{m}_5 h_5 - \dot{m}_3 h_3}{\dot{Q}_{in} + \dot{m}_w h_w} \quad \text{and} \quad \psi_{3-4-5} = \frac{\dot{m}_5 ex_5 - \dot{m}_3 ex_3}{\dot{Q}_{in}(1 - T_0/T) + \dot{m}_w ex_w}$$

whereas the energy and exergy efficiency of the same process when the total enthalpy of heated and humidified air is considered the output and the total enthalpy of of the stream in #3 is considered one of the system inputs, becomes

$$\eta_{3-4-5} = \frac{\dot{m}_5 h_5}{\dot{Q}_{in} + \dot{m}_w h_w + \dot{m}_3 h_3} \quad \text{and} \quad \psi_{3-4-5} = \frac{\dot{m}_5 ex_5}{\dot{Q}_{in} + \dot{m}_w ex_w + \dot{m}_3 ex_3}$$

The third psychrometric process given in Table 1.10 is cooling with dehumidification, process 6–x–7 (see also the process represented in Figure 1.18). In this process air is cooled to wet bulb temperature by putting it in contact with a colder surface with maintenance of a certain heat transfer rate to extract $\dot{Q}_{out}$ from the air stream. After water condenses, air temperature drops and the condensate is separated gravitationally. The useful effect in a cooling and de-humidification process must be equal to heat output plus the total enthalpy of air at the exit. The energy consumed may be considered to be the total enthalpy of the air stream at the input. Therefore one has

$$\eta_{6-x-7} = \frac{\dot{Q}_{out} + \dot{m}_7 h_7}{\dot{m}_6 h_6} \quad \text{and} \quad \psi_{6-x-7} = \frac{\dot{Q}_{out}(1 - T_0/T) + \dot{m}_7 ex_7}{\dot{m}_6 ex_6}$$

The evaporative cooling process is based on spraying water through a sprinkler system to produce cooling. Thus the humidity ratio and the relative humidity both increase. The useful output can considered the total enthalpy of the output stream, whereas the consumed energy is a summation of the total enthalpy of the input stream and the total enthalpy of injected water vapors. Based on this definition, the energy and exergy efficiencies of the evaporative cooling process are as follows:

$$\eta_{8-9} = \frac{\dot{m}_9 h_9}{\dot{m}_8 h_8 + \dot{m}_w h_w} \quad \text{and} \quad \psi_{8-9} = \frac{\dot{m}_9 ex_9}{\dot{m}_8 ex_8 + \dot{m}_w ex_w}$$

The last process included in Table 1.10 is that of adiabatic mixing of two streams of humid air. This process is denoted 10–11–12 and can be also observed on Molier diagram represented in Figure 1.18. In this adiabatic process the resulting stream (#12) carries the total enthalpy of the two mixing streams (#10 and #11). Thus, it appears logical to consider that the total enthalpy of the stream in #12 is the output, whereas the total enthalpy sum of the input streams $\dot{m}_{10} h_{10} + \dot{m}_{11} h_{11}$ represents the energy input. The energy efficiency is

$$\eta_{10-11-12} = \frac{\dot{m}_{12} h_{12}}{\dot{m}_{10} h_{10} + \dot{m}_{11} h_{11}} \equiv 1$$

As the process is adiabatic according to the EBE, the total enthalpy in the output stream is equal to the total enthalpy sum for the input streams. Therefore, the energy efficiency of this process, provided that there are no heat losses, must be 100%. Regarding the exergy efficiency, this is defined by

$$\psi_{10-11-12} = \frac{\dot{m}_{12} ex_{12}}{\dot{m}_{10} ex_{10} + \dot{m}_{11} ex_{11}}$$

which must always be smaller than 1 due to the inherent exergy destruction.

1.8 CONCLUDING REMARKS

In this chapter the fundamental concepts of thermodynamics are reviewed, with a focus on modeling power generation systems. In the modern world there is a continuously increasing demand for electric power as a basic commodity to satisfy many societal needs. Thermodynamics is the basic science concerned with conversion of thermal energy, fuel energy, nuclear energy, or other forms of energy into useful motive power. Because there are not endless resources it is important to generate power with highest efficiency. Another aspect regards the environment. Traditionally, major coal-based power plants intensely pollute the environment by emitting greenhouse gases and other pollutants. Novel, advanced power generating systems are in development, such as fuel cells, capable of working with less pollution and higher efficiency.

Regardless of the system type, thermodynamics analysis is the most basic method that can help in system assessment, modeling, design, irreversibility, and pollution reduction, etc. The most relevant thermodynamic concepts are presented at the beginning of the chapter. The equations of state, including that of ideal gas, Van der Waals, and the principle of corresponding states are extensively explained and illustrated with the help of diagrams calculated for real substances. The laws of thermodynamics are exposed in detail and the concepts of enthalpy, Gibbs free energy, and internal energy are defined rigorously. A section is dedicated to the concept of exergy and examples are given for calculation of the chemical exergy of compounds.

Thermodynamic analysis based on the first and second laws is a modern method used for thermal design and optimization which uses four types of balance equations: mass, energy, entropy, and exergy. These are explained in the text, as is the method of assessing exergy destruction during processes. One of the sections is dedicated to the notion of efficiency and its mathematic formulations for relevant cases. Both energy and exergy efficiencies are treated. Efficiency formulations are explained for most important engineered devices such as turbines, compressors, pumps, chemical reactors, and psychrometric processes.

Study Problems

1.1. Explain the difference between flow work and boundary work.

1.2. Explain the difference between closed and open thermodynamic systems.

1.3. What is the relationship between temperature and kinetic energy of molecules?

1.4. What is the difference between thermal and caloric equation of state?

1.5. Explain the principle of corresponding states? How much is the value of compressibility factor for ideal gas?

1.6. Derive the equation for mechanical work during an isothermal transformation of ideal gas.

1.7. Explain the difference between internal energy and enthalpy.

1.8. Define the notion of adiabatic process and explain the role of expansion coefficient for determining the work exchange during this type of process.

1.9. What is the difference between externally reversible and internally reversible processes?

1.10. What is the relationship between entropy change of a system, of its surroundings and of generated entropy?

1.11. Using thermodynamic tables or EES software calculate the thermomechanical exergy of saturated steam at 16 MPa. Assume standard values for temperature and pressure of the environment.

1.12. A cistern vehicle transports 20 t of propane and runs with 80 km/h, being exactly in the top of a hill of 100 m altitude. Calculate the total exergy stored in the cistern.

1.13. Stack gas of a power plant comprises 15% carbon dioxide by volume and the rest is made of other gaseous combustion products. The pressure of stack gas is 1.5 atm and the temperature 425 K. Calculate the amount of reversible work needed to separate carbon dioxide.

1.14. Calculate chemical exergy of octane based on standard exergy of the elements.

1.15. Write balance equations for a building wall assuming that exterior temperature is 5 °C, interior temperature is 25 °C, the wall thermal conductivity is 1 W/mK, the heat transfer coefficient is 10 W/m^{2}K on both sides. Calculate the entropy generation and exergy destruction per unit of wall surface.

1.16. Does an exergy analysis replace an energy analysis? Describe any advantages of exergy analysis over energy analysis.

1.17. Noting that heat transfer does not occur without a temperature difference and that heat transfer across a finite temperature difference is irreversible, is there such a thing as reversible heat transfer? Explain.

1.18. Write mass, energy, entropy and exergy balances for the following devices: (a) an adiabatic steam turbine, (b) an air compressor with heat loss from the air to the surroundings, (c) an adiabatic nozzle, and (d) a diffuser with heat loss to the surroundings

1.19. A turbine expands superheated steam at 10 MP having a superheating degree of 50 K. The expansion ratio is 2.5. The isentropic efficiency of the turbine is 0.8. Write balance equations and calculate exergy efficiency of the turbine.

1.20. In a combustion chamber natural gas is supplied at 10 atm while air comes in an excess ratio of 5. Assume a complete combustion of natural gas modeled as methane. Write balance equation and calculate the first and second law efficiency and exergy destruction of the combustion chamber.

1.21. Consider adiabatic mixing of 10 g/s moist air at 25 °C with relative humidity of 60% with 2 g/s of moist air at 10 °C with relative humidity of 40%. Write balance equation and determine the temperature and relative humidity of air for the mixed stream, entropy generation, exergy destruction and second law efficiency for the mixer. EES software or a Molier diagram can be used.

Nomenclature

A	area, m^2
a	acceleration, m/s^2
C	specific heat, kJ/kgK
COP	coefficient of performance
DOF	degree of freedom
$\dot{E}$	energy rate, kW
e	specific total energy, kJ/kg
$\dot{Ex}$	total specific exergy, kW
ex	specific exergy, kJ/kg
F	force, N
g	Gibbs free energy, kJ/kg
g	gravity acceleration, m/s^2
h	specific enthalpy, kJ/kg
KE	kinetic energy, kJ
k_B	Boltzmann constant
l	displacement, m
M	molecular mass, kg/kmol
m	mass, kg
N	number of particles

n	number of moles, mol
N_A	number of Avogadro, molecules/kmol
P	pressure, Pa
p	probability
PE	potential energy, kJ
Q	heat, kJ
$\dot{Q}$	heat rate, kW
q	mass specific heat, kJ/kg
$\mathcal{R}$	universal gas constant, kJ/kmolK
R	gas constant, kJ/kgK
S	entropy, J/K
s	specific entropy, kJ/kgK
$\dot{S}$	entropy rate, W/K
T	temperature, K
t	time, s
U	internal energy, kJ
u	specific internal energy, kJ/kg
V	volume, m^3
u	velocity, m/s
v	specific volume, m^3/kg
W	work, J
$\dot{W}$	work rate, kW
w	mass specific work, kJ/kg
x	vapor quality
Z	compressibility factor
z	elevation, m

Greek letters

η	energy efficiency
γ	adiabatic expansion coefficient
μ	chemical potential, kJ/kg
ω	humidity ratio
ϕ	total specific flow energy, kJ/kg
φ	relative humidity
Ψ	exergy efficiency
θ	total specific non-flow energy, kJ/kg

Subscripts

0	reference state
b	boundary work
C	cold
ch	chemical
conc	concentration
deliv	delivered
f	flow work
gen	generated
H	hot
in	input
ke	kinetic energy
L	low
loss	loss

nf	non-flow
out	output
p	constant pressure
pe	potential energy
ph	physical
r	reduced value
rev	reversible
surr	surroundings
sys	system
u	useful work
v	constant volume

Superscripts

$'$	saturated liquid
$''$	saturated vapor
ch	chemical

References

Cengel, Y.A., Boles, M.A., 2010. Thermodynamics: An Engineering Approach 7th edition. McGraw Hill, New York.

Carnot, S., 1824. Réflexions sur la puissance motrice du feu et sur les machines propres à développer cette puissance. Bachelier, Paris.

Dincer, I., Rosen, M.A., 2013. Exergy: Energy, Environment and Sustainable Development. Elsevier, Oxford, UK.

EES, 2013. Engineering Equation Solver. F-Chart Software, (developed by S.A. Klein), Middleton, WI.

Haynes, W.M., Lide, D.R., 2012. CRC Handbook of Chemistry and Physics, 92nd ed. CRC Press, New York Internet version.

ISU, 2006. The International System of Units (SI), eight ed. Inter-governmental Organization for the Metre Convention, Paris.

Rivero, R., Grafias, M., 2006. Standard chemical exergy of elements updated. Energy. 31, 3310–3326.

Wepfer, W.J., Gaggioli, R.A., Obert, E.F., 1979. Proper evaluation of available energy for HVAC. ASHRAE Trans. 85, 214–230.

$\dot{m}$ mass flow
out
p   hydraulic pressure
pe   potential energy
Ph   physical
r   reduced coordinate
rev   generated
surr   surroundings
sys   system
u   useful work
   total volume

Superscripts

   physical liquid
   saturated vapor
ch   chemical

References

Ghiaasiaan, S.M., 2011. Thermodynamics: An Engineering Approach. ... Oxford, UK.
Borgnakke, C., ... and Sonntag, R.E. Fundamentals of ... Oxford, UK.
Moran, M.J., ... 2012. Principles of Engineering Thermodynamics. ... Oxford, UK.
Smith, J.M., Van Ness, H.C., Abbott, M.M., 2005. Introduction to Chemical Engineering Thermodynamics, 2nd ed. CRC Press, New York.
Wark, K., Richards, D.E., ... 1999. Thermodynamics. McGraw-Hill, New York.
Rosen, M.J., Dincer, I., 1999. Exergy analysis of heating update. Energy, 24, 548-558.
Woudstra, N., Stougie, L.M., Dijkema, ... 1994. Exergy evaluation of available energy for HVAC. ASHRAE Trans. pp. 304-320.

Energy, Environment, and Sustainable Development

2.1 INTRODUCTION

Energy, environment, and sustainable development are three important and interrelated topics confronted by the modern world. World population continuously increases, with 9 billion people projected for 2050. Population increase necessarily implies increasing demand for commodities such as energy. Furthermore, higher energy consumption leads to an increased level of environmental pollution with negative consequences for present and future generations. Pollution related to anthropogenic activity creates intergenerational issues, as does the massive consumption of finite traditional resources such as fossil and nuclear fuels. It is ethical to assure access of future generations to clean air, land, and water and to food, energy, and knowledge with sustainable development.

Energy is extremely significant because it directly affects economic development and generation of wealth. National energy policies and strategies are required to maintain a balance between energy supply and demand. Local policies in the energy sector must consider environmental impact, technological development, economic performance, consumer preference, and integration with the world energy market. There are six essential factors that have to be addressed with sustainable energy solutions, namely, (1) efficiency, (2) cost effectiveness, (3) resources use, (4) design and analysis, (5) energy security, and (6) environment. These are detailed below:

(1) Better efficiency is a critical element. There is a strong connection between energy efficiency, environmental impact, and energy resource depletion because less consumption and less pollution are associated with increased energy efficiency.

(2) Better cost effectiveness appears to be one of the key tasks. The cost effectiveness of energy solutions is related directly and indirectly to sustainability. Cost effectiveness means more savings or reduced expenses for the same services or products. More cost effective solutions are necessary to ensure reduced environmental impact and increased sustainability.

(3) Better resource utilization is related to energy conservation and refers to the assembly of measures leading to the rational use of resources and energy conservation. Energy conservation leads to the stabilization of energy demand, reduced resource consumption, and decreased environmental impact, enhancing sustainability.

(4) Better design and analysis have a direct impact on technology development, innovation, and knowledge enhancement. This leads to a better understanding of processes, as well as identification of irreversibilities, energy loss, and potential system improvements.

(5) Better energy security implies development of energy policies and geopolitical strategies that will eventually assure equitable access to energy resources and therefore increased sustainability and a cleaner environment.

(6) Better environment is a basic human desire. Development of energy systems must take this desire into account. This stimulates the development of advanced power generation systems capable of producing multiple outputs with lower environmental impact than conventional processes.

In this chapter, Earth's main energy sources, the main problems related to environment pollution caused by power generation and consumption, and sustainable development of energy systems are analyzed. Some ways to improve and "greenize" energy systems are discussed, as are energy policy and systems assessment tools for determining environmental impact and sustainability.

2.2 ENERGY RESOURCES AVAILABLE ON EARTH

An *energy resource* is any form of energy available on Earth which can be converted into a useful form such as electrical power, mechanical power, or heat. Figure 2.1 provides a simple classification of energy resources as renewable energies, fossil fuels, and nuclear fuels. Energy resources include substances stored in the earth's crust (fossil fuels and nuclear fuels), flows such as wind or water currents, lakes storing the potential energy of water (hydro-energy), thermal energy from the earth's crust, electromagnetic radiation from the sun (sunlight), and gravitational forces from the interaction of the earth, the moon, and the sun.

The most important energy resource on earth is solar energy, which is generated by thermonuclear reactions in the sun. Solar energy impacts the earth's surface after traveling a

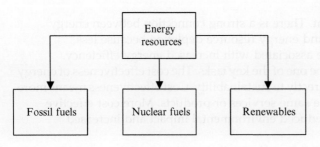

FIGURE 2.1 Classification of energy resources.

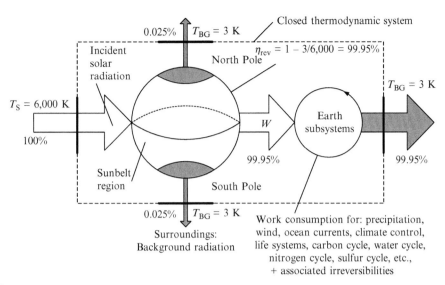

FIGURE 2.2   Thermodynamic model of the earth as a reversible heat engine coupled to various work-consuming earth subsystems.

distance of approximately 150×10^9 m. The sun's surface has a temperature close to 6,000 K and emits electromagnetic radiation in a broad spectrum, ranging from upper ultraviolet to far infrared. Solar radiation impacts the earth from the Arctic to Antarctica, reaching maximum intensity over the Sunbelt region. In polar regions solar radiation is present only about six months per year, at a lower intensity. Note that the earth is connected to the surrounding universe by two "heat exchangers": the Sunbelt on the "hot side" and the polar regions on the "cold side."

A thermodynamic model for converting solar energy on earth is presented in Figure 2.2. The earth is presented as a reversible heat engine that drives earth subsystems which consume entirely the useful work and generate irreversibility. According to the figure, the thermodynamic system involves solar radiation which crosses the system boundary at a temperature $T_S = 6,000$ K and the background radiation of the universe, which has a temperature of approximately $T_{BG} = 3$ K. The Sunbelt region acts as the source side heat exchanger, while the two polar regions act as sink side heat exchangers which reject heat to the background (see also Bejan and Lorente, 2006). According to Carnot, the efficiency of the reversible heat engine is $\eta_{rev} = 1 - 3/6000 = 99.95\%$. This means that 99.95% energy input received in the form of incident radiation can be converted into work. Because the earth's climate is steady, no energy is accumulated, and all produced work must be dissipated and rejected into the surroundings. In the model, the work is dissipated by the earth's subsystems for integral conversion of the work to heat. This heat is then rejected at the system boundary into the background radiation of the universe, which acts as a heat sink.

The generic "subsystems" block in the model accounts for the work consumption of all processes in the atmosphere, hydrosphere, biosphere, and lithosphere and the associated irreversibilities. Thus, solar energy drives the earth in the following ways:

- It maintains and regulates the earth's climate and temperature.
- It powers the earth's main cycles, including the water cycle, nitrogen cycle, carbon dioxide cycle, oxygen cycle, phosphorus cycle, sulfur cycle, etc.
- It propels the precipitation and hydrological system.
- It generates winds by producing thermal and pressure gradients in the atmosphere.
- It produces ocean currents by induced temperature and density differences.
- It supplies energy to all life systems in the lithosphere and hydrosphere.

A detail of solar radiation use by major subsystems of the terrestrial atmosphere, lithosphere, and hydrosphere is presented in Figure 2.3. There are two streams of radiation: incoming (gray arrows in the figure) and outgoing (black arrows). The incoming solar radiation at the outer border of the atmosphere represents 174 PW (Peta Watt $= 10^{15}$ W). This radiative energy enters the atmosphere. When entering, 7% of its energy is scattered back, 17% is reflected by clouds to extraterrestrial space, 19% is absorbed by the atmosphere and dissipated, 4% is absorbed by clouds, 6% is reflected back from the earth's surface, and 47% is absorbed by land and ocean.

The outgoing energy flux is due to the thermal radiation of the earth. A radiative heat exchange exists between the earth and extraterrestrial background radiation. The infrared radiation emitted by the earth's surface represents about 16% of the total incoming solar radiation. Of this 16%, about 6% is absorbed by clouds and atmospheric gases such as carbon dioxide, methane, ozone, and water vapor. The remaining 10% crosses the atmosphere and reaches extraterrestrial space. In addition, clouds emit infrared radiation toward extraterrestrial space; this amounts to 20% of the total incoming radiation. The atmospheric gases

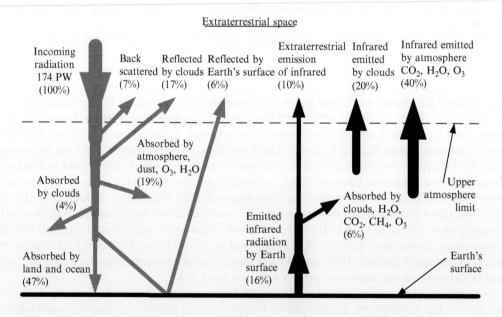

FIGURE 2.3   The energy inventory of incoming solar and outgoing thermal radiation on earth.

such as carbon dioxide, ozone, and water vapor emit infrared radiation to extraterrestrial space, representing 40% of the total incoming radiation. Note that the total outgoing radiation balances the total incoming radiation, and there is no net energy accumulation on the earth.

The energy forms derived from solar energy are hydro-energy, wind energy, ocean currents and waves, biomass energy, and solar radiation energy. In the subsequent paragraphs, each of these forms is detailed.

Forty-three percent of incoming extraterrestrial radiation is *beam radiation.* This is focalized (not dispersed) light that reaches the earth and sea surface. It is a source of energy for life systems, especially green plants that use photonic energy for photosynthesis process. In addition this energy heats the earth and contributes to climate control. The daily averaged beam radiation flux is 250 W/m^2. Note that the beam energy represents around 76 PW (i.e., 43% of 174 PW). If this is converted to electricity with a system having 10% efficiency then the generated power is 7.6 PW. Note that the world energy consumption rate is 0.015 PW; therefore, solar energy has the potential to produce 500 times more energy than the global demand.

Twenty-two percent of incoming radiation is retrieved in the form of hydro-energy, which is a form of potential energy caused by elevation of the water level. Hydro-energy is due to nature's water cycle (driven by solar energy), in which water evaporates from oceans and produces precipitation at higher altitudes, with formation of river basins and lakes.

Wind energy amounts to only 0.21% of the incoming solar radiation from which it is derived. Wind is the mechanical movement of air masses due to a gradient in pressure. The pressure difference between two adjacent locations is due the differential heating by solar radiation. Thus wind energy is eventually a form of solar energy dissipation. Wind energy can be harvested with wind turbines with good wind-to-electricity efficiency; however, the capacity factor for wind energy is only 0.3 because of intermittency.

Only 0.02% of beam radiation is used in the photosynthesis process of green plants on the globe's surface. Although the photosynthesis process has a low efficiency rate (1–4%) because of its massive production of sucrose, glucose, cellulose, and other chemical compounds, it represents a major source of energy and food worldwide. Biomass is derived from photosynthetic systems. Biomass in the form of wood, crops, grasses, straw, sugar cane, manure, dung, charcoal, municipal, and domestic waste, and paper mill residuals represents a form of energy with a calorific value range from 4 to 30 MJ/kg or 2 to 10 GJ/m^3.

In Figure 2.4 a breakdown of energy derived from incident solar radiation is presented. Recall that 34.77% of incident radiation at the outer shell of the terrestrial atmosphere is scattered or reflected back into extraterrestrial space. The remaining radiation is absorbed by the earth. A large percentage of absorbed solar radiation—43%—represents beam radiation impacting land and ocean. This is essentially converted into heat. Beam radiation is a valuable source of energy which can be harvested using a photovoltaic (PV) array, solar thermal panels, and concentrated solar radiation systems for power generation. Various energy conversion factors estimate the amount of power generation from solar radiation. For instance, beam radiation energy can be converted into shaft power using concentrated solar receivers and a heat engine. The practical conversion efficiency of such devices is about 25%, thus one assumes a power conversion factor of 0.25. Using this factor, the total shaft power generated if all beam radiation energy is converted to mechanical energy is 18.7 PW.

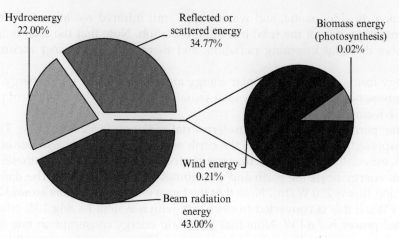

FIGURE 2.4   Breakdown of energy forms derived from incident solar radiation on earth.

The hydro-energy conversion factor is conservatively estimated at 0.8; thus the earth's potential hydropower is 30.6 PW.

Assuming a capacity factor of 0.3 and based on general efficiency data on wind turbines, the wind energy conversion factor is estimated at 0.6; thus the global power generated from wind energy is 220 TW. Biomass has a conservatively estimated conversion factor of 0.5, which leads to an estimated global power generation of 20 TW. Estimates of power generation potential from these energies derived from solar radiation are plotted on the bar chart in Figure 2.5.

Tidal energy is a form of energy essentially different from that of ocean currents. In contrast with ocean currents, a solar radiation-derived energy produced by ocean temperature

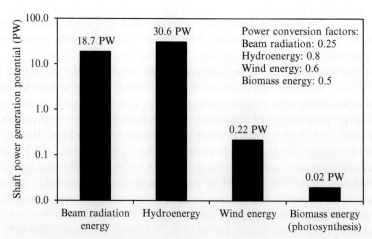

FIGURE 2.5   Estimated global power generation potential from energies derived from solar radiation.

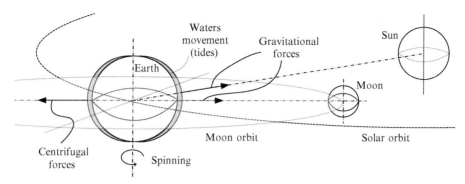

FIGURE 2.6 Tidal energy.

and density gradients, tidal energy is generated by gravitational and inertia forces caused by the planet turning. Tidal energy is illustrated in Figure 2.6. Gravitational forces from moon–earth and sun–earth interactions over ocean water lift massive quantities of water. Superimposed on gravitational forces is the inertia force caused by the planet's daily rotation, an angular velocity of 73 µrad/s. As the average radius of earth at the Equator is 6,000 km one can easily calculate the centrifugal acceleration: $a_{centrif} = r_{earth}\omega^2 = 0.032\,\text{m/s}^2$. Thus the net acceleration of ocean waters at the equatorial plane due to the compounding effect of the earth's gravity and centrifugal force is $a_{net} = g - a_{centrif} = 9.775\,\text{m/s}^2$. Along with the effect of gravitational forces from the moon and the sun that manifest preponderantly in the plane of their orbits, the compounding effect of the earth's gravity creates a net effect of rising waters.

According to Tiwari and Ghosal (2007), the energy from tides is estimated at 3 TW globally. Tidal energy can be harvested by impoundment systems that generate power either at ebb tide or at flood tide, or at both. The impoundment system is equipped with hydraulic turbines to generate power. The largest tidal energy power generation station operates in France at La Ranche and has a capacity of 240 MW. It operates in two ways using a hydrostatic head of 5 m.

Another of earth's major sources of energy is geothermal energy, derived from the thermal energy source generated at the earth's core. It is estimated that the earth's inner core temperature may reach 7000°C. Heat from the core propagates toward the earth's surface mainly by heat conduction. It is generally agreed that the temperature at 4,000 m below the earth's surface is relatively uniform at around 363 K. Geothermal energy at earth surface is around 2,000 times less intense than solar energy. According to Blackwell et al. (1991), geothermal energy intensity at the earth's surface is around 0.1 W/m^2, creating a total global amount of 44 TW, whereas the total amount of thermal energy stored at the earth's core is 10^{16} PJ. Based on these figures, one can easily calculate that at a rate of consumption of 44 TW geothermal energy will be exhausted after 10^{12} years. One concludes from here that geothermal energy is a renewable resource because it cannot be exhausted in humankind's foreseeable future.

Any energy resource that naturally occurs and is non-exhaustible is referred to as *renewable energy*. Three types of energy were discussed above: solar, tidal, and geothermal. All three are earth's fundamental renewable energy resources. A classification of renewable energy,

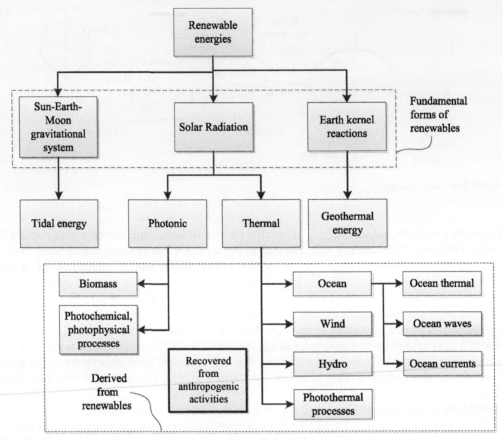

FIGURE 2.7   Classification of renewable energies.

including the forms derived from solar energy, is presented in Figure 2.7. One can divide renewable energy into fundamental forms and derived forms. Solar radiation is a fundamental form of energy that is manifested in many derived forms.

As suggested in the diagram in Figure 2.7, there are two modes of solar radiation energy: photonic energy and thermal energy. The energy of photons is harvested directly by green plants and cyanobacteria that use this energy for photosynthetic processes. This eventually generates biochemical energy used primarily by life systems of the vegetal kingdom. Indirectly, this energy is used by all life systems as a source of food, including the animal kingdom and humans, which also can use it for heating and power purposes. Photonic energy can be used directly for many photophysical processes, either natural or synthesized. An example of photonic energy in a natural system is animal and human vision. Sunlight photons are the energy source for vision. Another photophysical energy conversion of photons is used by PV (photovoltaic) arrays to produce electricity. In addition, many photochemical reactions conducted with the energy of photons produce various chemicals, including hydrogen and synthetic fuels.

The other form of solar radiation conversion uses thermal energy. Thermal energy in the ocean:

- heats the ocean surface and maintains a gradient temperature between the ocean bottom (typically 4 °C) the ocean surface (28 °C in the Sunbelt region),
- creates wave energy caused by the interaction between the ocean surface and the winds,
- creates ocean currents caused by water temperature and density difference between the Sunbelt and polar regions.

As discussed above, winds and hydro-energy are derived from solar thermal energy. In addition, solar thermal energy can be used to drive many photothermal energy conversion processes that generate heating, create chemical reactions, conduct physical processes, and generate power. For example, home heating with passive and active solar systems involves construction of South-facing buildings and the use of thermal solar panels for heating. Processes such as crop drying, laundry drying, and biomass drying are also good examples of the use solar thermal energy. In addition, concentrated solar radiation can be used to generate high temperatures. This heat can be used for thermochemical reactions that generate useful byproducts such as hydrogen and other synthetic materials, to generate high-temperature steam, or to produce power using externally driven heat engines.

Another source of energy is that derived using waste recovery from anthropogenic activity. Many combustible materials can be recovered from industrial and residential settings, including plastics, paper, wood materials, and garbage. Garbage at waste collection sites generates landfill gas, a combustible product. Recovered plastic materials can be converted into liquid fuels by catalytic reforming processes. Other waste materials can be combusted or incinerated, generating high temperatures. Many industrial processes generate waste heat that can be recovered and used as an energy source. Waste materials such as manure and certain forms of biomass can be converted into biogas by anaerobic digestion.

Certain biomasses recovered from crops and other plants can be converted into alcohols using aerobic fermentation processes. Various biomasses provide liquid fuels via esterification, Fischer–Tropsch synthesis, and other processes. Biomasses such as wood, crops, and other plants generate heat by direct combustion. Combustion of synthetic fuels such as the biogas of liquid biofuels produces high-temperature heat. Note that all heat derived from biomass sources represents anthropogenic renewable energy. Other heat derived from renewables is not anthropogenic as it occurs naturally via direct solar radiation, ocean heat, or geothermal energy. A general inventory of sustainable energies derived from renewable and anthropogenic sources is presented in Figure 2.8. The temperature level and Carnot factor $(1 - T_0/T_{\text{source}})$ of recoverable waste heat from industrial activity are listed in Table 2.1.

As seen in the table, there are various sources of waste heat due to human activity. For example, common steam cycle power plants reject heat at the condenser in the environment at temperature level in the range 30–60 °C. Gas turbine power plants reject exhaust gases at 600–800 °C. Usual fuel combustion processes for generation of low temperature heat (e.g., gas furnaces) generate flue gas at 150–200 °C. Automobile engines produce exhaust gases at 800–1000 °C. Based on the report by Latour et al. (1982) more than 65% of waste heat rejected by the industries in the United States is at low temperature level (below 200 °C). It can be inferred that low temperature heat is available in abundance to recover from industrial thermal waste. Table 2.1 suggests that most of the sustainable thermal energy sources are within this low

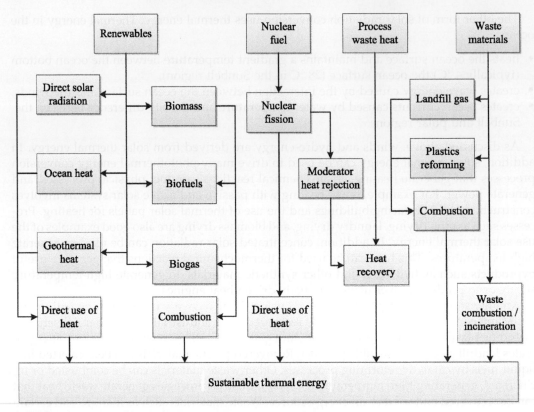

FIGURE 2.8   Sustainable thermal energy sources.

temperature range. It is of prime importance to develop methods and devices for harnessing thermal energy from such prevailing sources to meet a substantial portion of energy demand.

A comparison of renewable energy resources and the global daily energy demand is of importance because it reveals the relative magnitude of energy supply and energy demand. As indicated in Tiwari and Ghosal (2007), the average energy demand is 15 TW globally. The bar chart from Figure 2.9 compares the energy demand with energy potentially available from renewable sources.

As shown by the bar chart, the energy rate potentially derived from beam radiation, hydro-energy, wind energy and biomass energy is incommensurable higher (more than 3000 times) than the rate of energy demand globally. In addition, tidal energy is available at a rate five times smaller than world demand, while geothermal energy can generate approximately three times less power that global demand.

Other sources of energy available to humans are derived from fossil fuels and nuclear fuel. Conventional fossil fuels encompass three classes of materials: coal, petroleum and natural gas. In addition, there are several types of nonconventional fossil fuels that have started to be exploited more recently, such as shale oil, oils sands, and coal bed gas. Fossil fuels are mainly used for direct combustion with air in various systems such as furnaces, gas turbines, and internal combustion engines.

TABLE 2.1   List of Recoverable Waste Heat Sources of High Temperature

Thermal Source	Temperature (°C)	Carnot Factor
Steam power plants	230–250	0.40–0.43
Gas turbine power plants	600–800	0.65–0.72
Heating furnace exhaust	175–230	0.33–0.41
Automobile engine exhaust	400–700	0.55–0.69
Nitrogenous fertilizers plants	195–230	0.36–0.40
Pulp mills	140–200	0.27–0.37
Paper mills	140–200	0.27–0.37
Paperboard mills	140–200	0.27–0.37
Alkalines and chlorines production	170–220	0.32–0.40
Industrial inorganic chemical production	120–200	0.24–0.37
Industrial organic chemicals	120–300	0.24–0.47
Petroleum crudes and intermediate refining	150–200	0.29–0.37
Glass factories	400–500	0.55–0.61
Cement production	200–300	0.37–0.47
Blast furnaces	500–700	0.58–0.67
Iron foundries	425–650	0.57–0.67
Drying and baking ovens	93–230	0.18–0.40
Aluminum production	650–760	0.67–0.71
Copper production	760–815	0.71–0.72

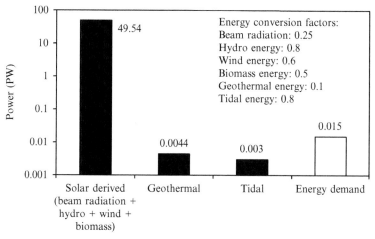

FIGURE 2.9   Energy rate comparison of three fundamental energy sources on earth and the average rate of global energy demand.

Coal is mainly formed from organic substances derived from fossilized plants with embedded mineral inclusions. The primary chemical element in coal is carbon, and coal has a carbon content of over 70% by weight. Coal has a calorific value of 25–35 MJ/kg, with the coal type lignite having the lowest calorific value and the coal type anthracite having the highest calorific value. According to BP (2010), the proven resources of coal globally are equivalent to the energy of 23.1×10^6 PJ and a resource over production ratio of $R/P = 122$ years. The R/P ratio is calculated based on the amount of the resource, preferably expressed in energy units, and the rate of coal production (or consumption) expressed in energy units per year. Coal is mainly used for major power plants and for industrial processes in the metallurgical and cement industries.

Petroleum is a naturally occurring hydrocarbon-based material found mainly as liquid. Nonconventional sources of petroleum are often present in solid forms such as bitumen or oil sands. Alkanes, cycloalkanes, aromatics, paraffin, and naphthalene are the main constituents of petroleum. The total world petroleum reserves include 30% of conventional oil, 25% extra heavy oil, 15% heavy oil and other petroleum forms such as bitumen. Shale oil or oil sands form a remaining 30% share. Proven fuel reserves of petroleum have an R/P ratio of ~50 years, and petroleum is primarily consumed by the transportation sector.

Natural gas containing methane as its combustible material occurs around the world, with the Russian Federation having the largest reserves. Natural gas is used in many industries, including fertilizer (ammonia, urea) production, and as fuel for heating, cooking, and in some cases for power generation. About 20% of world energy production is derived from natural gas combustion. Natural gas reserves are estimated to have an equivalent energy of 7×10^6 PJ with an R/P ratio of 60 years.

Conventional nuclear fuel is represented by fissile uranium ^{235}U, which naturally occurs in form of U_3O_8 ore. One ton of ore yields about 6 kg of fissile uranium, which is equivalent to 144 TJ of electrical energy (or 40 GWh), whereas 1 ton of coal can be used to generate 14,000 times less electricity. Conventional nuclear fuel reserves are estimated to have an equivalent of PJ with an R/P ratio of 50 years. Many more nonconventional nuclear fuel resources do exist, mainly in the form of thorium, as well as spent fuel material that might be used in the breeder reactors of next generation nuclear power plants. Figure 2.10 presents a bar chart comparison of reserves versus production ratio for conventional fuels, including coal, petroleum, natural gas, and uranium.

Projections of world energy demand by resource type are of crucial importance to energy planning and environmental management. Electricity demand represents the major component of energy consumption worldwide, and the demand is in such close correlation with human population that the projected increase of global population implies an unavoidable increase in consumption.

It is predicted that the increase of electrical energy consumption is not linearly related to the increase of population. Instead, the consumption of electricity per capita grows more quickly relative to population, given societal development and the diversification of tastes over time. In the twentieth century, the major use of residential electricity has been for lighting and appliances. Today, an important component of electricity demand is represented by communication devices, computers, Internet, and entertainment. The European Commission (2006) formulated some projections of world population increase and power demand until 2050. There are three scenarios proposed by the European Commission for determining

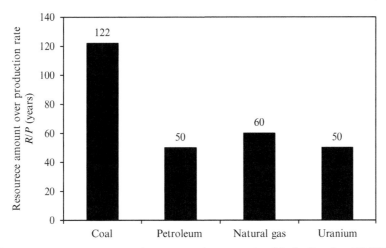

FIGURE 2.10 Proven reserves versus production ratio for conventional fuels. *Data from BP (2010).*

the projections. The scenarios account for the geopolitical context; petroleum, natural gas, and coal consumption profiles; development of hydrogen technologies; population growth; and other aspects. The three projection scenarios are as follows:

- Reference Case (RC) scenario: Nations exhibit a minimum degree of political initiative toward sustainable development.
- Carbon Constraint Case (CCC) scenario: Nations impose severe political limits on CO_2 emissions up to 2050.
- Hydrogen Case (H_2) scenario: Most countries implement policies to facilitate the development of a hydrogen economy.

Based on the projection by European Commission (2006), the plot from Figure 2.11 correlates the annual electric power consumption per capita with population growth. The curve has a knee by year 2030 after which the electricity consumption increases more sharply. Note that developing countries, with a population of around four billion, represent about 77% of the global population, but their electricity consumption represents only 25% of global consumption. Given the extra energy demand that should occur in developing countries due to population growth, diversification of services, and increases in standard of living, electricity consumption per capita will likely be 1.5–1.8 times higher than population growth rate after around 2030. Figure 2.12 illustrates a projection of population growth and global electricity consumption increase until 2050. It appears that electricity demand will be about three times higher in 2050 than in the present day.

A review of past literature reveals many studies concerned with global energy demand projections. Figure 2.13 illustrates some possible projections and past records of primary energy demand from conventional fuels, including coal, petroleum, natural gas, and uranium. The prediction shows that, in the next 40 years, fossil fuel consumption will tend to stabilize, while nuclear fuel demand will apparently continue to increase after 2050.

The projections of fossil fuel demand illustrated in Figure 2.13 are illustrated by Ermis et al. (2007) and are based on an artificial neural network (ANN) algorithm which was designed

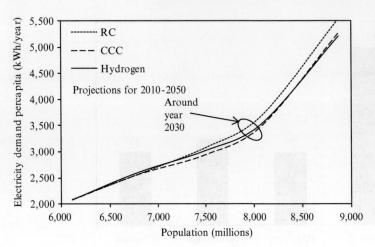

FIGURE 2.11 Correlation between projected population increase and electricity demand per capita [RC, reference case; CCC, carbon constraint case; Hydrogen, hydrogen economy case]. *Data from European Commission (2006).*

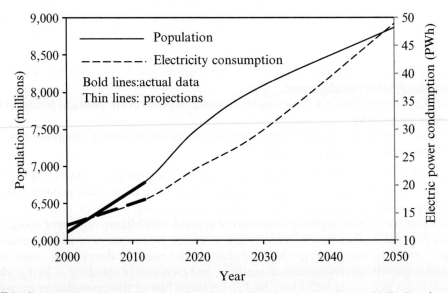

FIGURE 2.12 Projections of population growth and world electricity consumption until 2050. *Data from European Commission (2006), reference case.*

based on records of energy demand in the last 40 years. Table 2.2 gives the equations developed by Ermis et al. (2007) expressing the projected annual energy demand in exojoules as a function of year. The projection formulas for renewable energy and primary energies are indicated. Using the ANN method, Ermis et al. (2007) also developed projections of primary and sustainable energy until 2050. The corresponding prediction equations are presented in Table 2.2, with the projection plots in Figure 2.14.

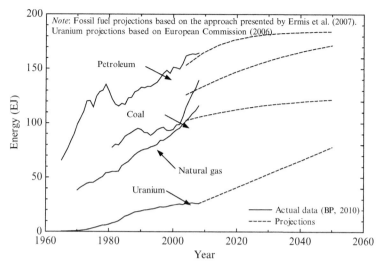

FIGURE 2.13 Historical records and future projections of fossil and nuclear fuels demand.

TABLE 2.2 Correlations for Energy Demand Projections Developed by Ermis et al. (2007)

Commodity	Annual Energy Demand Projection for 2010–2050 (Exojoules)
Coal	$E = 125.6667 \left\{ 1 + \exp\left[-\dfrac{Y - 1968.0551}{24.2314} \right] \right\}^{-1}$
Petroleum	$E = 186.5923 \left\{ 1 + \exp\left[-\dfrac{Y - 1984.2525}{27.0821} \right] \right\}^{-1}$
Natural gas	$E = 184.5575 \left\{ 1 + \exp\left[-\dfrac{Y - 1986.7137}{10.2282} \right] \right\}^{-1}$
Renewable energy	$E = 105.493 \left\{ 1 + \exp\left[\dfrac{Y - 1999.642}{26.387} \right] \right\}^{-1}$
Primary energy	$E = 589.024 \left\{ 1 + \exp\left[-\dfrac{Y - 1984.253}{22.721} \right] \right\}^{-1}$

The projections indicated in Figure 2.14 clearly indicate that the share of fossil fuel in global energy demand will decrease in the next 40 years, with a 5% drop in demand for fossil fuels allowing the sustainable energy share to grow by the same amount. According to this projection, the energy share by 2050 is 83% fossil and 17% sustainable. The projections of the European Commission (2006) are slightly different, as indicated in Figure 2.15 for the three scenarios mentioned above (RC, CCC, H₂). As the figure shows, the projected renewable

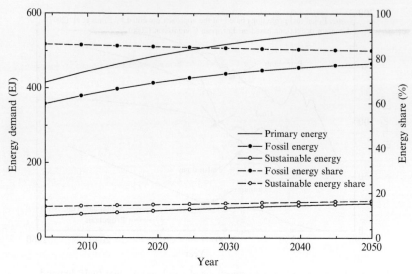

FIGURE 2.14   Global energy demand projection to 2050. *Data from Ermis et al. (2007).*

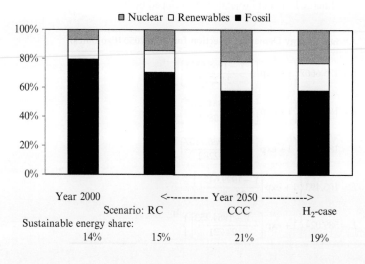

FIGURE 2.15   Share of fossil fuel, nuclear energy, and renewable energy in 2000 and projections for 2050. *Data from European Commission (2006).*

energy share by 2050 is 15% in the RC scenario, 21% in the CCC scenario, and 19% in the H_2 scenario. These predictions are close to that of the 17% renewable energy share projected by Ermis et al. (2007).

Figure 2.16 details the projections of renewable energy share for year 2050 with respect to energy shares in the year 2000. As shown, the share of wind and solar—which is very small today—will become more important by 2050, measuring ~20%. The plot in Figure 2.17

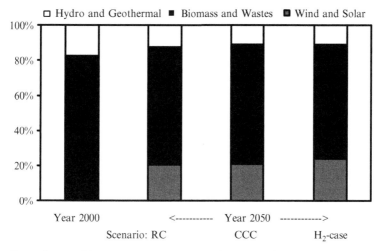

FIGURE 2.16 Share of renewable energy sources in 2000 and projection for 2050. *Data from European Commission (2006).*

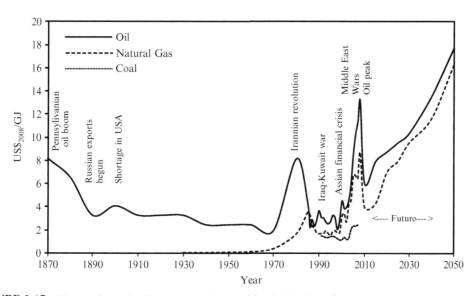

FIGURE 2.17 Historical trend and projection of price of fossil fuels. *Data from Dincer and Zamfirescu (2011).*

compares historic and projected prices for oil and natural gas. The continuous increase of fuel real value is mainly a consequence of fuel shortage, as indicated by the great influence of the Iranian revolution on the believed oil peak. The oil peak is the point in time when the maximum rate of global petroleum extraction is reached, after which the rate of production enters terminal decline.

2.3 ENVIRONMENTAL IMPACT OF POWER GENERATION SYSTEMS

It is generally agreed that any energy system impacts the environment. Over the past few decades, energy-related environmental concerns have expanded from primarily local or regional issues, encompassing the international and global nature of major energy-related environmental problems. Environmental problems are particularly apparent in developing or newly industrialized countries, where energy-consumption growth rates are typically extremely high and where environmental management has not yet been fully incorporated into the infrastructure. Nevertheless, at present, industrialized countries are mainly responsible for air pollution, ozone depletion, and the carbon emissions.

Figure 2.18 presents a generic power production system which consumes some fuels, generates useful work, and expels some pollutants in the environment. The system might cause pollutant emissions, accidents, hazards, ecosystem degradation through air and water pollution, animal poisoning, greenhouse gas emission, carbon monoxide leakages, stratospheric ozone depletion, and the emission of SO_2, NO_x, VOCs (volatile organic compounds), PM (particulate matter), and other aerosols.

The effluents expelled into the atmosphere by power generation systems can be divided into two categories: greenhouse gases (GHG) and aerosols. GHG are those chemicals which are released in the terrestrial atmosphere and produce the *greenhouse effect*. When released from natural and anthropogenic activities, GHG travel through the atmosphere and reach its upper layer, the *troposphere*. In the troposphere, GHG absorb an important part of the infrared radiation emitted by the earth surface. As a consequence, the earth's surface temperature tends to increase, and this process is called the *greenhouse effect*.

On the other hand, aerosols, such as VOCs, soot, and PM, are released continuously in the atmosphere and concentrate in its upper layers. Aerosols contribute to the earth's *albedo*. Due to their presence in the atmosphere, aerosols reflect a portion of the incident solar radiation back into space. As a result, the earth's temperature tends to decrease. This process can be denoted as the *albedo effect*. The balance between greenhouse and albedo effects establishes the earth's temperature and regulates the earth's climate. This mechanism of climate control is a natural process. However, since the industrial revolution, the anthropogenic impact on

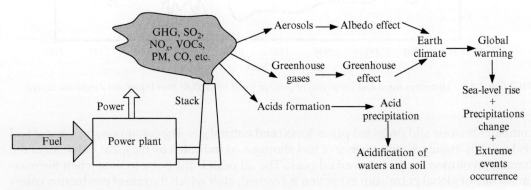

FIGURE 2.18   Environmental pollution from energy systems.

climate has become obvious due to accentuated emission of GHG by many activity sectors (energy, transportation, industry) which then induce *global warming*.

Global warming leads to drastic changes in natural systems. Permafrost ice is increasingly melting, and, consequently, the seal level tends to rise. Moreover, changes in global precipitation patterns have been observed (more dry areas and more regional flooding), along with a higher frequency of extreme events (tsunami, extreme winds, cyclones, earthquakes, tornadoes, etc.).

Another major environmental impact of energy systems results from acidic precipitation. Gaseous effluents expelled in the atmosphere by power generation systems can eventually form acids that return back to earth in precipitation, contributing to acidification of soil and seas. This acidification negatively affects all life systems. Other impacts are also possible, as it will be detailed subsequently. The main types of pollutant emissions due to energy systems are listed in Table 2.3, and their influence is explained.

The environmental impacts produced by pollutants can be classified in seven categories determined by life cycle assessment methodologies according to the norm ISO 14042 (2000). These impact categories are global warming, acidification, ozone depletion, toxicity,

TABLE 2.3 Atmospheric Pollutants Released by Power Generation Systems

Pollutant	Explanations
GHG	Greenhouse gases. These are gases that produce greenhouse effect. The main GHGs are CO_2, CH_4, N_2O. Greenhouse effect is the main cause of global warming
CO	Carbon monoxide. Arises mostly from the incomplete combustion of fuels, and poses a health and life of animals and humans at high risk upon inhalation. Intense emission of CO in open spaces may affect birds, whereas emissions in close spaces (residences, garages) may induce death of exposed persons
SO_2	Sulfur dioxide. It is a corrosive gas which is hazardous to human health and harmful to the natural environment. It results from combustion of coal and fuel oil, smelting of nonferrous metal ores, oil refining, electricity generation and pulp and paper manufacturing. It causes respiratory difficulties, damages green plants and is a precursor of acid precipitation
NO_x	Nitrogen oxides (NO and NO_2). It is produced by combustion of fuels at high temperature in all combustion facilities from large scale to small scale (including motor engines, furnaces etc). It can lead to respiratory problems of humans and animals. It can form acids in high altitude atmosphere
VOCs	Volatile organic compounds. Are volatile organic particles resulting from hydrocarbon combustion in engines. Have harmful effects in the atmosphere; it impede formation of stratospheric ozone
PM	Particulate matter. Particles in the air (fly ash, sea salt, dust, metals, liquid droplets, soot) come from a variety of natural and human-made sources. Particulates are emitted by factories, power plants, vehicles, etc., and are formed in the atmosphere by condensation or chemical transformation of emitted gases. PM cause of health and environmental effects including acid precipitation, damage to plant life and human structures, loss of visibility, toxic or mutagenic effects on people, and possibly non-accidental deaths
CFC, HCFC, HFC	Chlorofluorocarbons. These are mainly refrigerants or propellants. They can contribute to the destruction of the stratospheric ozone layer and to greenhouse effect

photooxidant formation, eutrophication, and depletion of abiotic resources. The norm ISO 14042 (2000) includes methodologies for quantitative assessment of each environmental impact category, using indicators. Table 2.3 lists the environmental impact categories and gives the impact indicators that are used to quantify each type of impact. The impact indicators are defined by convention and are based on some measurable quantities. For example, global warming is quantified by the *global warming potential* (GWP), measured in kilograms of CO_2, which is the equivalent of GHG emitted. The acidification potential is measured in SO_2 emitted, expressed in so-called Switzerland equivalent. Ozone depletion is quantified by *ozone depletion potential* (ODP) measured as the trichlorofluoromethane equivalent. The *toxicity* is measured with the 1,4-dichlorobenzene equivalent. Photooxidant formation is measured in kilograms of ethylene equivalent. Eutrophication is measured with the kilograms of PO_4 equivalent, and the depletion of abiotic resources is measured with the kilograms of antimony equivalent. The norm ISO 14042 (2000) establishes precise determination procedures for all these indicators.

The determination of GWP is detailed because greenhouse gas emissions represent the main environmental pollution produced by power generation systems. Water vapor is the most important greenhouse gas. However, the water vapor concentration in the atmosphere fluctuates rapidly, and water vapor absorbs radiation over a wide infrared spectrum. The principal GHG emitted in the atmosphere (except the water vapor) are listed in Table 2.4. Three of the gases shown in the table, namely carbon dioxide, methane, and nitrogen dioxide, are both part of natural cycles of carbon and nitrogen and part of anthropogenic emissions. Therefore, they were present in the atmosphere prior the industrial era, for which the year 1750 is considered the reference. As seen, the concentration of these gases in year 1750 is

TABLE 2.4 Environmental Impact Categories and Indicators for Impact Quantification

Impact Category	Explanation
Global warming	It is due to emission of greenhouse gases. It is quantified based on global warming potential GWP measured in kg CO_2 equivalent
Acidification	The impact of NH_3, NO_x, and SO_2 present in precipitations and their contribution to acidification of land and waters. It is expressed in SO_2 emitted in Switzerland equivalent
Ozone depletion	Depletion of stratospheric ozone layer, especially due to photoreactions involving chlorine. It is quantified by ozone depletion potential ODP which represents the amount of a substance that depletes the ozone layer relative to CFC11 (trichlorofluoromethane)
Toxicity	Refers to toxicity produced in air, freshwater, seawater, or terrestrial. It is quantified with respect to the toxicity of DCB (1,4-dichlorobenzene) and measured in kg DCB equivalent
Photooxidant formation	Depends on photochemical ozone creation potential expressed with respect to the reference substance: ethylene. Measured in kg of ethylene equivalent
Eutrophication	Covers all potential impacts of high levels of nutrients, the most important of which are nitrogen (N) and phosphorus (P). It is measured in kg PO_4 equivalent
Depletion of abiotic resources	Abiotic resources include all nonliving resources (coal, oil, iron ore, renewable energies, etc.). It is measured in kg antimony equivalent

higher than zero. Table 2.4 also identifies freons as manmade substances that are GHG. Here is a non-exhaustive list of freons in order of decreasing influence on greenhouse effect: 1,1,2-trichloro-1,2,2-trifluoroethane (CFC113), chlorodifluoromethane (HCFC22), CFC141b, CFC 142b, 1,1,1-trichloroethane (CH_3CCl_3), carbon tetrachloride (CCl_4), 1,1,1,2-tetrafluoroethane (HCFC134a). There are other GHG generated as a result of human activity, but they are emitted in lower quantity and consequently, their actual concentration in the atmosphere is low.

It is certain that atmospheric CO_2 levels will continue to increase significantly. The degree to which this occurs depends on the fixture levels of CO_2 production and the fraction of that production that remains in the atmosphere. Given plausible projections of CO_2 production and a reasonable estimate, which states that half the emitted amount will remain in the atmosphere, the concentration of CO_2 in the atmosphere will reach 600 ppm in sometime in the middle of the twenty-first century.

Anthropogenic activity induces an intensification of the greenhouse effect that leads to global warming. With greenhouse effect, there are two possibilities:

- Anthropogenic activity results in more GHG emissions. These gases will be more concentrated in the lower layers of the atmosphere where human activity occurs, and the troposphere then reradiates more energy toward the earth's surface. Therefore the global temperature increases.
- The GHG concentration reduces in the vicinity of the earth's surface and concentrates toward the upper layers of the atmosphere for some unknown reason. In this case, the troposphere reradiates less toward the earth, leading to a decrease of the earth's temperature.

Radiative forcing is defined as the net change in radiation balance at the tropopause produced by a specified cause. By convention, when radiative forcing is positive, it induces an increase of planetary temperature, and when it is negative, it leads to a decrease in temperature. The unit of measurement for radiative forcing is the unit radiation energy rate per square meter of earth surface, where the earth surface, by convention, is the area of the sphere having the average radius of the planet. The usual symbol for the radiative force is ΔF. Typical values for radiative forcing are 0–2 W/m^2. These values can be derived by determining how the concentration of GHG, aerosols, and atmospheric ozone induce radiative forcing. Depending on the concentration of the gas in the atmosphere, there are three regimes producing radiative forcing: low, moderate, and high concentration.

Quantifying the effect of a particular atmospheric gas on the climate is a multivariable problem. On one side, the radiative forcing induced by the respective gas is an indication of the direction in which the respective gas can influence the climate. If, for example, the gas absorbs more energy in the infrared spectrum, its greenhouse effect is accentuated, that is, it is associated with positive radiative forcing. However, the atmospheric lifetime of the gas is also important. This parameter indicates how long the gas is active in the atmosphere with respect to radiative balance control.

The net forcing produced during an infinitesimal time interval is given by the product $\Delta F_{CO_2}(t) f_{CO_2}(t) dt$, where the notation $f_{CO_2}(t)$ indcates that the fraction of gas mass existent in the atmosphere is a function of time. One can integrate the former quantity over a time horizon TH and obtain the total forcing produced by the respective amount of CO_2. If a greenhouse gas other than carbon dioxide is considered, its integrated forcing over the time horizon can be normalized with the integrated forcing of CO_2. This reasoning results in the following definition

$$GWP = \frac{\int_0^{TH} \Delta F_{GHG}(t) f_{GHG}(t) dt}{\int_0^{TH} \Delta F_{CO_2}(t) f_{CO_2}(t) dt}$$

where GWP is the global warming potential of the greenhouse gas (GHG). Table 2.5 gives the GWP of main GHGs for three time horizons, based on IPCC (2007).

The evolution of the atmospheric concentrations of the main GHGs based on historical records is presented in Figure 2.19. This figure shows a continuous increase of emissions from the industrial revolution to the present. Each of these gases has an effect on radiative forcing proportional to its GWP. Figure 2.20 presents the evolution of radiative forcing induced by the three main GHGs in the last 250 years. As shown, radiative forcing and thus the climate are mostly influenced by carbon dioxide emissions, followed by methane and nitrous oxide emissions. In turn, there is a correlation between increased radiative forcing due to GHG emissions and the global temperature and sea level.

Figure 2.21 shows how recorded global temperature and sea level increase with respect to year 1870. According to IPCC (2007), satellite data indicate that, in the last 35 years, the Arctic sea ice shrank by a factor of 2.7 times. In the last 100 years, the global surface temperature increased by about 0.74 K. The sea level increased by about 0.2 m in the last 150 years, and recent records indicate that global warming affects life systems on earth and in the seas.

From Figures 2.20 and 2.21, it can be noted that GHG emission from human activity increased sharply in the last 40 years. Figure 2.22 presents GHG emission by sectors of activity and by major GHG type. Since the industrial revolution, human activities related to fossil fuel combustion in industrial, commercial, residential settings and the transportation sector have dangerously impacted the global environment. The figures clearly show that the energy supply sector is responsible of the majority of GHG emissions in the atmosphere, with a 26% share, followed by industry with a 17% share. As mentioned above, carbon dioxide emissions

TABLE 2.5 Principal Greenhouse Gases and Their GWP

Gas	Absorption Spectrum (cm^{-1})	Atmospheric Concentration	Atmospheric Lifetime	GWP 20	GWP 100	GWP 500
CO_2	550–800	387 ppm	50–200 years	1	1	1
CH_4	950–1,650	1750 ppb	12 years	72	25	7.6
N_2O	1,200–1,350	314 ppb	120 years	289	298	153
$CFCl_3$	800–900	251 ppt	50 years	6,730	4,750	1,620
CF_2Cl_2	875–950	538 ppt	102 years	11,000	10,900	5200

Source: IPCC (2007).

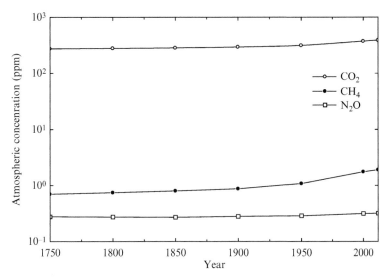

FIGURE 2.19 Evolution of atmospheric concentration of main GHGs starting with the industrial revolution. *Data from IPCC (2007).*

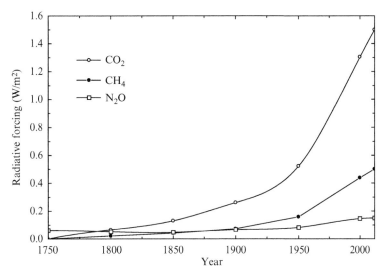

FIGURE 2.20 Evolution of radiative forcing due to main GHGs starting with the industrial revolution. *Data from IPCC (2007).*

are the most important anthropogenic impact on global warming. Figure 2.22b also reveals that combustion of fossil fuels is the most important source of carbon dioxide.

The albedo effect, which is the opposite of greenhouse effect, is caused by aerosols often generated by energy supply sector. According to IPCC (2007), black carbon, nitrates, soot, primary sulfates, organic carbon, VOCs, chlorides, ozone and trace metals generate a negative

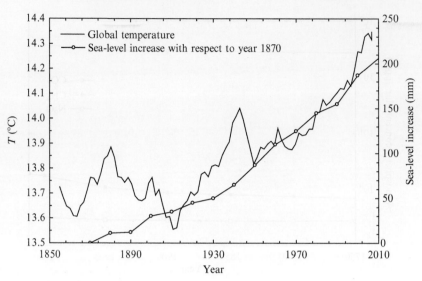

FIGURE 2.21   Recorded temperature and sea level changes during the last 250 years. *Data from IPCC (2007).*

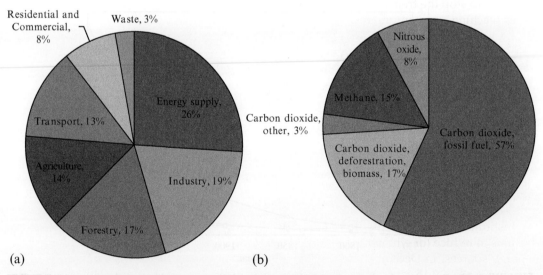

FIGURE 2.22   Anthropogenic GHG emissions by sectors (a) and major gas type (b). *Data from IPCC (2007).*

radiative forcing of at least—0.5 W/m^2, which is in not balanced with the radiative forcing generated by GHG emissions of appriximately +1.5 W/m^2. The inbalance is expected to increase in the future as GHG emissions are predicted to dramatically increase. Prediction of GHG emission in the next 100 years is of major importance in shaping the energy policy of today and promoting sustainable energy pathways. Depending on policy measures, it is

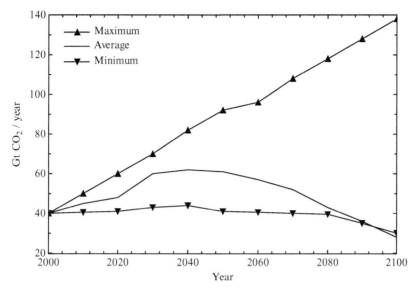

FIGURE 2.23   Projection of GHG emissions until the end of the current century. *Data from IPCC (2007).*

possible to stop the trend of increasing GHG emissions at certain point in time. In such case, after reaching a maximum, the concentration of GHG in the atmosphere may decrease.

IPCC (2007) considers a number of scenarios regarding GHG emissions. Figure 2.23 shows three possible projections of annual GHG emissions expressed in Gt CO_2 equivalent per year. The projection curves labeled "Maximum" and "Minimum" illustrate the 80th percentile range of emission scenarios published recently. The curve labeled "Average" represents one possible scenario, which assumes a world of very rapid economic growth, a population that reaches its maximum by 2050, and introduction of efficient technologies based on nuclear and renewable energy sources. In this scenario the global temperature will increase by approximately 2.4 K until 2011, as compared to an increase of 0.6 K, which is an estimate based on the assumption that, during the current century, all anthropogenic CO_2 emissions contribute to an GHG concentration equal to the concentration recorded in the year 2000. Moreover, the predicted 2.4 K temperature increase could contribute to a sea level rise in the range of 0.20–0.43 m.

Institutes and agencies in many developed and developing countries have started taking actions to reduce (or eliminate) the pollutant emissions and to attain a sustainable supply of energy sources. In December 1997, the International Kyoto Conference on Climate Change produced a list of 15 concrete proposals for curbing global greenhouse gas emissions. The list includes improving the fuel efficiency of automobiles, introducing solar power facilities, and planting forests to act as "green lungs" in densely populated areas.

Emission of SO_2 and NO_x, which characterizes the energy supply sector, has a direct environmental impact due to acidification effect. These gases may participate in the complex set of chemical transformations in the atmosphere resulting in acid precipitation. Road transport is also an important source of NO_x emissions. Most of the remaining NO_x emissions are due to fossil fuel combustion in stationary sources. Countries in which the mentioned energy-related activities occur widely are likely to be significant contributors to acid precipitation. A major problem with acid rain is that it often affects regions beyond the nations that produce it. A large variety of evidence shows the damages of acidic precipitation as follows:

- Acidification of lakes, streams, and ground waters.
- Toxicity to plants from excessive acid concentration.
- Corrosion to exposed structures.
- Resulting in damage to fish and aquatic life.
- Damage to forests and agricultural crops.
- Deterioration of buildings and fabrics.
- Influence of sulfate aerosols on physical and optical properties of clouds.

The acidification mechanism is outlined in the scheme from Figure 2.18. Figure 2.24 illustrates how acids form in the atmosphere and are deposited on land and seas. Acid precursors are mainly produced by the combustion of fossil fuels, especially coal and oil, as well as the smelting of nonferrous ores, and they can be transported long distances through the atmosphere and deposited on ecosystems.

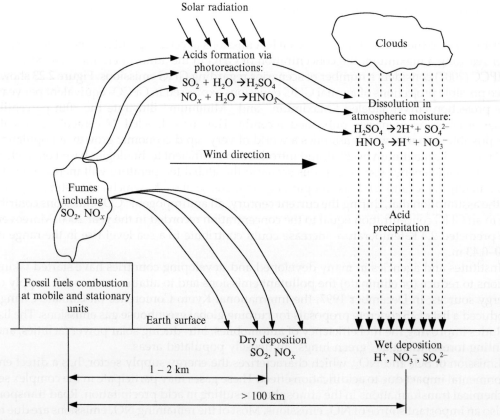

FIGURE 2.24 Mechanism of acid precipitation.

The majority of SO_2 emissions come from fossil fuel power plants, while the majority of NO_x emissions are emitted by the transportation sector. Another source of acidic precipitation is sour gas treatment. This treatment produces H_2S which then reacts with air to form SO_2. Dry deposition of SO_2 and NO_x creates an opportunity for acid formation at soil level or in seas. Direct deposition occurs at places 1–2 km from the emission source. Acid precursors that travel to high levels of the atmosphere enter into photoinduced reactions with water vapors and form acids such as sulfuric acid and nitric acid. These acids travel far from the source point of pollution and are dissociated in the atmospheric moisture. The dissociated acids then fall to the earth's surface with precipitation.

Acidification and other types of pollutions caused by energy systems may affect the quality of waters including groundwater used for drinking and irrigation. Efforts are continually made to control causes of energy-related pollution, such as geothermal fluids containing toxic chemicals, acid drainage from mines, coal wastes, effluents containing hazardous chemicals from power plants and refineries, and thermal pollution from the discharges of power plant cooling systems.

Economic priorities often cause the loss of land particularly suited for sustaining agriculture, housing, or natural ecosystems. In the energy sector, concern has focused on mining sites and hydroelectric reservoirs; the large land surfaces that might be needed for the large-scale exploitation of renewable energy forms such as solar power, wind power stations, or biomass production; the sites chosen for large, complex industrial processes such as fuel refining or electric power generation; and the disposal of solid wastes including radioactive wastes.

Hazardous wastes pose special health and environment threats and are mainly generated by the chemicals and metal industries. Non-hazardous wastes, including bottom ash from power plants and air-pollution control residues, pose disposal problems related to space and appropriate containment. The commercial use of some solid wastes, such as building industry products and transportation surfaces, is also limited by the size of the market. In addition to impacting water and air quality, the effects of energy generation can be observed at the level of ecosystems, food, coastal regions, industry, settlements, and health.

2.4 SUSTAINABILITY ASSESSMENT OF POWER GENERATION TECHNOLOGIES

In this period of transition from a fossil-fuel-based economy to an economy based on sustainable energy, it is essential to establish environmental impact criteria and indicators for power generation technologies, as they are major players affecting the pollution, economy, and wealth on a global scale. Apart from various existing ideologies, the status of technological development, theoretical assessment, and prediction tools play a major role in influencing policy making and high level decisions. Behind any policy and decision there must be a rationale to constrain the decision spectrum. Figure 2.25 illustrates how sustainable development strategies can be shaped with the help of assessment and advancements in technology.

Limited energy and resources have led in part to significant efforts toward energy-utilization efficiency improvement, as well as resource recycling and reuse. For example,

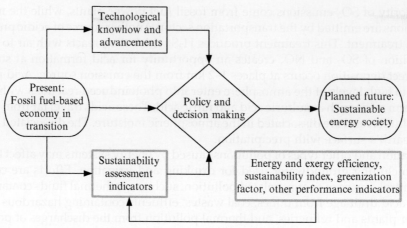

FIGURE 2.25　The role of sustainability assessment indicators in shaping policies for a planned sustainable energy society.

refuse and other solid wastes are now often used to supplement fuel supplies. Recycling (resource recovery) extends the lifetimes of many natural resources, and recycling is often profitable and usually beneficial to the environment. Energy efficiency and conservation can postpone shortages of energy resources, reduce environmental damage, and provide economic benefits. The efficient use of energy is of particular importance to developing countries, as it can forestall the need for very large capital investments. Enormous potential exists, through improvements in energy efficiency and conservation, for decreasing total world energy consumption and, thereby, the effects of energy consumption on the environment. Often such improvements require a myriad of small changes in consumption patterns.

Increased energy efficiency reduces energy-related environmental impacts such as those listed in the section "2.3 Environmental Impact of Power Generation Systems." These impacts include environmental damage due to extracting energy resources from the ground, as well as the competition for water between hydropower, agriculture and recreational activities. In addition, improved efficiency enhances the reliability of future energy supplies, and improves the longevity of energy supplies. The potential for energy efficiency is significant during both energy production and consumption (e.g., the 30% of oil in a reservoir that is extracted from onshore wells could be improved upon using secondary recovery techniques, such as water flooding, and thermal stimulation). The following measures are essential for reducing the environmental impact of energy systems:

- Increasing the efficiency of power generation systems.
- Energy conservation measures.
- Use of more environmentally benign energy sources.

When shaping sustainable energy policies and strategies, it is important to understand the benefit of clean energy technologies on one hand and the effects of fossil fuel based systems on the other. The kinds of energy technologies with zero or minimum negative environmental impact through associated pollution—more environmentally benign and sustainable technologies—are in general named renewable energy. Considering the benefits of renewable energy, developing and sustaining renewable energy supplies are assumed to be a key element in the interactions between nature and society. An essential increase in the scale and pace of effective policy development is required to change course toward a sustainable path. It is also important to possess theoretical tools to quantitatively assess renewable energy development and to:

- Help elaborate rationale strategies and policies.
- Help understand main concepts and issues related to renewable energy use and sustainability.
- Develop relations between renewable energy use and sustainability development.
- Encourage the strategic use and conservation of the renewable energy sources.
- Provide methods for developing and implementing energy security.
- Increase the motivation for implementing renewable energy strategies.
- Encourage the reduction of negative environmental impacts by considering the possible renewable energy strategies.
- Form a scientific platform to discuss the possible renewable energy strategies for sectoral use.

The need for new policies is justified by the fact that free markets must be required to meet the needs of vulnerable groups, reduce environmental pollution, and ensure the energy security. At the upper level, strategies are needed, with a strategy representing a plan or method for achieving a goal. (In our discussion "the goal" means progress toward a global society based on sustainable energy.) Once a strategy is established, the policy, which represents the course of actions to implement the strategy, must be elaborated. Any policy will specify "policy instruments" or specific measures taken to implement a policy. Examples of policy instruments are:

- Imposing efficiency standards.
- Setting public procurement policies.
- Imposing appliance labeling norms.
- Requiring the purchase or supply of energy from renewable sources.
- Supporting research and development in demonstration projects.

One of the most important aspects of policy implementation is the thorough evaluation of the costs of reducing CO_2 emissions. From a developing-country perspective, the discussion of costs and benefits has to take into account the need for policies promoting rapid economic growth. Achieving such a balance between economic development and emissions abatement requires the adoption of domestic policies aimed at improving the efficiency of energy use and facilitating fuel switching, as well as the implementation of international policies enabling easier access to advanced technologies and external resources.

It is expected that some countries will offer certain tax reductions for those businesses that promote renewable energy technologies, especially because these technologies are characterized by low or zero CO_2 emission.

Some possible solutions include cleaning fossil fuels before combustion, burning fuels more cleanly (i.e. using fluidized bed technology for coal), using renewable energies, switching to a hydrogen economy, implementing thermal energy storage technologies, promoting efficient public transport, and using more fuel-efficient vehicles.

Exergy analysis is of major importance in assessment of sustainability, because the exergy-based efficiency of systems and processes represents a true measure of imperfections. It also indicates the possible ways to improve the energy systems and to design better ones. Destruction of exergy must be reduced as much as possible. The assessment of exergy destruction offers the opportunity to quantify the environmental impact and the sustainability of any energy system. In the next paragraphs, we introduce the sustainability index (SI) and explain its influence in quantitative sustainability assessment and energy policies. Other sustainability assessment indicators are introduced as well. Some examples of energy policies and strategies include encouraging the expansion of energy- and exergy-efficient systems, expanding the use of renewable energy, widening clean fossil fuel combustion technologies. The environmental impact assessment (EIA) methods for energy systems can be classified in four categories:

- Environmental tools including impact assessment and ecological footprints.
- Thermodynamic tools, performance indicators, energy, exergy and material flux analyses.
- Sustainability tools including life cycle assessment, sustainable process index (SPI), exergetic SI, exergetic improvement potential, and greenization factor.
- Risk assessment.

The *EIA* is a tool used in assessing the potential environmental impact of a proposed activity. The derived information can assist decision making regarding whether or not the proposed activity will pose any adverse environmental impacts. The EIA process assesses the level of impacts and provides recommendations to minimize such impacts on the environment.

The *ecological footprint*s analysis is an accounting tool that enables the estimation of resource consumption and the waste assimilation requirements of a defined human population or economy in terms of corresponding productive land use.

The *SPI* is a means of measuring the sustainability of a process for producing goods. The unit of measure is m^2 of land. It is calculated from total land area required to provide raw materials, energy, infrastructure, production facilities, and disposal of wastes.

The *material flux analysis* is a materials-accounting tool that can be used to track the movement of elements of concern through a specified system boundary. The tool can be adapted further to perform a comparative study of alternatives for achieving environmentally sound options.

Risk assessment can estimate the likelihood of potential impacts, as well as the degree of uncertainty or likelihood that the impact will occur. Once management has been informed about the level of risk involved in an activity, the decision about whether such a risk is acceptable can be made.

Exergy may provide the basis for an effective measure of the potential impacts of a substance or energy form on the environment. It is important to mention that, in order to practice in the area of energy systems and the environment, engineers and researchers must have a thorough understanding of exergy and the insights that exergy can provide into the efficiency, environmental impact, and sustainability of energy systems. Furthermore, as energy policies increasingly play a role in addressing sustainability issues and the broad range of local, regional, and global environmental concerns, policy makers must appreciate the importance of exergy and its ties to these concerns. Thus, the need to understand the linkages between exergy, energy, and environmental impacts has become increasingly significant.

The resource depletion factor is based on the relative magnitude of exergy destruction within a system or process with respect to energy input. Connelly and Koshland (1997) proposed such a factor, denoting it the "depletion number." The depletion number is defined as follows:

$$D_p = \frac{Ex_d}{Ex_{in}}$$

The relationship between the depletion factor and exergy efficiency is

$$\psi = 1 - D_p$$

We express the sustainability of a fuel resource by a sustainability index (SI), which is the inverse of the depletion number:

$$SI = 1/D_p$$

Another criterion for assessing renewable energy systems is the improvement potential defined by Van Gool (1997):

$$IP = D_p \times \left(1 - \frac{Ex_{out}}{Ex_{in}}\right)$$

The greenization factor is then introduced by

$$GF = \frac{EI_{ref} - EI}{EI_{ref}}$$

where EI is the environmental impact factor.

It is important to note that the greenization factor varies from 0 to 1. A greenization factor of zero indicates that the system is not greenized. If the system is fully greenized, then the greenization factor is 1. Fully greenized systems depend on sustainable energy sources that zero or minimal environmental impact during the utilization stage (although some environmental impact is associated with system construction). In greenization factor equation, the environmental impact factor must be specified for two cases: the reference system and the greenized system. Depending on the specific problem analyzed, various types of environmental impact factors may be formulated. For many energy systems, specific GHG emissions can be used to determine the environmental impact factor.

In other cases, compounded environmental impact factors can be defined by accounting for several types of pollution. Defining an environmental impact factor in a comprehensive manner can be achieved by considering a weighted average of the indicators listed in Table 2.3. Some illustrative examples are given at the end of this section for better clarification.

Development of sustainable energy systems requires modification, adaptation, retrofitting, or replacement of the existing systems, which are normally based on conventional energy sources such as fossil fuels: natural gas, petroleum, or oil. Any energy system can be characterized by an environmental impact factor. The greenized system must have a lower

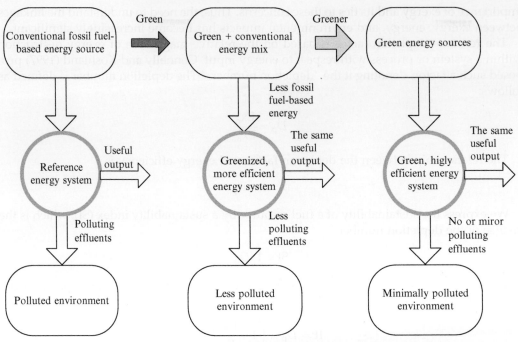

FIGURE 2.26   Greenization of energy systems for better sustainability. *Modified from Dincer and Zamfirescu (2012).*

environmental impact factor than the initial reference system. The greenization process of an exergy system is then illustrated schematically in Figure 2.26.

Pollution of the environment with GHG and other types of pollution produce reactions from society and natural ecosystems. The societal reaction to pollution can be of two kinds: *adaptation* and *mitigation*. Adaptation involves taking measures aimed at reducing the society's vulnerability to various pollution effects such as climate change (temperature rise, sea water level rise, droughts, storms, floods). A policy plan for increasing the efficiency of energy supply systems is a measure of adaptation to the pollution effects of energy systems and to the depletion of fuels. Other energy-related adaptation policies include diversification of energy sources by encouraging the development of renewable sources and improvement of electricity distribution infrastructure. An example of adaptation to the potentiality of sea level increase is in The Netherlands' national plan to prevent any future sea flood in their coastal area.

Another societal reaction to the environmental impact of energy systems is the *mitigation* of the systems' effects. GHGs are the most important target of mitigation. It appears that a good policy to encourage carbon mitigation entails imposing a carbon tax to encourage sustainable energy development. The carbon tax seems likely to be applied in most of the developed countries. By 2050, it is projected that developed countries will implement a carbon tax in the range of \$25–\$250 per ton of carbon dioxide equivalent emitted. If projected policy measures do not include a carbon tax, the mitigation of GHG will be decreased. Based on IPCC (2007), GHG mitigation with no carbon tax is estimated in the range of 5–7 Gt CO_2 equivalent. The GHG migitation with a carbon tax of \$100/t of CO_2 will increase the mitigation by

1.5–2 times, however. The global level of GHG mitigation ranges from a minimum of 5 Gt CO_2 equivalent per year to a maximum of 30 Gt CO_2 equivalent per year. Some mitigation measures are:

- Applying energy efficiency and energy conservation measures.
- Encouraging development of nuclear power generation systems.
- Switching fuels from coal to natural gas.
- Making use of renewable thermal energy from geothermal, biomass, or solar radiation.
- Making use of renewable electricity from solar radiation, hydro-energy, wind, biomass, geothermal, tidal, and ocean thermal (via ocean thermal energy conversion).
- Using combined heat and power systems and multi-generation systems.
- Applying carbon dioxide capture and storage.
- Applying waste heat recovery and energy extraction from waste materials.

Figure 2.27 provides an estimation of GHG mitigation according to energy supply and utilization by the year 2100, based on averaged data obtained from IPCC (2007). It can be remarked that the most of the mitigation is due to the use of renewable energies and most of the remaining mitigation is due to the application of energy efficiency and conservation measures.

The life-cycle GHG emission of electricity is a very important indicator. The estimated GHG emissions from power generation systems are presented in Figure 2.28 in terms of grams of CO_2 equivalent per kWh of electricity produced. Note that a power generation system using fresh biomass does not emit actual GHG in the atmosphere, since the GHG resulting from its combustion is sourced from the atmosphere, as the green plant in the biomass absorbed carbon dioxide during its life. Figure 2.27 indicates that the power generation technologies with smaller carbon footprints are hydropower and OTEC (ocean thermal energy conversion). It is interesting to note that PV arrays have a life-cycle GHG emission comparatively the same as nuclear power plants. Both nuclear and PV technologies emit negligible greenhouse gases during the power generation phase. However, in both cases, there are GHG emissions during the construction (manufacturing) and decommissioning

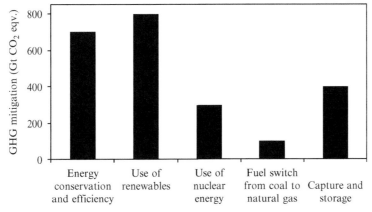

FIGURE 2.27   GHG mitigation potential by 2100. *Data from IPCC (2007), averaged.*

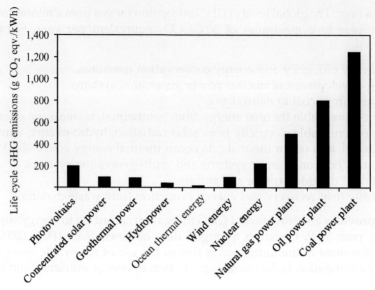

FIGURE 2.28   Lifecycle GHG emissions from various power generation technologies. *Data from Edenhofer et al. (2012).*

phases. Manufacturing PV arrays emits especially large quantities of GHG due to an energy intense process requiring high-tech procedures of silicone doping under vacuum. On the other hand, building nuclear facilities involves the construction of massive reinforced concrete structures, during which a large volume of GHG is emitted. Nevertheless, the highest level of GHG pollution comes from fossil fuel power plants, with 0.6 kg CO_2 for natural gas, 0.8 kg CO_2 for oil, and 1.25 kg CO_2 (in average) for coal-fired power plant per MWh of power generation.

Sustainability of carbon-based fuels can be also characterized by their *GHG emission factor* at the utilization phase, which can be calculated in a simplified manner, assuming a stoichiometric reaction of fuel combustion that goes to completion. If the fuel is used for heating purposes, then the stoichiometric GHG emission factor can be estimated based on the lower heating value (LHV) of the fuel (or net calorific value—NCV—which is applicable to solid fuels). Figure 2.29 suggests two types of energy conversion systems operated with carbon-based fuels. System (a) generates heat, while system (b) generates power. For both systems, it is assumed that a stoichiometric oxidation reaction is conducted in an adiabatic chamber such that the amount of heat generated is equivalent to the LHV (for gaseous and liquid fuels) or NCV (for solid fuels). Thus, $q = $ LHV, in kJ/mol.

If the system generates power (i.e., mechanical work), then the combustor is assumed to be ideally coupled to a reversible heat engine. As a result, the system in case (b) generates power in an measure equivalent to the specific chemical exergy of the fuel ($w = \mathrm{ex}^{ch}$, in kWh/mol). The stoichiometric reaction of oxidation determines the number of moles of carbon dioxide emission per number of moles of oxidized fuel. For example, for methane oxidation, one has $CH_4 + 2O_2 \rightarrow CO_2 + 2H_2O$. Therefore, 1 mol of GHG is emitted per mole of methane combusted.

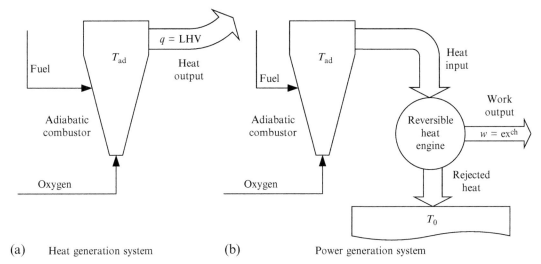

(a)   Heat generation system   (b)   Power generation system

FIGURE 2.29 Models of heat generation (a) and power generation (b) systems for determining the stoichiometric GHG emission factors of carbon-based fuels.

It can be easily observed that, for a stoichiometric oxidation process, the number of moles of carbon dioxide is equal to the number of atoms of carbon from fuel's molecule. One can denote with n_C the number of carbon atoms from fuel's molecule. With this notation, one can define a stoichiometric oxidation process for a saturated hydrocarbon with linear molecule. Consider the following example:

$$C_{n_C}H_{2n_C+2} + 0.5(3n_C+1)O_2 \rightarrow n_C CO_2 + (n_C+1)H_2O$$

We use EF_H to denote the stoichiometric GHG emission factor at heat generation of a fuel and define the factor according to

$$EF_H = \frac{44n_C}{LHV}$$

where the factor 44 in the numerator represents the molecular mass of carbon dioxide expressed in g/mol, and, the LHV of the assumed fuel is expressed in kJ/mol. Another factor of interest is the stoichiometric GHG emission factor at power generation of a fuel, denoted with EF_W and defined by

$$EF_W = \frac{44n_C}{ex^{ch}}$$

where ex^{ch} is the chemical exergy of the fuel expressed in kWh/mol.

Some numerical examples of calculating the emission factors of common fuels are given next. Take coal as a first example. The simplest assumption is to assimilate coal as graphite. Then the stoichiometric combustion reaction is

$$C + O_2 \rightarrow CO_2$$

which indicates that $n_C = 1$.

The LHV of coal can be estimated as the negative of formation enthalpy of carbon dioxide, which is LHV = 394 kJ/mol, and the chemical exergy of graphite (carbon) can be taken from Appendix B, as being $ex^{ch} = 0.114$ kJ/kWh. Given the above formulas, the GHG emission factors can be calculated as follows:

$$EF_H = \frac{44 \times 1}{394} = 0.112 \frac{gCO_2}{kJ} \text{ and } EF_W = \frac{44 \times 1}{0.114} = 385 \frac{gCO_2}{kWh}$$

A better estimate of emission factors can be obtained if one considers a specific type of coal such as anthracite or lignite. In general, any coal can be assimilated as a mixture of carbon, hydrogen, and sulfur (together with ash and moisture). A general chemical formula of complete coal combustion is as follows:

$$C_aH_{2b}S_c + (a + b + c)O_2 \rightarrow aCO_2 + bH_2O + cSO_2$$

where the stoichiometric factors for average anthracite are $a = 0.7, b = 0.2, c = 0.003$. These factors are from Dincer and Zamfirescu (2011) Chapter 6, assuming average values. Furthermore, the following values are approximated for the average anthracite: LHV = 300 kJ/mol, chemical exergy $ex^{ch} = 0.088$ kWh/mol. The above combustion equation then indicates that $n_C = 0.7$. It follows that

$$EF_H = 0.103 \frac{gCO_2}{kJ} \text{ and } EF_W = 351 \frac{gCO_2}{kWh} \text{ for anthracite}$$

Consider further gasoline fuel—assimilated as octane C_8H_{18}—with LHV = 4,959 kJ/mol and chemical exergy of $Ex^{ch} = 1.504$ kWh/mol. The stoichiometric combustion equation is

$$C_8H_{18} + 12.5O_2 \rightarrow 8CO_2 + 9H_2O$$

which indicates that $n_C = 8$. Therefore, for gasoline, one obtains

$$EF_H = 0.071 \frac{gCO_2}{kJ} \text{ and } EF_W = 234 \frac{gCO_2}{kWh} \text{ for gasoline}$$

Similarly, one may assume that diesel is formed of pure cetane (*n*-dodecane) with chemical formula $C_{16}H_{34}$, which gives $n_C = 16$. Diesal also has the following values: LHV = 9,972 kJ/mol and $Ex^{ch} = 2.861$ kWh/mol. Therefore,

$$EF_H = 0.071 \frac{gCO_2}{kJ} \text{ and } EF_W = 246 \frac{gCO_2}{kWh} \text{ for diesel}$$

Another fuel of interest is the liquefied petroleum gas, which can be represented by pure propane, C_3H_8. The stoichiometric number for carbon dioxide emissions is $n_C = 3$, the LHV = 2204 kJ/mol, and $ex^{ch} = 0.671$ kWh/mol. It results that

$$EF_H = 0.060 \frac{gCO_2}{kJ} \text{ and } EF_W = 197 \frac{gCO_2}{kWh} \text{ for propane}$$

Natural gas can be represented by methane, CH_4, with a stoichiometric number $n_C = 1$, LHV $= 811$ kJ/mol, and $ex^{ch} = 0.233$ kWh/mol. Therefore, it results that

$$EF_H = 0.054 \frac{gCO_2}{kJ} \text{ and } EF_W = 189 \frac{gCO_2}{kWh} \text{ for methane}$$

The above calculations indicate that the carbon dioxide emission factors for refined natural gas are the smallest while the factors for coal are the highest. This observation suggests that shifting from coal-based power generation to power generation with natural-gas-fired power stations may lead to environmental benefits. It is instructive to exemplify here the usefulness of the greenization factor in estimating the benefit of a greenization measure. Assume that a coal-fired power plant is replaced with a natural gas power plant producing the same amount of power. The greenization factor can be calculated in a simple manner, if one assumes that the environmental impact can be estimated by the emission factor given in terms of CO_2 emissions. Thus, taking the average anthracite coal as reference fuel, the environmental impact is estimated with (see above)

$$EI_{ref} = EF_W = 351 \frac{gCO_2}{kWh} \text{ for anthracite}$$

The environmental impact for the greenized power plant (fueled by natural gas) is then

$$EI = EF_W = 189 \frac{gCO_2}{kWh} \text{ for methane}$$

It results that the greenization factor obtained by shifting from coal-fired power plants to natural gas-fired power plants is

$$GF = \frac{EI_{ref} - EI}{EI_{ref}} = \frac{351 - 189}{351} = 0.46$$

which means close to 50% greenization, whereas a full greenization factor is always 1.

2.5 CONCLUDING REMARKS

This chapter provides an introduction to existing energy resources, types of energy generation, and their links to the environment and sustainable development. Predictions of energy consumption and population growth are correlated, and the environmental impact of energy use discussed. The main pollutants and the major types of environmental impact factors are introduced. The text also describes quantitative indicators used to characterize the environmental impact of involved processes. Several environmental criteria, including depletion number, improvement potential, sustainability factor, and greenzation factor, are also discussed.

Study Problems

2.1 How does the use of renewable energy resources improve sustainability?
2.2 What is the relationship between energy efficiency, environment, and sustainability?
2.3 What is the most abundant energy resource on earth? What is its derived forms?

2.4 Enumerate and describe briefly the fundamental energy sources on earth.

2.5 Explain the nature of tidal energy.

2.6 How many kilograms of green wood produce the same amount of energy as 1 kg of lignite? Compare the carbon dioxide emission per unit of power generated in each of the cases.

2.7 What are the main environmental impacts of energy generation and utilization?

2.8 Define radiative forcing and global warming potential.

2.9 How and why does acidic precipitation affect the environment?

2.10 Is it important to improve systems efficiency for better sustainability?

2.11 Quantify the environmental benefit of cogeneration of power and heat.

2.12 List the main sustainability indicators for power generation systems.

Nomenclature

C	concentration
D_p	depletion number
EI	environmental impact factor
Ex	exergy, kJ
ex	molar specific exergy, kJ/mol
F	radiative forcing
GF	greenization factor
GWP	global warming potential
IP	improvement potential
M	molecular mass, kg/kmol
m	mass, kg
m	mass flow rate, kg/s
n	net calorific value, MJ/kg
NCV	number of moles
Q	heat flux, kW
W	work rate, kW

Greek letters

α	stoichiometric factor
η	energy efficiency (coal utilization factor)
λ	excess air ratio
ψ	exergy efficiency

Subscripts

C	coal
in	inlet, input
net	net
out	output

Superscripts

ch	chemical

References

Bejan, A., Lorente, S., 2006. Constructal theory of generation of configuration in nature and engineering. J. Appl. Phys. 100, 041301.

Blackwell, D.B., Steele, J.L., Carter, L.S., 1991. Heat-flow patterns of the North American continent; a discussion of the geothermal map of North America. In: Selmons, D.B., Engdahl, E.R., Zoback, M.D., Blackwell, D.B. (Eds.), Neotectonics of North America. Geological Society of America, Boulder, CO, pp. 423–436.

BP, 2010. British Petroleum Statistical Review of World Energy, June. London, UK. Internet Source, http://www.bp.com/en/global/corporate/about-bp/energy-economics.html.

Connelly, L., Koshland, C.P., 1997. Two aspects of consumption: using an exergy-based measure of degradation to advance the theory and implementation of industrial ecology. Resour. Conserv. Recycl.. 19, 199–217.

Dincer, I., Zamfirescu, C., 2011. Sustainable Energy Systems and Applications. Springer, New York.

Dincer, I., Zamfirescu, C., 2012. Potential options to greenize energy systems. Energy. 46, 5–15.

Edenhofer, O., Madruga, R.P., Sokoma, Y., 2012. Renewable Energy Sources and Climate Change Mitigation. Special report of the Intergovernmental Panel of Climate Change, Cambridge University Press, Cambridge.

Ermis, K., Midilli, A., Dincer, I., Rosen, M.A., 2007. Artificial neural network analysis of world green energy use. Energ. Pol. 35, 1731–1743.

European Commission, 2006. World Energy Technology Outlook—2050. WETO-H_2. European Commission, Bruxelles.

IPCC, 2007. Climate change 2007: synthesis report. Intergovernmental Panel on Climate Change. IPCC Plenary XXVII, Valencia, Spain.

ISO International Standard 14042, 2000. Environmental Management—Life Cycle Assessment Life Cycle Impact Assessment. International Organisation for Standardisation (ISO), Geneva.

Latour, S.R., Menningmann, J.G., Blaney, B.L., 1982. Waste Heat Recovery Potential in Selected Industries.US Environmental Protection Agency—Industrial Environmental Research Laboratory report #EPA-600/S7-82-030.

Tiwari, G.N., Ghosal, M.K., 2007. Fundamentals of Renewable Energy Resources. Alpha Science International, Ltd., Oxford, UK

Van Gool, W., 1997. Energy policy: fairly tales and factualities. In: Soares, O., Cruz, A., Perreira, G., Soares, I., Reis, A. (Eds.), Innovation and Technology-Strategies and Policies, vol. 3. Springer, Amsterdam, pp. 93–105.

Fossil Fuels and Alternatives

3.1 INTRODUCTION

According to a general definition, a *fuel* represents a material which can be altered to release energy in a controlled manner. The "release of energy" supposes the existence of a mechanism of energy transfer. According to thermodynamics, the mechanism of energy transfer between a system and its surroundings must be either by work or by heat.

In fuel combustion systems, energy is released in the form of heat. In a fuel cell system, energy is released in the form of electrical work produced by electrochemical conversion of the energy embedded in the fuel.

In nuclear power reactor, the uranium fuel fissions and generate high temperature heat under controlled conditions. Various heat and work transfer mechanisms are possible in practical systems driven by fuels. Such fuel-driven systems involve purely radiative heat transfer between combustion gases and the walls of the combustion chamber maintained at a much lower temperature, heat transfer by mixed convection and radiation between hot flue gases and heat exchangers, heat transfer by convection between stack gases and heat recovery equipment, heat transfer between a nuclear reactor chamber and a high-pressure steam generator, work transfer between a steam turbine and an electric generator, work transfer between a piston and a crankshaft in reciprocating engines, and electrical work transfer between a fuel cell and its exterior circuit.

Fuels are very important for society. Most fuels are used in combustion processes. A combustion process represents a thermochemical reaction of oxidation that occurs with the production of a flame. In this process a fuel is combined with air from the atmosphere. A certain activation energy is required to initiate the oxidation reaction. Typically, the activation energy is obtained by increasing the temperature of the combustion reactants (fuel and air), and the electrochemical "combustion" process—more accurately denoted as electrochemical fuel oxidation—is used in devices such as fuel cells to generate electric power directly. Natural gas is the preferred fuel for solid oxide fuel cells—a commercially available technology to generate power and high-grade heat. Hydrogen is the preferred fuel for proton exchange fuel cells—also a commercially available technology for generating power, and possibly to recover low-grade heat.

Avoidance of flame combustion is advantageous in many instances because it limits the number of side products of the reaction process, it allows for better heat recovery from product gases, and it may facilitate the separation of carbon dioxide. Some advanced combustion technologies which operate flamelessly are catalytic combustion and chemical looping combustion.

Regardless of the types of combustion technology, fuels that contain carbon atoms generate carbon dioxide when reacting with air. If the source of fuel is an underground deposit such as an oil well, coal mine, natural gas field, tar sand bitumen, or oil shale, then, by using such fuels in the combustion process, additional carbon dioxide is released into the atmosphere. This may have negative consequences on the environment, as it affects the radiative balance of the earth and contributes to the global warming effect.

Fossil fuels are those fuels that can be mined from underground (or undersea) deposits and include high percentages of carbon. The main types of fossil fuels are coal, petroleum, and natural gas. The common theories consider that fossil fuels are formed from fossilized life systems, including any kind of vegetation or animal residuals that have been buried since historical times and decomposed biochemically by processes such as anaerobic digestion.

The world of today is dependent on the use of fossil fuels as a primary energy source, and this is especially obvious in all industrialized countries. Some major users of fossil fuels are the utility, industrial, residential, and transportation sectors. As mentioned in Chapter 2, for a sustainable development, it is important not to affect the natural concentration of gases in the atmosphere, and especially not to introduce additional carbon dioxide and other greenhouse gases (GHG) into it. Traditional combustion with fossil fuels may lead to slight incremental increase of GHG in the atmosphere, which in turn can induce radiative imbalance in the earth's atmosphere. Even if the radiative imbalance appears to be minor, it can have major consequences, including production of extreme events (droughts, floods, hurricanes, etc.).

There is an increased trend to promote alternatives to fossil fuels which are more environmentally benign. The leading candidate for alternative fuel is hydrogen, which is a carbon-free fuel and can be produced from sustainable energy sources such as renewables and nuclear, or from waste energy and materials. Another major fuel or fuel source—which is carbon-neutral—is biomass. According to the definition, *biomass* represents the biological material of any recently living organism. Biomass has relatively rich energy content and can be converted to heat and electricity in multiple ways: thermally, chemically, electrochemically, photo-chemically, or biochemically. Not only are biofuels better from an environmental standpoint than fossil fuels, but in some cases they are also more effective. For example, biodiesel—which is derived from plants—has a higher cetane rating than the usual fossil-based diesel (see Speight (2011) for details).

Use of fossil and biofuel blends represents a good choice for transition to a sustainable energy economy. Fissile materials such as uranium and thorium can be also considered as alternatives to fossil fuels since, at the utilization phase, a nuclear power plant emits only a negligible amount of GHG (due to indirect activities such as transportation, periodic running of emergency power generators for tests, etc.).

In this chapter, fossil fuels and alternative replacements are considered and reviewed. In the first part, general aspects are discussed and definitions of important parameters such as heating values, flammability, and autoignition temperature (AIT) are given. Next, fossil fuels—categorized as conventional and unconventional—are reviewed. Then the chapter focuses on biomass-derived fuels, fuel blends, and hydrogen. Some information regarding nuclear fuels and their role as a fossil fuel alternative are given in Chapter 5.

3.2 FUELS CLASSIFICATION AND MAIN PROPERTIES

Fuels can be categorized in two major classes, fossil fuels and alternatives to fossil fuels. Figure 3.1 illustrates a general classification of fuels. It is shown that conventional fossil fuels are coal, petroleum (oil), and natural gas as well as all petrochemical fuels obtained in refineries, such as gasoline, diesel, refined natural gas, etc. There are some unconventional fossil fuel sources which have started to be exploited more recently. These sources include tar sand bitumen, oil shale, and gas hydrates deposits. Synthetic crude oil and synthetic natural gas can be obtained from unconventional fossil fuel sources which can be further processed to obtain end user fuels.

As indicated in Figure 3.1, the first and the major alternative to fossil fuels is biomass. This is a very diverse fuel source containing organic substances, derived from recently living biological systems. Rough biomass can be combusted directly or it can be processed to obtain higher-quality fuels as solids, liquids, or gases. Waste materials and recovered waste energy can be also used to produce fuels. Some technologies of waste recycling for fuel are presently studied worldwide, e.g., conversion of waste plastic materials to diesel fuel. Hydrogen is also an alternative fuel if it is derived from sustainable sources using sustainable technologies with limited or no environmental pollution. Other synthetic fuels may be ammonia, urea, ethers, etc. Another alternative to fossil fuels is represented by fuel blends. Fuel blends emit lower carbon per unit of released energy because a part of emissions originates from a biofuel such as bioethanol, while another part is derived from the fossil fuel component.

Conventional nuclear fuel—uranium-235 (0.7% natural occurrence)—has been used in the last 60 years in nuclear power plants and represents a mature technology alternative to power generation from fossil fuels. Two unconventional fissile fuels which can be produced artificially from abundant resources are envisaged with the next generation of nuclear reactors.

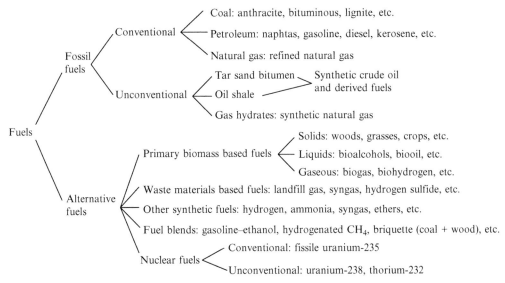

FIGURE 3.1 Classification of fuels.

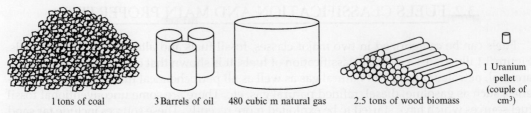

FIGURE 3.2   Calorific equivalents of main types of fuels.

These are plutonium ($^{239}_{94}$Pu), which can be obtained from non-fissile uranium-238 (99.3% natural occurrence) and uranium-238 ($^{233}_{92}$U), which is produced from thorium-232 (~100% natural occurrence).

The calorific value represents the essential technical parameter of a fuel. It expresses the heat which can be released per quantity of fuel utilized. Here, the quantity of fuel can generally be measured in kg, but other units are also used depending on the nature of the fuel. For fluids, the quantity is typically measured in volume units such as barrels (for petroleum) or normal cubic meters (for natural gas). Figure 3.2 illustrates graphically the calorific equivalents of various fuels. The calorific value of fuels is usually given in one of the following two forms:

- Gross calorific value (GCV) of solid fuels or higher heating value (HHV) for fluid fuels, which represents the heat of combustion when all combustion products are brought to the reactants' (fuel and oxidant) temperature, condensing all water vapor. Note that the gross heating value accounts for water existent in the fuel prior to combustion, which is of relevance for solid fuels such as coals and biomasses.
- Net calorific value (NCV) of solid fuels or lower heating value (LHV) of fluid fuels, which represents the heat of combustion case when all products are brought to the reactants' temperature but water remains in vapor phase; the NCV (LHV) is determined by subtracting the heat of evaporation of water from the GCV (HHV) value.

Fuels release energy due to the exothermic oxidation reaction with atmospheric air (except the nuclear fuels). Three atomic elements included in the molecular structure of any fuel are responsible for heat generation by oxidation, namely carbon, hydrogen, and sulfur. The heat released during oxidation of carbon, hydrogen, and sulfur is equivalent to the formation enthalpy of carbon dioxide, water, and sulfur dioxide. The oxidation reactions and their reaction enthalpies are given in Table 3.1. It can be observed that carbon has the highest reaction

TABLE 3.1   Reaction Enthalpies of Carbon, Hydrogen, and Sulfur Oxidation

Element	Oxidation reaction	Reaction enthalpy			Remark
		kJ/mol	kJ/kg	kJ/dm^3	
Carbon	$C + O_2 \rightarrow CO_2$	−393,486	32,790	65,580	Amorphous C at room temperature
Hydrogen	$H_2 + 0.5O_2 \rightarrow H_2O$	−241,811	120,905	9957	Standard pressure and temperature
Sulfur	$S + O_2 \rightarrow SO_2$	−296,792	9275	18,179	Beta phase S at room temperature

oxidation enthalpy per unit of volume, while hydrogen has the highest oxidation enthalpy per unit of mass.

The presence of carbon in a fuel molecule may lead to a high energy density (high heating value per unit of volume). Furthermore, as seen in Table 3.1, the presence of hydrogen in a fuel molecule may lead to a high HHV per unit of mass. In general, fuels containing carbon and hydrogen—e.g., hydrocarbons—represent the best option for the transportation sector or for remote applications: hydrocarbon fuel tanks are neither heavy nor voluminous.

The calorific value of fuels can be correlated to their hydrogen versus carbon ratio (H:C). There are two main theories regarding the influence of oxygen atoms present in some fuels (e.g., in alcohols such as CH_3OH, coals, and other fuels) on the HHV, which are discussed as follows. According to Ringen et al. (1979) the heating value of a general fuel containing hydrogen, carbon, sulfur, and oxygen atoms can be determined using the equation developed by Dulong in nineteenth century. The Dulong model is applicable for fuels which contain $<10\%$ oxygen by mass, and is given by

$$\text{HHV}(\text{kJ/kg}) = 337\,\mathcal{C} + 1442(\mathcal{H} - 0.125\,\mathcal{O}) + 93\,\mathcal{S} \tag{3.1}$$

where the mass fraction of constituent atoms from fuel's molecular composition is denoted with $\mathcal{C}$, $\mathcal{H}$, $\mathcal{S}$, and $\mathcal{O}$ for carbon, hydrogen, sulfur, and oxygen, respectively.

In Equation (3.1) it is assumed that oxygen atoms embedded in the fuel react with hydrogen only and form water, while the remaining hydrogen further reacts with combustion air. A more elaborate model for the heating value of fuels was developed by the middle of twentieth century by Boie at the University of Dresden. As indicated in Ringen et al. (1979), the model of Boie (1953) accounts for the possible existence of nitrogen atoms in fuels and the formation of NO_x. Note that NO_x formation is an endothermic process which reduces the heating value of fuels (formation enthalpy of NO is 91 kJ/mol and that of NO_2 is 34 kJ/mol). The equation of Boie is

$$\text{HHV}(\text{kJ/kg}) = 35.171\,\mathcal{C} + 1162.52\,\mathcal{H} + 104.67\,\mathcal{S} + 62.8\,\mathcal{N} - 110.95\,\mathcal{O} \tag{3.2}$$

The approximate formula for calculation of LCV (or NCV) is based on the mass fraction of moisture in the combustion (flue) gases, $\mathcal{M}$, and the latent heat of water, h_{fg}, according to

$$\text{LHV} = \text{HHV} - \mathcal{M}\,h_{fg} \tag{3.3}$$

A correlation between the HHV of fossil fuels and the H:C weight ratio can be obtained using Equations (3.2) and (3.3). For a first approximation, one assumes that the sulfur, nitrogen, and oxygen contents are negligible. As seen in Figure 3.3, the calorific value per unit of weight increases with the H:C ratio, a fact that suggests that lighter fuels have higher energy content. The average GCV of coals is roughly 18 MJ/kg, while the petroleum-derived fuels have a HHV, ranging from about 25 MJ/kg (for alcohols) to 55 MJ/kg (for methane).

In general, if a fuel is in contact with atmospheric air it does not ignite. The activation energy required for initiation of a combustion process can be obtained by increasing the fuel temperature. In a standard atmosphere (defined by 21% molar fraction of oxygen and a total pressure of 101.325 kPa), the lowest fuel temperature at which combustion is initiated without any exterior energy, such as a flame or a spark, is denoted as AIT. Simplified kinetic models

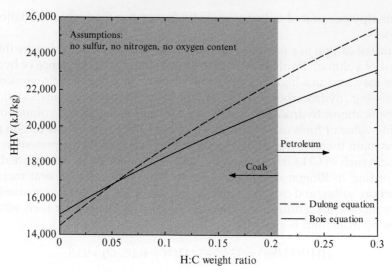

FIGURE 3.3   Approximate correlation between H:C ratio and higher heating value for fossil fuels.

are usually elaborated to predict AIT. The model must assume an air–fuel mixture at a certain temperature, and, based on mass and energy balance equations and chemical kinetics equations, a mathematical equation is formulated and solved in a transient mode to predict the evolution of species concentrations in time. If temperature is lower than AIT the mixture is stable at chemical equilibrium. If the temperature is equal to or higher than AIT, then the forward reaction proceeds at a high enough rate to produce autoignition.

The AIT depends strongly on the molecular structure of the fuel, such as the number of carbon atoms in the molecule. It also depends on the number and length of molecular bonds and on the level of branching. In recent years semi-empirical methods of AIT prediction based on the *structural contribution method* have been developed. In this method the molecular structure of the fuel is divided into functional groups which are assigned a weight factor determined from available experimental data. A regression formula for the AIT of hydrocarbons has been proposed by Albahri (2003) according to

$$AIT = 780.42 + 26.78f - 2.5887f^2 - 0.3195f^3 - 0.007825f^4 \tag{3.4}$$

where AIT results in K and f represents the so-called *overall structural group contribution* which is obtained as a superimposition of individual group contributions.

Any organic molecule can be decomposed into a number of structural groups, such as $-CH_3, =CH_2, >C<, =CH-, >CH-, >CH_2, >C=$ etc. Each functional group has an associated group contribution term denoted as f_i, where "i" is an indexation. Table 3.2 gives group contribution terms specific to paraffins, olefins, cyclic hydrocarbons, and aromatics. Many of these molecules are present in crude oil. The overall contribution is calculated with $f = \sum n_i f_i$, where n_i represents the number of "i" groups in the molecular structure.

Paraffins are one of the most encountered molecules in the constituency of crude oil. Paraffins have the general formula C_nH_{2n+2} and those with a linear molecule consist of two $-CH_3$ structural groups and $n-2$ structural groups of type $=CH_2$. The AIT is calculated with

TABLE 3.2 Group Contribution Factors for Autoignition Temperature of Hydrocarbons

Paraffins	f_i	Olefins	f_i	Cyclic	f_i	Aromatics	f_i
$-CH_3$	-0.8516	$=CH_2$	0.4682	$>CH_2$	-1.16	$=CH-$	0.4547
$>CH_2$	-1.402	$=CH-$	-1.9356	$>CH-$	0.0372	ortho $>C=$	0.9125
$>CH-$	0.0249	$\equiv CH$	-3.118	$>C<$	8.96	meta $>C=$	2.465
$>C<$	2.3226	$\equiv C-$	-1.136	$>C=$	-12.33	para $>C=$	2.097

Data from Albahri (2003).

Equation (3.4) for a wide series of paraffins, from $n=2$ (ethane) to $n=16$ (hexadecane), and the results are plotted in Figure 3.4. The behavior of AIT is nonlinear with molecular complexity (represented here by number of carbon atoms, n). In general, for lighter hydrocarbons, the AIT is higher. However, due to structural variations of the molecule there is not a monotonic decrease of AIT with n. Starting with decane ($n=10$) the AIT increases and decreases with n, but its value is around 200 °C.

In order for solid and liquid fuel to combust with air, a mixture of fuel vapors and ambient air (oxygen) has to be created. A non-gaseous fuel can emanate vapors due to evaporation or sublimation effect at the liquid–gas or solid–gas interface, depending on the case. Vapor generation is influenced by pressure and temperature conditions. At standard pressure the minimum temperature at which sufficient vapor is generated at the surface vicinity to form a flammable mixture with atmospheric air is denoted as *flash point temperature* (FPT). A source of ignition is required to inflate the flammable mixture formed at FPT. The FPT is a very important safety parameter of a fuel because it indicates the presence of combustible vapor and flammable mixtures of gases.

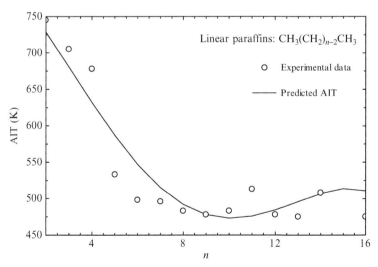

FIGURE 3.4 Predicted and experimental AITs for linear paraffins (predictions with Equation (3.4)). *Experimental data from Pan et al. (2008).*

The FPT can be correlated to the normal boiling point (NBP) of the fuel. According to Patil (1988) the following equation correlates the FPT with the NBP of organic compounds:

$$FTP = 4.656 + 0.844\,NBP - 0.000234\,NBP^2 \tag{3.5}$$

where the temperatures are expressed in K.

Figure 3.5 illustrates the variation of NBP and FPT with the molecular weight of linear paraffins. Here paraffins with a number of carbon atoms from 2 (ethane) to 20 (icosane) are considered. The FPT is below NBP and follows the same trend: it increases with the molecular mass of the hydrocarbon. It may be observed that combustion can be ignited in open air for all lighter paraffins, ranging from ethane to octane, because the FPT is below 13 °C. Heavier hydrocarbons, such as cetane (hexadecane) with an FPT of 135 °C, may require initial heating for ignition. This fact explains the need for fuel heating at cold startup of diesel engines, especially during the winter season. The conclusions derived from the results for linear paraffins illustrated in Figure 3.5 are valid for all types of petroleum fuels.

Ignition of any fuel can be initiated only under specific conditions of temperature, pressure, and oxidant (air) concentrations. In atmospheric air at standard pressure and temperature (101.325 kPa and 298.15 K), the fuel must be at a temperature higher than FPT (see above) but in addition the vapor concentration must be in a specified range denoted as *flammability limits*. The *lower flammability limit* (LFL) is defined as the volumetric concentration of vapor fuel in air below which there will be not enough fuel to allow for combustion to be self-sustained. If fuel concentration is too high, then another factor limits the ignition process: there is not enough oxygen for the combustion reaction to be maintained. The *upper flammability limit* (UFL) is defined as the volumetric concentration of fuel vapor in atmospheric air over which there is not enough oxygen to sustain combustion in the vicinity of the ignition point. For a combustion flame to be maintained, the fuel vapor concentration must be

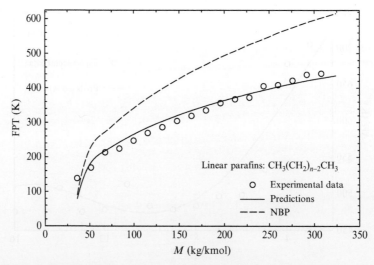

FIGURE 3.5 The normal boiling point and predicted and experimental values of flash point temperature for linear paraffins (predictions with Equation (3.5)). *Experimental data from Haynes and Lide (2012).*

between LFL and UFL. The flammability limits can be predicted based on a structural contribution method using polynomial relations similar to those in Equation (3.4). In Albahri (2003) correlations for LFL and UFL were developed based on a dataset comprising 500 combustible chemicals. The regression equations are as follows:

$$LFL = 4.174 + 0.8093f + 0.0689f^2 + 0.00265f^3 + 3.76 \times 10^{-5}f^4 \tag{3.6}$$

$$UFL = 18.14 + 3.4135f + 0.3587f^2 + 0.01747f^3 + 3.403 \times 10^{-4}f^4 \tag{3.7}$$

where the overall structural group contribution f can be determined based on individual group contributions listed in Table 3.3.

An example of utilizing the correlations for LFL and UFL according to Equations (3.6) and (3.7) is given in Figure 3.6, which represents the variation of flammability limits of linear paraffins with respect to the number of carbon atoms in the compound molecule. The LFL is predicted very well and decreases with increase of molecular complexity. For heavy hydrocarbons the LFL is below 1% by volume.

TABLE 3.3 Group Contribution Factors for Flammability Limits of Hydrocarbons

Paraff.	$f_{i,LFL}$	$f_{i,UFL}$	Oleof.	$f_{i,LFL}$	$f_{i,UFL}$	Cyclic	$f_{i,LFL}$	$f_{i,UFL}$	Aromatics	$f_{i,LFL}$	$f_{i,UFL}$
$-CH_3$	-0.8394	-1.4407	$=CH_2$	0.2479	-1.3126	$>CH_2$	-0.8386	-1.0035	$=CH-$	-1.2966	-0.8891
$>CH_2$	-1.1219	-0.8736	$=CH-$	-0.3016	-0.7679	$>CH-$	-0.9648	-0.4955	$>CH_2$	-1.6166	-1.0884
$>CH-$	-1.2598	-0.2925	$\equiv CH$	3.8518	-1.2849	$>C<$	-2.2754	0.1058	$>C=$ f.	-1.4722	-0.3694
$>C<$	-2.1941	0.2747	$\equiv C-$	1.3924	-0.4396	$>C=$	-0.1252	-0.5283	$>C=$ u.-f.	0.6649	-0.2847

Parraff., Paraffins; Oleof., Olefins; f., fused; u.-f., un-fused.
Data from Albahri (2003).

FIGURE 3.6 Predicted and experimental flammability limits for linear paraffins (predictions based on Equations (3.6) and (3.7)). *Experimental data from Haynes and Lide (2012).*

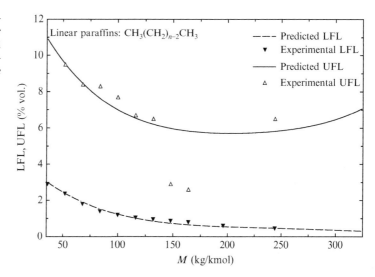

For prediction of UFL there are some discrepancies. The prediction is good for lighter hydrocarbons, in general, and one may conclude that the flash point tends to decrease with molecular complexity in this case. For heavier hydrocarbons the UFL dependency on molecular mass does not appear to be linear. In general, petrochemical fuels have a LFL below 5% and an UFL above 2.5–6%. Flammability limits of gases have higher values than those of liquids. Methane has a flammability range of 5–15% while *n*-octane has one of 0.96–6.5%.

3.3 FOSSIL FUELS

All fossil fuels resources are natural hydrocarbons as listed here with density in decreasing order and H:C ratio in increasing order: humic coals, sapropels, kerogen, bitumen, heavy and light oils, and natural gas. Here we review the main fossil fuel resources, namely coals, crude oil, natural gas, oil shale, tar sand bitumen, and gas hydrates. In the last part of this section refined fossil fuels are discussed in detail. Refined fossil fuels such as gasoline, LPG, diesel, kerosene, and jet fuel are obtained by crude oil processing in refineries.

3.3.1 Coal

Coal is a heterogeneous, combustible, sedimentary rock consisting of organic compounds formed by a complex process of conversion of fossilized biomass into solid material. From the beginning of the eighteenth to the mid-twentieth century, coal was the dominant fuel in industrialized societies. Coal can be used as fuel for heating, steam generation, or power generation (in coal-fired power plants) and as feedstock for various industries and processes such as the steel industry, production of synthetic transportation fuels, and production of a variety of chemicals.

Among other energy resources existent in the earth's crust (such as petroleum or uranium-235) it appears that coal is the most abundant. Moreover, because coal is the most readily available fuel source (it does not require refining) it appears that coal utilization in the energy sector will continue, although the conventional coal-fired power generation systems may no longer be used in the near future. It is expected that a new generation of coal-based power production plants will be developed using clean coal combustion technology, coal-biomass blending and/or carbon dioxide capture and sequestration. One of the very important strengths of coal as an energy resource is its high conversion efficiency in modern coal-fired power plants, which is in the range 37–40%, with the expectation of reaching 50% in very near future. This fact compensates for the GHG emission per unit of energy of fuel, which is the highest for coal with respect to petrochemical fuels.

According to the so-called *coal rank* there are four categories of coals: lignite, subbituminous, bituminous, and anthracite. Coal rank expresses the geological maturity of coal as a mineral. In underground deposits, coal transforms over time, starting with peat, then proceeding to lignite and up to anthracite. Peat—the precursor of coals—represents partially decayed vegetation of hemic, fibric, or sapric type that accumulates in wet soils under the absence of oxygen. Coals of low rank have high moisture and oxygen content and a soft friable texture, whereas higher rank coals are vitreous and hard, having low oxygen and moisture content and very good heating value compared to lower rank coals.

Apart from combustible constituents, coals contain two types of noncombustible compounds; these are *moisture* and *ash*. The presence of moisture in coal is either as physically or as chemically bonded molecules of water. Moisture is disadvantageous during coal combustion because a part of combustion heat must be used to heat, evaporate, and superheat the moisture. Consequently, the heating value of coals is degraded by their moisture content.

Ash represents the residue which remains after the complete combustion of coal and comprises various sulfates and oxides and other inorganic chemicals. Coal embeds in types of mineral matter that ultimately lead to formation of ash by combustion: extraneous mineral matter (gypsum, shale, sand, pyrite, clay, marcasite, calcium, magnesium) and inherent mineral matter (some of the constituents of the biomass from which coal was formed).

Lignite is the coal of lowest rank, having a dark brownish appearance, a moisture content higher than 60% by weight, ash content higher than 3%, and carbon content lower than 35%. Subbituminous coal is dull brown with a carbon content higher than 55% and moisture below 25%. Bituminous coal contains bitumen as its main combustible component and has a blackish, tarlike appearance, with a carbon content of over 65% and moisture below 15%. Anthracite, having a black-luster texture, is the coal with highest content of carbon at over 75% and with lowest moisture content at under 6%. Composition, density, and heating value of the main coal ranks are presented in Table 3.4.

Coal is comprised of the following main elements, which can be determined by a *proximate analysis*: volatile matter, fixed carbon, moisture, and ash. Proximate analysis identifies the main categories of compounds in coal using certain specific methods. *Volatile matter* is the percentage of products that are volatile—that is they can be released as gases during the pyrolysis process or by heating. Volatile matter refers to all components except water vapor that result from mixture evaporation. The loss of carbon dioxide from mineral embedded in coal structure is an example of volatile matter; other examples are hydrogen emanating from chloride minerals and sulfur from pyrite. If the mass of volatiles, ash, and moisture is extracted from the mass of coal, the remaining quantity is known as *fixed carbon*. As given in Table 3.4, for anthracite the fixed carbon is 75–85% by weight, while for bituminous, subbituminous, and lignite coal it is 50–70%, 30–57%, and 25–30%, respectively.

In general it is difficult to report coal analysis on an absolute basis. There are certain standard methods for proximate analysis which are based on physical and chemical processes

TABLE 3.4 Density, Calorific Value, and Composition of Coals by Rank

Rank	Density (kg/dm³)	GCV (MJ/kg)[a]	Composition[b]						
			C, %	H, %	O, %	S, %	N, %	Ash, %	Moisture, %
Anthracite	1.35–1.70	27.9–31.4	75–85	1.5–3.5	5.5–9.0	0.5–2.5	0.5–1.0	4.0–15.0	3.0–6.0
Bituminous	1.28–1.35	27.9–33.7	65–80	4.5–6.0	4.5–10.0	0.5–6.0	0.5–2.5	4.0–15.0	2.0–15.0
Subbituminous	1.35–1.40	17.4–23.3	55–70	5.5–6.5	15.0–30.0	0.3–1.5	0.8–1.5	3.0–10.0	10.0–25.0
Lignite	1.40–1.45	13.0–17.4	30–45	6.0–7.5	28.0–38.0	0.3–2.5	0.6–1.0	3.0–15.0	20.0–40.0

[a]*Ash-free basis.*
[b]*Percent by weight.*

such as drying, heating, pyrolysis, and combustion, but their actual application usually depends on specific laboratory equipment and practices. For example if coal is dried in air the moisture content in coal will be reduced until it reaches an equilibrium with the atmosphere in the laboratory where it is tested. Thus it is important to report in an unambiguous manner the results of proximate analysis. All proximate analyses will report the weight percentage of fixed carbon (C), volatile matter (VM), moisture (M), and ash (A). The following ways of reporting proximate analysis can be defined:

- *As-sampled*: the composition of C, VM, M, A are given as per weight of the coal sample.
- *As-received*: during transportation from sampling place to laboratory the moisture content may change; thus the as-received composition differs from the as-sampled analysis in an amount that is correlated with the moisture content change during transportation.
- *As-determined*: the proximate analysis is often made after drying the sample in the laboratory. Therefore, the composition of the received sample will differ from the composition of the analyzed sample due to the preliminary drying process.
- *Air-dried basis*: moisture is determined based on equilibrium water vapor concentration in laboratory conditions when the sample is dried in air. It is specified clearly in the analysis report whether hydrogen and oxygen in volatile matter are assumed to be part of moisture.
- *Dry basis*: the proximate analysis reports weight fractions of C, VM, A assuming that there is no moisture in the sample. In other words, after the content of moisture is determined, the weights are extracted from the sample weight.
- *Dry, ash-free basis*: in this case the moisture and ash content are subtracted from the sample's weight and only the weight fraction of fixed carbon and volatile matter are reported.

Denote with $C_{ar}, VM_{ar}, M_{ar}, A_{ar}$ the weight fractions of fixed carbon, volatile matter, moisture, and ash in the received coal sample (subscript "ar" stands for "as-received"), and M_{ad} is the moisture determined in the lab based on the air drying method. Due to the specificity of drying method M_{ad} as different than M_{ar}, in general $M_{ad} < M_{ar}$. The following conversion equation is valid between weight fractions on as-received (ar) and as-determined (ad) bases

$$\frac{C_{ar}}{C_{ad}} = \frac{VM_{ar}}{VM_{ad}} = \frac{A_{ar}}{A_{ad}} = \frac{100 - M_{ar}}{100 - M_{ad}}$$

Furthermore, if one denotes C_d, VM_d, A_d the fixed carbon, volatile matter, and ash fractions expressed on a dry basis, respectively, then the following conversion relations are valid

$$\frac{C_d}{C_{ad}} = \frac{VM_d}{VM_{ad}} = \frac{A_d}{A_{ad}} = \frac{100}{100 - M_{ad}}$$

The conversion between as-determined and dry, ash-free (subscript "daf") bases can be found with the help of the following equation

$$\frac{C_{daf}}{C_{ad}} = \frac{VM_{daf}}{VM_{ad}} = \frac{100}{100 - M_{ad} - A_{ad}}$$

Besides the proximate analysis, there is the so-called *ultimate analysis* which determines the elemental composition of coal and reports the weight fractions of carbon, hydrogen, sulfur, nitrogen, oxygen, and ash. The weight fractions of hydrogen and oxygen represent a special situation because these elements can be present in moisture (water, H_2O) and as well in the chemical constituency of volatile matter fraction or fixed carbon fraction. We denote $\mathcal{H}$ and $\mathcal{O}$ as the weight fractions of hydrogen and oxygen in the coal sample, respectively, on an as-received basis (this is hydrogen and oxygen atoms comprised in moisture and in the dry coal sample). Then, the mass fraction of oxygen and hydrogen atoms on a dry basis (subscript "d") can be determined based on the following equation

$$\frac{\mathcal{O}_d}{\mathcal{O} - 8/9\,\mathcal{M}_{ar}} = \frac{\mathcal{H}_d}{\mathcal{H} - 1/9\,\mathcal{M}_{ar}} = \frac{100}{100 - \mathcal{M}_{ar}}$$

As indicated in Dryden (1957), it is possible to correlate the volatile matter weight fractions of coals with the data resulting from ultimate analysis. Namely, the volatile matter weight fraction can be expressed with the following correlation:

$$\mathcal{VM} = 11.67\left(\frac{\mathcal{H}}{\mathcal{C}}\right)_{daf} + 1.4\left(1 - \frac{63}{\mathcal{C}_{daf}}\right)\left(\frac{\mathcal{O}}{\mathcal{C}}\right)_{daf} - 0.404$$

The GCV of coal can be made with equations similar to Equations (3.1) and (3.2) based on weight fractions of carbon, hydrogen, and sulfur and also the weight fraction of ash. Mason and Gandhi (1983) proposed the following regression equation which expresses the *GCV* of coal in MJ/kg as a function of weight fractions $\mathcal{C}, \mathcal{H}, \mathcal{S}, \mathcal{A}$ as follows:

$$\text{GCV} = 0.46\mathcal{C} + 1.44\mathcal{H} + 0.19\mathcal{S} + 0.1\mathcal{A} - 11.98 \tag{3.8}$$

More recently, Majumder et al. (2008) has developed a correlation to predict the GCV of coal on an as-received basis using data from the proximate analysis of coal. This correlation is

$$\text{GCV}_{ar} = 0.35\mathcal{C} + 0.33\mathcal{VM} - 0.11\mathcal{M} - 0.03\mathcal{A} \tag{3.9a}$$

where GCV_{ar} results in MJ/kg.

The NCV of coal on a wet basis (moisture included) can be estimated with the correlation of Van Loo and Koopejan (2008) as a function of *GCV* (given on dry basis) and weight fractions of moisture ($\mathcal{M}$) and hydrogen ($\mathcal{H}$), as given below:

$$\text{NCV} = \text{GCV}(1 - \mathcal{M}) - 2.444\mathcal{M} - 21.839\mathcal{H}(1 - \mathcal{M}) \tag{3.9b}$$

Design of coal furnaces must account for the thermodynamic properties of products of coal combustion. The combustion process of coal starts with a physical–chemical process of devolatilization during which the volatile matter releases from solid material while heating. Devolatilization is an initial stage of the pyrolysis process. At the same time *tar* is produced by destructive distillation of organic compounds such as hydrocarbons from coal (which are a combination of aromatic and aliphatic chains). The solid material remaining after devolatilization and tar release is denoted as *char*.

Determination of the standard enthalpy of coal (that is, the formation enthalpy at 298.15 K) is of high importance for design of coal combustion, gasification, liquefaction systems, etc. The standard enthalpy of coal can be determined based on a "black box" thermodynamic

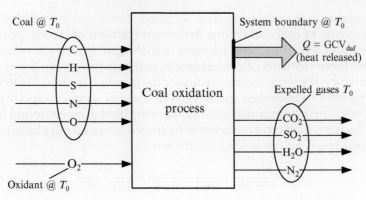

FIGURE 3.7   Thermodynamic model for determining enthalpy of formation values of coal.

model as illustrated in Figure 3.7. In standard condition coal (dry, ash-free) is fed at 298.15 K together with oxygen (O_2) in a stoichiometric amount. The resulting reaction products are only CO_2, H_2O, SO_2, and N_2. The atomic fractions of carbon, hydrogen, sulfur, nitrogen, and oxygen from the dry, ash-free coal can be calculated based on the weight fraction and molecular weight of the respective elements. Recalling that the molecular weights are 12, 2, 32, 28, and 32 kg/kmol for C, H_2, S, N_2, and O_2 respectively, the conversion of weight fractions into atomic "molar" fractions becomes

$$c = \frac{\mathcal{C}}{12v}, \; h = \frac{\mathcal{H}}{2v}, \; s = \frac{\mathcal{S}}{32v}, \; n = \frac{\mathcal{N}}{28v}, \; o = \frac{\mathcal{O}}{32v} \tag{3.10}$$

where $v = \frac{\mathcal{C}}{12} + \frac{\mathcal{H}}{2} + \frac{\mathcal{S}}{32} + \frac{\mathcal{N}}{28} + \frac{\mathcal{O}}{32}$.

Therefore, the chemical equation for coal combustion can be approximated with

$$cC + hH_2 + sS + oO_2 + (c + 0.5h + s - o)O_2 \rightarrow cCO_2 + hH_2O + sSO_2$$

From the above given equation, it results that the energy balance equation for the thermodynamic system represented in Figure 3.7, in which the enthalpy of the elements (O_2, N_2) in standard conditions is by definition zero, is written as follows:

$$\Delta H^0_{coal} = GCV_{daf} + c\Delta H^0_{CO_2} + h\Delta H^0_{H_2O} + s\Delta H^0_{SO_2}$$

The standard enthalpy of ash depends on ash composition, which in turn varies with the type of coal. Table 3.5 gives the main components of ash, their enthalpy and entropy of formation, their molecular mass, and the weight and molar fractions. The average molecular mass of ash ranges from 40 to 200 kg/kmol. The average formation enthalpy of ash ranges from about −900 to 1050 MJ/kmol. The average entropy of ash is about 50 kJ/kmol.K.

The specific heat of coal is calculated based on a dry, ash-free basis in the manner indicated in Table 3.6. It depends on the weight fraction of fixed carbon ($\mathcal{C}$) and the weight fractions of primary ($\mathcal{VM}_1$) and secondary ($\mathcal{VM}_2$) volatile matter. The total volatile matter is represented by the summation $\mathcal{VM} = \mathcal{VM}_1 + \mathcal{VM}_2$. If $\mathcal{VM} \leq 10\%$, then, as indicated in Eiserman et al. (1980), one has $\mathcal{VM}_1 = 0, \mathcal{VM}_2 = \mathcal{VM}$. If $\mathcal{VM} > 10\%$, and one has the following: $\mathcal{VM}_1 = \mathcal{VM} - 10\%$ and

TABLE 3.5 Average Composition of Coal Ash

Component	M	ΔH^0	ΔS^0	W.F. range	M.F. range
Sulfur trioxide (SO$_3$)	80	−437.9	132.7	0.1–12%	0.2–6.6
Alkali (K$_2$O, Na$_2$O)	62	−418.2	72.9	1–4%	3.1–2.8
Titanium oxide (TiO$_2$)	80	−945.2	50.2	0.5–2.5%	1.2–1.4
Magnesium oxide (MgO)	40	−601.5	26.8	0.3–4%	1.5–4.4
Calcium oxide (CaO)	56	−634.6	39.8	1–20%	3.5–15.8
Ferric oxide (Fe$_2$O$_3$)	160	−825.9	90.0	5–35%	6.1–9.6
Aluminum oxide (Al$_2$O$_3$)	102	−1674.4	51.1	10–35%	15.1–19.2
Silica (SiO$_2$)	60	−911.3	41.9	20–60%	44.1–65.1

M, molecular mass (kg/kmol); W.F. range, weight fraction range (kg component per kg ash); M.F. range, molar fraction range (moles component per moles of ash); ΔH^0, formation enthalpy (MJ/kmol); ΔS^0, formation entropy (kJ/kmol.K).
Source: Eiserman et al. (1980).

TABLE 3.6 Prediction Equations for Specific Heat of Coal, Ash, and Tar

Quantity	Equation
Specific heat of primary volatile matter	$C_{p,\,VM_1} = 0.728 + 3.391 \times 10^{-3}\,T$
Specific heat of secondary volatile matter	$C_{p,\,VM_2} = 2.273 + 2.554 \times 10^{-3}\,T$
Specific heat of fixed carbon matter	$C_{p,\,C} = -0.218 + 3.807 \times 10^{-3}\,T - 1.758 \times 10^{-6}\,T$
Specific heat of coal	$C_{p,\,coal} = CC_{p,C} + VM_1 C_{p,VM_1} + VM_2 C_{p,VM_2}$
Specific heat of ash	$C_{p,\,ash} = 0.594 + 5.86 \times 10^{-4}\,T$
Specific heat of tar	$C_{p,\,tar} = 4.22 \times 10^{-3}\,T$

Note: T, temperature in K; specific heat in kJ/kg.K.
Source: Eiserman et al. (1980).

$VM_2 = 10\%$. In Table 3.6, the predictive equations for the specific heat of ash and of tar are also given.

Entropy of coal is an important parameter for second-law-based thermodynamic analysis of coal-fired power generation systems. The entropy of coal can be calculated at a specified temperature based on specific heat and the formation entropy as follows:

$$s(T) = \Delta S^0_{coal} + \int_{T_0}^{T} \frac{C_p(T)}{T}\, dT$$

Eiserman et al. (1980) provide a correlation between formation entropy of coal and coal composition from ultimate analysis. The molar fractions defined according to Equation (3.10) are used to account for coal composition. The following factors for are introduced

$$f_1 = \frac{h}{c+n}, f_2 = \frac{o}{c+n}, f_3 = \frac{n}{c+n}, f_4 = \frac{s}{c+n}$$

Based on this formula, the Eiserman et al. (1980) correlation is further studied as given below:

$$s_{\text{daf}}^0 = 37.1653 - 31.4767 e^{-0.564682 f_1} + 20.1145 f_2 + 54.3111 f_3 + 44.6712 f_4 \qquad (3.11)$$

This correlation expresses the coal's formation entropy in kJ/kg.K on a dry, ash-free basis. Note that Equation (3.11) can be also used for predicting the formation entropy of char and tar.

The chemical exergy of good quality coals varies between 7 and 8.2 MJ per kg on a dry ash-free basis. The chemical exergy of coal, on a dry and ash-free basis, can be predicted with the correlation given by Kaygusuz (2009), which reads

$$\begin{aligned} \text{ex}_{\text{daf}}^{\text{ch}} = GCV - T_0 \left(s_{\text{daf}}^0 + f_1 s_{O_2} - c s_{CO_2} - \frac{1}{2} h s_{H_2O} - s s_{SO_2} - \frac{1}{2} n s_{N_2} \right) + c \text{ex}_{CO_2}^{\text{ch}} + \frac{1}{2} h \text{ex}_{H_2O}^{\text{ch}} \\ + s \text{ex}_{SO_2}^{\text{ch}} + \frac{1}{2} n \text{ex}_{N_2}^{\text{ch}} - f_2 \text{ex}_{O_2}^{\text{ch}} \end{aligned} \qquad (3.12)$$

where the exergy value results in MJ/kg_{DAF} and the f_i factors result from the stoichiometric equation for coal combustion and are as follows

$$f_1 = c + \frac{1}{4} h + s - \frac{1}{2} o, f_2 = c + \frac{1}{4} h + s - \frac{1}{2} o$$

and the molar fractions c, h, s, n, o are calculated based on weight fractions of carbon, hydrogen, sulfur, nitrogen, and oxygen according to Equation (3.10).

Note that the contribution of ash to chemical exergy is negligible; therefore the specific chemical exergy of coal, on a wet basis can be estimated with:

$$\text{ex}^{\text{ch}} = (1 - \mathcal{M} - \mathcal{A}) \text{ex}_{\text{daf}}^{\text{ch}} + \frac{\mathcal{M}}{M_{H_2O}} \text{ex}_{H_2O}^{\text{ch}} \qquad (3.13)$$

where $\mathcal{M}$, $\mathcal{A}$ are the weight fractions of moisture and ash, respectively, M_{H_2O} is the molecular mass of water and $\text{ex}_{\text{daf}}^{\text{ch}}$ can be predicted by Equation (3.12) (see Bilgen and Kaygusuz, 2008).

The specific emission factor at coal combustion is an important parameter for analyzing environmental impact. This factor represents the mass of carbon dioxide emitted at complete combustion per kg of coal. The *specific emissions factor* (*SEF*) can be calculated from the weight fraction of carbon in coal, C. The weight fraction must be divided by 12 kg/kmol to obtain the number of moles of carbon per kg of coal and then multiplied by 44 kg/kmol, which is the molecular mass of carbon dioxide:

$$SEF = \frac{44}{12} C \qquad (3.14)$$

Specific emission factors of coals by rank are given in Figure 3.8 for a range of moisture weight fractions, $\mathcal{M}$. The carbon content in each coal by rank corresponds to the range given in Table 3.4. Anthracite has the highest emissions because it has the highest content of carbon. For constant coal fraction (on a dry, ash-free basis) the specific emission factor decreases if the moisture increases. However, one has to note that for any increase of moisture in coal the NCV decreases.

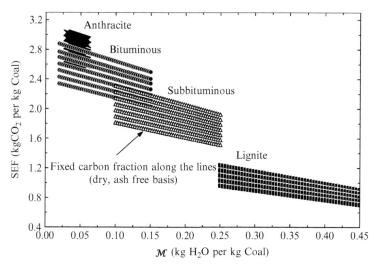

FIGURE 3.8 Specific emission factors of coals by rank as a function of moisture content.

The carbon emission per unit of net calorific heat has been introduced already in Chapter 2 as the ratio of the mass of carbon dioxide emitted per unit of heat generated (on an NCV basis) at stoichiometric combustion. This is referred to as the stoichiometric GHG emission factor at heat generation and denoted with EF_H.

The emission factor with respect to NCV is defined by the ratio of SEF to NCV, which is written as

$$EF_H = \frac{SEF}{NCV} = \frac{44}{12} \frac{\mathcal{C}}{NCV} \tag{3.15}$$

Figure 3.9 presents the stoichiometric emission factor for heat generation from coal by rank as a function of moisture content. This plot is obtained by applying Equation (3.15) and shows that, although anthracite has a high carbon content, the emissions of carbon according to the factor EF_H are rather small with respect to other coal ranks.

Another important emission factor is represented by the ratio of mass of carbon dioxide emission at stoichiometric combustion to the chemical exergy of the fuel. This kind of factor has been introduced in Chapter 2 and is denoted with EF_W. Here, the subscript W suggests that the fuel is utilized for work (or electrical or mechanical power) production. Based on this definition, one writes

$$EF_W = \frac{SEF}{ex^{ch}} = \frac{44}{12} \frac{\mathcal{C}}{ex^{ch}} \tag{3.16}$$

In the case that the chemical exergy of a coal is not specified, this factor can be calculated based on proximate and ultimate coal analyses, using Equations (3.11)–(3.13). The calculation process involves the preliminary determination of the specific exergy of coal as a function of rank and moisture content. Figure 3.10 illustrates the range of variation of specific exergy of coals by rank and moisture content. As seen, the specific entropy of lignite is the highest

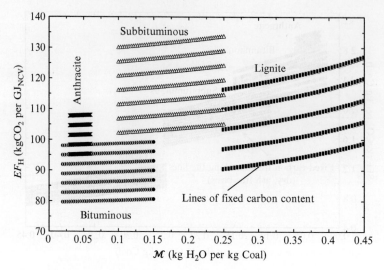

FIGURE 3.9   Emission factors with respect to NCV of coals by rank.

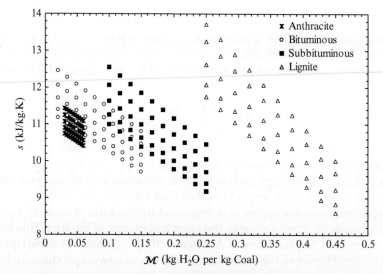

FIGURE 3.10   Specific entropy ranges of coals as a function of rank and moisture content.

among all coals, which indicates that the use of lignite as an energy source may induce large irreversibilities when compared to other coals. The chemical exergy of coals can be observed in Figure 3.11. The highest exergy content is that of anthracite, with values over 22 MJ/kg, whereas the chemical exergy of lignite is below 9 MJ/kg.

The specific emissions of GHG with respect to chemical exergy—as defined by Equation (3.16)—are presented for all ranks of coal in Figure 3.12. Bituminous coals and anthracite

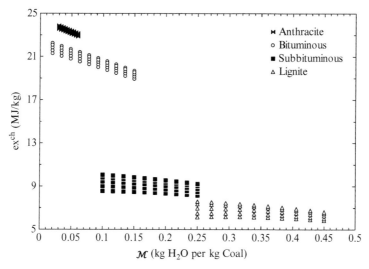

FIGURE 3.11 Chemical exergy ranges of coals as a function of rank and moisture content.

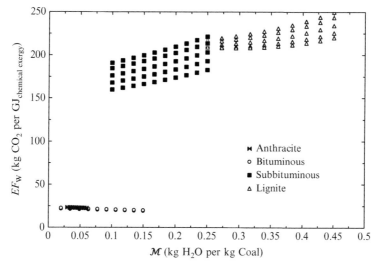

FIGURE 3.12 Emission factors with respect to chemical exergy of coals by rank.

have comparable emission factors and are the lowest among all coals, with ~25 kg CO_2 equivalent per GJ of chemical exergy. The emission factor of subbituminous coal is much higher, with over 150 kg CO_2 equivalent per GJ of chemical exergy. The highest emissions are those of lignite, with a range of 200–250 kg CO_2 equivalent per GJ of chemical exergy. Consequently, it is preferable to use anthracite or bituminous coals for power generation, whereas subbituminous and lignite can be used for heating purposes.

3.3.2 Petroleum

The word "petroleum" comes from Latin and is a combination of two words: "petra" which means rock and "oleum" meaning oil. Therefore, petroleum is a designation for oil found in rocks (sediments). Petroleum is a mineral resource comprising a range of hydrocarbon-based oils which can exist in an underground (or undersea) rock deposit as liquids, gases, or solids. Similar to coals—according to widely accepted theory—petroleum is formed from fossilized biological matter containing mostly lipids, amino-acids, carbohydrates, and lignins. Petroleum is a mixture of hydrocarbons that range from light ones (carbon number lower than five: methane, ethane, propane, butane, pentane) to heavier ones such as paraffins. Petroleum is toxic and flammable. Apart from carbon (83–87%) and hydrogen (10–14%), it contains sulfur (0.5–6%) and some metals. There is a large variety of petroleum reservoirs throughout the world—both conventional and nonconventional—including oil fields, natural gas fields, coal bed methane, oil shale, and tar sand bitumen reserves. In this section we refer to the main categories of conventional petroleum sources (crude oil, natural gas, oil shale, and tar sand bitumen) and to their properties.

3.3.2.1 Crude Oil

The main resource extracted from a conventional oil field is crude oil, consisting of a complex mixture of polymeric hydrocarbon chains of two categories: volatiles and resins. The lighter (low molecular mass) volatiles may be of naphthenic, aromatic, paraffinic, or combined nature. The more condense (heavier) hydrocarbons in crude oil consist of wax and asphaltenes. Crude oil may also contain sulfur, nitrogen, oxygen, metals, and other elements in smaller fractions.

One of the precursors of crude oil is "kerogen," which is a mixture of organic compounds and has a molecular weight on the order of 1000 kg/kmol. Different kerogens can release other fossil fuels such as natural gas, shale, or bitumen from tar sands. Bitumen represents the fraction of kerogen which is soluble in organic solvents. The conversion of kerogen into fossil hydrocarbons is referred to as catagenesis.

TABLE 3.7 General Classification of Crude Oil

Component of crude oil		Crude oil class (composition given in % by weight)				
		Paraffinic	Paraffinic/ naphthenic	Naphthenic	Paraffinic/naphthenic/ aromatic	Aromatic
Volatiles	Paraffins	46–61	42–45	15–26	27–35	0–8
	Naphthenes	22–32	38–39	61–76	36–47	57–78
	Aromatics	12–25	16–20	8–13	26–33	20–25
Resins	Waxes	1–10	1–51	Traces	0.05–1	0–0.5
	Asphaltenes	0–6	0–6	0–6	0–10	0–20

Source: Speight (2006).

According to Table 3.7, there are five general categories of crude oil, based on composition: paraffinic, paraffinic/naphthenic, naphthenic, paraffinic/naphthenic/aromatic, and aromatic. As seen in the table, the main components of crude oil in this classification are the weight fractions of paraffins, naphthenes, aromatics, waxes, and asphaltenes.

Naphthenes are cycloalkanes, a class of saturated hydrocarbons having at least one ring of carbon atoms in their molecular structure. Naphthenes have the chemical formula $C_nH_{2n-2r+2}$, with $r > 1$ representing the number of carbon rings. Aromatics are special types of hydrocarbons, such as toluene, xylene, and benzene, which produce polymers and have a planar molecular structure with at least one carbon ring and $4n+2$ electrons contributing to the covalent bonds. Waxes are polymerized hydrocarbons that are soluble in organic solvents and insoluble in water. They are of plastic aspect and are solid state at 25 °C, although their melting point is very low (starting at slightly above ~40 °C). Asphaltenes consist of a mixture of chemical compounds including hydrocarbons, nitrogen, oxygen, sulfur, and metals with a carbon to hydrogen atomic ratio of ~0.8 and average molecular mass of 800 kg/mol. Figure 3.13 presents the main volume fraction of various components in a typical crude oil composition. Pentane includes at least three molecular variations:, n-pentane, isopentane, and cyclopentane. Also, both n- and iso- structures are included in the diagram for hexane and pentane, respectively. In addition, cyclohexane is accounted for in the figure under the general label "hexane."

The general practice in petrochemistry is to measure the specific gravity of petrol (oil) in degrees Baumé or AIP degrees. The Baumé degree was defined in the eighteenth century as a scale for the hydrometer (an instrument used to measure the relative density of liquids) based on two reference points for calibration. For liquids lighter than water (viz. petroleum): mark 10 °Bé (Baumé) represents the specific gravity of pure water and mark 0 °Bé represents specific gravity of salted water with 10% NaCl by mass. The following equation shows the conversion between Baumé degree and specific gravity for liquids lighter than water at a temperature of 15 °C

FIGURE 3.13 Typical composition of crude oil. *Data from Berkowitz (1997), fraction by volume in liquid crude oil.*

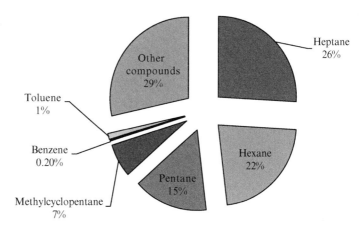

$$SG = \frac{140}{BSG + 130}$$

where BSG denotes the specific gravity expressed in Baumé degrees and SG represents the specific gravity as defined based on the ratio between the density of the liquid and the density of water:

$$SG = \frac{\rho}{\rho_{\text{water}}}$$

For historical reasons, at the beginning of twentieth century, the calibration of hydrometers in the United States had a 1.5 bias with respect to the Baumé scale. Due to petroleum trading issues the remedy for this situation—proposed by the American Petroleum Institute—has been to define an American measurement scale for the specific gravity of petroleum. The new American scale is denoted as the API scale and is defined with respect to the Baumé scale according to

$$\frac{APISG + 131.5}{BSG + 130} = \frac{141.5}{140}$$

where APISG represents the specific gravity measured in °API. At a temperature of 15 °C the specific gravity can be obtained from API scale gravity according to the conversion formula:

$$SG = \frac{141.5}{APISG + 131.5}$$

Specific gravity of crude oil varies with the exploitation site. Figure 3.14 presents the approximate distribution of API degree and specific gravity of crude oil at the main exploitation sites around the world. The composition and the properties of crude oil are of crucial importance for refinery processes. There are two main assays that are required to characterize crude oil, denoted as "inspection" and "comprehensive," respectively (these assays are analogous to proximate and ultimate analyses of coals composition). The inspection assay determines

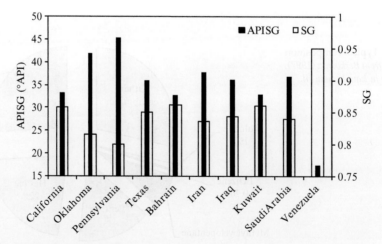

FIGURE 3.14 Specific gravity and API gravity degrees in crude oils. *Data from Speight (2006).*

the API gravity, sulfur content, pour point, viscosity, salt content, water content, wax content, trace metals, and distillation range of crude oil. The comprehensive assay determines the carbon residue yield, the volatility, the metallic constituents, the vapor pressure, and elemental analysis (i.e., determination of H:C, S:C, N:C, and O:C ratios). Typical ranges of weight fractions of carbon, hydrogen, sulfur, nitrogen, and oxygen in crude oil are represented in the chart in Figure 3.15.

The LCV of crude oil varies in the range of ~40–43 MJ/kg. There is an empirical correlation between specific gravity and the HHV of crude oil, which, according to Speight (2006), is given by

$$\text{HHV}(\text{MJ/kg}) = 52.7 - 8.8\text{SG}^2 \tag{3.17}$$

and is reported to have a deviation below 1%.

Crude oil is a mixture of components, and therefore, its standard chemical exergy must be calculated based on heuristic empirical expressions which account for the substance weight fractions of the constitutive chemical elements and compounds. The standard chemical exergy of crude oil, given in mass specific terms, can be calculated with the following formula adapted from Rivero et al. (2004):

$$\text{ex}^{\text{ch}} = \beta\,\text{HHV} + \sum_i \left(w_i \text{ex}_i^{\text{ch}}\right) + T_0 \sum_i R_i w_i \ln(w_i a_i) \tag{3.18}$$

where HHV represents the mass-based HHV of crude oil—which can be estimated with Equation (3.17)—and w_i and ex_i^{ch} represent the mass fractions and mass-based standard chemical exergies of water and the metallic elements such vanadium, iron, and nickel that enter into the constituency of crude oil, respectively.

The factor β in Equation (3.18) represents a correlation parameter between the HHV and chemical exergy of a mixture of fossil fuels with known weight fractions of $\mathcal{C}, \mathcal{H}, \mathcal{S}, \mathcal{N}$, and $\mathcal{O}$ (carbon, hydrogen, sulfur, nitrogen and oxygen, respectively). The factor β can be estimated using the equation of Szargut et al. (1988) as follows

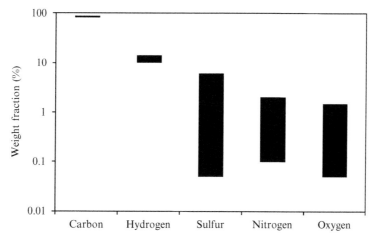

FIGURE 3.15   Range of weight fractions of main chemical elements present in crude oil. *Data from Speight (2006).*

$$\beta = 1.0401 + 0.1728f_1 + 0.0432f_2 + 0.2169f_3(1 - 2.0628f_1) + 0.0428f_4 \qquad (3.19)$$

where the factors f_i are defined by

$$f_1 = \frac{\mathcal{H}}{\mathcal{C}}, f_2 = \frac{\mathcal{O}}{\mathcal{C}}, f_3 = \frac{\mathcal{S}}{\mathcal{C}}, f_4 = \frac{\mathcal{N}}{\mathcal{C}}$$

The last term in Equation (3.18) is denoted as compositional exergy which accounts for the fact that crude oil is a mixture of hydrocarbon, water, and metals, each of them being present with a given mass fraction w_i and having an activity coefficient within the mixture, a_i; the factor R_i represents the ratio between the universal gas constant $\mathcal{R}$ and the molecular mass of the component i. In evaluation of Equation (3.18) there is negligible error induced by the assumption that the activity coefficient of all components is unity, a fact that facilitates calculations in cases where such activity coefficients are not known.

3.3.2.2 Natural Gas

Fossil natural gas consisting mainly of methane is a natural resource found mainly in either natural gas fields or in oil well fields (where is comes in association with oil) and coal beds (where methane is adsorbed in porous coals and is also known as coal bed methane or coal seams gas). Other important sources of natural gas are in the form of hydrates sediment present under permafrost regions. The typical composition of natural gas is given in Table 3.8.

The existence of noncombustible components as carbon dioxide, nitrogen, and oxygen in natural gas detracts from its heating value. One practical aspect is that natural gas is less flammable than other hydrocarbon-based fuels, even though methane—its principal constituent—is highly flammable. When confined, natural gas presents explosion danger. Methane is a chemical with a high greenhouse effect, having a potential seventy-two times higher than that of carbon dioxide for a time horizon of 20 years. Its lifetime in the atmosphere is 12 years. The main (average) properties of unrefined (or as-extracted) natural gas are given in Table 3.9.

The higher heating value of unrefined natural gas can be calculated based on the higher heating value of fuel components such as methane, ethane, propane, butane, etc. Assuming that w_i is the weight fraction of component i having the higher heating value HHV_i, then the natural gas higher heating value is

$$HHV = \sum w_i HHV_i$$

In the above equation, components such as moisture and carbon dioxide are not considered in this composition of natural gas. Therefore the weight fraction w_i refers only to components that can be combusted. Similarly, the chemical exergy of natural gas is calculated as a weighted average of the chemical exergy of all components (including carbon dioxide or traces of moisture)

TABLE 3.8 Typical Composition Ranges of Fossil Natural Gas

Component	Methane	Ethane	Propane	Butane	Pentane	Hexane	H_2	H_2S	O_2	CO_2
Weight fraction (%)	87.1–96.0	1.5–5.1	0.1–1.5	0.02–0.06	<0.18	<0.06	<0.02	1.9–2.6	0.01–0.1	0.1–1.0

TABLE 3.9 Principal Properties of Unrefined Natural Gas

Molecular mass (kg/kmol)	20
Density (kg/m^3)	1.832
Normal boiling point (°C)	N/A
Autoignition temperature (°C)	540
Octane number	120
LHV (MJ/kg)	18.2
HHV (MJ/kg)	20.2
ex^{ch} (MJ/kg)	18.8
Carbon fraction (% by weight)	73
CO$_2$ fraction (% by weight)	2.68
GHG emissions (kg CO$_2$/MJ$_{LHV}$)	147.2

$$ex^{ch} = \sum w_i ex_i^{ch}$$

According to Speight (2006) natural gas was discovered in Fredonia (New York, USA) by the first part of nineteenth century but started to be recognized as a valuable commercial commodity only in the second half of twentieth century. Nowadays, natural gas is extensively used for heating and cooking applications as well as power generation. Future expansion of natural gas use (after a refining process) in the transportation sector is envisioned because methane has the lowest carbon content of all fossil fuels. There have been many natural gas resources discovered in recent years, including unconventional wells such as gas shales, geo-pressured methane in saline brines, and tight gas sands.

3.3.2.3 Shale Oil

Solid sedimentary rocks containing kerogen are denoted as oil shale because unconventional crude oil can be extracted from them. Only a portion of the shale can be recovered as crude oil by thermochemical processes including pyrolysis. The remaining—unrecoverable portion—of oil shale processing must be deposited as residual materials, a fact that raises problems of land pollution and occupancy. Oil shale resources are abundant worldwide. In order to be economically profitable an exploitation of oil shale must be as large as millions of kilograms of mined material per annum. The processing of shale oil is essentially done by retorting, which consists of heating crushed shale in a closed vessel up to a temperature (~500 °C) at which the matter releases crude oil, gases, char, and steam. The fraction of extractable oil is 10–20% by volume with respect to the solid shale material.

In addition, the liquids resulting from oil shale retorting may require hydro-treating, that is, upgrading of fuel calorific value by reacting it with hydrogen to obtain a synthetic crude oil. The shale oil can also be retorted in an atmosphere of hydrogen, a process denoted as hydro-retorting. There are six mineral (rock) categories of oil shale resources, namely

TABLE 3.10 Typical Analysis Results for Oil Shale Given as Weight Fraction of Components

Proximate analysis		Ultimate analysis	
Component	Fraction	Parameter	Value
Fixed carbon	4%	H:C	0.14
Volatiles	41%	S:C	0.03
Moisture	3%	O:C	0.24
Ash	52%	N:C	0.02

Source: Hepbasli (2009).

tasmanite, marinate, kukersite, torbanite, lamosite, and cannel coal. As with coal, oil shale can be characterized via proximate and ultimate analysis. Some typical results for oil shale analyses are tabulated in Table 3.10.

The thermal conductivity of oil shale is rather low—between 0.2 and 2 W/mK—depending on the geological deposit, and the thermal diffusivity is an average of 5×10^{-7} m^2/s, with average heat capacity of 1.5 kJ/kg.K, average density of 2000 kg/m^3, retorting heat in the range of 250–850 kJ/kg, and LCV on the order of 8–10 MJ/kg, facts that must be accounted for at retorting equipment design. The chemical exergy for oil shale, as recommended in Hepbasli (2009), can be estimated based by an equation proposed by Kotas (1995) which was initially derived for solid fossil fuels. This relation is applied as follows:

$$\mathrm{ex}^{\mathrm{ch}} = (\mathrm{LHV} + 2442\,\mathcal{M})\,\beta + 9417\,\mathcal{S}$$

$$\beta = 1.0437 + 0.1882\frac{\mathcal{H}}{\mathcal{C}} + 0.061\frac{\mathcal{O}}{\mathcal{C}} + 0.0404\frac{\mathcal{N}}{\mathcal{C}}$$

where $\mathcal{M}$ is the mass fraction of moisture and $\mathrm{ex}^{\mathrm{ch}}$ and LHV are given in kJ/kg.

3.3.2.4 Oil Sands

This mineral resource of unconventional fossil fuel is also known as tar sand bitumen, bitumen-rocks oil, bituminous sand, impregnated rocks, oil sands, and rock asphalt and is mostly found in the Canadian province of Alberta (Athabasca Oil Sands). It consists of a sand-like rock (mainly quartz) impregnated with bitumen and moisture. Only a weight fraction of 10–20% from rough oil sand can be extracted as crude oil (denoted as synthetic crude oil in this context).

The molecular weight of bitumen is in the range of 540–800 kg/kmol. The valuable part of oil sand is bitumen, which represents a significant supply of energy and can be separated from the sand matrix by various methods and then converted into synthetic crude oil. In Table 3.11 the main properties of bitumen from oil sand in comparison to conventional crude oil are given. As observed, the hydrogen content in bitumen is ~3% less by weight with respect to conventional crude oil, which shows an important hydrogen deficit from a fuel utilization point of view. This illustrates that bitumen requires processing by hydro-treating (for hydrogen

TABLE 3.11 Average Properties of Bitumen Compared to Conventional Crude Oil

Property	Crude oil	Tar sand bitumen				
		World average	Athabasca	TST	PRS	NWAR
API gravity	35 °API	8 °API	11.6	11.1	10.3	14.4
Viscosity at 100 °C	10 cP	1700 cP	2000	4000	100,000	8000
Asphaltenes (wt.%)	5%	19%	16.4%	26.0%	16%	6.3%
Aromatics (wt.%)	25%	30%	31%	27%	31%	34%
Saturates (wt.%)	60%	19%	19%	17%	19%	21%
Resins (wt.%)	10%	32%	33.6%	30%	34%	37.7%
Carbon (wt.%)	86%	83%	82.5%	84%	84.4%	85.2%
Hydrogen (wt.%)	13.5%	10.6%	10.2%	10.1%	11%	11.7%
Sulfur (wt.%)	0.1%	4.8%	4.86%	4.38%	0.75%	0.59%
Nitrogen (wt.%)	0.2%	0.4%	0.47%	0.46%	1%	1.02%
Oxygen (wt.%)	0.2%	1%	1.7%	1.1%	2.2%	1.1%
Ash (wt.%)	N/A	0.2%	0.02%	0.12%	0.17%	0.04%
Vanadium	10 ppm	250 ppm	180	108	25	25
Nickel	5 ppm	100 ppm	112	53	98	120
LHV	45 MJ/kg	32–40 MJ/kg	41.2	41.6	48.1	43.7

Note: TST, Tar Sand Triangle (Utah); PRS, P.R. Springs (Utah); NWAR, Northwest Asphalt Ridge (Utah).
Source: Lee et al. (2007) and Bunger et al. (1979).

addition) and/or cocking (for carbon removal). Synthetic crude oil derived from bitumen has about 32 °API and 10 cP with sulfur and nitrogen weight fractions of 0.2% and 0.1%, respectively.

According to Bunger et al. (1979) the heating value of tar sand bitumen can be estimated based on the Boie equation—Equation (3.2)—as a function of the weight fractions of carbon, hydrogen, sulfur, oxygen, and nitrogen. The average composition of tar site bitumen at four major exploitation sites in North America is listed in Table 3.11. The chemical exergy of tar sand bitumen can be estimated using Equation (3.12) on a dry ash-free basis and Equation (3.13) on a wet basis.

3.3.3 Gas Hydrates

Clathrate hydrates (or gas hydrates) are assemblies of chemical compounds consisting in general of natural gas (or other organic materials or gases such as hydrogen, hydrogen sulfide, freons, hydrocarbons, noble gases, and carbon dioxide) entrapped in a crystalline matrix of water molecules. Clathrate hydrates are a major reserve of natural gas which can be found on sea floor and at permafrost regions and can be exploited to extract natural gas as fuel. It is noted that the volumetric storage of natural gas in hydrates is dense: 1 volume of hydrate

TABLE 3.12 Molar Fractions of Main Compounds of Some Naturally Occurring Hydrates

Compound	North America	Central America
Methane	99.93%	66%
Ethane	0.01%	2.9%
Propane	0.01%	14.7%
Isobutane	0.05%	3.7%

Source: Sloan and Koh (2007).

TABLE 3.13 Enthalpies of Dissociation in Function of Hydrate Composition

Hydrate	ΔH_d (kJ/mol)
$CH_4 \cdot 6H_2O$	18–54
$C_2H_6 \cdot 7.67H_2O$	25–71
$C_3H_8 \cdot 17H_2O$	27–129

Source: Sloan and Koh (2007).

stores more than 160 standard volumes of natural gas, and less than 15% of the recovered energy of methane is needed for clathrate dissociation and natural gas extraction (Sloan and Koh, 2007). In addition one notes that ~1% by volume of hydrates can be recovered from a reservoir and the volumetric heat of hydrate dissociation is ~370 MJ/kg.

In Siberia permafrost chlathrate hydrates were identified as a source of natural gas by 1965, when, as described in Sloan and Koh (2007), an extraction facility was in place at Messoyakha. Exxon started exploiting hydrates in Alaska by 1972. In the delta of Canada's MacKenzie River, there have been natural gas resources in the form of clathrate hydrates reported since 1974. In general it appears that the reserves of methane hydrates surpass those of other fossil fuels by about 10 times.

Table 3.12 provides the composition of clathrates from North and Central America. Clathrates contain mostly methane and some fractions of ethane, propane, and isobutane. Table 3.13 gives the dissociation enthalpies of some types of clathrates of methane, ethane, and propane. The dissociation enthalpy is very important for evaluating the gross and net calorific values of the clathrate fuel repository. Caloric values can be calculated as the difference between the heating value of resulting combustible gases and the dissociation enthalpy. Further, the heating value can be used to determine the chemical exergy of the clathrate.

3.3.4 Petrochemical Fuels

There are no (or very few) applications where crude oil is used directly, because its low calorific value, water content, metals, and pollutants may negatively affect any kind of combustion facility (or other types of devices using crude oil). Before use, crude oil and any other forms of petroleum must be refined and converted into more valuable fuels denoted as petrochemical fuels. These have low value in their original state and no practical applications. Examples of petrochemical fuels are liquefied petroleum gas (LPG), kerosene, diesel, gasoline, and jet fuels. Note that a refinery produces other valuable products such as waxes, asphalts, plastic materials, lubricants, and greases.

Even after separation of water, hydrogen sulfide, and other components, crude oil is still a mixture of many hydrocarbons ranging from very light to very heavy components. In a refinery, the crude oil has to go through successive chemical and physical processes such as desalination and cleaning, distillation, hydro-treating, etc. Distillation separates crude oil into different fractions depending on the difference of boiling temperatures. Atmospheric distillation is applied to separate the light compounds, whereas vacuum distillation extracts the heavier fuels.

Figure 3.16 is a simplified diagram of a petrochemical refinery indicating the main fuel and non-fuel products. Some of the main components of the diagram are the distillation columns,

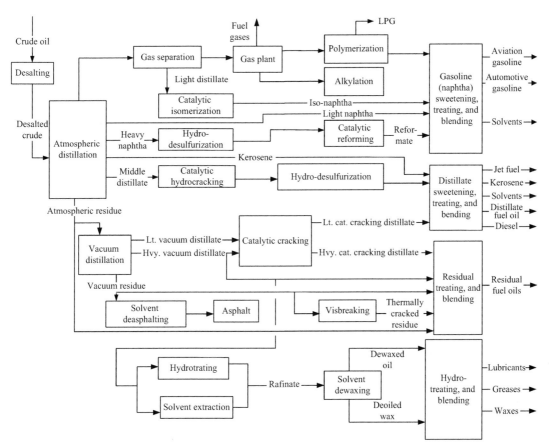

FIGURE 3.16 Simplified diagram of a petrochemical refinery.

the crude oil furnace, various catalytic reactors, separators and the heat exchanger network, the visbreaking unit, and the solvent dewaxing unit.

For the atmospheric distillation process the crude oil is heated to 350–380 °C and vaporized a pressure of about 200 kPa. About 25% atmospheric residue (liquid) is drawn at the bottom of the column. The following fuels result from the atmospheric distillation unit: heavy diesel, medium diesel, light diesel, and kerosene. The very light products, butane, in addition to light naphtha, leave as vapor at the top of the column. The atmospheric overhead is partially condensed in heat exchangers. When they are cooled, the naphtha condenses and leaves as liquid while the lighter products remain as gases. Uncondensed vapor flows to the fuel gas system, which is used as fuel for the furnaces.

The vacuum distillation unit operates at vacuum pressure (typically 18 kPa) to help in separating heavy hydrocarbons. The following products result from the vacuum distillation process: wash oil, heavy vacuum gas oil, and light vacuum gas oil. The residuum formed after the vacuum distillation process is eventually recovered as asphalt. The heavy vacuum distillate is partially converted into lubricants, greases, and waxes.

The range of petrochemical fuels obtained from crude oil refining is listed as follows:

- Naphthas—with 5–7 carbon atoms in the molecule (used mainly as solvents).
- Gasoline—with 8–11 carbon atoms, which is the basis of gasoline fuel for automobiles.
- Kerosene—with 12–15 carbon atoms, from which diesel fuel is extracted.
- Lubricating oils—with 16–19 carbon atoms; they show a high NBP.
- Solid hydrocarbons—with over 20 carbon atoms; they have a melting point at higher values than ambient temperature and are classified in order of increased molecular weight as paraffins, wax, tar, and asphaltic bitumen.

Table 3.14 gives the properties of the main petrochemical fuels. There are currently two main grades of jet fuel, Jet A-1 and Jet A, which are both kerosene-type fuels. There is another grade of jet fuel, Jet B, which is a wide cut kerosene (a mix of gasoline and kerosene), but it is

TABLE 3.14 Petroleum-Derived Fuels and Their Properties

Fuel name	Boiling point (°C)	Molec. weight (kg/kmol)	Density (kg/m³)	Carbon (% wt)	LHV (MJ/kg)	HHV (MJ/kg)	Chemical exergy (MJ/kg)	$EF_H \left(\dfrac{kg_{CO_2}}{GJ_{LHV}} \right)$
Light naphtha	0–150	100–150	~750	83.7	44.9	48.1	44.5	68.2
Gasoline	35–200	114	~745	85.1	43.5	46.5	47.5	71.9
Heavy naphtha	150–205	150–215	~850	85.4	43.0	46.1	49.0	64.2
Diesel fuel	150–370	233	~747	85.6	42.8	45.8	44.2	73.3
LPG[a]	−43	44	~580	81.8	46.0	52.0	54.9	65.3
Kerosene	205–260	170	~795	84.7	43.1	46.2	49.1	71.9
Jet Fuel	156–293	185	~800	76.0	43.2	46.9	45.3	63.4
Fuel oil	260–425	>200	~990	85.4	40.1	42.9	41.4	78.1

[a]LPG, liquid petroleum gas (assimilated as propane).

rarely used except in very cold climates. Jet A is a kerosene-based type of fuel, produced to an ASTM specification and normally only available in North America. It has the same flash point as Jet A-1, but a higher freeze point maximum ($-40\,°C$). Jet B is a distillate covering the naphtha and kerosene fractions. It can be used as an alternative to Jet A-1. Because it is more difficult to handle (with a higher flammability), there is only significant demand in very cold climates, where its better cold weather performance becomes important. Table 3.14 lists the average properties of Jet A and B fuels.

The fuel that is used mostly for road transportation is gasoline, which is a blend of aliphatic and aromatic chemicals obtained by distillation of petroleum in refineries. Diesel fuel is also very common for road transport—especially for large capacity vehicles—and marine ships and rail locomotives. Processing of petroleum to obtain diesel fuel is simpler than for gasoline fuel. The important aspect of diesel processing is reduction of sulfur content. Liquid petroleum gas is based mainly on propane, but often is mixed with butane in 60/40 proportion. The gases that compound LPG are all extracted from petroleum in refinery processes. The density of LPG is about the average of that of liquid propane and liquid butane at $25\,°C$. Alcohol-based fuels such ethanol and methanol can be produced using petroleum as fossil fuel source. However, these fuels are most often produced from biomass through fermentation and are categorized here as alternative fuels.

The chemical exergy of petrochemical fuels can be determined—as illustrated for example in Szargut (2005)—based on the chemical composition of the fuel, expressed with respect to the molar carbon in the molecule, namely

$$\left.\begin{array}{l} ex_{fuel} = LHV\,\beta \\ \beta = 1.0374 + 0.0159f_1 + 0.0567f_2 + 0.5985f_3[1 - 0.1737f_1] \end{array}\right\} \tag{3.20}$$

where factors f_i are defined in conjunction with Equation (3.19).

TABLE 3.15 Principal Properties of Refined Natural Gas

Parameter	Property
Molecular mass, g/mol	16 kg/kmol
Density	0.733 kg/m^3
Normal boiling point	$-162\,°C$
Autoignition temperature	560 °C
Octane number	130
LHV	50.7 MJ/kg
HHV	56.2 MJ/kg
ex^{ch}	52.4 MJ/kg
% C by weight	75%
% CO$_2$ by weight	2.75%
EF$_H$	54.2 kgCO$_2$/MJ$_{LHV}$

Refined natural gas is a special type of petrochemical fuel. Refinement of natural gas involves several processes aiming to remove the humidity (water), sulfur, carbon dioxide, helium, solid particle, and liquid hydrocarbons existent in the gas. Through refining the carbon content, the associated heating value is increased. Table 3.15 gives the main physical properties of natural gas in refined form.

3.4 ALTERNATIVE FUELS

As presented in the introduction, there is a large preoccupation worldwide about developing alternatives to fossil fuels which can mitigate the negative effect on global warming and environmental pollution but at the same time assure better energy security, especially with regard to non-depleting resources such as renewables. The main resource for producing alternative fuels is biomass, which itself can be combusted as a fuel or can be converted into multiple types of biofuels. In what follows the main types of biomass and their energy characteristics are reviewed, as are the main biofuels derived from biomass. Particular focus is placed on fuels derived from waste materials (including organic waste, municipal landfill waste, etc.) and on special fuels such as hydrogen, nitrogen-based fuels, and fuel blends.

3.4.1 Biomass

Biomass—which accounts for 15% of world energy consumption—can be derived from all kind of recently living matter on earth, including residual biological matter such as manure, dung, and crap, but also crops, herbs, trees, algae, etc. According to ultimate analysis, biomass consists of carbon, hydrogen, oxygen, nitrogen, and minute traces of sulfur. The energy content of biomass is relatively low compared with coal and other fossil fuels because of its high content of oxygen.

Biomass generates about the same amount of CO_2 as fossil fuels (when burned), but, from a chemical balance point of view, every time a new plant grows CO_2 is actually removed from the atmosphere. The net emission of CO_2 will be zero as long as plants continue to be replenished for biomass energy purposes. If the biomass is converted through gasification or pyrolysis, the net balance can even result in removal of CO_2. Energy crops such as fast-growing trees and grasses are called biomass feedstocks. The use of biomass feedstocks can help increase profits for the agricultural industry.

In principle one can categorize biomass sources as two kinds: energy crops and residual biomass materials. Here, the residual biomass can include all sorts of waste woods from demolition, furniture factories (sawdust), the building materials industry (sawdust, bark), residual fiberboards, and straw residuals from agriculture. Various residuals are also specific to forestry (tree branches, pruning, some trees) and to the food processing industry (kernels, seed shells, etc.), while residuals from paper mills, and any other kind of recoverable paper materials provide additional biomass stocks. Some examples of biomass are listed as follows:

- Wood: trees, tree stumps, dead trees, branches, wood residuals from forestry and wood processing industries.
- Agricultural residues: straw, sugarcane fibre, rice hulls, animal wastes, dried dung.

- Biodegradable wastes: from municipal and industrial sources.
- Vegetable oils: palm oil, corn oil, peanut oil, soy oil, canola oil.
- Energy crops: miscantus, sorghum, switchgrass, hemp, sugarcane, corn, poplar, eucalyptus, willow, aquatic plants.

There are two technological platforms for biomass conversion, biochemical and thermo-chemical, which are combined in a biorefinery. A biorefinery is a facility that processes bio-mass to generate multiple products like biodiesel, bioethanol, other liquid fuels, electric power, heat, hydrogen, and valuable bio-products or biochemicals. The concept of a biorefinery is illustrated schematically in Figure 3.17.

The upper branch comprises thermochemical processes (see Figure 3.17), including gasi-fication, gas separation, hydrogen production, Fischer–Tropsch synthesis to produce biodie-sel, and heat recovery to produce steam and process heat and electricity. The lower branch includes mostly biochemical processes, including lignin separation from biomass. Lignin is a valuable commercial product that can be used in many chemical processes, such as water formulation of dyes, production of humic acid, vanillin, leather tanning, and polyurethane foam. Through further enzymatic hydrolysis of cellulose one can produce a large variety of sugar intermediates. Further fermentation processes lead to production of ethanol and other biochemical byproducts.

Tables 3.16–3.18 give the compositions and heating values of some important types of bio-masses. In Table 3.16, the proximate analysis of eleven types of biomass is given. Proximate analysis determines the ash, moisture, volatiles and fixed carbon content in the biomass on a as-received basis. From the table it results that fixed carbon ranges from 5% to 20% by weight.

Table 3.17 presents the ultimate analysis data of 12 types of biomasses. The carbon content in biomass is high ranging from ~35% to 50% by weight. The content of hydrogen in biomass is relatively low with typical fractions of 3–6% by weight. Table 3.18 presents the heating

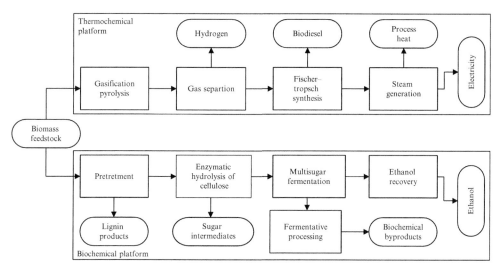

FIGURE 3.17 Conceptual diagram of an integrated biorefinery.

TABLE 3.16 Typical Proximate Analysis of Some Types of Biomass

Constituent	AS	WP	RS	WW	OW	SD	SL	NP	BB	SG	ST
Moisture	8	12	6	7	6.5	34.9	8	6	8.4	11.9	14.4
Ash	24	3	19	3	0.3	0.7	35	1	7.1	4.5	7.5
Volatiles	54	71	65	78	73.0	55.1	52	85	67.5	70.8	64.3
Fixed carbon	14	14	10	19	20.0	9.3	5	8	17.0	12.8	13.8

AS, almond shell; WP, walnut pruning; RS, rice straw; WW, whole tree wood chips; OW, oak wood; SD, saw dust; SL, sludge; NP, non-recyclable waste paper; BB, beech bark; SG, switch grass; ST, straw.
Source: Kalinci et al. (2010).

TABLE 3.17 Typical Ultimate Analysis of Some Types of Biomass

Constituent	AS	WP	RS	WW	OW	SD	SL	PW	NP	BB	SG	ST
C (wt.%)	36	48	38	51	47	40	36	48	49	48	43	40
H (wt.%)	4	4	3.5	3	6.5	6	4	6.7	5	6.0	6.1	5.6
N (wt.%)	1	0.6	0.5	0.35	0.3	0.5	6	0.10	0.1	0.7	0.7	1
S (wt.%)	0.05	0.03	0.06	0.05	0.2	0	1.1	~0.0	0.06	0.3	0.2	0.6
O (wt.%)	32	44	36	42	40	34.5	15	39.1	44	36	40	37.8
Ash (wt.%)	26.95	3.37	21.94	3.6	6	19	37.9	5.0	1.84	9	10	15

AS, almond shell; WP, walnut pruning; RS, rice straw; WW, whole tree wood chips; OW, oak wood; SD, saw dust; SL, sludge; NP, non-recyclable waste paper; BB, beech bark; SG, switch grass; ST, straw.
Source: Kalinci et al. (2010).

value and the specific chemical exergy of 15 types of biomasses. It is remarked that the exergy content of biomass sources is in the range of approx. 8–25 MJ/kg on wet basis.

The molecular mass of dry biomass can be calculated as indicated in Table 3.19. The GCV of biomass combustion can be calculated with the relation proposed by Van Loo and Koopejan (2008) according to

$$\text{GCV}(\text{M/kg}) = 34.91\mathcal{C} + 117.83\mathcal{H} + 10.05\mathcal{S} - 1.51\mathcal{N} - 1.034\mathcal{O} - 2.11\mathcal{A} \qquad (3.21)$$

where $\mathcal{A}$ is the weight fraction of ash in a dry (moisture-free) biomass. The NCV is calculated under the constraint that water in the exhaust gases is in vapor phase and all exhaust gases are at the same temperature as the biomass at the feed (standard temperature). The NCV can be calculated according to Van Loo and Koopejan (2008) with

$$\text{NCV}(\text{M/kg}) = \text{GCV}(1 - \mathcal{M}) - 2.444\mathcal{M} - 21.839\mathcal{H}(1 - \mathcal{M}) \qquad (3.22)$$

Szargut (2005) gives an equation for the chemical exergy of biomass (where the contribution of the sulfur to the combustion process is neglected) as follows:

$$\text{ex}^{\text{ch}} = \text{NCV}\beta, \text{ where } \beta = 1.0347 + 0.014 \, f_1 + 0.0968 \, f_2 + 0.0493 \, f_3 \qquad (3.23)$$

TABLE 3.18 Heating Values of Various Types of Biomass

Biomass	GCV, MJ/kg d.b	Moisture, w%, w.b.	NCV MJ/kg w.b.	NCV MJ/dm³ w.b.	ex^ch MJ/kg wb
Wood pellets	19.8	10	16.4	9.8	18–19
Pine wood	20.5	8	18.9	14	24.8
Woodchips	19.8	30–50	8.0–12.2	2.8–3.9	8–14
Grass	18.4	18	13.7	2.7	15–16
Cereals	18.7	15	14.5	2.5	16–17
Bark	20.2	50	8.2	2.6	9–10
Sawdust	19.8	50	8.0	1.9	8–9
Straw	18.7	15	14.5	1.7	16–17
Olive kernels	22.0	58	7.3	6.3	7–9
Almond shells	15.1	9	14.3	2.5	16.7
Walnut pruning	19	14.3	18.2	10.2	20.5
Rice straw	14.8	6	14.1	2.5	15.9
Whole tree woodchips	20.4	7.5	19.7	12	22.1
Sludge	15.4	8.6	14.5	2.5	15.9
Non-recyclable waste paper	18.8	6.3	17.7	10	20.1

Note: w%, percent by weight; d.b., dry basis; w.b., wet basis.
Source: Van Loo and Koopejan (2008), Dincer and Zamfirescu (2011), Kalinci et al. (2010).

TABLE 3.19 Equations for Estimating Molecular Mass of Biomass

Basis	Equation	Remark
Dry, ash free	$M_{daf} = 12c + 2h + 16o + 14n + 32s$	Molar fractions calculated with Equation (3.10)
Dry	$M_d = (1 - a)M_{daf} + aM_{ash}$	Molar fraction of ash a kmol ash/kmol dry
Wet	$M = (1 - m)M_d + 18m$	Molar fraction of moisture, m, kmol H_2O/kmol wet

An alternative expression for factor β in Equation (3.23) is proposed in Szargut and Styrylska (1964) for biomasses with low oxygen content O/C, as follows:

$$\beta(1 - 0.3035f_2) = 1.0412 + 0.216f_1 - 0.2499f_2(1 + 0.7884f_1 + 0.045f_4)$$

where factors f_i are defined above in connection to Equation (3.19).

3.4.2 Biofuels, Biogas, and Fuel Blends

Biomass sources can be used directly via combustion to generate thermal energy or can be converted by various methods to biofuels which can be either gases or liquids. The liquid

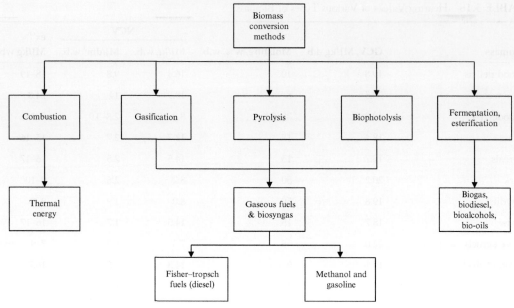

FIGURE 3.18   Pathways to convert biomass sources into biofuels or useful thermal energy.

fuels obtained from biomass can be alcohols or various types of oils, including ether, esters, and others. In Figure 3.18 are indicated graphically the main conversion pathways of biomass sources to biofuels and thermal energy. Apart from combustion (which generates thermal energy and/or power), biomass can be converted by gasification, pyrolysis, or biophotolysis. The process of biophotolysis uses bacteria (e.g., genetically modified cyanobacteria) to generate methane or hydrogen, and fermentation processes. Gasification, pyrolysis, and biophotolysis all produce combustible gaseous fuel mixtures.

Other processes to convert biomass are fermentation and esterification. Fermentation process results in liquid fuels of alcoholic type or in biogas. The majority content of biogas is methane (55–65% by volume) while the rest is mainly carbon dioxide (35–45% by volume) and other gases (hydrogen, nitrogen, oxygen, and ammonia). Biophotolysis, fermentation, and other methods like artificial photosynthesis and biological water–gas shift process can be classified as biological methods for biomass conversion.

Fermentation methods can be produced using anaerobic organisms and/or by the mechanism of photofermentation. Also, many phototropic organisms can produce hydrogen with the aid of solar energy, which is called photosynthesis conversion. Furthermore, some bacteria may easily perform water–gas shift reactions. Dark as well as photoheterotrophic (light fermentation) microorganisms can convert carbohydrate rich biomass into combustible fuels as hydrogen or biogas.

The primary source for biogas generated by biohydrogen is animal manure, which has the ability to deliver the necessary microorganisms for anaerobic digestion. Energy crops are also used for biogas production, like maize, sunflower, and grasses.

Landfills are deposits of solid municipal waste with a high content of organic materials and represent a major source of biogas. The first step is the acid or enzymatic hydrolysis of biomass to a highly concentrated sugar solution which is further fermented by anaerobic organisms to produce volatile fatty acids, methane or hydrogen, and CO_2. The organic acids are further fermented by the photoheterotrophic bacteria (Rhodobacter sp.) to produce CO_2 and methane or hydrogen, which is known as light fermentation. Combined utilization of dark and photofermentation was reported to improve the yield of hydrogen formation from carbohydrates. Many phototropic organisms, such as purple bacteria, green bacteria, cyanobacteria, and several algae, can be used to produce hydrogen with the aid of solar energy. Microalgae, such as green algae and cyanobacteria, absorb light energy and generate electrons. The electrons are then transferred to ferredoxin (FD) using the solar energy absorbed by the photosystem. The mechanism varies from organism to organism but the main steps are similar.

Alcohols are one of the most attractive classes of fuels derived from biomass sources. Practically all alcohols can be combusted to generate heat, but only few are suitable for common motor vehicles: ethanol (C_2H_5OH), methanol (CH_3OH), butanol (C_4H_9OH), and propanol (C_3H_7OH). Ethanol has qualities which makes it a competitor of gasoline in spark ignition engines because it allows for a higher compression ratio, higher flame speed, and leaner operation despite the fact that the energy density of ethanol is lower than that of gasoline and more toxic. The general equation for oxidation of alcohol is

$$C_xH_{2x+2}O + 1.5xO_2 \rightarrow xCO_2 + (x+1)H_2O$$

Alcohols are produced from biomass mainly via fermentation processes, in which case they are denoted as bioalcohols. However, it is also possible to synthesize the methanol form syngas. Simply, methanol can be produced by hydrogenation of carbon monoxide over a suitable catalyst—usually copper or zinc oxides:

$$CO + 2H_2 \rightarrow CH_3OH$$

This is an exothermic reaction favored by high pressure and low temperature.

In Table 3.20 the main properties of major alcohol fuels are given. Methanol (also denoted as methyl alcohol) is one of the cheapest fuels produced by crop or fruit residuals fermentation and also can be generated via wood pyrolysis. The main drawback of methanol as a fuel is that it is highly poisonous to humans, its ingestion in small quantities ($\sim$100 ml) being fatal.

TABLE 3.20 Properties of Some Alcohols Used as Fuels

Alcohol	Formula	Density	LHV (MJ/kg)	HHV (MJ/kg)	ex^{ch} (MJ/kg)
Methanol	CH_3OH	792 kg/m^3	19.9	22.7	22.4
Ethanol	C_2H_5OH	789 kg/m^3	28.8	29.7	29.5
Propanol	C_3H_7OH	786–803 kg/m^3	30.5	32.5	32.0
Butanol	C_4H_9OH	775–810 kg/m^3	33.0	35.7	35.1

Ethyl alcohol (or ethanol) can be produced by fermentation from sucrose, starch, or lingo-cellulosic biomasses. Currently, bioethanol is used in blends with gasoline (5–15% by volume) to reduce the specific emissions of road vehicles. Propanol and butanol with several isomers can be produced by fermentation of cellulosic biomasses (starch or sucrose).

Biodiesel is obtained by esterification of vegetable oils; the resulting fuel is clean of sulfur and can be used directly in diesel engines. An alternative route to biodiesel fabrication starts with bio-syngas and uses the Fischer–Tropsch method for conversion to diesel fuel. Biodiesel can be used in a blend with petroleum diesel in order to reduce the carbon and sulfur emissions. Also, alcohols like ethanol can be used to manufacture substitute diesel fuels. As shown in the next section, ethanol blends can fuel diesel engines.

Other types of fuels from biomass are those based on vegetable oils. They have a different structure than petroleum-based fuels. It appears that some vegetable oils can be used at diesel fuel replacement from an ecological and an economical point of view (see Demirbas, 1998). A list of common vegetable oils and their main properties is compiled in Table 3.21. It can be observed that the HHV of all vegetable oil fuel is about the same, at ~39 MJ/kg. The saponification value of oil represents the number of milligrams of sodium or potassium hydroxide required to saponify 1 g of oil, which is a measure of the molecular weight of the fatty acids that compose the oil. The iodine number gives the number of grams of iodine in 100 g of oil.

TABLE 3.21   Parameters of Main Vegetable Oil Fuels

Oil	Saponification value (mg KOH/g Oil)	Iodine value (g I/100 g oil)	HHV (MJ/kg)
Ailantus	206.34	107.18	39.38
Beech	202.16	105.15	39.59
Castor	202.71	88.72	37.41
Corn	194.14	119.41	39.64
Hazelnut kernel	197.63	98.62	39.83
Laurel	220.78	69.82	39.32
Linseed	188.71	156.74	39.33
Peanut	199.80	119.35	39.45
Poppy seed	196.82	116.83	39.59
Rapeseed	197.07	108.05	39.73
Sesame	210.34	91.76	39.42
Soybean	194.61	120.52	39.63
Spruce	207.09	96.08	39.44
Sunflower seed	191.70	132.32	39.57
Walnut kernel	190.82	135.24	39.56

Source: Demirbas (1997).

Demirbas (1998) correlated the HHV of vegetable oils with saponification and iodine values, according to

$$HHV = 49.43 - 0.041SV - 0.015IV$$

where SV and IV represent the saponification and iodine values; HHV is give in MJ/kg.

The typical biochemical route to biomass gasification leads to biogas generation. Biogas is a fuel consisting of a mixture of gases, especially methane, resulting from anaerobic fermentation of biomass. Composition of this biogas varies with the type of organic material used. The caloric power of biogas depends on the amount of methane in its composition, which could reach 20–25 MJ/m^3. Biogas can be used for stove heating, water heaters, torches, motors, and other equipment. Biogas typically contains 60–70% methane and 30–40% carbon dioxide with traces of nitrogen, ammonia, hydrogen, and hydrogen sulfide.

Several biodigester models are available. The basic digester is a reservoir built of bricks or concrete below ground level. A wall divides the biodigester into two semicylindrical parts for the purpose of retaining and providing circulation for the biomass loaded in a biofertilization process. The biodigester is loaded through the charge box, serving as a pre-fermenter. The load is typically a slurry containing biomass mixed with water. The charge box communicates with the digester through a pipe going down to the bottom. The output of the biofertilizer is through another pipe at a level that assures that the amount of biomass entering the biodigester is the same as that leaving it in biofertilizer form. It should also have a discharge box, tank, or dam to pump and/or deliver the biofertilizer directly to the consumer. For a production capacity of 5–6 m^3 biogas per day, the biodigester has a 3-m diameter and a 3-m height, consumes 250 l of biomass per day, and has a retention time of about 50 days. The biodigester is usually buried because underground temperatures are higher and more constant. The production rate of biogas depends on parameters such as the material temperature (preferably 30–35 °C), biodigestion acidity (i.e., the pH, preferably 6–8), nutrients (e.g., N$_2$) and their concentration, concentration of solids (preferably 7–9%).

The type of biomass used in the biodigester is diverse, including residues from industrial treatment of fruits, meats, cereals, and alcohol. Urban garbage can also be used as feedstock for biodigesters and thus transformed into sources of energy. The specific production of biogas is presented in Table 3.22 for various kinds of feedstock.

TABLE 3.22 Specific Biogas Production for Several Vegetable and Manure Feedstocks

Feedstock	Production (m^3 gas per ton of feedstock)	Feedstock	Production (m^3 gas per ton of feedstock)
Bean straws	380	Rice straws	300
Sunflower leaves	300	Wheat straws	300
Soy straws	300	Linen stem	360
Potatoes leaves	270	Dry tree leaves	250
Grapevine leaves	270	Birds	55
Bovines	40	Equines	48
Suidae	64	Oviparous	70

As in the case of biomass conversion into gas, its conversion into liquid fuels can be done by thermochemical or biochemical routes. Through liquefaction one obtains fuel oil. Liquefaction can be done by exposing a mixture of liquid water and solid biomass to high pressures and high temperature. The pressure is commonly on the order of 200 bar (close to the critical pressure of water). The biochemical route is through fermentation and leads to production of alcohols. In general biofuels are liquid fuels obtained from biomass (biodiesel, bioethanol, biomethanol, biogasoline, etc.). However, the term biofuel is general, referring to any kind of fuel derived from biomass, be it gas (biogas, hydrogen, carbon monoxide), liquid, or solid (the biomass itself is a solid fuel). Thus, in general, biofuels include wood, wood waste, wood liquors, peat, railroad ties, wood sludge, spent sulfite liquors, agricultural waste, straw, tires, fish oils, tall oil, sludge waste, waste alcohol, municipal solid waste, landfill gases, other waste, and ethanol blended into motor gasoline.

In the current context the use blends of biofuels and fossil fuels in the transportation sector is economically justifiable. One of the known fuel blends is M85, which contains 85% methanol and 15% gasoline. As methanol can be derived from biomass sources, M85 can achieve 85% carbon dioxide reduction compared to gasoline-only fuel. Pure methanol can be used as fuel instead of gasoline, in which case it is called M100. Similarly E85 fuel blend consists of 85% ethanol and 15% gasoline. Since ethanol is derived from biomass, wheel-to-wheel carbon dioxide reduction is at least 80%. Ethanol can be also used in diesel engines as a ED95 blend, containing 95% ethanol and 5% ignition agent.

Dimethyl ether (DME), having the chemical formula CH_3OCH_3, can be used as fuel in diesel engines or as a component of fuel blends for gasoline engines. One such fuel blend is formed by 30% DME and 70% LPG. DME can be produced from coal, biomass, or natural gas through various methods. The LCV of DME is 28.9 MJ/kg.

3.4.3 Hydrogen

Hydrogen is the simplest chemical element, with the atomic number $Z = 1$, and it is the most abundant chemical element in the universe. As a consequence of its simplicity, hydrogen can easily lose valence electrons and therefore is very reactive. For this reason, hydrogen cannot be found as an individual element on earth but rather is embedded in other material. Water is the most abundant resource of hydrogen on earth; hydrogen is also part of most fossil fuels and biomass. In nature, hydrogen can be also found in the form of hydrogen sulfide (H_2S) which is abundant in some springs, geothermal sites, and seas. Hydrogen has a high calorific value, the lowest molecular weight, the highest thermal conductivity of all gases, and the lowest viscosity. Table 3.23 gives the main thermophysical properties of hydrogen.

An important technical difficulty related to hydrogen storage is due to the existence of two isomers of the molecular hydrogen. In ortho-hydrogen the spin of the two protons of the hydrogen molecule are parallel. In the para-hydrogen configuration the spins are in opposite directions. At standard temperature, the hydrogen gas composition is of 25% ortho and the rest para. At cryogenic temperature there is practically no ortho hydrogen (<0.2%).

The heating value of hydrogen is much higher than that of conventional fuels by unit of weight. However, hydrogen cannot be kept in a condensed phase with current technology. Most often, hydrogen is stored in the form of compressed gas or as a cryogenic liquid. In any of these storage conditions, the heating value of hydrogen per unit of volume is lower than

TABLE 3.23 Thermophysical Properties of Hydrogen

Property	Description	Property	Description
Density	0.1 kg/m³ at 1 atm/298 K (gas) 1.34 kg/m³ at 1 atm/20.3 K (vapor) 70.79 kg/m³ at 1 atm and 20.3 K 11.69 kg/m³ at 350 atm/298 K (gas)	Critical point	32.97 K, 12.9 bar
		Normal boiling point	20.3 K (1 atm) 99.8% para-hydrogen + 0.02% ortho-hydrogen
Composition	75% ortho/25% para at 298.15 K Over 99.8% para in cryogenic H_2	Melting point	14.01 K (1 atm)
		Liquefaction	14.1 MJ/kg theoretical energy
LHV/HHV	119.9/141.8 MJ/kg	Diffusivity	In air, 0.61 cm/s
Exergy	118.05 MJ/kg (chemical)	Conductivity	0.177 W/mK
Auto-ignition temperature	850 K	Detonation	13–65% mix with air with 2 km/s, suppression of shock wave at 1.47 MPa
Flammability	4–18% and 59–75% mix with air	Flame speed	2.75 m/s

conventional fuels. Nevertheless, the efficiency of power generation systems fueled with hydrogen is much higher than that obtained with conventional fuel, and this fact compensates for the low storage density problem of hydrogen. Furthermore, hydrogen combustion is completely clean, producing only water vapor in the exhaust gas. Heat recovery can be applied intensively to hydrogen combustion to obtain liquid water which may be recycled to reproduce hydrogen via various methods for extracting hydrogen from water, including electrolysis and thermochemical splitting. Therefore, hydrogen can be viewed as an energy storage medium or energy carrier. The major problem with hydrogen as a fuel is the difficulty of storage in its pure form and the enormous cost of the hydrogen infrastructure that must be put in place in order to implement a hydrogen-energy-based economy. It is arguably believed that the storage and distribution problem is well balanced by the benefits of a hydrogen economy.

In order to identify pathways for sustainable hydrogen production one needs to inventory the natural resources of hydrogen, the available sources of energy which can be used to extract hydrogen from natural resources, and the applicable methods of hydrogen production. These items are summarized in Figure 3.19. The hydrogen-containing natural resources are water, fossil hydrocarbons, biomass, hydrogen sulfide, and anthropogenic wastes, as indicated in Figure 3.19a. Municipal sewage waters containing urea, farming wastes like manure, crops residues, etc. (which are sources of biogas), other wastes which generate landfill gas, and recycled plastic and cellulosic materials, etc., are all anthropogenic wastes from which hydrogen can be extracted.

Sustainable energy is required to extract hydrogen from any resource in a clean, non-polluting manner. Figure 3.19b lists the main energy sources that can be considered sustainable: solar, hydro, ocean thermal, tidal (including ocean currents), wind, biomass, geothermal, and nuclear. The careful use of any of these sources can generate electricity and/or high temperature heat and/or nuclear radiation with or without minor environmental impact. Such energy is used for hydrogen production via one of the methods listed in Figure 3.19c. These methods are categorized in six classes, namely. electrochemical,

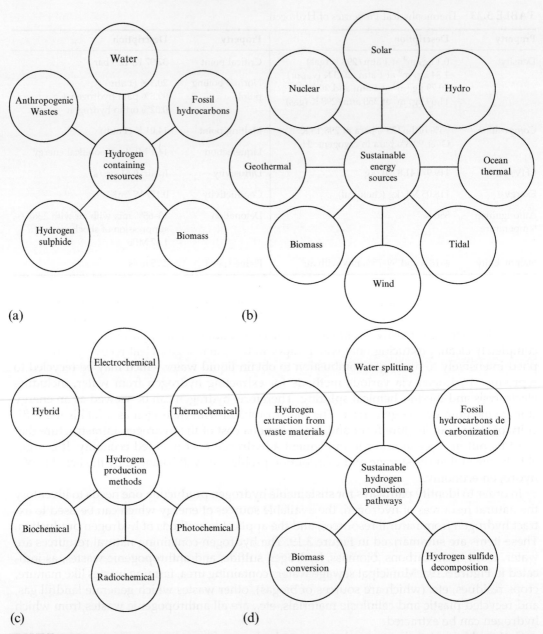

(a)

(b)

(c)

(d)

FIGURE 3.19 Natural hydrogen resources (a), sustainable energy sources (b), production methods (c), and sustainable pathways (d) for sustainable hydrogen production.

thermochemical, photochemical, radiochemical, biochemical, and hybrid (where hybrid refers to integrated systems that use any combination of the first five listed hydrogen production methods, e.g., electro-photochemical, photo-biochemical, electro-thermochemical, etc.).

Based on the above analysis we identified five possible pathways to generate hydrogen in a sustainable manner. As indicated in Figure 3.19d, these are water splitting, fossil hydrocarbon decarbonization, hydrogen sulfide decomposition, biomass conversion to hydrogen, and extraction of hydrogen from waste materials resulting from anthropogenic activity. Each pathway corresponds to a natural resource (including the anthropogenic waste) from which hydrogen can be extracted. For any of these pathways, the use of a specific combination of sustainable energy (Figure 3.19b) and hydrogen production method (Figure 3.19c) is possible. In the subsequent sections of this paper, the main options are addressed.

It is very difficult to store hydrogen in a satisfactorily dense phase such that the packed energy density is similar to that of the common fuels used today for vehicles. Based on HHV, 324 g of hydrogen has the same content of 1 kg of gasoline. The corresponding volume of 1 kg of gasoline is about 1.3 l; if 324 g of hydrogen is stored as gas under standard temperature and pressure conditions, the occupied volume is 3932 liters. At NBP (20.3 K) under atmospheric pressure, the density of cryogenic liquid hydrogen is 70.77 kg/m^3; the volume of 324 g of liquid hydrogen is 4.6 liters, which is 3.5 times larger than that of gasoline with the same energy content. Under standard temperature (298.15 K) and 400 bar pressure, hydrogen gas density is 25.98 kg/m^3, and the gas volume is 12.5 l or 9.6 times more than gasoline volume with same energy content. The following storage methods are considered for hydrogen:

- Compressed gas at standard temperature and very high pressure.
- Cryogenic liquid at standard pressure and 20 K.
- Physical binding of hydrogen molecule in a solid material matrix.
- Chemical binding to synthesize a denser chemical that can later release hydrogen.

Before storage through any of these methods, hydrogen must be purified to an acceptable degree; for compressed storage 4 ppm impurities are acceptable, while for cryogenic storage 1 ppm is recommended.

Some chemical elements like sodium, lithium, magnesium, and boron can create hydrides (called "chemical hydrides") that store hydrogen more densely than is possible in metal hydrides. There are two processes that can be applied for storing hydrogen in a solid matrix: physisorption (absorption of molecular hydrogen by a solid structure) and chemisorption (dissociation of the hydrogen molecule and bonding of the protons with the metal lattice). Many metal-hydride-based methods have been developed for both physisorption and chemisorption of hydrogen. In general they require some thermal energy at the discharge phase, and some cooling should be applied to enhance the efficiency of the hydrogen-charging process.

One interesting possibility for storing hydrogen in solid form is by first producing ammonia from it and then absorbing the ammonia gas in metal amines. About six molecules of ammonia can be bonded weakly by magnesium chloride to form $Mg(NH_3)_6Cl_2$, which embeds hydrogen of 109 g per liter and 92 g per kg of magnesium ammine. Metal amines are an attractive solution for hydrogen storage, but one has to account for the diminishing of the calorific value due to hydrogen extraction, or, alternatively, this method of storage should be applied when some combustion heat can be recovered to drive the ammonia-cracking reaction itself.

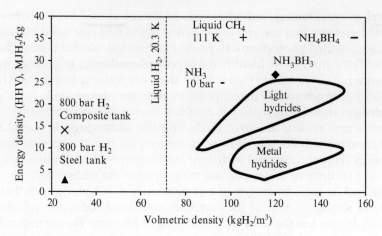

FIGURE 3.20 Density of hydrogen storage with various technologies. *Modified from Dincer and Zamfirescu (2011).*

Storing hydrogen chemically in the form of urea is also an attractive solution, as urea is very stable and can be stored for very long periods and safely transported. Urea as a commercial product is typically delivered in small-size prills. It can be combusted, having a low calorific value compared to that of wood, that is about 10 MJ/kg. Urea can be synthesized from biomass or other renewable energy sources that result in a CO_2-free fuel. Urea, with the chemical formula $CO(NH_2)_2$, is massively produced in industry and used as fertilizer. It is considered a nontoxic substance and several car manufacturers use it on their passenger vehicles for NO_x exhaust reduction.

Hydrogen can be stored chemically in the form of ammonia, which can be decomposed thermocatalytically to release hydrogen according to the reaction $NH_3 \rightarrow 1.5H_2 + 0.5N_2$. Ammonia can itself be used as fuel, as detailed in Zamfirescu and Dincer (2008, 2009).

Figure 3.20 presents the hydrogen storage density of various technologies. The storage density is presented in two different ways. On the horizontal axis it is the volumetric storage density in kg of hydrogen per unit of volume of storage facility. On the vertical axis it is presented as the energy embedded in the quantity of hydrogen that can be released by the storage; this figure is given per unit of mass (kg) of the storage facility (tank). The ammonia–boron compound with the chemical formula NH_4BH_4 appears to be the most "dense" hydrogen storage.

3.5 CONCLUDING REMARKS

We live in a transition period between the use of fossil fuels and the use of sustainable fuels as alternatives for the transportation, residential, energetic, and industrial sectors. During the last century the fuel sources for transportation and fuels for heating have been mainly fossil based: coal, petroleum, and natural gas. Presently other fuel sources are considered, with the general aim of achieving cleaner combustion and reduced carbon dioxide emissions. New alternatives to conventional fossil fuel include oil sands, oil shale, carbon hydrates, etc. These fuel sources are considered in conjunction with the production of synthetic fuels with reduced carbon. Such fuels may be hydrogen, ammonia, urea, methanol, ethanol or others. Another alternative consists of promoting fuel blends of fossil-based fuel and biofuels. The advantages of co-fueling are extremely relevant for both mobile and stationary applications. These include the opportunity to recover heat rejected by engines and upgrade the fuel value and thus improve the overall efficiency.

The chapter also details the fuel sources and the fuel properties, including lower and HHV and chemical exergy. Using biofuels is of major importance because this is a real means to reduce carbon dioxide pollution and promote cleaner options for the transportation and energy sectors. Many power generation systems based on fuel combustion reject heat into the environment. Recovering this heat for local hydrogen generation is a way of achieving improved fuel efficiency.

Study Problems

3.1 What are the features that differentiate alternative fuels from fossil fuels?

3.2 Define autoignition temperature and explain the difference between this parameter and the lower flammability limit.

3.3 What type of coal has the lowest emissions of GHG per unit of net calorific value?

3.4 Describe the process of coal combustion.

3.5 Which petrochemical fuel has the lowest emissions per unit of mass?

3.6 Why is crude oil not used directly as fuel but rather first refined?

3.7 Explain the difference between the thermochemical and biochemical platforms in a biorefinery.

3.8 Using Equations (3.21) and (3.22) study the influence of ash and moisture content on GCV, net calorific value, and chemical exergy of biomasses with low oxygen content. For study purposes assume fixed weight fractions of carbon, hydrogen, sulfur, oxygen, and nitrogen.

3.9 Compare wood and olive kernels from the point of view of specific emissions per unit of GCV and per unit of chemical exergy content.

3.10 Make a comparative chart of the main types of alcohol from the point of view of their net calorific value and chemical exergy.

3.11 Calculate the higher heating value of a vegetable oil fuel having the saponification value of 195 mg KOH/g oil and iodine value of 85 g I/100 g oil.

3.12 What would be the advantages of storing hydrogen in the form of ammonia compared to storing it in the form of ammonia borane onboard vehicles?

3.13 How much water needs to be carried onboard a vehicle fueled with methanol? Water is needed for methanol to re-form to hydrogen. Can this water be recovered from the methanol combustion process?

3.14 Compare the fuel volume per unit of chemical exergy for aqueous gasoline, aqueous methanol, ammonia, refined natural gas, and liquefied petroleum gas and comment on the results.

Nomenclature

A	weight fraction of ash
AIT	autoignition temperature, K
APISG	specific gravity in degrees API, °API
BSG	specific gravity in degrees Baumé, °B
C	weight fraction of carbon

c	molar fraction of carbon
C_p	specific heat, kJ/kg.K
EF	emission factor, kg_{CO_2}/GJ
ex	specific exergy, kJ/kg
FPT	flash point temperature, K
f	factor
GCV	gross calorific value, MJ/kg
H	molar enthalpy, kJ/kmol
$\mathcal{H}$	weight fraction of hydrogen
HHV	higher heating value, MJ/kg
h	molar fraction of hydrogen
h	specific enthalpy, kJ/kg
LFL	lower flammability limit
LHV	lower heating value, MJ/kg
M	molecular weight, kg/kmol
$\mathcal{M}$	weight fraction of moisture
$\mathcal{N}$	weight fraction of nitrogen
NBP	normal boiling point, K
NCV	net calorific value, MJ/kg
n	molar fraction of nitrogen
$\mathcal{O}$	weight fraction of oxygen
o	molar fraction of oxygen
$\mathcal{R}$	universal gas constant, kJ/kmol.K
$\mathcal{S}$	weight fraction of sulfur
SEF	specific emissions factor, kg_{CO2}/kg_{coal}
SG	specific gravity
s	molar fraction of sulfur
s	specific entropy, kJ/kg.K
T	temperature, K
UFL	upper flammability limit
$\mathcal{VM}$	weight fraction of volatile matter
w	weight fraction

Greek letters

β	correlation parameter between heating value and chemical exergy
Δ	difference
ρ	density, kg/m^3

Subscripts

ad	as-determined
ar	as-received
d	dry basis
daf	dry, ash free
fg	evaportation
H	with respect to NCV or LHV
W	with respect to chemical exergy

Superscripts

0	standard state (enthalpy of formation)
ch	chemical

References

Albahri, T.A., 2003. Flammability characteristics of pure hydrocarbons. Chem. Eng. Sci. 58, 3629–3641.

Berkowitz, N., 1997. Fossil hydrocarbons: Chemistry and Technology. Academic Press, San Diego, CA.

Bilgen, S., Kaygusuz, K., 2008. The calculation of the chemical exergies of coal-based fuels by using the higher heating values. Appl. Energ. 85, 776–785.

Boie, W., 1953. Fuel technology calculations. Energietechnik. 3, 309–316.

Bunger, J.W., Thomas, K.P., Dorrence, S.M., 1979. Compound types and properties of Utah and Athabasca tar sand bitumens. Fuel 58, 183–195.

Demirbas, A., 1998. Fuel properties and calculation of higher heating values of vegetable oils. Fuel 77, 1117–1120.

Dincer, I., Zamfirescu, C., 2011. Sustainable Energy Systems and Applications. Springer, New York.

Dryden, I.G.C., 1957. Chemistry of coal and its relationship to coal carbonization. J. Inst. Fuel. 30, 193–214.

Eiserman, W., Johnson, P., Conger, W.I., 1980. Estimating thermodynamic properties of coal, char, tar, and ash. Fuel Process. Technol. 3, 39–53.

Haynes, W.M., Lide, D.R., 2012. CRC Handbook of Chemistry and Physics, 92nd ed. CRC Press, Boca Raton, FL Internet Version.

Hepbasli, A., 2009. Exergetic modeling of oil shale-fired circulating fluidized bed systems. Energy Sources, Part A. 31, 325–337.

Kalinci, Y., Hepbasli, A., Dincer, I., 2010. Efficiency assessment of an integrated gasifier/boiler system for hydrogen production with different biomass types. Int. J. Hydrogen Energy. 35, 4991–5000.

Kaygusuz, K., 2009. Chemical exergies of some coals in Turkey. Energ. Sourc. A 31, 299–307.

Kotas, T.J., 1995. The Exergy Method of Thermal Plant Analysis. Krieger Publishing Company, Malabar, Florida.

Lee, S., Speight, J.G., Loyalka, S.K., 2007. Handbook of Alternative Fuel Technologies. CRC Press, New York.

Majumder, A.K., Jain, R., Banerjee, P., Baewal, J.P., 2008. Development of a new proximate analysis based correlation to predict calorific value of coal. Fuel. 87, 3077–3081.

Mason, D.M., Gandhi, K.N., 1983. Formulas for calculating the calorific value of coal and chars: development, tests, and uses. Fuel Process. Technol. 7, 11–22.

Pan, Y., Jiang, J., Wang, R., Cao, H., Zhao, J., 2008. Prediction of auto-ignition temperatures of hydrocarbons by neural network based on atom-type electro-topological-state indices. J. Hazard. Mater. 157, 510–517.

Patil, G.S., 1988. Estimation of flash point. Fire Mater. 12, 127–131.

Ringen, S., Lanum, J., Miknis, F.P., 1979. Calculating heating values from elemental compositions of fossil fuels. Fuel. 58, 69–71.

Rivero, R., Rendón, C., Galleos, S., 2004. Exergy and exergoeconomic analysis of a crude oil combined distillation unit. Energy. 29, 1909–1927.

Sloan, E.D., Koh, C.A., 2007. Clathrate Hydrates of Natural Gases, third ed. CRC Press, New York.

Speight, J.G., 2006. The Chemistry and Technology of Petroleum, fourth ed. CRC Press, New York.

Speight, J.G., 2011. The Biofuels Handbook. The Royal Society of Chemistry, Cambridge, UK.

Szargut, J., 2005. Exergy Method: Technical and Ecological Applications. WIT Press, Boston.

Szargut, J., Styrylska, T., 1964. Approximate evaluation of the exergy of fuels. Brennstoff Wärme Kraft. 16, 589–596.

Szargut, J., Morris, D.R., Steward, F.R., 1988. Exergy Analysis of Thermal, Chemical and Metallurgical Processes. Hemisphere Publishing Corporation, New York.

Van Loo, S., Koopejan, J., 2008. The Handbook of Biomass Combustion and Co-Firing. Earthscan, Sterling, VA.

Zamfirescu, C., Dincer, I., 2008. Using ammonia as a sustainable fuel. J. Power. Sources. 185, 459–465.

Zamfirescu, C., Dincer, I., 2009. Ammonia as a green fuel and hydrogen source for vehicular applications. Fuel Process. Technol. 90, 729–737.

4

Hydrogen and Fuel Cell Systems

4.1 INTRODUCTION

The idea of hydrogen utilization for power generation goes back to early 1970 in the context of the energy crisis that was caused by petroleum shortages. Hydrogen is scarcely available in its free form on earth, and hence its use as a fuel implies a hydrogen production step in which hydrogen is extracted from other materials. The sources of hydrogen are abundant: water, biomass, hydrogen sulfide, and even hydrogen-rich petroleum resources. Methods of hydrogen extraction from these kinds of sources were developed well before the 1970s (e.g., water electrolysis, coal gasification, natural gas reforming, etc.). The main issue that arises with applying these methods is related to several parameters, including environmental impact, cost effectiveness, commercial availability, reliability, etc.

In a *hydrogen economy*, environmentally sustainable hydrogen production methods must be implemented on a large scale. In this manner hydrogen in sufficient quantity can be generated and temporarily stored. Then, hydrogen is used on demand for power generation and any other uses. Of course, for generation of power hydrogen can be directly fed to fuel cells. The only exhaust is steam that is absorbed by nature with no or very limited environmental impact. Other major uses of hydrogen are for methanol and ammonia production. These two chemicals are crucial for the world economy because they are the feedstock for many other major products: formaldehyde, plywood, paints, textiles, fertilizers, and pharmaceutical substances. Besides these uses, ammonia and methanol can be directly used as fuels in fuel cells or engines. Other major uses of hydrogen do exist and will be briefly detailed later in this chapter. Nevertheless, the key element of a hydrogen economy is the development of sustainable hydrogen production on the one hand and effective fuel cell systems on the other.

From another perspective one notes that electrical power cannot be stored in large quantities because there are not any suitable devices to store electrons under fixed differences of potential. Supercapacitors cannot be applied on the scale of regional or national grids to satisfy all storage demand; at most they can dump partially the incidental demand peaks of very short duration. Other energy storage options for electricity grids have their disadvantage: batteries are heavy, costly and have a short lifetime, pumped hydro requires sites where dams can be built which are relatively expensive, compressed-air systems can be applied on an intermediate scale but not region wide; flywheels are applicable on a small scale only.

Hydrogen is storable directly in large volumes of compressed gas, absorbed in heavy masses of metal hydride, stored as a cryogenic liquid or be stored seasonally in amounts of tens of kilotons through reversible chemical conversion of ammonia or methanol or other chemical forms. Thus, stored hydrogen can be helpful in overcoming the fluctuating issues with renewable energy sources, such as solar and wind. Furthermore, the possibility of seasonal storage allows for alternating between the more efficient production during winter and the increased demand in summer. Again, fuel cells must be part of the system because they generate hydrogen power on demand in the most convenient manner.

Therefore, hydrogen has another major role besides that of being a fuel and a primary material for the worldwide economy. Namely, hydrogen is treated as a basic energy storage option that can moderate between the fluctuating and intermittent nature of the renewable energy availability (solar, wind, tidal, etc.) and the rather constant demand for electric power. Because of its roles as a synthetic fuel and energy storage capability, hydrogen is denoted as an *energy carrier*.

Power generation through hydrogen and fuel cells requires the development of a specific chain of technologies spanning from sustainable conversion of energy resources to hydrogen, to hydrogen storage, and finally to effective power generation with fuel cells. The focus of this chapter is placed on power generation using hydrogen and fuel cell systems. Therefore, the hydrogen role in power generation and the available methods of sustainable hydrogen production are discussed briefly. The most emphasis is given to fuel cell systems which are introduced in some detail, with their classification and their modeling equations presented thoroughly. Several important fuel cell systems, including integrated fuel cell and gas turbine systems, are introduced, and their efficiencies are defined through energy and exergy analysis methods with some examples and case studies.

4.2 HYDROGEN

Hydrogen is crucial for satisfying the power demand of the world in the future as the conventional energy resources deplete and by consequence there is a real need for conversion of fluctuating/intermittent renewable energies, especially solar, into storable forms; this necessarily requires hydrogen production. The need of solar-derived power generation through hydrogen can be clearly justified based on thorough predictions of worldwide energy demand increase. As will be made clear, development of sustainable hydrogen and fuel cell technology is not an optional choice, but rather it must be done!

According to the World Bank (2013), the predicted annual average power demand of the world is ~42 TW by 2050, an increase to three times today's consumption. But why is such a high increase in energy consumption predicted? and what will the solution to satisfy such a demand be? There are two closely related reasons: (i) the increasing tendency of the standard of living of humans which leads to higher and higher energy consumption per capita and (ii) the continuous increase of world's population. The current energy consumption worldwide per capita is on average about 2.5 kW. If this per capita consumption is maintained, with the prediction of a 9.6-billion world population by 2050, the energy demand will reach 24 TW.

However, allowing for the world to achieve the per capita usage of Europe and North America—that is, 4.5 and 9.5 kW, respectively—one can guess at a 2050 energy demand of

43 and 91 TW, respectively. Hence by rough estimation there will be a 50-TW energy demand by 2050. Asking from where the world will generate its energy demand beyond 2050 raises a dilemma because the conventional petroleum resources currently in exploitation and those that are proven will deplete by that time (30- to 40-year time span) due to the accelerated consumption (Rajeshwar et al., 2008).

The only energy resources available by 2050 will then be coal, nuclear, and renewables. Coal will be available for about an additional 70 years beyond 2050, with a total energy content of $\sim 10^{13}$ GJ. Using the data published in Rajeshwar et al. (2008) and Dincer and Zamfirescu (2011), one can find out the following:

- If coal power is generated at a rate of 7.5 TW then this resource will deplete in 100 years.
- If one new nuclear power station of 1 GW is commissioned every day from now until 2050 then the available power production from nuclear energy will become ~ 15 TW.
- If hydroelectric power stations are installed everywhere where it is technically feasible, the generated hydropower will be ~ 1.5 TW worldwide.
- If all geothermal sites produce power with 30% efficiency then 3 TW will be generated worldwide.
- If all available fertile land that is not destined for food production is used to grow energy, cultures then the available power generation will be ~ 10 TW worldwide.
- The wind power generation potential on land sites is ~ 2 TW worldwide.
- If 30% of the Sahara Desert is occupied by solar collectors to run a system that produces hydrogen as an energy carrier and uses fuel cells for continuous power production with an estimated overall efficiency of 10% then the power generated would be ~ 50 TW.

Deduced from the above is that there is no energy solution for the future without including concentrated solar power (CSP) production on a very large scale. Henceforth, the role of hydrogen and fuel cell systems becomes very important today for research and development and in the near future as an energy solution at planetary scale. In consequence, after one century of mechanization followed by another century of electrification, humankind is now on the verge of a new century that can be called the century of hydrogenation.

As shown in Naterer et al. (2013) (Chapter 1, Figure 1.1) about 60% of energy demand in 2050 will be satisfied from renewable sources. The hydrogen demand will increase even more because other sectors will benefit. For example, Dincer and Zamfirescu (2011) show that in Canada by 2025 there are predicted to be four major industrial sectors that demand hydrogen: crude oil upgrading (62% share), ammonia synthesis (25% share), methanol synthesis (6% share), and metallurgy (most the rest of the industrial share of hydrogen demand).

The implementation of hydrogen and fuel cells to satisfy a significant portion of energy demand and for load leveling at a regional level is illustrated schematically in Figure 4.1. Two manifolds exist for this grid system. First, hydrogen is produced from primary energy sources though sustainable methods. If fossil fuels are used for hydrogen production then ecological measures must be undertaken, such as carbon capture and sequestration. Hydrogen is then stored on a large scale and used on demand according to grid need through fuel cell generators. The second manifold implies the use of excess power when the demand is low to store energy in form of hydrogen. When demand is high again, hydrogen can be reconverted to power though fuel cell power generators.

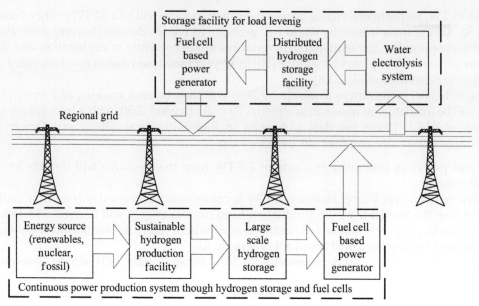

FIGURE 4.1   Regional grid with coupled hydrogen and fuel cell systems.

For an electrical grid the hourly average electricity price (HOEP) changes in pace with the generation capacity reserve: the price is higher when the reserve is lower. In addition, the HOEP tends to increase during peak demand. Therefore, in order to reduce the HOEP the capacity reserve must be increased. This can be easily done when energy storage is available, e.g., a hydrogen and fuel cell system is implemented. As shown in Naterer et al. (2013) load leveling through hydrogen and fuel cell systems can reduce the HOEP especially when hydrogen production is done at large scale. The HOEP is directly influenced by the cost of the stored hydrogen. As deduced from Naterer et al. (2008) and Naterer et al. (2013) (Figure 2.12), if the hydrogen production increases from 10 to 100 t/day the HOEP is expected to decrease by 10%.

The residential sector can also benefit from hydrogen and fuel cell systems. In a residence power can be generated locally using a fuel cell system fed with hydrogen. In addition, hydrogen can be fed into a furnace and combusted to generate heat for water and space heating on demand. Hydrogen can be distributed either through the existent network for natural gas (with some modifications or in some minor proportion can be blended with natural gas) or transported in cisterns and loaded in pressurized tanks at the residences. Figure 4.2 illustrates the hydrogen and fuel cell system for powering residences. There are many advantages of using such a system, one being the important reduction of greenhouse gas (GHG) emissions. Another advantage consists of the ability of local power generation with fuel cells which can reduce the peak loads on regional grids. A third advantage is that delocalized energy harvesting from renewable energy can be easily applied if hydrogen and fuel cell systems are in place. For example, power can be generated locally at residences or quarters from biomass/biofuel combustion, from wind energy, or from solar energy. The environmental benefit of hydrogen and fuel cell systems in residences connected to a distribution network for hydrogen can be exemplified by Canada, where in the vast majority of cases the residences are

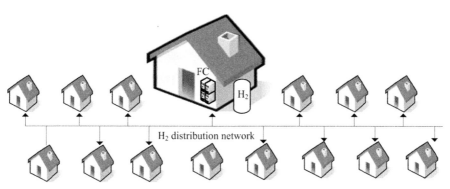

FIGURE 4.2 Powering and heating residences through hydrogen and fuel cells.

heated with natural gas from distribution networks. One can deduce from Naterer et al. (2013) that if Canadian residences were powered with hydrogen and fuel cells the reduction of GHG emission would be more than 48%. When hydrogen generated from solar and wind energy is distributed through pipeline networks, the generation sites for hydrogen can be delocalized. Table 4.1 presents the estimated GHG reduction for Canadian provinces. A bigger environmental benefit is obtained when hydrogen is used for cogeneration of power and heating (water heating and/or space heating). Considering that 20% of Canadian energy consumption is due to the residential sector it results that the environmental benefit of hydrogen and fuel cells for power generation or power and heat cogeneration is major.

The transportation sector is one of the most important consumers of liquid fuels worldwide and a major contribution to air pollution through GHG, NOx, VOCs, and SOx. The Canadian annual GHG emissions due to the transportation sector reach about 180 Mt CO_2 equivalent. Another indicator of the environmental impact of gas effluents is the *air pollution indicator*, which quantifies the integrated impact of airborne pollutants from automobile exhaust gases other than GHG which include CO, NOx, SOx, and VOCs. In a comprehensive lifecycle assessment by Dincer et al. (2010) the environmental impact of four passenger vehicle technologies, conventional, hybrid, electric, and hydrogen-fuel cell, were determined.

TABLE 4.1 GHG Reduction When Powering Canadian Residences with Hydrogen and Fuel Cells

Service	Province									
	NFL	**PEI**	**NS**	**NB**	**QUE**	**ON**	**MAN**	**SAS**	**AB**	**BC**
Appliances	17.7	7.362	18.4	18.1	18.4	17.1	20.6	16.5	15.9	19.3
Lighting	6.0	2.5	6.9	6.0	7.2	7.3	7.5	6.6	6.7	8.8
Space cooling	0.02	0.05	0.22	0.29	2.0	9.2	4.2	0.84	0.08	0.50
Total	23.72	9.912	25.52	24.39	27.6	33.6	32.3	23.94	22.68	28.6
GHG (kg/GJ)	234	235	243	221	1	251	3	334	143	104
GHG (t/year)	5.6	2.3	6.2	5.4	0.0	8.4	0.1	8.0	3.2	3.0
GHG reduction	58%	100%	50%	77%	100%	56%	100%	48%	100%	100%

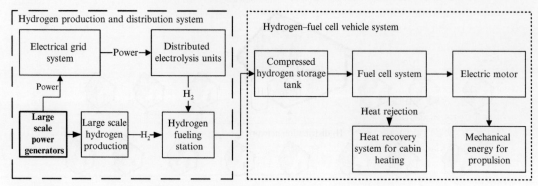

FIGURE 4.3   Hydrogen and fuel cell system for vehicles.

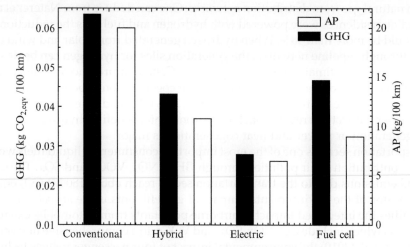

FIGURE 4.4   Life cycle GHG and AP emissions of four categories of road vehicles including the vehicle and fuel production and utilization phases. *Data from Dincer et al. (2010).*

Figure 4.3 shows the general layout of the hydrogen and fuel cell vehicle system. The GHG and air pollution indicators are determined for each category of the analyzed vehicles as summarized in Figure 4.4. The three studied alternative vehicle options reduce the pollution lifecycle by about 50% with respect to conventional power. The electric vehicle is characterized by air pollution during the utilization stage due to battery recycling. If the GHGs of hydrogen production could be decreased, the fuel cell vehicle has the best likelihood of having the lowest carbon footprint.

Figure 4.5 illustrates the specific GHG emission of the main classes of vehicles in Canada, given in g CO_2 equivalent per MJ with respect to the LHV of fuel. Based on the estimated GHG specific emission and the number of vehicles in operation the total GHG emissions can be determined. Further contribution of the Canadian transportation sector to the change in annual CO_2 concentration of earth atmosphere can be calculated.

Once the change of GHG concentration is determined for the next decade, the radiative forcing produced by each class of vehicles of the transportation sector can be estimated. Subsequently, the current estimated value of the climate sensitivity factor of 0.6 m^2K/W can be

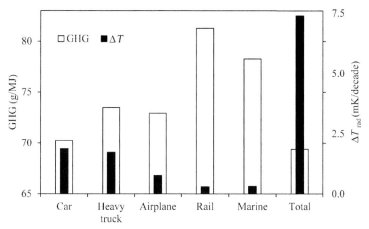

FIGURE 4.5 Energy-specific GHG emissions of vehicles in Canada and their contribution to earth temperature increase per decade. *Calculated from data from Naterer et al. (2013).*

used to determine the impact of the sectorial emissions on the earth temperature increase during the decade, due to the global warming effect. This type of study is summarized in Naterer et al. (2013). Based on their data, the predicted increase in earth temperature due to emissions from Canadian vehicles is given, as shown in Figure 4.5. As shown, these emissions lead to 7.5 mK temperature increased per decade. Once a hydrogen-based fuel cell system is put in practice, this global warming effect may effectively be mitigated.

In conclusion, hydrogen and fuel cell systems are considered an important part of the overall solution to sustainable power generation at global level. If hydrogen is produced using renewable energy sources a better air quality is obtained and damaging effects such as global warming, acid rains, and air pollution are diminished. Hydrogen and fuel cell systems become a key component for sustainable power generation due to the following:

- Compatibility with renewable energy systems strictly required in the future for better energy security and economic growth facilitation.
- Besides producing power hydrogen and fuel cell systems can produce useful heating.
- Fuel cells are very efficient, with better savings of primary energy.
- The variety of hydrogen and fuel cell technologies provide a flexible array of options for their use in various applications, with reduced environmental impact and increased efficiency.
- It favors the decentralization of power generation, providing more energy security.
- Even when fossil fuels are used to produce hydrogen it is almost certain that hydrogen and fuel cell systems are more effective than conventional systems.
- Hydrogen fuel is safer than conventional fuels for vehicles.

4.3 HYDROGEN PRODUCTION METHODS

In Section 3.4.4 of Chapter 3 hydrogen properties and storage methods are detailed. Storage of hydrogen appears to be very challenging because in normal atmospheric conditions it is a gas with a very low density of 40.8 g/m^3. Other possible methods of hydrogen storage are described in Figure 3.20, in terms of energy density and volumetric storage density. The most relevant physical properties of hydrogen are given in Table 3.4.4. Also, Figure 3.19 identifies the sustainable methods of hydrogen production. In this section, hydrogen production methods are detailed in some extent.

Figure 4.6 shows energy conversion pathways for hydrogen production methods. The following four forms of energy are extracted from basic energy sources: thermal, electrical, biochemical, and photonic (or radiation). Many hydrogen production methods exist, as shown in Figure 3.19c, that use four main forms of energy. As suggested in Figure 4.6, two main energy conversion pathways exists which lead to green or non-green hydrogen production.

Hydrogen production methods can be classified based on production scale in three categories. In Dincer and Zamfirescu (2012) the production scales are low (≤70 kg/day), medium (70–70,000 kg/day), and large (≥70,000 kg/day). Figure 4.7 shows the classification of hydrogen production methods as a function of production capacity. In what follows we will give some details on water electrolysis, thermochemical water splitting, gasification, hydrocarbon reforming, photocatalysis and photo-biochemical conversion.

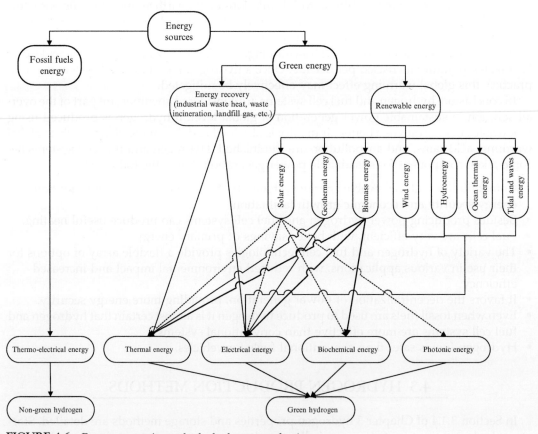

FIGURE 4.6   Energy conversion paths for hydrogen production.

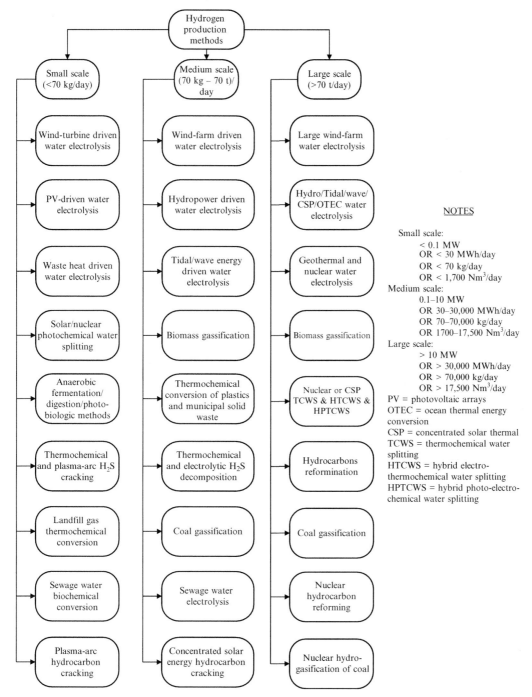

FIGURE 4.7 Hydrogen production methods according to production scale.

NOTES

Small scale:
    < 0.1 MW
    OR < 30 MWh/day
    OR < 70 kg/day
    OR < 1,700 Nm³/day
Medium scale:
    0.1–10 MW
    OR 30–30,000 MWh/day
    OR 70–70,000 kg/day
    OR 1700–17,500 Nm³/day
Large scale:
    > 10 MW
    OR > 30,000 MWh/day
    OR > 70,000 kg/day
    OR > 17,500 Nm³/day
PV = photovoltaic arrays
OTEC = ocean thermal energy conversion
CSP = concentrated solar thermal
TCWS = thermochemical water splitting
HTCWS = hybrid electro-thermochemical water splitting
HPTCWS = hybrid photo-electro-chemical water splitting

4.3.1 Water Electrolysis

The process of electrolysis was discovered in the early 1800s and formulated by 1834 according to Faraday's law, which states that the mass of matter that enters the reaction at the electrode is proportional to the electrical current. Today, water electrolysers are constructed for a wide range of power using three main technologies: alkaline electrolysers (the most mature technology), solid oxide high-temperature steam electrolysers (emergent technology), and proton-exchange-membrane electrolysers (PEMEs) (established technology). In principle, two options for water electrolysis systems exist for hydrogen production, and these are described in Table 4.2.

The most used electrolysers today are of alkaline type. They use an alkaline liquid electrolyte (pH >7), typically KOH or NaOH solutions, which facilitates the mobility of anions (hydroxyl, OH^-). In alkaline electrolysers, the following reactions occur

$$\begin{cases} \text{at cathode}: \ 2H_2O + 2e^- \rightarrow H_2(g) + 2OH^- \text{ with } E^0 = 0.408 \text{ V} \\ \text{at anode}: \ 2OH^- \rightarrow H_2O + 0.5O_2(g) + 2e^- \text{ with } E^0 = 0.822 \text{ V} \end{cases}$$

These electrolysers have two common configurations, monopolar and bipolar. In a monopolar configuration a number of alkaline cells are connected in parallel; therefore, the current consumed by the unit is a sum of the currents consumed by each individual cell, while all cells are subjected to the same potential difference. Figure 4.8 presents a monopolar alkaline electrolyser. A feed of aqueous electrolyte is provided to the electrolysis cell at the anodic compartment side. The typical concentration of base in the electrolyte is 30% (e.g., for KOH), which assures a good mobility of hydroxyls. The electrolyte must be recirculated and makeup water added such that mass transfer is enhanced at the electrodes while minimizing the concentration overpotentials.

A diaphragm separates the anodic and cathodic compartments. It allows passage of all ionic species but restricts the flow of gases; therefore, oxygen and hydrogen gases are completely separated and withdrawn from separate manifolds. Given that the cells are

TABLE 4.2   Water Electrolysis Systems for Hydrogen Production

Electrolysis technology	Liquid water electrolysis	High-temperature steam electrolysis
Configuration	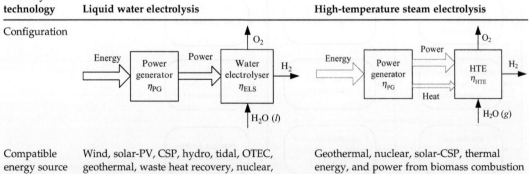	
Compatible energy source	Wind, solar-PV, CSP, hydro, tidal, OTEC, geothermal, waste heat recovery, nuclear, biomass, or fossil fuel energy through combustion and power plants	Geothermal, nuclear, solar-CSP, thermal energy, and power from biomass combustion or fossil fuel combustion

FIGURE 4.8 Configuration of an alkaline electrolyser of monopolar type.

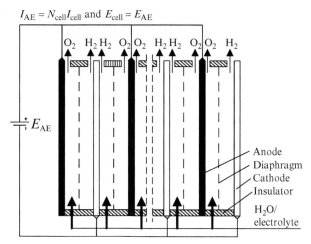

$$I_{AE} = N_{cell}I_{cell} \text{ and } E_{cell} = E_{AE}$$

electrically connected in parallel, the voltage on each cell must be the same, and the current absorbed by the unit is the sum of the currents absorbed by the cells as $I_{AE} = \sum I_{cell}$.

Due to constructive considerations, the ohmic losses are relatively high in monopolar electrolysers. If the cells are connected in series instead of parallel then fewer leads are required and as such ohmic losses can be reduced. Figure 4.9 illustrates a bipolar electrolyser; this has the cells connected in series. Therefore the voltage applied to the unit is the sum of voltages applied to the cells, while all the cells consume the same current as $E_{AE} = \sum E_{cell}$. For these cells, minimization of the gap between the electrodes becomes important because it reduces the overpotentials due to ionic transport. However, care must be taken to minimize the power surge between adjacent electrodes when these are placed too close to each other.

Alkaline electrolysers do not require expensive catalysts; rather steel electrodes coated with nickel can be used. Because of their planar electrodes the electrolysers can be assembled

FIGURE 4.9 Configuration of an alkaline electrolyser of bipolar type.

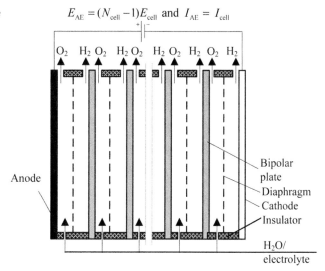

$$E_{AE} = (N_{cell} - 1)E_{cell} \text{ and } I_{AE} = I_{cell}$$

as stacks which confer excellent mechanical strength and allow the operation under pressure. Henceforth, the operation temperature can be raised over 373 K, the typical range being 353–475 K. The capacities of large-scale commercial alkaline electrolysers range from 50 MW to 100 MW, with hydrogen production up to 750,000 Nm^3/day.

The PEMEs, consisting of a stack of planar electrodes and an acidic electrolyte in solid phase, have a much smaller hydrogen production range—up to 1500 Nm^3 per day. A PEME is illustrated in Figure 4.10. Typically, a PEME stack uses a bipolar plate arrangement. The acidic electrolyte is a thin solid polymeric membrane—Nafion®—which is based on perfluorosulfonic acid. Hydronium (H_3O^+) permeates through Nafion® provided that the membrane is maintained wet. The operating temperature of PEME is 325–353 K. It requires ultrapure distilled water but generates hydrogen of 99.999% purity. The cost of a PEME stack is high because the bipolar plates require coating with very expensive platinum group materials, due to their exposure to the corrosive action of the acidic electrolyte. Moreover, the cathodes must be coated with expensive catalysts to promote proton reduction of such oxides as Ir or Ru. The anodic and cathodic reactions in a PEME are given as follows:

$$\begin{cases} \text{at anode}: H_2O \rightarrow 2H^+ + \frac{1}{2}O_2(g) + 2e^- \text{ with } E^0 = 1.23\,V \\ \text{at cathode}: 2H^+ + 2e^- \rightarrow H_2(g) \text{ with } E^0 = 0.0\,V \end{cases}$$

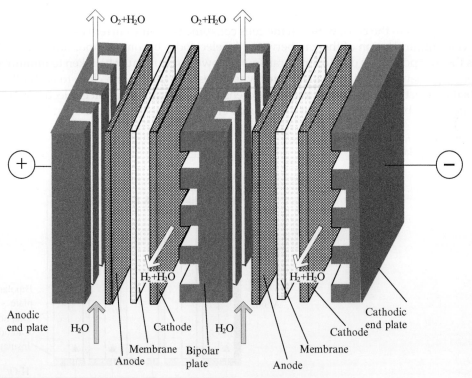

FIGURE 4.10   Stack arrangement of a proton-exchange-membrane electrolyser.

The high-temperature steam electrolysers represent an emerging technology. They include a solid oxide electrolyte, typically conductive to oxygen ions at high temperature, although there are some which allow for proton conduction at intermediate-temperature levels. The half-reactions at the electrodes of a solid oxide electrolyser cell (SOEC) with ionic oxygen conduction are written as

$$\begin{cases} \text{at anode}: 2O^{2-} \rightarrow O_2(g) + 4e^- \text{ with } E^0 = 0.226 \text{ V at } 1000 \text{ K} \\ \text{at cathode}: H_2O(g) + 2e^- \rightarrow H_2(g) + O^{2-} \text{ with } E^0 = 0.476 \text{ V at } 1000 \text{ K} \end{cases}$$

As it can be observed, the reversible cell voltage at the typical operating temperature of 1000 K is only 0.802 V, hence lower than the 1.23 V for low temperature cells. This indicates an advantage of SOEC, which requires less electrical power with respect to other cells. A solid oxide electrolysis cell stack is normally constructed in a bipolar configuration. Figure 4.11 represents the configuration of a single cell of a bipolar solid oxide electrolyser stack. It consists of bipolar end plates, gas channels, anode, cathode and electrolyte.

The electrolyte–electrode assembly consists of a thin layer of solid oxide packed between a porous cathode and a porous anode. Ceramic electrolytes based on scandium-stabilized zirconium or yttrium-stabilized zirconium are commonly used. These electrolytes allow the passage of oxygen ions O^{2-} under a driving force created by an electrical field and/or concentration gradient at temperatures of 1000–1300 K. Oxygen ions diffuse toward the positively charged anode under the influence of an electric field and concentration gradient.

The proton-conducting solid oxides use a solid electrolyte which becomes conductive to ions in a range of temperatures of 675–1025 K. The construction of these cells is very similar to that of PEME, consisting of bipolar plates with channels for the gas flow, porous electrodes, and electrolyte. The research work is in its early stage; no commercial systems are available.

The most important parameters for mathematical modeling of processes within electrolysers and the parameters that characterize performance are given in Table 4.3. Ultimately, using the information from the table, energy and exergy efficiencies of the electrolysis cell

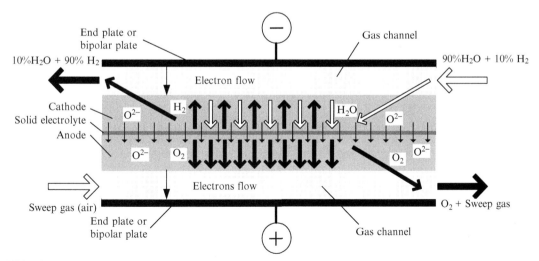

FIGURE 4.11 Configuration of a solid oxide electrolysis cell.

can be calculated. As the electrolysis system involves auxiliary components, its overall efficiency is downgraded.

The efficiency of hydrogen production through electrolysers also depends on the efficiency of power generation. For alkaline and PEM electrolysers which are supplied with power only, the overall hydrogen generation efficiency is given by $\eta_{H_2} = \eta_{PG}\eta_{ELZ}$ where subscripts refer to power generation (PG) and electrolysis (ELZ); a similar equation exists for exergy efficiency. Some cases are mentioned in Table 4.4.

Wind power generators or wind farms can be coupled to PEME or alkaline electrolysers (AE) depending on the hydrogen production scale. When doing so, one must account for the capacity factor of the wind power plant and the annually average wind speed. Using these parameters, the efficiency of hydrogen production becomes: $\eta_{H_2} = C(\bar{V})\,\eta_{ELZ}$. The capacity factor normally ranges from 20% to 50%, depending on the site. The energy from water

TABLE 4.3 Modeling and Performance Parameters for Electrolysers

Parameter and definition	Equations
Thermo-neutral potential (V): the potential applied to the adiabatic cell to transfer all reaction enthalpy, in which the term $T\,\Delta S$ is provided as heating via the Joule effect.	$E_{th} = \dfrac{\Delta H}{2F}$
Reversible cell potential (V): the potential applied to the cell in order to transfer the required free energy for the electrochemical reaction to proceed. It is also given by the Nernst equation as a function of the equilibrium constant and the standard reversible cell potential. The standard reversible cell potential is $E^0 = 1.229$ V. The equilibrium constant is $K_{eq} = \left(P_{H_2}P_{O_2}^{0.5}\right)/\left(a_{H_2O}P_0^{1.5}\right)$, where $P_0 = 101.325$ kPa and a_{H_2O} is the chemical activity of water in the electrolyte.	$E_{rev} = \dfrac{\Delta G}{2F}$ $E_{rev} = E^0 + \dfrac{RT}{2F}\ln\left(K_{eq}\right)$
Exchange current density (A/m^2): the current density corresponding to the reversible reaction. It expresses the fact that in order to proceed a minimum energy must be transmitted to the reaction which depends on the activation energy $\Delta G_{act} = 2FE_{act}$ and the pre-exponential factor $\mathcal{A}$.	$J_0 = zF\mathcal{A}\exp\left(-\dfrac{2FE_{act}}{RT}\right)$
Current density (A/m^2): the current in amperes divided by the area of the electrodes. It depends on the activation energy according to the Butler–Volmer equation, given here in simplified form, where the transfer coefficient $\alpha \in [0,1]$ expresses the proportion in which the applied electric field that reduces the activation energy.	$\dfrac{J}{J_0} = \exp(\alpha FE_{act}/RT) - \exp(-(1-\alpha)FE_{act}/RT)$
Activation overpotential (V): the potential required to overcome the activation energy for the forward reaction, defined by $\Delta G_{act} = 2FE_{act}$. For a small electrode polarization it is proportional to J/J_0. For polarized electrodes it is proportional to the logarithm of J according to the Tafel equation.	$E_{act} = (RT/F)(J/J_0)$, for $J \cong J_0$ $E_{act} = b\ln(J/J_0), J > J_0$ b Tafel constant.
Concentration overpotential (V): the overpotential required to transfer electrical energy to the electrode under an adverse concentration gradient of ionic species. It depends on the limiting current density at the electrodes (J_L), which is typically 10^6 times higher than the exchange current density.	$E_{conc} = \dfrac{RT}{zF}\ln\left(\dfrac{1+\frac{J}{J_L}}{1-\frac{J}{J_L}}\right)$
Energy efficiency (%): the ratio between the molar higher heating value of hydrogen and the electrical energy spent to generate it; note that the ratio $HHV/2F = 1.4696$.	$\eta = \dfrac{146.96}{E_{rev} + E_{act} + E_{conc}}$
Exergy efficiency (%): the ratio between the molar chemical exergy of hydrogen and the electrical energy spent to generate it. It can be expressed by a quality factor of 0.83.	$\psi = 0.83\,\eta$

$F = N_A\, e = 96,490$ C/mol (Faraday constant).
Additional modeling data and parameters can be found in Naterer et al. (2013), Chapter 4.

TABLE 4.4 Efficiency and Production Capacity of Green Electrolysis H_2 Production Methods

Technology	η (%)	Ψ (%)	$\dot{E}$ (MW)*
Wind turbine/farm + AE or PEME	22–40	21–39	0.01–400
Hydro/tidal power + AE	60–65	58–63	0.1–1000
OTEC power + AE	4–5	45–50	30–300
PV arrays/farm + PEME	4–6	3–5	0.001–0.1
CSP + AE or PEME	11–31	11–30	25–300
Geothermal Rankine + AE	10–15	20–30	10–300
Nuclear power + AE	19–32	17–29	100–1000
Waste heat advanced Rankine + AE or PEME	4–15	10–35	0.001–0.1
Geothermal + HTSE	11–16	22–32	10–300
CSP + HTSE	12–32	12–31	25–300
Nuclear + HTSE	22–35	20–33	100–1000

Data from Dincer and Zamfirescu (2012).

electrolysis can also be derived from hydro- or tidal power generators. If this is done, the efficiency of hydro-mechanical energy conversion to power is high, and consequently the overall hydrogen production efficiency is the highest among all existent methods using water electrolysis. Hydro/tidal OTEC power generation is of very low efficiency. Due to this fact, the energy efficiency of OTEC/electrolysis technology is the lowest. However, due to the use of advanced Rankine cycles (e.g., the Kalina cycle) the exergy efficiency becomes rather high, as seen in Table 4.4.

If solar radiation is harvested for power with the help of PV arrays the efficiency can reach 30%; however, electrical losses occur due to power conditioning. Due to the inherent losses the hydrogen production efficiency of practical systems with PV electrolysis is 5–7%, as detailed in Dincer and Zamfirescu (2012). Concentrated solar power (CSP) can also be coupled to electrolysis units. Although small-scale systems are viable (using power dishes and PEME) the more effective ones are at large scale, involving either solar through plants or heliostat fields and alkaline electrolyser units. Similarly, geothermal Rankine or nuclear Rankine plants can be coupled with alkaline electrolysers to generate hydrogen at a large scale of production.

An organic Rankine cycle can be used to recover waste heat sources and convert these to power output. There are many examples in industry where heat is wasted at intermediate-temperature level (~80–200 °C). Heat can be recovered from combustion gases in actual water heating, space heating, or power generation facilities. The power is used to drive PEM electrolysers which in turn produce sustainable hydrogen.

The systems that generate high-temperature heat, e.g., CSP, and next-generation nuclear reactors, can be used to produce hydrogen through high-temperature steam electrolysis (HTSE). In principle, using some heat upgrade and special heat exchanger networks, HTSE can be coupled to geothermal sources or conventional nuclear power plants. This offers another green solution for producing electrolytic hydrogen at scales up to hundreds of kilowatts or more.

EXAMPLE 4.1

The cathodic nickel electrode half-reaction in alkaline electrolyte is characterized by a current exchange density of $J_0=10\,\text{mA/m}^2$, Tafel slope of $b=120\,\text{mV}$, and transfer coefficient of $\alpha=0.2$. The cell operates at the limits between low and high polarization levels. Calculate the energy and exergy efficiency of the cell when the operating temperature is 60 °C.

- Table 4.3 gives the equation for activation overpotential for low and high polarization. At low polarization $E_{act}=(RT/F)(J/J_0)$. Replacing the values for R, T, F, J_0 the following relationship between current density and activation potential results: $E_{act}=2.61\,J$.
- For high polarization $E_{act}=b\ln(J/J_0)$; thus $E_{act}=0.5457\ln J$.
- When the cell operates at the limits between low and high polarization, $2.61\,J=0.5457\ln J$. Solving for J one obtains $J=0.1043\,\text{A/m}^2$.
- The activation overpotential for $J=0.1043\,\text{A/m}^2$ is $E_{act}=2.61\,J=0.2722\,\text{V}$.
- Assume that the cell operates at P_0 and water activity is 1, then the reversible cell potential is $E_{rev}=1.229\,\text{V}$.
- The cell operates at the limits of low and high polarization, so $E_{conc}\cong0\,\text{V}$.
- The total cell potential is $E_{tot}=E_{rev}+E_{act}+E_{conc}=1.501\,\text{V}$.
- The energy efficiency is $\eta=146.96/E_{tot}=97.9\%$ and $\psi=0.83\eta=81.26\%$.

4.3.2 Thermochemical Cycles

Water can be split into hydrogen and oxygen with the help of thermochemical cycles, which are processes that involve a closed sequence of chemical reactions. There is at least one endothermic reaction that is driven thermochemically at high temperature. The cycle can also include an electrochemical reaction in which case the name hybrid thermochemical cycle is used. The only consumed chemical is water as the rest of the intermediary chemical compounds are recycled. The hydrogen and oxygen product evolve as separate streams. Figure 4.12 illustrates the inputs and outputs specific to pure thermochemical and hybrid

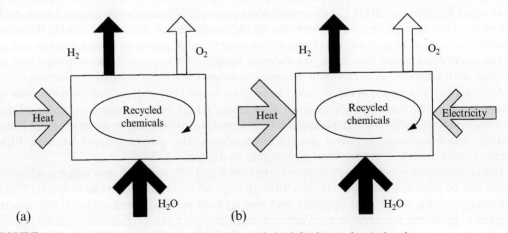

FIGURE 4.12   Black-box representation of pure (a) and hybrid (b) thermochemical cycles.

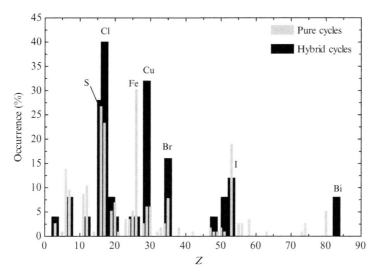

FIGURE 4.13 Occurrence of chemical elements in intermediary compounds of pure and hybrid thermochemical cycles. *Data from Naterer et al. (2013).*

thermochemical cycles. As seen, pure thermochemical cycles require only high-temperature heat as energy input, whereas hybrid cycles require bot power and heat.

Chemical compounds of sulfur, iodine, iron, and chlorine are often used in pure thermochemical cycles. For hybrid cycles the most common are the compounds of copper and chlorine. Figure 4.13 illustrates the use of chemical elements in thermochemical cycles.

Even when the cycle is pure thermochemical there is a need for power to drive the auxiliary equipment, such as mixers, pumps, conveyers, or compressors. According to the diagram in Figure 4.14 only a primary energy source in the form of high-temperature heat is used to provide both the required heat and electricity to the thermochemical cycle. If one denotes with η_{PG} the energy efficiency of power generation then the overall energy and exergy efficiencies of hydrogen production are

$$\eta_{H_2} = \frac{\dot{m}\,\mathrm{HHV}}{\dot{Q}_{TC} + \frac{\dot{W}_{TC}}{\eta_{PG}}} \quad \text{and} \quad \psi_{H_2} = \frac{\dot{m}\,\mathrm{ex}^{\mathrm{ch}}_{H_2}}{\left(\dot{Q}_{TC} + \frac{\dot{W}_{TC}}{\eta_{PG}}\right)\left(1 - \frac{T_0}{T_H}\right)}$$

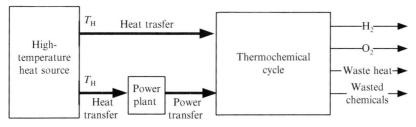

FIGURE 4.14 Hydrogen production system with thermochemical cycle.

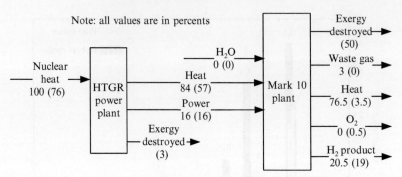

FIGURE 4.15   Energy and exergy flows in an integrated Mark 10—HTGR plant. *Data from Rosen (2008).*

where $\dot{Q}_{TC}$ and $\dot{W}_{TC}$ are the required heat and power for the cycle and T_H is the heat source temperature. An example of energy and exergy flows in a hybrid thermochemical cycle (Mark 10) is shown in Figure 4.15. The energy efficiency is 50% whereas the exergy efficiency is $\psi = 19/76 = 25\%$.

A comprehensive treatment of thermochemical cycles for water splitting is given in Naterer et al. (2013). Here we summarize the main issues, with a focus on production efficiency of the most important systems. The encountered reactions in thermochemical cycles can be categorized as five types: hydrolysis, hydrogen evolving, oxygen evolving, reagents recycling and electrochemical reactions. Some of the most encountered hydrolysis reactions are written as follows:

- $nH_2O + MO_m \rightarrow nH_2 + MO_{m+n}$, reaction of a metallic oxide MO_m with water to form a higher oxide MO_{m+n}, for example $4H_2O + 3Fe \rightarrow 4H_2 + Fe_3O_4$ at 773 K.
- $nH_2O + MX_m \rightarrow (n - m/2)H_2 + mHX + MO_n$, $2n \geq m$, reaction of a metallic halide with water to form hydrogen, hydrogen halide, and metallic oxide, for example $4H_2O + 3FeCl_2 \rightarrow H_2 + 6HCl + Fe_3O_4$ at 923 K.
- $H_2O + X_2 \rightarrow 0.5O_2 + 2HX$, hydrolysis with a halogen, for example $H_2O + Cl_2 \rightarrow 0.5O_2 + 2HCl$ at 873 K.
- $H_2O + X_2 + A \rightarrow 2HX + AO$, reaction of steam with a halogen in the presence of an oxygen acceptor, for example $H_2O + 0.5Br_2 + 0.5SO_2 \rightarrow HBr + 0.5H_2SO_4$ at 373 K.

Some relevant prototypical hydrogen and oxygen evolving and chemical recycling reactions schemes are given in Table 4.5. The most encountered electrochemical reactions are written as

$$2HCL \rightarrow H_2 + Cl_2, \ 363\,K, \ 0.99\,V$$

$$2HI \rightarrow H_2 + I_2, \ 373\,K, \ 0.5\,V$$

$$2H_2O + SO_2 \rightarrow H_2 + H_2SO_4, \ 310\,K, \ 22\,bar, \ 0.48\,V$$

$$2HBr \rightarrow H_2 + Br_2, \ 298\,K, \ 0.58\,V$$

$$2HBr \rightarrow H_2 + Br_2, \ 323\,K, \ 0.58\,V$$

To date, three main cycles are in the advanced stage of development for large-scale hydrogen production. The most advanced is the sulfur-iodine (SI) cycle, followed by the hybrid sulfur

TABLE 4.5 Prototypical Reactions for Thermochemical Cycles

Hydrogen evolving	Oxygen evolving	Chemicals recycling
$MX_m + HX_n \rightarrow H_2 + MX_{m+n}$, reaction of hydrogen halide with a metallic salt or a metal to form metallic halide	$MO_{m+1} \rightarrow 0.5O_2 + MO_m$, completed decomposition of an oxide or partial decomposition of a higher oxide	$M_3O_4 + 8HX \rightarrow 4H_2O + 3MX_2 + X_2$, (or $\rightarrow MX_2 + 2MX_3$), reaction of a metal oxide with a hydrogen halide
$2HX \rightarrow H_2 + X_2$, thermolysis of a hydrogen halide to form hydrogen and halogen	$AO \rightarrow 0.5O_2 + A$, thermolysis of oxyacid (or its salt e.g., chlorate, iodate, sulfate, nitrate) to form a halide (e.g., chloride, iodide, bromide) or oxide	$MX_{n+1} \rightarrow MX_n + 0.5X_2$, thermolysis of a metallic halide to form a lower halide and a halogen
$nH_2O + MO_m \rightarrow nH_2 + MO_{m+n}$, hydrolysis of a metallic oxide to generate hydrogen and a higher oxide MO_{m+n}		$MX_{n+1} + NX_n \rightarrow MX_n + NX_{n+1}$, reaction of two metallic halides that exchange one halogen atom
$nH_2O + MX_m \rightarrow (n - m/2)H_2 + mHX + MO_n$, $2n > m$, hydrolysis of a metallic halide to form H_2, hydrogen halide, and metallic oxide	$MO + X_2 \rightarrow 0.5O_2 + MX_2$, reaction of a metallic oxide with a halogen to form oxygen and halide $H_2O + Cl_2 \rightarrow 0.5O_2 + 2HCl$, reverse Deacon reaction of chlorine hydrolysis	$M_{n+1}O_{n+2} + 1/2nO_2 \rightarrow (1 + 1/n)M_nO_{n+1}$, oxidation of a metal oxide to form an oxide or oxidation of a metal sulfide to form a sulfate $M_2S + SO_2 + 3O_2 \rightarrow 2MSO_4$

(HyS) and hybrid copper chlorine-cycles. The simplified representation of the SI cycle is shown in Figure 4.16. This is a purely thermochemical cycle having all chemicals either in a liquid or a solid phase, which is very convenient for cycle construction because no solids must be manipulated. This cycle is developed mainly by General Atomic in the USA, the Japan Atomic Energy Agency, the Commissariat of Atomic Energy of France, and the Sandia National Laboratory. Several prototypes have been demonstrated with a capacity of up to 2 kg/day. According to predictions for the large-scale plant the cycle can achieve an energy efficiency of 47%.

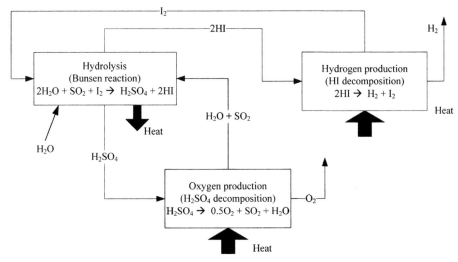

FIGURE 4.16 Simplified representation of the SI cycle.

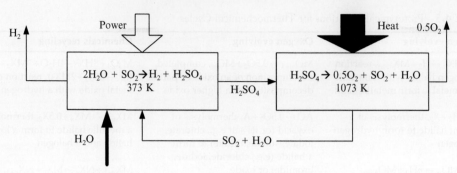

FIGURE 4.17   Simplified representation of the hybrid sulfur cycle.

The HyS cycle is presented in the simplified diagram in Figure 4.17. This cycle was proposed at the ISPRA research center in Italy in the 1970s and adopted for development at commercial scale by Westinghouse. The HyS cycle consists of a specialized electrochemical cell which is somewhat similar to a water electrolysis cell but has the anode depolarized due to the fact that sulfur dioxide is present in the electrolyte. The conceptual design of the Westinghouse electrolytic cell is shown in Figure 4.18. The process operates under 25 bar pressure with a slightly pressurized catholyte to impede sulfur dioxide migration.

The platinum-based electrodes are then used with various substrates. The electrolytic cell shows effective measured potentials of 0.6–1.05 V for current densities of 2–4 kA/m^2. The purity of hydrogen produced from the cell is 98.7%. The optimum concentration of sulfur dioxide in the feed stream is 30%. Based on flow-sheet analysis, under the assumption that electricity is generated with 45% efficiency, the net energy efficiency of the HyS cycle is about 42% (see the detailed description in Naterer et al., 2013).

The hybrid copper–chlorine cycle is in development mainly at Atomic Energy of Canada Ltd., the University of Ontario Institute of Technology, and the Argonne National Laboratory. There are various versions of this cycle having two up to five steps. The development status to date is reported in Naterer et al. (2014). A precursory of copper–chlorine cycles was proposed

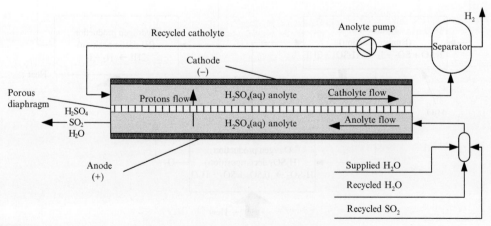

FIGURE 4.18   Conceptual diagram of the Westinghouse electrolytic cell.

by two researchers of General Electric, Wentorf and Hanneman (1974). They developed a physicochemical process of cuprous chlorine disproportionation. Dokyia and Kotera (1976) were the first to suggest an electrochemical step for the cycle. Carty et al. (1981) studied hydrolysis of cupric chloride with the coordination complex, namely $CuO \cdot CuCl_2$, which was later included in the five-step cycle. The cycle was confirmed experimentally in bench-top experiments by Lewis et al. (2003) at Argonne National Laboratory.

The conceptual flowchart of a three-step version of the Cu–Cl cycle is presented in Figure 4.19. Liquid water is supplied as stream #1 directly to the electrochemical cell, E. Wet cupric chloride $CuCl_2 \cdot nH_2O$(aq) is generated in #3 and heated in Hx1 to obtain a hydrated solid, $CuCl_2 \cdot nH_2O$(s). This stream, namely stream #4, is directed towards the oxychlorination reactor, R1, to generate solid copper oxychloride, gaseous hydrochloric acid, and steam. The gaseous phase in #5 is directed toward the heat exchanger system Hx2 where water condenses and hydrochloric acid completely dissolves in the aqueous solution which is fed to the electrochemical cell E as stream #6. The copper oxychloride #7 is heated in Hx3 and then fed into the thermolysis reactor R2 to release the oxygen #9. The molten cuprous chloride, collected from the bottom of the reactor as stream #11, is cooled and solidified in the heat exchanger system Hx5, then supplied to the electrolytic cell E as stream #12. The cycle and its variants are analyzed comprehensively in Chapter 6 of Naterer et al. (2013).

The analysis from Dincer and Zamfirescu (2012) shows that when thermochemical cycles are linked to CSP at large scale the expected range of hydrogen production efficiency is 36–40% for energy and 33–37% for exergy with a production range of 75–400 MW based on hydrogen's HHV equivalent. If nuclear power is used in conjunction with thermochemical cycles then efficiencies are 4–6% higher and the production scale is in the range of 400–1500 MW hydrogen HHV equivalent.

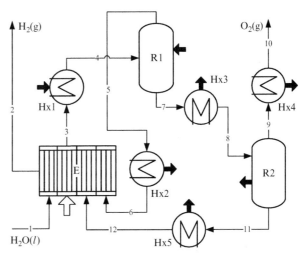

FIGURE 4.19 Simplified flow diagram of a three-step Cu–Cl cycle version.

E: $nH_2O(l) + 2CuCl(s) + 2HCl(aq) \rightarrow H_2(g) + 2CuCl_2 \cdot nH_2O(aq)$ at 350 K
R1: $2CuCl_2 \cdot nH_2O(s) \rightarrow CuO \cdot CuCl_2(s) + 2HCl(g) + (n-1)H_2O(g)$ at 673 K
R2: $CuO \cdot CuCl_2(s) \rightarrow 0.5O_2(g) + 2CuCl(l)$ at 800 K

➡ Heat input/output   ⇨ Work input

EXAMPLE 4.2

A hybrid thermochemical plant produces $200,000 \, Nm^3$ hydrogen per day while it consumes 500 MWh electricity and 2700 MWh heat at 1200 K. The energy efficiency for power generation is $\eta_{PG} = 30\%$. Calculate the energy and exergy efficiency of hydrogen production.

- The density of hydrogen in normal conditions is $\rho_0 = 82.35 \, g/m^3$ Thus, the mass of hydrogen produced daily is $\dot{m} = 200,000 \times 0.0485 = 16,470 \, kg/day$.
- Based on higher heating value, the energy in produced hydrogen is $E_{H_2} = 2330 \, GJ/day$.
- Based on chemical exergy, the exergy in hydrogen is $Ex_{H_2} = 118.1 \times 16,470 = 1944 \, GJ/day$.
- The required heat input is $\dot{Q}_{in} = 2700 \, (MWh/day) + \dfrac{500 \, (MWh/day)}{0.3} = 11,520 \, GJ/day$.
- The energy efficiency is $\eta = \dfrac{2330}{11,520} = 20.3\%$.
- The exergy efficiency is $\psi = \dfrac{1944}{11,520 \left(1 - \dfrac{298}{1200}\right)} = 22.4\%$.

4.3.3 Gasification and Hydrocarbon Reformation

Gasification can be applied to various types of solid combustible materials, such as coals, biomass, and municipal or industrial waste materials. Gasification involves combustion without oxygen and with steam addition. The final product of gasification is the synthesis gas which can be converted in hydrogen and carbon dioxide. Therefore, gasification is a method of hydrogen production, either green hydrogen if carbon neutral feedstock is use (e.g., biomass) or non-green hydrogen when coal is the feedstock.

Similar to gasification, liquid and gaseous hydrocarbons can be converted in synthesis gas by reforming and steam addition (water-gas shift reaction) and hydrogen is produced. In most of the cases, the sources are fossil fuels such as natural gas or oil, but various mineral oil and even biogas or bioalcohols (methanol, ethanol) can be reformed to obtain green hydrogen. In a simplified way, the gasification and reforming can both be described based on the following chemical equation:

$$C_aH_b + aH_2O + \text{Heat} \rightarrow (a + 0.5b)H_2 + aCO$$

Some examples of reactions that can occur during the gasification and reforming processes and their reaction heats are given in Table 4.6. Although the overall gasification and

TABLE 4.6 Main Reactions Involved in Reforming and Gasification

Process	Reaction	ΔH^0 (kJ/mol)
Methane reforming	$CH_4 + H_2O + \text{Heat} \rightarrow 3H_2 + CO$	206
Coal gasification	$C + H_2O \rightarrow H_2 + CO$	131
Coal hydro-gasification	$C + H_2 \rightarrow CH_4$	75
Boudouard reaction	$C + CO_2 \rightarrow 2CO$	172
Water-gas shift reaction	$CO + H_2O \rightarrow H_2 + CO_2$	−41

reforming processes look similar they require specific equipment and technology. Gasification requires oxygen, air, steam, or supercritical water as a gasification agent.

When oxygen is the agent the process temperature is over 1275 K. For this case the higher heating value of the obtained for syngas is the relative to cases when other gasification agents are used. The heating value is 10–15 MJ/m^3 but this. However, oxy-gasification is an expensive technology due to the need for pure oxygen agent, which plays the role of the main oxidant to provide heat to drive the overall endothermic gasification process. If instead of oxygen the employed agent is air then the cost is well reduced but the syngas will contain up to 60% nitrogen and the heating value is 4–6 MJ/m^3.

Steam is the most used gasification agent; it conveys the heat required by the process and it is also partially consumed by the process. The gasification temperature is superior to 1075 K and the heating value of generated syngas is 8–10 MJ/m^3. The syngas will be wet, containing moisture in a proportion of 18%, the rest being H_2 (45%) and CO (25%) and other gases. Supercritical gasification very much reduces the required temperature for the process; the range is 675–1000 K. Steam at high pressure (30 MPa) is fed into the reactor; the process suppresses the formation of tar and drying of coal or biomass becomes unnecessary.

The first step of gasification is pyrolysis, when volatiles are released and char and liquids are released. After the pyrolysis the combustible gases are reformed to hydrogen in the presence of steam. Several gasifiers exist, such as: (i) staged reformation with a fluidized-bed gasifier, (ii) staged reformation with a screw auger gasifier, (iii) entrained flow reformation, and (iv) partial oxidation. In staged steam reformation with a fluidized-bed reactor, the feed is first pyrolyzed in the absence of oxygen. Then the pyrolysis vapors are reformed to syngas with steam, providing added H_2 as well as the proper amount of oxygen and process heat, which comes from char oxidation. With a screw auger reactor, moisture and oxygen are introduced at the pyrolysis stage, and process heat comes from burning some of the produced gas. In entrained flow reformation, external steam and air are introduced in a single-stage gasification reactor. Partial oxidation gasification uses pure oxygen with no steam.

Assuming that the inputs in the gasifier or reformer are the energy embedded in steam and in feedstock, then the energy and exergy efficiencies can be defined as in Table 4.7. In the efficiency expressions, n represents the number of moles of feedstock (index f), steam (index s), tars (index T), and char (index C) needed to generate one mole of product. As mentioned in Dincer and Zamfirescu (2012), the overall efficiencies of hydrogen production by gasification and reforming methods become in the range of 60–70% for energy and 55–65% for exergy, with a range of production scale of 0.5–100 MW, respectively.

TABLE 4.7 Energy and Exergy Efficiency of Biomass Gasification Process

Case	Energy efficiency	Exergy efficiency
Only hydrogen is considered as output	$\eta = \dfrac{HHV_{H_2}}{n_f HHV_f + n_f \Delta h_f}$	$\psi = \dfrac{ex_{H_2}^{ch}}{n_f ex_f^{ch} + n_s ex_s}$
All product gas (PG) is considered as output	$\eta = \dfrac{HHV_{PG}}{n_f HHV_f + n_s \Delta h_s}$	$\psi = \dfrac{ex_{PG}^{ch}}{n_f ex_f^{ch} + n_s ex_s}$
The product gas (PG), tar (T), and char (C) are all considered as useful outputs	$\eta = \dfrac{HHV_{PG} + n_t HHV_t + n_c HHV_c}{n_f HHV_f + n_s \Delta h_s}$	$\psi = \dfrac{ex_{PG}^{ch} + n_t ex_t^{ch} + n_c ex_c^{ch}}{n_f ex_f^{ch} + n_s ex_s}$

4.3.4 Photochemical and Photo-Biochemical Methods

The photochemical methods purely use photocatalysts to split water by half-reactions (e.g., hydrogen-evolving reaction). A review of light-driven methods for hydrogen production (excluding biochemical based) is presented in Zamfirescu et al. (2011). Semiconductor nanoparticles that can act as photocatalysts are reviewed in Acar et al. (2013). Figure 4.20 presents the general mechanism of heterogeneous photocatalysis at the surface of a nanoparticle. In the presence of light, electrons are displaced from the valence to the conduction band of the semiconductor.

The electrons and the hole oxidize and/or reduce chemical species at the particle surface unless they recombine due to dissipative mechanisms. As indicated in the figure the water reduction reaction requires two electrons, which can be provided to the active center. If water reduction occurs the catalyst is depleted of electrons and no other reduction can occur until an oxidation process takes place at the catalyst surface to acquire electrons. To be sustainable, the photocatalysis requires continuous injection of electrons into the photocatalyst. This can be accomplished by dissolving in the electrolyte some chemical species that can play the role of electron donor. Examples of donors are sulfide or sulfite ions. There are many more examples of substances that can act as electron donors (see Acar et al., 2013). Figure 4.21 shows an

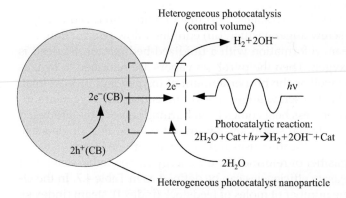

FIGURE 4.20 Representation of the heterogeneous photocatalysis process.

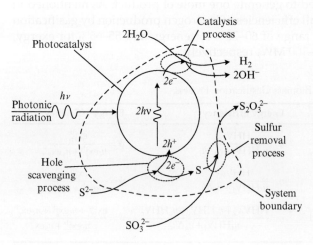

FIGURE 4.21 Representation of an actual heterogeneous photocatalytic system.

actual photocatalytic reaction which works with sulfide donors and generates hydrogen from water.

In this process, the whole scavenging process is performed by transferring two electrons from a sulfur generation reaction, $S^{2-} \rightarrow S + 2e^-$. In the electrolyte sulfite is also present, which facilitates a spontaneous reaction that consumes sulfur, $S + SO_3^{2-} \rightarrow S_2O_3^{2-}$ and avoids depositions on the catalyst. The wavelength of the photons must be such that their energy is higher than the band gap energy of the photocatalyst. In addition, the band gap energy of the photocatalyst must be higher than the reversible potential of the hydrogen-evolving reaction.

Homogeneous metallo-organic photocatalysts have also been studied in the past works. A typical homogeneous photocatalyst is ruthenium tris-2,2'-bipyridyne, $[Ru(bpy)_3]^{2+}$. This catalyst absorbs light at 450 nm and has a triplet exited state lifetime of 890 ns. If it decays at ground state the excited catalyst emits light at 610 nm. As shown in Figure 4.22, the molecular structure of the photocatalyst appears to be complex. In the ground state, the molecular orbital arrangement is of a singlet. The exited state acquires 2.75 eV and decays fast in a triplet state at 2.1 eV due to intersystem crossing. Once the electron donor is encountered, the catalyst is oxidized to a triplet at charge +1 and energy level of 1.08 eV. The catalyst is now able to transfer electrons to water at a potential higher than the reversible half-reaction potential of 0.822 V for water reduction. Once water is encountered hydrogen is evolved; if not, then the system decays to the ground state due to intersystem crossing.

Another light-driven process is biophotolysis which is a photonic-driven biochemical hydrogen production process from water. The photosynthesis process used by green plants to convert solar radiation into biochemical energy can be adapted to some extent for hydrogen

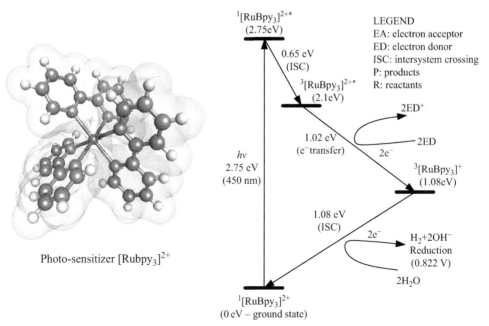

FIGURE 4.22   Photocatalytic cycle of water reduction with $[Ru(bpy)_3]^{2+}$.

generation and is the basis of biophotolysis. Production of hydrogen through biophotolysis can be classified into direct or indirect kinds and photo-fermentation kind.

In biophotolysis, some microorganisms sensible to light are used as biological converters in a specially designed photo-bioreactor. The most suitable microorganisms are microalgae, because they exhibit hydrogen evolution and can be cultured in closed systems that permit hydrogen capture. Micro-algal strains can be cultured that exhibit high hydrogen yields. One remarkable advantage of biophotolysis is to produce hydrogen from water in an aqueous environment at standard temperature and pressure. Biophotolysis is not yet developed for commercial use but is demonstrated at laboratory scale. Cyanobacteria (*Chlamydomonas reinhardtii*) generate and manipulate nitrogenase and hydrogenase enzymes in such a way that they may generate hydrogen and oxygen from water. The following two reactions describe the overall process of hydrogen and oxygen evolution from water:

$$C_6H_{12}O_6 + 12H_2O \xrightarrow{hv} 6CO_2 + 12H_2$$

$$12H_2O + 6CO_2 \xrightarrow{hv} C_6H_{12}O_6 + 6O_2$$

The pure biochemical route (with no light) occurs though the action of microbes which manipulate the energy stored in glucose, cellulose, sucrose, and other complex chemicals containing, for example, adenosine triphosphate (ATP) to extract hydrogen in the absence or presence of oxygen. When oxygen is completely absent or is present in very reduced quantities the biochemical conversion of organic matter to various forms of biochemical energy is called anaerobic digestion.

Anaerobic digestion is attractive as a hydrogen production method for two important reasons: (i) it can generate hydrogen from organic waste; and (ii) it stabilizes the waste which otherwise may become a source of uncontrolled microbial growth with potential danger of contamination of biological species. Moreover, the digester hardware is simple and the system is appropriate for mass production. Two types of bacteria are used in anaerobic digestion, namely the mesophilic (which is active at a temperature of 35–40 °C) and thermophilic kinds (active at 55–60 °C). Note also that hydrogen production is also possible by aerobic digestion, but this process is less productive and consequently less developed at the present stage. In this book the treatment is therefore limited to anaerobic digestion.

One major problem regarding reaction control in anaerobic digestion reactors is the inhibition of methanogens, which represent hydrogen-consuming bacteria. One possibility to eliminate these bacteria, which normally are present in the organic substrate, is to heat the substrate for short duration at about 100 °C. This will kill the bacterial population. Further, the biomass is seeded with hydrogen-producing bacteria like Clostridium and Bacillus species.

Some typical biochemical processes catalyzed by hydrogenase or nitrogenase enzymes to generate hydrogen are given as follows:

Acetic acid fermentation from sucrose $C_{12}H_{22}O_{11} + 2H_2O \rightarrow 4CH_3COOH + 4CO_2 + 8H_2$.
ATP reduction by hydrogenase $8H^+ + 8e^- + 16ATP \rightarrow 4H_2 + 16ADP + 16Pi$.
ATP reduction by nitrogenase $N_2 + 8H^+ + 8e^- + 16ATP \rightarrow 2NH_3 + H_2 + 16ADP + 16Pi$.

Note that when ATP is converted to adenosine diphosphate (ADP) it releases free energy of about 30.5 kJ/mol and inorganic phosphate (Pi).

Biochemical hydrogen production systems are used to generate biogas using various agricultural feedstocks or sewage waters or municipal wastes. The efficiency depends very much on the feedstock but can be assumed in the range of 5–15% whereas the production capacity is from 0.1 to 10 MW (Dincer and Zamfirescu, 2012). Photobiological methods can reach higher efficiency (up to 25%) but their capacity of production cannot be higher (1–100 kW, respectively).

4.4 FUEL CELLS

A fuel cell performs the opposite process of an electrolysis cell. Water is formed from hydrogen and oxygen gases and electricity is generated. Fuel cells are generally constructed in a stack arrangement with planar electrodes and hydrogen fed at the anode, where it loses electrons. An external circuit is attached to the anode to transfer electrons to the cathode, thus transferring power. The oxygen is fed at the cathode side where it receives electrons. Water is then formed from protons and oxygen ions at either the anode or cathode side, depending on the fuel cell type.

There are eight main types of fuel cells, as illustrated in Figure 4.23. Most fuel cell types are fed with hydrogen fuel; there are also direct ammonia fuel cells (DAFCs) and direct methanol

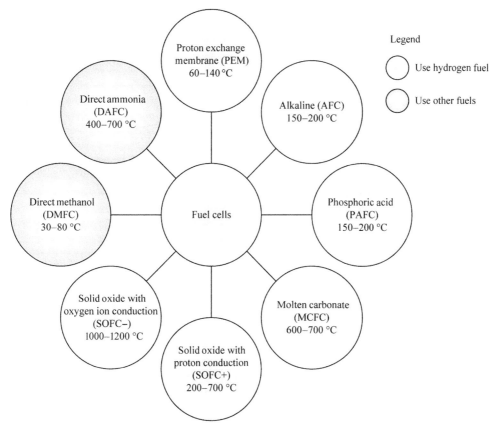

FIGURE 4.23   Types of fuel cells and their operating temperature range.

fuel cells (DMFCs); in addition, methane (natural gas) through reforming can be used as fuel in solid oxide fuel cells. The applications of fuel cell systems are mostly in power generation and propulsion for transportation vehicles. In special applications, fuel cells can be used to generate pure water or pure oxygen. Also, multigeneration of power, heating, water, and other products can be obtained. Figure 4.24 illustrates a classification of fuel cell applications. In order to generate extra-pure oxygen a larger capacity fuel cell is connected to a smaller capacity electrolyser which is fed with pure water produced by the fuel cell. The electrolyser generates pure oxygen for use and hydrogen which is fed back to the fuel cell.

The most important fuel cell application is for stationary power generation at intermediate scale either connected to or disconnected from the regional grid. Solid oxide and molten carbonate fuel cells (MCFCs) are suitable for cogeneration of high-temperature heat and power, thus they can be coupled with processes which require high grade heat. One of the processes that can be benefited by high grade heat is hydrocarbon fuel reforming. Due to this fact, SOFC and MCFC can be fuelled with natural gas or other fuels without the need of external reformers. Direct methanol and proton-exchange-membrane fuel cells (PEMFCs) can be used to cogenerate power and heating for residences (hot water and space heating).

The startup time is important in some applications. In vehicle propulsion and small power applications such as residences and portable devices, PEM, PAFC, AFC, and DMFC are the best choice because they require very small startup time. The alkaline fuel cell (AFC) is the preferred technology for space applications and is very attractive for road vehicles, although it requires a complicated system of electrolyte recirculation. For road vehicles, DMFC

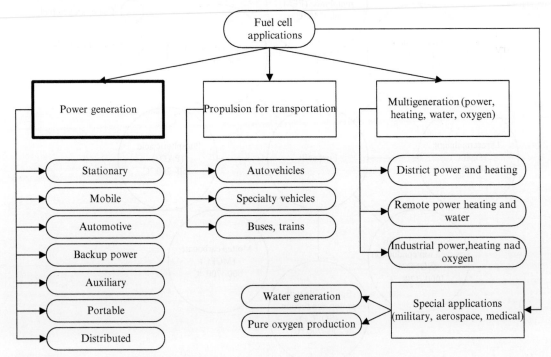

FIGURE 4.24 Applications of fuel cells.

represents a very reliable choice because it does not require any gas cleaning units, is simpler to implement, and requires a fuel tank of the same size as gasoline vehicles. Nevertheless, PEMFC are the most developed for vehicular application, although they require a compromise in driving range, as the fuel must be hydrogen. On the contrary, higher temperature fuel cells such as SOFC, MCFC, and DAFC are better suited to larger power stationary applications. In addition, DAFC appears to be a good choice for large vehicles that have long running times in steady mode such as railway locomotives, heavy truck road transporters, and ships. Also, SOFC can be implemented on vehicles as an auxiliary power unit which is in operation at steady running.

One of the most attractive fuel cell applications today is for powering laptops, mobile cells, and other small electronic devices. Here PEM and DMFC play a major role due to their reduced operating temperature.

4.4.1 Proton-Exchange-Membrane Fuel Cells

The simplified representation of a PEMFC is shown in Figure 4.25. The particular aspect of this type of fuel cell is represented by the solid polymer electrolyte, usually denoted as proton-conducting membrane. The electrolyte is very thin and allows for proton conduction through hydronium ions when the membrane is wetted. The half-reactions at the fuel cell electrodes are listed as follows:

$$\begin{cases} \text{at anode}: \ H_2(g) \rightarrow 2H^+(aq) + 2e^- \\ \text{at cathode}: \ 0.5O_2(g) + 2H^+ + 2e^- \rightarrow H_2O(l) \end{cases} \tag{4.1}$$

For the electrolyte a polystyrene sulfonate polymer (PSP) can be used; however, the proprietary membrane from DuPont—Nafion® consisting of a polytetrafluoroethylene (PTFE)—is proven to be more stable and has better conductivity than PSP membranes. Another proprietary membrane that is proven to be very effective is the perfluorocarbonic sulfonic acid Dow® from Dow Chemicals (Carrette et al., 2001).

Given that the protons are transported in hydronium form (H_3O^+), these acidic membranes must always be hydrated in order to operate. Therefore, water management in PEMFC is a major design issue. In order to assure good performance—with reduced ohmic losses—the

FIGURE 4.25   Simplified representation of a PEMFC.

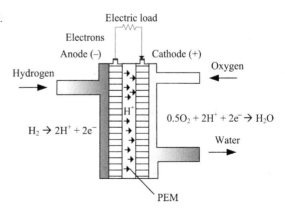

PEM is constructed in a sandwich-like architecture with porous planar electrodes forming a membrane electrode assembly. Noble metal catalysts must be used to coat the electrodes. The electrodes must be designed such that a tree-phase boundary is formed between the solid catalytic center, the aqueous ionic species, and the gaseous reactants. The catalytic center must be in direct contact with the electrode to ensure that electrons are supplied to or from the reaction site. Platinum is the preferred catalyst for the cathode, which is required to compensate for the slow kinetics. Also, for the anodic reaction a platinum catalyst is required when pure hydrogen is used as fuel. However, if CO impurities are present in hydrogen the platinum catalyst is immediately poisoned and becomes inactive. Due to the poisoning, complicated fuel purification systems may be required for PEMFC, depending on the case.

For construction of PEMFC stacks, a bipolar plate arrangement is used. Graphite is a good material to use for manufacture of the bipolar plates, although it is difficult on machinery. A bipolar stack arrangement for PEMFC is illustrated in Figure 4.26. The flow of gases is normally provided in cross-flow counter-current. Water must be circulated through the system in such a way that the membrane always remains wet. The efficiency of commercial PEMFC is superior to 55%.

4.4.2 Phosphoric Acid Fuel Cells

The phosphoric acid fuel cell (PAFC) runs the same electrode half-reactions as PEMFC. Protons are the charge carriers through an acidic electrolyte, and water is formed at the cathode. The difference between PAFC and PEMFC is due to the electrolyte. In PEMFC the

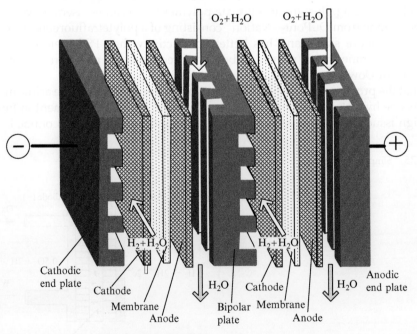

FIGURE 4.26 Bipolar stack arrangement for PEMFC.

electrolyte is a solid polymer, whereas for PAFC a liquid acidic electrolyte is used. The electrolyte is almost pure pressurized phosphoric acid ($\sim$100% concentration), which has a very low volatility at the specific temperatures of operation of $\sim$175 °C (the normal boiling point of the acid is 158 °C). The PAFC are the most developed at commercial level for megawatt range (up to 25 MW); it is very tolerant of CO_2 presence in the air stream (Carrette et al., 2001).

The liquid electrolyte is stabilized in a solid matrix of silicon carbide as shown in Figure 4.27. The arrangement is similar to that of PEMFC shown in Figure 4.25. Due to the stabilization in the silicon carbide the losses of electrolyte due to evaporation are very minimized, such that no acid must be replenished during the system lifetime. The SiC matrix comprises particles of micrometer size that are able to assure reduced ohmic losses. The electrodes are of planar geometry, made of PTFE with bonded platinum-black of 46 μmol/cm^2 load.

4.4.3 Solid Oxide Fuel Cells with Proton Conduction (SOFC+)

An emerging technology is that of solid oxide fuel cells with proton conduction. The electrode half-reactions are the same as the ones for PEMFC, Equation (4.1). However, the electrolyte is a proton-conducting metal oxide layer placed in between two porous, planar electrodes as shown in Figure 4.28.

The SOFC+ (solid oxide fuel cells with proton conduction) normally use a barium oxide. The proton-conducting solid oxide membranes bring the important advantage of letting the

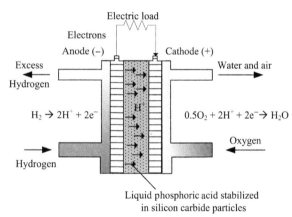

FIGURE 4.27 Simplified representation of a phosphoric acid fuel cell.

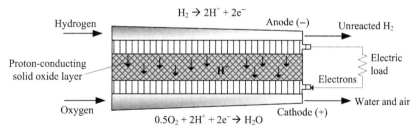

FIGURE 4.28 Simplified representation of a proton-conducting solid oxide fuel cell.

protons migrate from anode to cathode. As a consequence, the water formation reaction occurs at the cathode. Complete hydrogen utilization is therefore possible (in principle) in SOFC+, with direct implication for increasing system simplicity and compactness by eliminating the need of the afterburner. Moreover, because all the hydrogen is reacted electrochemically at the fuel cell cathode, practically no NOx is formed, and thus the fuel cell emission consists only of steam and nitrogen, i.e., it is clean. At present, certain efforts are devoted worldwide to developing proton-conducting membranes. In this respect barium cerate (BaCeO$_3$) based materials were identified as excellent solid oxide electrolytes because of their high proton-conducting capability over a wide range of temperatures (300–1000 °C).

The main problem with barium cerate results from the difficulty to sinter it in the form of a solid membrane. One option appears to be doping the barium cerate with samarium (Sm) which allows for sintering thin membranes, featuring a thickness as low as 50 μm and high-power densities in the range of 1300–3400 W/m^2.

4.4.4 Alkaline Fuel Cells

AFCs have slightly superior efficiency to PEMFC, over 60% (potentially they can reach 70%), taking advantage of hydroxyl as an excellent charge carrier with faster kinetics for oxygen reduction reaction. These systems were used in NASA space programs with great success as they provide an excellent stability. The catalyst requirements for AFC are less stringent than for PEMFC; in principle cheap nickel-based catalysts can be used. The electrolyte is a solution of potassium hydroxide (KOH, 30–40% by weight). The best configuration of AFC is with circulating liquid electrolyte, although options exist to embed the electrolyte in a matrix, and the half-reactions at the electrodes are given as follows:

$$\begin{cases} \text{at anode}: H_2(g) + 2OH^-(aq) \rightarrow 2H_2O(l) + 2e^- \\ \text{at cathode}: 0.5O_2(g) + H_2O(l) + 2e^- \rightarrow 2OH^-(aq) \end{cases} \quad (4.2)$$

One of the major problems with the KOH electrolyte is due to its spontaneous reaction with carbon dioxide from air. Air fed at the anode carries some amount of carbon dioxide; also hydrogen fuel can be altered with CO$_2$. If carbon dioxide is present in the electrolyte it reacts with hydroxyls and forms carbonate in aqueous solution (CO$_3^{2-}$). This fact severely degrades the electrolyte. Solutions exist to absorb CO$_2$, including swing absorption. Figure 4.29 shows the system configuration for an AFC system. Hydrogen penetrates the porous anode and water is generated at the electrode–electrolyte interface. The generated water is removed from the system with the electrolyte. The electrolyte is reconditioned in an external unit by eliminating the water and if needed fresh electrolyte may be added. Thereafter, the electrolyte is returned to the cell; fresh electrolyte enters into contact with the cathode where hydroxyls are generated. The AFCs can be constructed in a stack bipolar arrangement similar to that used for alkaline electrolysers.

4.4.5 Solid Oxide Fuel Cells with Oxygen Ion Conduction

Solid oxide fuel cells with oxygen ion conduction (SOFC−) are also an established technology. Their electrolyte is a solid layer of yttria-stabilized zirconia which operates at high

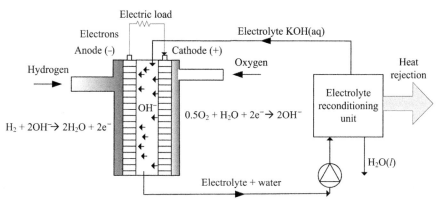

FIGURE 4.29 Alkaline fuel cell system configuration.

temperature (1000 °C). In fact, it leads to some important advantages for the SOFC−, namely:

- They are cheap and present for a long lifetime because no noble metal catalysts are needed for the electrodes.
- Internal reforming of alternative fuels (e.g., methane, syngas, methanol, ammonia) to hydrogen is facilitated, so they can use a reduced size fuel tank.
- The exhaust gases have high exergy, which can be converted into additional power and low-temperature heating.

During the operation, at the anode, the hydrogen is consumed and steam generated. Because of this the hydrogen's partial pressure decreases. As a consequence of the low partial pressure of hydrogen, the reaction kinetics are degraded and the only solution to compensate for this effect is to supply hydrogen in excess. The excess hydrogen must then be consumed. This can be done in multiple ways. One common method is to combust the excess hydrogen in an afterburner, with the released heat recovered or converted into work by a gas turbine. Thus, certain amounts of NOx are formed during the combustion of hydrogen with air. The half-reactions for SOFC− are given below:

The processes of SOFC− are somewhat similar to SOFC+ except that the mobile ions are different. Figure 4.30 illustrates the common configuration of an SOFC− in bipolar-plate stack arrangement. In any case, the fact that there is no liquid present is an advantage for the design because only two-phase solid–gas processes occur. A porous tubular wall design has been developed in the past, with fuel fed at the outer side and oxidant at the inner side. The planar design with bipolar plates is the dominant choice. The electrolyte thickness is about 20 μm, which is sandwiched between the porous planar electrodes.

4.4.6 Molten Carbonate Fuel Cells

MCFCs have efficiency higher than 60% while operating at high temperature. Moreover, MCFC can be used in cogeneration systems, in which case the fuel utilization efficiency can go over 85%. The electrodes do not need to be coated with expensive catalysts; simple carbon

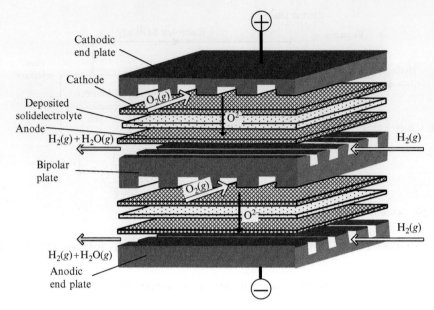

FIGURE 4.30   Configuration for an SOFC-bipolar-plate stack arrangement.

based, inexpensive electrodes operate very well at the high temperature specific to these cells. The electrolyte is a carbonate (e.g., LiK or LiNa carbonates) which is stabilized in an alumina-based porous matrix (e.g., $LiAlO_2$ with Al_2O_3 inserts). The cell efficiency can be over 55%; more efficiency can be obtained with an integrated system that uses the exergy of the expelled hot gases (CO_2 and steam). An advantage is that this fuel cell is insensitive to CO_2 or CO contamination. The durability of the fuel cell is low because the electrolyte is highly corrosive. The electrode reactions are written as follows:

$$\begin{cases} \text{at anode}: H_2(g) + CO_3^{2-} \rightarrow H_2O(g) + CO_2 + 2e^- \\ \text{at cathode}: 0.5O_2(g) + CO_2 + 2e^- \rightarrow CO_3^{2-} \end{cases} \tag{4.3}$$

An MCFC requires inexpensive nickel-based electrodes and can be constructed as a stack of bipolar plates. The system must include a system for recycling carbon dioxide in gaseous phase that includes a water separation process. The simplified configuration diagram of a MCFC is shown in Figure 4.31.

4.4.7 Direct Methanol Fuel Cells

DMFCs use a proton-conducting polymer electrolyte and operate with a 1:1 molar methanol–water mixture at low temperature (average 50 °C). These fuel cells are compact and store energy at high density; however, the rate of energy discharge (that is the power density) is rather low. The electrode reactions for DMFC are written as follows:

$$\begin{cases} \text{at anode}: CH_3OH(l) + H_2O(l) \rightarrow CO_2(g) + 6H^+ + 6e^- \\ \text{at cathode}: 1.5O_2(g) + 6H^+ + 6e^- \rightarrow 3H_2O(l) \end{cases} \tag{4.4}$$

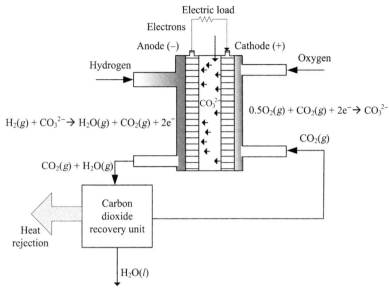

FIGURE 4.31   Molten carbonate fuel cell system configuration.

The technology of PEMFC is fully applicable to DMFC; the only major difference is the fuel, which comes in liquid phase. Because water is formed at the cathode and the fuel fed at anode requires water, the water recycling is very important. Figure 4.32 illustrates a simplified diagram of a DMFC system. Besides the water recycling requirement, there are other challenges with methanol fuel cells, the main one being the crossover of methanol. Due to the fact that it

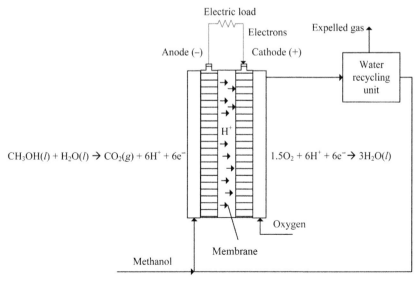

FIGURE 4.32   Direct methanol fuel cell system configuration.

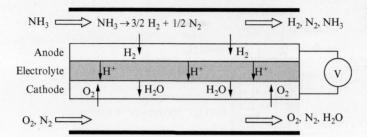

FIGURE 4.33 Process description for a direct ammonia fuel cell.

has a dipole-like molecule similar to that of water, methanol crosses relatively easily over Nafion® membranes. Reduction of crossover is obtained by using thicker membranes or modified membranes or composite membranes which use several types of polymers. As a consequence of these types of issues the efficiency of DMFC is rather low (~30%), as is their power generation per unit of volume.

4.4.8 Direct Ammonia Fuel Cells

DAFCs are of the SOFC+ type with selected catalysts at the anode where gaseous ammonia is fed as a source of hydrogen. The schematic of a DAFC is shown in Figure 4.33. Ammonia fed at the anode decomposes thermo-catalytically and generates protons that diffuse through the porous electrolyte. Water is formed at the cathode where protons encounter the oxygen. The achievable DAFC efficiency is on the order of SOFC+ fueled with hydrogen, i.e., over 55%. The system half-reactions are then given as follows:

$$\begin{cases} \text{at anode}: NH_3(g) \rightarrow N_2(g) + 3H^+ + 3e^- \\ \text{at cathode}: 1.5O_2(g) + 3H^+ + 3e^- \rightarrow 1.5H_2O(g) \end{cases} \tag{4.5}$$

4.5 FUEL CELL MODELING

Analysis and modeling of fuel cells and fuel cell systems is a complex problem, but very important for optimized design. Optimization of geometry and of the operating parameters is required for the fuel cell stack and the overall system. Thermodynamics, kinetics, and transport processes must be modeled. There are a number of fundamental equations from electrochemistry that must be applied for any fuel cell modeling attempt. Reaction kinetics and chemical equilibrium are very important. Fluid mechanics and heat transfer aspects may be accounted for. In this section we treat various aspects regarding fuel cell analysis and modeling.

An attempt to categorize fuel cell models is presented in Figure 4.34. It is rather difficult to classify models of fuel cells. In principle one can categorize the model based on modeling level, number of spatial dimensions taken in consideration, process time consideration, and the modeled processes taken into account. Modeling can be made at cell level (which is the deepest analysis), at stack level (which accounts for flow distribution, channeling, flow collection, temperature distribution, spatial component concentration, etc.). Also the

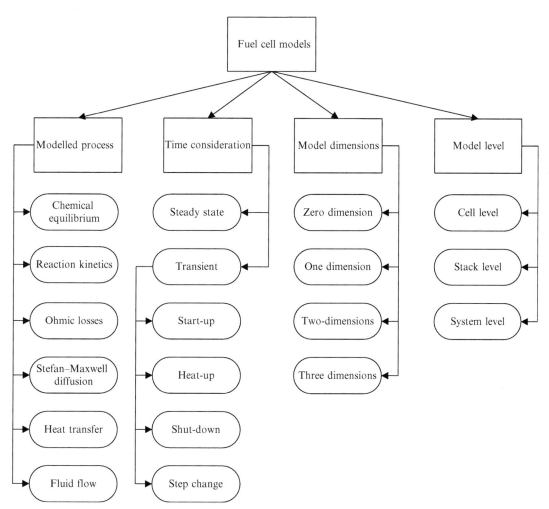

FIGURE 4.34 Categories of fuel cell models.

modeling can be done at system level, where the fuel cell stack is integrated in a larger system. General thermodynamic analysis through conservation laws and specific equations must be applied at system level.

Models can also categorized as a function of the physical–chemical processes considered. In general, reaction heats must be taken into account in modeling, as well as Gibbs free energy and entropy of the overall reaction. Moreover, chemical equilibrium should be included in models. Other aspects to be studied are reaction kinetics, electric charge distribution and ohmic losses of various kinds, diffusion of chemical species and of ionic species (Stefan–Maxwell–Knudsen), and heat transfer and fluid flow in channels.

Fuel cell models can be also classified as macro and micromodels. In macro-models, porous media models are applied to electrodes which are assumed electron-conducting media. The electrode–electrolyte interface is the place where electrochemical reactions occur. In micromodels, electrodes are considered porous structures made of electron-conducting and ion-conducting particles. The micromodels predict the electrochemical characteristics at the electrodes.

Models can be either steady-state or transient. Most of the published models in the literature refer to steady state operation, because it influences the design the most. However operation at startup, heat-up, shutdown and step change events must be analyzed too. Thus, transient models are applied for these situations.

The zero-dimension models consider the fuel cell as a black box, characterized by unique parameters such as temperature, generated voltage, current density, inlet and outlet flow rates, and concentrations. One-dimensional approach can be applied to a planar fuel cell or to tubular configurations with axial symmetry. Parameter variation is considered only in one relevant direction (that is, across the electrolyte) while the other directions are neglected. In two-dimensional modeling, one of the considered directions is across the electrolyte, and the other direction is parallel with stream flow directions in channels (which in general is the direction along which one observes relevant gradients of important quantities).

The most important equations for fuel cell modeling are the Nernst and Butler–Volmer equations which are given in Table 4.3 in the specific form used for electrolysis modeling. The processes are reversed in fuel cells with respect to the electrolysers. The Nernst equation for fuel cell modeling gives the open circuit voltage according to

$$E_{oc} = E_{cell}^0 - \left(\frac{RT}{zF}\right)\ln\left(\frac{P_{H_2O}}{P_{H_2}P_{O_2}^{0.5}}P_0^{0.5}\right) \tag{4.6}$$

where z are the number of charges per reaction and E_{cell}^0 is the standard cell potential (reversible cell voltage) at the given temperature (and standard pressure) which is given by

$$E_{cell}^0 = -\frac{\Delta H^0(T) - T\Delta S^0(T)}{zF} \tag{4.7}$$

where $\Delta H^0(T)$ and $\Delta S^0(T)$ are the standard enthalpy and entropy of the reaction.

In the Butler–Volmer equation the expression of activation energy must take account of the specific half-reaction and the number of charges which in the cases of specific fuel cells (e.g., direct methanol, molten carbonate, direct ammonia) can take very different forms than in the case of water electrolysis or hydrogen fuel cells. Here is the general Butler–Volmer equation:

$$\frac{J}{J_0} = \exp(\alpha F E_{act}/RT) - \exp(-(1-\alpha)F E_{act}/RT) \tag{4.8}$$

where the activation potential is given by

$$E_{act} = \frac{RT}{F}\sinh^{-1}\left(\frac{J}{zJ_0}\right) \tag{4.9}$$

In Equation (4.8) one must take into account that charges at the anode can be different than charges at the cathode. Some important performance and modeling parameters for fuel cells are given in Table 4.8. In a detailed modeling of the fuel cell, conservation laws take the form

TABLE 4.8 Modeling and Performance Parameters for Fuel Cells

Parameter	Equations
Oxygen utilization ratio: number of mols of oxygen reacted with respect to the quantity fed at the fuel cell inlet.	$U_a = \dfrac{n_{O_2,\text{utilized}}}{n_{O_2,\text{inlet}}}$
Fuel utilization ratio: similar to air utilization, number of mols of fuel reacted with respect to the quantity fed at the fuel cell inlet.	$U_f = \dfrac{n_{f,\text{utilized}}}{n_{f,\text{inlet}}}$
Excess air coefficient: amount of oxygen at the inlet versus stoichiometric oxygen.	$\lambda_a = U_f / U_a$
Open circuit cell voltage: voltage that exists between electrodes if there is no load connected. The actual load voltage is lower.	E_{oc} according to Equation (4.6)
Actual cell voltage: the open circuit voltage when the ohmic losses, activation, and concentration potential are extracted.	$E = E_{oc} - E_{ohm} - E_{act} - E_{conc}$
Area-specific ohmic resistance: represents the ohmic resistance of the electrolyte and depends on the contact resistance per unit of area, where ASR_{bulk} represents the summation of the resistivities of the materials transversed by ions and electrons, given as per unit of area of electrolyte.	$ASR = ASR_{contact} + ASR_{bulk}$ $E_{ohm} = J\, ASR$
Electric power output of the cell: the product of actual voltage and load.	$\dot{W} = E I$
Power density of fuel cell system: the ratio between net generated power and the volume of the power generating system (V_{system}).	$\dot{W}'' = \dfrac{\dot{W}}{V_{\text{system}}}$
Cell energy efficiency: the ratio between generated power and the energy from fuel based on LHV.	$\eta = \dfrac{\dot{W}}{n_f\, LHV}$
Cell energy efficiency: the ratio between generated power and the exergy supplied by the fuel through its chemical exergy.	$\psi = \dfrac{\dot{W}}{n_f\, ex^{ch}}$

of partial differential equations. In a fuel cell, one works with several chemical species, and as a result, the conservation equation of species becomes important. This equation is

$$\left[\frac{\partial C}{\partial t} + \nabla \cdot (\boldsymbol{u} \cdot C) = \nabla \cdot (D \cdot C) + S\right]_k$$

where t is time, C concentration, $\boldsymbol{u}$ velocity vector of the species k, and S is the source term that takes in the account of the rate of apparition or dis-apparition of chemicals.

One important characteristic for design and analysis of fuel cells is the voltage–current density curve. An example of this curve is shown in Figure 4.35. Once the curve is available it is possible to use it for fuel cell optimization purposes. On the abscissa is the current density. If the current density is nil, that circuit is open and the voltage read at the fuel cell electrodes is the open circuit voltage E_{oc}. At low-current density the voltage losses are dominated by the activation potential. In the region where the cell voltage decreases quasi-linearly the ohmic polarization is the dominant effect. It can be observed that at low-current density (no load) the power tends to zero (because $J \to 0$); similarly, at high-current density there is a limiting value over which the power tends to zero (because $E \to 0$). For current density higher than J_{Limiting} the fuel cell cannot generate power anymore, because of the dissipative effects. The dissipative effects can be observed in the form of heat $\dot{Q}'' = -T\Delta S$ and increase

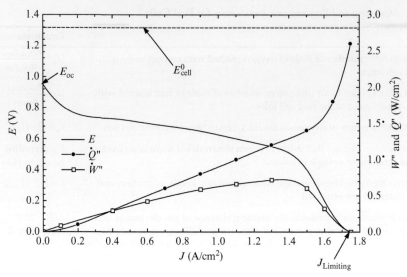

FIGURE 4.35 Typical voltage–current density curve of a fuel cell.

dramatically at high-current density. At a certain current density, the power delivered by the fuel cell presents a maximum.

The balance equations of a single cell take different forms depending on the type of fuel cell. We consider zero-dimensional geometry. The control volume is illustrated in Figure 4.36. The energy balance equation for the generic hydrogen-fueled fuel cell is written as follows:

$$\text{EBE}: (\dot{n}h)_{\text{H}_2\text{in}} + (\dot{n}h)_{\text{O}_2\text{in}} = (\dot{n}h)_{\text{H}_2\text{out}} + (\dot{n}h)_{\text{H}_2\text{Oout}} + (\dot{n}h)_{\text{O}_2\text{out}} + \dot{W} + \dot{Q} \qquad (4.10a)$$

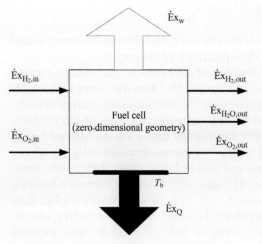

FIGURE 4.36 Control volume around a generic hydrogen-fueled fuel cell.

where $\dot{n}$ represents the molar rate of species. Here we assume that the fuel cell is fed with oxygen, although a similar equation can be written when the oxidant air consists of, e.g., 21% oxygen and 79% nitrogen. The entropy balance equation is:

$$\text{EnBE}: (\dot{n}s)_{H_2 in} + (\dot{n}s)_{O_2 in} + \dot{S}_g = (\dot{n}s)_{H_2 out} + (\dot{n}s)_{H_2 O out} + (\dot{n}s)_{O_2 out} + \frac{\dot{Q}}{T_b} \qquad (4.10b)$$

where T_b is the temperature of the system boundary. The exergy balance equation can be written as follows:

$$\text{ExBE}: \dot{E}x_{matter, in} = \dot{E}x_{matter, out} + \dot{E}x_Q + \dot{E}x_W + \dot{E}x_d \qquad (4.10c)$$

where $\dot{E}x_Q$ is the net exergy of heat release out of the control volume, $\dot{E}x_{matter, in}$ is basically formed from exergy associated with streams of fuel and oxidant input and $\dot{E}x_{matter, out}$ accounts for the exergy of the formed water and unreacted fuel and oxidant.

The useful exergy is at least the $\dot{E}x_W$ if no heat is cogenerated. This exergy is the same as the net generated power, $\dot{E}x_W = \dot{W} = E_{load}I$, where the net generated work by the fuel cell is given by the voltage across the load (E_{load}) and the current through the load. The voltage across the load results from the energy balance equation for the cell which is written in terms of voltages (E) as follows:

$$E_{rev}I = E_{load}I + E_{act}I + E_{conc}I + E_{ohm}I \qquad (4.11)$$

We now consider the case of a PEMFC for which we give the expression for voltage across the load and describe the relevant balance equations. The set of equations developed herein can be modified for other types of fuel cells, the equation derivation methodology being similar. Accounting for specific equations for activation, concentration, and ohmic overpotential in PEMFC, the following expression is given for the voltage across the load (as taken from Ay et al. (2006)):

$$E_{load} = 1.229 - 8.5 \times 10^{-4}(T - 298.15) + 4.3085 \times 10^{-5}T(\ln P_{H_2} + 0.5 \ln P_{O_2})$$
$$- \left(\frac{\alpha_a + \alpha_c}{\alpha_a \alpha_c}\right) \frac{RT}{zF} \ln\left(\frac{J}{J_0}\right) - J\left(\beta\frac{J}{J_L}\right)^\gamma - \frac{J\delta \exp\left(-4.185\frac{T-303}{T}\right)}{(0.005139\,\omega - 0.00326)} \qquad (4.12)$$

where the first term is the reversible standard open circuit voltage that is also mentioned in Table 4.3. The second term in Equation (4.12) accounts for the change in open circuit voltage due to the operation temperature, and the third term accounts for the combined effect of operating temperature and pressures in the output streams. Hence the first three terms in Equation (4.12) give the open circuit voltage at the operating conditions (pressures, temperature). The fourth term in the equation represents the activation overpotential. This depends on charge transfer coefficients α (which takes in general different values for the cathodic and the anodic reactions), the number of charges ($z=2$ for PEMFC), and the exchange current density J_0. The fifth term in Equation (4.12) represents the concentration overpotential which depends on the limiting current density (J_L, assumed the same—at an average value—for anode and cathode) and two empirical concentration overpotential constants denoted as β and γ. The sixth term is the ohmic overpotential. The energy corresponding to the ohmic overpotential dissipates internally as heat. The ohmic overpotential is related to the area-specific resistance (ASR) of the electrodes and of the membrane. This is related to the membrane thickness δ and the membrane humidity factor ω. Typical values and correlations for $\alpha, \beta, \gamma, J_0, J_L$ are given in Ay et al. (2006).

The charge transfer coefficients α can be assumed as 0.5 for anode and 1.0 for cathode, whereas the limiting current density can be taken as $J_L = 2000 \, A/m^2$. The exchange current density is taken as a function of temperature in the same manner as mentioned in Ay et al. (2006). J_0 is given by

$$J_0 \, (A/m^2) = 1.08 \times 10^{-17} \times \exp(0.086 \, T_b) \tag{4.13}$$

where the operation temperature can be assumed to be the same as the system boundary temperature.

The concentration overpotential exponent can be assumed to be $\gamma \cong 2.0$ as in Ay et al. (2006). The concentration overpotential factor β for PEMFC is documented in the same reference source as follows:

$$\beta = \begin{cases} (7.16 \times 10^{-4} T_b - 0.622) P_{cat} - 1.45 \times 10^{-3} T_b + 1.68 & \text{if } P_{cat} < 2 \, atm \\ (8.66 \times 10^{-4} T_b - 0.068) P_{cat} - 1.6 \times 10^{-4} T_b + 0.54 & \text{if } P_{cat} \geq 2 \, atm \end{cases} \tag{4.14}$$

where the cathode pressure for oxygen-only feed is related to the partial pressure of oxygen and that of water vapor, and it is $P_{cat} = 8.52 P_{O_2} + P_{sat}$ with P_{sat} being the saturation pressure of water vapor at the operating temperature of the cell. The membrane humidity factor is documented in Ay et al. (2006) according to the formula:

$$\omega = \begin{cases} 0.043 + 17.8 \, a - 39.85 \, a^2 + 39.85 \, a^3 & \text{for } 0 \leq a \leq 1 \\ 14 + 1.4(a - 1) & \text{for } 1 < a \leq 3 \end{cases} \tag{4.15}$$

where a is the membrane water activity coefficient defined by $a = y_{H_2O} P / P_{sat}$, with P the pressure in the membrane and y_{H_2O} the water molar fraction in the membrane, which can be assumed to be of linear variation from the molar fraction at anode $y_{H_2O,A} = (P_{sat}/P)_A$ to cathode $y_{H_2O,C} = (P_{sat}/P)_C$.

The partial pressure of hydrogen needed in Equation (4.12) is given by the pressure at the anode and molar fraction of hydrogen in the fed stream, namely $P_{H_2} = y_{H_2} P_A$. Similarly, the partial pressure of oxygen is given by $P_{O_2} = y_{O_2} P_A$. The molar fractions of hydrogen (at anode) and oxygen (at cathode) depend on the excess oxygen factor for the stoichiometric cathode λ_C and the excess hydrogen factor for the stoichiometric anode compartment λ_A used; commonly $\lambda_C = 3$ and $\lambda_A = 1.5$. It also depends on the dry molar fractions of oxidant and fuel in the fed streams at cathode and anode, y_C and y_A. Normally, the oxidant is atmospheric air and fuel is pure hydrogen, and therefore $y_C = 0.21$ and $y_A = 1$. The formula for molar fraction calculation of fuel and oxidant is documented in Ay et al. (2006) as follows:

$$\begin{cases} y_{H_2} = \dfrac{1 - y_{H_2O, A}}{1 + 0.5 y_A (1 + \lambda_A/(\lambda_A - 1))} \\ \\ y_{O_2} = \dfrac{1 - y_{H_2O, C}}{1 + 0.5 y_C (1 + \lambda_C/(\lambda_C - 1))} \end{cases} \tag{4.16}$$

EXAMPLE 4.3

In this example the energy and exergy analysis of a PEMFC is demonstrated. Equations (4.10)–(4.16) and efficiency definitions from Table 4.7 are used. The assumptions are listed as follows:

- Temperature of the control volume boundary is the same as the cell temperature, $T_b = 353$ K.

- Total pressure at anode is equal to total pressure at cathode and is $P = 4$ atm.
- Membrane thickness is $d = 0.18$ mm.
- α is assumed to 0.5 for anode and 1.0 for cathode.
- Concentration overpotential exponent is $\gamma = 2.0$.
- Limiting current density is 2000 A/m^2.
- Current density is 100 A/m^2.
- Oxygen utilization ratio is $U_a = 50\%$.
- Hydrogen utilization ratio is $U_f = 80\%$.

- We calculate first the stream parameters for oxygen. Given that $U_a = 50\%$ and $U_f = 80\%$ (from Table 4.7), the excess air coefficient $\lambda_a = 1.6$. This refers to the overall cell. Given that the cathode reaction rate must be two times faster than the anode reaction rate for steady operation, one deduce that $\lambda_C = 2\lambda_a = 3.2$, whereas $\lambda_A = \lambda_a = 1.6$. Also the air fuel ratio for the stoichiometric reaction is 0.5. Accounting for the excess air factor, the air fuel ratio becomes $AFR = 0.5\lambda_a/0.21 = 7.619$.
- Pure hydrogen and atmospheric air are used, thus $y_A = 1$ and $y_C = 0.21$ (oxygen molar fraction in air).
- The saturation pressure of water vapor at $T_b = 353$ K is $P_{sat} = 0.4647$ atm $\rightarrow y_{H_2O} = 0.1162$, the same for cathode and anode.
- The molar fraction of hydrogen from Equation (4.16) is $y_{H_2} = 0.3119 \rightarrow P_{H_2} = 1.248$ atm.
- The molar fraction of oxygen from Equation (4.16) is $y_{O_2} = 0.707 \rightarrow P_{O_2} = 2.811$ atm.
- As the molar fraction of water is the same for the anode and cathode, the activity of water inside the membrane becomes $a = 1$. Therefore, the membrane humidity factor from Equation (4.15) is $\omega = 17.84$.
- With $P_{cat} = 8.52P_{O_2} + P_{sat} = 24.41$ atm one calculates the concentration overpotential factor, which becomes $\beta = 6.29$.
- The current density results from Equation (4.14), namely $J_0 = 1.65 \times 10^{-6}$ A/m^2.
- The reversible cell potential corresponds to the first three terms in Equation (4.12), $E_{rev} = 1.194$ V.
- The activation overpotential is calculated from Equation (4.12), $E_{act} = 0.6075$ V.
- The concentration overpotential from Equation (4.12) is $E_{conc} = 0.0988$ V.
- The ohmic overpotential is calculated from Equation (4.12), $E_{ohm} = 0.112$ V.
- The voltage across the load is $E_{load} = 1.197 - 0.6075 - 0.083 - 0.112 = 0.3748$ V.
- To write the energy balance equation, the molar specific enthalpies must be estimated at inlet and outlet streams. This is done as listed below:
  - The molar specific enthalpy of hydrogen and that of air feed at $T_0 = 298.15$ K is nil.
  - The molar specific enthalpy of species in product stream at $T_b = 353$ K are 1591 J/mol for H$_2$, 1595 J/mol for air, and $-281,678$ J/mol for water.
  - Note that the number of mols of reacted hydrogen depends on the fuel utilization ratio according to $\dot{n}_{H_2r} = \dot{n}_{H_2} U_f$; also the rate of water formation is the same as the rate of hydrogen consumption $\dot{n}_{H_2O} = \dot{n}_{H_2r}$ and the rate of oxygen consumption is the half of the rate of hydrogen consumption.

Continued

EXAMPLE 4.3　　*(cont'd)*

- The energy balance equation becomes:

$$\text{EBE}: (\dot{n}_{H_2} h_{H_2})_{in} + \text{AFR}\,(\dot{n}_{H_2} h_{O_2})_{in} = \dot{W} + \dot{Q} + \left[(\dot{n}h)_{H_2} + (\dot{n}h)_{O_2} + (\dot{n}h)_{H_2O} + (\dot{n}h)_{N_2}\right]_{out}$$

  - If EBE is divided by $\dot{n}_{H_2,in}$ and one accounts for the fact that $\dot{W} = I\,E_{load}$ and $I = 2F\dot{n}_{H_2r}$ and one denotes $q = Q/\dot{n}_{H_2,in}$, then EBE becomes

$$h_{H_2,in} + \text{AFR}\,h_{air} = 2FU_f E_{load} + q + (1 - U_f)h_{H_2,out} + U_f h_{H_2O} + 3.76\,\text{AFR}\,h_{N2}$$
$$+ (\text{AFR} - 0.5U_f)h_{O_2}$$

  - When EBE is solved for q the value of $q = 109{,}972$ J/mol is obtained.

- Once q is determined the entropy balance equation can be solved for the generated entropy. EnBE is as follows:

$$\text{EnBE}: s_{H_2,in} + \text{AFR}\,s_{air} + s_{gen} = \frac{q}{T_b} + (1 - U_f)s_{H_2,out} + U_f s_{H_2O} + 3.76\,\text{AFR}\,s_{N_2} + (\text{AFR} - 0.5U_f)s_{O_2}$$

  - The following values for molar specific entropies of the species are obtained from thermodynamic data tables: $s_{H_2,in} = 119$ J/molK, $s_{air,in} = 209.8$ J/molK, $s_{H_2,out} = 123.9$ J/molK, $s_{O_2,out} = 137.3$ J/molK, $s_{N_2,out} = 186.7$ J/molK, $s_{H_2O} = 82.67$ J/molK.
  - The solution of EnBE gives $s_{gen} = 333.9$ J/molK.
- The destroyed exergy is given by $ex_d = T_0\,s_{gen} = 99.54$ kJ/mol.
- The exergy input in the system depends on the chemical exergy of the reactants.
  - The chemical exergies are extracted from tables as follows: $ex_{H_2}^{ch} = 236.1$ kJ/mol, $ex_{O_2}^{ch} = 3.97$ kJ/mol, $ex_{N_2}^{ch} = 0.72$ kJ/mol.
  - Because the reactants are provided at T_0 and P, the molar specific exergy input in the system is $ex_{in} = ex_{H_2}^{ch} + \text{AFR}\left(0.21\,ex_{O_2}^{ch} + 0.79\,ex_{N_2}^{ch}\right) + \sum n[h - h_0 - T_0(s - s_0)]$, where $h - h_0 = 0$ because the enthalpy of ideal gas does not depend on pressure. The molar exergy input is $ex_{in} = 279.554$ kJ/mol.
- The exergy efficiency is $\psi = 1 - (ex_d/ex_{in}) = 58.1\%$.
- The work generated for any kmol of hydrogen supplied is $w = 2\,F\,U_f\,E_{load} = 57.86$ kJ/mol.

4.6 OPTIMIZATION OF FUEL CELL SYSTEMS

Optimization of fuel cells and fuel cell systems can be done based on various objective functions. One possibility is to maximize the system efficiency (energy or exergy); also in some cases it is important to maximize the power generated per unit of system volume (power density). These two optimization options are in general divergent: when the system is optimized for the maximum power density its efficiency is not maximized.

Multi-objective optimization schemes can be applied for fuel cell system design. Multiple criteria such as efficiency, cost, and environmental impact can be considered to formulate the objective function(s). Setting constraints is one of the most important aspects of optimization. Constraints such as fixed size, imposed operating parameters, and imposed maximum cost can be applied, depending on the case. In many practical cases, multi-variable optimization methods must be applied.

In the case of multi-objective optimization two or more conflicting objective functions must be maximized (minimized). Different tradeoffs can be considered when selecting the best solution: e.g., maximizing profit and minimizing cost, minimizing energy consumption while maximizing performance, etc. The solution of multi-objective optimization problem consists of a Pareto set which can be represented on a map such that the tradeoffs can be easily observed.

With fuel cell systems there are many optimization opportunities. The most common example is the maximization of the power output by optimizing the current density and voltage across the load. The typical voltage–current diagram of a fuel cell shows that there is always an optimal current density for each fuel cell transfer of maximum power under certain operating conditions. Figure 4.35 shows an example of this optimum for maximum power generation.

In another optimization approach the parameter to be optimized can be the volumetric fraction occupied by the fuel cell itself in the overall volume of the fuel cell system. Note that the fuel cell system volume comprises two terms, the volume of the fuel cell stack V_{fc} and the volume of all auxiliary components V_{aux}. The volumetric fraction is defined by

$$f = \frac{V_{fc}}{V_{fc} + V_{aux}}$$

Two parameters can be optimized with respect to volumetric fraction of the fuel cell stack within the fuel cell system: the power density and the energy efficiency. The power density is defined as the net power generated by the system divided by the volume of the system. The energy efficiency is the net power generated divided by the energy carried by the fuel. There is a tradeoff between these two objectives and a multi-objective optimization problem can be formulated mathematically as follows:

$$\max\left\{ \dot{W}'''(f) = \frac{\dot{W}}{V}, \eta(f) = \frac{\dot{W}}{\dot{m}_f \, HHV} \middle| V = \text{constrained}, f = (0,1) \right\}$$

Note that a case study of this type of optimization is presented in Zamfirescu and Dincer (2009). It is shown that each of the two objective functions $\dot{W}'''(f)$ and $\eta(f)$ can be maximized with respect to f. In Figure 4.37 the parameters of the objective functions are normalized with respect to their maximums. Hence, $W^* = \dot{W}''' / \dot{W}'''_{max}$ and $\eta^* = \eta / \eta_{max}$.

These two parameters $\dot{W}'''$ and η are important especially in hydrogen-fueled vehicles that struggle with the problem of fitting a large hydrogen storage tank onboard and with the problem of maximizing the driving range. When the system is optimized for maximum efficiency the configuration is such that f is large, i.e., the fuel cell stack is large with respect to the other system components (the heat exchangers). For obtaining the maximum efficiency the stack occupies about 75% of the system volume.

The second option is to design the system for more compactness. In this case the system configuration is such that the stack occupies about 40% of the system's volume while the rest of the system is occupied by the heat exchangers. It can also be said that if the system is designed for maximum power generation, it loses efficiency (there is about 20% efficiency loss for the presented example). If the fuel cell is used for vehicle propulsion it is expected that driving range is reduced by a similar percent. In addition, if the system is designed for a

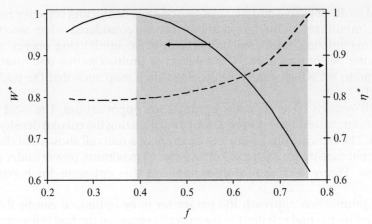

FIGURE 4.37   Example of fuel cell optimization curves for maximized power density and maximized efficiency. *Data from Zamfirescu and Dincer (2009).*

maximum energy efficiency to get the maximum driving range for the vehicle, the power density of the system is reduced by about 40% compared to the system optimized for a maximum power density. Therefore, if this option is chosen, one may end up with less useful space onboard for the system. This is even clearer in Figure 4.37, as the gray-shaded area for the range of the design parameter f. It apparently ranges between the two extreme points, i.e., 0.35 and 0.75. It is not feasible to go beyond this range. So, one can consider this as the optimum configuration domain for design.

4.7 INTEGRATED FUEL CELL SYSTEMS FOR POWER GENERATION

When used for any application a fuel cell stack must operate within a more complex system that has the function of supplying fuel and oxidant at appropriate pressure and temperature, recovering heat and work internally (when possible), and expelling heat, water, and reaction products. In many cases the fuel cell stack must be insulated and maintained to a certain optimum operating temperature. Also the balance of the plant has the role of assuring the startup and shutdown operations and also regulating the system to operate at part load and full load. In what follows we will present some typical fuel cell systems, including the fuel cell stack and the balance of the plant.

A typical PEM fuel cell power plant (including the fuel cell stack and the balance of the plant) is presented in Figure 4.38. The diagram comprises a hydrogen tank, pressure regulator, humidifiers, fuel cell stack, radiator, cooler, air compressor, and water recycle pump. Both the hydrogen gas and oxygen gas must be humidified before entering the fuel cell stack, because the Nafion membrane works only under humid conditions.

The system normally operates at a temperature under 100 °C and pressure slightly above the atmosphere. In these conditions, the formed water is in liquid state. The system continuously generates water when it runs. A part of the water is maintained in the system for

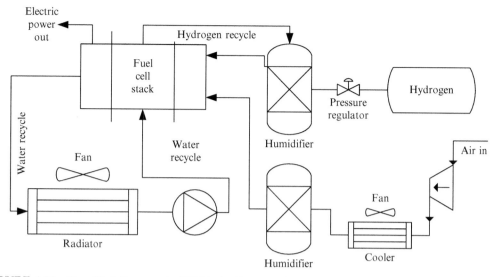

FIGURE 4.38   Simplified diagram of a PEM fuel cell power plant.

hydration purposes. The other part can be expelled (e.g., by spraying it out) or stored. However, storing water may raise problems when the system is applied to an autovehicle. The most important feature of the PEMFC power plant is its fast startup. Some electric power must be stored in a battery, as needed for running the air compressor and the pump at startup.

Figure 4.39 shows an SOFC+-based system for power generation which comprises the SOFC+ stack, a turbo-charger, four compact heat exchangers, a hydrogen fuel tank and a pressure-regulating device. Hydrogen is provided in state point #11 of the diagram.

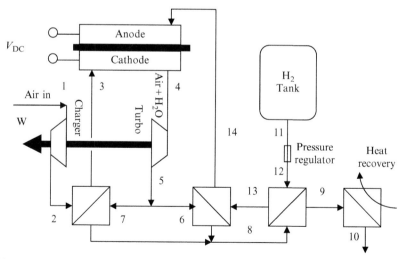

FIGURE 4.39   Fuel cell power plant based on proton-conducting SOFC+ technology.

After the pressure regulator, the hydrogen supplies the power generation system at #12 and then is preheated in two steps through #12–13 and #13–14. Air taken from the surroundings at #1 is compressed in the turbo-charger, delivered to the air-preheater at #2, and then to the fuel cell stack at #3. The exhaust of the fuel cell, consisting of oxygen-depleted air and steam, is directed toward the turbine inlet at #4 and expanded with work recovery up to state #5. The hot exhaust at #5 is used to preheat the two reactants: air (#7–8) and hydrogen (#6–8, #8–9) and then either released to the ambient environment or used for heat recovery in an additional heat exchanger (#9–10).

The conceptual design is made such that the two reacting streams, prior to being supplied to the fuel cell stack, are preheated to the same temperature in an equal number of two steps. In fact, the air temperature at #2 is same as the hydrogen temperature at #13, and the air temperature at #3 is same as the hydrogen temperature at #14.

Fuel cells that operate at higher temperature—like molten carbonate and solid oxide fuel cells—in general must be coupled with gas turbines in order to achieve satisfactory fuel utilization. Moreover, high-temperature heat is commonly recovered from the reaction product of these kinds of fuel cells and used for reforming some high-density fuels like compressed natural gas, alcohols, and others to hydrogen. An example is given in the diagram in Figure 4.40, consisting of a MCFC with reformed natural gas.

This is a simplified diagram of a system installed recently by the natural gas provider Embridge Inc. in Toronto, ON. The system is connected to the high-pressure natural gas network in #1. The high-pressure gas is first preheated (#1–2) and then expanded to generate about 1 MW electric power from a 6000 m³/h gas-flow rate. After expansion, a part of the

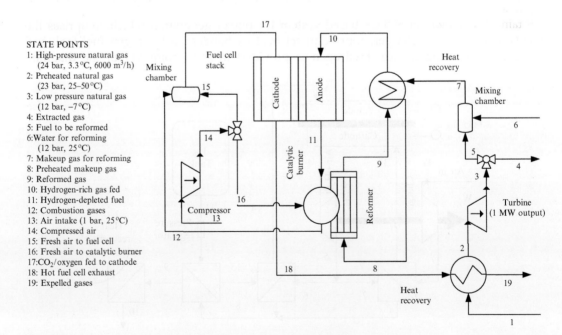

STATE POINTS
1: High-pressure natural gas
   (24 bar, 3.3 °C, 6000 m³/h)
2: Preheated natural gas
   (23 bar, 25–50 °C)
3: Low pressure natural gas
   (12 bar, −7 °C)
4: Extracted gas
5: Fuel to be reformed
6: Water for reforming
   (12 bar, 25 °C)
7: Makeup gas for reforming
8: Preheated makeup gas
9: Reformed gas
10: Hydrogen-rich gas fed
11: Hydrogen-depleted fuel
12: Combustion gases
13: Air intake (1 bar, 25 °C)
14: Compressed air
15: Fresh air to fuel cell
16: Fresh air to catalytic burner
17: CO₂/oxygen fed to cathode
18: Hot fuel cell exhaust
19: Expelled gases

FIGURE 4.40   Natural-gas-fueled system with molten carbonate fuel.

low-pressure gas is sent to the distribution network to be used for water/space heating in residences. The other part (#5) is mixed with water (#6), preheated (#7–8), and reformed (#8–9) and supplied to the MCFC in (#10). The resulting un-combusted anodic gases are further combusted with additional air (#16) in a catalytic burner (#11–12) which delivers its produced heat to the reformation process. The CO_2-rich combustion gases (#12) are mixed with air (#15) and fueled to the MCFC cathode in (#17). The heat carried by the cathode gas is recovered (#18–19) and used to preheat the high-pressure fuel. Fresh air from the ambient surroundings is used for combustion and the fuel cell process, as it is compressed first (#13–14) and then split (#15 and #16).

SOFC systems have the big advantage of operating at very high temperature, a fact that facilitates steam-assisted reforming of the primary fuel to hydrogen without the need of expensive catalysts. As explained above, owing to hydrogen consumption in the anode, partial pressure decreases, a fact that negatively affects the chemical equilibrium and reaction rate. Thus, hydrogen is fed in excess and only partially consumed in the fuel cell. One method of improving fuel utilization in the fuel cell is to recycle the product gases as illustrated in the diagram in Figure 4.41. With recirculation from $n_1^{CH_4}$ moles of fuel fed, it results that $(\zeta-1)\left(n_2^{CH_4}+n^{CO}+n^{H_2}\right)$ moles of un-used fuel, where $\zeta n_2^{CH_4} < n_1^{CH_4}$. If product gas recirculation is not applied, then the quantity of noncombustible gases necessary for the system to produce the same electric output would be higher $\zeta\left(n_2^{CH_4}+n^{CO}+n^{H_2}\right)$. The system with recirculation still requires an additional afterburner or another system to consume the remaining fuel, but is a fact that the fuel utilization within the fuel cell is better. Another aspect is that steam is needed for reforming the primary hydrocarbon fuel (in the above example, the methane). Through recirculation of product gases, steam is provided to the fuel feed side. Depending on the operating parameter, it is possible that this steam is sufficient to obtain enough fuel reformation. However, in many practical situations, the steam provided by product gas recirculation is not sufficient. Additional steam can be then provided through water condensation from exhaust gases, followed by water reheating and steam generation.

A complete diagram of an SOFC system is presented in Figure 4.42. The system uses natural gas from municipal distribution lines (assumed at 6–8 bar) which is reformed, with steam recovered from the product gas. Air from the ambient surroundings is aspired by the air compressor in #1 and discharged at high pressure in #2 with further preheating (#2–3); preheated air is fed into the fuel cell in #3. The oxygen-depleted air is discharged at the cathode side in #4. Natural gas enters in 5 and is preheated (#5–6) and mixed with superheated steam recycled from the fuel cell stack #28 and extracted from the exhaust gases #27. The make-up gas, consisting of natural gas, steam, uncombusted gases (hydrogen, methane, and carbon

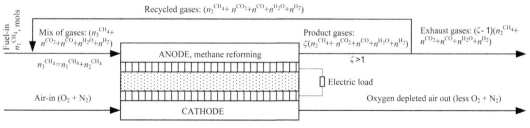

FIGURE 4.41 Product gas recycling in SOFC for steam reforming and better fuel utilization.

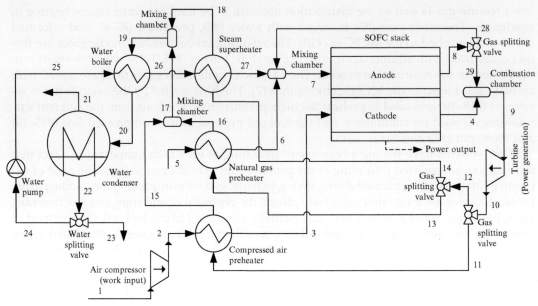

FIGURE 4.42   SOFC/gas turbine power generator with natural gas fuel. *Modified from Granovskii et al. (2008).*

monoxide) enters at the fuel cell anode in #7. The product gas exhaust at anode #8 is split in two streams—stream #28 (recycled) and stream #29 directed toward the combustion chamber, where it combusts with oxygen-depleted air from cathode #4, producing hot exhaust gas in #9.

The exhaust gas at high temperature and pressure in #9 is expanded with work generation in the gas turbine and retrieved downstream in #10, where it is split in three streams to be used for heat recovery (steams #11, #13, #14). The enthalpy of these gases is used for preheating air, preheating natural gas, and superheating steam. After this, all three streams are mixed in #19, and their enthalpy is used for steam generation from recovered water in the water boiler. Colder exhaust gases in #20 (assumed completely combusted) are further cooled in the water condenser which condensates water while releasing the un-condensable gases in #21. The resulting water is partially drained #23 (or used for other purposes) and partially recycled #24. The recycled water is first pressurized (#24–25), then boiled (#25–26), and the resulted steam superheated (#26–27).

Granovskii et al. (2008) calculated the energy and exergy efficiency of an SOFC system like that presented in Figure 4.42 and showed that depending on the operating condition it can achieve 73–85% (energy-based) and 72–85% (exergy-based). The energy and exergy efficiencies can then be defined for the SOFC system as follows:

$$\left.\begin{aligned} \eta &= \frac{W_{fc} + W_t - W_c - W_p}{LHV} \\ \psi &= \frac{W_{fc} + W_t - W_c - W_p}{ex_{ng}^{ch} + ex_{ng}^{thrm}} \end{aligned}\right\}$$

(4.17)

where the generated work W is given per molar unit of natural gas consumption, indices fc, t, c, p, ng represent fuel cell, turbine, compressor, pump, and compressed natural gas, respectively, ex is the molar chemical exergy, and the exponents "ch" and "thrm" mean chemical and thermomechanical, respectively.

The need for integration of fuel cell systems with other systems issues from other reasons in addition to that of achieving better fuel utilization. These reasons can be, for example, using other fuel than hydrogen and applying direct and indirect reforming—whichever is the case—or applying heat recovery for cogeneration of power and heat or to drive a bottoming cycle for additional power generation. In addition, fuel cells can be integrated in systems with other purposes than that of power generation. An example is the oxygen production system shown in Figure 4.43. These systems combine a fuel cell system with a water/steam electrolyser system. In this approach, a high-efficiency SOFC power plant is used to drive steam electrolysis. The SOFC system exhaust contains steam at very high temperature along with other combustion products (mainly nitrogen and carbon dioxide). A steam recovery unit is devised that basically cools the exhaust gases, condensates steam, separates water, boils it, and superheats the steam using extensive heat recovery from the gases.

Another fuel cell system—this time operating at intermediate temperature with an SOFC + device and gas turbine—is presented in Figure 4.44. This system operates with ammonia and cogenerates refrigeration and electric power. The cooling effect is produced in the ammonia storage tank, which features a cooling coil #10 and is thermally insulated. When ammonia vapor is drawn out of the tank #1, it removes enthalpy. Some liquid thus evaporates and generates refrigeration. At startup ammonia can be heated electrically through the heating element #2. At steady operation, heating is applied in steps (#2–3) and (#3–4) up to about 700 °C, the temperature at which ammonia is fed into the fuel cell. There ammonia is decomposed catalytically at the anode and the hydrogen consumed. Steam is formed at the cathode and expanded in #6 together with the resulting nitrogen to generate power for air compression. The element #7 is an electric-drive generator that during the startup drives the air compressor, while during the steady operation of the system generates some electricity. Heat recovery can be applied to the exhaust gases. The system emits only benign gases like steam and nitrogen.

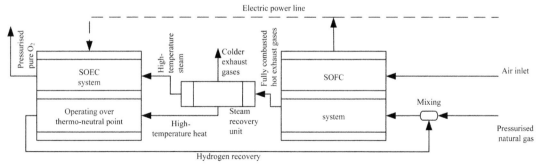

FIGURE 4.43   Integrated system with SOFC and SOEC for pure pressurized oxygen production.

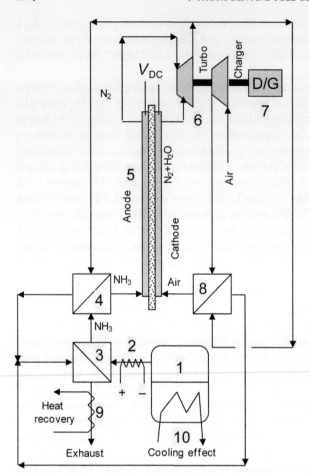

FIGURE 4.44   Direct ammonia fuel cell system with gas turbine for power and cooling.

AFCs also need improved fuel utilization, which can be achieved in principle by recirculation of the exhaust gases. However, their operating temperature is about 200 °C, which means a low enthalpy of the reaction products. Moreover, the electrodes of the AFCs are sensitive to poisoning, especially if the pressure is increased. Consequently it is appropriate neither to recycle the products nor to expand them in a gas turbine. An alternative is to couple AFC systems with an organic Rankine cycle operating at the bottoming side. The diagram from Figure 4.45 illustrates the concept: the exhaust gases from the AFC are combusted externally at constant pressure and the combustion heat from the exhaust gases is recovered through a heat exchanger. The heat is further transferred to a Rankine cycle which generates additional power. Preferably, for low to medium capacity application, the Rankine cycle should run with an organic fluid.

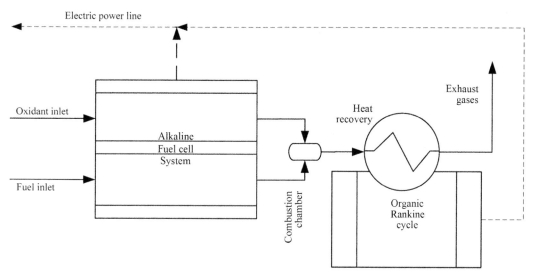

FIGURE 4.45   Alkaline fuel cell system with bottoming Rankine cycle for better fuel utilization.

4.8 CONCLUDING REMARKS

In this chapter hydrogen and fuel cell systems are introduced. The role of hydrogen and fuel cell systems in sustainable power generation at a global scale is discussed in detail. The energy demand per capita and the global population have an increasing trend. This is expected to be more drastic in the near future. More power production with the conventional systems means higher pollution and global warming effect as well as faster depletion of fossil fuel resources. It is shown that hydrogen and fuel cell system are the best choice for renewable energy utilization. This is important mostly for solar resources because this appears to be the most important for the future. Hydrogen production methods are described, with emphasis on electrolysis and thermochemical water-splitting cycles, which are very promising technologies. Fuel cell types are introduced and their specific applications presented. Modeling and optimization of fuel cells is very important for obtaining better designs with improved efficiency. So, a full section is dedicated to this topic. Several power generation systems with fuel cells are also presented.

Study Problems

4.1. In what way can a hydrogen economy influence the power generation sector?

4.2. Categorize the methods for hydrogen production.

4.3. Calculate the reaction enthalpy and the Gibbs energy of the water decomposition reaction at 25, 1000, and 2500 °C and compare the results.

4.4. Calculate the energy and exergy efficiency of an electrolysis cell when the operating temperature is 80 °C, the current exchange density $J_0 = 20$ mA/m^2, the Tafel slope $b = 100$ mV, the transfer coefficient $\alpha = 0.2$, and the cell operates at the limit between low and high polarization regimes.

4.5. Redo problem 4.4 for a range of current densities and determine the current-voltage characteristic.

4.6. Explain the concept of thermochemical water splitting for hydrogen production.

4.7. Describe the copper–chlorine hybrid cycle.

4.8. What is the difference between fuel reforming and gasification?

4.9. Explain the principle of photocatalytic water splitting.

4.10. Calculate the work needed to compress 1 kg of hydrogen from 1 bar pressure to 800 bar and compare it with hydrogen's higher heating value.

4.11. Describe the operation principle for fuel cells with a proton-conducting electrolyte.

4.12. What are the major advantages of alkaline fuel cell systems?

4.13. Is fuel cell operation benefited by high pressure and low temperature or by low pressure and high temperature?

4.14. Calculate the system in Figure 4.35 under reasonable assumptions and determine the efficiencies.

4.15. Calculate the system in Figure 4.36 under reasonable assumptions and determine the efficiencies.

4.16. Calculate the system in Figure 4.38 under reasonable assumptions and determine the efficiencies.

4.17. Calculate the system in Figure 4.39 under reasonable assumptions and determine the efficiencies.

4.18. Explain the equation of Nernst.

4.19. Describe the types of energy loss in fuel cells.

4.20. Do the case study in Figure 4.34 for a PEMFC.

Nomenclature

$\mathcal{A}$	pre-exponential factor
ASR	area-specific ohmic resistance (Ω)
E	electric potential (V)
ex	molar specific exergy (kJ/mol)
F	Faraday number (C/mol)
f	volumetric fraction
G	free energy (kJ)
H	enthalpy (kJ)
HHV	higher heating value (MJ/kg)
h	molar specific enthalpy (kJ/mol)
I	current (A)
J	current density (A/m^2)
K	equilibrium constant
LHV	lower heating value (MJ/kg)
m	mass flow rate (kg/s)
n	number of moles (mol)
P	pressure (kPa)
Q	heat (kJ)
$\dot{Q}$	heat rate (kW)
R	universal gas constant (kJ/mol K)
T	temperature (K)
U	utilization ratio

V volume (m^3)
W work (kJ)
$\dot{W}$ work rate (kW)
y molar fraction
z number of electrical charges

Greek letters

α transfer coefficient
β concentration overpotential factor
γ concentration overpotential exponent
δ membrane thickness (m)
η energy efficiency
ψ exergy efficiency
λ excess air
ω humidity factor

Subscripts

0   exchange current; reference state
act   activation
aux   auxiliary
c   compressor
conc   concentration
eq   equilibrium
f   fuel
fc   fuel cell
H   heat source
L   limiting
ng   natural gas
oc   open circuit
p   pump
PG   power generation
rev   reversible
s   steam
t   turbine
TC   thermochemical
tot   total

Superscripts

0   reversible
ch   chemical
thrm   thermomechanical

References

Acar, C., Dincer, I., Zamfirescu, C., 2014. A review on selected heterogeneous photocatalysts for hydrogen production from water. Chem. Rev. http://dx.doi.org/10.1002/er.3211.

Ay, M., Midilli, A., Dincer, I., 2006. Exergetic performance analysis of a PEM fuel cell. Int. J. Energy Res. 30, 307–321.

Carrette, L., Friedrich, K., Stimming, U., 2001. Fuel cells – fundamentals and applications. Fuel Cells 1, 5–39.

Carty, R.H., Mazumder, M.M., Schreiber, J.D., Pangborn, J.B., 1981. Thermochemical Production of Hydrogen. Institute of Gas Technology, Chicago, IL, Final report 30517, vols. 1–4.

Dincer, I., 2012. Green methods for hydrogen production. Int. J. Hydrogen Energy 37, 1954–1971.

Dincer, I., Zamfirescu, C., 2011. Sustainable Energy Systems and Applications. Springer, New York.

Dincer, I., Zamfirescu, C., 2012. Sustainable hydrogen production options and the role of IAHE. Int. J. Hydrogen Energy 37, 16266–16268.

Dincer, I., Rosen, M.A., Zamfirescu, C., 2010. Economic and environmental comparison of conventional, hybrid, electric and hydrogen fuel cell vehicles. In: Pistoia, G. (Ed.), Electric and Hybrid Vehicles: Power Sources, Models, Sustainability, Infrastructure and Market. Elsevier, Amsterdam.

Dokyia, M., Kotera, Y., 1976. Hybrid cycle with electrolysis using Cu–Cl system. Int. J. Hydrogen Energy 1, 117–121.

Granovskii, M., Dincer, I., Rosen, M.A., 2008. Exergy analysis of a gas turbine cycle with steam generation for methane conversion within solid oxide fuel cells. J. Fuel Cell Sci. Technol. 5 (3), 031005, 1–9.

Lewis, M.A., Serban, M., Basco, J., 2003. Hydrogen production at 550 °C using a low temperature thermochemical cycle. In: Proceedings of the OECD/NEA Meeting, Argonne National Laboratory.

Naterer, G.F., Fowler, M., Cotton, J., Gabriel, K., 2008. Synergistic roles of off-peak electrolysis and thermochemical production of hydrogen from nuclear energy in Canada. Int. J. Hydrogen Energ. 33, 6849–6857.

Naterer, G.F., Dincer, I., Zamfirescu, C., 2013. Hydrogen Production from Nuclear Energy. Springer, New York.

Naterer, G., Suppiah, S., Stolberg, L., Lewis, M., Wang, Z., Dincer, I., Rosen, M.A., Gabriel, K., Rosen, Secnik, E., Easton, E.B., Lvov, S., Odukoya, A., 2014. International program on hydrogen production with the copper–chlorine cycle. Int. J. Hydrogen Energy 39, 2431–2445.

Rajeshwar, K., McConnel, R., Licht, S., 2008. Solar Hydrogen Generation. Springer, New York.

Rosen, M.A., 2008. Exergy analysis of hydrogen production by thermochemical water decomposition using the Ispra Mark-10 cycle. Int. J. Hydrogen Energy 33, 6921–6933.

Wentorf, R.H., Hanneman, R.E., 1974. Thermochemical hydrogen generation. Science 185, 311–319.

World Bank, 2013. World Development Indicators (Spreadsheet). The World Bank, http://data.worldbank.org/data-catalog/world-development-indicators, accessed on July 2nd.

Zamfirescu, C., Dincer, I., 2009. Thermodynamic performance analysis and optimization of a SOFC-H+ system. Thermochim. Acta 486, 32–40.

Zamfirescu, C., Naterer, G.F., Dincer, I., 2011. Solar light-based hydrogen production systems. In: Anwar, S. (Ed.), Encyclopedia of Energy Engineering and Technology. Taylor & Francis, New York.

Conventional Power Generating Systems

5.1 INTRODUCTION

Power generating systems are generally treated as heat engines to convert heat input into work, hence to produce electricity at a sustained rate. Heat input is supplied by burning fossil fuels (coal, oil and natural) and biomass, or processing nuclear fuel, or harvesting thermal energy from renewable energy sources. For example, in a conventional coal-fired power plant (the term power station is also used), the energy of coal is eventually converted into power. In general, conventional power stations comprise multiple generating units which are designed to operate at their nominal load when they function optimally.

There are a number of well-known power generating systems denoted as conventional, namely the spark ignition engine, compression-ignition engine, steam Rankine or organic Rankine power plant, combustion turbine power plant, combined cycle power station, nuclear power station, and hydroelectric power station. All these conventional power generating systems (CPGSs) primarily produce mechanical work which is transferred to subsequent systems in the form of shaft rotation. In vehicles, shaft power developed by engines is transferred to the traction system for propulsion. In stationary power plants or generators the shaft power developed by the prime mover is used to rotate an electrical generator which converts the rotational mechanical power to electrical power.

The key component of a CPGS is the prime mover or the organ that produces shaft power. Two types of prime movers are used in CPGSs: positive displacement machines (e.g., reciprocating engines) and turbomachines. Reciprocating machines generally consist of piston-and-cylinder assemblies where the pressure force of an expanding gas is transformed in a reciprocating movement which subsequently is converted into shaft rotation. Turbomachines (turbines) convert kinetic energy of a fluid directly into shaft rotation.

Small-scale CPGS use in general reciprocating prime movers; these are the spark ignition engine and the compression-ignition engine. Large-scale CPGS use turbines as prime movers. The only CPGS which does not use heat as an energy source is the hydroelectric power plant, where hydraulic energy is the input. All other CPGS represent thermomechanical converters and operate based on a specific thermodynamic cycle. The steam Rankine cycle is used in coal-fired, gas-fired, and oil-fired power stations and conventional nuclear power plants. The Brayton cycle is used in gas turbine power plants. A diesel cycle is specific to compression-ignition engines, whereas the spark ignition engine operates based on the Otto cycle.

Any CPGS has its distinct type of equipment. As already mentioned, the most important equipment is the prime mover: steam power plants develop power with the help of steam turbines, gas turbine power plants develop power using specific turbomachinery as the prime mover (this is the gas turbine), hydropower plants use various types of hydraulic turbines, and internal combustion engines use reciprocating piston–cylinder systems for their admission, combustion, compression, and expansion processes, thus generating net work output.

In steam power plants the second major piece of equipment after the steam turbine is the steam generator. Conventional steam generators use to be fired with coal, oil, or natural gas. In a nuclear power plant the steam generator is more specialized, as it is heated using various types of systems aimed at transferring heat from the nuclear reactor to the boiling water in a controlled and safe manner. The specific nuclear-based power generating systems and their power cycles, conventional and advanced, are introduced in Chapter 6 of this book.

In this chapter, the CPGSs are presented in the following order: vapor cycle power plants, gas turbine cycle power plants, gas engines, and hydroelectric power stations. For steam power plants the thermodynamic cycle of steam Rankine type is presented first with various arrangements. Coal-fired power stations with their specific steam generators are then introduced. Organic Rankine cycle (ORC) systems are discussed as a variant of Rankine cycles using an organic working fluid instead of steam. The focus is then shifted to gas turbine cycle power plants with analyses of the air-standard Brayton cycle. The section on internal combustion power generating systems covers information about the Diesel, Otto, Stirling, and Ericson cycles. The last section before chapter's conclusion discusses hydro power plants. More importantly, the CPGSs and their components are analyzed thermodynamically by writing all balance equations for mass, energy, entropy and exergy, and the performance assessments of these systems and components are carried out by energy and exergy efficiencies as well as other energetic and exergetic performance evaluation criteria.

5.2 VAPOR POWER CYCLES

It is general knowledge that steam power emerged as a technology to generate motive force in eighteenth century. The steam Rankine cycles is the most used type among the power generating thermodynamic cycles for power generation worldwide. Another version of Rankine cycle is the organic Rankine cycle, a technology that has been developed to commercial level in the last few decades and is used now in some specific applications, especially in renewable energy systems.

In a first phase of their development (eighteenth to nineteenth centuries), steam power stations operated based on a non-condensing Rankine cycle using a reciprocating steam engine (also known as steam piston). The motive force generated by the non-condensing steam engine fueled with wood or coal has been extensively used in the nineteenth century in many industries, including textile, mining, and agriculture (for irrigation and crop processing), pumping stations, all sorts of mills, and for transport vehicles such as rail locomotives and ships.

Major progress with steam power generation was marked by the invention of the steam turbine by the end of the 1800s. Since then, the steam turbine has become a prime mover which is responsible today for more than 80% of electrical power generated worldwide. The operation of steam power plants is based on Rankine cycles that use steam as working fluid.

5.2.1 The Simple Rankine Cycle

The most basic vapor power cycle is the simple Rankine cycle. This thermodynamic cycle can be described as an ideal internally reversible cycle or as an actual cycle with four irreversible processes. The ideal cycle can be executed with a machine comprising the following components: an ideal pump that operates isentropically and adiabatically (no heat transfer, only work input), an ideal heat exchanger system which plays the role of vapor generator (with no pressure drops and heat losses), a turbine able to operate isentropically and adiabatically (no heat transfer, only work output), and a condenser with no pressure losses and an infinite heat transfer surface (or equivalently an infinite heat transfer coefficient with zero temperature difference between working fluid and heat sink). The vapor generator consists of a preheating section, boiling section, and vapor superheating section. When the working fluid is steam—which is the case for any conventional type of steam power plant—the system is denoted as "steam Rankine cycle." Note that when an organic fluid or another type of working fluid is used the allure of the T–s diagram might be quite different, depending on the type of the fluid.

The basic configuration (four-components) of the ideal Rankine cycle and the cycle's T–s diagram are presented in Figure 5.1. The cycle comprises four processes: (i) isentropic pressurization of the working fluid in saturated liquid state, process 1–2, (ii) isobaric vapor generation, process 2–3 (with sub-processes of heating 2–3a, boiling 3a–3b and superheating 3b–3), (iii) isentropic expansion of the superheated working fluid, and (iv) a condensation process of the working fluid until it reaches saturated liquid state, process 4–1.

The ideal Rankine cycle is internally reversible but it is not a totally reversible cycle because the heat addition process at the source side is not isothermal. Therefore, the cycle as shown in Figure 5.1 has external irreversibility due to heat transfer at the source side. The heat transfer irreversibilities in this case are determined by the temperature difference $T_3 - T$, where T represents the temperature of the working fluid which increases along the process 2–3a–3b–3.

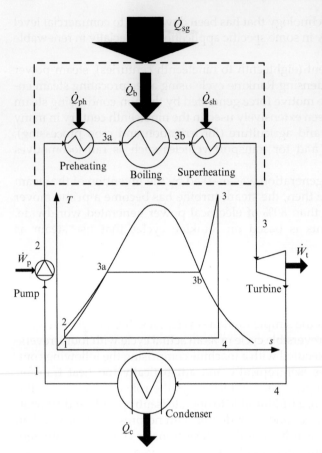

This involves a heat transfer with finite temperature difference between the heat source and working fluid. As the cycle is not externally reversible, an exergy efficiency smaller than 1 must be assigned to the cycle. The thermodynamic analysis can be pursued according to first and second laws.

The mass and energy balance equations for each of the components and the overall balances for the ideal cycle are given in Table 5.1. For the case of the ideal Rankine cycle there are no exergy destructions in condenser, pump, or turbine. However, provided that the heat source is at constant temperature, the externally irreversible Rankine cycle shows exergy destructions in the vapor generator which can be estimated based on the temperature difference between heat source and working fluid.

Let us denote the temperature of the heat source T_{so}. When the ideal Rankine cycle is analyzed, the heat source can be considered at a constant temperature which is equal to the highest temperature of the working fluid. The heat source T_{so} provides heat input to the processes of preheating, boiling and superheating.

In other words, the vapor generator is assumed with an "infinite surface for heat transfer" which assures that the working fluid reaches the heat source temperature level in the

TABLE 5.1 Mass and Energy Balance Equations for the Ideal Rankine Cycle

Comp.	Balance equations	Comp.		Balance equations
Pump (1–2)	MBE: $\dot{m}_1 = \dot{m}_2 = \dot{m}$ EBE: $\dot{W}_p = \dot{m}(h_2 - h_1)$	Vapor generator (2 – 3)	Preheating (2 – 3a)	MBE: $\dot{m}_2 = \dot{m}_{3a} = \dot{m}$ EBE: $\dot{Q}_{ph} = \dot{m}_{3a}(h_{3a} - h_2)$
Turbine (3–4)	MBE: $\dot{m}_3 = \dot{m}_4 = \dot{m}$ EBE: $\dot{W}_t = \dot{m}(h_4 - h_3)$		Boiling (3a-3b)	MBE: $\dot{m}_{3a} = \dot{m}_{3b} = \dot{m}$ EBE: $\dot{Q}_b = \dot{m}(h_{3b} - h_{3a})$
Condenser (4–1)	MBE: $\dot{m}_4 = \dot{m}_1 = \dot{m}$ EBE: $\dot{Q}_c = \dot{m}(h_4 - h_1)$		Superheating (3b – 3)	MBE: $\dot{m}_{3b} = \dot{m}_3 = \dot{m}$ EBE: $\dot{Q}_{sh} = \dot{m}(h_3 - h_{3b})$
Overall cycle	EBE: $\dot{W}_p + \dot{Q}_{sg} = \dot{Q}_c + \dot{W}_t$, where $\dot{Q}_{sg} = \dot{Q}_{sh} + \dot{Q}_b + \dot{Q}_{sh} = \dot{m}(h_3 - h_2)$			

thermodynamic state 3 (see the cycle in Figure 5.1); $T_3 = T_{so}$. On the other hand the condensation temperature ideally must be equal to the sink temperature; $T_1 = T_4 = T_{si}$.

In addition, the sink temperature is ideally the same as the reference temperature; $T_{si} = T_0$. The mass flow rate is the same for all system state points, hence it can be denoted generally with $\dot{m} = \dot{m}_1 = \dot{m}_2 = \ldots$. Furthermore, there is no entropy generation or exergy destruction in the condenser, pump or turbine. For the vapor generator the integrals in EnBE and ExBE can be solved directly because $T_{so} = \text{const}$. For the overall cycle the net power is:

$$\dot{W}_{net} = \dot{W}_t - \dot{W}_p \tag{5.1}$$

and total heat input is:

$$\dot{Q}_{vg} = \dot{Q}_{sh} + \dot{Q}_b + \dot{Q}_{sh} = \dot{m}(h_3 - h_2) \tag{5.2}$$

In Chapter 1 three general efficiency definitions that can be applied to Rankine cycles were given. The energy efficiency definition was given in Equation (1.67) and the exergy efficiency definition given either by Equation (1.68) or (1.69), which are equivalent. From Equations (5.1) and (5.2) the energy efficiency of the cycle can be defined in terms of net work output divided by the total heat input at source side (vapor generator), namely

$$\eta = \frac{\dot{W}_{net}}{\dot{Q}_{in}} = \frac{\dot{W}_t - \dot{W}_p}{\dot{Q}_{vg}} \tag{5.3}$$

The exergy efficiency in Equation (1.68) expresses the ratio between the net exergy delivered (net work output) and the actual exergy consumed. In the Rankine cycle the exergy consumed is the same as the exergy delivered by the heat source. Therefore, the exergy efficiency is given by

$$\psi = \frac{\dot{W}_{net}}{\dot{E}x_{in}} = \frac{\dot{W}_t - \dot{W}_p}{\dot{E}x_{so}} \tag{5.4}$$

The exergy efficiency when the heat source is of constant temperature is:

$$\psi = \frac{\dot{W}_{net}}{\left(1 - \dfrac{T_0}{T_{so}}\right)\dot{Q}_{so}} = \frac{\eta}{\eta_C} \tag{5.5}$$

where η_C is the Carnot factor of the constant temperature heat source, $\eta_C = 1 - T_0/T_{so}$.

The specific exergy of the flow represents a thermomechanical exergy calculated based on the specific enthalpy and entropy of the working fluid at a given thermodynamic state. According to its definition (see Chapter 1), the exergy is calculated with

$$ex = h - h_0 - T_0(s - s_0)$$

Here, index 0 represents the reference state which is assumed to be that of water at $T_0 = 25\,^\circ\text{C}$ and $P_0 = 101.325$ kPa. The variation of specific exergy Δex for any process is calculated with:

$$\Delta ex = \frac{\Delta \dot{E}x}{\dot{m}} = ex_{\text{out}} - ex_{\text{in}}$$

where $\dot{m}$ is a reference mass flow rate and subscripts "out" and "in" refer to outlet and inlet conditions, respectively. For each process within the ideal Rankine cycle either heat or work is exchanged between the working fluid and the surroundings. Processes that exchange work with the surroundings are the pumping and the expansion; the specific work is determined with:

$$w = \frac{\dot{W}}{\dot{m}} = h_{\text{out}} - h_{\text{in}}$$

Here, w is the mass-specific work. Similarly, the heat exchange processes that occur in the preheater, boiler, superheater, and condenser are described by the following equation, which expresses the mass-specific heat flux:

$$q = \frac{\dot{Q}}{\dot{m}} = h_{\text{out}} - h_{\text{in}}$$

Note that if heat or work is transferred out of the cycle then the sign is negative, whereas if heat or work is received by the working fluid the amount has a positive sign. Thus the work of the pump is negative, and that of the turbine is positive. Heat transfer at the preheater, boiler, and superheater is positive, whereas at the condenser it is negative.

In thermodynamic state 2 the working fluid must have the same specific entropy as in state 1 and the same pressure as in state 3. In thermodynamic state 3a the liquid is saturated at the specified boiling temperature T_b. In state 3b there is a saturated vapor at temperature T_b. State 3 is superheated vapor with the temperature equal to $T_b + \Delta T_{\text{sh}}$. State 4 represents the expanded working fluid at pressure equal to P_1, with the same specific entropy as in state 3. The vapor quality x (required for state 4) is related to specific enthalpy at the state point based on the equation

$$h = (1 - x)h_l + xh_v$$

where h_l and h_v are the specific enthalpies of saturated liquid and vapor, respectively, at the pressure corresponding to the thermodynamic state.

The mass, energy, entropy and exergy balance equations for the actual Rankine cycle are given in Table 5.2. Some important parameters are defined in Table 5.3. The back work

TABLE 5.2 Balance Equations for the actual Rankine Cycle of Basic Configuration

Component	Balance equations	Component	Balance equations
Pump (1–2)	MBE: $\dot{m}_1 = \dot{m}_2$ EBE: $\dot{m}_1 h_1 + \dot{W}_p = \dot{m}_2 h_2$ EnBE: $\dot{m}_1 s_1 + \dot{S}_g = \dot{m}_2 s_2$ ExBE: $\dot{m}_1 ex_1 + \dot{W}_p = \dot{m}_2 ex_2 + \dot{Ex}_d$	Vapor generator (2–3) Preheating section (2–3a)	MBE: $\dot{m}_2 = \dot{m}_{3a}$ EBE: $\dot{m}_2 h_2 + \dot{Q}_{ph} = \dot{m}_{3a} h_{3a}$ EnBE: $\dot{m}_2 s_2 + \int_{2}^{3a}\dfrac{d\dot{Q}_{ph}}{T_{so}} + \dot{S}_g = \dot{m}_{3a} s_{3a}$ ExBE: $\dot{m}_2 ex_2 + \int_{2}^{3a}\left(1-\dfrac{T_0}{T_{so}}\right)d\dot{Q}_{ph} = \dot{m}_{3a} ex_{3a} + \dot{Ex}_d$
		Boiling section (3a–3b)	MBE: $\dot{m}_{3a} = \dot{m}_{3b}$ EBE: $\dot{m}_{3a} h_{3a} + \dot{Q}_b = \dot{m}_{3b} h_{3b}$ EnBE: $\dot{m}_{3a} s_{3a} + \int_{3a}^{3b}\dfrac{d\dot{Q}_b}{T_{so}} + \dot{S}_g = \dot{m}_{3b} s_{3b}$ ExBE: $\dot{m}_{3a} ex_{3a} + \int_{3a}^{3b}\left(1-\dfrac{T_0}{T_{so}}\right)d\dot{Q}_b = \dot{m}_{3b} ex_{3b} + \dot{Ex}_d$
Turbine (3–4)	MBE: $\dot{m}_3 = \dot{m}_4$ EBE: $\dot{m}_3 h_3 = \dot{m}_4 h_4 + \dot{W}_t$ EnBE: $\dot{m}_3 s_3 + \dot{S}_g = \dot{m}_4 s_4$ ExBE: $\dot{m}_3 ex_3 = \dot{m}_4 ex_4 + \dot{W}_t + \dot{Ex}_d$	Superheating section (3b–3)	MBE: $\dot{m}_{3b} = \dot{m}_3$ EBE: $\dot{m}_{3b} h_{3b} + \dot{Q}_{sh} = \dot{m}_3 h_3$ EnBE: $\dot{m}_{3b} s_{3b} + \int_{3b}^{3}\dfrac{d\dot{Q}_{sh}}{T_{so}} + \dot{S}_g = \dot{m}_3 s_3$ ExBE: $\dot{m}_{3b} ex_{3b} + \int_{3b}^{3}\left(1-\dfrac{T_0}{T_{so}}\right)d\dot{Q}_{sh} = \dot{m}_3 ex_3 + \dot{Ex}_d$
Condenser (4–1)	MBE: $\dot{m}_4 = \dot{m}_1$ EBE: $\dot{m}_4 h_4 = \dot{m}_1 h_1 + \dot{Q}_c$ EnBE: $\dot{m}_4 s_4 + \dot{S}_g = \dot{m}_1 s_1 + \int_{4}^{1}\dfrac{d\dot{Q}_c}{T_{si}}$ ExBE: $\dot{m}_4 ex_4 = \dot{m}_1 ex_1 + \int_{4}^{1}\left(1-\dfrac{T_0}{T_{si}}\right)d\dot{Q}_c + \dot{Ex}_d$		
Overall cycle	MBE: there is no MBE because no mass crosses the boundary of the overall system. EBE: $\dot{W}_P + \dot{Q}_{sh} + \dot{Q}_b + \dot{Q}_{sh} = \dot{Q}_c + \dot{W}_t$; EnBE: $\int_{2}^{3a}\dfrac{d\dot{Q}_{ph}}{T_{so}} + \int_{3a}^{3b}\dfrac{d\dot{Q}_b}{T_{so}} + \int_{3b}^{3}\dfrac{d\dot{Q}_{sh}}{T_{so}} + \dot{S}_g = \int_{4}^{1}\dfrac{d\dot{Q}_c}{T_c}$ ExBE: $\dot{W}_P + \dot{Ex}_{so} = \dot{W}_t + \int_{4}^{1}\left(1-\dfrac{T_0}{T_{si}}\right)d\dot{Q}_c + \dot{Ex}_d$ where $\dot{Ex}_{so} = \int_{2}^{3a}\left(1-\dfrac{T_0}{T_{so}}\right)d\dot{Q}_{ph} + \int_{3a}^{3b}\left(1-\dfrac{T_0}{T_{so}}\right)d\dot{Q}_b + \int_{3b}^{3}\left(1-\dfrac{T_0}{T_{so}}\right)d\dot{Q}_{sh}$		

Note: MBE, mass balance equation; EBE, energy balance equation; EnBE, entropy balance equation; ExBE, exergy balance equation; T_{so} temperature at heat source [K]; T_{si} temperature at heat sink [K].

TABLE 5.3 Summary of Important Parameters for the Rankine Cycle

Quantity	Definition	Unit
Heat input	$\dot{Q}_{in} = \dot{Q}_{ph} + \dot{Q}_b + \dot{Q}_{sh}$	kW
Net work output	$\dot{W}_{net} = \dot{W}_t - \dot{W}_p$	kW
Energy efficiency, Equation (1.67)	$\eta = \dfrac{\dot{W}_{net}}{\dot{Q}_{in}}$	%
Exergy delivered	$\dot{Ex}_{deliv} = \dot{W}_{net}$	kW
Exergy efficiency, Equation (1.68)	$\psi = \dfrac{\dot{Ex}_{deliv}}{\dot{Ex}_{cons}}$	%
Carnot factor	$\eta_C = 1 - \dfrac{T_0}{T_3}$	%
Exergy efficiency (ideal maximum)	$\psi = \dfrac{\dot{W}_{net}}{\dot{W}_{rev}} = \dfrac{\eta}{\eta_C}$	%
Back work ratio	$BWR = \dfrac{\dot{W}_p}{\dot{W}_t}$	%
Pressure ratio	$PR = \dfrac{P_2}{P_1}$	-
Expansion ratio	$ER = \dfrac{v_4}{v_3}$	-

ratio (BWR) represents the ratio of the work required to turn the pump to the work delivered by the turbine. As observed, in the Rankine cycle the BWR is extremely low; below 1%. This is very favorable for high efficiency of power generation. The work necessary to pressurize liquid (pump work) is negligible with respect to the work generated by expansion of gas (turbine work). In this respect, it is an advantage of the Rankine cycle to operate with a working fluid that changes the phase from subcooled liquid to superheated vapor. Other types of thermodynamic cycles that operate with gas have a BWR higher than 30%.

The pressure ratio (PR) quantifies the ratio between high and low pressure of the Rankine cycle (see Table 5.3). In the steam Rankine cycle the value of PR is quite high due to the fact that steam expands in vacuum. Hence, it is necessary to use a turbine with multiple expansion stages to make the actual expansion process more efficient. Moreover, the parameter ER (expansion ratio) compares the vapor specific volume at turbine exit to the volume at turbine inlet (see Table 5.3). In the Rankine cycle the volume of expanded flow is more than one hundred times larger than the steam volume at the turbine inlet. The expansion ratio is a crucial parameter for turbine design and also quantifies the measure in which the size of low-pressure steam pipes at the turbine outlet must be larger than the high-pressure steam pipes at the turbine inlet.

EXAMPLE 5.1

We consider in internally reversible Rankine cycle of basic configuration with condensation temperature equal to the reference temperature of $T_c = 25\,°C$ and boiling occurring at $T_b = 200\,°C$ while the superheating degree is $\Delta T_{sh} = 200\,°C$. The cycle will be analysed thermodynamically according to energy and exergy.

Given that this is an ideal Rankine cycle, it is implicit that all internal processes and the heat transfer process at the condenser are reversible. The thermodynamic states 1–4 of the ideal Rankine cycle can be easily determined from the known parameters T_c, T_b, and ΔT_{sh}.

In Table 5.4 the calculation steps for the parameters in each of the thermodynamic states of the cycle are briefly explained. Because the mass flow rate is the same in all state points of the cycle all balance equations can be solved in a mass-specific form; intensive properties are therefore used, such as specific enthalpies, entropies, and volumes.

For the ideal steam Rankine cycle it is assumed that in state 1 the liquid at pump intake is in a saturated state. Thus, based on condensation temperature and a vapor quality $x_1 = 0$ all other parameters such as specific volume, specific enthalpy, specific entropy, and specific exergy of the working fluid can be determined with the help of a selected equation of the state for steam. A highly accurate equation of the state for water and steam is presented in Wagner and Pruss (1993); tabular steam data can be also found in many thermodynamic textbooks, such as Cengel and Boles (2010) or Moran et al. (2011). The Engineering Equation Solver software has implemented advanced equations of state for thermophysical and thermochemical properties of many other substances; this software was used to calculate numerical data within this book (see Klein, 2012).

TABLE 5.4 Thermodynamic States of Ideal Rankine Cycle for Example 5.1

State	State description	Exemplary case study parameters for cycle represented in Figure 5.1						
		P (kPa)	T(K)	v (m^3/kg)	x	h (kJ/kg)	s(J/ kg.K)	ex (kJ/kg)
1	Saturated liquid at T_c	3.169	298.2	0.001003	0	104.8	0.367	0
2	Subcooled liquid at $s_2 = s_1$ and $P_2 = P_3$	1554	298.2	0.001002	−0.38	106.3	0.367	1.555
3a	Saturated liquid at T_b	1554	473.2	0.001156	0	852.4	2.331	162.1
3b	Saturated vapor at T_b	1554	473.2	0.1273	1	2793	6.431	879.7
3	Vapor superheated with ΔT_{sh}	1554	673.2	0.1958	1.238	3255	7.252	1097
4	Two-phase fluid at T_c and $s_4 = s_3$	3.169	298.2	36.45	0.841	2157	7.252	0

Note that the vapor quality is by definition a quantity ranging from 0 (liquid) to 1 (vapor). However, in Table 5.4 the value of x goes beyond this limit in two cases. It is given a negative value for x at state 2; the negative sign indicates a subcooled liquid when $h < h_l$ ($x_2 = -0.38$; the magnitude of x_2 is proportional with the degree of subcooling). As well, a value of x superior to 1 at state 3 indicates a superheated vapor when $h > h_v$ ($x_3 = 1.24$; the magnitude of x_3 is proportional with the degree of superheating).

Continued

EXAMPLE 5.1 *(cont'd)*

The cycle diagram in *T–s* coordinates is given in Figure 5.2. As seen, when represented at scale, state 2 confuses with state 1 and the preheating line is superimposed over the saturated liquid line; this behavior is common for steam Rankine cycles. The processes specific to the studied cycle are indicated in Table 5.5. The summary of the calculated cycle parameters are given in Table 5.6. It is remarked that exergy efficiency is 62.5%, whereas the energy efficiency is only 34.8%.

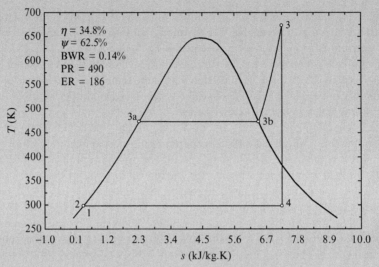

FIGURE 5.2 The *T–s* diagram of internally reversible Rankine cycle studied for Example 5.1.

TABLE 5.5 Processes of Ideal Rankine Cycle from Example 5.1

		Parameters for cycle case study in Example 5.1		
Process	Process description	q (kJ/kg)	w (kJ/kg)	Δex (kJ/kg)
1–2	Isentropic pumping $s_2 = s_1, P_2 = P_3, w_p = h_2 - h_1$	N/A	1.555	1.555
2–3a	Isobaric (pre)heating $x_{3a} = 0, T_{3a} = T_b, q_{ph} = h_{3a} - h_2$	746.1	N/A	160.6
3a–3b	Isobaric/isothermal boiling $x_{3b} = 1, q_b = h_{3b} - h_{3a}$	1940	N/A	717.6
3b–3	Isobaric superheating $P_3 = P_{3b}, q_{sh} = h_3 - h_{3b}$	462.3	N/A	217.7
3–4	Isentropic expansion $s_4 = s_3, P_4 = P_1, w_t = h_3 - h_4$	N/A	1097	1097
4–1	Isobaric/isothermal condensation $T_4 = T_1, q_c = h_4 - h_1$	2053	N/A	0.0

TABLE 5.6 Summary of Calculated Cycle Parameters for Example 5.1

Quantity	Value	Unit	Quantity	Value	Unit
Mass-specific heat input, q_{in}	3149	kJ/kg	Exergy efficiency, ψ	62.5	%
Net specific work output, w_{net}	1096	kJ/kg	Back work ratio, BWR	0.14	%
Energy efficiency, η	34.8	%	Pressure ratio, PR	490	–
Carnot factor, η_C	55.7	%	Expansion ratio, ER	186	–

5.2.2 Exergy Destructions in Rankine Power Plants

It is made clear from the above discussion and Example 5.1 that the Rankine cycle is externally irreversible. This is generally valid for any vapor power cycle which includes sensible heating processes such as liquid preheating and vapor superheating because these heat addition processes are non-isothermal. However, it is theoretically possible to imagine a totally reversible vapor power cycle if the heat addition is for a boiling process only.

In Figure 5.3 a totally reversible vapor power generator that operates based on a Carnot cycle (two isothermal and two adiabatic processes) is exemplified. At state 1 a two-phase water-steam mixture at 350 K having a vapor quality of ∼0.2 is enclosed in a cylinder and piston assembly. Due to presence of vapor, this mixture is compressible. The piston is moved upward slowly, performing an adiabatic–isentropic compression process (1–2) during which steam liquefies due to pressurization. In state 2 all vapors liquefies and only liquid is found in the cylinder. To perform the compression process a work input to the cycle is required. Further, during process 2–3 heat is added to the working fluid isothermally.

The piston moves freely in order to accommodate the larger volume of vapors formed during heat addition. No work is performed to the outside during process 2–3. In state 3 only saturated vapor is enclosed in the cylinder. There follows an isentropic expansion process (3–4) during which some moisture is formed until the liquid–vapor mixture reaches a vapor quality of 0.8 (as given in the particular example from Figure 5.3).

Work is generated during the piston stroke which compensates for the compression work, and thus a net work output is produced by the cycle. During the last process (4–1) heat is removed isothermally from the cycle and vapor condenses, thus the volume of the cylinder has to reduce. Although it appears theoretically realizable, this kind of engine has never been constructed because of some insurmountable technical obstacles. For example, during the isothermal heat addition the volume of the working fluid must increase about one thousand

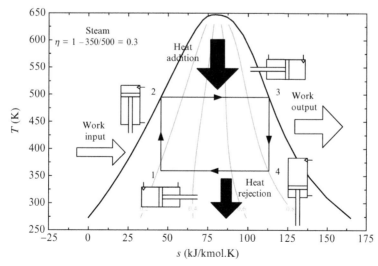

FIGURE 5.3 Totally reversible engine based on a two-phase cycle with steam. *Modified from Cengel and Boles (2010).*

times, which is unpractical. Also compressing and expanding in two phases must be performed extremely slowly in order to approach isentropic processes. In this section the irreversibilities and exergy destructions in actual vapor power plants are analyzed and methods for their quantification are given.

Each component of an actual power plant destroys exergy. The most important component of a Rankine power station is the steam turbine, which has the role of converting the energy carried by hot, high-pressure steam into useful shaft power which turns the electric generator. The other components of the plant—major and minor—have the sole role of working jointly to provide a continuous stream of steam at the turbine inlet. The major components of a Rankine power station besides the turbine are the steam generator, the pumps, and the condensers. Other important components are feedwater heaters and a preheater, superheater, reheater, and economizer. Minor components are many: separators, steam traps, drum, drains, valves, check valves, safety valves, etc. There are also components specific to various kinds of power stations—depending on the fuel source—coal, oil, natural gas, or nuclear.

More realistic analysis of the Rankine cycle is also necessary to determine the practically attainable efficiency of a given power plant. In this respect the exergy destruction due to water pumping, steam expansion, heat transfer, and fluid flow resistance must be accounted for. The common way in which irreversibilities are accounted for in power plant analysis is by the use of a number of parameters, as follows:

- Isentropic efficiency of turbines—that expresses the ratio of actual work to isentropic (i.e., reversible) work developed by the turbine
- Isentropic efficiency of pumps—that expresses the ratio of the isentropic (i.e., reversible) work to the actual work required by pumps
- Mechanical efficiency—which quantifies the difference between net amount of work developed by the working fluid and the mechanical work developed by the turbine shaft
- Electrical efficiency—which expresses the conversion efficiency of the mechanical power of the turbine shaft to the electrical power delivered to the consumer (e.g., electrical grid)
- Temperature difference at heat exchangers at source side—which quantifies the difference between the exergy extracted from the heat source and that delivered to the working fluid
- Heat losses at source-side heat exchangers and hot steam conduits—which induces exergy losses and a reduction of net exergy delivered by the heat source to the working fluid
- Temperature difference at condenser—which quantifies the exergy difference between that rejected by the working fluid and that received by the heat sink; there are two cases:
  - if the heat sink is a lake, then a minimal temperature difference of ~10 °C must be assumed between the condensation temperature and the sink temperature
  - if the heat sink is air and a cooling tower is used to reject heat then a temperature difference of a minimum 5 °C must be considered between air and cooling water and in addition a minimum temperature difference between cooling water and condensation has to be assumed; there must be at least a 15 °C temperature difference between condensation and the wet bulb temperature
- Subcooling degree—which is a technical requirement at pump inlet to prevent cavitation issues and malfunction; subcooling also leads to exergy losses with respect to ideal cycle

- Pressure drop in heat exchangers and conduits—which quantifies the loss of exergy due to the friction process between the fluid steam and walls (pipes, conduits, tubes, heat transfer surfaces) in which the fluid is confined or with which it comes in contact
- High-pressure steam leakage rate—which cannot be prevented because the turbine cannot be perfectly sealed and thus additional energy is required to supply additional fresh water as working fluid
- Air penetration rates at condenser—which cannot be stopped because the condenser operates in a vacuum; this increases the pressure in the condenser and reduces turbine work. Moreover, energy consumption is required to de-aerate the condenser and maintain a reasonable lower pressure
- Energy consumption for auxiliary equipment and power plant house—several pieces of auxiliary equipment, such as (depending on the case) conveyors, fuel injectors, fans, pumps, blowers, electric motors, and also lighting and requirements for the power plant house, will diminish the net power delivered to the consumer by a conventional power generation station

In Table 1.8 the balance equations and energy and exergy efficiency formulations for all important devices used in the power generation system are introduced. In this section the exergy analysis presented in Chapter 1 will be applied in order to exemplify how the irreversibilities in steam Rankine cycles can be assessed.

There are at least four additional processes that must be represented on an actual basic Rankine plant diagram that are not present on an ideal plant diagram: water subcooling at pump inlet, heat loss at hot steam conduits, high-pressure steam leak at the turbine, and fresh water addition at the condenser.

These processes are represented in Figure 5.4, which illustrates an actual steam Rankine cycle of basic configuration. The $\overline{T}$ notations in the figure represent the average system boundary temperature at locations where heat transfer processes occur. In subsequent paragraphs the thermodynamic analysis of the Rankine cycle with irreversibilities is presented and exergy destructions and entropy generations quantified. There are entropy transfer fluxes between surroundings and power plant which can be written mathematically as follows (see Figure 5.4):

$$\dot{S}_{ph} = +\frac{\dot{Q}_{ph}}{T_{ph}}; \; \dot{S}_b = +\frac{\dot{Q}_b}{T_b}; \; \dot{S}_{sh} = +\frac{\dot{Q}_{sh}}{T_{sh}}; \; \dot{S}_{loss} = -\frac{\dot{Q}_{loss}}{T_{loss}}; \; \dot{S}_c = -\frac{\dot{Q}_c}{T_c}; \; \dot{S}_{sc} = -\frac{\dot{Q}_{sc}}{T_{sc}}$$

where the sign indicates that entropy flux enters (+) or exits (−) the thermodynamic cycle.

The thermodynamic analysis of the cycle can start from thermodynamic state 1 where the working fluid is saturated. Three parameters must be specified in order to determine thermodynamic state 1; these are:

- The reference temperature T_0 (which can be the temperature of surrounding atmosphere if the condenser is cooled with air, or the lake temperature if the condenser is cooled with water; T_0 should be the lowest temperature of power plant surroundings)
- The assumed minimum temperature difference at the heat sink (condenser), ΔT_c
- The required subcooling degree ΔT_{sc} necessary for the pump to operate without cavitation.

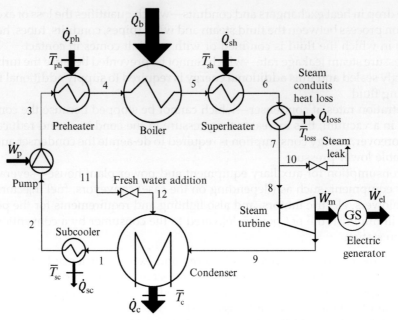

FIGURE 5.4   Representation of a Rankine power plant with multiple exergy destruction processes (finite difference heat transfer, heat losses, pressure drop, electromechanical losses, steam leak).

Thermodynamic state 1 is completely determined by specifying that the liquid is saturated ($x_1 = 0$) and that the temperature of working fluid is given by

$$T_1 = T_0 + \Delta T_{sc} + \Delta T_c$$

The degree of subcooling in practical applications is taken as $\Delta T_{sc} = 1 - 3\,^{\circ}C$, whereas $\Delta T_c = 10 - 15\,^{\circ}C$. The condensation pressure is the saturation pressure corresponding to T_1. There is negligible pressure drop in any practical subcooling process. However, some pressure drop must be accounted for across the condensation process. Hence, one can take $P_2 = P_1$. Further, the temperature of state 2 (subcooled liquid) is determined

$$T_2 = T_1 - \Delta T_{sc}$$

State 2 is completely specified because its pressure and temperature can be determined. Thermodynamic state 3 can be determined once the isentropic efficiency of the pump ($\eta_{s,p}$) and the pressure P_3 are specified. The isentropic efficiency allows for determination of specific flow enthalpy and is defined by the following equation (see also Chapter 1):

$$h_{(P_3,\, s_2)} - h_1 = \eta_{s,p}(h_2 - h_1)$$

The type of pump must be chosen in accordance with the head and the flow rate and economic criteria. For low head, centrifugal pumps of multistage configuration are generally preferred. For high head the preferred choice is of double-casing barrel pumps which are also tolerant to high water temperatures. The acceptable range of isentropic efficiency for pumps at power generating stations is 0.8–0.9.

The pressure in state 3 should be superior to the pressure is state 4 (boiling pressure) due to the existence of pressure drop in preheater. In general the preheater is considered as part of the vapor generator, and, depending on the case, it generally consists of an assembly of pipes heated from the exterior and working fluid circulating internally. In some typical cases (e.g., nuclear power stations) the preheater may be just a part of a shell and tube steam generator which has the hot heat transfer fluid circulating in tubes and water being preheated and boiled in the shell side. In typical power plants the pressure drop over preheaters leads to a decrease in saturation temperature of 0.1–0.3 °C. Denoting $\Delta T_{sat,ph}$ the decrease of saturation temperature due to pressure drop across preheaters, the pressure in state 3 results from the equation

$$P_3 = P_{sat}\left(T_b + \Delta T_{sat,\,ph}\right)$$

The pressure in state 4 corresponds to saturation at a specified boiling temperature that typically is in the range of $T_b = 100 - 330\,°C$, depending on the actual type of power generating station. In state 4 there is a saturated liquid, from which all state parameters can be determined based on temperature $T_4 = T_b$ and vapor quality $x_4 = 0$. There are many types of boilers, depending on the nature of the heat source. In a coal-fired power plant vertical tubular boilers are used, with boiling occurring in pipes heated by radiation and convection by combustion gases. In oil-fired plants the boilers may be of horizontal return tubular type, again with the boiling process occurring in tubes. Gas-heated power plants have boilers comprising a number of bare or extended-surface tubes forming a heat exchanger register and disposed in a horizontal configuration. In a nuclear power plant the steam generators a pool boiling process is the preferred one.

In any type of boiler there is an unavoidable pressure drop due to working fluid flow. The decrease in saturation temperature due to pressure drop over the boiler ($\Delta T_{sat,b}$) is on the order of 0.5–1.5 °C, depending on its actual type. Thus the thermodynamic state 5 at boiler exit is fully determined based on its temperature $T_5 = T_b - \Delta T_{sat,b}$ and the vapor quality $x_5 = 1$. The decrease in saturation temperature caused by pressure drop during the superheating process which occurs in superheater or reheater ($\Delta T_{sat,sh}$) is in a practical range of 0.3–0.7 °C. In fossil fuel power stations the superheaters are of tubular type (with or without extended surface), with steam circulated in tubes and heat provided from the exterior. In a nuclear power plant, the superheater may be the same unit as the steam generator. Using an assumed $\Delta T_{sat,sh}$, the thermodynamic state 6 (Figure 5.4) is determined based on pressure P_6 given by

$$P_6 = P_{sat}(T_5 - \Delta T_{sat,\,sh})$$

and on temperature $T_6 = T_{sh}$.

The pressure drop in steam conduits of power plants is in the range of $\Delta P_{drop} = 0.2 - 0.4\,MPa$. This pressure drop is due to the high complexity of the steam circuits, which include many types of fittings (screwed, flanged, compensators), headers and manifolds, valves, and check valves.

The steam conduits must be well insulated to minimize the heat losses. It is impossible to avoid heat losses at high-temperature steam conduits because there is generally a considerable distance between the steam generator and the turbine location. In addition, conventional power generating stations have massive steam conduits that must be supported and anchored at many points using special "heat transfer bridges" where heat losses are more accentuated. The temperature reduction of steam due to heat losses at hot steam conduits in power plants can be taken on the order of $\Delta T_{\text{loss}} = 10 - 30\,°\text{C}$. Hence the thermodynamic state 7 at the turbine inlet is determined based on its pressure and temperature.

The turbine's irreversibilities are accounted for by specifying the exergy efficiency ($\eta_{\text{s,T}}$). Hence, the specific enthalpy of steam at the turbine outlet results from:

$$h_7 - h_8 = \eta_{\text{s,T}}\left(h_7 - h_{(P_8, s_7)}\right)$$

The isentropic efficiency of the turbine has the highest influence on overall cycle efficiency, hence the correct choice of turbine is crucial for a power plant. After Parson's invention of the reaction steam turbine—in the last quarter of the nineteenth century—steam turbine technology developed fast in the twentieth century. In the first quarter of the nineteenth century there were many types of proprietary steam turbine designs, such as Parson, Westinghouse-Parson, De Laval, Terry, Kerr, Curtis, Hamilton-Holtzworth, and Allis Chalmers. Steam turbines are generally considered highly reliable equipment in conventional power generation stations. They can be classified as impulse and reaction turbines or as condensing, extraction, reheat, and back pressure (non-condensing) turbines. Reheat and condensing turbines are those which are normally used in conventional Rankine power stations. Modern turbines are conceived as a succession of stages of impulse and reaction blades, with impulse-type of stages at higher pressures and reaction-type stages at lower pressures. The isentropic efficiency of modern steam turbines for conventional power stations spans from 0.8 to 0.9.

The pressure at the turbine outlet (state 8) is practically the same as pressure at the condenser inlet. The number of stages of the turbine can be approximated based on the isentropic nozzle formula that correlates the pressure in the throat (P^*) with pressure at the nozzle inlet and the specific heat ratio of the steam, γ. For superheated steam $\gamma = 1.14$, while saturated steam is generally approximated with $\gamma = 1.3$. The general formula for throat pressure is

$$P^* = \left(\frac{2}{\gamma + 1}\right)^{\frac{\gamma}{\gamma - 1}} P_{\text{in}}$$

from which, using the values of γ, it results that

$$P^* = \begin{cases} 0.546\,P_{\text{in}}, \text{ for superheated steam} \\ 0.576\,P_{\text{in}}, \text{ for saturated steam} \end{cases} \tag{5.6}$$

where P_{in} stands for pressure at the turbine stage inlet.

Moisture does not actually form inside the steam turbine if the vapor quality is higher than 96% because of the character of the flow, which is a fast flow. Hence, the thermodynamic states in the turbine are metastable: there is not sufficient time for heat transfer or for steam condensation to occur. The thermodynamic locus connecting states with steam quality of 96% (or 4% moisture) is known as the Wilson line. If the thermodynamic state in a turbine stage is above the Wilson line, then the steam is dry. If the thermodynamic state is below Wilson line, in that particular stage steam condensation occurs.

The magnitude of pressure drop on the low-pressure steam line and condenser altogether is of 1–2 kPa, which leads to a saturation temperature drop of $\Delta T_{\text{sat, c}} = 2 - 4\,°\text{C}$. The steam condenser is a massive shell-and-tube heat exchanger with steam condensing at the shell side and with water circulating in pipes.

The mechanical efficiency of the power plant is a parameter that accounts for losses caused by turbine friction, steam leak, and any other mechanical losses at the turbine and is defined by

$$\eta_m = \frac{w_m}{w_{net}} = 0.88 - 0.94$$

In power generation stations the turbines are optimized to rotate synchronously with the power grid—at 3600 RPM in North America, Japan, and South Korea, where the electrical grid is at 60 Hz and at 3000 RPM and where the grid frequency is 50 Hz. The generator, which is connected in line with the turbine shaft is of synchronous type with a copper-coiled rotor and stator. Both the rotor and the stator generate an important amount of heat. For the stator, heat is removed by cooling from the exterior, but for the rotor it is difficult to reject the generated heat because of the small free space between stator and rotor. The current technology for rotor cooling uses hydrogen as a heat transfer fluid. Hence, an enclosure is built around rotor. The generator is linked to a transformer that raises the electric voltage up to over 100 kW in compatibility with the electrical grid. The electrical efficiency, representing the ratio between electrical energy delivered by the generator–transformer assembly and the mechanical energy transferred by the shaft to the generator is defined by

$$\eta_{el} = \frac{w_{el}}{w_m} = 0.9 - 0.96$$

There are many auxiliary devices in power plants that consume power, and hence diminish, even if in small measure, the net generation. Here are listed some relatively important auxiliary power-consuming processes and devices of conventional power generation stations:

- If condenser is cooled with water from a lake, there must be installed water circulation pumps and water purification systems which require electrical motors to drive pumps
- If condenser is cooled with hyperboloid cooling tower there must be installed water recirculation pumps and water-feeding pumps driven by electrical motors
- If condenser is cooled by mechanical-draft cooling tower, besides the use of electrically driven pumps there must be installed electrically driven fans for air circulation in forced convection
- In coal-fired power stations there are coal conveyers, ash conveyers, coal mills, and pulverizers that consume power
- In natural gas and oil-fired power station the burners must have blowers driven by electrical motors
- Chimneys when installed require electrically driven fans and blowers
- In fossil fuel plants rotary wheel economizers turned by electric motors are installed at exhaust gases
- Some boilers require additional feed pumps
- Turbine auxiliary parts, such as lubrication system and generator cooling system, require some power consumption
- The power station house requires power for lighting, appliances, alarms, and data acquisition, etc.

For a conventional power generating station, the power consumption for all auxiliary devices and processes, including that for the steam power station house, represents a small fraction of the generated electric power, usually 0.5–1%. Other losses such as those caused by steam leakage and air penetration are small and can be neglected with respect to other more important energy losses.

In Table 5.7 the main parameters which quantify the irreversibilities of steam power stations are summarized. The turbine's isentropic efficiency influences the cycle efficiency to a very large measure. A change of isentropic efficiency of 10% leads to a change of cycle energy efficiency of ~3–4% while exergy efficiency change is 8–9%. Temperature difference between the working fluid and the heat source affects the exergy efficiency. As an order of magnitude, a change of temperature of 50 °C may induce a reverse change of exergy efficiency of 1–2%. Heat losses at hot steam conduit inlets influence the cycle's energy efficiency, but in very small measure, ~0.2%. All other irreversibilities have even smaller influence (independently) on energy efficiency of the cycle.

TABLE 5.7   Parameters That Quantify the Irreversibility of the Rankine Cycle

Parameter	Name and definition[a]	Typical range
$\eta_{s,p}$	Isentropic efficiency of pump $h_{3s} - h_2 = \eta_{s,p}(h_3 - h_2)$	0.8–0.9
$\eta_{s,t}$	Isentropic efficiency of turbine $h_7 - h_8 = \eta_{s,t}(h_7 - h_{8s})$	0.8–0.9
η_m	Mechanical efficiency: $\eta_m w_{net} = w_m$	0.88–0.94
η_{el}	Electrical efficiency: $\eta_{el} w_m = w_{el}$	0.9–0.96
ΔT_c	Required temperature difference between condensation and heat sink	10–15 °C
ΔT_{sc}	Required subcooling degree at pumps inlet $\Delta T_{sc} = T_1 - T_2$	1–3 °C
$\Delta T_{sat, ph}$	Saturation temperature decrease due to pressure drop across preheater $\Delta T_{sat,ph} = T_{sat,3} - T_{sat,4}$	0.1–0.3 °C
$\Delta T_{sat, b}$	Saturation temperature decrease due to pressure drop across boiler $\Delta T_{sat,b} = T_4 - T_5$	0.5–1.5 °C
$\Delta T_{sat, sh}$	Saturation temperature decrease due to pressure drop across superheater/reheater $\Delta T_{sat,sh} = T_5 - T_{sat,6}$	0.3–0.7
ΔP_{drop}	Pressure drop in hot steam conduits before the turbine $\Delta P_{drop} = P_6 - P_7$	0.2–0.4 MPa
ΔT_{loss}	Temperature drop over hot steam conduits due to heat loses $\Delta T_{loss} = T_6 - T_7$	10–30 °C
$\Delta T_{sat, c}$	Saturation temperature decrease due to pressure drop at sink side $\Delta T_{sat,c} = T_8 - T_1$	2–4 °C

[a]See Figure 5.4 for notations.

Certain inequality-type relations must be satisfied between state parameters in order comply with the second law of thermodynamics. More specifically, the energy and entropy balance equations for the subcooler (1–2) require

$$h_1 = h_2 + q_{sc} \quad \text{and} \quad s_1 + s_{gen} = s_2 + \frac{q_{sc}}{T_{sc}}$$

As the generated entropy must be positive, from the above equations it results that the following inequality must be satisfied

$$\overline{T}_{sc} < \frac{h_1 - h_2}{s_1 - s_2}$$

which reflects the fact that the cooling medium of the subcooler must have an average temperature $\overline{T}_{sc}$ below a certain value; in addition it is known that $\overline{T}_{sc} > T_0$.

The energy and entropy balance on the preheater are

$$h_3 + q_{ph} = h_4 \quad \text{and} \quad s_3 + \frac{q_{ph}}{T_{ph}} + s_{gen} = s_4$$

From the above equation it results that the following inequality must be satisfied

$$\overline{T}_{sh} > \frac{h_4 - h_3}{s_4 - s_3}$$

which states that the average temperature of the heat source temperature at the preheater must be higher than a threshold value that insures that the generated entropy is positive.

Similarly inequalities are derived from the energy and entropy balances on the boiler and superheater as follows:

$$\overline{T}_b > \frac{h_5 - h_4}{s_5 - s_4} \quad \text{and} \quad \overline{T}_{sc} > \frac{h_6 - h_5}{s_6 - s_5}$$

Furthermore, because the preheater, boiler, and superheater are all heated from the same heat source, it results that the following inequalities must be satisfied

$$\overline{T}_{sh} > \overline{T}_b > \overline{T}_{ph}$$

The irreversibilities due to heat losses and pressure drop which occurs at conduits that transport hot, high-pressure steam to the turbine are accounted for by specifying the parameters ΔP_{drop} and ΔT_{loss} defined in Table 5.7. The energy and entropy balance equations for process 6–7 (see Figure 5.4), which models these types of losses, require that

$$\overline{T}_{loss} < \frac{h_6 - h_7}{s_6 - s_7}$$

The energy and entropy balances for the condenser can be combined to eventually derive the following inequality that must be satisfied

$$\overline{T}_{sc} < \overline{T}_c < \frac{h_8 - h_1}{s_8 - s_1}$$

As a conclusion of this section it is clearly noted that the effect of irreversibilities is very important for the steam Rankine cycle and must be quantified accurately in order to predict the actual plant performance. Furthermore, exergy efficiency reveals that the organ which destroys most of the steam exergy is the turbine. Hence, the selection of turbine and its design improvement are crucial issues when designing steam power plants.

The increase of cycle efficiency is very important in practice because it leads to important savings of thermal energy supply (or fuels) and reduction of environmental impact. The engineering goal is to "impinge" the design toward the limits imposed by the second law of thermodynamics. In other words, the design engineer must find ways to reduce the cycle irreversibilities.

In many engineering thermodynamics textbooks some methods to increase the efficiency of the ideal Rankine cycle are discussed. Basic thermodynamic analysis shows that increase of source temperature and/or decrease of sink temperature leads to higher cycle efficiency of any thermodynamic cycle. This is confirmed by a simple analysis of the Carnot factor $1 - T_0/T$ with a decreasing T_0 and increasing T. Let us assume that T_0 is replaced by $T_0 - \Delta T$ and T is replaced with $T + \Delta T$. Figure 5.5 reports Carnot factor η when ΔT is varied from 0 to 50 K. For the reference case ($\Delta T = 0$) it is taken that $T_0 = 25\,°C$ and $T = 200\,°C$, in which case $\eta_{ref} = 0.557$. Three types of Carnot factors are reported, corresponding to three cases:

- Sink temperature (condensation) is decreased with ΔT, hence $\eta = 1 - (T_0 - \Delta T)/T$.
- Source temperature is increased with ΔT, hence $\eta = 1 - T_0/(T + \Delta T)$.
- Both source and sink temperatures are changed, hence $\eta = 1 - (T_0 - \Delta T)/(T + \Delta T)$.

The results show that it is relatively more important to reduce the sink temperature than to increase the source temperature. If sink temperature decreases by 50 K then efficiency increases by ∼8%, from 55.7–63.1%. If source temperature is increased with 50 K then the

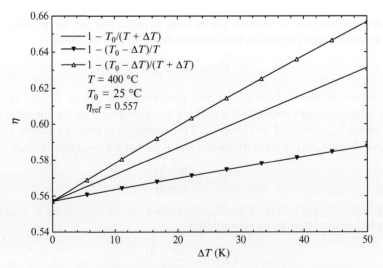

FIGURE 5.5 Improvement of Carnot factor by changing source or/and sink temperatures.

Carnot factor will increase only by 3%. If both temperatures are changed with 50 K then the Carnot factor increases by 10%.

The same trend is observed in the ideal Rankine cycle. In the steam Rankine cycle the condensation temperature is limited by the freezing point of water when working fluid freezes; however, for practical reasons one cannot run water as a working fluid at temperatures lower than $4-5\,^\circ C$ (note that at the bottom of deep lakes and with seawater there is a steady temperature of $\sim 4\,^\circ C$. A simple parametric study is reported in Figure 5.6 which shows the variation of energy efficiency of an ideal steam Rankine cycle when the condensation temperature is decreased by $\Delta T=0 \ldots 50\,K$ with respect to a reference value of $55\,^\circ C$, and when boiling temperature is increased with same ΔT with respect to a reference value of $175\,^\circ C$, and when the temperature of superheated vapor increases from the reference value of $400\,^\circ C$ with by the amount ΔT; only one parameter is changed at a time while all other cycle parameters remain unchanged.

As seen, the reference cycle efficiency is 27%. If condensation temperature is decreased from 55 to $5\,^\circ C$ the cycle efficiency increases to 36.2% (or a 9.2% increase). If boiling temperature increases from 175 to $225\,^\circ C$ the cycle efficiency reaches 31.9% (that is, a 4.9% increase). Note that in the last case the pressure of the boiler increases from 890 to 2550 kPa, while the highest temperature of the cycle is assumed to be constant, $T_5=400\,^\circ C$. The increase of boiler pressure due to increase of boiling temperature is illustrated in Figure 5.6. If the boiling temperature is $225\,^\circ C$ and condensation is at $5\,^\circ C$ then the cycle efficiency becomes 40% (or a 13% increase with respect to the reference case). The last curve plotted on the graph in Figure 5.6 shows the variation of energy efficiency when the temperature of superheated vapor, T_5, increases from 400 to $450\,^\circ C$. In this case the cycle efficiency increases relatively less, from 27% to 27.9%.

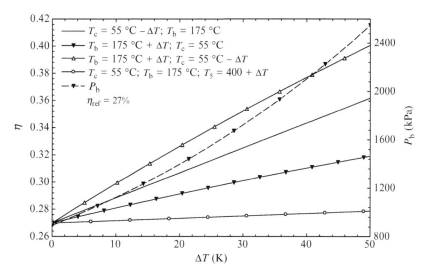

FIGURE 5.6 Influence of condensation and boiling temperatures and of the degree of superheating on ideal steam Rankine cycle efficiency.

This simple study suggests a strategy of efficiency improvement for the steam Rankine cycle: first one must set the condensation temperature at the lowest value possible, then one needs to increase the boiling temperature (and pressure) to the highest technical-economic value that is advantageous, and, as a last measure, one must set the temperature of superheated vapor as high as possible based on the practical temperature level that can obtained from the specific heat source considered in the application. Note also that the boiler pressure is limited by technical–economical considerations related to materials selection. For example, in conventional nuclear power plant the boiling pressure is ~5 MPa and the superheated vapor temperature is limited to ~350 °C for safety reasons. Furthermore, in a conventional Rankine cycle—which includes a boiling process—the highest pressure should be lower than the critical pressure of steam, which is 22.06 MPa. Newly emerging non-conventional Rankine cycles can operate at supercritical pressure and hence have better efficiency than the conventional cycles.

EXAMPLE 5.2

We give an example of the thermodynamic analysis of a Rankine cycle with exergy destructions. The cycle configuration and state points are as shown in Figure 5.4. The assumed parameters for the cycle are given in Table 5.8. It is required to determine the entropy generated and exergy destroyed for each process, the exergy destroyed by the cycle, the exergy wasted, the cycle efficiencies, the BWR and expansion ratio, and the number of turbine stages.

The mass balance equations require that the mass flow rate is the same at each state point, $\dot{m}_i = \dot{m}$. The energy balance equations can be written for each process as indicated in Example 5.1. From energy balances the temperature, pressure, specific volume, specific entropy, and specific enthalpy at each state point can be calculated. From these the specific exergy at each state point can be determined. It is reasonable to assume that the reference state is liquid water at $T_0 = 25\,°C$, $P_0 = 101.325\,kPa$. Entropy and exergy balance equations give the entropy generated and exergy destroyed for each component, respectively. In Table 5.8 the relevant equations are given together with the results.

It is remarked that the maximum exergy destruction is at the turbine at 41.2%, which is assumed to operate with an isentropic efficiency of 85% per stages (see Table 5.8). The number of stages of the turbine is determined based on Equation (5.6). As seen in the cycle representation in Figure 5.7 the expansion process in the turbine proceeds in seven stages. The first four stages operate above the Wilson line, hence condensation does not occur and impulse blades can be used. The subsequent three stages operate under condensation, when the flow volume increases more prominently across the stage. For this reason the turbine stages must be of reaction type (with expansion both in stator and rotor). The exergy destruction due to internal irreversibilities is denoted with ex_d in the table. The waste exery represents also a form of exergy destruction, namely, exergy destruction due to system interaction with the surroundings, $ex_{waste} = ex_{d,surr}$.

Furthermore, the energy efficiency of the non-ideal cycle is as low as 30.1% and the exergy efficiency with respect to the maximum reversible work output (Carnot factor of the heat source) is $\psi = \eta/\eta_C = 55.6\%$. The exergy efficiency relative to the actual exergy input into the cycle is $\psi = w_{net}/ex_{in} = 68\%$. If the cycle will operate reversibly under the same conditions, then the energy efficiency is 38.5% and exergy efficiency is 87.2%.

EXAMPLE 5.2 (cont'd)
TABLE 5.8 Performance Parameters of a Non-Ideal Rankine Cycle of Basic Configuration

Assumed parameters		Cycle performance parameters			
Parameter	**Value**	**Irreversibilities**			
T_0	25 °C	Component	s_{gen} (J/kg.K)	ex_d (kJ/kg)	(%)
T_b	250 °C	Subcooler (1–2)	0.193	0.06	0.0
T_{sh}	400 °C	Pump (2–3)	2.311	0.689	0.2
$\eta_{s,T}$	0.85	Preheater (3–4)	222.8	66.4	17.3
$\eta_{s,P}$	0.85	Boiler (4–5)	276.0	82.3	21.5
ΔT_c	7 °C	Superheater (5–6)	71.0	21.2	5.5
ΔT_{sc}	2 °C	Heat losses[a] (6–7)	70.0	20.6	5.4
$\Delta T_{sat,ph}$	0.2 °C	Turbine (7–8)	529.7	157.9	41.2
$\Delta T_{sat,b}$	1 °C	Condenser (8–1)	113.9	33.9	8.9
$\Delta T_{sat,sh}$	0.7 °C	Specific energy and exergy flows (kJ/kg)			
ΔP_{drop}	0.3 MPa	$q_{in} = q_{ph} + q_b + q_{sh}$			3077
ΔT_{loss}	20 °C	$w_{net} = w_t - w_p$			926.3
$\Delta T_{sat,c}$	3 °C	$q_{waste} = q_{in} - w_{net}$			2151
T_{sc}	304 K	$ex_{in} = q_{ph}\left(1 - \dfrac{T_0}{T_{ph}}\right) + q_b\left(1 - \dfrac{T_0}{T_{source,b}}\right) + q_{sh}\left(1 - \dfrac{T_0}{T_{sh}}\right)$			1362
T_{ph}	500 K	$ex_d = \sum\limits_i ex_{d,i}$			383.1
$T_{source.b}$	550 K	$ex_{waste} = ex_{d,surr} = ex_{in} - w_{net} - ex_d$			52.8
T_{sh}	600 K	Cycle efficiencies, BWR and ER			
T_{loss}	338 K	$\eta = w_{net}/q_{in} = 30.1\%$ $\eta_C = 1 - T_0/T_{sh}^- = 54.1\%$			BWR = 0.5%
$T_{sink,c}$	305K	$\psi = \eta/\eta_C = 55.6\%$ $\psi = w_{net}/ex_{in} = 68\%$			ER = 267

[a]Heat losses from hot steam conduits (losses due to pressure drop are included).

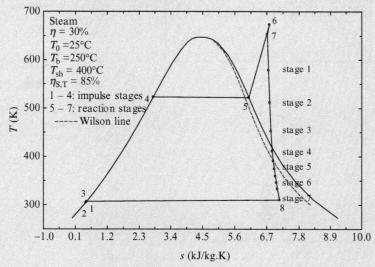

FIGURE 5.7 Non-ideal steam Rankine cycle of basic configuration (Example 5.2).

5.2.3 Ideal Reheat Rankine Cycle

Reheating is a method of improving Rankine cycle efficiency which consists of inter-stage heating of the expanding steam. After the first stage of expansion which typically reduces initial steam pressure by one-fourth, the steam is heated up (or close) to the maximum heat source temperature. After the second expansion stage the steam reaches condensation pressure. Although more heat input is required for reheating, the efficiency of a reheating cycle is higher because the two-stage turbine develops relatively more work.

Double-reheating cycles do exist and have been used in conventional power generation plants since the 1950s. With respect to the basic Rankine cycle, the efficiency improvement of using a single reheat is on the order of 5% and using double reheat it is 8%. Hence, using more than two reheat stages does not appear economically justifiable for conventional power generation systems (Cengel and Boles, 2010). The use of a reheating cycle is also justified because in practical applications the maximum temperature of the cycle is constrained for one of two reasons: (i) the nature of the heat source limits the temperature at which heat is available (e.g., geothermal heat, waste heat recovered, concentrated solar receiver with associated temperature of 200–800 °C) or (ii) the safety and material resistance issues (which limits boiler temperature to a maximum of 800 °C for conventional fossil-fuel-based power generation plants and 350 °C for conventional nuclear power plants).

In Figure 5.8 the diagram of a single-reheat Rankine power plant is presented. As seen, after the high-pressure turbine (HPT) the expanded steam is reheated in a special heat exchanger and then further expanded in the low-pressure turbine (LPT). The thermodynamic cycle for the ideal single-reheat Rankine cycle in a T–s diagram is presented in Figure 5.9.

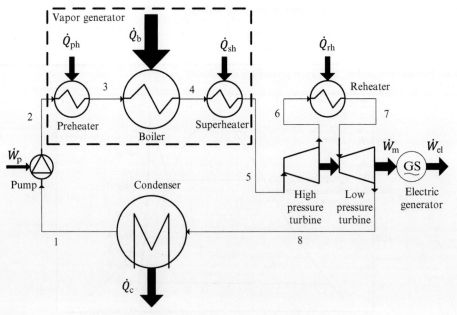

FIGURE 5.8   Single-reheating Rankine power plant.

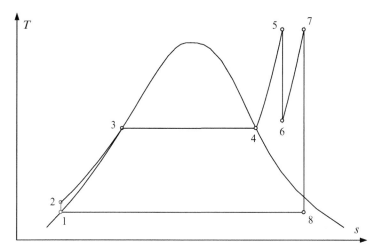

FIGURE 5.9   Ideal single-reheat Rankine cycle.

One of the features which is important for reheat cycle design is that there is an optimum pressure for the reheating process at which the efficiency of the cycle is maximized for any fixed boiling temperature. This optimum can be observed if the cycle efficiency is plotted against the pressure ratio across the HPT. Figure 5.10 illustrates the existence of the optimum pressure ratio for a relevant case study. Assume that the working fluid is steam and that the highest temperature of the working fluid is fixed to $T_5 = T_7 = 400\,°C$. The variation of cycle efficiency with pressure ratio $PR = P_5/P_6$ is shown for five values of boiling temperature (see Figure 5.10). The optimum pressure ratio is in the range of 3 to 4.6. The vapor quality

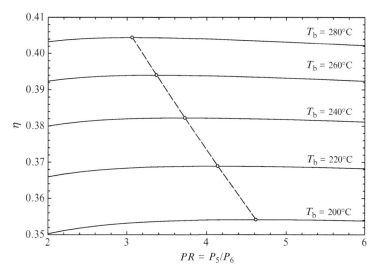

FIGURE 5.10   Variation of energy efficiency of single-reheat cycle with pressure ratio ($PR = P_5/P_6$) and boiling temperature (T_b) for fixed superheated and condensation temperatures ($T_{sh} = 400\,°C$, $T_c = 30\,°C$).

at the turbine exit decreases for optimum expansion from $x_8 = 0.94$ when $T_b = 200°C$ to $x_8 = 0.83$ and boiling temperature is $280°C$.

EXAMPLE 5.3

This example considers a single-reheat ideal Rankine cycle with a configuration similar to that indicated in Figure 5.7. The condensation temperature is $T_c = 30°C$, boiling is at $280°C$, and steam is superheated up to $400°C$. Determine the optimum pressure ratio for the HPT, the maximum cycle efficiency, and the mass-specific work for the HPT and LPT and show the T–s diagram of the cycle.

The mass and energy balances for each component of the cycle are written as indicated in Table 5.1. The mass flow rate is the same for all state points, hence the mass-specific quantities can be used. The pressure ratio over the HPT is varied in the range of 3 to 4 and a maximum efficiency of the cycle in the given conditions is determined to be 40.44% for $PR_{opt} = 3.066$. For this case the mass-specific work developed by the HPT is 280.1 kJ/kg, and the LPT develops 1099 kJ/kg. The cycle diagram is shown in Figure 5.11.

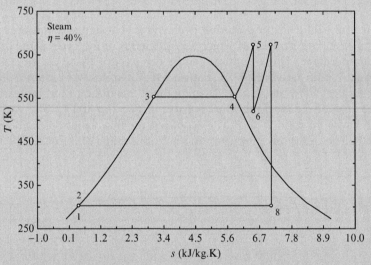

FIGURE 5.11   Ideal single-reheat Rankine cycle for $T_c = 30°C$, $T_b = 280°C$, $T_{sh} = 400°C$.

5.2.4 Ideal Regenerative Rankine Cycle

At the beginning of the twentieth century, the regenerative Rankine cycle started to be used for power generation with improved efficiency. This cycle is a modification of the basic Rankine cycle. It consists of applying internal regeneration within the cycle aimed at preheating the working fluid after the liquid pressurization process. Regeneration increases the average boiler temperature and makes the ideal Rankine cycle approach better than the Carnot cycle.

Let us compare the ideal Rankine cycle represented in Figure 5.1 with the Carnot cycle. As observed, the process path 3–4–1–2 resembles a part of the Carnot cycle. Also the boiling process 3a–3b is an isothermal heat input such as that in the Carnot cycle. However, compared to the Carnot processes, processes 2–3a and 3b–3 show large abatement because these are non-isothermal heat additions. In particular, process 2–3a has a negative effect on cycle efficiency because it dramatically reduces the temperature at which heat input occurs. The idea of regeneration consists of transferring heat internally from expansion process 3–4 to preheating process 2–3a such that the need for heat input at a lower temperature (below boiling) is reduced or eliminated.

The schematic of the ideal regenerative Rankine cycle is presented in Figure 5.12a. Imagine that the turbine is "surrounded" by a stream of working fluid in liquid phase which receives heat by heat transfer from the expanding flow to increase its temperature according to process 2–3. The rate of heat is indicated in the figure with $\dot{Q}_{ph}$. Because there is a heat transfer process superimposed over the isentropic expansion process in the ideal regenerative Rankine cycle the turbine will operate non-adiabatically (and isentropically, that is with no irreversibilities due to expansion) according to a process represented by the line 5–7 in Figure 5.12b.

The energy balance equation for the regeneration process (see Figure 5.12) can be written with mass-specific quantities as shown below:

$$q_{ph} + w_t = h_5 - h_7 \quad \text{and} \quad q_{ph} = h_3 - h_2$$

One also imposes that the ideal regenerative Rankine cycle develops the same turbine work as the internally reversible Rankine cycle of basic configuration operating under the same condition ($w_t = h_5 - h_6$); hence one has:

$$h_3 - h_2 + h_5 - h_6 = h_5 - h_7 \rightarrow h_6 - h_7 = h_3 - h_2$$

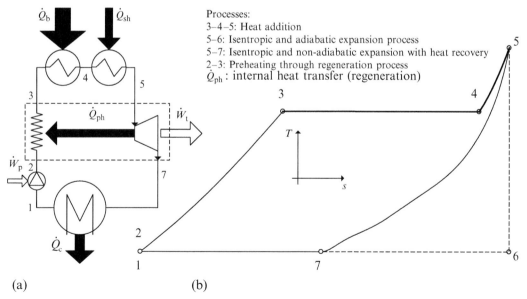

Processes:
3–4–5: Heat addition
5–6: Isentropic and adiabatic expansion process
5–7: Isentropic and non-adiabatic expansion with heat recovery
2–3: Preheating through regeneration process
$\dot{Q}_{ph}$: internal heat transfer (regeneration)

(a)　　　　(b)

FIGURE 5.12 Ideal regenerative Rankine cycle with non-adiabatic turbine—(a) power plant schematics, (b) T–s diagram (represented at scale for $T_b = 250\,°C$, $T_c = 30\,°C$, $\Delta T_{sh} = 150\,°C$).

The efficiency of the basic, internally reversible Rankine cycle operating under the same conditions as the ideal regenerative cycle (T_c, T_b, T_5 are the same) is expressed as follows:

$$\eta = \frac{w_t - w_p}{q_{in}} = \frac{h_5 - h_6 - (h_2 - h_1)}{h_5 - h_2} = 1 - \frac{h_6 - h_1}{h_5 - h_2}$$

The efficiency of the ideal regenerative Rankine cycle can be expressed as follows:

$$\eta_R = \frac{w_t - w_p}{q_{in}} = \frac{h_5 - h_7 - (h_3 - h_2) - (h_2 - h_1)}{h_5 - h_3} = 1 - \frac{h_7 - h_1}{h_5 - h_3}$$

Comparing η and η_R expressed with the help of above three equations, one has that

$$\eta_R = 1 - \frac{h_6 - h_1}{h_5 - h_3} + \frac{h_3 - h_2}{h_5 - h_3} > 1 - \frac{h_6 - h_1}{h_5 - h_2} + \frac{h_3 - h_2}{h_5 - h_3} > \eta$$

where one notices that $h_5 - h_2 > h_5 - h_3 > 0$ (see Figure 5.12). Thus, it is demonstrated that the efficiency of the ideal regenerative Rankine cycle is necessarily higher than that of the ideal Rankine cycle of basic configuration operating between the same temperature limits. This increase in efficiency can be as much as 18–20% when the working fluid is steam and the boiling temperature in the range of 200–300 °C.

According to Figure 5.12a, the regenerative Rankine cycle cannot be implemented in practice because it is not convenient to extract heat from an expanding flow; the turbine construction would be too complicated and not feasible. In addition, due to heat extraction, the moisture at the turbine outlet is far too high (it can reach more than 60%).

What can be done to overcome these construction obstacles is to extract some steam from the turbine at intermediate pressures and condense it to further extract heat for water preheating. This process will diminish turbine work but it will increase the cycle efficiency because the average temperature at source side becomes higher. There are several known ways to transfer heat from the condensing steam extracted from the turbine to the water preheating process. The simplest system is the use of a closed feedwater heater that represents, in fact, a condenser.

In Figure 5.13 is presented a regenerative Rankine cycle with a closed feedwater heater. After it condenses in the regenerator (process 7–8), the working fluid is throttled to lower pressure using a device called a "steam trap" (process 8–9). A fraction f of vapor is extracted from the turbine at intermediate pressure (in state 7) and directed toward the regenerator. The remaining vapor (that is the fraction $1 - f$) is expanded up to condenser pressure (state 10). In a first section of the condenser (10–11) the steam reduces the vapor quality to an intermediate value that corresponds to state 11. The vapor quality of the throttled working fluid in 9 is the same as that in state 11, hence $x_9 = x_{11} = x_{12}$. The condensation process continues with full flow until saturated liquid is obtained in 1. The relevant mass balance equations and the extracted vapor fraction ($f \in [0,1]$) for the system are:

$$\dot{m}_6 = \dot{m}_7 + \dot{m}_{10}; \quad f\dot{m}_6 = \dot{m}_7; \quad \dot{m}_9 + \dot{m}_{11} = \dot{m}_{12} \tag{5.7}$$

The energy balance of the regenerator also imposes that

$$(1 - f)(h_7 - h_8) = h_3 - h_2 \tag{5.8}$$

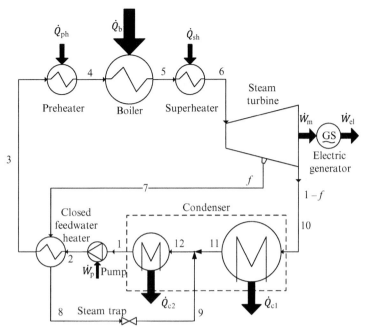

FIGURE 5.13 Regenerative Rankine power plant with closed feedwater heater and steam trap.

The exergy balance of the regenerator will determine the irreversibility due to heat transfer, owing to the fact that $T_7 = T_8 > T_3 > T_2$. The exergy balance is as follows:

$$ex_2 + f\,ex_7 = ex_3 + f\,ex_8 + ex_{d,\ reg} \qquad (5.9)$$

where ex_d is the exergy destroyed by the regenerator. In addition, there is exergy destruction due to throttling process 8–9. The exergy balance for the steam trap can be written as follows to determine the exergy destruction:

$$f\,ex_8 = f\,ex_9 + ex_{d,\ trap} \qquad (5.10)$$

The energy balance on the condenser requires

$$f\,h_9 + (1 - f)h_{10} = h_1 + q_c \qquad (5.11)$$

where q_c represents the mass-specific heat flux rejected by the condenser (in kJ/kg). If $\dot{m}$ is the mass flow rate of working fluid in state 1, then, with reference to Figure 5.13, it results that the heat rate rejected by the condenser is

$$\dot{Q}_c = \dot{m}q_c = \dot{Q}_{c1} + \dot{Q}_{c2} = \dot{m}\left(q_{c1} + q_{c2}\right)$$

where $q_{c1} = (1 - f)(h_{10} - h_{11})$ and $q_{c2} = h_{12} - h_1$, and one also notes that $h_9 = h_{11} = h_{12}$.

The vapor quality of the extracted steam in regenerative Rankine schemes is slightly below unity. The efficiency of the regenerative cycle is slightly influenced by the quality of extracted steam. For higher quality, the energy efficiency tends to be better, whereas the exergy efficiency

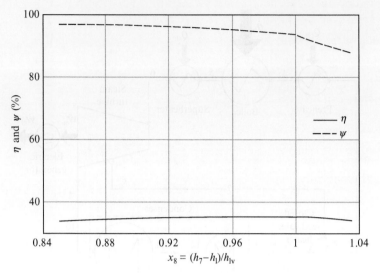

FIGURE 5.14 Energy and exergy efficiency of regenerative steam Rankine cycle as a function of vapor quality of extracted steam [assumptions: cycle configuration as in Figure 5.13; $T_c = 30\,°C, T_b = 200\,°C, T_6 = 400\,°C, T_7 - T_3 = 10\,°C$].

is not influenced. This evolution can be observed in Figure 5.14, which indicates that the best extraction is obtained when vapor quality is around 0.9. It also results that it is detrimental to extract superheated steam because both the energy and exergy efficiencies decrease sharply.

The exergy destructions within the cycle can be reduced and the energy efficiency can be improved if the steam trap is replaced with a pump followed by a mixer. Such a scheme is denoted as a regenerative Rankine cycle with a closed feedwater heater and pump and is represented in Figure 5.15. The system has the advantage of reducing the size of the condenser, and it also reduces the pumping power. However, there are still exergy destructions related to the regenerator and the mixer (the mixer is in fact a direct contact heat exchanger, which, in this case, mixes streams that have about the same temperature). The energy balance on the regenerator becomes in this case:

$$(1-f)h_2 + fh_8 = (1-f)h_3 + fh_9 \tag{5.12}$$

whereas the exergy balance becomes

$$(1-f)ex_2 + fex_8 = (1-f)ex_3 + fex_9 + ex_{d,\,reg} \tag{5.13}$$

The energy and exergy balances for the mixer are written as

$$(1-f)h_3 + fh_{10} = h_4 \quad \text{and} \quad (1-f)ex_3 + fex_{10} = ex_4 + ex_{d,mix} \tag{5.14}$$

For the ideal cycle exemplified in Figure 5.16 the exergy destruction at the regenerator is 22.86 kJ/kg while the exergy destruction at the mixer is negligible (0.046 kJ/kg). The extracted steam fraction is 11.3% and the energy efficiency of the cycle is slightly improved with respect to the energy efficiency of the one with the steam trap analyzed previously: 35.4% versus 35.1%. This cycle is also illustrated in the P–h diagram in Figure 5.17 where the non-isobaric processes are made apparent.

FIGURE 5.15 Regenerative Rankine power plant with closed feedwater heater, high-pressure pump, and flow mixer.

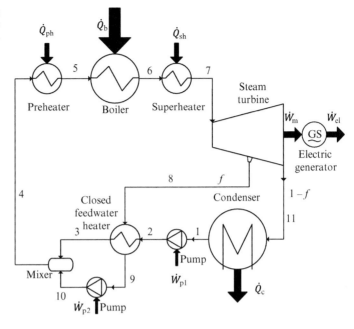

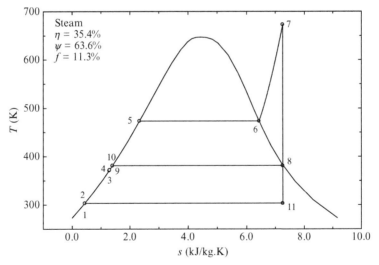

FIGURE 5.16 Ideal regenerative Rankine cycle with closed feedwater heater of configuration in Figure 5.15 for $T_c = 30\,°C$, $T_b = 200\,°C$, $\Delta T_{sh} = 200\,°C$, $\Delta T_{reg} = 10\,°C$.

The exergy destruction in the regenerator–mixer system can be further reduced if the closed feedwater heater is replaced with an open feedwater heater (OFW). This is in fact a direct contact heat exchanger. An improved Rankine plant design of regenerative type is introduced in Figure 5.18. The exergy destruction in the OFW must be smaller with respect to the closed feedwater heater because the temperature of extracted steam can be equal to the

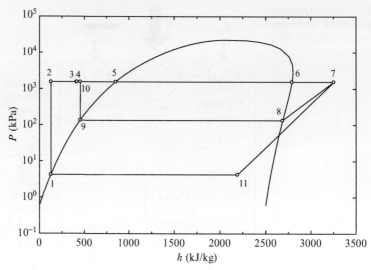

FIGURE 5.17   Ideal regenerative Rankine cycle with closed feedwater heater of configuration in Figure 5.15 for $T_c = 30\,°C$, $T_b = 200\,°C$, $\Delta T_{sh} = 200\,°C$, $\Delta T_{reg} = 10\,°C$ represented in a $T–h$ diagram (non-isobaric processes are made apparent).

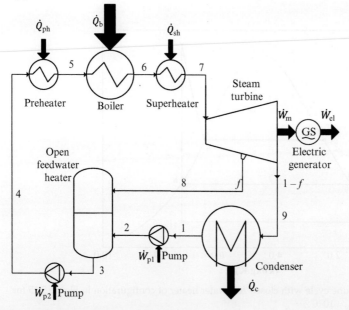

FIGURE 5.18   Regenerative Rankine power plant with open feedwater heater.

temperature of the condensate. In addition, the OFW has a simpler and cheaper construction than the closed one. An example of an ideal cycle with this configuration is illustrated in Figure 5.19.

A combination of open and closed feedwater heaters with multiple steam extraction pressures can be used to increase cycle efficiency even more. This approach is used in practice as a technical and economical trade-off for steam Rankine power stations. The OFW is cheaper

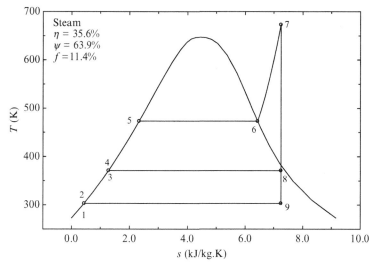

FIGURE 5.19   Ideal steam Rankine cycle with open feedwater heater for $T_c = 30\,°C$, $T_b = 200\,°C$, $\Delta T_{sh} = 200\,°C$.

and more efficient than the closed feedwater heater, but requires additional pumps, which involve more costs. On the contrary, the closed feedwater heaters are more costly and less efficient, but these heaters do not necessitate additional pumps. Moreover, multiple feedwater heaters of closed type can be cascaded using steam traps and a single pump, although this solution destroys more exergy.

EXAMPLE 5.4

Let us consider a regenerative steam Rankine cycle of the configuration illustrated in Figure 5.13. Assume that $T_c = 30\,°C$, $T_b = 200\,°C$, $\Delta T_{sh} = 200\,°C$ and that all internal processes are reversible except the feedwater heater, which operates with a temperature difference for heat transfer defined by $\Delta T_{reg} = T_7 - T_3 = 10\,°C$ (see Figure 5.13 for the state point numbers). We want to determine the extracted steam ratio and the energy and exergy efficiencies and to represent the cycle in the T–s diagram.

In order to solve this problem mass, energy, entropy, and exergy balance equations must be written for each system component; hence for the nine components of the system, there will be in total 27 equations. Three equations come from the fact that the pump and the turbines operate isentropically, which means that $s_2 = s_1$; $s_6 = s_7 = s_{10}$. In addition, the extracted steam fraction is defined by Equation (5.8), and because all internal processes are reversible the mechanical and electrical efficiency are $\eta_m = \eta_{el} = 1$. In total one has 33 equations that form a closed system that can be solved to determine thermodynamic parameters at all of the state points.

A simple code can be written to solve the algebraic system numerically. Here we give the solution obtained with EES as shown in Table 5.9, where the thermodynamic parameters for each of the state points is given. For the specific exergy calculation a reference state must be assumed. As in previous examples, the reference state for the steam Rankine cycle is taken to be liquid water at the standard T_0, P_0. Once the thermodynamic states are determined, the extracted steam fraction results directly from

Continued

EXAMPLE 5.4   (cont'd)

TABLE 5.9   State Point Parameters Calculated for Example 5.4

State	P (kPa)	T (K)	v (m³/kg)	x	h (kJ/kg)	s (kJ/kgK)	ex (kJ/kg)
1	4.246	303.2	0.001004	0	125.7	0.4365	0.1745
2	1554	303.2	0.001004	-0.3738	127.2	0.4365	1.73
3	1554	371.2	0.001041	-0.2271	411.8	1.283	33.84
4	1554	473.2	0.001156	0	852.4	2.331	162.1
5	1554	473.2	0.1273	1	2793	6.431	879.7
6	1554	673.2	0.1958	1.238	3255	7.252	1097
7	134	381.2	1.287	0.9984	2685	7.252	527.3
8	134	381.2	0.00105	0	453	1.397	41.2
9	4.246	303.2	4.432	0.1347	453	1.516	5.573
10	4.246	303.2	27.97	0.8503	2192	7.252	34.25
11	4.246	303.2	4.432	0.1347	453	1.516	5.573
12	4.246	303.2	4.432	0.1347	453	1.516	5.573

Equation (5.8), which gives the solution $f = 12.7\%$. The energy efficiency is 35.1%. To calculate the exergy efficiency, one assumes that the heat source is at a constant temperature level equal to the temperature of the thermodynamic state at the turbine inlet, $T_{so} = T_6 = T_b + \Delta T_{sh} = 673\,\text{K}$. Therefore the Carnot factor is 0.5571 and the exergy efficiency is 63.06%. The cycle diagram is presented at scale in Figure 5.20.

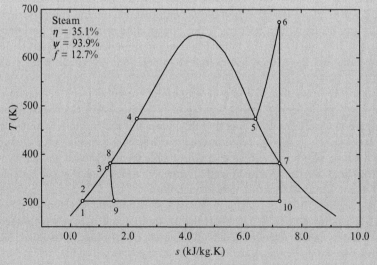

FIGURE 5.20   Ideal regenerative Rankine cycle for steam power plant with closed feedwater heater and steam trap for $T_c = 30\,°\text{C}$, $T_b = 200\,°\text{C}$, $\Delta T_{sh} = 200\,°\text{C}$, $\Delta T_{reg} = 10\,°\text{C}$ (Example 5.4).

EXAMPLE 5.5

In Figure 5.21 is exemplified a steam Rankine power plant with multiple feedwater heaters and steam extraction points. The cycle operates according to the following main parameters: $T_c = 30\,^\circ\text{C}$, $T_b = 200\,^\circ\text{C}, \Delta T_{sh} = 200\,^\circ\text{C}$.

Mass balance equations for this system become significantly more complex with respect to the other systems discussed previously. Table 5.10 gives the mass balance equations for all components of the system and also introduces the steam extraction fractions $f_i, i = 11, 14, 17, 18$. Using the steam fractions the energy and exergy balances can be written for each plant component according to Equations (5.8)–(5.14).

Optimal flow extraction fractions can be determined based on various multivariable optimization methods to maximize the energy efficiency of the cycle. Here the genetic algorithm optimization implemented in EES is used to maximize cycle efficiency; the optimum extraction fractions are determined as follows: $f_{11} = 5.4\%, f_{14} = 4.7\%, f_{17} = 5.8\%, f_{18} = 3.8\%$. The maximized energy efficiency for the ideal Rankine cycle case study is $\eta = 36.5\%$. The exergy efficiency of the cycle is 65.5%; exergy is destroyed in feedwater heaters, steam traps, the condenser, and the steam generator. The thermodynamic cycle is represented in the T–s diagram shown in Figure 5.22.

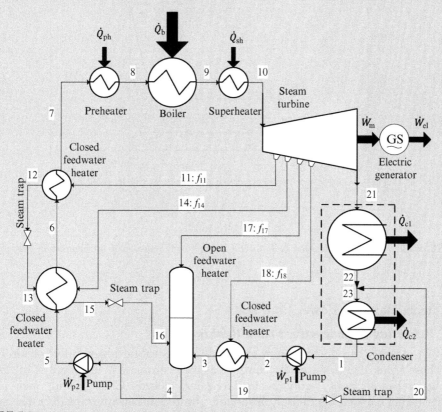

FIGURE 5.21 Regenerative Rankine power plant configuration with multiple feedwater heaters.

Continued

EXAMPLE 5.5 *(cont'd)*

TABLE 5.10 Mass Balance Equations for Rankine Cycle Represented in Figure 5.21

Component	MBE	Component	MBE	f definitions
LPP	$\dot{m}_2 = \dot{m}_1$	CFWH$_{2-3}$	$\dot{m}_3 = \dot{m}_2$; $\dot{m}_{18} = \dot{m}_{19} = \dot{m}_{20}$	$\dot{m}_{11} = f_{11}\dot{m}_{10}$
HPP	$\dot{m}_5 = \dot{m}_4$	OFWH	$\dot{m}_3 + \dot{m}_{16} + \dot{m}_{17} = \dot{m}_4$	$\dot{m}_{14} = f_{14}\dot{m}_{10}$
ST$_{15-16}$	$\dot{m}_{16} = \dot{m}_{15}$	CFWH$_{5-6}$	$\dot{m}_6 = \dot{m}_5$; $\dot{m}_{13} + \dot{m}_{14} = \dot{m}_{15}$	$\dot{m}_{17} = f_{17}\dot{m}_{10}$
PH	$\dot{m}_8 = \dot{m}_7$	CFWH$_{6-7}$	$\dot{m}_7 = \dot{m}_6$; $\dot{m}_{11} = \dot{m}_{12}$	$\dot{m}_{18} = f_{18}\dot{m}_{10}$
B	$\dot{m}_9 = \dot{m}_8$	STB	$f_{11} + f_{14} + f_{17} + + f_{18} + f_{21} = 1$	$\dot{m}_{21} = f_{21}\dot{m}_{10}$
SH	$\dot{m}_{10} = \dot{m}_9$	COND	$\dot{m}_{20} + \dot{m}_{21} = \dot{m}_1$	$\dot{m}_1 = (f_{18} + f_{21})\dot{m}_{10}$

MBE, Mass balance equation; LPP, low-pressure pump; CFWH, closed feedwater heater; OFWH, open feedwater heater; HPP, high-pressure pump; ST, steam trap; PH, preheater; B, boiler; SH, superheater; STB, steam turbine; f, steam extraction fraction.

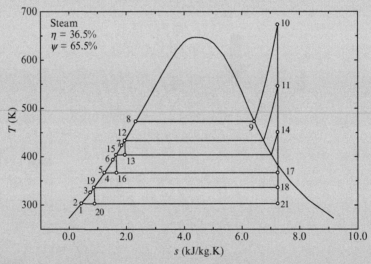

FIGURE 5.22 Ideal steam Rankine cycle for the power plant represented in Figure 5.21 [assumptions: $T_c = 30\,°C, T_b = 200\,°C, \Delta T_{sh} = 200\,°C$].

5.2.5 Steam Rankine Power Stations

5.2.5.1 Reheating-Regenerative Steam Rankine Cycle

The vast majority of power stations are based on reheating-regenerative steam Rankine cycles. These types of cycles are very complex as they comprise many types of components, including preheater, boiler, superheater, reheater, closed (CFWH) and open (OFWH) feedwater heaters, mixer, condenser, deaerator, pumps, and steam traps (ST). The cycle design depends on the type of heat source and heat sink. For most cases of large power generation the heat source is derived from coal combustion.

The heat sink is in general a lake; however, there are many cases in which power plants use air as a heat sink. In this case the plant is equipped with cooling towers—either with natural convection or with forced convection. In some situations (e.g., in colder regions) the heat sink is represented by a district heating system for cogeneration. In this section we give an illustrative example of a reheating-regenerative steam Rankine cycle as presented in Figure 5.23. This power plant comprises eighteen main components. There are also four steam extraction ports with corresponding steam extraction fractions denoted with $f_{12}, f_{17}, f_{20}, f_{21}$. The thermodynamic analysis of the cycle becomes complex because multiple variables and equations are involved and optimization is required for energy efficiency maximization. We will illustrate in detail how the cycle can be analyzed using balance equations and how the optimization problem for maximum efficiency can be formulated.

Mass, energy, entropy and exergy balance equations must be written for each component of the power plant in order to form a system of equations that can be solved to determine thermodynamic parameters defining each state. The following observation is very useful for simplifying the mass balance equations: in state 11 (turbine inlet) the mass flow rate is

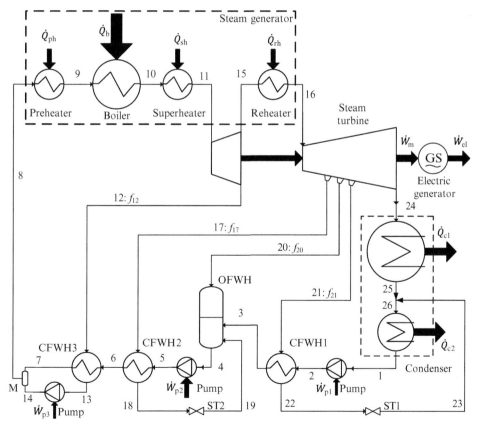

FIGURE 5.23 Rankine power plant with single reheating and multiple feedwater heaters (M, mixer; CFWH, closed feedwater heater; OWFH, open feedwater heater; ST, steam trap).

the maximum for the plant; also $\dot{m}_{11} = \dot{m}_{10} = \dot{m}_9 = \dot{m}_8$. Hence, the steam injection fractions are defined with respect to $\dot{m}_{11}$ as follows:

$$f_{12}\dot{m}_{11} = \dot{m}_{12}, f_{17}\dot{m}_{11} = \dot{m}_{17}, f_{20}\dot{m}_{11} = \dot{m}_{20}, f_{21}\dot{m}_{11} = \dot{m}_{21}$$

In addition, in Figure 5.23 it can be observed that mass balance equations can be reduced to the following set of equations expressed as a function of flow fractions in other state points of the power plant, defined by $f_i\dot{m}_{11} = \dot{m}_i$, namely:

$$f_1 = f_2 = f_3 = f_{26}; \ f_4 = f_5 = f_6 = f_7; \ f_{12} = f_{13} = f_{14}; \ f_{17} = f_{18} = f_{19}; \ f_{21} = f_{22} = f_{23}$$

$$f_{15} = f_{16}; \ f_{24} = f_{25}; \ f_{25} + f_{23} = f_{26}; \ f_{12} + f_{15} = 1; \ f_4 = f_{15}; \ f_{16} = f_{17} + f_{20} + f_{21} + f_{24}$$

Using the flow fractions f_i, the energy, entropy, and exergy balance equations are written for each component as indicated in Table 5.11. The balance equations together with the parameters that quantify exergy destructions (e.g., isentropic efficiencies of pumps and turbines, temperature differences due to heat transfer) form a closed system of equations which can be solved for temperatures, pressures, and specific enthalpy, entropy, and volume at each state point provided that all extraction fractions are specified ($f_{12}, f_{17}, f_{20}, f_{21}$) and the condensation, boiling, and superheating temperatures (T_1, T_9, T_{11}) are given.

TABLE 5.11   Balance Equations for Ideal Rankine Cycle of Power Plant in Figure 5.23

Component	Equations	Component	Equations
Pump 1 (states 1–2)	MBE: $\dot{m}_1 = \dot{m}_2$ EBE: $f_1h_1 + w_{p1} = f_1h_2$ EnBE: $s_1 + s_g = s_2$ ExBE: $f_1ex_1 + w_{p1} = f_1ex_2 + ex_d$	CFWH1 (states 2–3)	MBE: $\dot{m}_2 = \dot{m}_3$ EBE: $f_1h_2 + f_{21}h_{21} = f_1h_3 + f_{21}h_{22}$ EnBE: $f_1s_2 + f_{21}s_{21} + s_g = f_1s_3 + f_{21}s_{22}$ ExBE: $f_1ex_2 + f_{21}ex_{21} = f_1ex_3 + f_{21}ex_{22} + ex_d$
Pump 2 (states 4–5)	MBE: $\dot{m}_4 = \dot{m}_5$ EBE: $f_4h_4 + w_{p2} = f_4h_5$ EnBE: $s_4 + s_g = s_5$ ExBE: $f_4ex_4 + w_{p2} = f_4ex_5 + ex_d$	OFWH (states 3–4–19–20)	MBE: $f_1\dot{m}_{11} + f_{17}\dot{m}_{11} + f_{20}\dot{m}_{11} = f_4\dot{m}_{11}$ EBE:$f_1h_3 + f_{17}h_{18} + f_{20}h_{20} = f_4h_4$ EnBE: $f_1s_3 + f_{17}s_{18} + f_{20}s_{20} + s_g = f_4s_4$ ExBe: $f_1ex_3 + f_{17}ex_{18} + f_{20}ex_{20} = f_4ex_4 + ex_d$
Pump 3 (states 13–14)	MBE: $\dot{m}_{13} = \dot{m}_{14}$ EBE: $f_{12}h_{13} + w_{p3} = f_{12}h_5$ EnBE: $s_{13} + s_g = s_{14}$ ExBE: $f_{12}ex_{13} + w_{p3} = f_{12}ex_{13} + ex_d$	CFWH2 (states 5–6–17–18)	MBE: $f_4\dot{m}_{11} + f_{17}\dot{m}_{11} = f_4\dot{m}_{11} + f_{17}\dot{m}_{11}$ EBE: $f_4h_5 + f_{17}h_{17} = f_4h_6 + f_{17}h_{18}$ EnBE: $f_4s_5 + f_{17}s_{17} + s_g = f_4s_6 + f_{17}s_{18}$ ExBE: $f_4ex_5 + f_{17}ex_{17} = f_4ex_6 + f_{17}ex_{18} + ex_d$
Mixer (states 7–8–14)	MBE: $f_4\dot{m}_{11} + f_{12}\dot{m}_{11} = f_8\dot{m}_{11}$ EBE: $f_4h_7 + f_{12}h_{14} = h_8$ EnBE: $f_4s_7 + f_{12}s_{14} + s_g = s_8$ ExBE: $f_4ex_7 + f_{12}ex_{14} = ex_8 + ex_d$	CFWH3 (states 6–7–12–13)	MBE: $f_4\dot{m}_{11} + f_{12}\dot{m}_{11} = f_4\dot{m}_{11} + f_{12}\dot{m}_{11}$ EBE: $f_4h_6 + f_{12}h_{12} = f_4h_7 + f_{12}h_{13}$ EnBE: $f_4s_6 + f_{12}s_{12} + s_g = f_4s_7 + f_{12}s_{13}$ ExBE: $f_4ex_6 + f_{12}ex_{12} = f_4ex_7 + f_{12}ex_{13} + ex_d$
Preheater (states 8–9)	MBE: $\dot{m}_8 = \dot{m}_9$ EBE: $h_8 + q_{ph} = h_9$ EnBE: $s_8 + q_{ph}/T_{ph} + s_g = s_9$ ExBE: $ex_8 + q_{ph}(1 - T_0/T_{ph}) =$ $ex_9 + ex_d$	Boiler (states 9–10)	MBE: $\dot{m}_9 = \dot{m}_{10}$ EBE: $h_9 + q_b = h_{10}$ EnBE: $s_9 + q_b/T_b + s_g = s_{10}$ ExBE: $ex_9 + q_b(1 - T_0/T_b) = ex_{10} + ex_d$

TABLE 5.11   Balance Equations for Ideal Rankine Cycle of Power Plant in Figure 5.23—Cont'd

Component	Equations	Component	Equations
Superheater (states 10–11)	MBE: $\dot{m}_{10}=\dot{m}_{11}$ EBE: $h_{10}+q_{sh}=h_{11}$ EnBE: $s_{10}+q_{sh}/T_{sh}+s_g=s_{11}$ ExBE: $ex_{10}+q_{sh}(1-T_0/T_{sh})=$ $\quad ex_{11}+ex_d$	Reheater (states 15–16)	MBE: $\dot{m}_{15}=\dot{m}_{16}$ EBE: $f_{15}h_{15}+q_{rh}=f_{15}h_{16}$ EnBE: $f_{15}s_{15}+q_{rh}/T_{sh}+s_g=f_{15}s_{16}$ ExBE: $f_{15}ex_{15}+q_{rh}(1-T_0/T_{rh})=f_{15}ex_{16}+ex_d$
Turbine 1 (states 11–12–15)	MBE: $\dot{m}_{11}=\dot{m}_{12}+\dot{m}_{15}$ EBE: $h_{11}=f_{12}h_{12}+f_{15}h_{15}+w_{t1}$ EnBE: $s_{11}+s_g=f_{12}s_{12}+f_{15}s_{15}$ ExBE: $ex_{11}=f_{12}ex_{12}+f_{15}ex_{15}$ $\quad +w_{t1}+ex_d$	Turbine 2 (states 16–17–20–21–24)	MBE: $f_{15}=f_{17}+f_{20}+f_{21}+f_{24}$; EBE: $f_{15}h_{16}=f_{17}h_{17}+f_{20}h_{20}+f_{21}h_{21}+f_{24}h_{24}+w_{t2}$; EnBE: $f_{15}s_{15}+s_g=f_{17}s_{17}+f_{20}s_{20}+f_{21}s_{21}+f_{24}s_{24}$; ExBE: $f_{15}ex_{15}=f_{17}ex_{17}+f_{20}ex_{20}$ $\quad +f_{21}ex_{21}+f_{24}ex_{24}+w_{t2}+ex_d$
ST 1 (states 22–23)	MBE: $\dot{m}_{22}=\dot{m}_{23}$ EBE: $h_{22}=h_{23}$ EnBE: $f_{21}s_{22}+s_g=f_{21}s_{23}$ ExBE: $f_{21}ex_{22}=f_{21}ex_{23}+ex_d$	ST 2 (states 18–19)	MBE: $\dot{m}_{18}=\dot{m}_{19}$ EBE: $h_{18}=h_{19}$ EnBE: $f_{17}s_{18}+s_g=f_{17}s_{19}$ ExBE: $f_{17}ex_{18}=f_{17}ex_{19}+ex_d$

As seen in Table 5.11 the balance equations can be written using mass-specific quantities. These quantities are defined with respect to the mass flow rate at the HPT inlet; thus: $w=\dot{W}/\dot{m}_{11}$ and $q=\dot{Q}/\dot{m}_{11}$. Furthermore, the energy efficiency can be maximized according to the following objective function:

$$\max\left\{\eta=\frac{w_{net}}{q_{in}}=F(f_{12},f_{17},f_{20},f_{21})\text{ with given }T_1,T_9,T_{11}\text{ and }f_i\in(0,1)\right\}$$

where $F(f_{12},f_{17},f_{20},f_{21})$ represents the functional relation between steam extraction fractions and the energy efficiency of the cycle, a relation which is established by the nonlinear system of equations mentioned above.

Note that the certain additional assumptions must be made in order for the optimization problem to be completely specified. These assumptions are as follows:

- at pump suction the liquid is saturated, hence $x_1=x_4=x_{13}=0$
- at the steam traps inlet the liquid is saturated, hence $x_{18}=x_{22}=0$
- at the OFW outlet there is a saturated liquid, hence $x_4=0$
- there is a minimum temperature difference at the closed feedwater heater outlet denoted with ΔT_{hx}; it is assumed that $T_{22}-T_3=T_{18}-T_6=T_{13}-T_7=\Delta T_{hx}$.

EXAMPLE 5.6

Let us consider a steam cycle of the configuration described in Figure 5.23. The following parameters are assumed: $T_c=30°C$; $T_b=293°C$; $T_{sh}=T_{rh}=400°C$; $\Delta T_{hx}=10°C$, where T_{sh} and T_{rh} are the superheating and reheating temperatures, respectively. We want to determine the optimum steam extraction fractions that maximize cycle energy efficiency for two cases: (i) all pumps and turbines

Continued

EXAMPLE 5.6   (cont'd)

operating reversibly and (ii) the isentropic efficiency of the pumps and steam turbines is 0.85. Exergy destructions for each of the components will be determined for case (ii).

The results of the optimization problem (solved with EES software) are presented in Table 5.12 for case (i). We assume that water is saturated liquid at the pump inlet and at closed feedwater heater outlets (states 13, 18, 22). There are optimal values for steam extraction fractions and optimum extraction pressures. It can be observed that the largest extraction is required after the first stage of expansion with $f_{12} = 14.4\%$ and the extraction fraction decreases when decreasing the pressure. The mass-specific net generated work is 1069 kJ/kg while heat input is 2350 kJ/kg.

TABLE 5.12   Performance Parameters of Optimized Ideal Rankine Cycle of Figure 5.23

Parameter	Value
Parameters with assumed values	$T_c = 30\ °C;\ T_b = 293\ °C, T_{sh} = T_{rh} = 400\ °C;\ \Delta T_{hx} = 10\ °C$
Optimum extraction fractions	$f_{12} = 14.4\%,\ f_{17} = 6.7\%,\ f_{20} = 8.3\%,\ f_{21} = 5.1\%$
Optimized extraction pressures	$P_{12} = 3{,}381\ kPa;\ P_{17} = 988\ kPa;\ P_{20} = 266\ kPa;\ P_{21} = 39.3\ kPa$
Energy and exergy efficiencies and Carnot based efficiency	$\eta = \dfrac{w_{net}}{q_{in}} = 45.5\%,\ \psi = \dfrac{\eta}{1 - T_0/T_{sh}} = 81.6\%$
Net generated work	$w_{net} = w_{t1} + w_{t2} - w_{p1} - w_{p2} - w_{p3} = 1{,}069\ kJ/kg$
Heat input and exergy consumed	$q_{in} = h_{11} - h_8 = 2{,}350\,kJ/kg;\ ex_{cons} = ex_{11} - ex_8 = 885\,kJ/kg$
Exergy destroyed	$ex_d = \sum ex_d = 62.18\ kJ/kg;$ for each component the exergy destruction in percents with respect to total is as follows: $ex_{d,c} = 34\%;$ $ex_{d,CFWH1} = 12.6\%$ $ex_{d,CFWH2} = 11.2\%;\ ex_{d,CFWH3} = 19.9\%;\ ex_{d,M} \cong 0.2\%;$ $ex_{d,OFWH} = 20.1\%;\ ex_{d,ST1} = 1.0\%;\ ex_{d,ST2} = 1.0\%$
Pressure ratio over turbines	$PR_1 = P_{11}/P_{12} = 2.3;\ PR_2 = P_{16}/P_{24} = 796$
Expansion ratio over turbine	$ER_1 = \dfrac{v_{12}}{v_{11}} = 1.91;\ ER_2 = \dfrac{v_{24}}{v_{16}} = 300$
Back work ratio	$BWR = \dfrac{w_{p1} + w_{p2} + w_{p3}}{w_{t1} + w_{t2}} = 0.7\%$

Note: $T_c = T_1;\ T_b = T_9;\ T_{sh} = T_{rh} = T_{11}.$

The energy efficiency (which is maximized) reaches 45.5%. Moreover, one notes that vapor quality at the isentropic turbine outlet is 80%, which is acceptable. The exergy efficiency of the system is exceptionally good, namely $\psi = 81.6\%$, due to the assumed reversible operation of pumps and turbines. Note that the average temperature at the steam generator is approximated to 603.2 K (Carnot factor 50.6%); the Carnot factor associated with the turbine inlet temperature is 55.7%.

The total exergy destruction for case (i) is 61.18 kJ/kg. The highest exergy destruction occurs in the condenser at 34% of total, followed by the OFW and the high-pressure closed feedwater heater CFWH3, each with ~20% of total exergy destruction. The exergy destruction in the mixer is negligible.

EXAMPLE 5.6 (cont'd)

To solve the cycle in case (ii) the assumption that water is saturated liquid at the pump inlet and at closed feedwater heater outlets is maintained. In addition, the assumption that there is negligible pressure drop in all devices with heat exchange and in pipelines is maintained. Furthermore, for case (ii) temperature differences are assumed for the steam generator as follows: 75 °C for the preheating section, 50 °C for the boiling section, and 25 °C for the superheating and reheating sections.

In these conditions the optimum extraction rates are found to have relatively similar values as for case (i), as follows: $f_{12} = 13.3\%$, $f_{17} = 6.1\%$, $f_{20} = 7.7\%$, $f_{21} = 4.8\%$. The pressure ratio over the HPT is 2.8 and over the LPT is 660. All state point parameters of the optimized cycle are given in Table 5.13. The energy efficiency for case (ii) is degraded to 39.2%, as compared to the case (i) with 45.5%.

TABLE 5.13 Cycle Parameters of Reheating-Regenerative Steam Rankine Cycle for Case (ii)

State	P (kPa)	T (K)	v (m³/kg)	x (kg/kg)	h (kJ/kg)	s (kJ/kgK)	ex (kJ/kg)	Mass-specific parameters (kJ/kg)		
								Component	w	q	ex$_d$
1	4.246	303.2	0.001004	0	125.7	0.4365	0.1745	Pump 1	0.3	N/A	0.04
2	244.6	303.2	0.001004	−0.1862	126	0.4367	0.4166	CFWH1	N/A	108	8.05
3	244.6	338.6	0.00102	−0.1183	274.1	0.8988	10.77	OFWH	N/A	262	12.0
4	244.6	399.9	0.001067	0	532.4	1.6	60.07	Pump 2	8.2	N/A	0.91
5	7769	400.9	0.001063	−0.5245	541.8	1.603	68.44	CFWH2	N/A	136	6.83
6	7769	437.4	1.977	−0.417	698.4	1.977	113.6	CFWH3	N/A	256	11.2
7	7769	503.6	0.001203	−0.2144	993.3	2.604	221.5	M	N/A	N/A	∼0.0
8	7769	503.8	0.001203	−0.2138	994.1	2.606	221.8	Preheat.	N/A	312	21.3
9	7769	566.2	0.001377	0	1305	3.187	359.7	Boiler	N/A	1456	62.2
10	7769	566.2	0.02433	1	2761	5.759	1049	Superheat.	N/A	381	7.3
11	7769	673.2	0.03549	1.262	3143	6.382	1245	HPT	213	N/A	20.8
12	2820	546.7	0.08119	100	2930	6.451	1011	Reheater	N/A	264	5.1
13	2820	503.6	0.00121	0	992.3	2.614	217.5	LPT	743	N/A	125.3
14	7769	504.9	0.001205	−0.2102	999.3	2.616	223.9	Condenser	N/A	1466	24.2
15	2820	546.7	0.08119	1.07	2930	6.451	1011	ST1	N/A	0.0	0.62
16	2820	673.2	0.106	1.238	3234	6.953	1165	ST2	N/A	0.0	0.54
17	877	535.3	0.2736	1.098	2973	7.041	878.4	Cycle parameters			
18	877	447.4	0.00112	0	738.1	2.084	121.4	Heat input	$q_{in} = 2,413$ kJ/kg		
19	244.6	399.9	0.0701	0.09422	738.1	2.114	112.4	Net work	$w_{net} = 946$ kJ/kg		

Continued

EXAMPLE 5.6   *(cont'd)*

TABLE 5.13   Cycle Parameters of Reheating-Regenerative Steam Rankine Cycle for Case (ii)—Cont'd

State	P (kPa)	T(K)	v (m³/kg)	x (kg/kg)	h (kJ/kg)	s (kJ/kgK)	ex (kJ/kg)	Mass-specific parameters (kJ/kg)			
								Component	w	q	ex_d
20	244.6	419.3	0.7737	1.019	2757	7.161	626.7	Efficiency	$\eta = 39.2\%$		
21	39.28	348.6	3.845	0.9465	2511	7.319	333.8	Exergy input	$ex_{in} = 1,253$ kJ/kg		
22	39.28	348.6	0.001026	0	315.8	1.021	16.1	Efficiency	$\psi = 75.5\%$		
23	4.246	303.2	2.575	0.07825	315.8	1.064	3.31	Exergy destruction	$ex_d = 306.6$ kJ/kg		
24	4.246	303.2	29.05	0.883	2271	7.513	35.56	Pressure ratio	HPT: 2.8; LPT: 660		

This example underlines the importance of irreversibilities in the steam generator and the turbines. Due to the additional exergy destruction the cycle exergy efficiency for case ii) reduces to 75.5% (for case (i)) the exergy efficiency is 81.6%). The relative exergy destructions for cycle components are represented in the chart in Figure 5.24. Under the assumed conditions the most exergy destruction occurs in the LPT (41%), followed by the steam generator (20%).

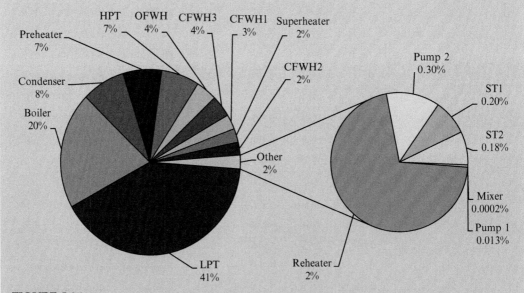

FIGURE 5.24   Exergy destructions for reheating-regenerative cycle in Example 5.6 (case ii).

5.2.5.2 Coal-Fired Power Stations

Although their flue gas stacks highly pollute the atmosphere with carbon dioxide, sulfur dioxide, nitrous oxide emissions, and other pollutants, coal-fired power stations are used in nearly every country of the world. These types of power generating stations are responsible for the majority of global electricity production. They operate based on steam Rankine cycles (generally on reheating-regenerative schemes) and comprise a coal-fired furnace as steam generator.

Two main combustion technologies are used currently in coal-fired power stations, namely pulverized coal boilers (the most predominant) and fluidized bed boilers (used when the coal rank is lower). Although the two technologies are different with regard to transport phenomena, the transient heat and mass transfer from a combusting coal particle to the surroundings are similar in both cases: combustion occurs at the coal particle surface, which gradually consumes while generates incandescent, hot gases. Coal is supplied at the lower part of the furnace where it devolatilizes and ignites in an oxidative atmosphere. The oxygen comes from two sources: fresh air injection and recirculated flue gases. In the first part of the combustion process, the volatiles are oxidized while the temperature of coal particle increases. Thereafter, fixed carbon oxidizes at very fast rate.

In Figure 5.25 the typical configuration of a pulverized coal-fired furnace at a conventional power station is illustrated. It has the roles of preheating, boiling, superheating, and reheating high-pressure steam. The furnace normally is in the form of a tall brick structure of a 10m to 50m height, depending on the type of coal and the residence time of coal particles required for complete combustion. Pulverized coal-fired furnaces are crucial to the utility industry. Typically, pulverized coal-fired furnaces comprise a coal mill which grinds coal to form small powder-like particles conveyed with primary air to the pulverization point, where ignition occurs.

The average temperature of gases at the lower part of the pulverized coal furnace (below point A in Figure 5.25) is typically around 650–700 K. A calandria boiler system with vertical tubes is installed at the lower half of the furnace. Between the hot gases inside the furnace duct and the calandria tubes placed at the periphery an intense radiative heat transfer occurs. Under exposure to intense thermal radiation, water boils.

The calandria system comprises a steam drum (see Figure 5.25) which accumulates liquid water at the lower part. Water flows gravitationally downward and feeds into a system formed of multiple headers arranged such that water is distributed uniformly into calandria tubes. The boiling process in the calandira system is incomplete and is facilitated by the fact that the water is close to its saturation state. Given that the boiling water is lighter than the feed water, in the calandria system is formed a natural circulation process due to gravity. In some cases circulation pumps are also used to enhance the water circulation rate.

In later stages of the combustion process, carbon oxidizes and the flue gas temperature increases further. This process occurs approximately between locations A and B in Figure 5.25. At point B, coal particles are completely consumed and the flue gas temperature reaches its maximum value (around 1200 K). In between points B and C the heat of the flue gases is transferred (mainly radiatively) to the superheater and reheaters of the steam Rankine plant. At points C and D there is a heat transfer process by forced convection and radiation between flue gases and the boiling water prior its return to the steam drum. Saturated vapor exists

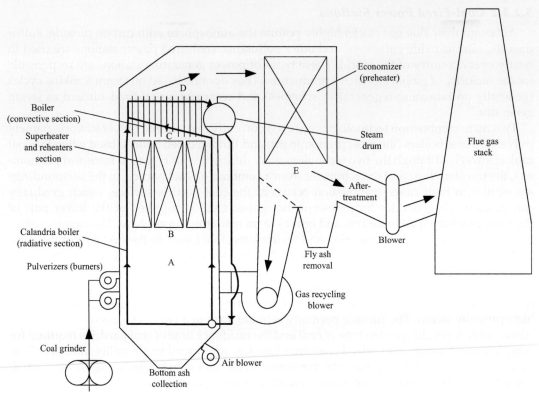

FIGURE 5.25   Typical coal-fired steam generator configuration for conventional power stations.

above the liquid in the steam drum. The next heat transfer process in the furnace occurs between D and E where the economizer is installed, which preheats water prior to feeding it into the steam drum. The heat transfer in the economizer is mostly caused by a forced convection process. After point E, a part of the oxygen-containing hot flue gases is returned to the combustion zone in the furnace. Fly ash is removed and electrostatic devices are typically used to remove particulate matter. Further, the flue gases are directed to the stack using a mechanical draft system.

An example of Rankine cycle configuration for a coal-fired steam Rankine power station is shown in Figure 5.26. The system is a double reheating-regenerative Rankine cycle. Liquid water in 1 is pumped by a low-pressure pump (LPP), preheated in a low-pressure closed water-feeding heater (LPFW), further heated in the OFW, and then is pressurized by the high-pressure pump (HPP).

After state 5, pressurized water is the preheated in closed feedwater heater (HPPFW) and in the economizer, and then it is fed as saturated liquid in 7 to the steam drum. Saturated vapor is extracted from the steam drum in 8 and superheated and expanded in the HPT. The expanded steam in 10 is then reheated (RH$_1$) and expanded in the intermediate-pressure turbine (IPT). The expanded steam in 12 is again reheated (RH$_2$) and expanded in the LPT. Steam is extracted in 15, 17, 18, and 19 to be used in the water preheating process. In the

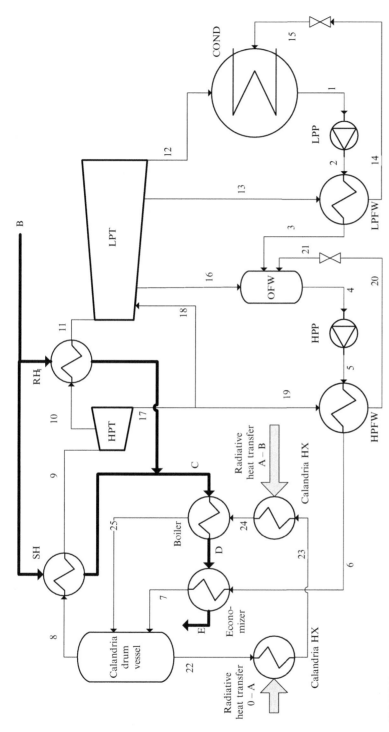

FIGURE 5.26 Diagram of a coal-fired steam Rankine cycle.

calandria system, saturated liquid water from the steam drum bottom is extracted in 22 and heated radiatively in the "Calandria HXs" and then further heated by combined convection–radiation heat transfer in the "BOILER" section. The figure illustrates the heat transfer between hot flue gases and steam as shown with bold line. The process B–C represents an approximation of the heat transfer process between the flue gases and the steam at the superheater and the two reheaters. The process C–D represents the heat transfer in the "BOILER" heat exchanger (see Figure 5.26), whereas process D–E represents the heat transfer in the preheater.

The coal type and its specific energy and exergy contents play a role mostly in the design of the furnace and the efficiency of the power plant as well as its air pollution characteristics. The gross calorific value of coal varies from 13 to $\sim$34 MJ/kg; chemical exergy (which is sensitive to moisture content) varies from about 5 to $\sim$25 MJ/kg; and the specific GHG emissions of coal combustion are in the range of 70–140 kg CO_2 equivalent per GJ on a net calorific value basis (see Chapter 3).

For thermodynamic analysis the mass, energy, entropy, and exergy balance must be written for each component of the system. A logical sequence of computation steps requires solving the equations related to coal combustion in the furnace first.

In the actual coal furnace system oxidant in excess is used. Furthermore, additional oxygen is supplied to the combustion process from recirculated flue gases. Therefore, the combustion process can be represented by the following equation

$$\mathcal{R} + y\mathcal{P} + \lambda x_{st}(O_2 + 3.76N_2) \rightarrow (1+y)\mathcal{P} + a\mathcal{A} \tag{5.15}$$

where $\mathcal{R}$ stands for reactant (referring to coal only) and $\mathcal{P}$ for products (except the ash), respectively, x_{st} represents the stoichiometric molar fraction of fresh oxygen, $\lambda > 1$ is the excess air factor, $y \geq 0$ is the flue gas recycling factor, and $\mathcal{A}$ stands for the ash. It is useful to describe coal using atomic fractions expressed per mole of coal on an as-received basis. Hence, the reactants $\mathcal{R}$ can be represented by the following equation

$$\mathcal{R} = cC + 0.5hH_2 + sS + 0.5oO_2 + 0.5nN_2 + aA + mH_2O \tag{5.16}$$

where c, h, s, o, n, a, m are determined from weight fractions as-received by dividing with molecular masses.

The reaction products, assuming that complete consumption of coal generates only the products CO_2, H_2O, SO_2, O_2, N_2, and ash, is described by the following equation:

$$\mathcal{P} = cCO_2 + (0.5h + m)H_2O + sSO_2 + (\lambda - 1)x_{st}O_2 + (0.5n + 3.76\lambda x_{st})N_2 \tag{5.17}$$

The stoichiometric molar fraction of fresh air, x_{st}, results from atomic species balance:

$$x_{st} = c + 0.25h + s - 0.5o \tag{5.18}$$

As mentioned above, the first phase of the combustion process is represented by the process of pyrolysis coal and volatile combustion and occurs in a spatial region that extends approximately between the furnace bottom and the coal's pulverization location. Figure 5.27 gives a sketch representing the matter and heat fluxes in the ignition and volatile combustion zone of the furnace. During the pyrolysis, a series of combustible materials such as long- and short-chain hydrocarbons (e.g., C_nH_{2n+2}) and sulfur and some amount of carbon dioxide are released. Hence, a part of fixed carbon is released during pyrolysis as hydrocarbons with a carbon-to-hydrogen ratio of $\sim$0.25–0.47. If one denotes with $z \in (0,1)$ the fraction of fixed

FIGURE 5.27 Representation of ignition zone process and involved streams (furnace bottom).

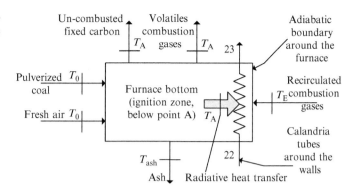

carbon released during pyrolysis, then the volatile combustion phase can be described by the following chemical equation

$$\mathcal{R} + \lambda x_{st}(O_2 + 3.76N_2) + y\mathcal{P} \rightarrow c(1-z)C + \mathcal{P}_{vc} + aA \tag{5.19}$$

where $\mathcal{P}_{vc}$ represents the combusted volatile gases defined according to

$$\mathcal{P}_{vc} = n_{H_2}H_2 + n_{CO_2}CO_2 + n_{H_2O}H_2O + n_{SO_2}SO_2 + n_{O_2}O_2 + n_{N_2}N_2 \tag{5.20}$$

with

$$n_{H_2} = 0.5h(1-z); \; n_{CO_2} = c(z+y); \; n_{H_2O} = 0.5h(z+y) + m(1+y); \; n_{SO_2} = s(1+y)$$

$$n_{O_2} = (1-z)(c+0.25h) + (1+y)(\lambda-1)x_{st}; \; n_{N_2} = (1+y)(0.5n + 3.76\lambda x_{st})$$

standing for the number of moles of components per mole of combusted coal.

In the second phase of combustion (see Figure 5.28) the remaining un-combusted fixed carbon from the first phase oxidizes and the temperature of flue gases increases. Recirculated gases are added to the process as a means of adjusting the temperature at the end of combustion phase. The combustion process (see A–B in Figure 5.25) is described by the following equation

$$\mathcal{P}_{vc} + c(1-z)C \rightarrow (1+y)\mathcal{P} \tag{5.21}$$

FIGURE 5.28 Representation of fixed carbon combustion process and involved streams.

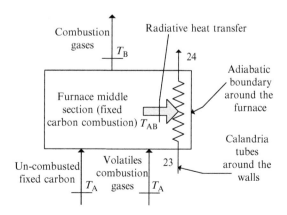

Notice that the summation of Equation (5.19) and Equation (5.21) gives Equation (5.15), which describes the overall process of coal combustion.

The boiling process continues in the heat exchanger placed between points C and D as indicated in Figure 5.25. Once the thermodynamic states in the furnace (0, A, B, C, D, E, F, 6, 7, 8, 9, 10, 11, 22, 23, 24, 25) are completely determined the analysis proceeds further with balance equations of all other components of the Rankine steam power plant. According to energy balance, the mass flow rate in states 4, 5, 6, 7, 8, 9, and 10 must be the same. According to the overall energy balance on the boiler one has

$$\dot{Q}_{\text{boiler}} = \dot{H}_{25} - \dot{H}_{22} = \dot{m}_7(h_8 - h_7) \tag{5.22}$$

where $\dot{H} = \dot{m}h$ is the enthalpy rate of the flow.

In the above equation one accounts for the fact that a saturated liquid enters in state 7 whereas a saturated vapor exits out of the drum in 8. The equation is then solved for $\dot{m}_7$. The enthalpy in 25 results from the enthalpy in state 24 and the enthalpies of states C and D on the flue gas side, according to the energy balance

$$\dot{H}_{25} - \dot{H}_{24} = \dot{H}_{\text{C}} - \dot{H}_{\text{D}}$$

In the example presented above the NOx emissions are neglected, which is a reasonable assumption for first hand calculations. More detailed models must include chemical equilibrium analysis of the combustion process and chemical kinetics in order to determine the NOx emissions. As a rule of thumb, NOx emissions represent half of SO$_2$ emissions per unit of kW net power. On average, for a coal-fired power plant the specific emissions are as follows: 900 g/kWh CO$_2$, 7 g/kWh SO$_2$, and 4 g/kWh NOx.

In the subsequent part of this section the summary of two case studies of energy and exergy analyses of two actual power plants is given. A larger-scale power plant is selected, as Nanticoke Power Generation Station (NPGS) in Ontario, with 500-MW installed capacity per power generating unit. The exergy analysis of the NPGS power plant is summarized here based on the work of Dincer and Rosen (2013). For comparison purposes, the second case study refers to a lower-capacity power plant with a 32-MWgenerator output, operated by Tecro Power Systems Ltd. (TPSL) in India which was previously analyzed in Regualagadda et al. (2010).

The NPGS has steam-reheating power generating units with three stages of turbines (high, intermediate, and low pressure) and a water-cooled condenser. The steam generator operates on the coal pulverization principle. The feedwater system comprises LPP and HPP, two closed feedwater heaters with steam traps, one OFW, and five steam extraction points from the turbines.

The TPSL power plant has a circulating fluidized bed furnace as a steam generator with a nominal capacity of 140 t/h superheated steam. The ultimate coal analysis for TPSL is as follows: moisture 25%, ash 0.88%, hydrogen 4.06%, nitrogen 1.1%, sulfur 0.075%, oxygen 7.935%, carbon 60.95%. The feedwater system comprises a LPP, a HPP, three closed feedwater heaters, and one OFW. The steam cycle operates with no reheating and the turbine has three steam extraction points. The condenser is aircooled.

The general technical parameters of both power stations are compared in Table 5.14. The layout of a Rankine plant is described by the simplified diagram in Figure 5.29, which is applicable to both the NPGS and TPSL plants and consists of four main subsystems: feedwater system, steam generator, power production system, and condenser. The state points indicated in the diagram are described in Table 5.15, and the energy and exergy are given for

TABLE 5.14 General Technical Parameters of NPGS and TPSL Power Stations

Parameter	NPGS	TPSL
Cycle type	Single reheating	No reheating
Nameplate power	523 MW	32 MW
Coal GCV	32.8 MJ/kg	22.0 MJ/kg
Coal exergy	34.2 MJ/kg	26.0 MJ/kg
Steam generator	Pulverized coal	Circulating fluidized bed
Condenser type	Water cooled	Air cooled
Sink temperature	287 K	314 K
Energy efficiency	37%	25%
Exergy efficiency	36%	30%
Specific emissions	1.262 kg GHG/kWh	1.218 kg GHG/kWh

FIGURE 5.29 Simplified layout of steam Rankine power generating station (state point descriptions in Table 5.14).

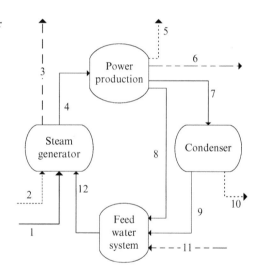

each stream. In addition, the exergy destructions are given in the Table 5.16. As remarked in Table 5.14, the coal quality is much better at NPGS, hence the efficiencies are higher.

Due to a lower quality of coal the exergy destructions of the TPSL steam generator are relatively more important than at NPGS (87% vs. 75% with respect to overall exergy destructions). Comparatively, the feedwater system destroys 3% of exergy for both the NPGS and TPSL stations. The destruction of exergy in the power production subsystem is more important at NPGS than at TPSL (15% vs. 8%).

Conventional power generators based on a the steam Rankine cycle using fossil fuels other than coal (e.g., natural gas or fuel oil) do exist. However, due to economic advantages and efficiency improvements, oil- and gas-fired power plants are combined types, including a combustion turbine and a steam turbine. These types of combined plants will be discussed later in this chapter, after the introduction of gas turbine power plants below.

TABLE 5.15 Streams Parameters for NPGS (523 MW) and TPSL (32 MW) Generating Units

		NPGS			TPSL		
Stream	Description	T (K)	$\dot{E}$ (MW)	$\dot{Ex}$ (MW)	T (K)	$\dot{E}$ (MW)	$\dot{Ex}$ (MW)
1	Coal fed	300	1368	1427	300	120	124
2	Air	300	0	0	300	0	0
3	Flue gas	393	74	62	500	26	9
4	High-pressure steam	710	1780	838	793	125	46
5	Heat loses from high-temperature steam	298	10	0	298	4	0
6	Net power	n/a	524	524	n/a	32	32
7	Saturated vapor	308	775	54	331	67	1.6
8	Extracted steam	550	449	142	497	7	5.6
9	Condensed vapors	308	26	0	331	7	0
10	Heat rejected by the condenser	308	746	0	331	60	0
11	Pump power input	n/a	13	13	n/a	1	1
12	Preheated water	500	487	132	473	31	4

TABLE 5.16 Exergy Destruction of Power Plant Subsystems for NPGS and TPSL

Component	NPGS	TPSL
Feedwater system	23 MW (3%)	2.6 MW (3%)
Steam generator	600 MW (75%)	73 MW (87%)
Power production	118 MW (15%)	6.9 MW (8%)
Condenser	54 MW (7%)	1.6 MW (2%)
Total	795 MW (100%)	84.1 MW (100%)

EXAMPLE 5.7

We give an example of thermodynamic analysis of a coal-fired power plant having a furnace arranged according to the sketch in Figure 5.25 and a steam Rankine cycle according to Figure 5.26. Assume that the coal power plant operates with good quality coal, anthracite, with ultimate analysis as indicated in Table 5.17.

The chemical exergy, formation enthalpy, and formation entropy for coal are calculated first, based on equations indicated in Chapter 3. In Table 5.18 the calculation sequence for coal composition

TABLE 5.17 Coal and Ash Composition from Ultimate Analysis, By Weight, As Received

Ultimate analysis (% by weight AR)								Ash analysis (% by weight)								
C_{ar}	$\mathcal{H}_{ar}$	S_{ar}	O_{ar}	N_{ar}	A_{ar}	$\mathcal{M}_{ar}$	M_{coal}	SO_3	K_2O	TiO_2	MgO	CaO	Fe_2O_3	Al_2O_3	SiO_2	M_{ash}
75	7.5	1	5	1	2	8.5	6.82	4	3	2	1	10	30	25	25	85.02

M_{coal}, molecular mass of coal in kg/kmol; M_{ash}, molecular mass of ash in kg/kmol; molecular mass of compound (coal or ash) is calculated based on component weight fractions w_i and molecular weight of components M_i with the equation: $M = (\sum w_i/M_i)^{-1}$; AR, as-received.

EXAMPLE 5.7 (cont'd)

TABLE 5.18 Coal Compositions and Estimated Chemical Exergy, Formation Enthalpy and Entropy

Parameters	Equation	Value
Dry basis ultimate analysis (kg/kg$_{db}$)	$\dfrac{C_{db}}{C_{ar}}=\dfrac{\mathcal{H}_{db}}{\mathcal{H}_{ar}}=\dfrac{S_{db}}{S_{ar}}=\dfrac{O_{db}}{O_{ar}}=\dfrac{N_{db}}{N_{ar}}=\dfrac{A_{db}}{A_{ar}}=\dfrac{100}{100-\mathcal{M}_{ar}}$	82.0, 8.2, 1.1, 5.5, 1.1, 2.2[a]
Dry ash-free ultimate analysis (kg/kg$_{daf}$)	$\dfrac{C_{daf}}{C_{db}}=\dfrac{\mathcal{H}_{daf}}{\mathcal{H}_{db}}=\dfrac{S_{daf}}{S_{db}}=\dfrac{O_{daf}}{O_{db}}=\dfrac{N_{daf}}{N_{db}}=\dfrac{100}{100-\mathcal{A}_{db}}$	83.8, 8.4, 1.1, 5.6, 1.1[b]
Atomic fractions in daf basis (kmol/kg$_{daf}$)	$c_{daf}=\dfrac{C_{daf}}{1200};\ h_{daf}=\dfrac{\mathcal{H}_{daf}}{100};\ s_{daf}=\dfrac{S_{daf}}{3200};\ o_{daf}=\dfrac{O_{daf}}{1600};\ n_{daf}=\dfrac{N_{daf}}{1400}$	69.8, 83.8, 0.3, 3.5, 0.8[c]
Atomic fractions in as-received basis (kmol/kg$_{ar}$)	$c_{ar}=\dfrac{C_{ar}}{1200};\ h_{ar}=\dfrac{\mathcal{H}_{ar}}{100};\ s_{ar}=\dfrac{S_{ar}}{3200};\ o_{ar}=\dfrac{O_{ar}}{1600};\ n_{ar}=\dfrac{N_{ar}}{1400};$ $a_{ar}=\dfrac{A_{ar}}{101,200};\ m_{ar}=\dfrac{\mathcal{M}_{ar}}{1800}$	62.5, 75, 0.31, 3.12, 0.71, 0.23, 4.7[d]
Atomic fractions per kmol of coal as received (kmol/kmol$_{ar}$)	$c=c_{ar}M_{coal};\ h=h_{ar}M_{coal};\ s=s_{ar}M_{coal};\ o=o_{ar}M_{coal};\ n=n_{ar}M_{coal};$ $a=a_{ar}M_{coal};\ m=m_{ar}M_{coal}$	0.426, 0.512, 0.002, 0.021, 0.005, 0.002, 0.032[e]
Formation entropy of ash	$\Delta S^0_{ash}=\sum w_i\Delta S^0_i$ in kJ/kg.K, Table 3.5 for formation entropies of components. $S^0_{ash}=\Delta S^0_{ash}M_{ash}$ in kJ/kmol.K	0.66 kJ/kg.K 56 kJ/kmol.K
Formation enthalpy of ash	$\Delta H^0_{ash}=\sum w_i\Delta H^0_i$, in MJ/kg, Table 3.5 for formation enthalpies of components. $H^0_{ash}=\Delta H^0_{ash}M_{ash}$ in kJ/kmol	−11.4 MJ/kg −968 MJ/kmol
Chemical exergy of ash	$ex^{ch}_{ash}=\sum w_i ex^{ch}_i$ in MJ/kg, Eq. (1.41), Table 1.6, Table 3.4 for $\Delta G^0=\Delta H^0+T_0\Delta S^0$ $ex^{ch,mol}_{ash}=ex^{ch}_{ash}M_{ash}$ in MJ/kmol	0.4878 MJ/kg 41.5 MJ/kmol
Coal formation entropy	ΔS^0_{daf} is calculated according to Equation (3.11) in dry, ash-free basis	55.1 kJ/kg.K
Coal formation entropy in as-received basis	$\Delta S^0_{ar}[\text{kJ/kg.K}]=\dfrac{100-\mathcal{A}_{ar}-\mathcal{M}_{ar}}{100}\Delta S^0_{daf}+\dfrac{\mathcal{A}_{ar}}{100}\Delta S^0_{ash}+\dfrac{\mathcal{M}_{ar}}{100}\Delta S^0_{H_2O};$ $S^0_{ar}[\text{kJ/kmol.K}]=\Delta S^0_{ar}M_{coal}$	49.66 kJ/kg.K 338.7 kJ/kmol.K
Gross calorific value	GCV_{db} is calculated based on Equation (3.8); $GCV_{daf}=100/(100-\mathcal{A}_{db})GCV_{db}$	37.9 MJ/kg$_{db}$ 38.8 MJ/kg$_{daf}$
Coal formation enthalpy	$\Delta H^0_{daf}=GCV_{daf}+c\Delta H^0_{CO_2}+h\Delta H^0_{H_2O}+s\Delta H^0_{SO_2};\ \Delta H^0_{ar}[\text{kJ/kg}_{ar}]=$ $\dfrac{100-\mathcal{A}_{ar}-\mathcal{M}_{ar}}{100}\Delta H^0_{daf}+\dfrac{\mathcal{A}_{ar}}{100}\Delta H^0_{ash}+\dfrac{\mathcal{M}_{ar}}{100}\Delta H^0_{H_2O};\ H^0_{ar}[\text{kJ/kmol}]=\Delta H^0_{ar}M_{coal}$	−0.75 MJ/kg$_{daf}$ −2.25 MJ/kg$_{ar}$ −15 MJ/kmol
Coal chemical exergy	ex^{ch}_{daf} is calculated with Equation (3.12); $ex^{ch}_{ar}=\dfrac{100-\mathcal{A}_{ar}-\mathcal{M}_{ar}}{100}ex^{ch}_{daf}+\dfrac{\mathcal{A}_{ar}}{100}ex^{ch}_{ash}+\dfrac{\mathcal{M}_{ar}}{100}ex^{ch}_{H_2O};$ $ex^{ch,mol}_{ar}=ex^{ch}_{ar}M_{coal}$	22.3 MJ/kg$_{daf}$ 19.9 MJ/kg$_{ar}$ 136.1 MJ/kmol

[a]Dry basis weight percentages are listed in order for $C_{db},\mathcal{H}_{db},S_{db},O_{db},N_{db},\mathcal{A}_{db}$.
[b]Dry ash-free weight percentages are listed in order for $C_{daf},\mathcal{H}_{daf},S_{daf},O_{daf},N_{daf}$.
[c]Atomic fractions for DAF coal are listed in order for c,h,s,o,n and are given in mol/kg$_{daf}$.
[d]Atomic fractions for as-received coal listed in order for c,h,s,o,n,a,m and given in mol/kg$_{ar}$.
[e]Atomic fractions per kmol of coal as-received listed in order for c,h,s,o,n,a,m and given in kmol/kmol$_{coal}$.
db, dry basis; ar, as received; daf, dry, ash free.

Continued

EXAMPLE 5.7   (cont'd)

(such as dry-basis, dry, ash-free basis, and atomic fractions of constituents) and the main thermo-dynamic parameters of coal and ash (such as chemical exergy, calorific values, formation entropy, and enthalpy) are given. Based on the assumed coal and ash composition the value of 99 or 13.2 MJ/kg_{ar} is obtained for chemical exergy on an as-received basis.

Table 5.19 gives the calculated stream concentrations, enthalpies, entropies, and chemical exergies for the first phase of the combustion process. For 1 kmol of coal supplied to the combustion process 5.64 kmol of water is circulated in the calandria tubes. Water boils partially up to a quality of

TABLE 5.19   Stream Parameters and Balance Equations for Ignition Zone

| Stream | Composition | | T (K) | H (MJ) | S (kJ/K) | Ex (MJ) |
	Compound	Amount[a]				
Pulverized coal	C	0.426	298.15	−15.4	338.7	136.1
	H	0.512				
	S	0.002				
	O	0.021				
	N	0.005				
	A	0.002				
	H_2O	0.032				
Fresh air	O_2	0.655	298.15	0.0	619	0.004
	N_2	2.462				
Water inlet (#22)	H_2O $(x=0)$ [b]	5.64	620.5	167.6	380.5	54.6
Fixed carbon	C	0.098	773.0	0.697	1.877	40.2
Combusted volatile matter	CO_2	1.054	773.0	−453.7	2067	101.5
	H_2O	0.719				
	SO_2	0.006				
	O_2	0.422				
	N_2	6.661				
Ash	A	0.002	500.0	−4.15	0.401	0.195
Boiling water exit (#23)	H_2O $(x=0.25)$	5.64	620.5	222.1	468.4	82.9
Energy balance equation	$H_{in}=H_{out}+Q_{gen}^{steam}=-414.6$ MJ; $Q_{gen}^{steam}=H_{23}-H_{22}=40.0$ MJ					
Entropy balance equation	$S_{in}+S_g=S_{out}=2,191$ kJ/K; $S_g=24.9$ kJ/K $=1.14\%$ of S_{out}					
Exergy balance equation	$Ex_{in}=Ex_{out}+Ex_d=153.0$ MJ; $Ex_d=7.4$ MJ $=4.9\%$ of Ex_{in}					

[a]Given in kmol component per kmol coal (as-received basis).
[b]Formation enthalpy, formation entropy, and chemical exergy not included.

EXAMPLE 5.7 (cont'd)

25% in the lower part of the furnace. The combusted volatile matter consisting of CO_2, H_2O, SO_2, and N_2 reaches a temperature of 773 K at point A. The bottom ash leaves the furnace at 500 K. Based on the energy balance equation it results that the enthalpy transferred to boiling water in region 0–A of the furnace is 40 MJ/kmol of supplied coal. The generated entropy represents 1.14% of output entropy whereas the destroyed exergy is 4.9% of the exergy at the input. Table 5.20 gives the streams parameters and the balance equations for the fixed carbon combustion zone of the furnace.

TABLE 5.20 Streams Parameters and Balance Equations for Carbon Combustion Zone

Stream	Composition Compound	Amount[a]	T (K)	H (MJ)	S (kJ/K)	Ex (MJ)
Boiling water outlet (#24)[b]	H_2O ($q=0.35$)	5.64	620.5	234.8	488.9	89.5
Combustion gases in B	CO_2	1.152	1073	−467.9	2167	136.5
	H_2O	0.778				
	SO_2	0.006				
	O_2	0.295				
	N_2	6.661				
Energy balance equation	$H_{in}=H_{out}+Q_{gen}^{steam}=-454.6$ MJ; $Q_{gen}^{steam}=H_{24}-H_{23}=13.3$ MJ					
Entropy balance equation	$S_{in}+S_g=S_{out}=2,218$ kJ/K; $S_g=26.8$kJ/K$=1.23\%$ of S_{out}					
Exergy balance equation	$Ex_{in}=Ex_{out}+Ex_d=145.5$ MJ; $Ex_d=8.0$ MJ$=5.5\%$ of Ex_{in}					

Note: balances are given per unit of kmol of coal. The input streams are outputs of previous unit (Table 5.16).
[a]kmol component per kmol coal.
[b]Formation enthalpy, entropy, and chemical exergy not included for water.

From the table it results that in the calandria tubes of region A–B steam augments its vapor quality from 0.25 to 0.35. The heat input to steam is 13.3 MJ/kmol of coal fed; the entropy generation represents 1.23% of entropy in B while exergy destruction is 5.5% of exergy input in A. The vapor quality of steam at its return to the calandria drums in state 25 (Figure 5.26) is 75%.

The flue gas parameters in the non-combustion zone of the furnace comprising states C, D, E, and F (Figure 5.25) can be observed in Table 5.21. From this table it results that the exergy destruction at the superheater is 7.53 MJ/kmol of coal, whereas exergy destruction of the reheater is 2.18MJ/kmol. The exergy destruction of the economizer is reported as 2.38 MJ/kmol of coal.

The balance equations are written for all components of the plant. A system of algebraic equations is formed which can be solved for thermodynamic parameters (temperature, pressure, enthalpy, entropy, exergy) at each state point. Values for the isentropic efficiency of the pumps and turbines must be assumed.

The imposed known parameters for modeling are the water condensation temperature at the condenser inlet and outlet, temperature of coal at feed (taken as T_0), boiling temperature (T_8), and superheating (T_9) and reheating (T_{10}) temperatures. Furthermore, the minimum temperature difference for heat exchangers is taken as $\sim$10 K.

Continued

EXAMPLE 5.7   (cont'd)

TABLE 5.21 Stream Parameters and Irreversibilities for Furnace's Non-Combustion Zones

Stream	T (K)	H (MJ)	S (kJ/K)	Ex (MJ)	Component	Irreversibilities S_g (kJ/K)	Ex_d (MJ)
C	644	−504.3	1932	76.3	Superheater	25.2	7.53
					Reheater	7.33	2.18
D	630	−507.9	1926	74.4	Calandria boiler	0.15	0.05
E	473	−550.1	1784	38.4	Economizer	8.00	2.38
F[a]	473	220	713.6	15.4	Total	40.7	12.13

[a]*Flue gas recirculation ratio: 0.6.*
Note: the values of enthalpies, entropies, and exergies are given for 1 kmol of coal combusted.

With these assumptions the algebraic equation system is closed and can be solved. Once thermodynamic state points are determined the irreversibilities in the form of entropy generation and exergy destruction are determined for each system component.

The system components are grouped into four major subsystems: furnace, turbines, condenser, and feedwater heating system. The results for each thermodynamic state are given in Table 5.22. The exergy destruction for each of the subsystems is shown graphically in Figure 5.30. The maximum exergy destruction is in the furnace, followed by the turbines. The HPT destroys 8.2% exergy, whereas the LPT destroys 31.7%.

TABLE 5.22 Working Fluid Stream Parameters (Example 5.7)

Stream	n	P	T	H	S	Ex	Stream	n	P	T	H	S	Ex
0[a]	1	298.15	0.101	0.002	6.6	-0.082	13	0.4787	115.9	377	22,238	60.68	4187
1	3.84	4.246	303.2	8695	30.2	5.264	14	0.4787	115.9	377	3753	11.64	321.6
2	3.84	260.6	303.2	8716	30.21	23.13	15	0.4787	4.246	303.2	3753	12.57	44.68
3	3.84	260.6	367	27,200	85.54	2011	16	0.2577	260.6	402	12,496	32.37	2867
4	4.229	260.6	402	41,248	123.6	4745	17	1.752	551.1	428.7	83,845	207.2	22,223
5	4.229	16,000	404.1	42,750	124.2	6081	18	1.62	551.1	428.7	77,564	191.6	20,558
6	4.229	16,000	418.7	47,479	135.7	7383	19	0.1312	551.1	428.7	6281	15.52	1665
7	4.229	16,000	620.5	125,678	285.4	40,948	20	0.1312	551.1	428.7	1552	4.488	224.4
8	4.229	16,000	620.5	196,597	399.6	77,792	21	0.1312	260.6	402	1552	4.511	217.5
9	4.229	16,000	773.1	251,108	480.1	108,298	22	5.639	16,000	620.5	167,571	380.5	54,597
10	2.478	5530	620.5	136,008	284.5	51,394	23	5.639	16,000	620.5	222,106	468.4	82,930
11	2.478	5530	773.1	152,999	309	60,862	24	5.639	16,000	620.5	234,847	488.9	89,550
12	3.362	4.246	303.2	136,627	452	2133	25	5.639	16,000	620.5	238,490	494.8	91,442

[a]*Reference state.*
n, *number of moles (kmol/kmol coal); P, pressure (kPa); T, temperature (K); H, enthalpy (kJ/kmol coal); S, enthalpy (kJ/K for 1 kmol coal); Ex, exergy (kJ/kmol coal).*

EXAMPLE 5.7 *(cont'd)*

FIGURE 5.30 Exergy destructions and efficiencies of the studied coal-fired Rankine plant.

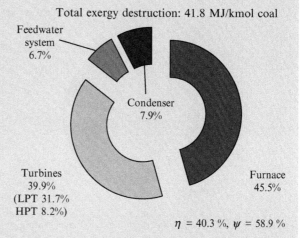

Total exergy destruction: 41.8 MJ/kmol coal

Feedwater
system
6.7%

Condenser
7.9%

Turbines
39.9%
(LPT 31.7%
HPT 8.2%)

Furnace
45.5%

$\eta = 40.3\,\%,\ \psi = 58.9\,\%$

Figure 5.31 illustrates graphically the amounts of exergy destruction for each of the components of the furnace: calandria boiler (segments 0–A, A–B, C–D), superheater, reheater, and economizer. The subsystem destroys 19 MJ/kmol of coal (or 45.5% of input exergy). Proportionally, the super-heater destroys the most exergy among the furnace components at 39.6%, followed by the first calandria segment (0–A) with 30.2% and the reheater with 11.5%.

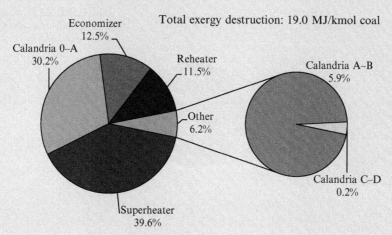

Total exergy destruction: 19.0 MJ/kmol coal

Economizer
12.5%

Calandria 0–A
30.2%

Reheater
11.5%

Calandria A–B
5.9%

Other
6.2%

Calandria C–D
0.2%

Superheater
39.6%

FIGURE 5.31 Exergy destruction in the furnace.

A similar pie chart is presented in Figure 5.32 to show the exergy destruction of the components of the feedwater heating subsystem. Among the seven components (low-pressure feedwater heater, OFW, throttling valve 14, HPP, high-pressure feedwater heater, throttling valve 2, and LPP) the most exergy destruction (66.6%) happens in the low-pressure feedwater heater.

Continued

EXAMPLE 5.7   (cont'd)

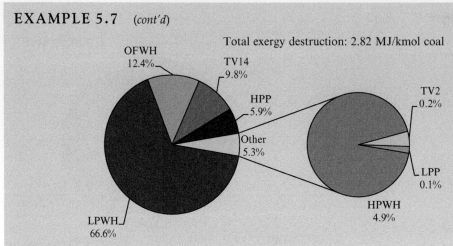

FIGURE 5.32   Exergy destruction in feedwater heating subsystem.

The energy and exergy efficiencies of the overall system result from Figure 4.24 and are 40.3% (energy) and 58.9% (exergy). Also the specific emissions of CO_2 and SO_2 can be calculated from Equation (5.15). Accordingly, $c = 0.426$ moles of CO_2 are generated for mole of coal combusted; moreover, $s = 0.002$ moles of SO_2 are produced for one mole of coal combusted. There is 756 g CO_2/kWh net power and 5.04 g SO_2/kWh net power. Figure 5.33 indicates the molar composition of flue gas as well as the specific CO_2 and SO_2 emissions.

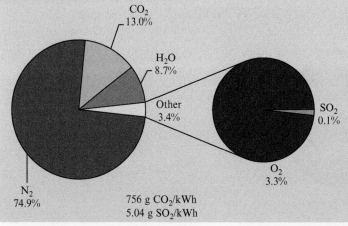

FIGURE 5.33   Molar composition of flue gases and specific CO_2, SO_2 emissions.

5.2.6 Organic Rankine Cycles

Unlike the more common steam Rankine cycles, an organic working fluid is used in the ORCs. Local and small-scale power generation as well as renewable energy systems and low-grade heat recovery systems are the best applications for ORC technology. Some examples of low-temperature heat sources from renewable sources are solar radiation, biomass combustion systems, heat recovery systems from engine exhaust gases, geothermal energy, and ocean thermal energy.

TABLE 5.23 List of some ORC Manufacturers and Technology Descriptions

Manufacturer	Power range	Heat source temperature	Technology description	Applications
ORMAT, US	200 KW–72 MW	150°–300 °C	Fluid : n-pentane	Geothermal, WHR, solar
Turboden, Italy	200 KW–2 MW	100°–300°C	Fluids: OMTS, Solkatherm Axial turbines	Geothermal, CHP
Adoratec, Germany	350 KW–1600 KW	300 °C	Fluid : OMTS	CHP
GMK, Germany	50 K–2 MW	120°–350 °C	Fluid: GL160 (GMK proprietary) 3000 rpm multistage axial turbines	Geothermal, CHP, WHR
Koehler-Ziegler, Germany	70–200 KW	150–270 °C	Fluid: Hydrocarbons, screw expander	CHP
UTC, US	280 KW	>93 °C	N/A	Geothermal, WHR
Cryostar	N/A	100–400 °C	Fluids: R245fa, R134a Radial inflow turbine	WHR Geothermal
Freepower, UK	6 KW–120 KW	180–225 °C	N/A	WHR
Tri-o-gen, NL	160 KW	>350 °C	Turbo-expander	WHR
	50 KW	>93 °C	Twin screw expander	WHR
Infinity turbine	250 KW	>80 °C	Fluid: R134a Radial turbo expander	WHR

WHR, waste heat recovery; CHP, combined heating and power.

ORC technology has received increased attention in the last decades. The manufacturers of ORC provide a wide range of solutions, based on the temperature level and sources. Table 5.23 gives a list of some existing ORC manufacturers and briefly describes their technology.

Working fluid characteristics have great influence in determining the cycle configuration. Selection of the working fluid must be done in compliance with the ORC application, which is influenced by the type of heat source, level of temperature, modes of heat transfer, and scale of the application. Pure fluids and fluid mixtures can both be considered. Also, the retrograde characteristic of the working fluid plays a crucial role in determining the cycle configuration. For example, if a retrograde fluid is used, then the expansion of saturated vapor occurs in the superheated region. This is due to the fact that the slope of the vapor saturation curve on a T–s diagram for retrograde fluids is always positive (e.g., toluene). On the other hand, if the fluid is regular then the expansion of saturated vapor occurs in the two-phase region. The slope of the vapor saturation curve of regular fluids is negative in the T–s diagram (e.g., R134a).

One important feature related to organic working fluid refers to the fact that a large pressure ratio over the turbine can be obtained for relatively small pressure differences. For example, with the organic fluid R123 the ORC operates at a boiling pressure of 5.87 bar and a condensation pressure of 0.85 bar. The pressure ratio is thus 6.9 and the pressure difference is 5.02 bar. This point signifies the advantage of using organic fluids as the working fluid in Rankine cycles. The most important fluid properties for organic working fluid selection are:

- Thermodynamic and physical properties such as density, boiling enthalpy, liquid heat capacity, viscosity, thermal conductivity, melting point temperature, normal boiling point, boiling latent heat (must be high), critical temperature, critical pressure, retrograde or regular behavior, and zeotropic or azeotropic characteristic in the case of mixtures.
- Environmental impact and safety factors such as ozone depletion potential (ODP), global warming potential (GWP), atmospheric lifetime (e.g., R11 45 years, R152 0.6 years), toxicity, flammability.
- Availability, stability at high temperature, and cost effectiveness.

Table 5.24 gives the main categories of working fluids for ORCs and the typical applications for each category. Some proposed working fluids are propylene, propane, R227ea, R236fa, RC318, cyclohexane, R245fa, R141b, R365mfc, R11, R113, R114, toluene, fluorinol (CF_3CH_2OH), R41, R123, R245ca, isobutene, propane–ethane, and siloxanes (MM, MD4M, D4, D5). Table 5.24 lists the working fluids in order of their molecular masses and gives some main parameters, namely normal boiling point, critical pressure and temperature, ODP, and GWP.

Both turbo-expanders and positive displacement expanders are considered prime movers for ORCs. The choice depends on the scale of the application, the working fluid, pressure differentials, and pressure and volume ratios. For large-scale applications turbo-expanders will be the preferred choice because of their higher efficiency. At lower-scale applications, when pressure drop across the prime mover must be high, the preferred choice is a positive displacement expander because it has contact seals which reduce the bypass leakages. When pressure difference per stage is high, turbo-expanders (turbines) have bypass leakage flow at blade tips which is accentuated for small-scale applications.

TABLE 5.24 Classes of Working Fluids for ORC and Typical ORC Applications

Class of working fluid	Condensation	Boiling	Applications
Alkanes, fluorinated alkanes, ethers and fluorinated ethers	30 °C	100 °C	Geothermal
R123, iso-pentane, HFE7100, Benzene, Toluene, p-xylene	30 °C	150–200 °C	Waste heat recovery
R123, isopentane, R245ca, R245fa, butane, isobutene, and R-152a	55 °C	60–150 °C	Waste heat recovery; cascading with internal combustion engines
Butyl-benzene, propyl-benzene, ethyl-benzene, toluene	90°C	250–350 °C	Biomass combustion, combined heating and power; concentrated solar power
R245fa, R123, R134a, n-pentane	35 °C	60–100 °C	Waste heat recovery
R123, n-pentane, PF5050	90 °C	70–90 °C	Geothermal
R227ea, propylene, RC318, R236fa, isobutane, R245fa	25 °C	80–120 °C	Geothermal

There are many configurations of the ORC which differ from the typical reheat-regenerative steam cycle configuration that normally uses steam extraction. For example, in ORC with retrograde fluid it is possible to use a counter-current heat exchanger as a regenerator. In addition, ORC with retrograde fluids can operate without superheating, which makes them better suited for applications that involve heat recovery or solar energy. Other configurations are the reheat-regeneration cycle and supercritical cycle.

An ORC of simple configuration is shown in Figure 5.34. The cycle operates with the organic fluid R134a and is heated with a hot fluid. This cycle configuration can be applied for waste heat recovery when the heat source is low grade. Notice that superheating is not required because the moisture at the end of the expansion process is very low when saturated vapors are expanded (Table 5.25).

The regenerative schemes for ORC do not require (in general) vapor extraction (as in the case of steam cycles). Rather an additional heat exchanger is inserted between the turbine exit and the pump exit, which allows for a transfer of heat between the hotter working fluid at the turbine exhaust and the colder liquid at the pump discharge. The configuration of a regenerative ORC is presented in Figure 5.35. An example of ORC with regeneration that uses R134a as working fluid is presented in Figure 5.36. In this cycle, boiling occurs at 353 K; it starts with saturated liquid in state 4a and reaches saturated vapor in state 4b (pressure drop is neglected). In the condenser, there is a de-superheating process according to line 6–1a, where 1a represents saturated vapor (see Figure 5.36). Further, the condensation process continues until reaching saturated liquid in state 1.

The exergy destruction in the regenerator is accounted for with the help of regenerator effectiveness. This parameter is defined based on the sketch in Figure 5.37, which illustrates a regenerator in the form of a heat exchanger that facilitates a heat transfer process between a hotter stream and a colder stream. Both streams exchange sensible heat. The effectiveness is defined by the ratio between the heat transfer rate that ideally can be extracted from the hotter

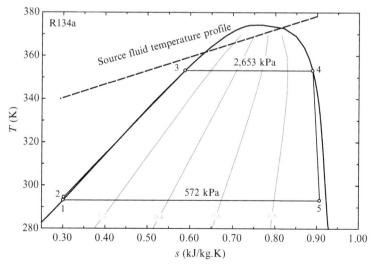

FIGURE 5.34 Simple four-component ORC with R134a for low-grade waste heat recovery.

TABLE 5.25   Working Fluids Commonly Used in ORC

Fluid	M (kg/kmol)	NBP (°C)	T_c (°C)	P_c (MPa)	ODP	GWP
Methanol	32.04	64.4	240.2	8.104	N/A	N/A
R290	44.1	−42.1	96.68	4.247	0	~20
Ethanol	46.07	78.4	240.8	6.148	N/A	N/A
R32	52.02	−51.7	78.11	5.784	0	675
R600a	58.12	−11.7	135	3.647	0	~20
R600	58.12	−0.5	152	3.796	0	~20
R152a	66.05	−24	113.3	4.52	0	124
R601	72.15	36.1	196.5	3.364	0	~20
Cyclohexane	84.16	80.7	280.5	4.075	N/A	N/A
R407c	86.2	−43.6	86.79	4.597	0	1800
Toluene	92.14	110.6	318.6	4.126	N/A	N/A
R500	99.3	−33.6	105.5	4.455	0.738	8100
R134a	102.03	−26.1	101	4.059	0	1430
R141b	116.95	32	204.2	4.249	0.12	725
R12	120.91	−29.8	112	4.114	1	10,890
R123	152.93	27.8	183.7	3.668	0.02	77
Biphenyl	154.2	255	547	3.25	N/A	N/A
R114	170.92	3.6	145.7	3.289	1	10,040
R113	187.38	47.6	214.1	3.439	1	6130
RC318	200.03	-6	115.2	2.778	0	10,250

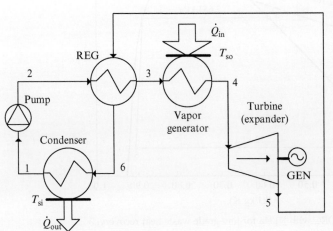

FIGURE 5.35 Regenerative ORC configuration.

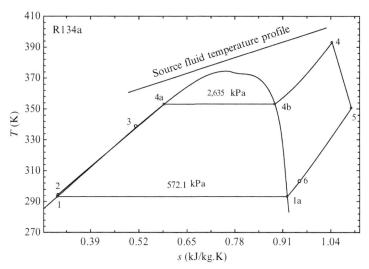

FIGURE 5.36   Regenerative ORC with R134a for low-grade heat sources.

FIGURE 5.37   Sketch for the definition of regenerator effectiveness.

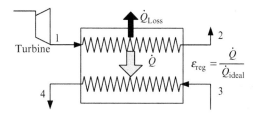

stream ($\dot{Q}_{\text{ideal}}$) and the heat which is actually received by the colder stream ($\dot{Q} = \dot{m}_{\text{colder}}(h_4 - h_3)$). Note that the ideal, maximum heat rate that the hotter fluid can transfer occurs when $T_2 = T_3$ (see Figure 5.37); therefore the ideal maximum heat transfer in the regenerator occurs when $h_2 = h_2(T_3)$ and is given by $\dot{Q}_{\text{ideal}} = \dot{m}_{\text{hotter}}(h_1 - h_2(T_3))$. According to this definition, the regenerator effectiveness is

$$\varepsilon_{\text{reg}} = \frac{\dot{Q}}{\dot{Q}_{\text{ideal}}} = \frac{\dot{m}_{\text{colder}}(h_4 - h_3)}{\dot{m}_{\text{hotter}}(h_1 - h_2(T_3))} \qquad (5.23)$$

An interesting type of regenerative ORC is that using retrograde working fluids where saturated vapor is fed into the turbine while the expanded fluid is in superheated state. An example of such a cycle is shown in Figure 5.38, with toluene as the working fluid. As seen, the boiling temperature is 573 K and the boiling pressure is 1665 kPa. For a condensation at 303 K the saturation pressure is 2.911 kP (vacuum). Therefore the pressure ratio over the turbine is 395; at the same time the volumetric expansion ratio is 285. Figure 5.38 is suggested the temperature profile at the source when a hot fluid which exchanges sensible heat is used as the heat source (e.g., hot combustion gases, high-temperature geothermal brine, etc.).

An emerging type of ORC is the supercritical cycle. This cycle works generally based on a scheme with regeneration such as that presented in Figure 5.35. In supercritical cycles the

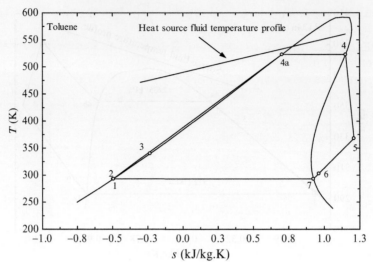

FIGURE 5.38   ORC with a retrograde working fluid (toluene) and expansion of saturated vapor.

working fluid is pressurized to supercritical pressure and heated. As the pressure is supercritical, there is no boiling during heating. Rather, when the temperature becomes superior to critical point the working fluid enters into a supercritical state where the thermodynamic properties are neither of a liquid nor of a gas but a combination of these. Given that there is no boiling, the typical pinch point problem that occurs in vapor generators for Rankine cycles is minimized. Therefore, a gain on exergy efficiency can be obtained with respect to cycles with boiling. However, due to the high pressures associated with supercritical cycles, one must exercise care when selecting a working fluid.

An example of a supercritical ORC operating with R404A working fluid (which is a blend of refrigerants) is shown in Figure 5.39. The high pressure in this cycle is 44.82 bar while

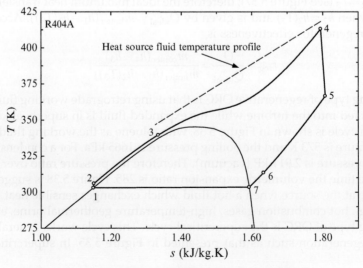

FIGURE 5.39   Supercritical-regenerative ORC with R404a (refrigerants blend).

the low pressure is 14.28 bar, making a pressure ratio of 3.1, which is applicable to positive displacement expanders. The temperature profiles match very well, proving that there is reduced exergy destruction in a supercritical ORC. In the regenerative heat exchanger a pinch point appears due to the nonlinear temperature profile of the supercritical fluid.

EXAMPLE 5.8

In this example the thermodynamic analysis of a regenerative cycle with the retrograde fluid toluene is demonstrated. The cycle configuration is according to Figure 5.35. Assume that the boiling temperature is 525 K, condensation is at 305 K cooled with water, the isentropic efficiency of the pump is 0.8, and isentropic efficiency of the turbine is 0.9. The heat source is a hot flue gas of 650 K that is assumed as ideal gas air. The minimum temperature difference in the heat exchangers is 10 K and the log mean temperature difference (LMTD) at the condenser is 18 K. We will determine the efficiencies, the exergy destructions, and the effectiveness of the regenerator and the mass flow rates for 100 kW power generated.

The balance equations in conjunction with the algebraic equations for pump and turbine isentropic efficiency and the temperature assumptions form a closed system that can be solved to determine the thermodynamic parameters for each state point. Table 5.26 gives the assumptions and balance equations for each of the cycle components.

TABLE 5.26 Balance Equations for Example 5.8

Component	Process	Assumptions	Balance equations
Pump	1–2	$h_{2s} - h_1 = \eta_{S,p}(h_2 - h_1)$ with $\eta_{S,p} = 0.8$ adiabatic; $x_1 = 0$	MBE: $\dot{m}_1 = \dot{m}_2 = \dot{m}$ EBE: $\dot{m}_1 h_1 + \dot{W}_p = \dot{m}_2 h_2$ EnBE: $\dot{m}_1 s_1 + \dot{S}_g = \dot{m}_2 s_2$ ExBE: $\dot{m}_1 ex_1 + \dot{W}_p = \dot{m}_2 ex_2$
Regenerator	2–3 and 5–6	$T_6 = T_2 - 10$ $\varepsilon_{reg} = \dfrac{h_3 - h_2}{h_5 - h_{(p_5, T_2)}}$, adiabatic	MBE: $\dot{m}_2 = \dot{m}_3 = \dot{m}_5 = \dot{m}_6 = \dot{m}$ EBE: $\dot{m}_2 h_2 + \dot{m}_5 h_5 = \dot{m}_3 h_3 + \dot{m}_6 h_6$ EnBE: $\dot{m}_2 s_2 + \dot{m}_5 s_5 + \dot{S}_g = \dot{m}_3 s_3 + \dot{m}_6 s_6$ ExBE: $\dot{m}_2 ex_2 + \dot{m}_5 ex_5 = \dot{m}_3 ex_3 + \dot{m}_6 ex_6 + \dot{E}x_d$
Vapor generator	3–4	$T_{so,1} = T_4 + 10$ $T_{so,pinch} = T_{4a} + 10$, adiabatic, isobaric	MBE: $\dot{m}_3 = \dot{m}_4 = \dot{m}$; $\dot{m}_{so,1} = \dot{m}_{so,2} = \dot{m}_{so}$ EBE: $\dot{m}_3 h_3 + \dot{m}_{so,1} h_{so,1} = \dot{m}_4 h_4 + \dot{m}_{so,2} h_{so,2}$ EnBE: $\dot{m}_3 s_3 + \dot{m}_{so,1} s_{so,1} + \dot{S}_g = \dot{m}_4 s_4 + \dot{m}_{so,2} s_{so,2}$ ExBE: $\dot{m}_3 ex_3 + \dot{m}_{so,1} ex_{so,1} = \dot{m}_4 ex_4 + \dot{m}_{so,2} ex_{so,2} + \dot{E}x_d$
Turbine	4–5	$h_4 - h_5 = \eta_{S,t}(h_4 - h_{5s})$ with $\eta_{S,t} = 0.9$, adiabatic	MBE: $\dot{m}_4 = \dot{m}_5 = \dot{m}$ EBE: $\dot{m}_4 h_4 = \dot{m}_5 h_5 + \dot{W}_t$ EnBE: $\dot{m}_4 s_4 + \dot{S}_g = \dot{m}_5 s_5$ ExBE: $\dot{m}_4 ex_4 = \dot{m}_5 ex_5 + \dot{W}_t + \dot{E}x_d$
Condenser	6–1	$\Delta T_{pinch} = T_{1a} - T_{si} = 10$ K $LMTD_{si} = 10$ K, adiabatic, isobaric	MBE: $\dot{m}_6 = \dot{m}_1 = \dot{m}$; $\dot{m}_{si,1} = \dot{m}_{si,2} = \dot{m}_{si}$ EBE: $\dot{m}_6 h_6 + \dot{m}_{si,1} h_{si,1} = \dot{m}_1 h_1 + \dot{m}_{si,2} h_{si,2}$ EnBE: $\dot{m}_6 s_6 + \dot{m}_{si,1} s_{si,1} + \dot{S}_g = \dot{m}_1 s_1 + \dot{m}_{si,2} s_{si,2}$ ExBE: $\dot{m}_6 ex_6 + \dot{m}_{si,1} ex_{si,1} = \dot{m}_1 ex_1 + \dot{m}_{si,2} ex_{si,2} + \dot{E}x_d$

Continued

EXAMPLE 5.8 *(cont'd)*

The chemical exergies do not need to be considered because there is no chemical reaction. For the thermomechanical exergy calculation the standard reference state of $T_0=298.15\,\text{K}$ and $P_0=101.325\,\text{kPa}$ is considered. For toluene, which is always enclosed inside the heat engine circuits, the reference state is saturated liquid at T_0.

A pinch point calculation procedure must be applied for both the vapor generator and the condenser. In this respect, the requirement of a 10 K-minimum temperature difference is called. Hence, the temperature of the gas at source side must be superior by 10 K with respect to the working fluid temperature at state 4a, which corresponds to the pinch point. In addition we assume that the temperature difference between the source inlet temperature and the working fluid temperature at the turbine entrance is 10 K. Regarding the condenser, a temperature difference between the saturated vapor (state 1a) and the water coolant of 10 K is imposed (this corresponds to the pinch point). Further, the temperature of water at inlet must be iterated until $LMTD$ at the condenser becomes 18 K. The $LMTD$ must be calculated with $\Delta T_1=T_1-T_{\text{si},1}$ where $T_{\text{si},1}$ is the water temperature at sink inlet and $\Delta T_2=T_6-T_{\text{si},2}$ where $T_{\text{si},2}$ is the water temperature at sink outlet. The known equation for $LMTD$ is: $LMTD=(\Delta T_{\max}-\Delta T_{\min})/\ln(\Delta T_{\max}/\Delta T_{\min})$ where $\Delta T_{\min}=\min(\Delta T_1,\Delta T_2)$ and $\Delta T_{\max}=\max(\Delta T_1,\Delta T_2)$.

The temperatures, pressures, and enthalpy, entropy, and exergy rates for each thermodynamic state of the cycle are given in Table 5.27. In the same table the rates of heat and work transfer and the rate of exergy destruction for each cycle component are also given. The energy efficiency of the cycle is 38% while the exergy efficiency of 50%. Based on 100 kW net power generation the mass flow rate of the working fluid must be 0.3786 kg/s.

TABLE 5.27   Thermodynamic State Points, Components, and Cycle Parameters for Example 5.8

State	T (K)	P (kPa)	$\dot{H}$ (kW)	$\dot{S}$ (kW/ K)	$\dot{Ex}$ (kW)	Component	$\dot{Q}$ (kW)	$\dot{W}$ (kW)	$\dot{Ex}_d$ (kW)
1	305	5.362	−56.78	−0.165	0.05225	Pump	N/A	0.96	0.188
2	305.7	1710	−55.82	−0.1644	0.8287	Regenerator	119.7	N/A	6.613
3	459.9	1710	63.84	0.1494	26.92	Vapor generator	263.3	N/A	8.668
4a	525	1710	126.4	0.2765	51.63	Turbine	N/A	100.96	6.615
4b	525	1710	216.2	0.4476	90.4	Condenser	163.3	N/A	7.187
4	650	1710	327.2	0.6371	144.9	Cycle parameters			
5	513.3	5.362	226.2	0.6593	37.29	Energy efficiency		$\eta=38\%$	
6	315.7	5.362	106.6	0.3703	3.804	Exergy efficiency		$\psi=50\%$	
1a	305	5.362	101.7	0.3526	4.182	Mass flow rate	Working fluid	$\dot{m}=0.3786$ kg/s	
1so	660	101.3	1025	11.73	198		Source fluid	$\dot{m}_{\text{so}}=1.528$ kg/s	
2so	535	101.3	824.6	11.39	97.69		Sink fluid	$\dot{m}_{\text{si}}=6.159$ kg/s	
1si	288.9	101.3	406.1	1.443	−24.24	Turbine	Pressure ratio	$PR=319$	
2si	295	101.3	564.6	1.986	−27.63		Expansion ratio	$ER=285$	

EXAMPLE 5.8   *(cont'd)*

In Figure 5.40a the thermodynamic cycle is illustrated in the T–s diagram. On the same diagram the temperature profiles of the heat source and heat sink fluids are suggested. This illustrates the irreversibility due to the finite temperature difference for heat transfer at heat source and heat sink heat exchangers. The exergy destructions by the cycle components are represented in the chart in Figure 5.40b. As seen, the exergy destructions are well balanced among the cycle components, except for the pump, which destroys negligible exergy.

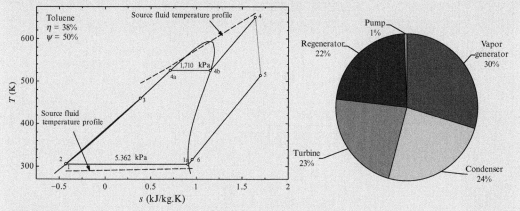

FIGURE 5.40   ORC with toluene for Example 5.8; (a) thermodynamic cycle, (b) exergy destruction percentages.

5.3 GAS POWER CYCLES

In gas power cycles the working fluid does not experience any liquid–vapor transformation process (boiling, condensation) as in vapor power cycles. Rather, the working fluid in gas power cycles remains in a gaseous state of aggregation during all cycle processes. There are three types of totally reversible power cycles, the Carnot, the Stirling, and the Ericson. There are three main types of internally reversible power cycles, namely Otto, Diesel and Brayton.

5.3.1 Totally Reversible Gas Power Cycles

The condition for a cycle to be totally reversible is that its heat addition and rejection processes be isothermal. If the heat addition and rejection processes are not isothermal then the cycle must be externally irreversible because a finite temperature difference must occur between a thermal reservoir at fixed temperature and the working fluid. The Carnot, Stirling and Ericson cycles include two isothermal heat addition processes. The other two processes (from a total of four) are adiabatic for the Carnot cycle, isochoric for the Stirling cycle and isobaric for the Ericson cycle. The efficiency of any totally reversible heat engine (e.g., Carnot, Stirling, Ericson) connected to a heat source of temperature T_H and to a heat sink of temperature T_L is given by the Carnot factor.

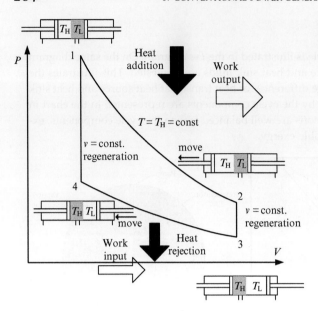

FIGURE 5.41 *P–V* diagram of the ideal Stirling cycle.

$$\eta_{\text{tot, rev}} = 1 - \frac{T_L}{T_H}$$

The *P–v* diagram of the Carnot cycle is given in Figure 1.12a. This cycle is realizable by a heat engine working with an ideal gas and comprising a heat exchanger connected to the heat sink followed by an adiabatic–isentropic compressor, followed by a heat exchanger connected to the heat source, followed by an adiabatic–isentropic turbine.

Regarding the Stirling cycle, this can be performed in a double-effect piston and cylinder set-up which contains a regenerator porous matrix placed inside as shown schematically in Figure 5.41. The working medium can be an ideal gas. Obviously, the processes of heat transfer, regeneration, and work exchange must be reversible. The descriptions of the processes are given below, as follows:

- Isothermal expansion with heat addition, process 1–2: The expansion space is heated externally, and the gas undergoes isothermal expansion. As the left piston moves leftward it generates useful work output.
- Isochoric cooling with heat regeneration, process 2–3: Both pistons are moved slowly rightward at the same pace so that the enclosed volume remains constant. The gas passes through the porous matrix of the regenerator and reaches the other side.
  This process is assumed to be frictionless (reversible). During the process a heat transfer occurs between the gas and the regenerator matrix. At the beginning of the process the gas is at temperature T_H and at the end of it the temperature of the gas reaches $T_L < T_H$.
- Isothermal compression and heat rejection, process 3–4: Active work is given to the piston on the right from the outside so that this moves and compresses the gas. At the same time the cylinder is in thermal contact with the heat sink. Therefore, it continuously removes heat from the gas to the heat sink such that the compression process remains isothermal.

The work input required by process 3–4 is smaller than the work generated by process 1–2 such that the net generated work is positive.

- Isochoric heating with heat regeneration, process 4–1: Both pistons displace leftward, keeping a constant space between them such that the enclosed volume remains constant. During the process, heat is transferred from the matrix to the gas such that at the end of the process the gas increases its temperature to T_H.

The practical implementation of an ideal Stirling engine is impossible because there is much unavoidable irreversibility, especially for heat transfer under finite temperature difference. For example, heat must be transferred from the regenerator matrix to the gas under infinitesimal temperature difference such that at the end of the process the matrix is at the initial gas temperature and the gas is at the initial matrix temperature. This process is practically impossible. Although it is in principle possible to provide heat to the Stirling engine by internal combustion, doing this will induce very high irreversibilities to the regeneration process because combustion is fast while regeneration must be slow. Therefore, practical Stirling engines are implemented either with external combustion or using heat sources such as concentrated solar radiation or waste heat recovery. Hence the cylinders of the Stirling engine are heated and cooled by external sources which need additional temperature differences and also some thermal response time. Despite these disadvantages the Stirling engine has been used successfully with a low-temperature heat source using helium or air as the working fluid. However, designing a Stirling engine is a very complicated task because the dynamic behavior of the mechanism and performance of the heat exchangers highly influence the efficient operation.

The Ericson cycle comprises two isothermal and two isobaric processes. The processes are described by the P–V diagram shown in Figure 5.42. The Ericson heat engine consists of a compressor, an expander, and one heat exchanger with the role of regenerator (the configuration looks similar to that of a closed-loop Brayton cycle; the difference is that the

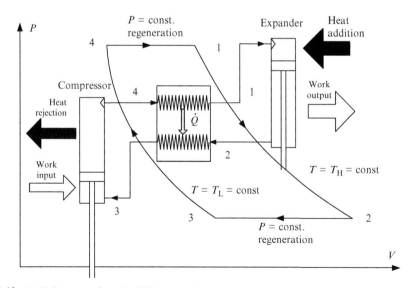

FIGURE 5.42 P–V diagram of an ideal Ericson cycle.

compression and expansion processes for the Ericson cycle are not adiabatic but rather are accompanied by heat transfer such that, instead of being isentropic, these processes are isothermal). The cycle processes are (see Figure 5.42):

- Isothermal expansion and heat addition, process 1–2: In this process the working fluid is heated by an external heat source and expansion occurs simultaneously with the addition of heat at constant temperature. During this process useful work output is obtained.
- Isobaric heat removal with regeneration, process 2–3: The working fluid passes through the regenerator, where its temperature reduces at constant pressure while simultaneously the regenerator transfers the heat to process 4–1.
- Isothermal compression with heat removal, process 3–4: In this process the working fluid is compressed with constant temperature maintained due to a continuous heat rejection process. The work input required for compression is covered by a part of the work generated by the expansion.
- Isobaric heat absorption from the regenerator, process 4–1: The compressed working fluid flows through the regenerator where heat is transferred and the temperature increases. For a reversible process the regenerator effectiveness is 1.

The difficulty of implementing an ideal Ericson cycle arises from the irreversible nature of the isothermal compression, isothermal expansion and, regeneration processes. Maintaining constant temperature during the compression imposes an extremely slow process and an "infinitely" large heat transfer area of the cylinder such that the temperature differences are minimized. The same issues stand for the expander. Regarding the regenerator, due to the finite temperature differences required for a heat transfer to occur, the effectiveness cannot be 1. A regenerator with an effectiveness of 1 requires an infinite heat transfer surface which at least leads to finite pressure drop along the flow, hence the generation of irreversibility.

5.3.2 Otto and Diesel Power Cycles

The fundamental thermodynamic cycles that are internally reversible are, as mentioned above, Brayton, Otto, and Diesel. All these cycles are a good choice for internal combustion engines and thus for engine-driven power generators. The Brayton cycle has been extensively analyzed in the two previous sections. This cycle is better suited to large-scale power generation in combined cycle power plants (CCPPs). In this section the Otto and Diesel cycles are analyzed in detail. These cycles are the most common for engine-driven generators, which consist of an internal combustion engine installed on the same chassis as an electrical generator and a power regulation block. Applications of engine-driven power generators can be both stationary and portable (mobile).

The Otto cycle is the ideal cycle for spark ignition engines. This cycle was established in the 1870s after successful demonstration of the four-stroke spark-ignition engine by Nikolaus Otto. The cycle and the processes are presented in Figure 5.43, where the piston–cylinder operation and the P–v diagram are illustrated. The cycle consists of two isentropic and two isochoric processes of a gas. The numerical example in the diagram is for a volume ratio of $r_v = v_1/v_2 = 8$ and a pressure ratio of $r = P_3/P_1 = 90$. As will be shown, for these conditions the energy efficiency with real air as the working fluid is 47.3%, while the exergy efficiency is 51.9% and the Carnot factor is 0.911.

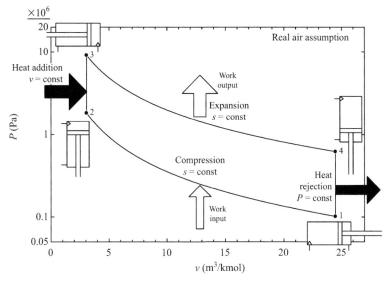

FIGURE 5.43   Ideal Otto cycle under real air assumption for $r_v = 8, r = 90$.

In thermodynamic state 1 there is a gas enclosed in a cylinder in thermal equilibrium with the low-temperature reservoir at $T_1 = T_0$. The processes are given as follows:

- Isentropic compression, process 1–2: The piston is moved leftward and compresses the gas while at the same time the cycle receives work from the outside. The process is isentropic and adiabatic.
- Isochoric heat addition, process 2–3: The piston remains in fixed position while heat is added to the working fluid. In an internal combustion engine this heat is due to a combustion process (with spark ignition). The temperature and pressure of the gas increase considerably during the isochoric heat addition process.
- Isentropic expansion, process 3–4. The piston moves down (see Figure 5.43) and produces a motor stroke that generates usable work. The expansion continues until the volume reaches the maximum stroke, when $v_4 = v_1$.
- Isochoric heat removal, process 4–1: The piston remains in fixed position while heat is removed from the gas by placing the gas in thermal contact with the heat sink. In practical applications with the internal combustion engine, during process 4–1 the heat is removed by expelling the gas into the atmosphere while allowing fresh air in.

The Otto cycle is externally irreversible because its heat addition and removal processes are not isothermal. Therefore, the efficiency of this cycle is lower than that given by the Carnot factor. Under the standard air assumption an analytical expression for the cycle efficiency can be derived. During the isochoric processes there is no work exchange with the surroundings, therefore one must have

$$Q_{in} = C_v(T_3 - T_2) \quad \text{and} \quad Q_{out} = C_v(T_4 - T_1)$$

Using the above equation and the isentropic equations $T_2 = r_v^{\gamma-1} T_1$ and $T_3 = r_v^{\gamma-1} T_4$ it results that the efficiency of the internally reversible Otto cycle is

$$\eta = 1 - \frac{Q_{out}}{Q_{in}} = 1 - \frac{T_4 - T_1}{T_3 - T_2} = 1 - r_v^{1-\gamma} \tag{5.24}$$

The exergy efficiency of the cycle results from an exergy balance equation which can be written as follows:

$$\text{ExBE}: \dot{Q}_{in}\left(1 - \frac{T_0}{T_{so}}\right) + \dot{W}_{in} = \dot{Q}_{out}\left(1 - \frac{T_0}{T_0}\right) + \dot{W}_{out}$$

From the above balance equation, in which it is assumed that the sink temperature is the same as the reference temperature T_0, the following equation for the exergy efficiency results:

$$\psi = \frac{\dot{W}_{net}}{\dot{Ex}_{in}} = \frac{\dot{W}_{net}}{\dot{Q}_n\left(1 - \frac{T_0}{T_{so}}\right)} = \frac{\eta}{1 - \frac{T_0}{T_{so}}}$$

In Figure 5.44 the variation of energy and exergy efficiency of the Otto cycle against the volume ratio is given for three pressure ratios. The typical range of volume ratio during the compression process for spark ignition engines is ~7 to 11. In this range the energy efficiency of the ideal Otto cycle is of 40–50%. The exergy efficiency depends on an additional parameter besides r_v, which is the pressure ratio between the maximum and minimum pressures in the cycle. Observe that the exergy efficiency for $r_v = 7\ldots11$ and $r = 70\ldots120$ spans from 44% to ~68%; the exergy efficiency of the internally reversible Otto cycle gives an indication of the magnitude of exergy destructions due to the heat transfer at source and sink side; the exergy destructions are therefore on the order of 32–66% of exergy input.

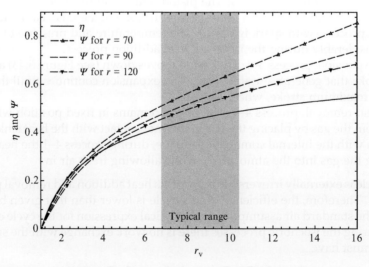

FIGURE 5.44   Efficiency variation with r_v for ideal Otto cycle under standard air assumption.

The Diesel cycle is used in many applications of industrial scale power generation where large-engine-generator groups can be installed. The engines that are based on the Diesel cycle are denoted as compression-ignition internal combustion engines. This thermodynamic cycle and the compression-ignition engine were proposed in Germany by Rudolf Diesel around 1895. In this type of engine the ignition of the combustion process is obtained by compressing the air up to a temperature superior to the auto-ignition of the fuel. In the moment when the air reaches sufficiently high temperature, pressurized liquid fuel is injected into the cylinder, and the ignition process occurs instantly.

Because the air temperature must be sufficiently high (over 800 K) the volume ratio (also called the compression ratio) for the Diesel cycle must typically be in the range of $r_v = 12$ to 24. No spark is given to initiate the combustion process in a compression-ignition engine. So, the duration of the combustion is relatively slow. Consequently, the pressure can be kept almost steady, provided that the injection occurs at top dead center (the upper most position of the piston). As the volume increases during combustion, the temperature of the gases will follow this increase. Figure 5.45 presents an ideal Diesel cycle. It comprises the following processes:

- Isentropic compression, process 1–2: The gas entrapped in the cylinder is compressed adiabatically by moving the piston leftward. Work consumption is required.
- Isobaric heat addition, process 2–3: During the heat addition process the pressure is maintained constant by increasing the cylinder volume. The temperature of the gas increases due to the heat addition process.
- Isentropic expansion, process 3–4. This is the motor stroke when the piston produces useful work. The expansion stroke stops when $v_4 = v_1$.
- Isochoric heat removal, process 4–1: Heat is rejected while the cylinder volume remains constant. This process is similar to that described for the Otto cycle above.

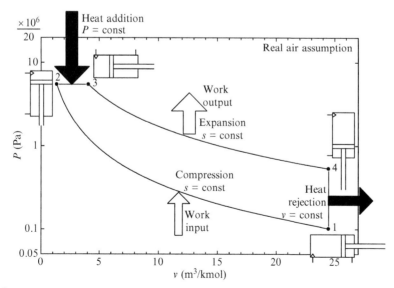

FIGURE 5.45   Ideal Diesel cycle under real air assumption for $r_v = 18, r_c = 3$.

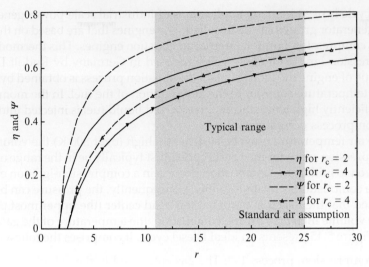

FIGURE 5.46   Efficiency variation with r_v for ideal Diesel cycle under standard air assumption.

It is easy to demonstrate (see Cengel and Boles, 2010) that the efficiency of the standard air Diesel cycle is given by the following equation:

$$\eta = 1 - \frac{1}{\gamma r^{\gamma-1}} \frac{r_c^{\gamma} - 1}{r_c - 1} \tag{5.25}$$

Figure 5.46 illustrates the variation of energy and exergy efficiency of the Diesel cycle for a range of typical compression ratios (r_v). As seen, the ideal cycle energy efficiency can reach values of over 60%, while the exergy efficiency approaches 80%.

5.3.3 Gas Turbine (Brayton) Power Cycles

Combustion turbine power plants started to be developed commercially in the 1930s based on a simple Brayton cycle of open type that includes a turbo-compressor, a combustion chamber, and a gas turbine. The combustion turbine appeared to be the most successful technology for conversion of the chemical exergy of gaseous and liquid fuels into electric power. A large pallet of fuels could be used, starting with natural gas, fuel oil, and coal gas. Further technological developments led to commercialization of advanced combustion turbine power generators with enhanced efficiency due to use of a regenerator, gas reheater, and compressor intercooler and turbine-blade cooling, which allowed for a higher operating temperature. With respect to a steam Rankine power plant, the combustion turbine offered an essentially shorter startup time (below 1 min) which recommended their use for peak load compensation in regional power grids. The efficiency of combustion turbine power plants reached remarkable values of over 40% in conditions when the expelled gases were still at an elevated temperature (over 625 K). Hence, conventional power generation systems that coupled a combustion turbine with a Rankine cycle were developed starting in the 1950s. These were known as combined cycles and their efficiency of power generation surpassed 55%. Specific details of CCPPs are given in Section 5.4, and Section 5.3 focuses on standard-air Brayton cycle configurations and exergy destruction in combustion turbine power stations.

5.3.3.1 Air-Standard Brayton Cycle

In an actual combustion turbine the working fluid is represented by pressurized combustion gases which expand and generate power. In order to make fuel utilization efficient, the excess air provided to the combustion chamber is in the range from a minimum 4 up to over 50. Hence, air is well present in combustion gases, and consequently approximating the properties of combustion gases with those of air is generally accepted.

When the working fluid is modeled as ideal gas air with specific heats corresponding to 298 K ($c_p = 1.005$ kJ/kg.K, $\gamma = 1.4$), then the combustion turbine cycle is denoted an "air-standard Brayton cycle." If all internal processes are reversible, then the cycle is called and "ideal, standard-air Brayton cycle." We will study next the ideal Brayton cycles of various configurations aiming to determine the influence of the main parameters on the upper limit of efficiency according to energy and exergy. Therefore, for this section's analyses the working fluid is standard air and the compressors and turbines and heat exchangers operate with no internal irreversibility.

The most basic Brayton cycle comprises four processes, namely isentropic compression, isobaric heat addition, isentropic expansion, and isobaric heat rejection. The mechanical devices that perform these processes are the compressor, heater, turbine, and cooler indicated schematically in the plant diagram from Figure 5.47. The fact that in the Brayton cycle the heat addition and rejection processes are isobaric rather than isothermal means that this cycle is externally irreversible. Recall that only the cycles that have isothermal heat addition and rejection (e.g., Carnot, Stirling, and Ericson) can be externally reversible. Another remark is that the Brayton cycle can be implemented both as an internal combustion engine system and as an

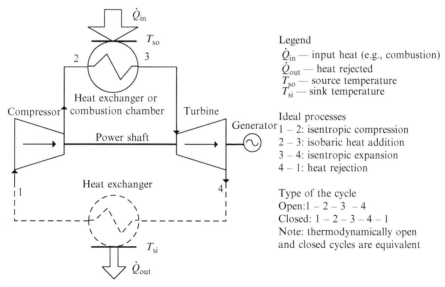

FIGURE 5.47   Schematics of a simple Brayton-cycle-based power plant.

external combustion system, depending on the manner in which the isobaric heat transfer occurs at the heat source side.

The overwhelming majority of combustion turbine power plants operate based on an open Brayton cycle: 1–2–3–4 in Figure 5.47. Due to the direct connection with the atmosphere, the intake pressure and gas-expelling pressure are equal to the atmospheric pressure, which represents a limitation of the turbine's working conditions. This limitation can be overcome if the cycle is closed using a heat exchanger having the role of a cooler. Hence, the turbine can be made to discharge in a vacuum, which means that more work is generated and an enhanced efficiency is obtained. The turbine discharge pressure can be adjusted indirectly by varying the heat sink temperature. A closed Brayton cycle (1–2–3–4–1 in Figure 5.47) must operate with an inert working fluid, either dry air or helium, etc. The heat input can be derived from a combustion process but is applied externally—by heat transfer—to the heater process 2–3. The combustion gases never mix with the working fluid in this case, and the Brayton-cycle-based system becomes an external combustion engine. It is important to note that the open and closed Brayton cycles are equivalent from a thermodynamic point of view. Namely, in the open cycle the atmosphere plays the role of cooler, where the isobaric cooling process 4–1 occurs, whereas in a closed cycle the same process is conducted in an engineered heat exchanger.

The T–s diagram of an ideal air-standard Brayton cycle of basic configuration is shown in Figure 5.48. The energy and exergy balance equations can be written for each of the cycle processes. Assume that the surrounding atmosphere is at standard conditions defined by T_0, P_0, which define the reference state for exergy calculations. The energy (EBE), entropy (EnBE), and exergy (ExBE) balance equation for each process are given in Table 5.28. The ideal cycle is internally reversible, which means that the compression and expansion processes evolve isentropically (no entropy is generated). However, the heat transfer processes at the heat sink and heat source are assumed to be internally reversible but externally irreversible, which means that the system boundary at the source is set to T_3 (highest temperature within the

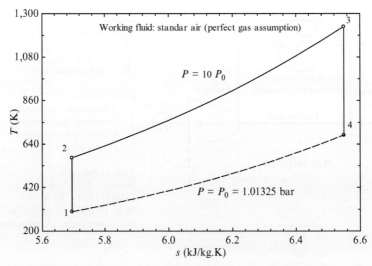

FIGURE 5.48 Representation of a basic air-standard Brayton cycle in T–s diagram for $r=10$ and $T_1/T_3=0.24$.

TABLE 5.28 Energy, Entropy, and Exergy Balance Equations for the Ideal Basic Brayton Cycle

Process	MBE	EBE	EnBE	ExBE
Compression (1–2)	$\dot{m}_1 = \dot{m}_2$	$h_1 + w_c = h_2$	$s_1 + s_{g,c} = s_2$ $s_{g,c} = 0$	$ex_1 + w_c = ex_2 + ex_{d,c}$ $ex_{d,c} = 0$
Heat addition (2–3)	$\dot{m}_2 = \dot{m}_3$	$h_2 + q_{so} = h_3$	$s_2 + \frac{q_{so}}{T_3} + s_{g,so} = s_3$	$ex_2 + q_{so}\left(1 - \frac{T_0}{T_3}\right) = ex_3 + ex_{d,so}$
Expansion (3–4)	$\dot{m}_3 = \dot{m}_4$	$h_3 = h_4 + w_t$	$s_3 + s_{g,t} = s_4$ $s_{g,t} = 0$	$ex_3 = ex_4 + w_t + ex_{d,t}$ $ex_{d,t} = 0$
Heat removal	$\dot{m}_4 = \dot{m}_1$	$h_4 = h_1 + q_{si}$	$s_4 + s_{g,si} = s_1 + \frac{q_{si}}{T_1}$	$ex_4 = ex_1 + q_{si}\left(1 - \frac{T_0}{T_1}\right) + ex_{d,si}$
Overall	N/A	$w_c + q_{so} = w_t + q_{si}$	$\frac{q_{so}}{T_3} + s_g = \frac{q_{si}}{T_1}$	$w_c + q_{so}\left(1 - \frac{T_0}{T_3}\right) = w_t + q_{si}\left(1 - \frac{T_0}{T_1}\right) + ex_d$

cycle) and the sink temperature is set to T_1 (lowest temperature within the cycle). Consequently, there will be entropy generation and exergy destruction within the cycle.

The net power generated by the cycle is represented by the difference $w_{net} = w_t - w_c$. Also, the net generated power can be expressed based on BWR, a parameter introduced also for the Rankine cycle. The BWR represents the ratio between the work consumed by the compressor (s) and the work generated by the turbine(s). Mathematically it is written as

$$BWR = w_c / w_t$$

Therefore one has $w_{net} = w_t (1 - BWR)$. Another way of expressing the net power output results from the overall energy balance given in Table 5.28: $w_{net} = q_{so} - q_{si}$. Furthermore, the energy efficiency of power generation (η) is represented by the ratio of net power production and the rate of heat input; this can be mathematically written as the ratio of mass-specific quantities as follows:

$$\eta = 1 - \frac{q_{si}}{q_{so}}, \text{ furthermore } \psi_1 = \frac{q_{so} - q_{si}}{q_{so}\left(1 - \frac{T_0}{T_3}\right)} = \frac{\eta}{1 - \frac{T_0}{T_3}} > \eta$$

Note that in the above equation the exergy efficiency (ψ_1) of the cycle has been introduced as the ratio of useful work output to the exergy input at the heat source. An alternative way of defining the exergy efficiency is with respect to the exergy received by the cycle, which is represented by the exergy input at source (ex_{in}) and the exergy output at sink (ex_{out}). From exergy balance equations given in Table 5.28 one has

$$\psi_2 = \frac{w_{net}}{ex_{in} - ex_{out}} = 1 - \frac{ex_d}{q_{so}\left(1 - \frac{T_0}{T_3}\right) - q_{si}\left(1 - \frac{T_0}{T_1}\right)}$$

There is third way of expressing the exergy efficiency of the cycle, namely as the ratio of its actual energy efficiency and the efficiency of a reversible heat engine operating between the T_1 and T_3 heat reservoirs; hence

$$\psi_3 = \frac{\eta}{1 - \frac{T_1}{T_3}} = \frac{1 - \frac{q_{si}}{q_{so}}}{1 - \frac{T_1}{T_3}}$$

One remarks from the above three exergy equations that if $T_1 = T_0$ then $\psi_1 = \psi_2 = \psi_3$. This identity is valid for an open-type Brayton cycle, which naturally assumes that the reference temperature is that of the surrounding air. For special cases with closed Brayton cycles, the temperature T_1 may be different than T_0, hence $\psi_1 \neq \psi_2 \neq \psi_3$.

For the standard-air Brayton cycle, a perfect gas equation of state can be applied to further derive analytical expressions for cycle efficiency. The perfect gas equation of state assumes constant specific heats (independent of temperature). As mentioned above, for standard air, the specific heats are estimated at 298.15 K. Using the general equation $\Delta h = c_p \Delta T$ and the energy balance equation, the energy efficiency becomes

$$\eta = 1 - \frac{c_p(T_4 - T_1)}{c_p(T_3 - T_2)} = 1 - \left(\frac{T_1}{T_2}\right) \frac{\left(\dfrac{T_4}{T_1}\right) - 1}{\left(\dfrac{T_3}{T_2}\right) - 1}$$

Another general equation for perfect gas describes the interrelation between pressure and temperature along an isentropic process 1–2 according to $T_1 P_2^\kappa = T_2 P_1^\kappa$, with

$$\kappa = (\gamma - 1)/\gamma$$

and γ as the specific heat ratio. Given that processes 1–2 and 3–4 are isentropic and evolve between the same pressures, $P_1 = P_4$ and $P_3 = P_2$, one has that

$$\frac{T_3}{T_4} = \left(\frac{P_3}{P_4}\right)^\kappa = \left(\frac{P_2}{P_1}\right)^\kappa = \frac{T_2}{T_1} \rightarrow \frac{T_4}{T_1} = \frac{T_3}{T_2}$$

Therefore, the energy efficiency of the standard-air Brayton cycle becomes

$$\eta = 1 - r^{-\kappa}, \text{ with pressure ratio defined by } r = \frac{P_2}{P_1} > 1 \tag{5.26}$$

From the above equation it results that the efficiency of the air-standard Brayton cycle depends only on pressure ratio. In other words, it does not depend directly on the cycle's highest and lowest temperatures (T_3 and T_1). However, due to the cycle configuration, some restrictions must be imposed on the sink and source temperatures for the cycle to function. Hence, the temperature of the heat source (T_3) must be greater than the temperature at the compressor discharge (T_2). Given that $T_2 = r^\kappa T_1$, it turns out that for the cycle to function one must have

$$T_3 > r^\kappa T_1$$

Differently than for energy efficiency, the magnitude of the heat sink temperature influences the cycle's exergy efficiency because the exergy input depends on this temperature. The exergy efficiency of the cycle decreases when the source temperature (T_3) becomes higher and the pressure ratio is fixed because η will not vary and the Carnot factor grows.

According to the efficiency equation, the cycle efficiency increases monotonically with the pressure ratio, r. Also, with pressure ratio increase the temperature at the turbine entrance increases. Because of material-related issues both the pressure and temperature at the turbine inlet must be limited to maximum values. Typically, the pressure at the gas turbine inlet is limited to 20–30 bar. The maximum temperature is limited to about 1200–1500 K. These constructive limitations justify the cycle design for a maximum acceptable heat source temperature. If this strategy is taken, some design parameters trade-offs can be found, as discussed next.

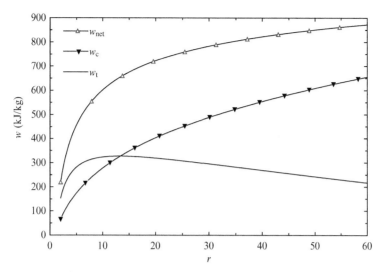

FIGURE 5.49 Mass-specific work variation with pressure ratio for the basic standard-air Brayton cycle when heat sink and source temperatures are restricted to 298 and 1200 K, respectively.

As in Figure 5.49, when the temperatures at the heat source (T_3) and heat sink (T_1) are restricted to fixed values the net generated work has a maximum for a certain pressure ratio. It is observed that for small pressure ratios the specific work generated by the turbine is relatively small with respect to the work consumed by the compressor. In another extreme situation, when the pressure ratio has a high value, the work generated by the compressor is relatively high as compared to the magnitude of work generated by the turbine. Hence, in these two extreme situations (low- and high-pressure ratios) the net generated work represented by the difference between w_t and w_c is small. However, for intermediate-pressure ratios the net generated work is high and reaches a maximum at a certain pressure ratio.

In Figure 5.49 it is clearly observed that the net, mass-specific work output w_{net} has a maximum for pressure ratio $r_{opt} \cong 14$. This is very relevant for power plant design because if the mass-specific work is high, then the power plant is more compact. In order to assess the system compactness for an imposed net power output one can introduce a system specific size parameter (SSP) defined as the reciprocal of the net mass-specific work output, namely

$$SSP = \frac{1}{w_{net}} \left[\frac{kg}{kJ} \right] \tag{5.27}$$

The specific size parameter is plotted in Figure 5.50 against the pressure ratio. On the same plot are superimposed the energy and exergy efficiency curves for the cycle. As observed, the minimum SSP (which corresponds to the most compact system) corresponds to the "knee" of the efficiency curves. If the pressure ratio is higher than r_{opt} then the rate of efficiency decreases fast, and from the economic point of view it does not make sense to enhance the efficiency because the plant size and the equipment cost will increase substantially.

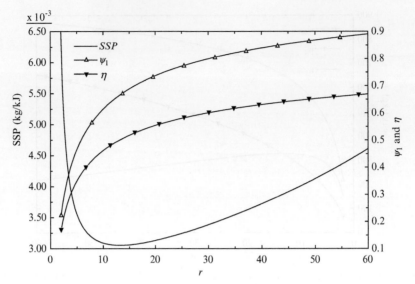

FIGURE 5.50 Minimization of the specific size parameter of the system (*SSP*) for the standard-air Brayton cycle with sink/source temperatures restricted to 298 and 1200 K, respectively.

The efficiency of the Brayton cycle can be improved if some modifications to the basic configuration are applied, such that the external irreversibility due to the finite temperature difference at sink and source is reduced. The improvement can be obtained in three ways:

- By regeneration, in which case the air is preheated after compression using the heat recovered from the cot gas expelled by the turbine. Hence, the heat input is reduced for similar net work output and efficiency increases.
- By reheating, in which case two (or more) subsequent turbines are used with inter-stage reheating. Typically in an actual combustion turbine only a high-pressure and a LPT are installed on the same shaft, with a secondary combustion chamber inserted in between stages such that the working gas is reheated prior to the inlet into the lower-stage turbine. The effect of reheating is represented by a substantial increase in net work output which leads to an increase in cycle efficiency.
- By intercooling, in which case the air compression is done in stages, with inter-stage cooling and heat rejection into the environment. The intercooling leads to a decrease in compression work, thus the BWR decreases and the overall cycle efficiency increases.

EXAMPLE 5.9

In order to explain some important features of the Brayton engine consider now as a numerical example the air-standard cycle represented in Figure 5.48 for which the compression ratio is $r = 10$ and the temperature ratio of $\tau = T_1/T_3$ is 0.24. The parameters for each state point are to be calculated, as are specific work, heat, entropy generated, and exergy destructions for each component and cycle efficiencies. For this case the cycle parameters resulting from energy and exergy analyses are given in Table 5.29. These parameters were calculated based on balance equations listed in Table 5.28.

EXAMPLE 5.9 *(cont'd)*

TABLE 5.29 Calculated Parameters for Brayton Cycle Presented in Figure 5.48

State	T (K)	P (kPa)	h (J/g)	s (J/gK)	ex (J/g)	Process	w (J/g)	q (J/g)	s_g (J/gK)	ex_d (J/g)
1	298.15	101.325	298.6	5.69	0	Compression	278.3	n/a	0	0
2	571	1013.25	576.9	5.69	278.3	Heating	n/a	740.9	0.254	75.67
3	1234	1013.25	1318	6.55	764.5	Expansion	619.7	n/a	0	0
4	685	101.325	698.1	6.55	144.8	Cooling	n/a	399.5	0.486	144.8

Cycle parameters: $T_0 = 298.15$ K; $P_0 = 101.325$ kPa; BWR = 44.9%; $\eta_C = 75.8\%$; $w_{net} = 341.4$ J/g; $\eta = 46.1\%$; $ex_d = 220.5$ J/g; $\psi = 60.8\%$; *Note:* η_C = Carnot factor

5.3.3.2 *Regenerative Brayton Cycle*

We introduce the improved Brayton cycle with regeneration as shown in Figure 5.51. The regenerator will preheat the pressurized air delivered by the compressor in 2 and hence will increase the air temperature prior to entry into the main heater at source in 3. The stream of hot low-pressure gas 5 is further cooled in the regenerator while transferring the heat to the working fluid. After the regenerator, the colder stream 6 is either expelled into the environment or recycled through a cooler (6–1), depending on the cycle type (open or closed).

Due to the fact that the heat transfer in the regenerator must be from stream 5–6 to stream 2–3 the temperature at the turbine outlet must be higher than the temperature at the compressor exit; otherwise, the application of regeneration is not possible. Hence, using the notations in Figure 5.51, one must have $T_5 > T_2$. The T–s diagram of the Brayton cycle with regeneration is given for a case study in Figure 5.52. On the diagram, the reduction of heat input in conditions in which the net work output is constant can be clearly observed.

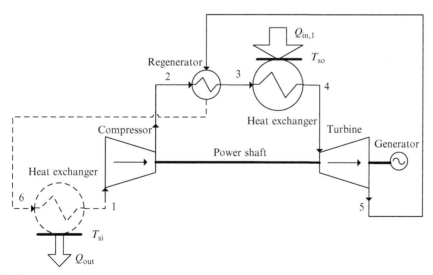

FIGURE 5.51 Regenerative Brayton cycle power plant.

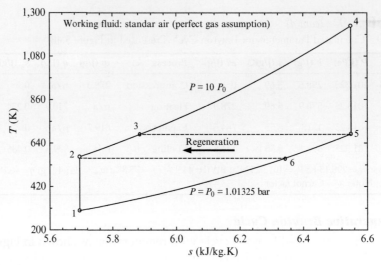

FIGURE 5.52   The T–s diagram of an air-standard Brayton cycle with regeneration.

Under the air standard and no internal irreversibility assumptions one can derive a simple expression for exergy efficiency of the Brayton cycle with regenerator. The assumption of no internal irreversibility implies that in the regenerator there is an ideal match between temperature profiles of the hot and cold streams. That is, the temperature in state 3 (regenerator outlet) reaches the temperature level of state 5 (turbine outlet); hence $T_3 = T_5$. Furthermore, due to the standard air assumption (constant specific heat), the energy balance on the regenerator involves

$$h_3 - h_2 = h_5 - h_6 \rightarrow T_3 - T_2 = T_5 - T_6 \rightarrow T_2 = T_6$$

Moreover, the "no internal irreversibility assumption" implies isentropic expansion and compression processes which can be modeled based on an ideal gas equation of state in a similar manner to that indicated above for the simple configuration Brayton cycle; one has

$$\frac{T_2}{T_1} = \left(\frac{P_2}{P_1}\right)^\kappa = \left(\frac{P_4}{P_5}\right)^\kappa = \frac{T_4}{T_5}$$

The energy efficiency results from an energy balance for the overall cycle that states that the net work output must be equal to heat input minus heat rejected; hence, the energy efficiency becomes "one minus heat rejected by heat input." Or, using the above two equations, the energy efficiency becomes

$$\eta = 1 - \frac{c_p(T_6 - T_1)}{c_p(T_4 - T_3)} = 1 - \frac{T_1 T_2}{T_4 T_1} \frac{1 - \left(\dfrac{T_1}{T_2}\right)}{1 - \left(\dfrac{T_5}{T_4}\right)} = 1 - \tau r^\kappa, \text{ where } \tau = \frac{T_1}{T_4} \tag{5.28}$$

In the above equation, the ratio τ of the lowest and highest temperatures in the cycle is introduced. Given that $T_4 > T_2 = r^\kappa T_1$ one notes that $\tau < r^{-\kappa}$ for the cycle to be possible. Note that for an extreme case of $\tau = r^{-\kappa}$ the energy efficiency is nil. Furthermore, if one imposes the

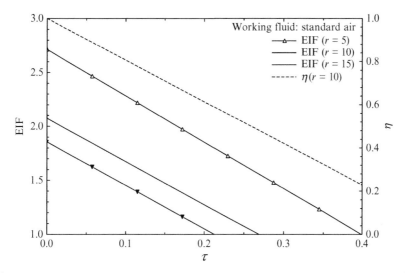

FIGURE 5.53   Efficiency improvement factor (*EIF*) and regenerative Brayton cycle efficiency (η) for a range of temperature ratios (τ) and three pressure ratios (*r*).

condition that the cycle efficiency with regeneration must be higher than that without regeneration one has to satisfy the inequality $1 - r^{-\kappa} < 1 - \tau r^{\kappa}$, or $\tau < r^{-2\kappa}$, which is more restrictive than the previous condition. It is also remarked that the efficiency of the regenerative Brayton cycle reaches the Carnot limit regardless of the pressure ratio when the source temperature is infinity, because then $\tau = 0$ and hence $\eta = 1$. Furthermore, one can introduce an *efficiency improvement factor* (*EIF*) for the regenerative Brayton cycle with respect to basic configuration of the cycle defined as an efficiency ratio according to: $EIF = \eta_{regeneration}/\eta_{basic}$. For standard air assumption the EIF expression becomes

$$EIF = \frac{1 - \tau r^{\kappa}}{1 - r^{-\kappa}} \tag{5.29}$$

Figure 5.53 illustrates the dependence of the efficiency improvement factor on the sink-source temperature ratio (τ) and various pressure ratios (*r*) for standard air. Low-pressure ratios and high source temperatures facilitate better EIF.

5.3.3.3 Reheat-Regenerative Brayton Cycle

An improved diagram for the Brayton cycle which includes a reheater in addition to the regenerator is presented in Figure 5.54. The application of reheating leads not only to a higher work generation but also to higher temperature of the gas exhausted by the turbine. Consequently, there is a higher potential of heat recovery and regeneration when reheating is applied. A thermodynamic *T–s* diagram of the Brayton cycle with reheating and regeneration is shown in Figure 5.55. The mathematical description of all involved processes is given in Table 5.30 for the air-standard case. As remarked, the standard reference state is assumed at compressor suction (state 1). All internal processes are reversible. Recall that the cycle is externally reversible due to temperature differences at heat source and heat sink. A heat source reservoir of a temperature equal to the highest temperature in the cycle (T_4) is

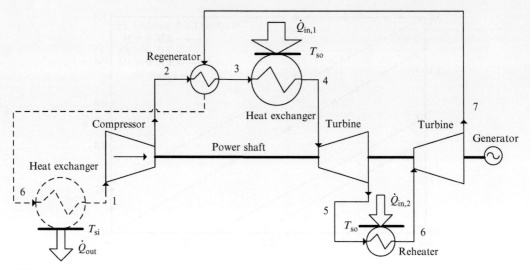

FIGURE 5.54 Improved Brayton cycle with reheating and regeneration [legend as in Figure 5.51].

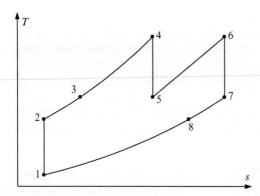

FIGURE 5.55 Brayton cycle with reheating and regeneration (see Figure 5.54).

assumed. The heat sink is at the standard reference level of temperature $T_0 = T_1 = 298.15\,\mathrm{K}$ for open cycle configurations.

The exergy destructions which occur in the heater (3–4), reheater (5–6), and cooler (8–1) are calculated based on exergy balance equations. The inter-stage pressure for expansion ($P_5 = P_6$) must be selected such that the work generated jointly by the turbines is maximized. Recall that the overall pressure ratio $r = P_2/P_1$ is an imposed parameter, but the first stage pressure ratio $r_1 = P_4/P_5$ is taken as a variable. $T_6 = T_4$. So, the specific work generated by the two turbines is given by

$$w = \kappa R T_4 \left(r_1^\kappa + r^\kappa r_1^{-\kappa} - 2 \right) \text{ with } \kappa = \frac{\gamma - 1}{\gamma} \text{ and } \gamma = \frac{C_p}{C_v}$$

The above expression can be differentiated with respect to r_1 and the result equated to zero such that the optimum value is obtained, which maximizes w. It is simple to show that the following equation is obtained $r_1^\kappa + r^\kappa r_1^\kappa = 0$, which can be solved for r_1 to get $r_{1,\mathrm{opt}} = r^{0.5}$. This type of result is also valid for inter-stage compression and is applied in order to reduce the compression work.

TABLE 5.30 Process Descriptions for Standard-Air Reheating-Regeneration Cycle (Figure 5.54)

Process	Description	Equations
1–2	Isentropic compression	$ds=0$; $Pv^{\gamma}=$ct.; assume $\gamma=1.4$ (standard air), $T_1=298.15$ K, $P_1=101.325$ kPa $P_1v_1=RT_1$, $s_1=165$ kJ/kmol.K, $h_1=8{,}649$ kJ/kmol; $s_2=s_1$, assume r, e.g., $r=15$ $P_2=r\,P_1$, $v_1^{\gamma}=r\,v_2^{\gamma}$, $T_1v_1^{\gamma-1}=T_2v_2^{\gamma-1}$, $h_2-h_1=C_p(T_2-T_1)$; $h_1+w_c=h_2$
2–3	Isobaric heat addition	$dP=0$, $P_3=P_2$; energy balance on regenerator: $h_3-h_2=h_7-h_8$; also $dh=C_pdT\rightarrow$ $h_3-h_2=C_p(T_3-T_2)$; $P_3v_3=RT_3$; $s=C_p\frac{dT}{T}\rightarrow s_3-s_2=C_p\ln\left(\frac{T_3}{T_2}\right)$
3–4	Isobaric heat addition	$dP=0$, $P_4=P_3$, assume $\tau=0.2\rightarrow T_1=\tau\,T_4$; $P_4v_4=R\,T_4$; $dh=C_pdT$; $C_p=7/2R\rightarrow$ $h_4-h_2=C_p(T_4-T_2)$, $ds=C_p\frac{dT}{T}\rightarrow s_4-s_2=C_p\ln\left(\frac{T_4}{T_2}\right)$
4–5	Isentropic expansion	$ds=0$, $s_5=s_4$, $P_4=r^{0.5}P_5$, $P_4v_4^{\gamma}=P_5v_5^{\gamma}$, $T_4v_4^{\gamma-1}=T_5v_5^{\gamma-1}$, $h_5=h_1+C_p(T_5-T_1)$, $h_4=h_5+w_{t,1}$
5–6	Isobaric heat addition	$dP=0$, $P_6=P_5$, assume reheating to T_4 (max. temperature), $T_6=T_4$, $P_6v_6=R\,T_6$, $h_6-h_5=C_p(T_6-T_5)$, $s_6-s_5=C_p\ln\left(\frac{T_6}{T_5}\right)$
6–7	Isentropic expansion	$ds=0\rightarrow s_7=s_6$; $P_7=P_1$; $P_7v_7^{\gamma}=P_6v_6^{\gamma}$; $P_7v_7=RT_7$, $h_7=h_1+C_p(T_7-T_1)$; $h_6=h_7+w_{t,2}$,
7–8	Isobaric heat rejection	$dP=0$, $P_8=P_7$, assume heat rejection down to $T_8=T_2$; $P_8v_8=R\,T_8$ $h_7-h_8=C_p(T_7-T_8)$, $s_7-s_8=C_p\ln\left(\frac{T_7}{T_8}\right)$
8–1	Isobaric heat rejection	$dP=0$, $P_8=P_1$, $q_{out}=h_8-h_1$

EXAMPLE 5.10

Consider a case study of the reheat-regenerative cycle based on the cycle diagram in Figure 5.54. The compression ratio is $r=15$ and the temperature ratio is $\tau=0.2$. The exergy and the cycle are internally reversible.

The thermodynamic analysis is based on balance equations given in Table 5.30. The system of equations formed by the balance equations is solved and the thermodynamic parameters at each of the cycle's state points are determined. Table 5.31 gives the thermodynamic properties for each state point. The efficiency of the cycle is 0.79; the energy efficiency is 0.64 and the Carnot factor of the heat source of 1491 K is 0.8. The T–s diagram of the cycle is presented in Figure 5.56.

TABLE 5.31 Thermodynamic Properties at State Points for the Cycle in Example 5.10

P (kPa)	T (K)	h (kJ/kmol)	s (kJ/kmolK)	ex (kJ/kmol)
101.325	298.2	8649	164.978	0
1520	646.3	18,780	164.978	10,130
1.520	1012	29,440	178.040	16,890
1520	1491	43,350	189.298	27,450
392.43	1012	29,440	189.298	13,540
392.43	1491	43,350	200.555	24,100
101.325	1012	29,440	200.555	10,180
101.325	646.3	18,780	187.493	3.419

Continued

EXAMPLE 5.10   *(cont'd)*

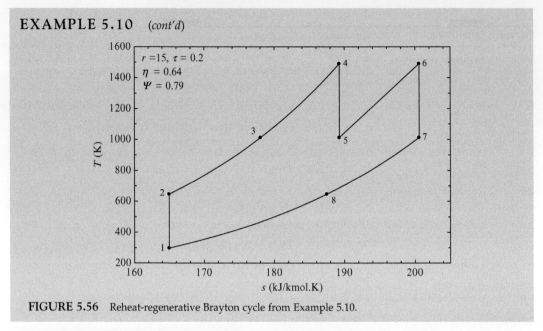

FIGURE 5.56   Reheat-regenerative Brayton cycle from Example 5.10.

5.3.3.4 *Brayton Cycle with Intercooler*

When inter-stage cooling is applied for the compression process, the energy and exergy efficiency of the Brayton cycle is increased for two main reasons: the work consumed by the compressor is reduced, and relatively more heat can be recovered by regeneration. A Brayton cycle with intercooling, reheating, and regeneration is shown in Figure 5.57. The cycle diagram in $T–s$ coordinates is shown in Figure 5.58 as a case study.

For the case study the same compression ratio ($r = 15$) and temperature ratio ($\tau = 0.2$) as for the cycle studied above, with reheating and regeneration only, is taken. It becomes clear that intercooling leads to important improvement of the cycle's energy and exergy efficiency, which both increase significantly (by about 8%). The exergy destructions due to external irreversibilities for the case study can be appreciated in percentages of the total exergy destruction as follows: 22% at the heater and reheater and 28% at the intercooler and cooler.

5.3.3.5 *Exergy Destructions in Brayton Cycle Power Plants*

Estimation of exergy destructions in Brayton cycles is of major importance because they affect in significant measure the efficiency of power generators. One special issue results from the fact that the backwork ratio of Brayton cycles is relatively high. Consequently, the isentropic efficiency of turbomachinery dramatically affects the cycle performance. In a Brayton cycle, the turbo-compressors and the turbines must have isentropic efficiency of over 80% for reasonable economic viability of the power plant. Besides the irreversibility of turbomachinery, there are important irreversibilities related to heat transfer across finite temperature differences and due to pressure drop in ducts and heat exchangers. In this section, the types and magnitudes of irreversibilities in Brayton cycles are investigated, and the impact of those on exergy destruction is analyzed for a relevant case study.

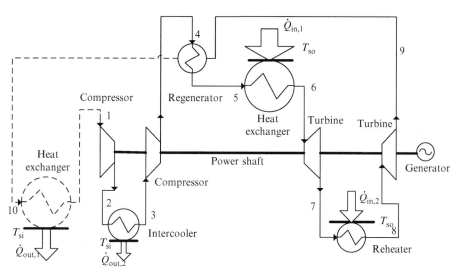

FIGURE 5.57 Brayton cycle with inter-cooling, reheating, and regeneration.

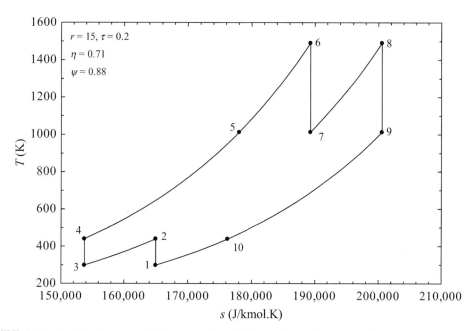

FIGURE 5.58 The T–s diagram of a Brayton cycle with intercooling, reheating, and regeneration.

The isentropic efficiency is probably the most important parameter that quantifies the irreversibility in Brayton cycle. Typical isentropic efficiency of a compressor that equips a Brayton-cycle-based power plants is in the range of $\eta_{S,c} = 0.8 - 0.85$. The isentropic efficiency of a compressor has been introduced (in Table 1.8) as the ratio of the isentropic work to actual work consumed by the compressor when performing the task of increasing the pressure of a working fluid. Note that the isentropic efficiency has an integrative nature as it accounts for multiple types of irreversibility that occur inside a compressor. Specifically, the irreversibility during the compression process is higher in compressors than in turbines because of intrinsic dissipative phenomena, such as weak compression waves that occur mostly in compressors.

The nature of the working fluid impacts in certain measure the magnitude of dissipations, specifically due to the specific heat: if the specific heat is higher, the capacity of energy accumulation as internal energy within the fluid is higher, and this directly affects the temperature and the exergy of the stream discharged by the compressor. Because in combustion turbine power plants the air excess ratio is set rather high, the specific heat of combustion products is very similar to that of air with the same temperature and pressure conditions. This is why it becomes reasonable to assume that the working fluid of a Brayton cycle is air. However, the variation of specific heat with temperature should be taken into account to get an accurate estimate of exergy destructions in turbomachinery.

In order to illustrate the influence of working fluid nature on exergy destruction in a compressor let us consider a case study of air compression under standard conditions with a compression ratio $r = 15$. Three types of air—as the working fluid—are considered: standard air (ideal gas with a constant specific heat), real gas air (dry air modeled with an accurate equation of state that accounts for specific heat variations with temperature and for other real gas effects), and humid air (with an assumed relative humidity of 60% at compressor intake). For all cases the same isentropic efficiency of 0.8 is assumed.

Figure 5.59 illustrates the compression processes in the T–s diagram for the three cases considered. When the working fluid is standard air, the temperature corresponding to isentropic discharge is 646.3 K. This values is determined from the equation $TP^\kappa = \text{const.}$ with the exponent $k = (1 - \gamma)/\gamma$. The temperature for actual discharge is determined by taking into account a constant specific heat and the isentropic efficiency which allows for deriving the following relationship

$$C_p(T_{2,s} - T_1) = \eta_S C_p(T_2 - T_1)$$

where subscript 2,s refer to isentropic discharge and subscript 2 to the actual discharge.

The compressor suction in the case study is assumed to have standard conditions, and as a result, the thermomechanical exergy at intake is set to zero. The work consumed by the compressor is given by $w_c = h_2 - h_1 = C_p(T_2 - T_1)$. The exergy balance equation for the compression process with standard-air working fluid can be used to determine the exergy destructions as follows:

$$0 + w_c = h_2 - h_1 - T_0(s_2 - s_1) + ex_d \rightarrow ex_d = T_0(s_2 - s_1)$$

The discharge temperature of standard-air working fluid is $T_{2,i} = 733.4\,\text{K}$ and the exergy destruction is 34.884 kJ/kg. In a similar manner, the exergy destruction is calculated for

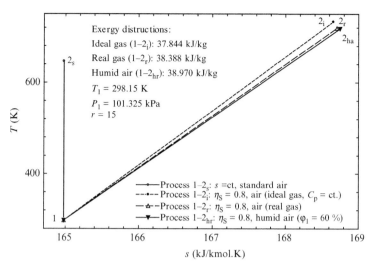

Exergy distructions:
Ideal gas (1–2$_i$): 37.844 kJ/kg
Real gas (1–2$_r$): 38.388 kJ/kg
Humid air (1–2$_{hr}$): 38.970 kJ/kg
T_1 = 298.15 K
P_1 = 101.325 kPa
r = 15

—•—Process 1–2$_s$: s =ct, standard air
---•---Process 1–2$_i$: η_S = 0.8, air (ideal gas, C_p = ct.)
—▲--Process 1–2$_r$: η_S = 0.8, air (real gas)
—▼—Process 1–2$_{hr}$: η_S = 0.8, humid air (φ_1 = 60 %)

FIGURE 5.59 Comparison of actual compression process and exergy destruction for three working fluids—standard air, real gas air, and humid air.

the process 1–2r shown in Figure 5.59, which represents the actual compression of real air working fluid. The discharge temperature in this case becomes $T_{2,r}$=719.2 K and the exergy destruction is higher, 38.388 kJ/kg. In the case of humid air (process 1–2$_{ha}$ in Figure 5.59) the discharge temperature is even lower $T_{2,ha}$=719.1 K and the exergy destruction is 38.970 kJ/kg. As seen in this example, the exergy destruction in the compressor is influenced by the working fluid used. Although standard air is a reasonable approximation for compressor modeling, for improved accuracy the use of a real gas model of air (that accounts for specific heat variation with temperature) is recommended.

Regarding the turbines, their isentropic efficiency is in an approximate range of $\eta_{S,t}$= 0.8 – 0.9. In order to observe the influence of working fluid on the exergy destruction in the turbine let us analyze comparatively, as a case study, the exergy destruction when one operates with (i) standard air, (ii) ideal gas air with variable specific heat, and (iii) combustion gases. In all cases the isentropic efficiency of the turbine is taken as $\eta_{S,t}$=0.85, the inlet temperature is T_1=1273 K, and the inlet pressure is P_1=rP_0, where the compression ratio r=15 and P_0 is the standard pressure. The calculations for case (i) are given in Table 5.32. The final result shows that the molar specific exergy destroyed amounts to ex_d=1.4 kJ/mol when the exergy input to the turbine is ex_{in}=18.4 kJ/mol; hence, the fraction of exergy destruction with respect to input exergy can be expressed based on an *exergy destruction fraction* parameter which can be defined by $EDF=ex_d/ex_{in}$.

The second case considers air with variable specific heat $C_p=C_p(T)$. The common method of calculating entropy and enthalpy changes with an ideal gas with variable specific heat requires tabulation of C_p values against temperature. Common thermodynamic tables give $C_p(T)$ and in addition they contain the tabulated data for the integral $h^0-h_0^0 = \int_{T_0}^{T} C_p(T)dT$,

TABLE 5.32 Case Study of Destroyed Exergy During Air-Standard Expansion in Brayton Cycle

Calculation step	Assumptions	Equations	Calculated values
Reference state	$T_0 = 298.15$ K, $P_0 = 101.325$ kPa $h_0 = 8,649$ kJ/kmol, $s_0 = 165$ kJ/kmol.K	$\phi_0 = h_0 - T_0 s_0$ $ex_0 = 0$ $P_0 v_0 = R\,T_0$	$\phi_0 = -40.5$ kJ/mol $v_0 = 24.47$ m^3/kmol
Parameters of state 1 (inlet)	$T_1 = 1,273$ K, $P_1 = r\,P_0$, $r = 15$, standard air, $C_p = $ ct. $C_p = 29.101$ kJ/mol.K	$P_1 v_1 = R\,T_1$ $h_1 - h_0 = C_p\,(T_1 - T_0)$ $s_1 - s_0 = C_p \ln\left(\dfrac{T_1}{T_0}\right) - R \ln\left(\dfrac{P_1}{P_0}\right)$ $ex_1 = h_1 - T_0 s_1 - \phi_0$	$v_1 = 6.964$ m^3/kmol $h_1 = 37.0$ kJ/mol $s_1 = 185$ J/mol.K $ex_1 = 22.5$ kJ/mol
Parameters of state 2s	Process 1–2s: isentropic expansion, $\gamma = 1.4$	$s_{2s} = s_1, P_{2s} = P_0, P_{2s} v_{2s}^{\gamma} = P_1 v_1^{\gamma},$ $v_1 T_1^{\gamma-1} = v_{2s} T_{2s}^{\gamma-1},$ $h_{2s} - h_0 = C_p(T_{2s} - T_0),$ $ex_{2s} = h_{2s} - T_0 s_{2s} - \phi_0$	$v_{2s} = 48.19$ m^3/kmol $T_{2s} = 587.2$ K $h_{2s} = 17.1$ kJ/mol $s_{2s} = 185$ J/mol.K $ex_{2s} = 2.5$ kJ/mol
Parameters of state 2	$\eta_{S,t} = 0.85$	$P_2 = P_0, (h_1 - h_2) = \eta_{S,t}\,(h_1 - h_{2s})$ $(T_1 - T_2) = \eta_{S,t}(T_1 - T_{2s}), P_2 v_2 = R\,T_2$ $s_2 - s_0 = C_p \ln\left(\dfrac{T_2}{T_0}\right) - R \ln\left(\dfrac{P_2}{P_0}\right)$ $ex_2 = h_2 - T_0 s_2 - \phi_0$	$v_2 = 56.6$ m^3/kmol $T_2 = 690.1$ K $h_2 = 20.0$ kJ/mol $s_2 = 189$ J/mol.K $ex_2 = 4.1$ kJ/mol
Exergy balance	Process 1–2: adiabatic expansion	$ex_1 = ex_2 + w_t + ex_d, ex_{in} = ex_1 - ex_2$ $w_t = h_2 - h_1, \psi = \dfrac{w_t}{ex_{in}}, \text{EDF} = \dfrac{ex_d}{ex_{in}} = 1 - \psi$	$ex_d = 1.4$ kJ/mol $ex_{in} = 18.4$ kJ/mol EDF $= 7.63\%$

where h^0 is the enthalpy at T and h_0^0 is the enthalpy at the reference state temperature T_0. Furthermore, the following integral is also tabulated in common thermodynamic tables

$$\int_{T_0}^{T} \frac{C_p(T)}{T} dT = s^0 - s_0^0, \quad \text{where} \quad s_0^0 = s(T_0, P_0) \quad \text{and} \quad s^0 = s(T, P_0)$$

Hence, the enthalpy at any temperature T is expressed with $h(T) = h^0(T)$ and the entropy at any T and P is given by $s(T, P) = s^0(T) - R \ln(P/P_0)$. Using these considerations the calculations for case (ii) are presented in Table 5.33.

For calculation of temperature variation along the isentrope when the working fluid is ideal gas with variable specific heat the procedure presented in Chapter 7 of Cengel and Boles (2010) is used. This method consists of using the *reduced pressure*, which can be derived as follows. Say that 1 and 2 are two thermodynamic states along the isentrope. Then, because $s_1 = s_2$ one can write:

$$s^0(T_1) - R \ln\left(\frac{P_1}{P_0}\right) = s^0(T_2) - R \ln\left(\frac{P_2}{P_0}\right)$$

TABLE 5.33 Case Study of Destroyed Exergy During Real Air[a] Expansion within Brayton Cycle

Calculation step	Assumptions	Equations	Calculated values
Reference state	$T_0 = 298.15$ K, $P_0 = 101.325$ kPa	$\phi_0 = h_0 - T_0 s_0$ $ex_0 = 0$ $P_0 v_0 = R\,T_0$	$\phi_0 = -40.5$ kJ/mol $v_0 = 24.5$ m^3/kmol $h_0^0 = 8,649$ kJ/kmol $s_0^0 = 165$ kJ/kmol.K
Parameters of state 1 (inlet)	$T_1 = 1,273$ K, $P_1 = r\,P_0, r = 15$	$h^0(T_1) = 39.5$ kJ/mol, $s^0(T_1) = 209.9$ J/mol.K $h_1 = h^0(T_1), s_1 = s^0(T_1) - R\ln\left(\frac{P_1}{P_0}\right), P_1 v_1 = R\,T_1$ $ex_1 = h_1 - T_0 s_1 - \phi_0$	$v_1 = 6.964$ m^3/kmol $h_1 = 39.5$ kJ/mol $s_1 = 187.4$ J/mol.K $ex_1 = 24.2$ kJ/mol
Parameters of state 2s	Variable $C_p; C_p = f(T, \text{only})$	$s_{2s} = s_1, P_{2s} = P_0, h_{2s} = h^0(T_{2s}), P_{r,1} = \exp\left(\frac{s^0(T_1)}{R}\right)$ $P_{r,2s}(T_{2s}) = P_{r,1}\frac{P_{2s}}{P_1} = \exp\left(\frac{s^0(T_{2s})}{R}\right); P_{2s} v_{2s} = RT_{2s}$ $ex_{2s} = h_{2s} - T_0 s_{2s} - \phi_0$	$v_{2s} = 52.2$ m^3/kmol $h_{2s} = 18.7$ kJ/mol $s_{2s} = 187.4$ J/mol.K $ex_{2s} = 3.36$ kJ/mol
Parameters of state 2	$\eta_{S,t} = 0.85$	$P_2 = P_0, (h_1 - h_2) = \eta_{S,t}(h_1 - h_{2s}), h_2 = h^0(T_2)$ $P_2 v_2 = RT_2, s_2 = s^0(T_2) - R\ln\left(\frac{P_2}{P_0}\right)$ $ex_2 = h_2 - T_0 s_2 - \phi_0$	$v_2 = 60.46$ m^3/kmol $h_2 = 21.82$ kJ/mol $s_2 = 191.98$ J/mol.K $ex_2 = 5.12$ kJ/mol
Exergy balance	Process 1–2: adiabatic expansion	$ex_1 = ex_2 + w_t + ex_d, ex_{in} = ex_1 - ex_2,$ $w_t = h_2 - h_1, \psi = \frac{w_t}{ex_{in}}, EDF = \frac{ex_d}{ex_{in}} = 1 - \psi$	$ex_d = 1.36$ kJ/mol $ex_{in} = 19.0$ kJ/mol $EDF = 7.13\%$

[a]Here "real air" represents air modeled as ideal gas with variable specific heats.

Rearranging the above equation it results that

$$\frac{P_2}{P_1} = \frac{\exp\left(s^0(T_2)/R\right)}{\exp\left(s^0(T_1)/R\right)} = \frac{P_{r,2}}{P_{r,1}} \tag{5.30}$$

where the reduced pressure is introduced as $P_r = \exp(s^0(T)/R)$, which is a function of temperature only.

The above equation allows for correlating pressure variation with temperature variation during an isentropic process. As seen in Table 5.33 the entropy destruction fraction in case (ii) is 7.13%, hence it is slightly lower than for standard air.

Let us assume for case (iii) the complete combustion of CH_4 with an excess air fraction $\lambda = 5$. The composition of combustion gases results from:

$$CH_4 + 2\lambda(O_2 + 3.76N_2) \rightarrow CO_2 + 2H_2O + 7.52\lambda N_2 + 2(\lambda - 1)O_2$$

Hence, the molar fractions of compounds in the product gases result according to the total number of moles per mole of reacted methane: $1 + 2 + 7.52\lambda + 2(\lambda - 1) = 48.6$. The molar fractions in the product gas are $y_{CO_2} = 2.0\%$, $y_{H_2O} = 4.1\%$, $y_{N_2} = 77.4\%$, $y_{O_2} = 16.5\%$. As the temperature of gases at the turbine inlet is assumed to be above $T_1 = 1273$ K, the adiabatic flame temperature in any given condition must be estimated to verify that $AFT > T_1$. The AFT results from the energy balance over the adiabatic combustion chamber, that is $H_R = H_P$, where H is the total enthalpy and subscripts refer to reactants (R) and products (P), respectively. The solution of this equation will give $AFT = 1544$ K, well above the working temperature T_1.

TABLE 5.34   Case Study of Destroyed Exergy During Combustion Gas Expansion

Calculation step	Assumptions	Equations	Calculated values
Reference state	$T_0 = 298.15$ K, $P_0 = 101.325$ kPa	$\phi_0 = h_0 - T_0 s_0;\ h_0 = \sum y_i h_i^0(T_0);\ s_0 = \sum y_i s_i^0(T_0)$ $ex^{ch} = \sum y_i ex_i^{ch},\ i = CO_2,\ H_2O,\ N_2,\ O_2$ $P_0 v_0 = R\,T_0$	$\phi_0 = -77.7$ kJ/mol $v_0 = 24.5$ m^3/kmol $ex^{ch} = 1.998$ kJ/mol
Parameters of state 1 (inlet)	$T_1 = 1{,}273$ K, $P_1 = r\,P_0$, $r = 15$	$P_1 v_1 = R\,T_1, h_1 = \sum y_i h_i^0(T_1),$ $s_1 = \sum y_i s_i^0(T_1) - R\ln\left(\frac{P_1}{P_0}\Pi y_i^{y_i}\right),$ $ex_1 = ex^{ch} + h_1 - T_0 s_1 - \phi_0$	$v_1 = 6.964$ m^3/kmol $h_1 = 13.5$ kJ/mol $s_1 = 223$ J/mol.K $ex_1 = 26.5$ kJ/mol
Parameters of state 2s	Variable C_p; $C_p = f(T, \text{only})$	$s_{2s} = s_1, P_{2s} = P_0, h_{2s} = \sum y_i h_i^0(T_{2s}),$ $P_{2s} v_{2s} = R\,T_{2s},$ $P_1 \exp\left(\dfrac{\sum y_i s_i^0(T_1)}{R}\right) = P_2 \exp\left(\dfrac{\sum y_i s_i^0(T_{2s})}{R}\right),$ $ex_{2s} = ex^{ch} + h_{2s} - T_0 s_{2s} - \phi_0$	$v_{2s} = 53.05$ m^3/kmol $h_{2s} = -7.5$ kJ/mol $s_{2s} = 223$ J/mol.K $ex_{2s} = 5.58$ kJ/mol
Parameters of state 2	$\eta_{S,t} = 0.85$	$P_2 = P_0, (h_1 - h_2) = \eta_{S,t}(h_1 - h_{2s}), h_2 = \sum y_i h_i^0(T_2)$ $P_2 v_2 = R\,T_2, s_2 = \sum y_i s_i^0(T_2) - R\ln\left(\frac{P_2}{P_0}\Pi y_i^{y_i}\right),$ $ex_2 = ex^{ch} + h_2 - T_0 s_2 - \phi_0$	$v_2 = 61.16$ m^3/kmol $h_2 = -4.35$ kJ/mol $s_2 = 227.8$ J/molK $ex_2 = 7.37$ kJ/mol
Exergy balance	Process 1–2: adiabatic expansion	$ex_1 = ex_2 + w_t + ex_d, ex_{in} = ex_1 - ex_2,$ $w_t = h_2 - h_1, \psi = \dfrac{w_t}{ex_{in}},\ EDF = \dfrac{ex_d}{ex_{in}} = 1 - \psi$	$ex_d = 1.35$ kJ/mol $ex_{in} = 19.17$ kJ/mol $EDF = 7.03\%$

Because the magnitude of exergy in combustion gases is affected by the chemical exergy of components, the calculation steps given in Table 5.34 take this into account. However, the magnitude of chemical exergy is not high because no combustible substances are present. Of particular interest is the calculation of entropy when a mixture of ideal gases is involved. If one denotes y_i as the molar fraction of the components $i = 1 \ldots n$ then the partial pressures of each of the components is given by $P_i = y_i P$, where P is the total pressure. Therefore, the pressure term for the ideal gas mixture entropy is determined accounting for the identity $\sum_{i=1}^n y_i = 1$ as follows:

$$\sum_{i=1}^n \left[y_i \ln\left(\frac{P_i}{P_0}\right) \right] = \ln\left[\prod_{i=1}^n \left(\frac{y_i P}{P_0}\right)^{y_i} \right] = \ln\left[\left(\frac{P}{P_0}\right)^{\sum_{i=1}^n y_i} \prod_{i=1}^n y_i^{y_i} \right] = \ln\left(\frac{P}{P_0} \prod_{i=1}^n y_i^{y_i} \right)$$

Therefore, the entropy of the gas mixture at temperature T and pressure P becomes

$$s(T, P) = \sum_{i=1}^n y_i s_i^0(T) - R\ln\left(\frac{P}{P_0} \prod_{i=1}^n y_i^{y_i} \right) \tag{5.31}$$

Another required relation is pressure variation along an isentropic process. Assume an isentropic process evolving from state 1 to state 2. Given that $s_1(T_1, P_1) = s_2(T_2, P_2)$, the above equation can be used to obtain

$$\sum_{i=1}^n y_i s_i^0(T_1) - R\ln\left(\frac{P_1}{P_0} \prod_{i=1}^n y_i^{y_i} \right) = \sum_{i=1}^n y_i s_i^0(T_2) - R\ln\left(\frac{P_2}{P_0} \prod_{i=1}^n y_i^{y_i} \right)$$

By rearranging the equation one obtains a relationship between P_1 and P_2 expressed as a function of the reduced pressures. In the case of gas mixtures, the reduced pressure is given in a similar manner as for a single gas, as follows: $P_r = \exp(\sum_{i=1}^{n} y_i s_i^0 / R)$. Therefore, one has

$$\frac{P_2}{P_1} = \frac{P_{r,2}}{P_{r,1}} = \frac{\exp\left(\sum_{i=1}^{n} y_i s_i^0 (T_2)/R\right)}{\exp\left(\sum_{i=1}^{n} y_i s_i^0 (T_1)/R\right)}$$

In Table 5.34 the temperature at the end of isentropic expansion T_{2s} results from iterative solving of the equation $P_{r,2s} = \exp(\sum_{i=1}^{n} y_i s_i^0 (T_{2s})/R)$. Another iterative solving process is required to determine the temperature T_2 at the end of the non-isentropic expansion process; this is done by solving the equation $h_2 = \sum y_i h_i^0 (T_2)$ for T_2.

The results of the case study are comparatively analyzed in Figure 5.60. As seen, the temperature at the end of the expansion is predicted reasonably well when air with variable specific heats is the working fluid because the result is very close to that obtained with combustion gases. The standard-air prediction is significantly different. However, with regard to the exergy destruction fraction, the standard-air model gives a very reasonable prediction (i.e., 7.6% vs. 7.05 with combustion gases). In conclusion, this brief comparative study demonstrates the usefulness of standard air assumption in preliminary calculations for Brayton cycle and emphasizes that if the specific heat variation with temperature is accounted for then the predictions become sufficiently accurate. Note in addition that for turbines the pressure ratio per stage assuming an air-standard model ($\gamma = 1.4$) is given by the relation

$$\frac{P^*}{P_{in}} = 0.528$$

where P^* is the pressure in the throat and P_{in} is the pressure inlet per stage; the number 0.528 results from Equation (5.6).

Another source of irreversibility besides that caused by the non-isentropic process in the turbomachinery is due to the existence of friction in bearings and other mechanical

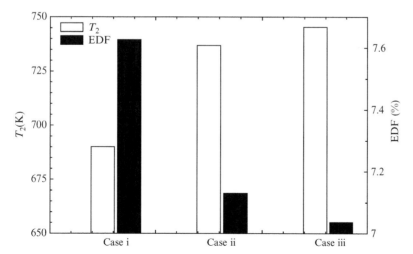

FIGURE 5.60 Comparison of turbine exit temperatures (T_2) and exergy destruction fractions (EDF) for three cases (see text for description).

components. The mechanical energy conversion efficiency is similar to that for a steam turbine system and has values in the approximate range of 0.9 to 0.95. The mechanical to electrical conversion efficiency is in the range of about 0.94 to 0.97.

Source side heat exchangers such as a heater and reheater have irreversibility caused by a finite temperature difference across the heat transfer surface. This situation occurs when the heater and reheater are heated from external sources for heat transfer. Let 1 and 2 be the thermodynamic states of working fluid at the entrance and exit of the heat source heat exchanger respectively. Let T_{so} be the average temperature of the heat source. With these notations the energy and entropy balances for standard air assumption are written in mass (or molar) specific form as follows:

$$C_p T_1 + q = C_p T_2 \text{ and } s_1 + \frac{q}{T_{so}} + s_{gen} = s_2$$

Therefore, because the generated entropy must be positive, the following restrictive relationship must exist between temperatures

$$T_{so} > \frac{T_2 - T_1}{\ln\left(\frac{T_2}{T_1}\right)}$$

A temperature difference of $\Delta T_{so} = T_{so} - T_2$ or reasonable amount must be considered between the average source temperature and average working fluid temperature. In most of practical situations heat is provided to the Brayton cycle via internal combustion. In such a case, it is logical to assume that the source temperature is equal to the temperature of combustion gases at the exit of combustion, $T_{so} = T_2$. If this is the case, the equality from above is automatically satisfied because if one denotes $\theta = T_2/T_1$ the condition $s_{gen} \geq 0$ is equivalent to $\theta \ln \theta - \theta + 1 \geq 0$ which is true for any $\theta > 0$; note that if $\theta = 0$ or $T_2 = T_1$ there is no entropy generated, $s_{gen} = 0$.

The exergy destruction at the combustion chamber can be approximated with better accuracy when the chemical exergy of product gases and C_p variation with temperature are considered. Let us assume an adiabatic combustion chamber as in Figure 5.61a. For this case, state 1 is the preheated air inlet, state 2 is the fuel inlet and state 3 is the temperature of combustion gases. The molar fraction of combustion gases is denoted with y_i.

The total exergy of fuel assumed in gaseous phase is

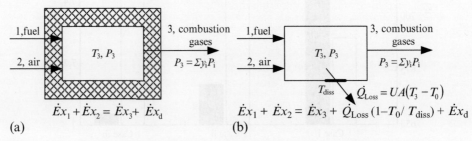

(a)   (b)

FIGURE 5.61 Exergy destruction in combustion chamber: (a) adiabatic combustion chamber and (b) non-adiabatic combustion chamber.

$$ex_1 = ex_f^{ch} + h_f^0(T_1) - h_f^0(T_0) - T_0\left[s_f^0(T_1) - R\ln\left(\frac{P_1}{P_0}\right) - s_f^0(T_0)\right] \tag{5.32}$$

If the fuel is in liquid phase, then the pressure term in the above equation is zero. The total exergy of air at state 2 is

$$ex_2 = ex_{air}^{ch} + h_{air}^0(T_2) - h_{air}^0(T_0) - T_0\left[s_{air}^0(T_2) - R\ln\left(\frac{P_2}{P_0}\right) - s_{air}^0(T_0)\right]$$

The total exergy expression for state 3 is the following

$$ex_3 = \sum y_i ex_i^{ch} + \sum y_i\left[h_i^0(T_3) - h_i^0(T_0)\right] - T_0\left\{\sum y_i\left[s_i^0(T_3) - s_i^0(T_0) - R\ln\left(\frac{P_3}{P_0}\prod y_i^{y_i}\right)\right]\right\}$$

The exergy destruction in the adiabatic combustion chamber results from the above equation and the exergy balance equation ($\dot{n}_1 ex_1 + \dot{n}_2 ex_2 = \dot{n}_3 ex_3 + \dot{Ex}_d$), while the generated entropy per mole of working fluid is given by the equation $\dot{Ex}_d = T_0\dot{S}_{gen}$. Here, the molar flow rate is denoted with $\dot{n}_i$, $i = 1, 2, 3$. In addition one notes that the temperature of combustion gases must be equal to the adiabatic flame temperature, $T_3 = AFT$, which results from the energy balance equation

$$\dot{n}_1 h_f^0(T_1) + \dot{n}_2 h_{air}^0(T_2) = \dot{n}_3 \sum y_i h_i^0(T_{aft})$$

Combustion chambers operate at a fairly high temperature with respect to the surroundings. Consequently, heat losses to outside are unavoidable. Because actual combustion chambers work non-adiabatically the exergy destruction is higher than that of an adiabatic chamber. Figure 5.61b illustrates a non-adiabatic combustion chamber.

An effectiveness parameter (ε_{cc}) for the combustion chamber can be introduced which expresses the relative magnitude of gas temperature increase for a non-adiabatic vs. adiabatic combustion chamber. If one denotes the adiabatic flame temperature T_{aft}, the air temperature at the combustion chamber entrance T_2 (see Figure 5.49b), and the combustion gases temperature T_3, then the combustion chamber effectiveness is by definition

$$\varepsilon_{cc} = \frac{T_3 - T_2}{T_{aft} - T_2} \tag{5.33}$$

Once an estimate of the combustion chamber effectiveness is made available, the temperature of combustion gases can be immediately determined from T_{aft} and T_2. Moreover, the heat losses are determined according to the energy balance equation:

$$\dot{n}_1 h_f^0(T_1) + \dot{n}_2 h_{air}^0(T_2) = \dot{n}_3 \sum y_i h_i^0(\varepsilon_{cc}(T_{aft} - T_2) + T_2) + \dot{Q}_{Loss}$$

After calculation of the heat loss rate ($\dot{Q}_{Loss}$), the exergy destruction of the non-adiabatic combustion chamber can be determined with the help the of exergy balance equation. As shown in Figure 5.61b the exergy balance equation for the non-adiabatic combustion chamber is

$$\dot{n}_1 ex_1 + \dot{n}_2 ex_2 = \dot{n}_3 ex_3 + \dot{Q}_{Loss}(1 - T_0/T_{diss}) + \dot{Ex}_d$$

where the temperature at which the heat losses are dissipated in the environment (T_{diss}) is the same as the temperature of the warm surface of the casing of the gas turbine assembly.

The key in preliminary thermodynamic analysis of the combustion chamber is to assume a reasonable value of the effectiveness, ε_{cc}. Extracting the energy balance equation for an adiabatic combustion chamber from the similar equation for a non-adiabatic chamber one obtains

$$\dot{Q}_{Loss} = \dot{n}_3 \sum y_i \left[h_i^0(T_{aft}) - h_i^0(\varepsilon_{cc}(T_{aft} - T_2) + T_2) \right]$$

With the assumption that at high temperature the combustion gases can be modeled as ideal gases with constant specific heat and that heat loss can be written as $\dot{Q}_{Loss} = UA(T_3 - T_0)$, where U is the overall heat transfer coefficient and A is the heat transfer area for heat leakages, the above equation becomes

$$UA[\varepsilon_{cc}(T_{aft} - T_2) + T_2 - T_0] \cong \dot{n}_3 C_p(T_{aft} - \varepsilon_{cc}(T_{aft} - T_2) + T_2 - T_0)$$

The above equation can be solved for ε_{cc} provided that all other parameters are specified. There is an analytical solution which is given as a function of the number of transfer units for the combustion chamber (NTU) and a dimensionless adiabatic flame temperature (θ_{aft}) according to:

$$\varepsilon_{cc} = \frac{NTU\, \theta_{aft} - 1}{(NTU + 1)(\theta_{aft} - 1)} \tag{5.34}$$

where the NTU and θ_{aft} are defined by

$$NTU = \frac{\dot{n}_3 C_p}{UA} \text{ and } \theta_{aft} = \frac{T_{aft} - T_0}{T_2 - T_0}$$

In the preliminary thermodynamic analysis of combustion turbines it is very important to have a reasonable assumption for the combustion chamber effectiveness; this will allow for estimation of exergy destruction. Therefore, knowing a practical range for combustion chamber effectiveness will facilitate the analyses, parametric studies, and optimization. A simple analysis can be performed to determine a practical range, as shown subsequently.

Assume that the surrounding air is at reference state $T_0 = 298$ K and the working fluid of the Brayton cycle is standard air. The typical compression ratio is $r = 16$; if two-stage compression is used then the compression ratio is about $r = 4$ per stage. The temperature at the end of compression can be determined as for an isentropic process with $T/P^\kappa = $ ct. with $\kappa = (\gamma - 1)/\gamma$ and $\gamma = 1.4$. Thus, for $r = 4 \ldots 16$ the air temperature at the end of compression is in the range of 450–650 K. The adiabatic flame temperature in the context when the combustion gases temperature at combustion turbine inlet is at least of 1000 K can be assumed to 1500–2500 K. Hence, the practical range for θ_{aft} can be estimated to 4–10. Using this range of values the combustion chamber effectiveness is plotted against the NTU as shown in Figure 5.62.

The number of thermal units (NTU) of the combustion chamber is better when thermal insulation is thicker. If one denoted k the thermal conductivity of the insulation, then the overall heat transfer coefficient for heat leakage from the combustion chamber can be approximated with $U = k/\delta$, where δ is the thickness of the insulation. Further, from the NTU definition it results that the thickness of the insulation is given by $\delta = kA\, NTU/\dot{n}_3 C_p$. The thermal capacity of the combustion chamber is higher when the flow rate of combustion gases $\dot{n}_3$ is higher, and so are the size and the "exposed" area of the chamber. Consequently one can take $A = $ ct.$\dot{n}_3$ and thus $\delta = $ ct.$k\, NTU/C_p$. Using some typical values for insulation thickness, NTU, k, C_p,

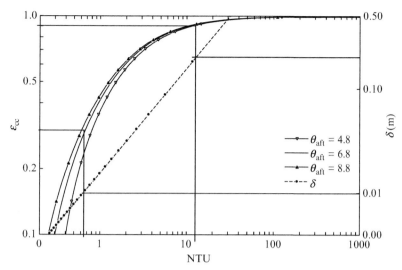

FIGURE 5.62   Effectiveness of combustion chamber versus *NTU*.

from practice one can finally approximate the insulation thickness with $\delta \cong \text{const.} NTU$ with const. $= 0.015 \, [\text{m}^{-1}]$. The insulation thickness is superimposed over the plot in Figure 5.62. It shows that for typical values of thermal insulation (say from 1 cm to 20 cm assuming a combustion chamber capacity from kW to hundreds of MW range) the range of *NTU* should be 0.6 to 11 and the range of combustion chamber effectiveness from 0.3 to 0.9.

For a numerical case study of exergy destruction let us assume $r = P_3/P_0 = 15$, excess air $\lambda = 2.0$, and methane combustion. From above analysis we know that the combustion gases in state 3 are of following molar fractions: $y_{CO_2} = 2.0\%, y_{H_2O} = 4.1\%, y_{N2} = 77.4\%, y_{O_2} = 16.5\%$. In state 1 we have $n_1 = 1$ mole of natural gas preheated at a temperature of $T_1 = 400 \, \text{K}$. In state 2 we must have $n_2 = 2 \lambda n_1 = 10 n_1$ moles of air preheated at $T_2 = 800 \, \text{K}$. In state 3 we must have the following amount of combustion gases: $n_3 = [3 + 7.52\lambda + 2(\lambda - 1)]n_1 = 48.6 n_1$ (see chemical equation above for the combustion turbine example). The process in the combustion chamber is isobaric: $P_1 = P_2 = P_3$. The temperature of combustion gases (T_3) depends on the assumed effectiveness of the combustion chamber, which varies from $\varepsilon_{cc} = 0.4$ to 1.0 (Figure 5.63).

One important parameter for exergy analysis is the temperature at which heat leaking out of the combustion chamber is dissipated to surroundings. If the system boundary is set far, then $T_{diss} = T_0$, otherwise, the dissipation temperature is somewhat higher. For the case study we assumed three values for T_{diss}, as seen in Figure 5.51. The exergy destruction fraction for the combustion chamber can go up to 40–50% when the effectiveness is low and becomes ~20% when the operation is adiabatic. For $\varepsilon_{cc} = 90\%$, the *EDF* is around 25%.

Another important source of exergy destruction at the combustion chamber is due to the fact that the chemical reaction cannot go to completion (the reactants do not fully transform into products). This kind of irreversibility is related to the design of the combustion chamber and some crucial operational parameters such as excess air, adiabatic flame temperature, and combustion chamber effectiveness. The combustion reactions are spontaneous and have a

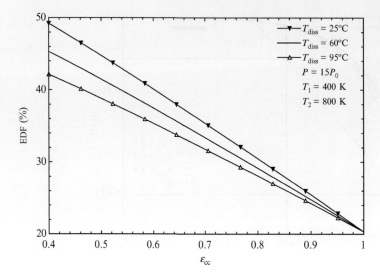

FIGURE 5.63 Exergy destruction fraction versus combustion chamber effectiveness for a case study.

highly negative free energy, which is an indication that the reaction evolves at non-equilibrium. In a gas turbine there is always some non-combusted fuel in flue gases.

This is one of the reasons why in practice gas turbines are used in conjunction with heat recovery steam generators in CCPPs. The degree of completion of the combustion reaction is given by the "extent of reaction," a parameter that varies from 0 to 1; the value 1 corresponds to a complete reaction, while 0 corresponds to a "no reaction" case. A general chemical reaction can be described with

$$\sum \mathcal{R} \rightarrow \zeta \sum \mathcal{P} + (1 - \zeta) \sum \mathcal{R}$$

where $\mathcal{R}$ and $\mathcal{P}$ are generic names for reactants and products, respectively, and $\zeta \in [0, 1]$ represents the extent of reaction.

Note that the maximum value of the extent of reaction is 1. However, if the reaction is at equilibrium, then the extent of reaction satisfies the inequality $\zeta \leq \zeta_{eq} \leq 1$. The free energy of the reaction is given by $\Delta G = G_P - G_R$, where G represents the free energy of the products (P) and reactants (R). Here $G_{P,R} = H_{P,R} - T S_{P,R}$, where T is the reaction temperature, which normally is taken as the temperature of the products. If the combustion process is adiabatic, then $\Delta H = H_P - H_R = 0$, thus $\Delta G = \Delta H - T \Delta S = T \Delta S$. The adiabatic flame temperature and the entropy of the reaction for adiabatic combustion can be determined from the following system of equations

$$\begin{cases} \sum_i n_i h_i(T_i) \big|_R = \zeta \sum_j n_j h_j(T_{aft}) \big|_P + (1 - \zeta) \sum_k n_k h_k(T_{aft}) \big|_R \\ \Delta S = \zeta \sum_j n_j s_j \left(T_{aft}, y_j P\right) \big|_P + (1 - \zeta) \sum_k n_k h_k \left(T_{aft}, y_k P\right) \big|_R - \sum_i n_i s_i \left(T_i, y_i P\right) \big|_R \end{cases}$$

If the reaction entropy is negative then the reaction is spontaneous. This is the case with the majority of combustion reactions. A numerical example is given next to illustrate how the extent of the reaction affects the adiabatic flame temperature and hence the exergy destruction.

Assume an adiabatic combustion chamber of methane combustion operating at parameters similar to the previous example. The only difference is that this time the assumption that the reaction goes to completion is relaxed. Therefore, the actual reaction for the adiabatic combustion is

$$CH_4 + 2\lambda(O_2 + 3.76N_2) \rightarrow$$

$$\rightarrow \zeta(CO_2 + 2H_2O + 2(\lambda - 1)O_2 + 7.52\lambda N_2) + (1 - \zeta)(CH_4 + 2\lambda(O_2 + 3.76N_2))$$

The total number of moles of products per mole of CH_4 depends on the extent of reaction and can be determined from the above equation: $n = \zeta(1 + 3 + 2(\lambda - 1) + 7.52\lambda) + (1 - \zeta)(1 + 2\lambda + 7.52\lambda) = 9.52\lambda + 1$. The following amount of moles for each species in product gases exist, for one mole of CH_4 reactant: $n_{CO_2} = \zeta$, $n_{H_2O} = 2\zeta$, $n_{CH_4} = 1 - \zeta$, $n_{O_2} = 2(\lambda - \zeta)$, $n_{N_2} = 7.52\lambda$. For our case study with the assumed $\lambda = 2$ the following molar fractions exist in product gases as functions of the extent of reaction: $y_{CO_2} = 0.0499\zeta$, $y_{H_2O} = 0.0998\zeta$, $y_{CH_4} = 0.0499(1 - \zeta)$, $y_{O_2} = 0.1996 - 0.0998\zeta$, $y_{N_2} = 0.7505$. The enthalpy and entropy of reactants can be directly calculated for $T_1 = 400$ K (temperature of CH_4) and $T_2 = 800$ K (for air) and total pressure $P = r P_0$, with $r = 15$. This is shown below:

$$\begin{cases} H_R = n_1 h_f^0(T_1) + n_2 \left[0.21 h_{O_2}^0(T_2) + 0.79 h_{N_2}^0(T_2)\right] = -9.764\,\text{MJ/kmol} \\ S_R = n_1 s_f(T_1, P) + n_2 \left[0.21 s_{O_2}^0(T_2, 0.21P) + 0.79 s_{N_2}^0(T_2, 0.79P)\right] = 998.3\,\text{kJ/kmol.K} \end{cases}$$

The enthalpy and entropy of the reaction products can be calculated provided that the extent of reaction and the combustion chamber effectiveness are specified. Figure 5.64 shows a parametric study that gives the evolution of exergy destructions with the extent of reaction and the combustion chamber effectiveness. The adiabatic flame temperature is maximum when the reaction is complete (1617 K) and degrades to 822 K for $\zeta = 0.3$. Also the exergy

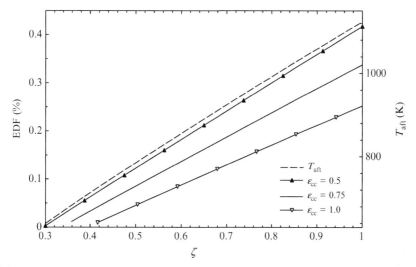

FIGURE 5.64   The influence of the extent of combustion reaction on exergy destructions.

destruction fraction is higher when combustion chamber effectiveness is lower. In addition it is remarked that the less reaction occurs (the extent of reaction is lower) the less exergy destruction is obtained. This type of behavior is the logical consequence of the second law of thermodynamics: more exergy must be destroyed when more moles of reactants enter into reaction. In combustion the extent of reaction can be assumed in the range of $\zeta \in [0.9, 1]$.

Another type of exergy destruction in Brayton-cycle-based power plants occurs in the regenerator. This exergy destruction can be accounted for by the effectiveness of the regenerator ε_{reg}, which is defined by the actual heat recovered at the ideal maximum heat recovery potential. Figure 5.65 illustrates a regenerator which recovers heat from hot gases expelled by the turbine in 1 to preheat combustion air according to process 3–4. Note that for a general case the actual heat recovered is given by $\dot{Q} = \dot{m}_{colder}(h_4 - h_3)$.

What is the ideal heat transfer rate that can be transferred to the colder fluid? This is obtained when the colder fluid is heated up to the maximum temperature of the hotter fluid, T_1. At the same time, the hotter fluid temperature must decrease to the minimum possible value, which is T_3, thus: $\dot{Q}_{ideal} = \dot{m}_{hotter}(h_1 - h_2(T_3))$.

For most practical situations in a Brayton cycle $\dot{m}_{colder} = \dot{m}_{hotter}$ and the working fluid is ideal gas; therefore, the enthalpy depends on temperature only. For standard air assumption, the molar specific heat is $C_p = 7/2R$; therefore, $h_2(T_3) = h_3(T_3)$. Thus, the mathematical expression for regenerator effectiveness definition is (see Figure 5.65):

$$\varepsilon_{reg} = \frac{\dot{Q}}{\dot{Q}_{ideal}} = \frac{h_4 - h_3}{h_1 - h_3} \cong \frac{T_4 - T_3}{T_1 - T_3} \tag{5.35}$$

The regenerator effectiveness is in general in the range of 0.8 to 0.9; a value of 1 signifies an ideal regenerator for which there is no entropy generation or exergy destruction. This is possible only if between the hotter and colder streams there is no temperature difference at any time. In the above equation, the sign "$\cong$" becomes "$=$" if the cycle operates with standard air.

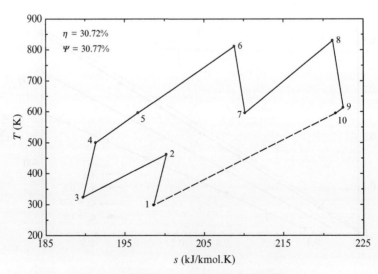

FIGURE 5.65 Case Study of Brayton cycle with intercooling and reheating.

The heat exchangers at the sink side (cooler and intercooler) introduce additional irreversibility to the cycle and thus produce exergy destruction. With reference to the notations in Figures 5.57 and 5.58 one must state that in an actual system cooling processes 2–3 and 10–1 cannot occur down to the reference temperature, T_0. The irreversibility due to finite temperature difference can be accounted for by specifying a minimum temperature in the cycle which is superior to T_0. In general, one can assume $\Delta T_{si} = 10 - 30$ K as the minimum temperature difference for the sink side heat exchangers, where ΔT_{si} is defined by the equation

$$\Delta T_{si} = T_3 - T_0$$

If the cycle operates in open loop then one can set $T_1 = T_0$; otherwise, for closed loop operation, one must set $T_1 = T_0 + \Delta T_{si}$. For the intercooler one must satisfy the following inequality, which guarantees that the generated entropy is not negative:

$$T_0 + \Delta T_{si} < \overline{T}_{si} \leq \frac{T_2 - T_3}{\ln\left(\dfrac{T_2}{T_3}\right)}$$

where the state notations are as indicated in Figure 5.57. Here, $\overline{T}_{si}$ is the average temperature of the colder fluid at the sink side heat exchanger. A similar inequality must exist for the cooler (process 10–1).

Additional irreversibility in Brayton-cycle-based plants are caused by the pressure drop in heat exchangers and other flow elements. The additional exergy destruction due to flow circulation must be a small percent of the work developed by the turbine. A parameter denoted as "flow work ratio" FWR can be introduced to help take into account the additional exergy destruction owing to the pressure drop at source and sink side heat exchangers. The flow work ratio is defined by

$$\text{FWR} = \frac{\Delta ex_{d,\Delta P}}{w_{rev}} \tag{5.36}$$

where $\Delta ex_{d,\Delta P}$ is the additional exergy destruction due to the existence of pressure drop, and w_{rev} is the reversible work developed by the turbine; both quantities are expressed in mass (or molar) specific terms.

The reversible work of the turbine under r pressure ratio and air standard assumption is $w_{rev} = \kappa R T_0 / \tau (r^\kappa - 1)$, with $\kappa = (\gamma - 1)/\gamma$ and $T_0 = \tau T_{tit}$ (T_{tit} being the temperature at turbine inlet). The exergy destruction due to pressure drop with air standard assumption will be given by

$$\Delta ex_{d,\Delta P} = -R \ln\left(1 - \frac{\Delta P}{P}\right) \tag{5.37}$$

where P is P_0 for sink side pressure drop and rP_0 for source side pressure drop. Using the above two equations it results that the pressure drop for the sink side heat exchanger can be approximated with

$$\Delta P_{si} = P_0\left\{1 - \exp\left[-\kappa\frac{\text{FWR}}{\tau}(r^\kappa - 1)\right]\right\} \cong P_0[1 - \exp(-1.7\,\text{FWR})] \tag{5.38}$$

TABLE 5.35 Parameters that Quantify the Irreversibilities in a Brayton Cycle

Parameter	Name and definition[a]	Typical range
$\eta_{s,c}$	Isentropic efficiency of compressor $h_{2s} - h_1 = \eta_{s,c}(h_2 - h_1); \; h_{2s} = h(P_2, s_1)$	0.8–0.85
$\eta_{s,t}$	Isentropic efficiency of the turbine $h_6 - h_7 = \eta_{s,t}(h_6 - h_{7s})$	0.8–0.9
η_m	Mechanical efficiency: $\dot{W}_{shaft} = \eta_m \dot{W}_{net}; \; \dot{W}_{net} = \dot{W}_t - \dot{W}_c$	0.9–0.95
η_{el}	Electro-mechanical energy conversion efficiency: $\dot{W}_{el} = \eta_{el} \dot{W}_{shaft}$	0.94–0.97
ΔT_{so}	Minimum temperature difference at source: $\Delta T_{so} = T_{so} - T_6 > \dfrac{T_6 - T_5}{\ln\left(\dfrac{T_6}{T_5}\right)} - T_6$	10–100 K
ε_{cc}	Effectiveness of combustion chamber $\varepsilon_{cc} = \dfrac{T_6 - T_5}{T_{aft} - T_5}$	0.3–0.9
ζ	Extent of reaction of combustion $\sum \to \zeta \sum \mathcal{P} + (1 - \zeta) \sum$	0.9–1
ε_{reg}	Effectiveness of the regenerator $\varepsilon_{reg} = \dfrac{\dot{Q}}{\dot{Q}_{ideal}} = \dfrac{h_5 - h_4}{h_9 - h_4} \cong \dfrac{T_5 - T_4}{T_9 - T_4}$	0.8–0.9
ΔT_{si}	Minimum temperature difference at sink: $\Delta T_{si} = T_3 - T_0$	10–30 K
FWR	Flow work ratio $\mathrm{FWR} = \dfrac{\Delta ex_{d,\Delta P}}{w_{rev}}; \; \Delta P_{si} \cong P_0[1 - \exp(-1.7\,\mathrm{FWR})]$ $P_{so} \cong \Delta P_{si}/r$	0.001–0.005

[a]Definitions are given based on diagram in Figure 5.57.

In the above equation is it approximated that $\tau \cong 0.2, k \cong 1.4, r \cong 16$, which are average values for combustion turbine applications. The pressure drop at the source side for this pressure drop will be $\Delta P_{so} \cong \Delta P_{si}/r$. The value range for the *FWR* parameter can be assumed as 0.001–0.005. A summary of the parameters discussed above that quantify the exergy destructions in Brayton cycles is given in Table 5.35.

EXAMPLE 5.11

A case study is next presented of a Brayton cycle of the same configuration as in Figure 5.57, this time with the assumption of no internal irreversibilities being relaxed. The assumed parameters that quantify the exergy destruction and the relevant equations for each process are given in Table 5.36. The working fluid is assumed to be ideal gas air consisting of a mixture of 21% oxygen and 79% nitrogen.

EXAMPLE 5.11 (cont'd)

TABLE 5.36 Assumptions and Balance Equations for Brayton Cycle Case Study

Component	Assumptions	Balance equations
Stage I compressor (process 1–2)	Adiabatic, $\eta_{S,c1} = 0.85$, $r = 15$ $T_1 = T_0 = 298.15$ K, $P_2 = r^{0.5}P_1$ $P_1 = P_0 = 101.325$ kPa Air: $0.21\%\ O_2 + 0.79\%\ N_2$	MBE: $\dot{m}_1 = \dot{m}_2$ EBE: $\dot{m}_1 h_1 + \dot{W}_{c,1} = \dot{m}_2 h_2$ EnBE: $\dot{m}_1 s_1 + \dot{S}_{g,c1} = \dot{m}_2 s_2$ ExBE: $\dot{m}_1 ex_1 + \dot{W}_{c,1} = \dot{m}_2 ex_2 + \dot{Ex}_{d,c1}$
Intercooler (process 2–3)	$\Delta T_{si} = T_3 - T_0 = 25$ K Pressure drop: FWR $= 0.0025$ $P_2 - P_3 = 0.0042\,P_2$ $\overline{T}_{si} = \dfrac{T_2 - T_3}{\ln\left(\dfrac{T_2}{T_3}\right)} - \Delta T_{si}$	MBE: $\dot{m}_2 = \dot{m}_3$ EBE: $\dot{m}_2 h_2 = \dot{m}_3 h_3 + \dot{Q}_{ic}$ EnBE: $\dot{m}_2 s_2 + \dot{S}_{g,ic} = \dot{m}_3 s_3 + \dfrac{\dot{Q}_{ic}}{\overline{T}_{si}}$ ExBE: $\dot{m}_2 ex_2 = \dot{m}_3 ex_3 + \dot{Q}_{ic}\left(1 - \dfrac{T_0}{\overline{T}_{si}}\right) + \dot{Ex}_{d,c1}$
Stage II compressor (process 3–4)	Adiabatic, $\eta_{S,c2} = 0.85$, $r = 15$, $P_4 = r^{0.5}P_3$	MBE: $\dot{m}_3 = \dot{m}_4$ EBE: $\dot{m}_3 h_3 + \dot{W}_{c,2} = \dot{m}_4 h_4$ EnBE: $\dot{m}_3 s_3 + \dot{S}_{g,c2} = \dot{m}_4 s_4$ ExBE: $\dot{m}_3 ex_3 + \dot{W}_{c,2} = \dot{m}_4 ex_4 + \dot{Ex}_{d,c2}$
Regenerator (processes 4–5 and 9–10)	$\varepsilon_{reg} = \dfrac{T_5 - T_4}{T_9 - T_4} = 0.85$, $\dot{Q}_{Loss} = 0$ Pressure drop: FWR $= 0.0025$ $P_5 - P_4 = 0.0042\,P_4$ $P_9 - P_{10} = 0.0042\,P_{10}$	MBE: $\dot{m}_4 + \dot{m}_9 = \dot{m}_5 + \dot{m}_{10}\ \dot{m}_4 = \dot{m}_5, \dot{m}_9 = \dot{m}_{10}$ EBE: $\dot{m}_4 h_4 + \dot{m}_9 h_9 = \dot{m}_5 h_5 + \dot{m}_{10} h_{10}$ EnBE: $\dot{m}_4 s_4 + \dot{m}_9 s_9 + \dot{S}_{g,reg} = \dot{m}_5 s_5 + \dot{m}_{10} s_{10}$ ExBE: $\dot{m}_4 ex_4 + \dot{m}_9 ex_9 = = \dot{m}_5 ex_5 + \dot{m}_{10} ex_{10} + \dot{Ex}_{d,reg}$
Combustion chamber (process 5, 11–6, fuel CH_4)	Isobaric, non-adiabatic: $P_6 = P_5$, $\varepsilon_{cc,1} = \dfrac{T_6 - T_5}{T_{aft,1} - T_5} = 0.85$, Dissipated heat at $T_{diss} = 368$ K Complete combustion (no CO), but non-reacted $CH_4, \zeta = 0.95$, Excess air: $\lambda = 3$	MBE: $\dot{m}_5 + \dot{m}_{11} = \dot{m}_6$ EBE: $\dot{m}_5 h_5 + \dot{m}_{11} h_{11} = \dot{m}_6 h_6 + \dot{Q}_{Loss,1}$ EnBE: $\dot{m}_5 s_5 + \dot{m}_{11} s_{11} + \dot{S}_{g,cc,1} = \dot{m}_6 s_6 + \dfrac{\dot{Q}_{Loss,1}}{T_{diss}}$ ExBE: $\dot{m}_5 ex_5 + \dot{m}_{11} ex_{11}$ $\quad = \dot{m}_6 s_6 + \dot{Q}_{Loss,1}\left(1 - \dfrac{T_0}{T_{diss}}\right) + \dot{Ex}_{d,cc,1}$
High-pressure turbine (process 6–7)	Adiabatic, $\eta_{S,t1} = 0.9$ $r^{0.5}P_7 = P_6, r = 15$ Working fluid: ideal gas mixture of combustion gases	MBE: $\dot{m}_6 = \dot{m}_7$ EBE: $\dot{m}_6 h_6 = \dot{m}_7 h_7 + \dot{W}_{t,1}$ EnBE: $\dot{m}_6 s_6 + \dot{S}_{g,t1} = \dot{m}_7 s_7$ ExBE: $\dot{m}_6 ex_6 = \dot{m}_7 s_7 + \dot{W}_{t,1} + \dot{Ex}_{d,t1}$
Secondary combustion chamber (process 7–8)	Isobaric, non-adiabatic: $P_8 = P_7$, $\varepsilon_{cc,2} = \dfrac{T_8 - T_7}{T_{aft,2} - T_7} = 0.85$ Complete combustion (no CO), but non-reacted CH_4, $\zeta = 0.995$, Excess air: $\lambda = 6$	MBE: $\dot{m}_7 + \dot{m}_{12} = \dot{m}_8$ EBE: $\dot{m}_7 h_7 + \dot{m}_{12} h_{12} = \dot{m}_8 h_8 + \dot{Q}_{Loss,2}$ EnBE: $\dot{m}_7 s_7 + \dot{m}_{12} s_{12} + \dot{S}_{g,cc,2} = \dot{m}_8 s_8 + \dfrac{\dot{Q}_{Loss,2}}{T_{diss}}$ ExBE: $\dot{m}_7 ex_7 + \dot{m}_{12} ex_{12}$ $\quad = \dot{m}_8 s_8 + \dot{Q}_{Loss,2}\left(1 - \dfrac{T_0}{T_{diss}}\right) + \dot{Ex}_{d,cc,2}$
Low-pressure turbine (process 8–9)	Adiabatic, $\eta_{S,t2} = 0.9$ $r^{0.5}P_7 = P_6, r = 15$ Working fluid: ideal gas mixture of combustion gases	MBE: $\dot{m}_8 = \dot{m}_9$ EBE: $\dot{m}_8 h_8 = \dot{m}_9 h_9 + \dot{W}_{t,2}$ EnBE: $\dot{m}_8 s_8 + \dot{S}_{g,t2} = \dot{m}_9 s_9$ ExBE: $\dot{m}_8 ex_8 = \dot{m}_9 s_9 + \dot{W}_{t,2} + \dot{Ex}_{d,t2}$

Continued

EXAMPLE 5.11   (cont'd)

In addition it is assumed that the excess air is $\lambda=3$ for the primary combustion chamber and $\lambda=5$ for the secondary combustion chamber. These values can be selected by trial and error aiming for a similar exergy destruction in both turbines. The pressure drop in the intercooler and regenerator are accounted for by assuming a flow work ratio of FWR$=0.0025$. Compressors are both assumed to have an isentropic efficiency of 0.85, while the turbines have isentropic efficiencies of 0.9. In addition an electromechanical energy conversion efficiency is considered as the product $\eta_m \eta_{el}$, which is taken to equal 0.89. The combustion chambers operate non-adiabatically, with an effectiveness of 0.85 for both.

The gas temperature at the exit of the combustion chamber is determined based on the adiabatic flame temperature and the combustion chamber effectiveness. The molar fractions of gaseous species must be determined as a prerequisite for calculating the adiabatic flame temperature. These molar fractions result from an assumed chemical reaction and the assumed extent of reaction (Figure 5.66).

In this case study it is assumed that the combustion is complete, that is only CO_2 and H_2O are produced and no other compounds are formed. Furthermore, a reasonable value is assumed for the extent of reaction; this allows for determination of the molar fractions of species in the combustion gases and also of the adiabatic flame temperature. Note that in the secondary combustion chamber only additional fuel is injected because the molar fraction of oxygen is sufficiently high to assure a combustion process with the desired excess ratio.

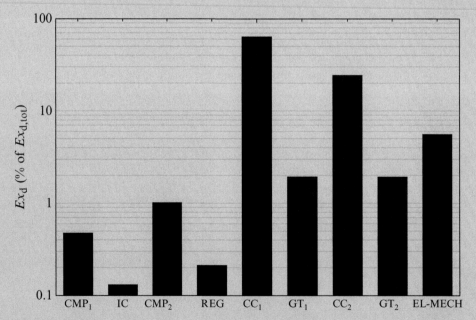

FIGURE 5.66   Exergy destructions for each component of Brayton cycle analyzed in case study.

EXAMPLE 5.11 *(cont'd)*

The exergy and energy efficiencies for the cycle analyzed in the case study are 30.7% (energy) and 30.8% (exergy). The BWR of the cycle is rather good (BWR = 0.15) because of the application of intercooling. The impact of internal irreversibilities on exergy efficiency is obvious.

The exergy destructions are the highest in the primary combustion chamber with respect to total exergy destruction. About 64% of the exergy destruction within the cycle occurs in the primary combustion chamber and about 24% is caused by the processes in the secondary combustion chamber. Also, the electromechanical exergy destruction is relatively high (5.6%) due to the fact that a relatively low electromechanical efficiency has been assumed.

5.4 COMBINED CYCLE POWER PLANTS

From the example above, it results that the temperature of the combustion gases at their discharge into the atmosphere is rather high (>595 K). Although the gas has a high temperature its pressure is low ($\sim P_0$), and therefore no further expansion is possible. The still high energy of the expelled gases can be recovered if the heat is transferred to a Rankine cycle. CCPPs are attractive power generation options due to a higher thermal efficiency than that found in individual steam or gas turbine cycles. In these systems, the expelled low-pressure hot gases from the gas turbine are directed to a heat recovery steam generation (HRSG) unit where additional fuel is sprayed. There is sufficient oxygen in the gases to maintain a combustion process that leads to temperature increase. Further, the hotter gases are used in a sequence of heat exchangers to generate steam for a Rankine cycle.

The main challenge in designing a combined cycle is proper utilization of gas turbine exhaust heat in the steam cycle in order to achieve optimum steam turbine output. According to the benefits of CCPP, the number and output power of such cycles have increased recently. Combined cycles have higher thermal efficiency as well as higher output power when compared to gas turbine and steam cycles. Higher efficiencies of CCPPs compared to Brayton and Rankine cycles have made them quite attractive for power generation. Based on these advantages and less specific emissions, CCPPs have widely been used all around the world.

The main design parameters of CCPPs are: pressure ratio (r), compressor isentropic efficiency ($\eta_{s,c}$), gas turbine isentropic efficiency ($\eta_{s,gt}$), steam turbine isentropic efficiency ($\eta_{s,st}$), gas turbine inlet temperature (T_{TIT}), duct burner mass flow rate ($\dot{m}_{db}$), pressure and temperature of superheated steam, condenser temperature, temperature difference at the pinch point, isentropic efficiency of the pumps (η_p).

A typical diagram of a CCPP is illustrated in Figure 5.67. In this diagram a Brayton cycle of basic configuration is considered. The hot gases expelled in D (see Figure 5.67) have a relatively large flow rate and are directed into the duct burner. In the first section of the duct burner fuel is sprayed. The gases reach the highest temperature in state 11.

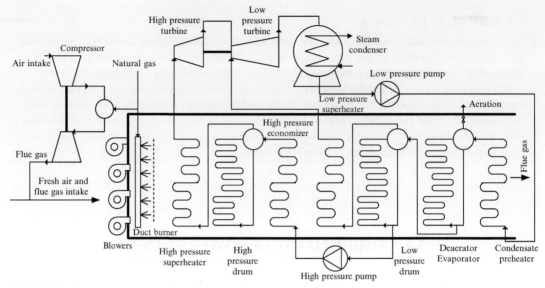

FIGURE 5.67 Schematic flow diagram of a dual-pressure combined cycle power plant. *Adapted from Ahmadi and Dincer (2011).*

The hot gases exchange heat sequentially with the steam superheater, high-pressure calandria boiler, economizer, low-pressure steam superheater, low-pressure calandria boiler, deaerator-evaporator (DEA-EVP), and condensate preheater. Eventually, the cooler gases exit the drum boiler at temperature T_{out} (see Figure 5.67).

In the Rankine cycle the condensate is pressurized to an intermediate pressure (process 20–1) and then directed to the condensate preheater. A DEA-EVP follows, where steam is partially boiled in a calandria system. Low-pressure steam is extracted from the evaporator drum and further directed to the low-pressure calandria boiler where the full amount of saturated low-pressure steam is generated in 5. The steam is then superheated (process 5–6) and injected into the turbine system. Liquid drained from the evaporator drum is further pressurized and directed to the high-pressure economizer (process 7–8). Next, saturated steam is generated in the high-pressure calandria boiler and superheated according to process 9–10.

The efficiency of the CCPP can be expressed in terms of net work output (the sum of net work generated by the gas turbine and by the steam turbine minus the work consumed by the compressor and the pumps) and total fuel energy input. The exergy efficiencies and the exergy balance equations for the CCPP shown in Figure 5.67 and are given in Table 5.37.

Some common assumptions for modeling these type of plants are as follows: all the processes are steady state and steady flow, the principle of ideal gas mixture is applied for the air and combustion products, the fuel injected into the combustion chamber is assumed to be natural gas, heat loss from the combustion chamber is considered to be 3% of the fuel's lower heating value, and all other components are considered adiabatic. In Figure 5.68 the typical temperatures of the main components of the Rankine plant in a CCPP are presented, from Ahmadi and Dincer (2011).

TABLE 5.37 Exergy Balance Equations and Exergy Efficiency Definitions for a CCPP

Component	Exergy balance equation	Exergy efficiency definition
Compressor	$\dot{Ex}_A + \dot{W}_{CMP} = \dot{Ex}_B + \dot{Ex}_{d,CMP}$	$\psi_{CMP} = 1 - \dot{Ex}_{d,CMP}/\dot{W}_{CMP}$
Combustion chamber	$\dot{Ex}_B + \dot{Ex}_{fuel,CC} = \dot{Ex}_C + \dot{Ex}_{d,CC}$	$\psi_{CC} = 1 - \dot{Ex}_{d,CC}/(\dot{Ex}_B + \dot{Ex}_{fuel,CC})$
Gas turbine	$\dot{Ex}_C = \dot{Ex}_D + \dot{W}_{GT} + \dot{Ex}_{d,GT}$	$\psi_{GT} = 1 - \dot{Ex}_{d,GT}/(\dot{Ex}_C - \dot{Ex}_D)$
Duct burner	$\dot{Ex}_D + \dot{Ex}_{fuel,DB} = \dot{Ex}_{11} + \dot{Ex}_{d,DB}$	$\psi_{DB} = 1 - \dot{Ex}_{d,DB}/(\dot{Ex}_D + \dot{Ex}_{fuel,DB})$
HRSG	$\dot{Ex}_1 + \dot{Ex}_{11} = \dot{Ex}_{10} + \dot{Ex}_6 + \dot{Ex}_{18} + \dot{Ex}_{d,HRSG}$	$\psi_{HRSG} = 1 - \dfrac{\dot{Ex}_{d,HRSG}}{\dot{Ex}_1 + \dot{Ex}_{11} - \dot{Ex}_{d,HRSG}}$
Steam turbine	$\dot{Ex}_{10} + \dot{Ex}_6 = \dot{Ex}_{19} + \dot{W}_{ST} + \dot{Ex}_{d,ST}$	$\psi_{ST} = 1 - \dot{Ex}_{d,ST}/(\dot{Ex}_{10} + \dot{Ex}_6 - \dot{Ex}_{19})$
Condenser	$\dot{Ex}_{19} + \dot{Ex}_{21} = \dot{Ex}_{20} + \dot{Ex}_{d,CND}$	$\psi_{CND} = 1 - \dfrac{\dot{Ex}_{d,CND}}{\dot{Ex}_{19} + \dot{Ex}_{21} - \dot{Ex}_{22}}$
Pump	$\dot{Ex}_{20} + \dot{W}_P = \dot{Ex}_1 + \dot{Ex}_{d,P}$	$\psi_P = 1 - \dot{Ex}_{d,P}/\dot{W}_P$
CCPP (overall)	$\dot{Ex}_A + \dot{W}_{CMP} + \dot{Ex}_{fuel,CC} + \dot{Ex}_{fuel,DB} + \dot{Ex}_{21} + \dot{W}_P$ $= \dot{W}_{GT} + \dot{W}_{ST} + \dot{Ex}_{18} + \dot{Ex}_{22} + \dot{Ex}_{d,CCPP}$	$\psi_{CCPP} = \dfrac{\dot{W}_{GT} + \dot{W}_{ST} - (\dot{W}_{CMP} + \dot{W}_P)}{\dot{Ex}_{fuel,CC} + \dot{Ex}_{fuel,DB}}$

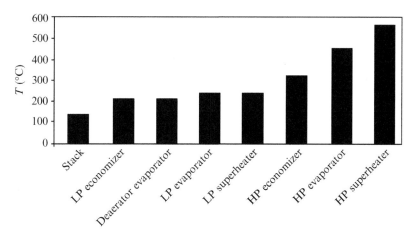

FIGURE 5.68 Variation of hot gas temperature for various heat transfer elements of HRSG at Neka (Iran) CCPP (comparison of modeling results with measured values). *Adapted from Ahmadi and Dincer (2011).*

5.5 HYDROPOWER PLANTS

Hydro-energy is derived from solar energy. The difference in elevation of water level—upstream and downstream of a dam—represents an accumulation of potential energy which eventually can be transformed into mechanical work and thus into electricity. There are a number of water turbines and waterwheels which can generate rotation work when exposed to water flow. This is a potential worldwide method for electric power production and has gained significance because the majority of rural properties have rivers with water streams and small water heads that can be used as primary energy.

Energy is commonly used for stationary machine drivers, generation of electricity, and water storage in elevated reservoirs. Small electric power plants can still be used with waterwheels or, depending on the flow or the available head, with turbines. One important characteristic of a hydropower plant is the head, which represents the difference in water level elevation. Based on this, hydro-energy plants are classified in two categories:

- Low head power plants: head = 5–20 m;
- High heat power plants: heat = 20–1000 m.

As a function of the heat and the volumetric flow rate of water, specific turbines or waterwheels can be selected. The known hydro-turbines are Pelton, Francis, Michel-Banki, Kaplan, and Deriaz; also some water pumps can work efficiently in reverse, as turbines.

The conversion of hydro-energy into work is governed by the Bernoulli equation, a particular form of energy conservation equation. Accordingly, the thermodynamic limit of hydro-energy conversion into work is given by

$$W = \dot{V}\left(0.5\rho v^2 + \rho g \Delta z + P\right)$$

where $\dot{V}$ is the volumetric water flow rate, ρ water density, v water velocity, Δz the elevation difference, and P water pressure.

The thermodynamic limit is never reached in practice due to friction in ducts and irreversibilities in the turbine. Thus the efficiency of hydro-energy conversion is commonly 80% with respect to water head $\rho g \Delta z$. The waterwheels are the least efficient devices for converting the water energy due to losses by friction, turbidity, and incomplete filling of the buckets. The water pushes the shovels tangentially around the wheel. The water does not exert thrust action or shock on the shovels as is the case with turbines. The advantage of waterwheels is that they can operate in dirty water or water with a suspension of solids.

A Pelton machine is a turbine of free flow (action). The potential energy of the water becomes kinetic energy through injectors and control of the needles that direct and adjust the water jet on the shovels of the motive wheel. They work under approximate atmospheric pressure with a typical head in the range 400–2000 m. Compared with the Francis turbine, the Pelton head has a better efficiency curve.

The Francis turbine is used in small hydroelectric power plants with a head in the range of 3 to 600 m and hundreds of dm^3/s flow rate of water. This turbine is very sensitive to cavitation and works well only close to the design point. Its operation becomes unstable at powers 40% or more lower than nominal.

The Michel-Banki turbine works with radial thrust. The range of power goes up to 800 kW per unit and the flow rate varies from 25 to 700 dm^3/s, with a head in the range of 1–200 m. The number of slats installed around the rotor varies from 26 to 30, according to the wheel circumference, whose diameter is from 200 to 600 mm. This multi-cell turbine can be operated at one- or two-thirds of its capacity (in the presence of low or average flows) or at full capacity (in the presence of design flows: that is, three-thirds). The turbine can be operated even at 20% of its full power.

The Kaplan turbine is a hydraulic propeller turbine adapted to low heads, from 0.8 to ~5 m. In addition, this turbine has the advantage of maintaining its electromechanical parts out of the water. This feature eases routine inspection and maintenance and adds safety in the case of floods.

Deriaz Turbines were developed in the 1960s and can reach a capacity of up to 200 MW, with flow rates in a broad range (from 1.5 to 250 m^3/s) and with heads of 5 to 1000 m. The runner diameter may be up to 7000 mm with six to eight runner blades. Diagonal turbines operate very economically as either turbines or pumps. The thrusting water follows an approximately conic surface around the runner.

The use of water pumps in reverse, as turbines, for small hydroelectric power plants has become quite popular because of the appreciable reduction in facility costs. These pumps, usually of small capacity, have been used for many years in industrial applications to recover energy that would otherwise be lost. They present the following advantages:

- They cost less because they are mass produced for other purposes (as water pumps for buildings and residences).
- Their acquisition time is minimal because they have a wide variety of commercial standards and are available in hardware stores and related shops.

However, there are few disadvantages: slightly reduced efficiency compared to the same head height used for water pumping, and, high sensitivity to the cavitation and operating parameters change.

EXAMPLE 5.12

A typical hydropower setting is illustrated in Figure 5.69 and has a $\Delta z = 100$ m level difference between the lake's surface and downstream waters. On the water stream, with a volume flow of $\dot{V} = 6000 \, \text{dm}^3/\text{s}$, is installed a turbine coupled to an electric generator that produces $\dot{W} = 4.5 \, \text{MW}$ power with $\eta_g = 95\%$ efficiency. We want to calculate the efficiency of the water turbine itself, the overall efficiency, and the power generated by the turbine.

FIGURE 5.69 Hydropower setting for the numerical example.

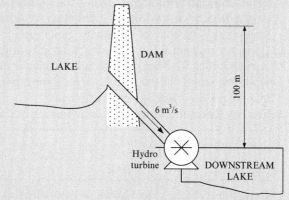

The energy rate supplied to the turbine-generator system is $\dot{E} = \rho \dot{V} \times g \times \Delta z$; assuming a density of water of $1 \, \text{kg/dm}^3$, the energy rate is $\dot{E} = 5.886 \, \text{MW}$. Thus the overall conversion efficiency is $\eta = \dot{W}/\dot{E} = 4.5/5.886 = 76\%$. One can assume that the overall conversion efficiency is the product of the turbine and the generator efficiencies $\eta = \eta_t \times \eta_g$; thus the turbine efficiency is $\eta_t = \eta / \eta_g = 0.76/0.95 = 80\%$. Thus the mechanical power generated by the turbine for shaft rotation is $\dot{W}_{\text{shaft}} = \eta_t \times \dot{E} = 0.8 \times 5.886 = 4.71 \, \text{MW}$.

5.6 CONCLUDING REMARKS

This chapter provides a comprehensive presentation of conventional power generating cycles and systems and their energy and exergy analyses with emphasis on irreversibility quantifying through exergy destruction. Economic and environmental impact of these systems is briefly discussed. The overwhelming majority of power stations in use today are based on steam Rankine cycles operated with steam turbines. Besides steam Rankine power plants, other types of conventional power generation systems such as combustion turbines, (based on the Brayton cycle and its variants), internal combustion engines (based on Otto, Diesel, Stirling, and Ericson cycles), and hydropower stations are introduced gradually, starting with the most basic ones (e.g., the four-component Rankine cycle) and continuing with improved ones, such as reheat and regenerative cycles. The efficiency and other relevant indicators are presented for each type of conventional power generation station, including a coal-fired steam Rankine power plant, combustion turbines and combined cycles, and internal-combustion-engine-based generators. For large-scale hydropower stations an illustrative example is presented.

Study Problems

1. Consider a vapor generator for a steam Rankine cycle that operates with no pressure drop. The boiling temperature is 200 °C; subcooled liquid enters the vapor generator at 160 °C and superheated vapor leaves at 350 °C. The steam generator is heated with hot combustion gases which are assimilated as ideal gas with a constant specific heat of 1050 J/kg.K. There are no heat losses; the temperature difference at pinch point is 10 K and the gas temperature at the hot end is 390 °C. Write the balance equations for this system and determine the LMTD, the mass-specific exergy destructions, and the exergy destruction fractions for the preheating, boiling, and superheating regions. Determine the exergy efficiency of the steam generator.

2. Consider a steam Rankine cycle of basic configuration which operates with no internal irreversibility. The heat source has a constant temperature with a Carnot factor of 0.6 while the sink is at the standard T_0. Determine the optimum boiling temperature that maximizes the cycle's energy efficiency.

3. Rework Example 5.2 for the following input data set:

T_0 298 K	T_{sh}	500 °C	$\eta_{s,P}$	0.8	ΔT_{sc}	2 K	$\Delta T_{sat,b}$ 1 K	ΔP_{drop} 0.2 MPa	$\Delta T_{sat,c}$ 3 K	T_{ph}	500 K
T_b 575 K	$\eta_{s,T}$	0.9	ΔT_c	7 °C	$\Delta T_{sat,ph}$ 0.2 K	$\Delta T_{sat,sh}$ 0.7 K	ΔT_{loss}	20 K	T_{sc}	304 K	$T_{source.b}$ 625 K
T_{sh} 775 K	$T_{sink,c}$ 305 K	T_{loss} 338 K									

4. Consider a single reheat steam power plant as shown in Figure 5.8. The heat source is hot ideal gas with a constant specific heat of 1005 J/kg.K and hot end temperature of 775 K. The minimum temperature difference at heat exchangers is 10 K. The heat sink temperature is 300 K. The turbines and pump operate with an isentropic efficiency of 0.85 and the electro-mechanical efficiency is 0.9. There are no pressure drops in conduits or heat exchangers. Perform the following: (i) determine the Rankine cycle by specifying the best boiling temperature and thermodynamic parameters at all state points; (ii) analyze the cycle based on energy and exergy methods and

determine exergy destructions for each component, cycle efficiencies, and *BWR*; and (iii) determine the mass flow rate of working fluid when the net power generated is 25 MW.

5. Consider the regenerative Rankine cycle in Figure 5.13. Assume that the reference temperature is that of lake water (condenser is cooled with water) at 283 K. The cycle is heated with ideal gas consisting of 12% CO_2, 20% H_2O, and the rest N_2. The hot end temperature is 475 K. The minimum temperature difference for heat transfer processes is 10 K. The isentropic efficiency of turbines and pumps is 0.85. The pressure drop at the steam generator corresponds to a 0.2 K temperature drop during boiling while the heat losses are 2% of steam generator duty. Determine the cycle diagram in *T–s* coordinates. Analyze the cycle with energy and exergy methods and determine the efficiencies and exergy destructions for each component.

6. Repeat the Study Problem 5 for the cycle represented in Figure 5.23. Determine the optimum extraction fractions and extraction pressures such that the cycle energy efficiency is maximized.

7. A regenerative ORC operates with siloxane D5 (available in EES). Determine the cycle diagram provided that the boiling temperature is 600 K and saturated vapor is expanded. Assume that the condenser is cooled with air having the annual average temperature of 303 K. The minimum temperature at the heat exchangers is 10 K. The cycle is supplied with heat from a thermal oil with a constant specific heat of 2.5 kJ/kg.K. Determine the exergy destructions for each of the cycle components, the efficiencies, and the regenerator effectiveness.

8. Consider three gas power cycles as shown in Figure 5.70. The first cycle is Otto $1-2-3_{\text{Otto}}-4-1$ with compression ratio $r_v = 16$. State 1 is at T_0, P_0. The second is Diesel $1-2-3_{\text{Diesel}}-4-1$ for which the cut-off ratio is not given but can be deduced from

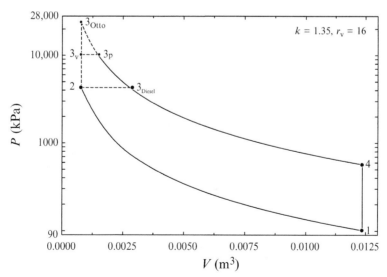

FIGURE 5.70   Gas power cycles for Study Problem 8.

the diagram because states 3_{Otto}, 3_p, 3_{Diesel} and 4 are laying on the same isentrope. The third cycle, 1–2–3v–3p–4–1, is a dual cycle (in between Otto and Diesel). Calculate the efficiency of each cycle under (i) air standard assumption and (ii) air as ideal gas with variable specific heat assumption (Hint: use the reduced pressure method according to Equation (5.30)).

9. Consider a closed loop regenerative air-standard Brayton cycle which is heated from solar energy with 20 kW of power. Assume that the hot end temperature is 500 K and the cold end temperature is 393 K as the cycle is cooled with water from a lake ($T_{lake} = 383$ K). The Carnot factor for solar radiation is 0.95 and the regenerator effectiveness is 0.85, turbine isentropic efficiency is 0.9, and compressor isentropic efficiency is 0.8. Analyze the cycle with energy and exergy methods and determine the BWR, efficiencies, exergy destructions, and the T–s diagram.

10. A reheat-regenerative open loop Brayton cycle as in Figure 5.54 is heated through methane combustion. The combustion reaction is complete and the combustion chamber is adiabatic with an excess air of $\lambda = 5$. The pressure in the combustion chamber is 10 bar. The pressure drop on the circuit connecting the compressor with the turbine is 2 bar. The regenerator effectiveness is 0.85, turbine isentropic efficiency is 0.95, and compressor isentropic efficiency is 0.85. There are electromechanical loses of 3%. Determine the cycle parameters, the efficiencies, and exergy destructions.

11. Perform energy and exergy analysis for the cycle given as a case study in Figures 5.57 and 5.58, provided that the isentropic efficiency of the compressors is 0.8 and the isentropic efficiency of the turbines is 0.9.

12. Rework Example 5.11 with the only difference that the methane combustion is not complete; the molar ratio between CO and CO_2 in the combustion product is 1:5.

13. Perform a detailed energy and exergy analysis of the combined cycle power plant in Figure 5.67; make reasonable assumptions.

14. Rework Example 5.12 for $\Delta z = 200$ m and $\dot{V} = 12{,}000 \, dm^3/s$, with turbine efficiency of 80% and electromechanical losses of 4%.

Nomenclature

BWR	back work ratio
EDF	exergy destruction fraction
ER	expansion ratio
$\dot{E}x$	exergy rate, W
ex	specific exergy, kJ/kg
FWR	flow work ratio
f	flow fraction
h	specific enthalpy, kJ/kg
$\dot{m}$	mass flow rate, kg/s
NTU	number of thermal units
$\dot{n}$	molar flow rate, kmol/s
P	pressure, kPa
PR	pressure ratio
$\dot{Q}$	heat rate, W
q	mass-specific heat flux, kJ/kg
$\dot{S}$	entropy rate, W/K

SSP	specific size parameter, kg/kJ
s	mass-specific entropy, kJ/kg.k
T	temperature, K
v	specific volume, m^3/kg
$\dot{W}$	work rate, W
w	mass-specific work, kJ/kg
x	vapor quality

Greek letters

γ	specific heat ratio
η	energy efficiency
ψ	exergy efficiency
θ	dimensionless temperature

Subscripts

0	reference state
b	boiling
c	condensation
C	Carnot (factor)
cons	consumed
d	destroyed
deliv	delivered
drop	pressure drop
el	electrical
gen	generated
in	input
loss	losses
m	mechanical
mix	mixer
net	net generation
l	liquid
out	output
p	pump
ph	preheating
R	regenerative
rev	reversible
rh	reheating
sat	saturation
sc	subcooling
si	sink
sh	superheating
so	source
t	turbine
trap	steam trap
v	vapor
vg	vapor generation

Superscripts

*	throat condition
(–)	average value

References

Ahmadi, P., Dincer, I., 2011. Thermodynamic analysis and thermoeconomic optimization of a dual pressure combined cycle power plant with a supplementary firing unit. Energy Conver. Manage. 52, 2296–2308.

Cengel, Y.A., Boles, M.A., 2010. Thermodynamics: An Engineering Approach. McGraw-Hill, New York.

Dincer, I., Rosen, M.A., 2013. Exergy: Energy, Environment and Sustainable Development. Elsevier, London.

Dunbar, W.R., Moody, S.D., Lior, N., 1995. Exergy analysis of an operating boiling-water-reactor nuclear power station. Energy Conver. Manage. 36, 149–159.

Klein, S.A., 2012. Engineering Equation Software, Internet source http://www.fchart.com/assets/downloads/ees_manual.pdf, accessed in February 2013.

Moran, M.J., Shapiro, H.N., Boettner, D.D., Bailey, M.B., 2011. Fundamentals of Engineering Thermodynamics, seventh ed. John Wiley and Sons, Hoboken, NJ.

Regualagadda, P., Dincer, I., Naterer, G.F., 2010. Exergy analysis of a thermal power plant with measured boiler and turbine losses. Appl. Therm. Eng. 30, 970–976.

Wagner, W., Pruss, A., 1993. International equations for the saturation properties of ordinary water substance. J. Phys. Chem. Ref. Data. 22, 783–787, Revised according to the International Temperature Scale 1990. Addendum to Journal of Physical and Chemical Reference Data 6, 893 (1987).

6

Nuclear Power Generation

6.1 INTRODUCTION

Nuclear power stations are used in thirty-one countries or 15% of the total number of countries in the world. Moreover, the generated nuclear power covers 13% of world energy consumption (WNA, 2013), which represents ~2.5 PW. Currently, the most nuclear power is generated in the United States with a 31.3% share, followed by France with a 16.8% share; some other countries with an important share of nuclear power generation are the Russian Federation (6.4%), Japan (6.2%), South Korea (5.8%), Germany (4%), Canada (3.5%), Ukraine (3.4%), and China (3.3%).

All nuclear power stations worldwide use uranium as nuclear fuel extracted from uranite mineral resources. The richest country in uranite is Australia, with a 23% share, followed by Kazakhstan with 15%, Russia with 10%, South Africa and Canada with 8%, United States of America with 6%, Brazil, Namibia, and Niger with 5%, Ukraine with 4%, Jordan and Uzbekistan with 2%, and India with 1%.

In a nuclear power station a controlled nuclear reaction is maintained in the reactor. The nuclear reaction generates mainly neutron radiation that is immediately converted to high-temperature heat. A steam generator with special design—depending on the reactor type—is then used to extract the thermal energy produced by the reaction and to obtain pressurized superheated steam at 525–625 K. The steam generator is part of a steam Rankine power plant which produces power with a typical energy efficiency of 30%.

The key to achieving the current level of technology of nuclear power stations has been the establishment of a new body of science in the last quarter of the nineteenth and the first quarter of the twentieth centuries: nuclear physics. The first laboratory demonstration proving the possibility of generating nuclear radiation in a controlled manner was realized in 1924 by the research group of the well-known scientists Frederic and Irene Joliot Curie in Paris and their PhD student Stefania Maracineanu, which showed that lead, being activated with radioactive polonium, starts emitting radiation (Neues Wiener Journal, 1934). Further experiments of the research group led to the formulation of a theoretical model for artificial radio-activity in aluminum bombarded with alpha particles, a discovery which was of utmost importance for the progress of nuclear energy engineering; this discovery was acknowledged by the 1935 Nobel Prize for chemistry awarded to Frederic and Irene Joliot Curie.

The proof of the concept for a nuclear reactor was achieved only 7 years later in Chicago by Enrico Fermi, who demonstrated the first nuclear reactor with a controlled chained fission reaction of uranium. Later a nuclear reactor of 1 MW was built and tested at Oak Ridge Laboratories (1943) and the first large-size nuclear reactor (of 250 MW installed capacity), called the B reactor, began to operate near Richland Village, WA (1944). The first nuclear power plant was commissioned in Idaho, USA, in 1951, with an installed electricity of 100 kW. CANada Deuterium Uranium (CANDU) technology achieved a commercial level in 1962. During the last 40 years, many kinds of nuclear reactors have been developed for the purpose of electric power production and marine and space mission applications.

Nuclear power generation is a crucial player in the generation of wealth, economic development, and energy security. In this chapter, nuclear power generation is discussed starting from the most basic concepts of nuclear reaction, to the engineering of controlled generation of nuclear heat and power cycles, to the advanced concepts of a nuclear reactor. As cogeneration brings a better recognition and wider acceptance of nuclear energy for daily applications in various sectors, this topic is also discussed, with emphasis on producing district heating, process heat, potable water from desalination, hydrogen, and other synthetic fuels.

6.2 NUCLEAR REACTIONS

It is well known that some chemical elements are radioactive; they can transform themselves into other chemical elements and emit or absorb radiation through nuclear processes such as radioactive decay, or they can participate in fusion or fission reactions. How do nuclear reactions occur and how can they be controlled for steady generation of high-temperature heat? This is the main question that we want to answer in this chapter, by providing the basic elements of nuclear energy and power generation applications.

Let us recall that the atom is the smallest unit of a chemical element (radius of about 1 Å) and that there are 118 known atoms corresponding exactly to the periodic table of elements. The atom is made up of a nucleus and a number of electrons. The nucleus is formed of protons and neutrons. In normal condition the atom is neutral with respect to electric charge; thus, the number of electrons is the same as the number of protons. The nucleus cannot be broken down by chemical reactions, but it can be affected in its constituency by nuclear reactions. Table 6.1 gives the formal definition of the elementary particles which are considered the most significant ones for nuclear power generation.

The nucleus has a rest mass in the range of 10^{-24}–10^{-22} g. In some specific nuclear reactions, a portion of the nucleus mass may disappear, as it can be converted into radiation energy according to the well-known Einstein formula, which is written here $\Delta E = \Delta m \, c^2$ to emphasize that a change of nucleus mass of Δm corresponds to an emitted (or absorbed) energy amount noted with ΔE. In order to have an idea of the energy generated by nuclear reactions, let's assume that a mass of 10^{-25} g is lost by a nucleus which represents the thousandth part of the average atomic mass of the chemical elements. The generated energy caused by the conversion of this amount of mass into radiation is calculated to be 56 MeV.

TABLE 6.1 A List of Significant Particles in Nuclear Physics

Particle	Symbol	Definition
Electron	e^-	Subatomic particle having one negative elementary electric charge and a very small rest mass, negligible with respect to the nucleus mass
Positron	e^+	The antimatter that annihilates an electron having the same rest mass but positive charge
Proton	p^+	Subatomic particle having one positive elementary electric charge and rest mass. Proton is categorized as a nuclide because it enters into the constituency of the atomic nucleus
Neutron	n	Subatomic particle having no electric charge and a rest mass approximately equal to that of proton. The neutron is categorized as a nuclide
Neutrino	v	This particle has zero electric charge and non-zero rest mass; it is smaller than an electron and travels with a speed close to that of light
Antineutrino	$\bar{v}$	The antimatter that can annihilate a neutrino
Photon	hv	The elementary quanta of electromagnetic radiation (light). Photon has a dualistic behavior as wave and particle. Its speed of propagation is the speed of light; therefore it has no rest mass. It has no electric charge, but it has dynamic mass, momentum, and spin. Photons can be annihilated by absorption when light interacts with matter

If 5 kmol of atoms (~300 kg radioactive matter, on average) would react in the same way, then the energy released is 27 EJ, or the same as the annual nuclear power generated worldwide with the installed capacity; this gives an idea of the immense potential of using nuclear reactions for power generation.

The balance of atomic masses is a very important issue for analysis of nuclear reactions. At the same time, the atomic mass (denoted with A, also known as mass number) is a parameter that identifies chemical elements. Any chemical element is identified by the number of protons in its nucleus, called the atomic number (or proton number) and denoted with Z. However, for a chemical element the number of neutrons may differ from the number of protons within the nucleus. A chemical element may have some degree of variety in its nucleus corresponding to a variation in the number of neutrons.

Therefore, although the electric charge of the nucleus of a chemical element is fixed, the nucleus mass varies depending on the number of neutrons generating a range of subspecies. Nuclei having the same number of protons and a different number of neutrons are called *isotopes*. The abundance of an isotope represents the probability of identifying the particular isotope in nature among all known isotopes of the same chemical element.

Note that one isotope of particular interest is carbon-12 ($^{12}_{6}C$), which is the most abundant isotope of carbon in the universe with an abundance of 98.89%. This isotope has $Z = 6$ protons and $N = 6$ neutrons and has a mass of 1.992×10^{-26} kg. By convention, $^{12}_{6}C$ is used to define the *atomic mass unit* (AMU) according to the following statement:

Carbon-12 has 12 AMU.

One can simply calculate 1 AMU $= 1.992 \times 10^{-26}/12 = 1.660538782 \times 10^{-27}$ kg. In Table 6.2, the abundance and the atomic masses of some selected isotopes relevant to nuclear-power-generating reactors are given. There are a number of universal constants that characterize the subatomic processes. These constants and other relevant constants for nuclear processes are

TABLE 6.2 Abundance and Atomic Mass of Selected Isotopes Relevant to Nuclear Power

Symbol	Name	Abundance	Atomic mass (AMU)
^1_1H	Hydrogen	>99%	1.00727646661
^2_1H	Deuterium	<1%	2.0141078
^3_1H	Tritium	Traces	3.0160492
^3_2He	Helium-3	<0.0002%	3.0160293
^4_2He	Helium-4	>99.999%	4.002602
$^{11}_6C$	Carbon-11	Traces	11.002035
$^{12}_6C$	Carbon-12	98.9%	12.0000
$^{13}_6C$	Carbon-13	1.1%	13.00335
$^{14}_6C$	Carbon-14	Traces	14.003241
$^{56}_{26}Fe$	Iron-56	91.754%	55.93493758
$^{222}_{86}Rn$	Radon-222	~100%	222.0175777
$^{232}_{90}Th$	Thorium-232	~100%	232.03806
$^{234}_{92}U$	Uranium-234	0.0054%	234.035265
$^{235}_{92}U$	Uranium-235	0.7204%	235.0439299
$^{238}_{92}U$	Uranium-238	99.2742%	238.0507826
$^{239}_{94}U$	Plutonium-239	Negligible [a]	239.0521634

[a]Artificially created; negligible natural occurrence.

given in Table 6.3. Nuclear reactions result in emission or absorption of various types of nuclear radiation. According to quantum mechanics, the matter absorbs and emits radiative energy in fixed packages (quanta). In general, nuclear radiation consists of a flux of unstable particles. The basic types of radiation encountered in nuclear reactions are listed and described in Table 6.4.

The release and absorption of energy during the nuclear reaction can be determined based on the conservation of energy principle which in quantum mechanics must account for the possibility of mass-to-energy conversion. The reconfiguration of the nucleus after the reaction is responsible for the released or absorbed energy. Both attractive and repulsive forces manifest among nuclides that form the nucleus. Coulomb forces produce proton repulsion and the Yukawa forces attract the nuclides. The equilibrium between attractive and repulsive nuclear forces can be stable, unstable, or metastable depending on the exact situation. If the equilibrium in the nucleus is stable, then the isotope is stable; if the nuclear equilibrium is unstable or metastable, then the isotope is radioactive—that is, it will spontaneously disintegrate to form a system with stable equilibrium.

One can determine the binding energy for any nucleus if one assumes that the nucleus discomposes in its protons and neutrons. Consider an element E with Z protons and A nuclides (A is also denoted as the atomic number). The reaction $^A_Z E \rightarrow Zp^+ + (A-Z)n + \Delta E_b$ describes the

TABLE 6.3 Universal Constants in Nuclear Physics

Universal constant	Symbol	Value	Description
Speed of light in a vacuum	c	299,792,458 m/s	The maximum speed at which matter, energy, and information can be transported in the universe
Plank constant	h	$6.626,069,573 \times 10^{-34}$ Js or $4.135,667,517 \times 10^{-15}$ eVs	Energy carried by a photon of 1 Hz or the ratio between photon energy and its associated frequency
Elementary electric charge	e	$1.602,176,487 \times 10^{-19}$ C	Electric charge carried by a single proton
Mass of electron	m_e	0.000,548,600 AMU	The rest mass of the electron
Radius of electron	r_e	2.817,940,326,700 fm	Electron's radius according to classical theory
Mass of proton	m_p	1.007,276,466,610 AMU	The rest mass of the proton
Mass of neutron	m_n	1.008,664,916,566 AMU	The rest mass of the neutron
Electric constant	ε_0	$8.854,188 \times 10^{-12}$ F/m	Dielectric permittivity of vacuum

TABLE 6.4 Basic Types of Nuclear Radiation

Radiation	Symbol	Definition
Alpha	α	This radiation is a fast-moving flux of ^{4}He nuclei (helium nucleus consisting of two protons and two neutrons). Hence this is an ionized radiation of which particles have two elementary charges (+2e). The typical kinetic energy is of 5 MeV per particle, corresponding to a speed of 54×10^6 km/h
Beta	β^-	Negatively ionized particle radiation consisting of a flux of fast-moving electrons
	β^+	Positively ionized particle radiation consisting of a flux of fast-moving positrons
Gamma	γ	Electromagnetic radiation in the wavelength spectrum of picometers carrying an energy of hundreds of keV per photon
Röntgen	X	Electromagnetic radiation in the spectrum of 0.1–10 nm carrying an energy of approx. 1–100 keV per photon
Neutron	n	This radiation consists of a beam of free neutrons. This radiation is therefore non-ionized. However, when interacting with atoms it produces strong ionizing effects because photon emission occurs which displaces electrons at higher energy levels and ionizes the atom. The neutron radiation can be categorized based on the kinetic energy of the neutrons. Some subcategories of neutron radiation are thermal neutrons (25 meV), slow neutrons (1–10 eV), and fast neutrons (1–20 MeV)

un-binding process of the nucleus with the release of the binding energy ΔE_b. This energy is caused by the difference in mass between the protons and neutrons forming the nucleus and the mass of nucleus, which is heavier. Therefore, the binding energy of any nucleus can be calculated based on $\Delta E_b = c^2 \Delta m$ according to the formula:

$$\Delta E_b = -\left[Z \times m_{p^+} + (A - Z)m_n - m\left(^A_Z E\right) \right] c^2 \tag{6.1}$$

Note that for a mass change of 1 AMU the binding energy is $\Delta E_{b1} = 931.4$ MeV. A factor that quantifies the nuclear stability of an isotope is given by the ratio of binding energy to the atomic mass, denoted with the nucleus stability factor (NSF)

$$NSF = \frac{\Delta E_b}{m_{nucleus}} \left[\frac{MeV}{AMU} \right]$$

Because the mass of proton and the mass of neutron as well are both of the order of ~1 AMU (see Table 6.3) it results that the mass of nucleus is approximately equal to the sum of protons and neutrons. Therefore, the NSF gives the binding energy per number of nuclides. When the binding energy per number of nuclides is higher the nucleus is more stable. In Figure 6.1 one can observe that ^{56}Fe has about the maximum binding energy per nucleon (the peak is very close to ^{56}Fe and corresponds to less abundant isotopes of ^{62}Ni followed by ^{58}Fe). The nucleus of ^{4}He is also very stable, showing a peak with respect to its neighbors. This high stability is the reason this isotope results from many nuclear re-actions, including fusion that is believed to occur in the sun's core, where hydrogen is converted into ^{4}He, or the alpha decay where a fast moving nucleus of ^{4}He is generated.

Nuclei that have a small atomic mass (smaller than that of magnesium, atomic number $Z = 24$) tend to associate and form heavier nuclei through nuclear-fusion-type reactions. Atoms naturally repel each other, so fusion is easiest with these lightest atoms. However, to force the atoms together takes extreme pressure and temperature.

Elements with atomic mass in the range between magnesium and xenon (atomic number $Z = 131$) are relatively stable as the stability factor is relatively flat. For elements heavier than xenon the nucleus size becomes large, a fact that affects the balance of forces: attractive forces become weaker and the repulsive forces stronger. Therefore, heavy elements have the ten-dency to disintegrate into lighter elements through a nuclear fission reaction. The most easily fissionable elements are the isotopes are uranium-235 and plutonium-239.

Fissionable elements are flooded with neutrons causing the elements to split. When these radioactive isotopes split, they form new radioactive chemicals and release extra neutrons that create a chain reaction if other fissionable material is present. While uranium (atomic number 92), is the heaviest naturally occurring element, many other elements can be made by adding protons and neutrons to nuclei with the help of particle accelerators or nuclear reactors (Figure 6.1).

Besides fusion and fission reactions there is another nuclear process specific to unstable nuclei, namely the *radioactive decay*. The non-stable isotopes are denoted as radionuclides. In a radioactive decay a radionuclide discomposes by emitting nuclear radiation (alpha, beta, or gamma). Assume that a batch of matter contains N_0 radionuclide at an initial moment. Due to decay reaction the number of radionuclides reduces after time t to $N(t)$. The decay law ex-presses the time required for the number of radionuclides to reduce to half. The *half-life time* $t_{1/2}$ is defined according to the following equation:

$$\frac{N(t)}{N_0} = \left(\frac{1}{2} \right)^{t/t_{1/2}}$$

The half-life time of some radionuclides are given in Table 6.5. The half-life time is a very important parameter for designing nuclear fuel cycles. Nuclear fuels after use in a nuclear

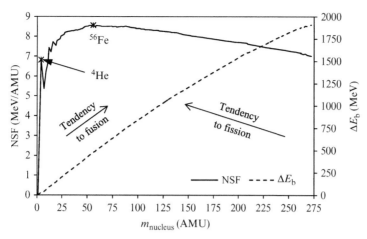

FIGURE 6.1 Nuclear stability factor and binding energy as a function of nucleus weight.

TABLE 6.5 Half-Life Time Values for Some Selected Radionuclides

Radionuclide	Half-life time	Radionuclide	Half-life time
Tritium, $_1^3$H	12.32 years	Carbon-14, $_6^{14}$C	5730 years
Radon-222, $_{86}^{222}$Rn	3.8235 days	Thorium-232, $_{90}^{232}$Th	1.4×10^{10} years
Uranium-234, $_{92}^{234}$U	245,500 years	Uranium-235, $_{92}^{235}$U	7.038×10^{8} years
Uranium-238, $_{92}^{238}$U	4.468×10^9 years	Plutonium-239, $_{94}^{239}$Pu	24,100 years

reactor are still radioactive in some measure. The half-life time allows for predicting the radiation decay of spent fuel in fuels repositories. The nuclear fuel cycle is detailed in the next section.

Here we will focus on two main nuclear reactions that are relevant for power generation: thermonuclear fusion and fission. A basic reaction for thermonuclear fusion is

$$_1^2H + _1^2H \rightarrow _2^4He + 28.4 \text{ MeV} \tag{6.2}$$

where the thermal energy emitted by the reaction is mainly gamma radiation. The most difficult problem with this reaction consists of making two deuterium nuclei collide.

There is considerable activation energy of ~0.02 MeV per collision required by reaction (6.2). However, the activation energy is negligible with respect to the amount of generated energy, which is 28.4 MeV for each collision. In order to transfer the activation energy for reaction initiation the temperature of the deuterons (deuterium nuclei) must be raised to 144×10^6 K, which is probably possible on the sun but definitely not on earth. Another thermonuclear reaction, practically demonstrated on earth through an explosive process, is the fusion of deuteron with triton (the tritium nucleus) according to

$$_1^2H + _1^3H \rightarrow _2^4He + n + 17.6 \text{ MeV} \tag{6.3}$$

The activation energy for reaction (6.3) is ~10 keV, with an associated temperature of 77×10^{6} K. In an explosive process (the hydrogen bomb) this temperature is achieved via a pilot reaction of uranium-235 fission. However, with current technology steady, controlled generation of thermal energy from the fusion reaction is not possible, especially because the process has an explosive nature with an extremely short duration of 1 μs. Research is underway to develop power generation technology from nuclear fission using various options such as electromagnetic fields and plasma and thermally induced shock waves to maintain the reaction confinement space at high temperature.

Nevertheless the most important reaction for nuclear power technology is currently the nuclear fission of uranium, a naturally occurring fuel resource. In a fission reaction a heavy nucleus is "bombarded" with a "slow" neutron. By capturing a neutron the nucleus enters into a metastable state. In order to reach equilibrium, the nucleus splits in two parts of about half weight and generates neutron radiation and energy.

The only known naturally occurring fissionable isotope is $^{235}_{92}U$. However, the abundance of this isotope on earth is only 0.7%. Only two other fissionable artificial isotopes are known: the uranium isotope $^{233}_{92}U$ and the plutonium isotope $^{239}_{94}Pu$. Both these isotopes can be produced starting from more abundant resources through neutron bombardment. Plutonium-239 is obtained from $^{238}_{92}U$ (~99.3% natural occurrence) through the following overall reaction

$$^{238}_{92}U + n \rightarrow {}^{239}_{94}Pu + 2\beta^{-}$$

The artificial fissionable uranium isotope (uranium-233) is produced from thorium-232 (abundance ~100%) through the overall reaction

$$^{232}_{90}Th + n \rightarrow {}^{233}_{92}U + 2\beta^{-}$$

The nuclear fission process of uranium-235 requires one slow thermal neutron n_{th} and generates a number $\kappa > 1$ of fast neutrons and heat (in the form of gamma radiation). The first step of the reaction is the bombardment of a nucleus of $^{235}_{92}U$ with one thermal neutron. As an intermediary product the process generates the metastable uranium-236, $^{236m}_{92}U$. The fission reaction can be written in a simplified manner as follows:

$$^{235}_{92}U + n_{th} \rightarrow {}^{236m}_{92}U \rightarrow {}^{A_1}_{Z_1}E_1 + {}^{236-\kappa-A_1}_{92-Z_1}E_2 + \kappa n + Q_{release} \tag{6.4}$$

The metastable $^{236m}_{92}U$ is formed, which further splits into two lighter nuclei, namely $^{A_1}_{Z_1}E_1$ and $^{236-\kappa-A_1}_{92-Z_1}E_2$, and emits κ neutrons. The mass number A_1 (i.e., the number of nucleons) of the first element $\left({}^{A_1}_{Z_1}E_1 \right)$ can be found in the range of 75–160, with a higher occurrence between 92 and 144. The resulting elements E_1 and E_2 have metastable nuclei and therefore will break apart, emitting gamma rays.

For reaction (6.4) the mean energy released by one mole of fissionable nuclear fuel (any of $^{235}_{92}U$, $^{233}_{92}U$, $^{239}_{94}Pu$) is $Q_{released} = 200$ MeV $= 19.3 \times 10^{12}$ J. With this, one calculates 82.13 TJ per kg of $^{235}_{92}U$, 82.83 TJ per kg of $^{233}_{92}U$, and 80.75 TJ per kg of $^{239}_{94}Pu$. For natural uranium fuel, accounting for the occurrence of 0.7%, the associated exergy is ~584 GJ/kg.

EXAMPLE 6.1

Calculate the binding energy of tritium using the data from Tables 6.2–6.4.

- For tritium the following data is extracted from the tables:
  - $Z = 1$
  - $A = 3$
  - $m\left(^3_1H\right) = 3.016,049\,\text{AMU}$
- For protons and neutrons the following masses are found in the tables
  - $m_p = 1.007,276,466,610\,\text{AMU}$
  - $m_n = 1.008,664,916,566\,\text{AMU}$
- Using Equation (6.1) we obtain:
  - first $\Delta m = m_{p^+} + 2m_n - m\left(^3_1H\right) = 0.008,557,100\,\text{AMU}$
  - and thereafter $\Delta E_b = c^2\,\Delta m = 931.4 \times 0.0085571 = 7.9700827\,\text{MeV}$.

6.3 NUCLEAR FUEL

As mentioned in the introduction the fuel for nuclear power stations is extracted from uranium ore—uranite—which may contain up to 0.3% uranium oxide. Most uranite resources are found in Australia, including 23% of the global share. The map shown in Figure 6.2

FIGURE 6.2   Worldwide distribution of uranium-fuel resources. *Data from NEA (2008).*

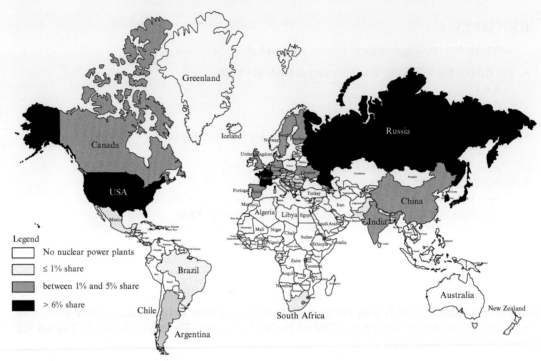

FIGURE 6.3   Worldwide nuclear fuel utilization for power generation. *Data from WNA (2013).*

illustrates the worldwide distribution of uranium fuel resources. The total known resources are on the order of 5 Tg (teragrams, 10^{12} g), with estimated theoretical resources (not yet proven) of 16 Tg. In addition to uranium there are about 20 Tt resources of fertile Thorium-232 fuel.

Once the fuel is extracted from natural deposits it follows a specific fuel processing chain until it is fed to the nuclear reactors. Practically all primary nuclear material used throughout the world is uranium, even though some limited applications may use thorium. The worldwide consumption of uranium fuel for power generation is represented on the map in Figure 6.3. For the future, thorium-232 is foreseen as the principal material to be extracted by mining for producing nuclear fuel.

After use at nuclear power stations the spent fuel—still radioactive—is further processed and eventually disposed. Various types of repositories for nuclear spent fuel exist, as will be presented later in this section, when the *back-end* of the fuel cycle will be discussed.

The nuclear fuel cycle comprises a front-end phase, a fuel utilization phase, and a back-end phase. The front-end covers all the processing, starting with mining and ending with fuel fabrication. The back-end includes fuel recycling, materials recovery, and disposal.

Figure 6.4 shows the main processing steps of a uranium fuel cycle and gives the approximate energy balances for producing 100 Mt of reactor fuel. At the mining site the uranium mineral is mined, milled, and extracted using chemical leaching. The resulting product is the uranium cake, which is the rough commercialized form of uranium fuel containing

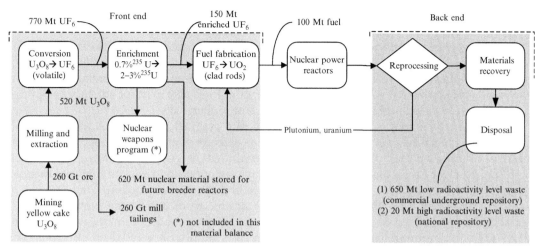

FIGURE 6.4 Simplified diagram for nuclear fuel cycle. *Modified from Bodansky (2004).*

triuranium octaoxide (U_3O_8). About 260 Gt of ore is required to eventually produce 0.1 Gt of fuel. If enrichment is needed then U_3O_8 is converted to UF_6 and processed through isotope separation to increase the amount of ^{235}U to 3.5%. If enrichment is not needed the triuranium octaoxide is converted into uranium dioxide (UO_2) which can be fed directly to reactors (e.g., CANDU). At the extraction phase of triuranium octaoxide about 520 Mt of product is generated for 100 Mt of fuel while close to 260 Gt of mill tailings are deposited.

The conversion of U_3O_8 to UO_2 is through a thermal process (heating is required). If enrichment is required the uranium dioxide must be further converted to uranium hexafluoride. The steps for the entire conversion process from U_3O_8 to UO_2 are illustrated in Figure 6.5. After uranium dioxide is obtained a hydro-fluorination process follows in which uranium dioxide is reacted with anhydrous hydrofluoric acid (HF). In a further conversion step the uranium tetrafluoride is combined with fluorine in gas phase to obtain uranium hexafluoride. The product extraction is done through distillation and crystallization using special steel

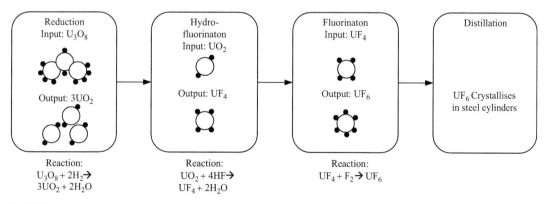

FIGURE 6.5 The conversion process to obtain uranium hexafluoride.

drums. From 770 Mt of UF_6 only 150 Mt of enriched fuel can be obtained; the residual material is stored for future use with the next generation of breeder reactor.

Further processing includes molding the uranium-fuel into a pellet form. There are various ways in which the fuel pellets can be assembled and clad. Recycled fertile materials are also embedded in fuel pellets (e.g., plutonium and depleted uranium). In commercial reactors for power generation the fuel is used in solid form. Liquid fuels also exist for advanced applications and research, including liquid uranyl salts and molten salts.

A typical fuel element is sintered and formed into cylindrical cartridges with a core of uranium which can be of a metallic material (the uranium), of an alloy (e.g., with aluminum), or of uranium dioxide (UO_2) or uranium carbide (UC). One important aspect is fuel cladding. Nuclear fuel must be embedded in a layer of corrosion resistant material that allows for good heat transfer between the fuel core and the reactor coolant. Cladding has a reduced cross-section for the absorption of thermal neutrons and at the same time does not allow escape of radioactive materials. The cladding is made normally in zirconium alloys (2.5% niobium and 97.5% zirconium, which is highly penetrative to neutrons) or, as in more recent technologies, in silicon carbide. One of the problems with fuel cladding is its catalytic effect of splitting water molecules and hydrogen production; this effect is more pronounced in case of accidents when the temperature of the cladding increases and the catalytic activity enhances. Silicon carbide cladding shows very reduced water-splitting activity.

Recent advances in nuclear fuel manufacturing led to the development of tristructural isotropic-coated fuel particles (TRISO) which have improved safety and versatility. These fuel particles are designed to retain fission fragments under any possible circumstances, as they have a strong negative coefficient of reactivity. This means that TRISO fuel is intrinsically safe because there is no need for active cooling in case of accident and the reaction stops immediately when the neutron rate is reduced.

TRISO comprises a 0.5 mm spherical fuel particle of uranium dioxide which is coated in four protective shells, as shown in Figure 6.6. The outer layer acts a pressure vessel, not allowing the passage of any radioactive particles up to temperatures as high as 1900 K. TRISO particles have a diameter of 1 mm and can be formed as pebbles (of ~6 cm in diameter) or as hollow cylinders of 25 mm in diameter and 4 mm in length.

After fuel is used, it is discharged from the reactor and stored in interim locations. Further, the used uranium is transported to disposal places for permanent storage, or reused, depending on the case. For breeding reactors, the uranium fuel is always recycled. In this case,

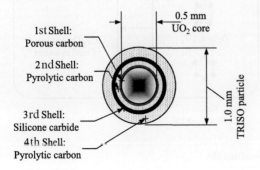

1st Shell: Porous carbon

2nd Shell: Pyrolytic carbon

3rd Shell: Silicone carbide

4th Shell: Pyrolytic carbon

0.5 mm UO_2 core

1.0 mm TRISO particle

FIGURE 6.6   TRISO fuel particle description.

the fuel is reprocessed to extract the fertile and the fissionable isotopes which then are transported to production places to obtain new fuel pellets. The materials remaining after reprocessing can be processed even more by recovering valuable isotopes for other uses.

The nuclear waste comprises many radioactive isotopes and transuranic elements which have an extremely long half-life time. Some of these elements are neptunium, plutonium, tectenium, and iodine. Keeping such elements for a long time implies vitrification or calcination, which embeds them in an amorphous matter with which they do not react for a long period of time.

An interesting aspect is that the radiation emitted by nuclear waste generates heat which is on the order of about 1.5 kW per ton of uranium in 10 years. This heat can in principle be used as a heat source either to generate power with low-temperature differential heat engines or for heating applications. However, there is a lack of confidence regarding the safety of using waste cooling systems as heat sources and such solutions have not been implemented to date.

After appropriate packing, the waste is deposited in long-term geological storage locations excavated in the form of tunnels in pits at more than 1000 m below the earth surface (see IAEA, 2011). Nuclear wastes can be categorized in three classes, namely:

- Low-level waste—which comprises the residual radioactive material from medical and industrial facilities which use radionuclides for various processes involving, for example, gamma- or X-ray generation. Some of the radioactive material disposed by other nuclear facilities is also low-level waste.
- High-level waste—generated during the nuclear fuel cycle, consisting of solid and liquid material which is highly radioactive.
- Transuranic waste—comprising radioactive materials with a decay time of longer than 20 years which emit alpha particles with sensible intensity.

Among all categories the safe disposal of high-level waste is the most important. These wastes are stored in underground repositories built in several places throughout the world, one of them being at Yucca Mountain in the United States, planned to accommodate 70 kt of nuclear waste. The countries that have national plans for nuclear waste repositories are the following:

- United States—storing the waste in tuff host rock in Yucca Mountain
- Finland—to be completed in 2020, storing the waste in granite host rock, at Olkiluoto location
- Germany—to be completed in 2020, with storage in salt, at Gorleben location
- France—to be completed by 2020, with storage in clay and granite
- Canada—to be completed in 2025, with storage in granite
- Japan—to be completed by 2030, with storage in sedimentary rock and granite
- Switzerland—with opening target in 2050, and storage in granite and clay
- Sweden—with completion target in 2020, and storage in granite host rock

There are various safety measures that must be taken to avoid direct exposure of humans and environment to radiation emitted by nuclear waste. Shielding is applied wherever possible. Also the spent fuel assemblies are placed in cooling pools and transferred into shielded protective canisters. The hazard that is possibly created by the nuclear waste is the dose that can be received through inhalation of escaped radionuclides. The maximum dose rate that a person can take is of 20–50 mSv per annum, where "mSv" stands for millisievert, which is the

TABLE 6.6 Comparison of Nuclear Fuel with Fossil Fuels and Biomass

Fuel	Heat transfer medium	Temperature (°C)	Carnot factor	Energy density (approximate values)		GHG emissions
				MJ/t	MJ/m³	
Nuclear fuel (fission)	Liquid	300–500	0.5–0.6	83,000	1,550,000	Indirect: associated with reactor construction
	Gas	700–950	0.7–0.75	16,000	300,000	
Fossil fuel	Flue gas	1000–1500	0.76–0.83	0.045	0.035	CO_2 in flue gas; About 0.5–1 t CO_2 emitted per kWh thermal
Biomass	Flue gas	500–1500	0.62–0.83	0.017	0.015	Neutral

equivalent dose to a tissue, a calculation based on the absorbed radiation dose and factors dependent on the radiation type and other relevant parameters, such that the quantity represents a quantitative measure of radiation effects; note that $1\,Sv = 1\,m^2 s^{-2}$.

A brief comparison of nuclear fuel, fossil fuels, and biomass is given in Table 6.6. Definitively, the nuclear technology produces the most power per unit of hardware volume or mass. Coal and biomass combustion facilities are massive because they must include fuel handling facilities and large stacks; this decreases the energy density of the generated power. However, the Carnot factor, and therefore the exergy efficiency, associated with combustion technologies is the highest.

6.4 NUCLEAR REACTORS

6.4.1 Reactivity Control

The steady and safe operation of a nuclear reactor requires the maintaining of a balance between the energy generated by the fission reaction and the energy removed by heat transfer from the fuel cladding to the coolant. Both the heat generation rate and the heat transfer rate can be adjusted using specific mechanisms. The heat generation rate depends on the reaction rate.

The nuclear reaction rate is related to the rate of neutrons that are able to initiate new reactions. Moreover, the stable operation of the reactor depends on the balance of neutrons generated by the reaction versus the neutrons absorbed by the surrounding medium. Once it is produced, the fission reaction of produces κ fast neutrons that, in principle, can be slowed down and further used to initiate new fission reactions. This observation suggests that it is possible to obtain a self-sustained nuclear fission reaction (chain reaction). Two parameters influence the balance between the generated and absorbed neutrons: the *neutron cross-section* and the *critical mass*; these parameters are very important in the design of fission nuclear reactors.

The neutron cross-section expresses the likelihood of interaction between a neutron and a target nucleus. This probabilistic quantity is measured in units of area (e.g., cm^2). The typical values are on the order of $10^{-24}\,cm^2$, therefore a special submultiple of the unit for area is used, namely *the barn*, where 1 barn $= 10^{-24}$ cm^2. The cross-section denoted with σ is defined implicitly from the following equation

$$r\left[\frac{\text{interactions}}{\text{time} \times \text{target atom}}\right] = \sigma\left[\frac{\text{area}}{\text{target atom}} \times \frac{\text{interactions}}{\text{neutron}}\right]\phi\left[\frac{\text{neutrons}}{\text{time} \times \text{area}}\right] \qquad (6.5)$$

where r is the rate of interaction [nuclear interactions per second per target atom] and ϕ is the neutron flux [neutrons per second and cross-section perpendicular to the neutron beam] and σ is measured in barn of cross-section times the number of interactions per neutrons per target atom.

The cross-section values for fission fuel encounter by neutrons vary in the range of 0.05–300 barn. The cross-section of heavy water encounter by neutrons is 3–4 barn when the neutrons are scattered. In a nuclear reactor a certain amount of uranium fuel is placed in a special vessel surrounded by a coolant and a moderator, which have the role of heat transfer and neutron deceleration respectively. Neurons are emitted by the radioisotope material due to its decay. When atoms of uranium are hit by the neutrons some of them produce fission and generate many other neutrons. Some of these hit other fuel atoms and generate new fission reactions. If more atoms of fuel (or more mass) are brought together then there is a higher probability of initiating a chain reaction.

The critical mass represents the smallest amount of fissile fuel which must be placed in a reaction confinement to maintain a chained nuclear reaction. If the mass of fissile material is higher than critical mass then higher chances exist that the reaction rate will increase, whereas when the mass is less than critical the reaction will cease as insufficient neutrons are emitted to maintain the chain reaction.

The following three cases of fission reaction are possible, depending on the parameter κ—denoted as the *neutron multiplication factor*—which represents the number of neutrons producing a new fission per neutron that initiates the reaction:

$$\kappa = \begin{cases} > 1, \text{ supercritical} \\ = 1, \text{ critical} \\ < 1, \text{ subcritical} \end{cases}$$

When $\kappa = 1$ the reaction proceeds at steady rate and the reactor operates at "criticality." When $\kappa > 1$ the reactor is in supercritical ranges, therefore the reaction and heat release rates amplify. The rate can grow extremely fast and explosion becomes possible. When the reactor operates in subcritical range the process slows down. Note that the critical mass for uranium-235 is $\sim$50 kg (the corresponding volume is $\sim$3 l).

The nuclear reaction rate can be controlled (and maintained stable) in two ways, corresponding to two different reactor technologies: (i) by slowing the neutrons as in "thermal reactor" technology and (ii) by enriching the fissile material in a proper proportion as in "fast neutron reactor" technology. The cross-section parameter plays the most important role in design for both options.

The fast neutron reactors use nuclear fuel blended with enriched fissile material. The enrichment is done such the probability of fast neutrons colliding with fissile atoms becomes high enough to maintain the chain reaction. Henceforth these reactors do not need any moderator because the there is no need to slow down the neutrons. When bombarded with neutrons, the fertile materials transform into fissionable fuel. Uranium-238 can be made fissile in

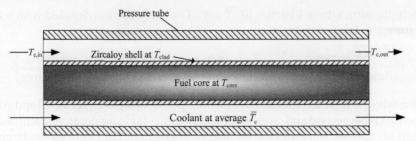

FIGURE 6.7   Fuel-coolant arrangement in the pressure tube of a pressurized water-cooled reactor.

about two and one-half days, thus natural-uranium-based reactors are called "Fast Breeding Reactors" (FBR). In an FBR, ~2.7 neutrons are generated for one absorbed neutron to produce fission, while 1.7 neutrons produce fuel breeding. Thorium breeds slower than natural uranium, often after about 21 days. Thus, thorium-based reactors are called "Thermal Breeding Reactors" (TBR).

Thermal reactors use a neutron moderator, which contains a material that slows down the fast neutrons that result from each individual fission reaction. The kinetic energy of the slow neutron—which is named the thermal neutron—is on the order of magnitude of the surroundings medium. Being slow, the thermal neutrons have a higher cross-section than the fast neutrons; in other words, the probability of colliding with fissionable material is higher.

Regardless of the specific type of nuclear reactor the thermal energy transfer process is similar: heat is transferred from a hot core by thermal conduction to the cladding surface; further heat is transferred by convection to a heat transfer fluid. Figure 6.7 illustrates a typical fuel-coolant arrangement for pressurized water-cooled reactors. The core temperature is about 2200 °C, whereas the temperature of the Zircaloy cladding is ~340 °C. The average temperature of water coolant is 305 °C, with water at the entrance being at $T_{c,in} = 290°C$ and at the exit $T_{c,out} = 315 °C$.

Of major importance in fission reactor design is the evaluation of heat generation rate per unit of volume, q'''. This parameter varies across the reactor core volume and is influenced primarily by the intensity of the neutron and gamma radiation produced by the nuclear reaction. In turn the radiation intensity is influenced by the kinetic energy of the emitted particles, the flux density energy spectrum, the density of various atomic constituents, and the atomic cross-sections.

In the design process of the reactor the heat conduction equation must be solved in a region around the fissionable fuel where heat is generated. The medium is non-homogenous, therefore the general heat conduction equation in differential form is

$$-\nabla\cdot(k\nabla T) + \rho c_{\mathrm{p}}\frac{\partial T}{\partial t} = \dot{q}''' \tag{6.6}$$

where $\dot{q}''' = \dot{q}'''(r,t)$ is the volumetric heat generation rate which varies in time and space (axial symmetry is assumed), r is the radial coordinate, and k is the thermal conductivity.

The heat conduction problem can be approximated as one-dimensional axisymmetric. The volumetric heat generation rate can be determined with the help of the parameter ϵ defined as the average energy dissipated by a single interaction and with the help of parameter N''', representing the number of fissile nuclei per unit of volume. Note that the average energy dissipated, the cross-section, the particle flux, and the numbers of atoms per unit of volume vary spatially and temporarily. Moreover, the nature of the radiation varies and secondary effects can occur (e.g., scattering, low energy X-ray emission, ionization, photon emission, beta emission, Compton scattering, etc.). The rate of nuclear interactions given by Equation (6.5) must be multiplied with the number of fissile nuclei per unit of volume and with the energy dissipation parameter so that the rate of heat generation is obtained; hence one has

$$\dot{q}''' = \sigma \phi N''' \epsilon \tag{6.7}$$

In Equation (6.6) one must account for thermal conductivity variation with the temperature. The average thermal conductivity of UO_2 can be taken to be about 6 W/mK. However, one must note that for certain operating conditions when the UO_2 fuel temperature reaches 2800 °C the fuel melts and its thermal conductivity varies sharply. The melting temperature of Zircaloy is around 1900 °C, whereas its average thermal conductivity is 10 W/mK.

For a common case in practice, fuel rods have a diameter of 5 mm and generate a linear heat flux of 15 kW/m; therefore at the rod surface the heat flux density q''' is superior to 750 MW/m^3. The temperature difference between the core and the rod surface is about 2000 °C. The temperature also varies along the fuel rod.

For preliminary design calculation purposes it is customary to assume that the fuel assemblies are cylindrical of radius R_f and the height z; they are also assumed to be uniformly distributed in the cross-section of the reactor. Then the heat generated by a fuel assembly located at the radius r in the reactor cross-section (the radial distance from the center of the reactor to the center of the rod) is given by

$$\dot{q}(r) = \pi R_f^2 \int_0^z \dot{q}'''(r,1) dl \tag{6.8}$$

If one denotes with $\dot{Q}_{out}$ the heat rate delivered to the heat transfer fluid (coolant), n_{fuel} the number of fuel assemblies, $e_{fuel} = 200$ MeV the average energy generated by the fission reaction of a single fuel atom, and $e_{diss} < e_{fuel}$ the actual energy dissipated by the fuel rods for a single interaction, then the maximum heat that could be generated by the reactor if all fuel elements would generate at the same rate q_{max} can be approximated with the simplified equation given by Lamarsh and Baratta (2001):

$$\dot{q}_{max} = \dot{q}(0) = 2.32 \frac{\dot{Q}_{out}}{n_{fuel}} \frac{e_{diss}}{e_{fuel}} \tag{6.9}$$

Therefore, the maximum volumetric heat generation for a fuel rod of radius R_f and length z is obtained by dividing $\dot{q}_{max}$ by the rod volume as follows:

$$\dot{q}'''_{max} = \frac{\dot{q}_{max}}{\pi R_f^2 z} \tag{6.10}$$

The volumetric heat generation at any radius r in a reactor cross-section is obtained by integration of Equation (6.8); according to Lamarsh and Baratta (2001) this integration leads to

$$\dot{q}'''(r) = \dot{q}'''_{max} J_0 \left(2.405 \frac{r}{R_r} \right) \tag{6.11}$$

where R_r is the reactor radius and J_0 is the zero order Bessel function of first kind.

EXAMPLE 6.2

A cylindrical nuclear reactor of the radius $R_r = 2$ m is loaded with $n_{fuel} = 50,000$ fuel elements of cylindrical shape. It is given that the energy dissipated by nuclear interaction is $e_{diss} = 185$ MeV and the output heat by the reactor is 1 GW. Calculate the maximum heat rate and the heat rate at middle radius and at 5% from the reactor periphery.

- First one calculates the maximum heat rate according to Equation (6.9)

$$\dot{q}_{max} = 2.32 \frac{1\,GW}{50,000} \frac{185\,MeV}{200\,MeV} = 42.9\ kW$$

- With Equation (6.11) and $\dot{q}_{max} = 42.9$ kW the heat rate at any radius r from the reactor center is given by

$$\dot{q}(r) = 42.9\, J_0(1.2025\,r)\ [kW]$$

- Therefore, at half the reactor radius $R_r = 1$ m one gets

$$\dot{q}_{middle} = 42.9 \cdot 0.669 = 28.74$$

- Further one calculates the radius corresponding to a location at 5% from reactor periphery to be $r = 0.95, R_r = 1.9$ m; hence one has

$$\dot{q}_{periphery} \cong \dot{q}(r = 1.9\,m) = 42.9 \cdot 0.06379 = 2.737\ kW$$

6.4.2 Exergy Destructions in a Nuclear Reactor

Nuclear radiation has very high exergy content, practically 100%. Based on the kinetic theory of ideal gases, the energy associated with a particle is $e = 3/2 k_B T$ where $k_B = 1.38 \times 10^{-23}$ J/K is the Boltzmann constant. The neutrons emitted by the fission reaction are fast neutrons which are immediately thermalized by the moderator. The energy spectrum of fast neutrons is 1–20 MeV. Therefore, the particle temperature associated with this kinetic energy is extremely high; for 1 MeV it corresponds to 7.7E9 K, while for 20 MeV one has 1.5E11 K. Therefore, the Carnot factor of the neutron radiation can be assumed to be unity when the reference temperature is on the order of 300 K. This means that the nuclear energy is assimilated to an exergy.

Nevertheless, the exergy of the neutron radiation is destroyed in the reactor by multiple processes. Figure 6.8 represents a simplified reactor model which is useful for identifying the main exergy destruction processes within a general nuclear reactor. In Figure 6.8, the system components are assimilated with boxes. In each box the specific process that take place is indicated. The arrows that connect the boxes represent transfer processes such as neutron radiation, conduction heat transfer, combined heat transfer, and fluid flow, according to the legend. The nuclear fission reaction occurs in the fuel pellet core and emits neutron radiation (mainly) and gamma radiation or other minor particles. The main amount of nuclear radiation interacts with fuel pellets (#1) and generates heating. Some neutron radiation escapes to the moderator (#2) where it is scattered back or partially converts to heating. Other neutron radiation (#3) reaches auxiliary components of the reactor (e.g., the shielding structure) and is converted to heat. The hot fuel pellets transfer heat though the heat conduction mechanism towards the pellet surface (#4) and further to the cladding surface (#5).

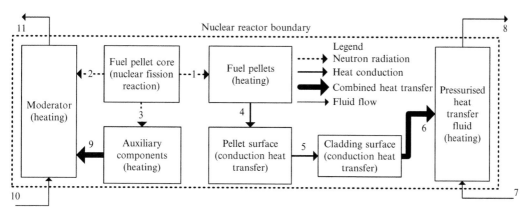

FIGURE 6.8 Simplified thermodynamic model for a generic nuclear reactor.

The cladding surface interacts through combined heat transfer mechanisms (mainly con-duction and convection) with the heat transfer fluid (#6) which is heated from its entrance temperature (at #7) to its exit temperature (at #8). The heat generated within the auxiliary components of the reactor is transferred by convection (or convection conduction) heat trans-fer to the surroundings moderator (#9). The moderator is typically circulated to a cooling sys-tem to extract the moderator heat. It enters in the reactor colder at (#10) and exists warmed at (#11). Simple energy and exergy balance equations for the reactor allow for defining energy and exergy efficiencies. Table 6.7 gives these equations and definitions.

TABLE 6.7 Balance Equation and Efficiency Definition Equations for Generic Nuclear Reactor

Item	Equation	Description
Energy balance equation	$\dot{E}_{nuc} = \dot{E}_{htf} + \dot{E}_{mod}$	• $\dot{E}_{nuc} = \dot{E}_1 + \dot{E}_2 + \dot{E}_3$ is the nuclear radiation rate emitted by the fission reaction • $\dot{E}_{htf} = \dot{m}_{htf}(h_8 - h_7)$ is the thermal energy rate transferred to the heat transfer fluid of mass flow rate $\dot{m}_{htf}$ and specific enthalpies $h_{7,8}$ • $\dot{E}_{mod} = \dot{m}_{mod}(h_{11} - h_{10})$ is the thermal energy rate transmitted to the moderator fluid, energy that is rejected into the environment
Exergy balance equation	$\dot{E}x_{nuc} = \dot{E}x_{htf} + \dot{E}x_{mod} + \dot{E}x_d$	• $\dot{E}x_{nuc} = \dot{E}_{nuc}$ exergy rate of nuclear radiation; Carnot factor is unity; associated temperature is infinity • $\dot{E}x_{htf}$ and $\dot{E}x_{mod}$ are the exergy rates transmitted to the heat transfer fluid and moderator, respectively. Moderator exergy is a lost exergy • $\dot{E}x_d$ is the exergy destroyed by the nuclear reactor
Energy efficiency	$\eta = \dfrac{\dot{E}_{htf}}{\dot{E}_{nuc}}$	The useful effect is the energy rate transmitted to the heat transfer fluid; the expense is the energy rate delivered by the fission
Exergy efficiency	$\psi = \dfrac{\dot{E}x_{htf}}{\dot{E}x_{nuc}}$	The useful effect is the exergy rate transmitted to the heat transfer fluid; the expense is the exergy rate delivered by the fission

EXAMPLE 6.3

Consider a pressurized light water reactor modeled according to the model in Figure 6.8 where stream 10–11 is replaced with a heat leak flux from the reactor vessel to the environment. Calculate the energy and exergy efficiency of the reactor as well as overall exergy destruction and exergy destruction for each of the processes. It is given that: the nuclear radiation transfer to the pellet core is 1 GW (for all reactor), temperature of the pellet core is 1900 K, temperature at the pellet surface is 675 K, temperature at cladding surface is 600 K, temperature of the heat transfer fluid at inlet is 500 K and at outlet 550 K, nuclear radiation dissipated into the moderator is 8 MW, and radiation dissipated into auxiliary components is 2 MW.

- First one calculates the energy (and exergy) input from the radiation

$$\dot{E}_{nuc} = \dot{E}x_{nuc} = \dot{E}_1 + \dot{E}_2 + \dot{E}_3 = 1000 + 8 + 2 = 1010\,\text{MW}$$

- The energy transferred to the heat transfer fluid is $\dot{E}_1 = 1000$ MW. Therefore the energy efficiency is $\eta = 1000/1010 = 99\%$.
- The mass flow rate of heat transfer fluid results from its energy balance

$$\dot{E}_6 = \dot{m}_{htf} c_p (T_8 - T_7) \rightarrow 1000 = \dot{m}_{htf} 4.185(550 - 500) \rightarrow \dot{m}_{htf} = 4.78\,\text{t/s}$$

- The exergy lost by the moderator is determined based on the assumed average temperature of the moderator; as this is a pressurized light water moderator the average moderator temperature is the same as the average temperature of the heat transfer fluid, thus $\bar{T}_{htf} = 525$ K. The reference temperature is assumed to be the temperature of a lake used for condenser cooling and is taken as $T_0 = 15\,°\text{C}$. Therefore the exergy lost by the reactor is

$$\dot{E}x_{loss} = (\dot{E}_2 + \dot{E}_3)\left(1 - \frac{T_0}{\bar{T}_{htf}}\right) = 10\left(1 - \frac{288}{525}\right) = 4.5\,\text{MW}$$

- The exergy transferred to the heat transfer fluid is $\dot{E}x_{htf} = \dot{E}x_8 - \dot{E}x_7$; noting that the enthalpy rate difference for heat transfer fluid is $\dot{H}_8 - \dot{H}_7 = \dot{E}_6$, the $\dot{E}x_{htf}$ becomes

$$\dot{E}x_{htf} = (\dot{H}_8 - \dot{H}_7) - T_0(\dot{S}_8 - \dot{S}_7)$$

where the entropy rate difference can be calculated from:

$$\dot{S}_8 - \dot{S}_7 = \dot{m}_{htf} c_p \ln\left(\frac{T_8}{T_7}\right) = 1.9\frac{\text{MW}}{\text{K}}$$

- Therefore, $\dot{E}x_{htf} = 1000 - 288\cdot1.9 = 452.8\,\text{MW}$ and the exergy efficiency becomes

$$\psi = \frac{\dot{E}x_{htf}}{\dot{E}x_{nuc}} = \frac{452.8}{1000} = 45.28\%$$

- The exergy destructions then become
  - Overall: $\dot{E}x_{nuc} = \dot{E}x_{htf} + \dot{E}x_{loss} + \dot{E}x_d \rightarrow \dot{E}x_d = 552.7$ MW
  - Fuel pellet heating: $\dot{E}x_1 = \dot{E}_1(1 - T_0/T_4) + \dot{E}x_d \rightarrow \dot{E}x_d = 151.6$ MW
  - Conduction through pellets: $\dot{E}_1(1 - T_0/T_4) = \dot{E}_1(1 - T_0/T_5) + \dot{E}x_d \rightarrow \dot{E}x_d = 275.1$ MW
  - Conduction through cladding: $\dot{E}_1(1 - T_0/T_5) = \dot{E}_1(1 - T_0/T_6) + \dot{E}x_d \rightarrow \dot{E}x_d = 53.3$ MW
  - Convection to heat transfer fluid: $\dot{E}_1(1 - T_0/T_6) + \dot{E}x_7 = \dot{E}x_8 + \dot{E}x_d \rightarrow \dot{E}x_d = 67.2$ MW
  - Auxiliaries heating: $\dot{E}x_3 = \dot{E}_9(1 - T_0/\bar{T}_{htf}) + \dot{E}x_d \rightarrow \dot{E}x_d = 1.1$ MW
  - Moderator heating and heat loss: $\dot{E}x_2 + \dot{E}_9(1 - T_0/\bar{T}_{htf}) = \dot{E}x_{loss} + \dot{E}x_d \rightarrow \dot{E}x_d = 4.4$ MW

6.4.3 Conventional Reactors

Reactor technology has evolved toward a set of established concepts, most of which are employed today in nuclear power stations. In this section we will discuss the past and established concepts of nuclear reactors, with a focus on pressurized water and boiling water concepts, which are the most used today. A classification of established concepts of nuclear reactors is shown in Figure 6.9. The first generation of nuclear reactors comprises four major concepts; from this first generation none of the power plants are in use today. The concepts

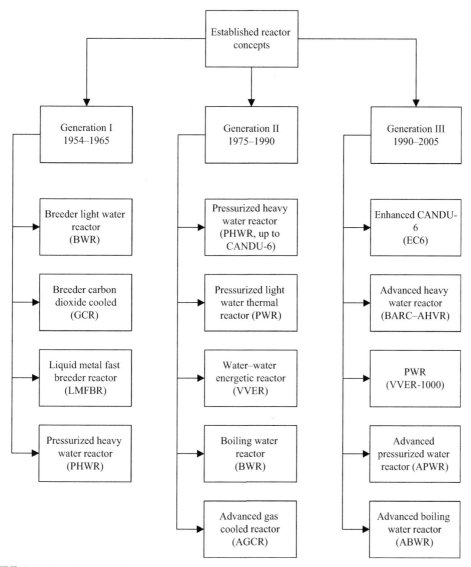

FIGURE 6.9   Classification of the established concepts of nuclear reactors.

developed in the first generation are: a boiling water reactor (BWR), gas-cooled reactor, liquid metal fast breeder reactor (LMFBR), and pressurized heavy water reactor (PHWR). Some typical examples of first generation reactors are listed as follows:

- BWR—DGS 1 (Dresden Generating Station 1, Illinois), with 210 MW installed power generation, commissioned 1960, decommissioned 1990.
- LMFBR—Fermi 1 from Monroe USA with 100 MW installed power generation; commissioned 1956, decommissioned 1972.
- Gas-cooled reactor—Mangox 1 (Clader Hall, UK) with 60 MW installed power generation, commissioned 1956, decommissioned 2003.
- PHWR—NPD Rolphton (Ontario), with 19.5 MW installed power generation, commissioned 1961, decommissioned 1987.

Other examples are the Shippingport Atomic Power Station commissioned in 1958 and decommissioned in 1982, with 60 MW installed power generation and a thermal breeder light water moderator, and the Obninsk Nuclear Power Station, built in the former Soviet Union (FSU), with 5 MWe installed power generation, commissioned in 1954 and decommissioned in 2002.

One of the most used types of reactors is the pressurized water reactor. The earliest design option for a pressurized water reactor is the PHWR, which uses heavy water as moderator. The operating principle of the PHWR is shown in Figure 6.10 as it was developed by AECL (Atomic Energy Canada Ltd) under the name CANDU (CANada Deuterium Uranium). The fuel is placed in tubes through which PHWR coolant flows. The coolant transfers the nuclear heat to a heat exchanger, which is typically a steam generator; the produced heat comes to about 350 °C. Outside the pressure tubes there is the moderator, consisting of a heavy water bath in which the tubes are submerged.

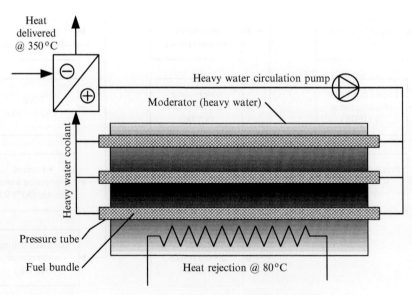

FIGURE 6.10 Pressurized heavy water reactor, based on the CANDU concept.

The pressure in the moderator is around 1 bar. The moderator is continuously cooled and the heat is rejected at about 80 °C. The fuel rods can be replaced or extracted independently at any time though special mechanisms. In the moderator bath there are a series of graphite rods (not shown) with the role of attenuating the neutron flux and thus controlling the reaction rate and criticality. The CANDU reactor evolved with improved options and increased capacity based on the same pressure tube concept and the use of natural uranium and heavy water. The model CANDU-6 corresponds to the second generation of nuclear reactors and has a 500–600-MWe capacity. The Enhanced CANDU reactor EC6 has an output of 740 MWe and has the ability to operate at part-load conditions.

The core concept of the BWR appertains to the first generation of nuclear reactors (see Figure 6.9). The BWR uses water as a moderator and coolant at the same time, with the difference that the reactor produces saturated steam. Figure 6.11 shows a simplified schematic of a BWR. Typically the pressure in the BWR is of 75 bar and the corresponding temperature is 285 °C. The reactor has a special construction which allows for a stable boiling process. When it enters in the reactor, water is first guided into a downcomer where it is preheated. It then rises and flows around the vertical fuel rods though the reactor core region. The resulting twophase flow, having ~15% vapor quality, is directed toward a cyclone separator and steam dryer at the top of the reactor, where the saturated steam is collected. In

FIGURE 6.11 Simplified schematic of the boiling water reactor.

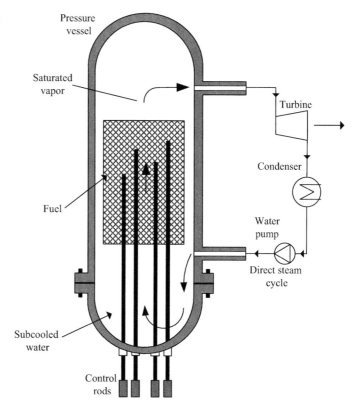

general layout, the reactor is kept in a containment structure, while the steam turbine, the condenser, and the pumps are placed on the exterior.

Gas-cooled reactors (GCR) were first developed in the United Kingdom (MAGOX reactor, see above). In the original MAGOX the coolant was carbon dioxide pressurized at ~20 bar. The fuel is natural uranium and the reactor is of fast breeding type. The moderator is graphite. The temperature level of the generated heat is compatible with linkage to a steam power plant. Figure 6.12 shows the simplified diagram of a gas cooled reactor which uses carbon dioxide coolant.

The most used in the United States is the pressurized water reactor (PWR), which has found applications in power generation and propulsion of marine vessels like submarines, aircraft carriers, and ice breakers. The simplified diagram of the PWR is illustrated in Figure 6.13 and comprises the fuel bundles, the moderator vessel, the pressurizer vessel, the heat exchanger (steam generator), and the coolant pump. The PWR uses slightly enriched uranium in the form of UO_2 containing about 3% $^{235}_{92}U$ placed in tubes of Zircaloy and sunk in light water, which plays the role of moderator. The fuel and the moderator are kept in a pressure vessel designed to operate at about 155 bar pressure; water boils at 344 °C.

However, in the reactor (pressure) vessel water is heated from 275–315 °C and always remains in liquid state. In fact, in order to enhance the heat transfer between the fuel rods and the water, the flow is set such that subcooled nucleate boiling occurs at the surface of the rods; the small vapor bubbles formed are immediately absorbed into the subcooled liquid water. Overall, the operation is very stable.

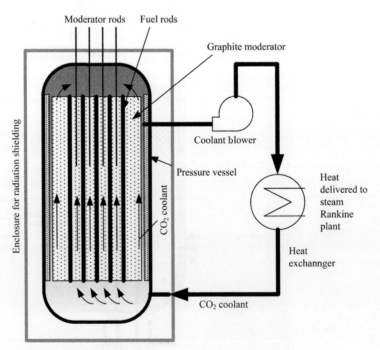

FIGURE 6.12   Simplified diagram of gas-cooled reactor (GCR) of first generation.

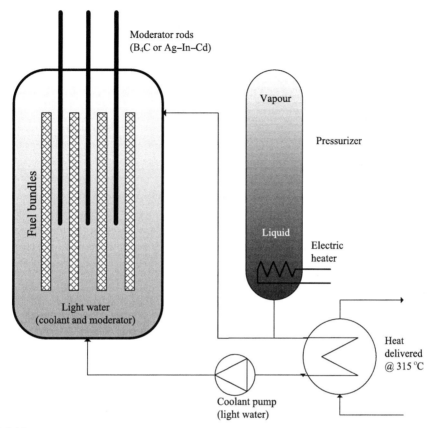

FIGURE 6.13 Pressurized light water reactor (PWR).

The light water is an excellent moderator. In order to control the reaction rate, additional moderators are used in the form of movable bars made from boron carbide or Ag–In–Cd. The pressure is maintained through a pressure vessel equipped with submerged electric heaters meant to maintain a vapor pressure of 155 bar; the vapor is located at the top of the pressurizer, while its bottom is always full of subcooled liquid. A heat exchanger is used to deliver the nuclear heat from the primary circuit to a secondary circuit, which may be a steam generator aimed to run a turbine which commonly turns a generator or propels a marine vehicle. The generated steam is typically 275 °C and 60 bar.

The VVER 440, which refers to a second generation nuclear reactor, was first commissioned before 1970, with the most common design version known as V230, with a 440-MWe output. It is cooled and moderated with light water; therefore the Russian VVER is a version of the PWR. There are some design features in the VVER that distinguish it from the PWR, namely its use of a horizontal boiler instead of a vertical arrangement.

The Generation III nuclear reactors are based on former design principles but with improved efficiency, modularity, safety, standardization of components, 1.5 times increased lifetime (~60 years), and reduced capital and maintenance costs. The reactors of Generation

III were commissioned between about 1990 and 2005. They include the Enhanced CANDU-6 (EC6) reactor, the VVER-1000/392 pressurized water reactor, the BARC–AHWR (Indian advanced heavy water reactor), the advanced pressurized water reactor (APWR), and the advanced boiling water reactor (ABWR). Note that the ABWR was developed by a GE–HNE consortium, which included General Electric (GE) and Hitachi Nuclear Energy (HNE). The APWR was developed by Mitsubishi Heavy Industries (MHI). The VVER-1000 has about 33% efficiency with a net power generation of 1000 MWe and a hottest coolant temperature of 321 °C. The APWR capacity is ∼1500 MWe and units are in operation in Japan.

EXAMPLE 6.4

Consider a pressurized water reactor (of thermal type) having a fuel rod radius of $R_f = 5$ mm, clad with a Zircaloy shell of $t = 0.5$ mm thickness, and with a rod height of $z = 4$ m. One can calculate the heat flux and the temperature of the cladding outer surface provided that the temperature in the core is given; assume that the core temperature is $T_c = 2200$ °C. Calculate the volumetric heat flux rate, the heat flux rate, and the temperature at the cladding surface. The thermal conductivity of the fuel is given as $k_f = 2$ W/mK and that of the cladding as $k_c = 20$ W/mK.

- First one can calculate the volumetric heat generation at the axis; one obtains

$$\dot{q}'''_{max} = \frac{96}{2 \times 4 \times (5 \times 10^{-3})^2} = 480 \text{ MW/m}^3$$

- The energy balance over the fuel rod is written as

$$\dot{q} = \dot{q}'''_{max} \times \pi R_f^2 z = \dot{q} \times 2\pi (R_f + t)z$$

where the LHS represents the heat generated in the rod and the RHS represent the heat flux exiting the lateral rod surface; $\dot{q}''$ is the heat flux at the outer cladding.
- One obtains $\dot{q}'' = 109$ W/cm^2.
- If one denotes $T_{f,s}$ the temperature at the fuel rod surface, and $\Delta T_f = T_c - T_{f,s}$ is the difference of temperature between the rod core and the rod surface, then, from the equation for heat transfer by conduction for the cylinders, one can easily obtain

$$\dot{q}'''_{max} \times \pi R_f^2 h = 4\pi h k_f \Delta T_f$$

- It results in $\Delta T_f = 1500$ °C. On the other hand, the heat transfer equation though the cladding shell is

$$\dot{q}'''_{max} \times \pi R_f^2 h = \frac{2\pi h k_c \Delta T_c}{\ln(1 + t/R_f)}$$

- From this equation, one obtains $\Delta T_c = 300$ °C.
- From these results it follows that:
  - $T_{f,s} = 2200 - 1500 = 700$ °C
  - The temperature at the cladding surface is $T_{c,s} = 400$ °C.
- Therefore, the temperature at which the heat transfer fluid is delivered by the nuclear reactor is estimated at $T_{out} = 350$ °C.

EXAMPLE 6.4 *(cont'd)*

- As the reactor is of thermal type, 5–10% of the nuclear heat is transmitted to the moderator and is considered to be lost. Therefore, the energy efficiency is $\eta = 90$–95%. The Carnot factor in this example is calculated as

$$1 - T_0/T_{out} = 1 - 300/650 = 0.54$$

- The exergy efficiency of the reactor becomes

$$\psi = \eta \times (1 - T_0/T_{out}) = 0.54\,\eta = 49 - 51\%$$

6.4.4 Advanced Nuclear Reactors

There are a number of advanced nuclear reactors currently in the design stage. This means that the conceptual phase of development is ended and the detailed design is underway. These reactors are sometimes denoted as Generation III+ and are the following:

- Advanced CANDU (ACR-1000) reactor.
- European pressurized reactor (EPR).
- VVER-1200.
- APWR-IV.
- ESBWR—economic simplified BWR.
- AP-1000 reactor.
- APR-1400.
- EU–APWR.
- IFR—integral fast reactor.
- PBR—pebble bed reactor.
- HTGCR—high-temperature gas-cooled reactor.
- SSTAR—small sealed transportable autonomous reactor.
- CAESAR—clean, environmentally safe advanced reactor.
- SR—subcritical reactor.
- TBR—thorium-based reactor.
- AHWR—advanced heavy water reactor.
- KAMINI—Uranium-233 based reactor.

The ACR-1000 is a hybrid design that uses heavy water as a moderator and light water as a coolant to generate 1200 MWe. It includes enhanced safety features with respect to former CANDU versions and it is to be commissioned by 2016. The AP-1000 is the Westinghouse Generation III+ design with $\sim$1150 MWe output. The EPR was designed by an international consortium of Siemens AG, Areva NP, and Electricité de France (EDF). It is a PWR-type reactor with 37% efficiency, generating 1650 MWe. In Finland, construction of the first EPR power plant started in 2005 with a plan to be commissioned in 2013. The ESBWR is a BWR designed by GE–HNE with a capacity of 1520 MWe and lifetime of 60 years. APR-1400 is an APWR adopted by South Korea, with 1455 MWe and first commissioning expected in 2013. The VVER-1200 of Generation III+ is an evolution of the VVER-1000 with a lifetime expectancy of 50 years and power range of 1200–1300 MWe. The EU–APWR is a version of APWR by MHI to be commissioned in Europe at a power generation capacity of ~1600 MWe.

In South Africa, the pebble bed modular reactor (PBMR) was under development for almost 16 years under the funding of Eskom, its partners, and the South African government (Koster et al., 2003). No equipment failure or human error can produce an accident with the PBMR because its design and construction are inherently safe. This greatly simplifies the safety design of the system. Its design evolved from the high-temperature reactor of Siemens to a scheme that uses a Brayton cycle turbine with a closed loop. The hot outlet temperature is 900 °C, with around 400 MW of thermal output per unit. Currently, the PBMR technology is pending as it has not found the necessary investors.

India is the sole country that has developed a research program to study the thorium fuel cycle for nuclear power reactors. The government typically has policy programs formulated as 5-year plans. Five "nuclear energy parks" were set up by the Indian government. The Bhabha Atomic Research Centre (BARC) of India leads the research activities on the country's nuclear reactors. Two of the Indian nuclear power stations are coupled with desalination plants (WNA, 2013). In the vicinity of Madras, a prototype of a fast breeder reactor (FBR) of 500 MWe is under construction by an Indian consortium that includes the government and Bharatiya Nabhikiya Vidyut Nigam Ltd.

The FBR reactor is a thorium-based reactor project in India. It will be fueled with uranium–plutonium fuel and it will have a blanket of uranium/thorium to breed fissile ^{233}U. The thorium-based breeder may become one of the most remarkable non-GIF next generation commercial reactors, as it is able to use the large amount of thorium resources in India. Six more FBRs are planned by 2020. In a third phase, the fissile ^{233}U and plutonium from FBR will be transferred to existing Indian advanced heavy water reactors where they will generate about one-third of the power, the rest being supplied by additional thorium fuel.

The Russian Federation is very advanced in fast breeder reactor research. Russian federal funding for novel nuclear reactors is around 2 billion USD for a multi-year period (until ~2020, see Table 6.8). The modular SVBR 100 concept is a low power lead–bismuth-cooled fast breeder reactor for pressurized process steam or power generation. This small reactor is very versatile and can be configured for many applications. Russia led the IAEA INPRO project in 2001.

TABLE 6.8 Russian Next Generation of Commercial Nuclear Reactors

Reactor	Characteristics	Commissioning	Applications
SVBR-100	Lead–bismuth cooled fast breeder reactor 265–280 MW thermal or ~100 MWe Steam generation @: 440–482 °C, 4.7–9.5 MPa, 460–580 t/h	2017	– Modular power or process steam generation – Coastal or floating power generators – Desalination – Process heat
BN 600/800	Sodium-cooled fast neutron reactor	2016	Power generation
BREST 300	Lead-cooled fast reactor 300 MWe	2020	Power generation
MBIR 150	150 MW thermal fast neutron reactor of IV generation	2020	Multipurpose research

Source: WNA (2013).

6.4.5 Generation IV Nuclear Reactors

The US DOE promoted an international initiative on the development of advanced nuclear reactors for power and process heat which was established in 2001 as the Generation IV International Forum (GIF). The concept "Generation IV" has been proposed by the DOE as a method of facilitating international cooperation in the development of advanced nuclear systems by the 2030s and beyond. Generation IV reactors are categorized as theoretical designs currently under research.

The US DOE implemented a Generation IV national program starting in 2002.The terminology "nuclear energy systems" is used instead of nuclear reactors to indicate that the fourth generation is also aimed also at power and process heat production. In January 2000, the Office of Nuclear Energy, Science, and Technology of the DOE facilitated discussions of senior representatives from nine countries (Argentina, Brazil, Canada, France, Japan, Republic of Korea, South Africa, the United Kingdom and the United States) aiming to establish an international collaboration in the development of Generation IV nuclear energy systems. Three other countries later adhered to the charter, including Switzerland (2002), China, and the Russian Federation (2006). The European Union adhered to GIF in 2003 with its atomic energy agency, Euratom. The International Atomic Energy Agency (IAEA) and the Nuclear Energy Agency of the Organisation for Economic Co-operation and Development (OECD) participate in the charter as permanent observers.

The fourth generation of nuclear reactors is represented by six reactor concepts selected by the GIF. These concepts will be discussed in more detail in Chapter 7. Here they are briefly introduced with the purpose of facilitating the discussion of the technology roadmap, research issues, and existing research programs. GIF aims at the development of safer, sustainable, more economical, physically secure, and proliferation-resistant nuclear reactors. Generation IV reactors are intended to become commercially available by 2030. According to GIF (2013), the Generation IV nuclear reactors include the following designs.

- Gas-cooled fast reactor (GFR)—it features a fast-neutron-spectrum, helium-cooled reactor and closed fuel cycle (spent fuel is reprocessed and reused); it generates 1200 MWe and processes heat at 850 °C.
- Very-high-temperature reactor (VHTR)—it is a graphite-moderated, helium-cooled reactor with a once-through uranium fuel cycle (spent fuel is not reprocessed); it has a projected core outlet temperature of 1000 °C and the units have a 600 MW output.
- Supercritical-water-cooled reactor (SCWR)—this is a high-temperature, high-pressure water-cooled reactor that operates above the thermodynamic critical point of water; it is planned for a 1700-MWe output and a core outlet temperature in the range of ~500–650 °C.
- Sodium-cooled fast reactor (SFR)—it features a fast-spectrum, sodium-cooled reactor and closed fuel cycle for efficient management of actinides and conversion of fertile uranium; its core outlet temperature is 550 °C, and it is planned for three capacities: low (50–150 MWe), intermediate (300–600 MWe) and large (600–1500 MWe).
- Lead-cooled fast reactor (LFR)—it features a fast-spectrum lead/bismuth eutectic liquid-metal-cooled reactor and a closed fuel cycle for efficient conversion of fertile uranium and management of actinides. It includes two versions: (i) small size at 19.8 MWe and 567 °C core outlet temperature and (ii) high power at 600 MWe and 480 °C.

- Molten salt reactor (MSR)—it produces fission power in a circulating molten salt fuel mixture with an epithermal-spectrum reactor and a full actinide recycling fuel cycle. The reference capacity is 1000 MWe with a coolant exit from the core at 700–800 °C.

The SCWR represents an evolution of older reactor concepts such as the BWR, PWR, and PHWR, which use water as a coolant. In an SWCR, the pressure of the water coolant in the reactor is maintained over a critical value of 22.1 MPa such that water does not boil. This simplifies the design, increases the efficiency up to 50%, and addresses the issues related to safety, although more stress occurs on materials. There are two design options of SWCR, either with pressure tubes or a pressure vessel. Variants of SWCR concepts with pressure tubes are under development in Canada (SCWR–CANDU) and Russia, while variants with pressure vessels are under development in France, Japan, and Korea.

The planned electricity generation range with SWCR reaches up to 1.5 GW electric (no SCWR has been built yet to date). The system uses light water coolant at an operational pressure of 25 MPa. It has design flexibility, with an inlet temperature up to ∼625 K and outlet up to 900 K, with both thermal (up to 60 GW days per ton of heavy metal) and fast (up to 120 GW days per ton of heavy metal) neutron spectra, three types of fuel (uranium dioxide, thorium, and mixed oxide fuel), two types of fuel cycles (once-through or closed) and three variants of moderators (light water, heavy water, or zirconium hydride solid moderator).

Figure 6.14 illustrates the two main design concepts of the SWCR, namely the pressure vessel configuration and the pressure tube configuration. The CANDU–SWCR configuration

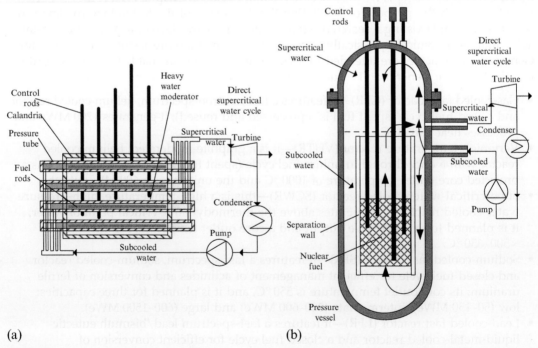

FIGURE 6.14   Supercritical water cooled reactor configurations: (a) Pressure tube (CANDU–SWCR), (b) Pressure vessel. *Modified from GIF (2013).*

uses pressure tubes as illustrated in Figure 6.14a. The CANDU–SWCR, described also in Naterer et al. (2013), has a similar geometrical arrangement to traditional heavy water reactors developed by AECL (Atomic Energy of Canada Limited). However, in the supercritical version, light water is circulated in tubes and the nuclear fuel is based on a thorium fuel cycle. The moderator is heavy water which surrounds the pressure tubes in a calandria vessel with enhanced safety functions.

The pressure vessel configuration—illustrated in Figure 6.14b—represents an evolution of the BWR and PWR concepts where the nuclear fuel is placed inside a pressure vessel surrounded by water which has the role of coolant and moderator. As SWCR operates at supercritical pressure, no boiling occurs inside the pressure vessel, but rather water reaches the state of a supercritical fluid, characterized by both the temperature and pressure, with values above the critical point. Supercritical water simplifies the design because there is no need for a primary or a secondary circuit for heat transfer as typically used in current subcritical water reactors. Moreover, the need for a pressurizer vessel is also eliminated because water in a supercritical state is compressible.

The next generation of SWCR will likely operate at about 25 MPa. The pressure vessel variant of the reactor under development in Japan and Europe is part of the GIF. In Japan, both fast and thermal SCWR development is projected. The pressure tube variant of the SCWR is under development in Canada, for an inlet water temperature in the reactor of ∼625 K, with the outlet at ∼900 K. Besides the development of the Canadian reactor, which is a GIF-adopted concept, there has been a pressure tube version developed in Russia as a non-GIF design, namely by the Research and Development Institute of Power Engineering (RDIPE). The reactor planned by RDIPE is a thermal reactor which operates with water at 25 MP and with a 543 K inlet temperature and 818 K outlet temperature.

Another GIF concept is the GFR, with variants in development at Euratom (Joint Research Centre), France (CEA—Commissariat of Atomic Energy), Japan (JAEA—Japan Atomic Energy Agency and ANRE—Agency for Natural Resources and Energy), and Switzerland (PSI—Paul Scherrer Institute). This is a fast-neutron-spectrum reactor with a closed fuel cycle and helium gas used as a coolant, as shown in Figure 6.15. It is envisioned that the GFR will have on-site fuel processing for actinide recycling with a minimal requirement for nuclear material transportation.

The helium coolant enters the reactor at ∼733 K and leaves at 1023 K and 6 MPa. The planned thermal power of this reactor is 2.4 GW. The active core of a nuclear reactor is about 2 m diameter with a height of 6 m, and it uses a mixed nitride fuel with a breeding ratio of 1.2.

Another Generation IV fast reactor is the sodium-cooled fast reactor (SFR), as schematically illustrated in Figure 6.16. Note that the pool-type variant, although different from the point of view of operation, has a similar loop-type design. The reactor core comprising the nuclear fuel is submerged in molten sodium. The arrangement comprises guiding walls for the flow such that forced convection is promoted and eventually the molten sodium is heated and further forced to flow through a primary heat exchanger. The heat is transferred to a secondary loop with molten sodium. A pump is used to circulate the molten sodium through the secondary circuit. Eventually the heat is transferred to the steam generator. Variants of SFR have been developed by Euratom, France, Japan, China, Korea, Russia, and USA. The reactor features a closed fuel cycle (spent fuel is reprocessed). The concept of the reactor has been verified by systems operated during the period of 1967–1985 on smaller scale reactors in the

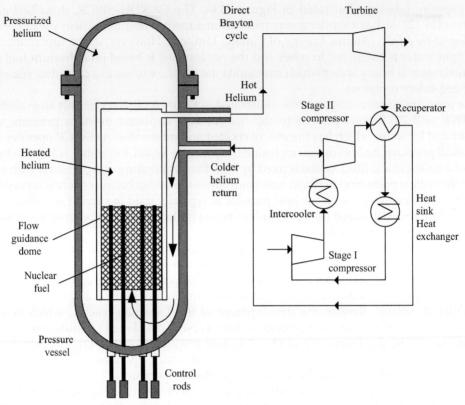

FIGURE 6.15   Conceptual schematic of Generation IV gas-Cooled fast reactor. *Modified from GIF (2013).*

United States, Japan, India, Germany, France, UK, and Russia. There are two main variants of SFR, namely a loop type and a pool type. Both systems consist of primary and secondary sodium loops that eventually transfer heat to a steam generator which produces superheated steam at 19.2 MPa and 793 K, while the water inlet is at ~512 K. The temperature of the secondary sodium loop at the inlet/outlet of the steam generator is 793/608 K; the envisaged thermal power is ~3.5 GW.

The VHTR is planned for a thermal generation capacity of around 600 MW per unit to supply process heat or generate electricity at high efficiency. The reactor uses a thermal neutron spectrum moderated with graphite and cooled with helium. Either pebble bed or prismatic graphite block cores are considered for VHTR. The temperature of helium at the core outlet is 1273 K. The VHTR technology is built on previous practical experience of many research groups and industries, with the most relevant examples of test reactors currently under operation in Japan—the HTTR (high-temperature test reactor) with a 30 MW thermal capacity, and in China—the HTR-10 (high-temperature reactor with a 10 MW thermal output). Other variants of VHTR concepts are: (i) the PBMR proposed in South Africa, (ii) GTHTR-300C proposed in Japan, (iii) ANTARES proposed in France (a reactor design adopted by AREVA), (iv) NHDD proposed in Korea (nuclear hydrogen development and demonstration

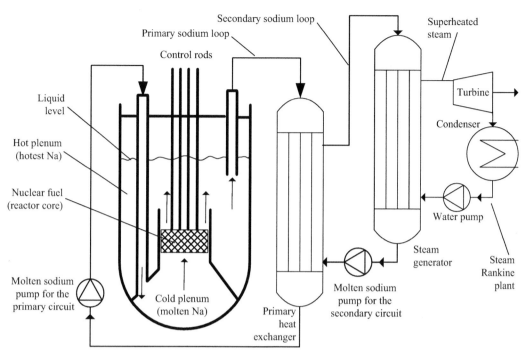

FIGURE 6.16 Conceptual schematic of Generation IV sodium fast-cooled reactor (loop type). *Modified from GIF (2013).*

reactor), (v) GT–MHR proposed in the United States (gas turbine modular helium reactor developed by General Atomics), and (vi) NGNP proposed in the United States (next generation nuclear reactor for process heat and electricity developed at the Idaho National Laboratory with funding support by the Department of Energy). Past test units and prototypes of high-temperature gas reactors have been operated in the United States, Germany and the United Kingdom. The test reactor of Japan—the HTTR—uses helium at 4 MPa with a 668 K inlet temperature and 1223 K outlet temperature.

The typical design of a VHTR system is illustrated schematically in Figure 6.17. The reactor's core consists either as a prismatic block or a pebble bed. Helium gas is circulated through the reactor core in a top-down direction such that hot helium accumulates at a bottom plenum. An intermediate heat exchanger is used to transfer the heat for downstream processing—either for power generation or for use as process heat at high temperature. The test reactor of Japan—the HTTR—uses helium at 4 MPa with a 668 K inlet temperature and 1223 K outlet temperature. The Chinese HTR-10 test reactor works with helium coolant at 3 MPa and a 523 K inlet temperature and 973 K outlet temperature. The GTHTR-300C reactor design for full commercial use is planned to operate at 5–7 MPa helium pressure with an inlet temperature in the range of 860–936 K and outlet temperature of 1123–1223 K.

Both Brayton and Rankine power cycles are considered for electricity generation with VHTR systems. Brayton gas turbine cycles can operate either in a direct configuration coupled to the reactor (where hot helium is expanded in a turbine, then further cooled in a regenerator

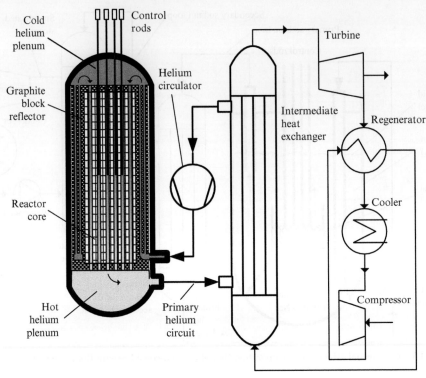

FIGURE 6.17   Typical configuration of a VHTR plant. *Modified from GIF (2013).*

heat exchanger, and then compressed, reheated in the regenerator, and then returned at the reactor inlet) or an indirect configuration (which uses an indirect heat exchanger and a secondary helium or helium–nitrogen loop). When a secondary heat exchanger is used, the temperature decreases according to the log-mean temperature difference. At a capacity for full-scale plants (hundreds of MW thermal), it is around 100–150 K. For example, the intermediate-temperature heat exchanger in the GTHTR-300C reactor is about 750 K at the inlet and 1173 K at the outlet. The Chinese full-scale VHTR reactor is projected to have a ∼500 MW thermal output with a helium gas pressure of 7 MPa, delivered at 1023 K from an input temperature of 523 K. It generates superheated steam in a secondary loop at ∼13 MPa and ∼840 K from subcooled water at ∼475 K.

The LFR is another concept of Generation IV systems that uses a fast-spectrum nuclear process with a closed fuel cycle. The core is cooled with a liquid metal consisting of lead or lead–bismuth eutectic. As opposed to all other reactors (such as water reactors, gas reactors), lead-cooled (or liquid-metal-cooled) reactors do not require a moderator. The LFR design is currently in a development phase in the United States (SSTAR variant—small secure transportable autonomous reactor) and at Euratom (ELSY variant—European lead-cooled system). The planned capacity is about ∼20 MW electric for SSTAR and 600 MWe for ELSY. The ELSY generates superheated steam at 18 MPa and 723 K from the water inlet at 610 K.

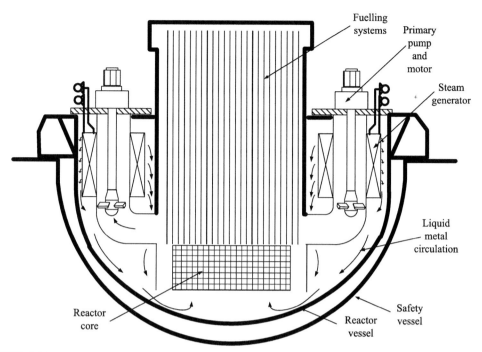

FIGURE 6.18 Conceptual schematic of the European ELSY lead fast reactor (LFR). *Modified from GIF (2013).*

The SSTAR delivers supercritical CO_2 at 20 MPa and 825 K. The ELSY is an LFR version adaptable for large-scale nuclear hydrogen production. In its first planned implementation, the system can be coupled to a high-temperature alkaline electrolyzer to split water or a steam-methane reforming plant to extract hydrogen from natural gas with reduced pollution, compared to other methods.

The reactor configuration for the ELSY concept is illustrated in Figure 6.18. It includes a cylindrical housing with a hemispherical bottom inside which the liquid metal is recirculated. In the liquid metal, there are eight submerged steam generators with a unit thermal power of 175 MW each. Future advancement of the LFR is underway in the United States after a first phase of testing with the SSTAR reactor; a hydrogen generation (or process heat) dedicated LFR will be deployed in the 2020s—denoted STAR-H_2. This will generate 400 MW thermal power at 1073 K.

The MSR operates with dissolved nuclear fuel in a molten fluoride salt. The initial MSR concept was developed in the 1950s at the Oak Ridge National Laboratory (Fermi-1 concept mentioned above). This reactor is suitable for a thorium fuel cycle. Recently a non-moderated thorium molten salt reactor (TMSR–NM) was developed at CNRS-Grenoble in France. The reactor is designed for a 2.5 MW thermal output and uses a fluoride-based molten salt mixture which contains uranium and thorium. The operating temperature of the reactor is 900 K. The general layout of the reactor concept for the Generation IV MSR is illustrated in Figure 6.19. The summary of the main parameters of Generation IV nuclear reactors is given in Table 6.9.

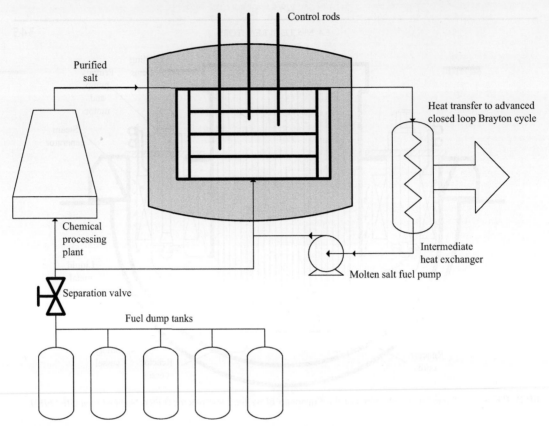

FIGURE 6.19 Conceptual diagram of the molten salt reactor. *Modified from GIF (2013).*

TABLE 6.9 Main Parameter of the Planned Generation IV Nuclear Reactors

Parameter	GFR	LFR	MSR	SFR	SCWR	VHTR
Installed power (MW thermal)	600	400–3600	1000	1000–5000	3850	600
Power density (MWth/m³)	100	100	22	350	100	6–10
Power cycle	Closed Brayton	Steam Rankine	Advanced closed Brayton	Steam Rankine	Supercritical Rankine	Closed Brayton
Net energy efficiency	48%		44–50%	44%	44%	>50%
Coolant	Helium at 90 atm 763/1123 K	Pb–Bi 825 K	Molten fuel salt 825/975 K	825 K	Sc Water 550/780 K, 25 MPa	Helium 910/1273 K
Reactor type	Fast breeder	Fast breeder	Thermal spectrum	Fast spectrum	Thermal or fast spectrum	Thermal spectrum
Fuel cycle	Closed (U, Pu)	Closed	Multiple options: once-through, actinide recycling, etc. (dissolved U and Pu fluorides in molten salts)	Closed (Pu, U, actinides)	Closed	Once through

CASE STUDY 6.1—LASALLE COUNTY BWR

Figure 6.20 represents a simplified diagram of the LaSalle County Nuclear Power Station (LCNPS). The BWR of the LCNPS was developed by GE. The power station, which is located in the vicinity of Chicago, has two generating units of ~1140 MW each and is operated by Commonwealth Edison Company.

The Rankine cycle for BWR represented in Figure 6.20 is a single reheat cycle with no vapor superheating at the high-pressure turbine.

The plant uses a moisture separator for steam reheating before the low-pressure turbine. Simultaneously, in the moisture separator, saturated steam is partially condensed along path 9–14. The corresponding condensation heat is transferred to stream 6–7 for reheating. The actual LaSalle plant has an assembly of steam turbines comprising a high-pressure turbine and three low-pressure turbines, all installed on the same shaft as the main power generator.

The main turbines have in total ~11 steam extraction points to supply the feedwater heating system. In the simplified diagram, the extraction points with similar parameters are joined such that only three essential extractions can be represented, as #12, #13, and #14. There is an additional auxiliary turbine installed (FWT in Figure 6.20) which runs a secondary power generator (SG). The power produced by this turbine is used to run the pumps of the feedwater system (see the dashed line in the figure, representing electric power). The plant also comprises a complex feedwater heating system that includes four pumps and ten feedwater heaters. This system is represented in a simplified manner with a single unit in the diagram in Figure 6.20.

A part of the power generated by the main generator is diverted to the reactor recirculation pumps required to assure the stability of the boiling process. The reactor has two water recirculation

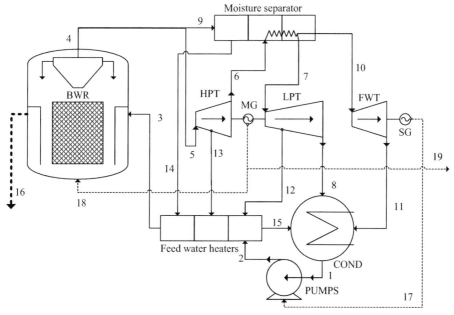

FIGURE 6.20 Simplified diagram of LaSalle County BWR nuclear power station. *Modified from Dunbar et al. (1995).*

Continued

TABLE 6.10   Streams Parameters for LCNPS Generating Unit

Stream	Description	T (K)	P (kPa)	E (MW)	Ėx (MW)
1	Water, condensate out	306	5.03	132.5	6.3
2	Water, pressurized	316	6653	146.7	20.5
3	Water, preheated	489	7308	1530.8	381.2
4	Steam, saturated	555	6653	4830.5	1945.7
5	Steam, to high-pressure turbine	552	6329	4646.0	1871.3
6	Moist steam, to reheater/moisture separator	359	1138	3715.7	1221.1
7	Steam, superheated to low-pressure turbine	340	1082	3532.6	1175.0
8	Moist steam, to condenser	314	8	2192.1	173.2
9	Saturated steam, to moisture separator	555	6604	184.5	74.2
10	Saturated steam, to FWT[a]	540	1096	64.5	21.3
11	Moist steam, to condenser, throttled	406	276	49.7	4.7
12	Extracted steam, to feedwater heaters	401	251	590.5	178.8
13	Extracted steam, to feedwater heaters	468	1442	516.7	178.8
14	Extracted hot water, saturated	510	3764	303.1	77.1
15	Condensate return, via steam trap	311	14	26.2	2.0
16	Heat loss from nuclear reactor	298	n/a	5.5	0
17	Electric power to feedwater pumps	n/a	n/a	14.8	14.8
18	Electric power to reactor circulation pumps	n/a	n/a	12.2	12.2
19	Net power output	n/a	n/a	1132.0	1132.0

[a]FWT, turbine/generator assembly for feedwater pump powering.
Source: Dunbar et al. (1995).

loops served by twenty internal jet pumps, a condensate flow pumping station, and a pumping station related to the water demineralization system.

The main thermodynamic parameters for each stream are given in Table 6.10. Based on those parameters, energy and exergy balances can be written for each component, as indicated in Table 6.11. From the energy balances the energy transferred or consumed or provided by each functional unit is determined, as are the energy losses. From the exergy balance equation, all exergy destructions by the functional units are determined. The pie chart in Figure 6.21 illustrates the distribution of exergy destruction among the system components. Remarkably, the overwhelming majority of exergy destruction (~83%) occurs in the reactor itself. In order to determine the exergy associated with nuclear fuel it is assumed that nuclear energy is a form of heat with an extremely high-temperature, such that the Carnot factor can be approximated with one. As a consequence of this approximation, the energy and exergy efficiency of the nuclear power station become coincident. It is determined that the total exergy destruction is 2104.4 MW, whereas the exergy efficiency is 36.1%.

Continued

TABLE 6.11 Energy and Exergy Balances on LCNPS Subsystems

Subsystem	Energy balance	Exergy balance
Condenser	$\dot{E}_8 + \dot{E}_{11} + \dot{E}_{15} = \dot{E}_1 + \dot{Q}_{cond}$ $\dot{Q}_{cond} = 2136\,\mathrm{MW}$	$\dot{Ex}_8 + \dot{Ex}_{11} + \dot{Ex}_{15} = \dot{Ex}_1 + \dot{Q}_{cond}(1 - T_0/T_1) + \dot{Ex}_{d,cond}$ $\dot{Ex}_{d,cond} = 116.8\,\mathrm{MW}$
Feedwater pumps[a]	$\dot{E}_1 + \dot{E}_{17} = \dot{E}_2$ $\dot{W}_{pump} = \dot{E}_{17} = 14.2\,\mathrm{MW}$	$\dot{Ex}_1 + \dot{Ex}_{17} = \dot{Ex}_2 + \dot{Ex}_{d,pump}$ $\dot{Ex}_{d,pump} = 0.4\,\mathrm{MW}$
Feedwater heaters[a]	$\dot{E}_2 + \dot{E}_{12} + \dot{E}_{13} + \dot{E}_{14} = \dot{E}_3 + \dot{E}_{15}$ $\dot{Q}_{fwh} = \dot{E}_3 + \dot{E}_{15} = 1557\,\mathrm{MW}$	$\dot{Ex}_2 + \dot{Ex}_{12} + \dot{Ex}_{13} + \dot{Ex}_{14} = \dot{Ex}_3 + \dot{Ex}_{15} + \dot{Ex}_{d,fwh}$ $\dot{Ex}_{d,fwh} = 72\,\mathrm{MW}$
Nuclear reactor	$\dot{E}_3 + \dot{E}_{18} + \dot{Q}_{BWR} = \dot{E}_4 + \dot{E}_{16}$ $\dot{Q}_{BWR} = 3293\,\mathrm{MW}$	$\dot{Ex}_3 + \dot{Ex}_{18} + \dot{Q}_{BWR}(1 - T_0/T_\infty) = \dot{Ex}_4 + \dot{Ex}_{16} + \dot{Ex}_{d,BWR}$ $\dot{Ex}_{d,BWR} = 1741\,\mathrm{MW}$, where $T_\infty = \infty$
Moisture separator	$\dot{E}_6 + \dot{E}_9 = \dot{E}_7 + \dot{E}_{10} + \dot{E}_{14}$ $\dot{Q}_{msep} = \dot{E}_6 + \dot{E}_9 = 3900\,\mathrm{MW}$	$\dot{Ex}_6 + \dot{Ex}_9 = \dot{Ex}_7 + \dot{Ex}_{10} + \dot{Ex}_{14} + \dot{Ex}_{d,msep}$ $\dot{Ex}_{d,msep} = 21.9\,\mathrm{MW}$
High-pressure turbine (HPT)	$\dot{E}_5 = \dot{E}_6 + \dot{E}_{13} + \dot{W}_{HPT}$ $\dot{W}_{HPT} = 413.6\,\mathrm{MW}$	$\dot{Ex}_5 = \dot{Ex}_6 + \dot{Ex}_{13} + \dot{W}_{HPT} + \dot{Ex}_{d,HPT}$ $\dot{Ex}_{d,HPT} = 57.8\,\mathrm{MW}$
Low-pressure turbine (LPT)	$\dot{E}_7 = \dot{E}_8 + \dot{E}_{12} + \dot{W}_{LPT}$ $\dot{W}_{LPT} = 750.0\,\mathrm{MW}$	$\dot{Ex}_7 = \dot{Ex}_8 + \dot{Ex}_{12} + \dot{W}_{LPT} + \dot{Ex}_{d,LPT}$ $\dot{Ex}_{d,LPT} = 73\,\mathrm{MW}$
Main generator (MG)	$\dot{E}_{18} + \dot{E}_{19} + \dot{Q}_{loss} = \dot{W}_{HPT} + \dot{W}_{LPT};$ $\dot{Q}_{loss} = 19.1\,\mathrm{MW}$	$\dot{W}_{HPT} + \dot{W}_{LPT} = \dot{E}_{18} + \dot{E}_{19} + \dot{Q}_{loss}(1 - T_0/T_{SG}) + \dot{Ex}_{d,gen}$ $\dot{Ex}_{d,gen} = 19.1\,\mathrm{MW}$, $\dot{W}_{net} = \dot{E}_{19} = 1132\,\mathrm{MW}$, $T_{MG} = T_0$
Auxiliary turbine (FWT)	$\dot{E}_{10} = \dot{E}_{11} + \dot{W}_{FWT}$ $\dot{W}_{FWT} = 14.8\,\mathrm{MW}$	$\dot{Ex}_{10} = \dot{Ex}_{11} + \dot{W}_{FWT} + \dot{Ex}_{d,FWT}$ $\dot{Ex}_{d,FWT} = 1.8\,\mathrm{MW}$
Secondary generator (SG)	$\dot{W}_{FWT} = \dot{E}_{17} + \dot{Q}_{loss}$ $\dot{Q}_{loss} = 0.6\,\mathrm{MW}$	$\dot{W}_{FWT} = \dot{E}_{17} + \dot{Q}_{loss}(1 - T_0/T_{SG}) + \dot{Ex}_{d,FWT}$ $\dot{Ex}_{d,FWT} = 0.6\,\mathrm{MW}$; $T_{SG} = T_0$

[a]*The actual (detailed) diagram of LCNPS comprises four feedwater pumps and ten feedwater heaters, which are represented in the simplified diagram as only one pump and one feedwater heater.*

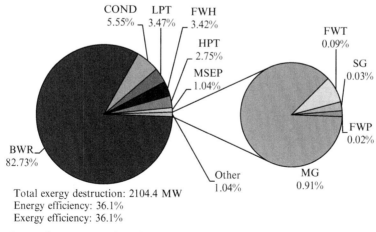

Total exergy destruction: 2104.4 MW
Energy efficiency: 36.1%
Exergy efficiency: 36.1%

COND 5.55% · LPT 3.47% · FWH 3.42% · HPT 2.75% · MSEP 1.04% · BWR 82.73% · Other 1.04% · MG 0.91% · FWP 0.02% · FWT 0.09% · SG 0.03%

FIGURE 6.21 Exergy destruction in the subsystems of a LCNPS power generation unit.

CASE STUDY 6.2—PICKERING NUCLEAR POWER STATION

A simplified diagram of CANDU Pickering Nuclear Power Station (PNPS) is given in Figure 6.22. The nuclear reactor generates superheated pressurized steam in #6 at 527 K and 4250 kPa. The Rankine cycle is of single reheat type. Table 6.12 gives the thermodynamic parameters of each state point within the Rankine power plant, including exergy and energy. A detail of the main heat transfer processes inside the nuclear reactor is explained with the help of Figure 6.23. Heat generated by nuclear reaction is transferred via a neutronic radiation to the fuel pellets (#1) and to the low-pressure heavy water moderator (#7) that surrounds the pressure tubes. The temperature associated with neutronic radiation is assumed to be infinity.

Hence, the Carnot factor associated with neutronic radiation is equal to one. The pellets will heat up under the radiation; a part of thermal energy is transferred as heat to the pellet surface (#2) and then to the cladding surface (#3). From the cladding surface, heat is transferred to the high-pressure heavy water coolant (#4) which is circulated inside the tubes. The fluid enters with a lower temperature in (#5) and exits at a higher temperature in (#6). Also, hot moderator is extracted in (#9), then cooled and returned in (#8). The heat transfer process parameters within the CANDU PNPS nuclear reactor are given in Table 6.13.

The remaining thermal energy is completely transferred to the pressurized heat transfer fluid (heavy water circulated through the pressure tubes). The heat losses within the reactor are very small compared to the generated heat; thus the energy efficiency of thermal power generation is rather high, namely 94.8%. Regarding the exergy analysis, there are irreversibilities associated with each heat transfer process. Figure 6.24 gives a representation of irreversibility distribution among processes inside the reactor.

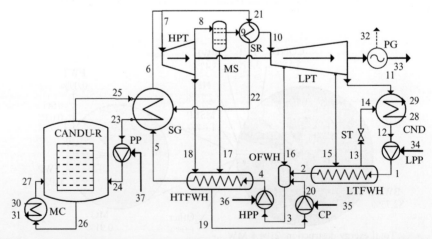

FIGURE 6.22 Simplified diagram of CANDU Pickering Nuclear Power Station (PNPS) (modified from Dincer and Rosen (2007)). LPP, low-pressure pump; LTFWH, low-temperature feedwater heater; CP, condensate pump; OFWH, open feedwater heater; HTFWH, high-temperature feedwater heater; HPP, high-pressure pump; PP, primary circuit pump; SG, steam generator; MC, moderator cooler; CANDU-R, nuclear reactor; HPT, high-pressure turbine; MS, moisture separator; SR, steam reheater; LPP, low pressure turbine; PG, electric power generator; CND, condenser; ST, steam trap.

Continued

TABLE 6.12   Streams Parameters for PNPS Generating Unit

Stream	Description	T (K)	P (kPa)	E (MW)	$\dot{E}x$ (MW)
1	Subcooled water	296.5	1480	211.55	1.13
2	Preheated water	373.2	1400	207.88	26.5
3	Condensed water	396.9	1400	344.21	53.16
4	Subcooled water	397.4	5400	347.93	56.53
5	Preheated water	437.1	5350	476.02	96.07
6	Superheated steam	527.2	4250	2226.9	862.32
7	Superheated steam	527.2	4250	2060.02	797.7
8	Moist steam	425	500	1705.5	500.4
9	Saturated steam	433.2	500	1629.8	476.5
10	Superheated steam	511.2	450	1733.2	508.3
11	Low-pressure steam	296.5	2.86	1125.1	44.4
12	Condensed water	296.5	2.86	20.1	0.2
13	Saturated liquid	334	20.7	15.9	1.1
14	Moist steam	296.5	2.86	15.9	1.1
15	Extracted steam	334	20.7	204.0	28.1
16	Extracted steam	459.2	255.0	61.1	16.0
17	Extracted liquid	433.2	618.0	75.7	23.7
18	Extracted steam	449.9	928	138.7	44.6
19	Saturated liquid	407.2	304	75.04	12.3
20	Subcooled liquid	407.4	1480	75.3	12.5
21	Superheated steam	527.2	4250	166.88	64.62
22	Condenser water	527.2	4250	63.6	17.8
23	Cold heavy water	522.2	8320	7861.2	2188.6
24	Heavy water input	522.6	9600	7875.4	2201.6
25	Hot heavy water	565.1	8820	9548.2	2984.2
26	Warm heavy water	337.7	101.0	207.0	16.0
27	Cooled heavy water	316.2	101.0	117.0	5.3
28	Cooling water in	288.2	101.0	0.0	0.0
29	Cooling water out	299.2	101.0	1107.2	20.6
30	Cooling water in	288.2	101.0	0.0	0.0
31	Cooling water out	299.2	101.0	90	1.7

Continued

TABLE 6.12   Streams Parameters for PNPS Generating Unit—Cont'd

Stream	Description	T (K)	P (kPa)	E (MW)	Ėx (MW)
32	Heat loss, generator	n/a	n/a	11.1	0.0
33	Brut power output	n/a	n/a	544.8	544.8
34	LPP power in	n/a	n/a	1.0	1.0
35	CP power in	n/a	n/a	0.2	0.2
36	HPP power in	n/a	n/a	3.7	3.7
37	PP power in	n/a	n/a	14.3	14.3

Source: Dincer and Rosen (2007).

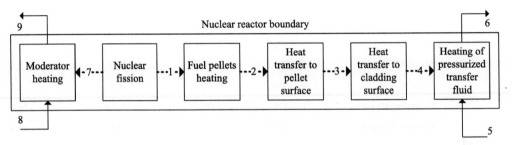

FIGURE 6.23   Representation of main heat transfer processes within CANDU reactor. *Modified from Dincer and Rosen (2007).*

TABLE 6.13   Heat Transfer Process Parameters in CANDU PNPS Nuclear Reactor

Stream	Transport process description	T (K)	E (MW)	Ėx (MW)
1	Radiative heat transfer (neutrons)	∞	1673	1673
2	Conduction heat transfer	2000	1673	1455
3	Combined heat transfer	675	1673	956
4	Convection heat transfer	575	1673	881
5	Pressurized heavy water	523	7875	2202
6	Pressurized heavy water	565	9548	2984
7	Radiative heat transfer (neutrons)	∞	90	90
8	Cold moderator fluid in	316	117	5
9	Warmed moderator fluid out	338	207	16

Source: Dincer and Rosen (2007).

Continued

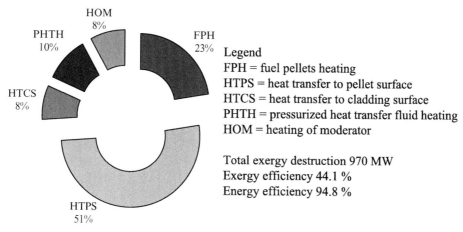

FIGURE 6.24 Exergy destruction within the CANDU nuclear reactor of PNPS.

Based on these parameters, the exergy destructions of each of the main processes inside the reactor can be determined. From the detailed calculations presented in Table 6.14, the energy generated by the nuclear reaction is 1673 MW. From this energy 90 MW are lost as $\dot{E}_7$, namely the energy transferred to the low-pressure moderator.

The maximum exergy destruction occurs at heat transfer from the core to the surface of fuel pellet, which represents 51%. Exergy destruction associated to fuel pellet heating by neutronic radiation is also high, representing 23%. The total exergy destruction in the reactor is 970 MW, which corresponds to 44.1% exergy efficiency. The 8% exergy destruction due to moderator heating is caused 90.0% by heat dissipation from neutron radiation, with 2.8% due to pressure tube heating, 2.7% due to calandria tube heating, 2.6% due to calandria tank heating, 1.2% due to reactor shield heating, and 0.7% due to dump tank heating.

Exergy destruction for each component of the PNPS can be calculated based on exergy balance equations using the data from Table 6.14. The results are represented in the chart in Figure 6.25. The maximum exergy destruction is in the nuclear reactor, with 79.8% of the total exergy destruction, amounting to 1125.1 MW. Based on exergy destruction and nuclear exergy input, the power generation exergy efficiency of the plant is 31.1%.

TABLE 6.14 Energy and Exergy Balances for Main Processes in CANDU PNPS Reactor

Process	Energy balance	Exergy balance
Fuel pellets heating	$\dot{E}_1 = \dot{E}_2 = 1673\,\text{MW}$	$\dot{E}x_1 = \dot{E}x_2 + Ex_d; \dot{E}x_d = 218$ MW
Heat transfer to pellets surface	$\dot{E}_2 = \dot{E}_3 = 1673$ MW	$\dot{E}x_2 = \dot{E}x_3 + Ex_d; \dot{E}x_d = 499\,\text{MW}$
Heat transfer to cladding surface	$\dot{E}_3 = \dot{E}_4 = 1673$ MW	$\dot{E}x_3 = \dot{E}x_4 + Ex_d; \dot{E}x_d = 75$ MW
Pressurized heat transfer fluid heating	$\dot{E}_4 + \dot{E}_5 = \dot{E}_6 = 9548$ MW	$\dot{E}x_4 + \dot{E}x_5 = \dot{E}x_6 + \dot{E}x_d; \dot{E}x_d = 99\,\text{MW}$
Heating of moderator	$\dot{E}_7 + \dot{E}_8 = \dot{E}_9 = 207$ MW	$\dot{E}x_7 + \dot{E}x_9 = \dot{E}x_9 + \dot{E}x_d; \dot{E}x_d = 79\,\text{MW}$
Overall	$\dot{E}_5 + \dot{E}_8 + \dot{E}_{\text{nuc}} = \dot{E}_6 + \dot{E}_9$ $\dot{E}_{\text{nuc}} = \dot{E}_1 + \dot{E}_7 = 1763$ MW	$\dot{E}x_5 + \dot{E}x_8 + \dot{E}x_{\text{nuc}} = \dot{E}x_6 + \dot{E}x_9 + \dot{E}x_d$ $\dot{E}x_{\text{nuc}} = \dot{E}x_{\text{nuc}}; \, \dot{E}x_d = 970$ MW
Efficiency	$\eta = \frac{1673}{1763} = 94.8\%$	$\psi = \frac{Ex_6 - Ex_5}{Ex_{\text{nuc}}} = 44.1\%$

Continued

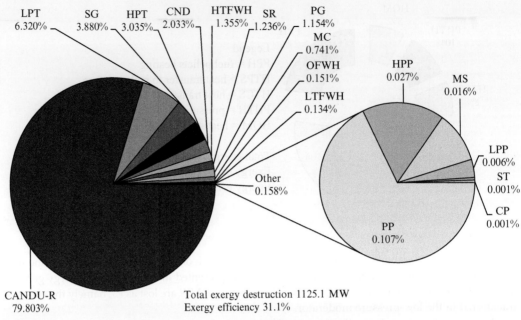

FIGURE 6.25   Exergy destructions diagram for CANDU Pickering Nuclear Power Station.

6.5 NUCLEAR-BASED COGENERATION SYSTEMS

Cogeneration, which is the simultaneous production of thermal and electrical energy, has significant potential for transforming present energy supply systems so that they become more sustainable. In any type of thermal power plant a portion, usually 20–45%, of the input thermal energy is converted to electricity, and the remainder is rejected to the environment as waste heat. As nuclear power generation units can produce a high-pressure and high-temperature vapor, they can be utilized to generate heating, cooling, or freshwater with a topping cycle. In this chapter, we try to address the main nuclear-based cogeneration plants briefly. Besides power these cogeneration plants co-produce heating, freshwater, cooling, hydrogen, process heat, synthetic fuels or other commodities with market value. In Chapter 9 another perspective on the cogeneration system will be provided, when discussing multigeneration systems.

When cogeneration is applied better resource utilization is obtained. This is illustrated in Figure 6.25. In a cogeneration system the thermodynamic cycle can be connected to two heat sinks. One heat sink is the reference environment, typically air or a lake or river. The heat engine rejects heat at a temperature higher than the environment depending on the specific demand. In district heating or desalination applications the required temperature level is low (60–120 °C). In other applications heat can be required at intermediate or very high values. Obviously, the applications with a low-grade heat demand are the most favorable economically because valorizing the low grade heat will increase the return on investment in the nuclear power plant and make it more economically attractive.

A short numerical example can be given related to options (a) and (b) presented in Figure 6.26. Assume that there is a demand of 300 MW power and 350 MW heat. If heat and power are produced with separate technologies (for example, power is derived from a nuclear station and heat is generated by a natural-gas-burning furnace) then the following situation is valid: for 300 MW power the required thermal energy is 900 MW, and for 350 MW heating one needs to combust 410 MW equivalent natural gas for 85% furnace efficiency. Hence the overall efficiency is 650/1310 MW = 49.6%. With a cogeneration approach there is no need for natural gas, but rather 350 MW of heat is recovered from the nuclear power plant; note that in order to produce 300 MW power the power plant must reject into the environment about 600 MW of waste heat. Hence cogeneration is motivated by an efficiency of 650/900 MW = 72%.

Although there are some challenges, such as coping with the danger of radioactive contamination, there is obviously great potential for nuclear heat. Cogeneration with nuclear energy has been used in the past and is used today in a limited number of applications. One of the most common uses of nuclear-based cogeneration is for district heating. The worldwide experience with nuclear district heating measures 500 reactor-years, which represents only 4.1% of the global total of 12,193 reactor-years, according to IAEA (2006). Nuclear desalination has

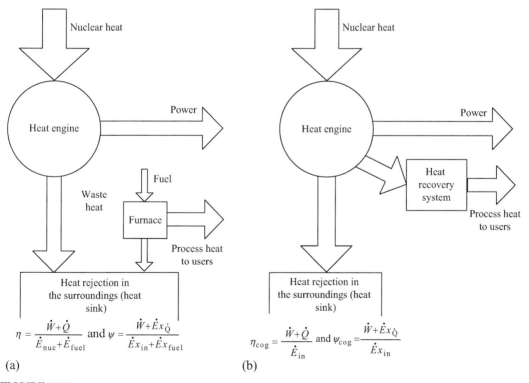

$$\eta = \frac{\dot{W}+\dot{Q}}{\dot{E}_{nuc}+\dot{E}_{fuel}} \text{ and } \psi = \frac{\dot{W}+\dot{E}x_{\dot{Q}}}{\dot{E}x_{in}+\dot{E}x_{fuel}}$$

(a)

$$\eta_{cog} = \frac{\dot{W}+\dot{Q}}{\dot{E}_{in}} \text{ and } \psi_{cog} = \frac{\dot{W}+\dot{E}x_{\dot{Q}}}{\dot{E}x_{in}}$$

(b)

FIGURE 6.26   Value added by cogeneration with respect to power generation: (a) power and heat production with separate technologies and (b) cogeneration-integrated technologies for heat and power production from the same source.

also been used in some countries with a cumulated experience of 250 reactor-years (~2% of global nuclear power station experience). Some limited experience exists for industrial heat cogeneration with nuclear reactors, namely for heavy water production, industrial steam cogeneration, salt refining, and cardboard production. Two main categories of heat demand can be satisfied through cogeneration:

- Tertiary sector, more specifically including heat demands in residential, commercial, governmental building settings (water and space heating or air conditioning).
- Primary and secondary sectors that require heat at a wide temperature range and capacities for water heating, space heating, drying processes, manufacturing, mining, agriculture, chemical processing, etc.

Figure 6.27 shows the temperature level with Generation IV nuclear reactors versus requirements of some relevant heat demanding processes. When heat is transported remotely from the nuclear reactor to the user site measures must be taken to avoid any possibility of radioactive contamination through the heat transfer fluid. Two configurations are suggested for reactor radioactive isolation, as shown in Figure 6.28. In the HLH configuration there is a higher pressure in the reactor loop and process loop. Possibly, radioactive contaminants may leak into the isolation loop; however, due to the pressure, contamination by radioactive particles remains confined in the isolation loop and does not pass further. The isolation loops may be treated for cleaning with specific methods. With the LHL configuration the isolation loop has the highest pressure and therefore will not allow penetration of particles or fluids from the other two loops.

Figure 6.29 suggests a possible linkage between a nuclear Rankine steam power plant and a heat-demanding system for cogeneration. The steam generator which produces superheated steam is part of the multistage Rankine power plant. A reheating circuit can be integrated with the steam generator. When cogeneration is required, a part of the superheated steam

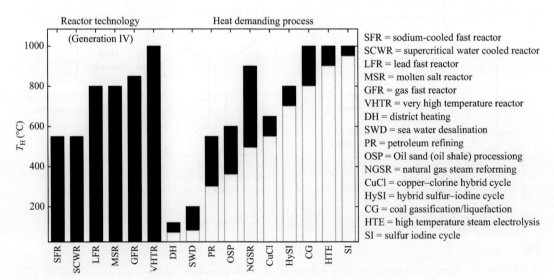

FIGURE 6.27   Temperature level with Generation IV nuclear reactors versus requirements of some relevant heat-demanding processes.

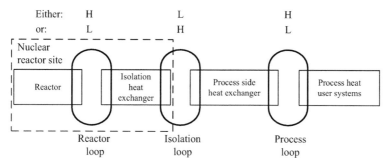

FIGURE 6.28   Reactor isolation configuration to avoid radioactive contamination.

from the steam reheating circuit is directed toward a secondary low-pressure turbine (stream #27). Steam is expanded to an intermediate pressure in accordance with the temperature level required for cogeneration. After a heat transfer process with the cogeneration loop steam fully condenses and the liquid water is returned at the steam trap of the open heat water feeder. For applications where a higher temperature is required for the cogeneration loop the low-pressure turbine is not used, rather the cogeneration heat exchanger is supplied directly with high-temperature steam (Figure 6.29).

The power and heat products of a nuclear reactor station can be converted in industrial parks into multiple commodities such as steam, potable water, and hydrogen and used further for many processes, as suggested in Figure 6.30. Some examples of heat-demanding industrial processes are as follows:

- Aluminum production: the alumina-rich bauxite ore (Al_2O_3) must be extracted in slurry form mixed with sodium hydroxide at 110°–270 °C, with the intermediary products being sodium aluminate and aluminum hydroxide. Further, a rotating kiln is used for calcination of aluminum hydroxide at 1100 °C.
- Steam-coal gasification: this process can be conducted in two ways (i) steam-coal gasification and (ii) hydrogasification. In hydrogasification coal is combined with hydrogen in an exothermic reaction to produce synthetic natural gas. Nuclear heat is used to produce synthesis gas and extract hydrogen through steam reforming and water-gas shift reactions. The processes can be coupled with the VHTR for a required heat of 950 °C.
- Steam-assisted gravity drainage: this process is used to extract bituminous oil sands where steam at 10–15 MPa and 200–240 °C is required. In addition this process requires hydrogen for synthetic crude oil production.

As the population is growing, many regions are experiencing increased freshwater demands that greatly exceed the supply capability of existing infrastructures. The problem is compounded by increases in both the pollution and salinity of freshwater resources. Lack of freshwater is a prime factor inhibiting regional economic development. Seawater desalination is an important option for satisfying current and future demands for freshwater in arid regions in close proximity to the sea.

Desalination units need energy to separate salt from saline waters. The energy can be either heat for seawater distillation or mechanical energy to drive pumps for pressurization of seawater across membranes using reverse osmosis (RO). The major source of heat energy comes from fossil fuels, including coal, oil, and gas. Most of the large-size plants based on thermal and

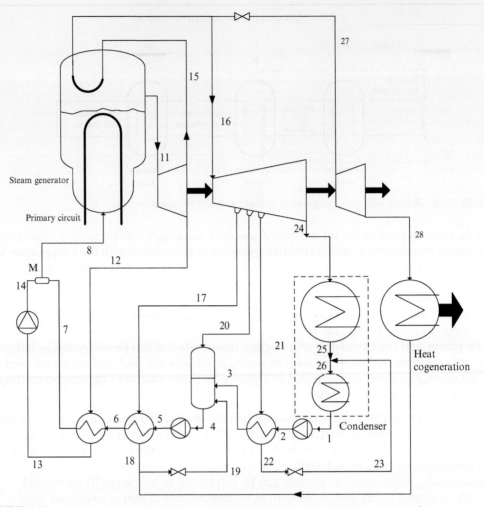

FIGURE 6.29 Integration example of the cogeneration loop with a nuclear Rankine power plant.

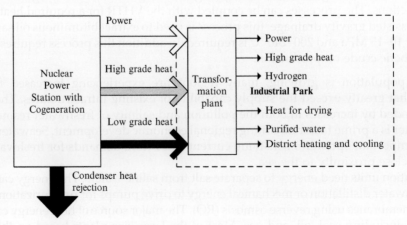

FIGURE 6.30 Linkage of a nuclear cogeneration station with an industrial park with multiple commodity demands.

membrane processes are located near thermal power stations, which utilize fossil fuels to supply both steam and electrical power for desalination. However, the use of fossil fuel to produce freshwater has several side effects. As the combustion of fossil fuels emits greenhouse gases to the environment, it will lead to global warming, one of the main concerns of this century. In addition, these fuels are non-renewable and will be exhausted in next 50 years.

Therefore, because of their versatility as an energy source and hydrocarbon feedstock, there is a keen interest in conserving fossil fuels for other industrial applications, especially in countries with inadequate sources of energy. Over the long term it is not practical to establish a desalting economy based solely on fossil fuels because they have limited availability, whereas freshwater must continue to be available to sustain humankind. Long-term reliance on desalting should only occur if the availability of the energy supply is comparable to the required availability of the product. This criterion can be satisfied by nuclear energy. Interest in using nuclear energy for producing potable water has been growing worldwide in the past decade. This has been motivated by a wide variety of factors, from the economic competitiveness of nuclear energy to energy supply diversification; from conservation of limited fossil fuel resources to environmental protection; and by the spin-off effects of nuclear technology in industrial development. One of the main cogeneration units using nuclear power as a topping cycle is used to produce freshwater by multistageflash desalination techniques.

Hydrogen is expected to play a significant role as energy carrier in the future. Hydrogen can be used as fuel in almost every application where fossil fuels are utilized today. In contrast to fossil fuel, its combustion is without harmful emissions, disregarding NO_x emissions that can be effectively controlled. In addition, hydrogen can be transformed into useful forms of energy more efficiently than fossil fuels. Hydrogen is as safe as other common fuels, despite its common perception. Hydrogen is not an energy source and hence it does not exist in nature in its elemental form.

Therefore, hydrogen must be produced from water, the most abundant source of hydrogen, or from other sources. However, splitting of water for hydrogen production necessitates energy that is higher than the energy that can be obtained from the produced hydrogen. Therefore, hydrogen is considered an energy carrier for a suitable form of energy like electricity. It is commonly accepted that hydrogen is one of the promising energy carriers and that the demand for it will increase greatly in the near future, for it can be utilized as a clean fuel in diverse energy end-use sectors, including the conversion to electricity with no CO_2 emission. In addition, the nuclear fuel supply is estimated to be readily available even on a once-through fuel cycle for 50 to 100 years. Using breeder reactors and a closed fuel cycle, it is virtually inexhaustible. Because of its relative abundance, nuclear fuel is relatively cheap compared to fossil-based fuels.

Possible routes to generate hydrogen from water using nuclear energy are also depicted in Figure 6.31, with additional paths which assume processes other than water decomposition, such as coal gasification, natural gas reforming, petroleum naphtha reforming, or hydrogen sulfide cracking. In all such processes, high-temperature nuclear heat is used to conduct the required chemical reactions. Coal gasification and natural gas reforming methods include the water-gas shift reaction to generate additional hydrogen and reduce the carbon monoxide to CO_2. Nuclear-based reforming of fossil fuels is a promising method of high viability in the transition toward a fully implemented hydrogen economy which will reduce the world dependence on fossil fuel energy.

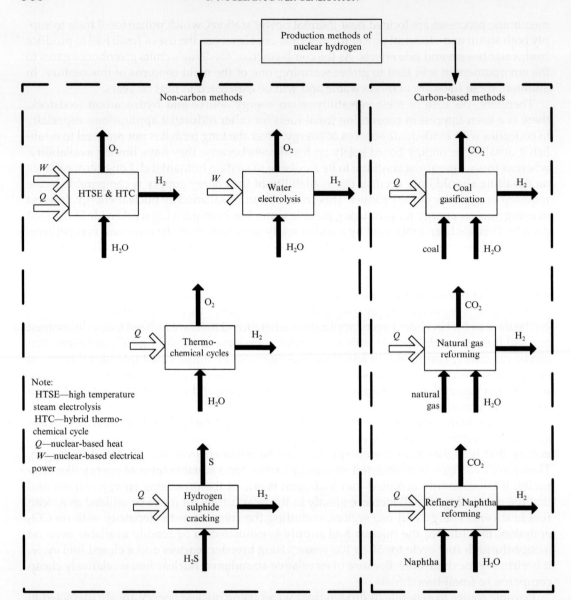

FIGURE 6.31 Production methods of nuclear hydrogen which use heat and/or electricity inputs.

Although fossil fuel reforming can be used to generate hydrogen, it is attractive to use nuclear heat to synthesize Fischer–Tropsch fuels or methanol. Such fuels can also be used for transportation vehicles. With this method, nuclear heat is used to generate synthesis gas, and then to generate either diesel or methanol. Another process of great interest is fertilizer production, such as ammonia and urea, with nuclear energy. A nuclear ammonia/urea production facility would comprise the nuclear reactor, a heat transfer system for high-temperature nuclear process heat, a nuclear power plant, a nuclear hydrogen generation unit, an air separation unit, H_2/CO_2 separators, a Haber–Bosch synthesis unit, and an NH_3/CO_2 reactor to generate urea.

Hydrogen sulfide is a potential source of hydrogen. Hydrogen sulfide can be extracted from some geothermal wells, oil wells, volcano sites, and seas. The Black Sea represents the highest reserve of hydrogen sulfide in the world. In the Black Sea, the hydrogen sulfide can be viewed as a renewable resource because it is generated by microorganisms under the influence of solar radiation. Typically, thermo-catalytic cracking of H_2S can be applied to generate hydrogen and sulfur, as two valuable products using nuclear energy.

CASE STUDY 6.3: INTEGRATED DESALINATION AND HYDROGEN PRODUCTION WITH NUCLEAR REACTOR

In this example, which is based on the work by Orhan et al. (2010), we analyze a coupling of the Cu–Cl cycle with a desalination plant for hydrogen production from nuclear energy and seawater. To produce hydrogen by splitting the water molecule one needs to supply freshwater to the hydrogen production facility. Thus, to avoid causing one problem while solving another problem, hydrogen could be produced from seawater rather than limited freshwater sources. Nuclear energy is used to drive the process. Here we consider five configurations, in hopes of determining or helping to determine an optimum option to couple the Cu–Cl cycle with a desalination plant.

Configuration 1: the Cu–Cl cycle is coupled to a desalination plant using nuclear energy, as shown in Figure 6.32. Salty water is put into the desalination plant and salt is removed from the water using waste energy from the nuclear reactor moderator at 70–80 °C. Freshwater supplied by the

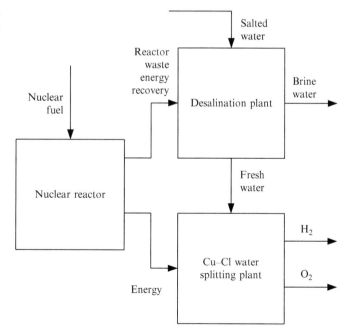

FIGURE 6.32 Using waste energy from nuclear reactor for desalination process. *Adapted from Orhan et al. (2010).*

Continued

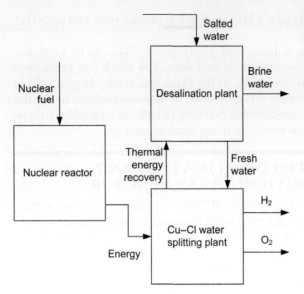

FIGURE 6.33 Using recovered energy from Cu–Cl plant. *Adapted from Orhan et al. (2010).*

desalination plant is decomposed into hydrogen and oxygen by the Cu–Cl cycle driven by nuclear energy. humidification–dehumidification (HD) technology is used for desalination.

Configuration 2: the thermal energy recovered from the Cu–Cl cycle is transferred to the desalination plant to remove salt from freshwater. As illustrated in Figure 6.33, the desalination plant operates as a subsystem of the Cu–Cl cycle. Considering the overall system, only process/waste energy from the nuclear reactor and salty water enter the system. Hydrogen is produced, and oxygen and salt are byproducts. A drawback to this configuration is the efficiency decrease (~3–5%) incurred by the Cu–Cl cycle as the recovered energy is used for the desalination process rather than within the cycle itself. But the Cu–Cu cycle is a subsection of the plant and, when the desalination plant and the Cu–Cl cycle are considered as a combined system, the efficiency is not affected significantly as the recovered energy is used within the overall system. The efficiency of the combined system (desalination plant and copper chlorine cycle) is about 0.4. Multiple effect desalination technology is used.

Configuration 3: the nuclear thermal energy is used directly in the desalination plant and to drive the hydrogen plant as indicated in Figure 6.34. The thermal energy recovered in the Cu–Cl cycle is used within that cycle. A desalination method with a high capacity and low production cost is used as high-grade energy is available. Multistage flash desalination method is used.

Configuration 4: in addition to nuclear thermal energy, solar energy is used to complete the energy requirement for the process. The solar energy drives directly and completely the desalination plant which supplies the hydrogen production plant with freshwater. Process and waste energy from a nuclear plant are used in the Cu–Cl cycle as indicated in Figure 6.35. Vapor compression (VC) desalination technology is applied.

A drawback of this configuration is its dependence on the availability of solar energy, which is intermittent. As the capacity of the desalination plant and hence the Cu–Cl cycle determines the required capacity of the solar collectors, the site for the Cu–Cl cycle and desalination plant is chosen carefully. However, as constraints likely exist regarding the location of the nuclear reactor, this configuration is likely advantageous if the nuclear reactor is located in an area with high solar

Continued

FIGURE 6.34 Direct use of energy generated by the nuclear reactor for desalination process. *Adapted from Orhan et al. (2010).*

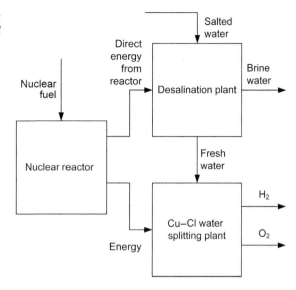

FIGURE 6.35 Using solar energy for desalination process. *Adapted from Orhan et al. (2010).*

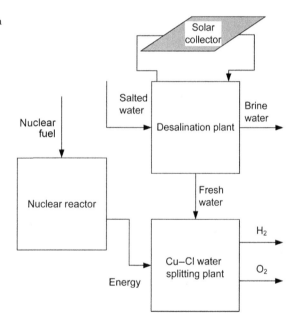

insolation. Otherwise, a desalination process with low capacity and low inlet energy requirements is used. RO technology is applied.

Configuration 5: off-peak electricity from a nuclear reactor is used to operate desalination plant. When off-peak electricity is available, it is usually less expensive than peak electricity and thus is used by many industrial processes. Off-peak electricity can also be used to desalinate seawater and to produce hydrogen, and could be beneficial for the same reason. This configuration includes

Continued

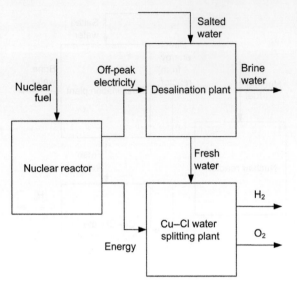

FIGURE 6.36 Using off-peak electricity for desalination process. *Adapted from Orhan et al. (2010).*

a Cu–Cl cycle coupled with a desalination plant driven by off-peak electricity (as in Figure 6.36). In this case, efficient membrane processes are used as electrical energy is supplied.

The overall energy efficiency of the coupled system, $\eta_{overall}$, represents the fraction of energy supplied to produce hydrogen from salted water that is recovered in the energy content of H_2 based on its lower heating value. The total energy required to produce hydrogen from salty water can be written as

$$Q_{in,\,total} = Q_{in,\,Cu-Cl} + Q_{in,\,Desalination} \tag{6.12}$$

and the overall efficiency is

$$\eta_{overall} = LHV/Q_{in,\,total}, \tag{6.13}$$

where LHV is the lower heating value of hydrogen.

The variation of costs for hydrogen production capacity at a Cu–Cl thermochemical plant is shown in Figure 6.36. These costs are based on producing hydrogen from freshwater and do not include costs associated with desalination. As can be observed, the capital cost of the Cu–Cl cycle varies from 1.8 to 0.3 $/kg H_2, based on the capacity of the cycle. The capital cost of the cycle per unit of hydrogen output is less for a larger capacity plant while the production cost (energy cost) remains constant at around 0.9 $/kg H_2, mainly because the reaction energy (per unit hydrogen produced) of any chemical or physical reaction occurring in the Cu–Cl cycle does not change based on plant capacity. The total cost of hydrogen production is the sum of the capital cost, energy cost, and also some additional costs such as operating, storage, and distributions costs. In Figure 6.37, two main cost items (capital and energy costs) are given, along with the total cost to build a pilot plant using the Cu–Cl cycle, presumably at the commercialization stage. The total cost varies from 3.5 to 2 $/kg H_2, in an inversely proportional relationship with plant capacity.

The total cost of the desalination plant is shown in Figure 6.38. The least expensive desalination method is HD at 7.5 $/(kg H_2/day), which is considered in Configuration 1, and the technology with the highest initial cost (11.9 $/(kg H_2/day)) is multistage flash desalination, which is used

Continued

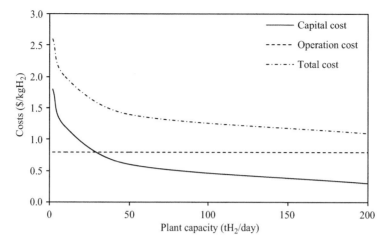

FIGURE 6.37 Variation of costs with hydrogen production capacity of a Cu–Cl thermochemical water decomposition plant using freshwater. *Adapted from Orhan et al. (2010).*

FIGURE 6.38 Total cost of desalination. *Adapted from Orhan et al. (2010).*

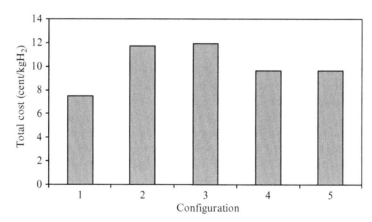

in Configuration 3. The total costs of VC and RO appear to be the same at 9.62 $/(kg H_2/day), while the capital cost of multiple effect desalination is 11.7 $/(kg H_2/day).

The energy efficiencies of the Cu–Cl cycle, desalination plant, and the overall system, including the Cu–Cl cycle and the desalination plant, are shown in Figure 6.39. The effect of the Cu–Cl cycle on the overall system is dominant, as the desalination plant uses much less energy than the Cu–Cl cycle. Thus, the efficiency of the Cu–Cl cycle and the overall system are very similar for each case. Configuration 1 exhibits a higher efficiency, as waste heat from the nuclear reactor (which is assumed to be "free") is utilized in this option. In Configuration 2, recovered heat from the Cu–Cl cycle is used for desalination instead of internally, within the cycle. Thus, the Cu–Cl cycle and the overall system operate at lower efficiencies, as the effect of the cycle is very important on the overall system.

Continued

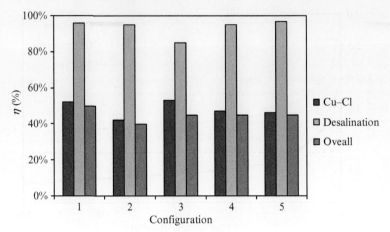

FIGURE 6.39   Efficiency of nuclear-driven hydrogen production and water desalination.

6.6 CONCLUDING REMARKS

In this chapter we discuss nuclear energy generation from a sustainable perspective. The nuclear theory is revisited along large lines, with a focus on derivation of binding energy. The nuclear fuel cycle is introduced. The method of producing nuclear heat is explained in detail, as is the possibility of quantifying the efficiency of nuclear heat generation on an energy and an exergy basis. The main nuclear reactors are revisited and various applications of nuclear energy discussed. Some examples demonstrate briefly the calculation of nuclear reactor heat generation and efficiency in terms of useful heat delivered with the primary nuclear energy used. Both energy and exergy efficiency are introduced. Apart from the most common use of nuclear energy—for electric power production—there is major interest in nuclear energy as a source of process heat. Nuclear process heat can be used for hydrogen production, desalination, or other processes. Three case studies are also presented in this chapter to provide better coverage and understanding.

Study Problems

6.1. Compare uranium fuel and gasoline in terms of energy content per unit of volume and mass.

6.2. Define and explain binding energy.

6.3. Calculate the binding energy of uranium isotopes.

6.4. Determine the decay time of carbon-14 if $N/N_0 = 0.1$.

6.5. How one can estimate the available energy from nuclear fusion?

6.6. Explain how a nuclear reaction can be controlled.

6.7. Define the energy and exergy efficiency of nuclear thermal energy generation.

6.8. Define the term "nuclear cross-section."

6.9. A reactor comprises a $n=40{,}000$ cylindrical fuel assembly distributed over a circular cross-section of the reactor with the radius of $R_r=1.8\,\text{m}$. The average energy generated of fissionable material in this configuration is $e_g=175\,\text{MeV}$ per interaction. The generated heat is 3 GW. Calculate the heat generation at a distance (r) from the fuel rod axis and the maximum heat flux.

6.10. The fuel rod radius in the example above is $R_f=7\,\text{mm}$ and is clad with a Zircaloy shell of $t=0.7\,\text{mm}$ thickness and the rod height is $z=5.7\,\text{m}$. Calculate the heat flux and the temperature of the cladding outer surface provided that the core temperature is $T_c=2600\,°\text{C}$.

6.11. Describe the principle of a pressurized water reactor and that of a boiling water reactor and emphasize the differences between them.

6.12. Describe the CANDU reactor and explain the role of the heavy water.

6.13. List the three top countries having uranium fuel reserves.

6.14. What do you understand by the "process of conversion?"

6.15. Define the unit of measure "Sievert."

6.16. What is the advantage of nuclear hydrogen production?

6.17. Do a case study for a district heating cogeneration plant with nuclear energy based on the diagram in Figure 6.32.

6.18. Rework Example 6.10 with other reasonable assumed values for data input.

6.19. Rework Example 6.9 for data corresponding to a CANDU reactor of 600 MW.

6.20. Design a system using nuclear heat cogeneration for aluminum production. Make a brief documentary study to determine the heat demands. Use the very high nuclear reactor of Generation VI for heat generation.

Nomenclature

A	atomic mass (AMU)
c	speed of light (m/s)
E	energy (J)
ΔE_b	binding energy (MeV)
e	specific energy (J/kg)
e_x	specific energy (J/kg)
N	number of neutrons
N	number of atoms or particle density
q	heat amount per unit of mass (J/kg)
R	radius, m
r	radial coordinate, m
T	temperature, K
t	time, s
Z	number of protons

Greek letters

α	transfer coefficient
η	energy efficiency
ψ	exergy efficiency
λ	excess air
ϵ	dissipation parameter
σ	cross-section, cm^2
ϕ	radiation particle flux

Subscripts

0	exchange current; reference state
act	activation
aux	auxiliary
c	compressor
conc	concentration
eq	equilibrium
f	fuel
fc	fuel cell
H	heat source
L	limiting
ng	natural gas
oc	open circuit
p	pump
PG	power generation
rev	reversible
s	steam
t	turbine
TC	thermochemical
tot	total

Superscripts

0	reversible
ch	chemical
thrm	thermomechanical

References

Bodansky, D., 2004. Nuclear Energy: Principles, Practices and Prospects, second ed. Springer, New York.

Dincer, I., 2012. Green methods for hydrogen production. Int. J. Hydrogen Energy. 37, 1954–1971.

Dunbar, W.R., Moody, S.D., Lior, N., 1995. Exergy analysis of an operating boiling-water-reactor nuclear power station. Energy Convers. Manage.. 36, 149–159.

GIF, 2013. Generation IV International Forum. Internet site http://www.gen-4.org, accessed in October 2013.

IAEA, 2006. Nuclear power reactors in the world, IAEA-RDS-/26.

IAEA, 2011. The management system for the development of disposal facilities for radioactive waste. NW-T-1.2.

Koster, A., Matzner, H.D., Nicholsi, D.R., 2003. PBMR design for the future. Nuclear Engineering and Design. 222, 231–245.

Lamarsh, J.R., Baratta, A.J., 2001. Introduction to Nuclear Engineering, third ed. Prentice Hall, Upper Saddle River, NJ.

Naterer, G.F., Dincer, I., Zamfirescu, C., 2013. Hydrogen Production from Nuclear Energy. Springer, New York.

NEA, 2008. Uranium, 2007: Resources, Production and Demand, Report NEA 6345: OECD.

Orhan, M.F., Dincer, I., Naterer, G.F., Rosen, M.A., 2010. Coupling of copper-chloride hybrid thermochemical water splitting cycle with a desalination plant for hydrogen production from nuclear energy. Int. J. Hydrogen Energy. 35, 1560–1574.

Neues Wiener Journal. 1943 June 5th http://anno.onb.ac.at/cgi-content/anno?aid=nwj.

WNA, 2013. World Nuclear Association Website, accessed in November 2013.

Renewable-Energy-Based Power Generating Systems

7.1 INTRODUCTION

Power generation from renewable energy has become increasingly important because it confers better energy security and smaller environmental impact with respect to conventional energy sources. Renewable energies are generally available at any location on the earth's surface. In some places, such as shorelines, there are wind resources or tides, in other places there is geothermal energy available, in agricultural areas there is biomass, and there is solar radiation falling on the earth's surface, everywhere.

The success of renewable energy power generation depends a lot on the economics because the payback period must be reasonably short. In this context the engineering challenge is to design systems with high energy conversion efficiency but reasonably low cost. This is a difficult trade-off problem for the designer because when efficiency is higher, the system cost tends to be higher. Harvesting these renewable energy sources and converting them to power is challenging due to the following facts:

- In general, the resources are fluctuating and intermittent.
- The resource intensity is relatively low, therefore it requires a large land area for energy harvesting. Consequently, the power generation per power plant footprint is very reduced (e.g., one hectare of solar power plant generates 100 kW continuous power, or using one hectare of land, one generates an average of 10 kW from wind power; however, with a nuclear power plant, one generates about 2.2 MW per hectare).
- Although renewable energy is freely available, the infrastructure required for its harvesting is rather costly as a consequence of low resource availability per unit of land area and, for most cases, modest energy conversion efficiency.

As introduced in Chapter 2, these energies are named as follows: solar, wind, geothermal, biomass, hydroenergy, and ocean energy (e.g., ocean thermal, tidal, wave, and ocean currents). Their classification is presented in Figure 2.7 in Chapter 2. The energy recovered from anthropogenic waste is also considered a form of renewable energy. In general, energy is wasted from anthropogenic activities in two forms: materials that have energy value, and heat rejected by industrial and other processes at a temperature essentially higher than the ambient temperature.

There are many opportunities to implement power generation systems from renewable energy sources. A common aspect of most renewable-energy-based power generation systems is the requirement for energy storage. The energy storage system has the role of mediating between power generation and power demand.

Let us consider solar energy. During daytime the system is able to generate power or to convert light into a more usable form of energy. However, power (or heating) is demanded not only during daylight, but also during nighttime. Therefore, the system must be sized such that it generates power on demand. An energy storage unit must be integrated with the power generation system for this purpose. The storage can be in batteries (electrochemical), in hydrogen (chemical), in hydrostatic pressure (mechanical), or in many other ways.

Some types of renewable energy have a high availability factor and do not fluctuate. Examples are geothermal and ocean thermal energy. At most of the geothermal sites, there is no need for energy storage because the heat is generated steadily. Therefore, geothermal power plants can be directly connected to the grid to generate base-load power. An ocean thermal energy conversion (OTEC) platform operating in Sunbelt seawater will work steadily because its heat source and heat sink temperature are quasi-constant during the whole year. However, for OTEC case, there is a need for energy storage for transportation purposes, and this is normally done by converting solar energy into a synthetic fuel (e.g., ammonia or hydrogen).

In this chapter, power generation methods from renewable energy sources are presented. Emphasis is given to power conversion methods that are applicable to any specific form of renewable energy. Renewable energies are available in the form of light, heat, mechanical, and chemical energy. If, for example, the renewable energy is light, then photovoltaic panels are used to directly produce power. If the energy is available in mechanical form, then special turbines can be used (wind or hydro turbines, depending on the case). If the energy is available in the form of heat, then special thermodynamic cycles can be applied to generate power. If the energy is available in chemical form, such as biomass, biofuel, biogas, or a salinity gradient in oceans, then several specific methods exist; for example, biomass, biofuel, and biogas can be combusted and the generated heat can be used for a specific power cycle, or specific fuel cell systems; also, membrane technology can be applied to extract power from salinity gradients on a large scale.

7.2 SOLAR POWER GENERATION SYSTEMS

In this section solar power generation systems are discussed in detail. Solar radiation and its spectrum, intensity, and exergy are introduced. Photovoltaic systems (PV) are explained. Concentrated solar collectors that generate high-temperature heat are discussed, and some methods to store solar thermal energy at high temperature are briefly mentioned. The following systems are detailed, focusing mostly on their specific thermodynamic cycles for power generation: solar tower with heliostat field, solar through power plants (including imaging and non-imaging concentrators), and solar dish heat engine.

7.2.1 Solar Radiation

Solar light is an electromagnetic radiation of wavelength spectrum ranging from ultraviolet (100 nm) to far infrared (1 mm). Actually, the ultraviolet-C (UVC) spectrum, comprising shortest wavelengths from 100 to 280 nm, is almost negligible in the total amount of sunlight falling on the light surface. With respect to energy rate, the most sunlight is in the visible light spectrum (380–780 nm), with a 44% energy fraction. The rest of the energy content of sunlight on the earth is comprised of 4% ultraviolet A and B (UVA and UVB), from 380 to 280 nm, and 52% infrared (IR), covering the long wavelength range from 780 to 1 mm. However, the light intensity in the infrared-C (IRC) sub-spectrum from 3 μm to 1 mm is negligible. The solar light spectrum is shown in Figure 6.1, based on the reference standard.

Adopted by the American Society for Testing and Materials (ASTM), the reference solar spectrum ASTM AM 1.5 (2008) has been given as a chart that expresses the variation of spectral irradiance (measured in W/m^2 nm) with the wavelength. The spectrum has four irradiation components (see Figure 6.1): extraterrestrial $I_{\lambda,etr}$, global on a $37°$-tilt south-facing surface $I_{\lambda,gt}$, direct normal and circumsolar $I_{\lambda,dc}$, and diffuse radiation $I_{\lambda,diff}$. This reference spectrum corresponds to a zenith angle of $\theta_z = 48.2°$ (the solar zenith angle is the angle between the local vertical and the direction of the sun).

The extraterrestrial radiation intensity (measured in W/m^2) corresponds to the light intensity at the upper edge of the atmosphere and is obtained by integration of the extraterrestrial spectral irradiance over the whole spectrum of wavelengths. According to Duffie and Beckman (2013), the extraterrestrial solar radiation varies according to the day of the year (counted from the 1st of January) as described by the following equation:

$$I_{ext} = \int_0^\infty I_{\lambda,ext} \, d\lambda = I_{sc}\left(1 + 0.033 \, \cos\left(\frac{360}{365}n\right)\right) \tag{7.1}$$

where I_{sc} is commonly called the *solar constant* and represents the annual average of extraterrestrial solar radiation intensity, for which the value is $I_{sc} = 1367 \, W/m^2$. Note that the extraterrestrial radiation varies throughout the year because it depends on the astronomical distance between the sun and the earth. This distance is maximum at aphelion (on the 3rd of July) at 152×10^6 km and minimum at perihelion (on the 3rd of January) at 147×10^6 km.

The presence of the atmosphere attenuates the solar radiation intensity due to various aspects such as albedo, atmospheric absorption, and scattering. An energy inventory of incoming and outgoing solar radiation on the earth is given in Figure 2.3. Due to scattering, an observer on the surface of the earth sees sunlight from two sources: the direct and circumsolar radiation with a blue-yellow color, showing that the peak spectrum is at ~550 nm, and the sky radiation (or diffuse radiation) with blue color, with a peak around 450 nm (see Figure 7.1).

The most important parameter that influences both the intensity of solar radiation at the earth's surface and the spectrum is the air mass. The air mass is defined as the ratio between the path length of sunrays through the atmosphere and the effective atmosphere thickness at the local zenith. The sketch in Figure 7.2 illustrates the air mass definition with $AM = L_0/L$. Air mass depends on the zenith angle, the day of the year, and the geographical latitude. At sea level when the sun is at zenith, the air mass is $AM = 1$, while if the sun is at horizon then $AM = 38.2$. Several types of spectral distributions of solar light can be obtained as a function of air mass, as follows:

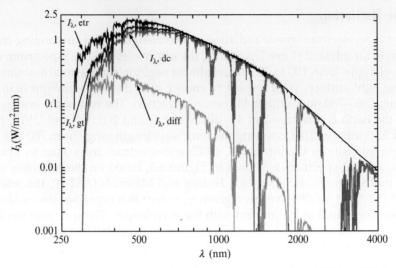

FIGURE 7.1   The ASTM 1.5 reference spectrum for solar irradiation. *Data from ASTM 1.5 (2008).*

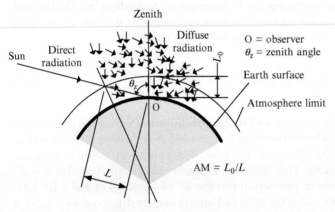

FIGURE 7.2   Sketch for definition of air mass parameter.

- AM0—the extraterrestrial spectrum at the edge of the atmosphere.
- AM1—the extraterrestrial spectrum with sun at the zenith (equator, sunbelt regions).
- AM1.1—obtained when zenith angle is ~25° and applicable to tropical regions.
- AM1.5—the most widely adopted case ($\theta_s = 48.2°$, discussed above).
- AM2—for zenith angle of 60° at high latitudes (>45°).
- AM3—for zenith angle of 70° at even higher latitudes (>60°).
- AM38.2—the air mass spectrum corresponding to horizon ($\theta_z = 90°$).

The air spectra can be predicted according to the methodology adopted by NREL (National Renewable Energy Laboratory), which is based on the paper of Gueymard et al. (2002) and can be calculated with the help of the software SMARTS, described in Gueymard (1995). Besides the air mass, the solar spectrum depends on the water content and ozone in the atmosphere, and also on turbidity, aerosol types and concentration, cloudiness, haziness, and optical thickness of the atmosphere. In Figure 7.3, the most important AM spectra are presented in detail.

Direct radiation is obtained directly from the solar disk which is viewed at an angle of 5°. The direct radiation has two subcomponents: the direct beam that is the radiation of the solar disc itself viewed at an angle of 0.53° and the circumsolar radiation that refers to the ring around the sun which covers the angular region from 0.53° to 5°. The diffuse component, which is the light coming indirectly to the observer, is more intense when air mass is higher. In general the diffuse radiation is only 10% of global radiation, the rest being the direct and circumsolar components.

The AM spectra from Figure 7.3 shifts very slightly toward shorter wavelengths when the air mass increases. Also, the intensity of the radiation decreases with the air mass. For AM1 the global normal radiation is ~1040 W/m², for AM 1.1 ~1015 W/m², for AM1.5 1000 W/m², for AM2 840 W/m², for AM3 ~715 W/m², while for AM38 it is only 20 W/m². An important parameter for practical solar radiation estimation is the global horizontal intensity, which can be calculated with the following equation from Singh and Tiwari (2005):

$$\frac{I_{gh}}{f_{gd}I_{ext}\cos\theta_z} = \exp\left[-\left(AM\left(4.5\times10^{-4}AM^2 - 9.67\times10^{-3}AM + 0.108\right)f_{ch} + f_{at}\right)\right] \qquad (7.2)$$

Here, the extraterrestrial factor is given in Equation (7.1) as a function of the day of the year. The zenith angle varies depending on the hour and geographical location. The factor f_{ch} represents the cloudiness and haziness of the atmosphere. This factor is dimensionless and takes

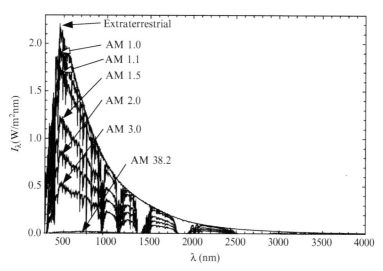

FIGURE 7.3 Global spectral radiation on a horizontal surface as a function of air mass. *Data calculated with NREL SMARTS software Gueymard (1995).*

values in the range of 2–6 depending on the geographical location on the earth's surface. The term f_{at} represents the direct radiation depletion factor due to compounded effects of aerosol level and thermal disturbances in the atmosphere. The range of practical values for this factor is 0.2–0.5. The factor f_{gd} represents the ratio between the intensity of global radiation and direct radiation. This factor can be correlated with the number of sunshine hours. From Singh and Tiwari (2005) for 5 h of sunshine the daily diffuse radiation is 75% of the global radiation and for nine hours of sunshine it reduces to only 25%. All factors, f_{ch}, f_{at}, and f_{gd}, are dimensionless.

A key question is how to convert light into other forms of energy which are more easily accessible. In principle, solar energy can be converted into thermal energy, electricity, and chemical energy. We will subsequently give some insights into the mechanisms of light interaction with matter, which are encountered in light energy conversion processes. To begin with, recall that in physics the notion of "matter" refers to any substance that occupies space and possesses rest mass. However, light has no rest mass. In a vacuum it propagates with a speed equal to the speed of light, given by the constant $c = 299,792,458$ m/s.

When light passes through transparent matter (e.g., the earth's atmosphere), the speed of light is reduced according to the refractive index defined with $n = c/v$, where v is the speed of light propagation through matter. As the refractive index of the atmosphere is ~1.0003, the speed of light in the terrestrial atmosphere is ~299,700,000 m/s. This is a simple example, showing that matter interacts with light through refraction. Other known matter-light interactions are reflection, absorption, and emission.

Max Planck discovered in 1901 that light (or very high-frequency electromagnetic radiation) can be absorbed or emitted in quanta of energy which are proportional to frequency. Through this discovery Planck solved the "ultraviolet catastrophe," that is, the impossibility of explaining the blackbody radiation spectrum using the classical theory expectation based on a continuous mode oscillator model. The blackbody concept emerged in 1860, with Kirchhoff, Stefan, Boltzmann, and Wien being the main contributors. The blackbody model consists of an ideal absorber in the form of a cavity with a pinhole. Light that enters through the pinhole is entrapped in the cavity and never escapes; rather, it is thermalized. Similarly, light is emitted from the cavity. Early discoveries showed that the emitted energy rate is proportional to the fourth power of the temperature. Spectroscopes were invented at that time and light irradiance measured at every wavelength. It was observed that spectral irradiance has a maximum that, according to the Wien law, depends on temperature only.

The schematics of the blackbody cavity receiver and the spectral distribution of the emissive power are shown in Figure 7.4. One observes the displacement toward shorter wavelengths (ultraviolet) when temperature increases.

The Wien law is considered very important because it shows that the emissive power of the blackbody does not depend on wavelength, but on the product λT. A form of the Wien displacement law is $I_{\lambda,b} = T^5 f(\lambda T)$. Experimentally, it has been determined that when I_λ is maximum then the product $(\lambda T)_{max} = c_2/5$, where $c_2 = 0.014388$ mK is denoted as the second radiation constant.

The total emissive power of the blackbody is obtained by integration of the spectral emissivity for all spectrum of wavelengths, namely: $I_b = \int_0^\infty T^5 f(\lambda T) d\lambda = T^4 \int_0^\infty f(\lambda T) d(\lambda T)$. Furthermore, using the Stefan–Boltzmann law, the emissive power of the blackbody in a vacuum is:

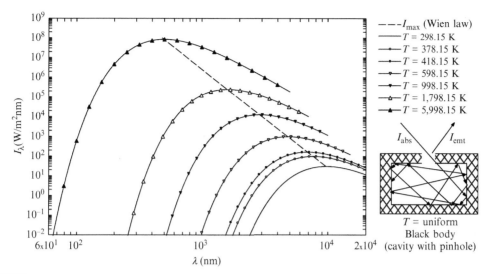

FIGURE 7.4 Emission spectrum of blackbody radiation [chart prepared with Engineering Equation Solver (Klein, 2013)].

$$I_{b} = \sigma T^4 \qquad (7.3)$$

with the Stefan–Boltzmann constant $\sigma = 5.67 \times 10^{-8}$ W/m²K⁴.

Based on the hypothesis of quantization of the electromagnetic energy, Planck derived the form of the function $f(\lambda T)$ for blackbody spectral emissive power. This determines the distribution of blackbody radiation, which is given by

$$I_{\lambda,b} = T^5 \frac{c_1(\lambda T)^{-5}}{\exp(c_2/\lambda T) - 1} \qquad (7.4)$$

where $c_1 = 3.7418 \times 10^{-16}$ Wm² is the first radiation constant.

Based on Equation (7.4) Planck introduced a new constant—the Planck constant—denoted with h and defined by $h = c_2 k_B/c = 6.55 \times 10^{-34}$ Js. The following relations can be easily derived, namely: $c_1 = 2\pi h c^2$ and $\sigma = (2\pi^5 k_B^4)/(15c^2 h^3)$.

The blackbody concept is a helpful idealization that allows for estimation as a first approximation of the emissive power of materials that are not blackbody. The common way to account for the discrepancy between blackbody and an actual material is by using the total emissivity ε, which is smaller than unity because the emissive power of actual materials is smaller than that of the blackbody at the same temperature. Therefore, the value of emissivity can be determined for each material and the emissive power will be expressed with $I = \varepsilon \sigma T^4$.

More intriguing is predicting the spectral distribution of an actual (gray) body heated at a given temperature. Furthermore, it is important to be able to assign to any polychromatic or monochromatic radiation a temperature. This can be done by recalling that the quanta of light—denoted photon—has an energy proportional with the reciprocal of wavelength, whereas the proportionality factor is the Planck constant multiplied with the speed of light constant

$$E_\lambda = h\frac{c}{\lambda} \tag{7.5}$$

As viewed at microscopic level, light carries kinetic energy ($KE_\lambda = 0.5\,E_\lambda$) but also vibronic energy. Although at microscopic level the photon can transfer work, from a macroscopic viewpoint, it transfers heat when it interacts with matter. Hence, the blackbody radiation transfers entropy at the temperature of the blackbody. From Equation (7.3) the entropy becomes $\dot{S}_b'' = \sigma T^3$. In a recent work Chen et al. (2008) suggested that the spectral entropy of light should be quantized by analogy with photon energy: $\dot{S}_b'' = \int_0^\infty \dot{n}_\lambda S_\lambda d\lambda$, where $\dot{n}_\lambda$ is the rate of photons of the same wavelength in the blackbody spectrum. The factor S_λ is the entropic contribution of individual photons which should be assumed in analogy with energy quanta as a function of wavelength only. Therefore, one has: $hc\sigma = \int_0^\infty (\lambda T)f(\lambda T)S_\lambda d(\lambda T)$. Analogy is claimed again with the spectral distribution of blackbody radiative energy. This means that the integrant must be a function of λT only. However, as S_λ must also be a function of λ only, it results that S_λ is a constant.

Chen et al. (2008) determined that $S_\lambda = 2.69952\,k_B = 3.7268 \times 10^{-23}$ J/K and denoted this parameter as the "entropy constant of a photon." As S_λ is a constant, the entropy of any radiation of a given spectral distribution I_λ can be determined from $hc\dot{S}'' = S_\lambda \int_0^\infty \lambda I_\lambda d\lambda$. Moreover, when interacting with matter, light can be viewed as a heat source of temperature T_{rad} (temperature of the radiation) which is easily derived as follows:

$$T_{rad} = h\frac{c}{S_\lambda}\frac{\int_0^\infty I_\lambda d\lambda}{\int_0^\infty \lambda I_\lambda d\lambda} \tag{7.6}$$

As a consequence of Equation (7.6), when light interacts with matter at a reference temperature T_0, it transfers exergy according to the Carnot factor for the temperature of the radiation, namely

$$\dot{Ex}'' = \left(1 - \frac{T_0}{T_{rad}}\right)I \tag{7.7}$$

where $I = \int_0^\infty I_\lambda d\lambda$ is the total irradiance.

EXAMPLE 7.1

In this example the exergy and entropy content of an AM 1.5 spectrum and the temperature of radiation components are calculated and compared with the extraterrestrial radiation spectrum. The global horizontal, direct, and diffuse radiation components are considered.

- The AM 1.5 spectrum is given in a tabular form with the spectral irradiance components as a function of wavelength. A particular spectrum is considered here, not the standard ASTM AM 1.5. For the particular spectrum taken as an example, the relevant quantities are shown in Table 7.1. Note that SMARTS is free software provided by NREL for calculation of solar radiation spectra.

EXAMPLE 7.1   (cont'd)

TABLE 7.1   Parameters for AM 1.5 Spectrum Considered in Example 7.1

Parameter	Extraterrestrial	Horizontal Diffuse	Horizontal Direct
$I = \int_0^\infty I_\lambda d\lambda$	1349 W/m^2	94.32 W/m^2	600.7 W/m^2
$S = \frac{S_\lambda}{hc} \int_0^\infty I_\lambda d\lambda$	0.2286 W/K	0.010 W/K	0.0987 W/K

TABLE 7.2   Radiation Analysis Results for Example 7.1

Parameter	Extraterrestrial	Horizontal Diffuse	Horizontal Direct	Global Horizontal
T_{rad}	5900 K	9411 K	6086 K	6394 K
Carnot factor	0.9495	0.9683	0.951	0.9534
$\dot{E}x''$	1281 W/m^2	91.33 W/m^2	571.3 W/m^2	662.63 W/m^2

- The calculation of radiation temperature can be done with Equation (7.6). This equation is equivalent to $\dot{S}'' T_{rad} = I$
- Once T_{rad} is determined for each radiation component the Carnot factors can be calculated for $T_0 = 298.15$ K
- Further, the exergy of radiation components is calculated with Equation (7.7). The results are presented in Table 7.2.
- Note that, for calculating the global horizontal component, the entropies are added for the diffuse and direct radiation $(0.01 + 0.0987)T_{rad} = 94.32 + 600.7 \rightarrow T_{rad} = 6394$ K
- It is noted that the temperature of diffuse radiation is the highest, and this is in agreement with the blue color of the sky. Furthermore, direct radiation has a higher temperature than the extraterrestrial radiation; this is mostly due to radiation absorption in the IR by water vapor, making the spectrum richer in shortwaves relative to the extraterrestrial.

7.2.2 Classification of Solar Power Generators

Light interacts with matter mainly in three ways: (i) it displaces electrons producing photovoltaic electricity, (ii) it displaces electrons that release vibronic energy that dissipates into heat, generating heat, and (iii) it displaces electrons, which eventually generates electrochemical reactions to produce synthetic fuels that store chemical energy convertible into power on-demand. Hence, power (i.e., electric or mechanical power) can be obtained from solar energy in many ways. Figure 7.5 gives a classification of solar power generation systems. Only the systems that directly produce electricity or mechanical shaft power are included in the classification. These are of two types: photovoltaic systems (which directly produce electrical power) and photothermal systems that generate heat that is used as a heat source

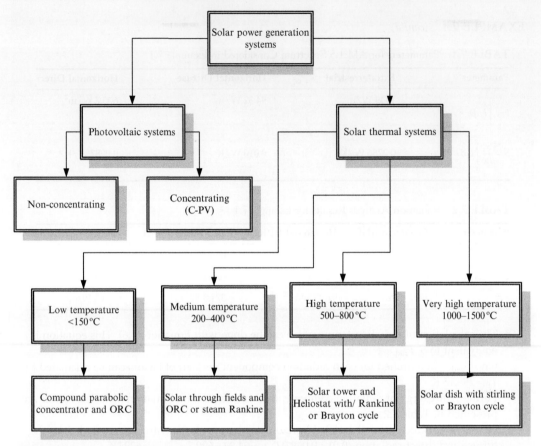

FIGURE 7.5   Classification of solar power generation systems.

for a specific heat engine to run a turbine/expander and to generate mechanical shaft power; this is eventually converted into electrical power using an electrical generator.

As the solar energy is intermittent, suitable means of energy storage are required. Three types of storage methods are mainly used for solar energy. These are: (i) storage of generated power in electrical batteries, (ii) thermal storage of heat generated photo-thermally, and (iii) conversion of solar energy to hydrogen for storage and on-demand power generation with the help of fuel cell systems.

The classification of solar power generation systems can also be made according to the scale of application. Figure 7.6 suggests a classification of solar power plant systems that use a heat engine driven by solar thermal energy. The systems are classified in two categories, namely of small-scale production capacities (on the order of kW) and large-scale ones, from a few MW to hundreds of MW installed capacity. The figure indicates the type of solar concentrator and the type of heat engine for each category.

The low-power generation systems use Rankine power generators operated with organic fluids (e.g., toluene), that is using organic Rankine cycles (ORCs), which are believed to be the most effective. Supercritical CO_2 cycles (ScCO$_2$) or transcritical cycles are also considered.

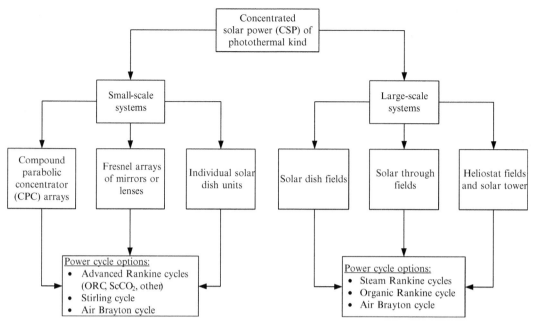

FIGURE 7.6 Classification of solar power generation systems according to the scale of application.

Ammonia-water, ammonia, and other refrigerants are also worth considering in Rankine engines of low capacity that can use low-cost refrigeration compressors of scroll or screw type in reverse, namely as expanders. This feature increases the marketability of independent low-power solar driven generators.

Other heat engine options are Stirling engines which operate at very high pressures, on the order of 200 bar and temperatures in the range of 700–800 °C, working with helium or hydrogen. Hydrogen is highly flammable, which imposes severe safety issues. A drawback with these systems is that hydrogen and helium leak easily, which raises maintenance problems. However, Stirling systems are very compact and reach high engine efficiency of around 40%, leading to overall electricity production efficiency of 22–23% for 10 h/day operation and installed capacity of 10–25 kW. One of the main drawbacks of using Stirling engines in solar applications is related to the long warm-up time needed, which is in contradiction with the reality of solar energy's fluctuating nature.

The air Brayton cycle—used for both small- and large-scale systems—appears to be an excellent choice due to the fact that the system has low thermal inertia and hence it can adapt to solar radiation fluctuations.

The solar power generation systems may be made more economically attractive when heat is cogenerated (e.g., water heating, space heating, district heating, process heating). The solar cogeneration system can be classified as:

- Photovoltaic-thermal: non-concentrating (PV/T) and concentrating (C-PV/T).
- Photothermal based on Rankine or Brayton cycle with cogeneration.
- Hybrid system which combines PV and thermal cycles, as in the ORC.

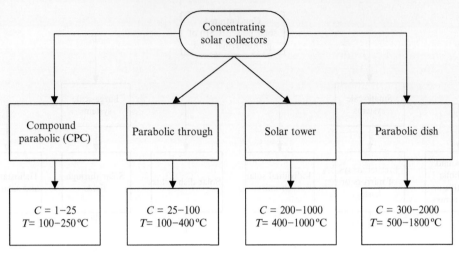

FIGURE 7.7   Solar thermal collectors.

The solar collector types that concentrate solar radiation for electricity generation through heat engines are classified in Figure 7.7. For temperatures higher than about 200 °C, a non-imaging concentrator shows major heat losses, and thus the preferred concentrators are of imaging type. The imaging concentrators use optical systems to create an image of the sun on a small hot spot (the solar receiver). The imaging concentrator can concentrate the radiation on a line or on a spherical point. The imaging concentrators must track the sun in order to be able to focus the sun on the receiver. Parabolic through collectors are those that concentrate the solar radiation along a line.

7.2.3 Photovoltaic Systems

7.2.3.1 Photovoltaic Cells

The photovoltaic cell was discovered at the end of the 1800s and evolved into a mass-produced commercial product—the PV-array—during the last half of the twentieth century. The main drive for development of photovoltaic arrays was their application to powering satellites. The cost of PV-arrays has dropped 250 times the cost at their first commercial application period in the early 1950s to the current price of around 60¢/W.

PV-arrays consist of a number of silicon wafers that form a p–n junction. When illuminated, the energy of light transforms into movement of electrons which can generate an electric current though an external circuit, thus transferring power. Two main semi-conductor technologies are in use today for solar cell manufacturing: silicon and thin film.

Silicon technology is the most evolved, with more than 90% of the market share and higher efficiency. Thin film technology has high potential for the future. The construction of a PV-array is illustrated in Figure 7.8. In essence, a p–n junction is formed, which is exposed to light. Electrical leads are attached to collect the photo-generated current. A substrate is used as support and a transparent cover is attached. An optical coating can be applied to allow only the desired light spectrum to pass and to prevent light reflection from the useful spectrum.

FIGURE 7.8   Construction of a PV-cell.

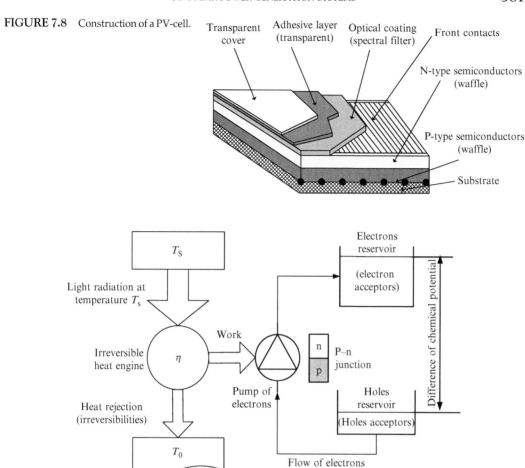

FIGURE 7.9   Schematic model of solar energy conversion in photovoltaic cells.

The light energy conversion process in a photovoltaic cell can be contemplated as suggested in Figure 7.9. Only the photons having energy higher than the energy gap between the conduction and valence bands of the semiconductors is useful. This leads to a first category of irreversibility because the photons with low energy either are reflected or ultimately transform in heat. Other irreversibilities are due to reflection and scattering of useful photons. Also there is certain probability that a photon will encounter an electron from the valence band. This probability is related to electron density in both the valence and conduction bands. Therefore, photon absorption followed by useful displacement of the electron is a stochastic process from a macro-scale viewpoint. Hence, it generates entropy. Therefore, the light to electrical energy conversion is similar to a heat engine process operating irreversibly between the temperature of radiation and the temperature of the surroundings. The work generated by this engine is equal to the generated photocurrent times the electrical potential difference across the p–n junction.

One of the most important parameters of a solar cell is the spectral internal quantum efficiency—denoted with $\Phi_{i,\lambda}$ (or sometimes IQE). This parameter is defined as the rate of electrons displaced across the semiconductor junction $\dot{N}_{e,\lambda}$vs vs. the rate of photons absorbed by the wafer at a certain wavelength $\dot{N}_{ph,abs,\lambda}$. Mathematically, $\Phi_{i,\lambda}$is defined by

$$\Phi_{i,\lambda} = \frac{\dot{N}_{e,\lambda}}{\dot{N}_{ph,\,abs,\lambda}} = \frac{J_{ph,\lambda}}{e} \frac{hc}{\lambda I_{abs,\lambda}(\lambda)} \tag{7.8}$$

where $J_{ph,\lambda}[A/m^2]$ is the electric current density generated by the absorbed photons on the wafer and $I_{abs,\lambda}(\lambda)[W/m^2\,nm]$ is the spectral irradiance at wavelength λ of the absorbed photons in a direction perpendicular to the PV-cell plane.

Not all photons on a PV-cell can be absorbed by the wafer because a part of them are reflected by the glassing and transparent coatings and another part are transmitted through the wafer. Therefore, one defines the external spectral quantum efficiency $\Phi_{e,\lambda}$ (ofen denoted EQE) as follows:

$$\Phi_{e,\lambda} = \frac{\dot{N}_{e,\lambda}}{\dot{N}_{ph,\lambda}} = \Phi_{i,\lambda}(1 - \mathcal{R}_\lambda - \mathcal{T}_\lambda) = \frac{J_{ph,\lambda}}{e} \frac{hc}{\lambda I_\lambda(\lambda)} \tag{7.9}$$

where $\dot{N}_{ph,\lambda}$ is the rate of incident photons on the surface of the cell, $\mathcal{R}_\lambda$ is the reflectance of the cell, $\mathcal{T}$ is the transmittance of the wafer, and $I_\lambda(\lambda)$ is the spectral irradiance at wavelength λ of the incident photons.

The equivalent electric circuit of a PV-cell is presented in Figure 7.10. It consists of a source of photovoltaic current in parallel with a diode and a shunt resistor (R_{sh}). In series with the external load is the internal resistance (R_s). The photovoltaic current density J_{ph} splits into three currents such that $J_{ph} = J_{Load} + J_D + J_{sh}$, where J represents the electric current per arc of PV-array. When the cell operates in short circuit ($R_{Load} \to 0$) practically all electrons generated by the photovoltaic effect pass through the external circuit, that is $J_{sc} \cong J_{ph}$.

Therefore, the short-circuit current density is obtained if one solves Equation (7.9) for the photonic current and one integrates for all spectrum wavelengths. The prerequisite for doing this calculation is the knowledge of the spectral distribution of external quantum efficiency. The EQE can be determined by specific measurements. The following equation results for short-circuit current density

$$J_{sc} \cong J_{ph} = \frac{e}{hc} \int_0^\infty \lambda \Phi_{e,\lambda} I_\lambda d\lambda \tag{7.10}$$

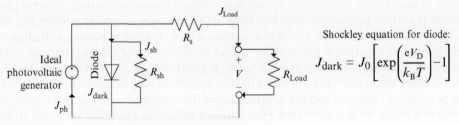

FIGURE 7.10 Equivalent electrical circuit of a PV-cell.

The problem with operation at short circuit is that the cell does not transfer power to the exterior. Indeed, the power transferred to the load is $\dot{W}''_{\text{Load}} = J_{\text{Load}} V_{\text{Load}} = J_{\text{Load}} R^2_{\text{Load}}$ and tends to zero when $R_{\text{Load}} \rightarrow 0$. In another extreme case, $\dot{W}''_{\text{Load}} \rightarrow 0$ when $R_{\text{Load}} \rightarrow \infty$, that is the cell operates in open-circuit mode. In this case the current through the load must be zero. Consequently, the excited electrons accumulate at the n-junction whereas the holes (or electron depletion) accumulate at the p-junction. Due to charge accumulation the voltage across the p–n junction increases until it reaches a maximum value, known as open-circuit voltage.

If the open-circuit cell is placed in dark conditions then there is no photocurrent, $J_{\text{ph}} = 0$. However, there is background radiation received at the p–n junction at the temperature of the cell, T_c. This radiation is sufficient to allow some of the electrons to jump over the potential barrier of the junction, E_g. This creates a current (or a current density J_0), and therefore, a current J_D will discharge though the shunt resistance R_{sh} (see Figure 7.10). Here the current density J_0 is commonly known as the diode saturation current (that is, the current generated by the thermal radiation at the temperature of the cell). Therefore, a voltage can be measured at the leads of the cell, which is the open-circuit voltage $V_{\text{oc}} = R_{\text{sh}} J_D A_c$.

The saturation current is proportional with the number of photons generated by blackbody radiation at the temperature of the cell that have energy higher than the band-gap energy (E_g) of the p–n junction. For any type of p–n junction the band gap energy can be determined. Denote the band gap energy with $E_g = k_B T_g$, where T_g is called the effective temperature of the band gap. Then the number of photons of blackbody radiation at cell temperature having energy higher than the band gap is given by the spectral distribution of blackbody energy according to:

$$\dot{N}_g = \frac{2\pi k_B T_c^3}{h^3 c} \int\limits_{T_g/T_c}^{\infty} \frac{\chi^2}{e^\chi - 1} d\chi$$

where χ is a dummy variable and $\dot{N}_g$ is measured in photons/m^2 of exposed surface.

Following Landsberg (2003), the saturation current density generated by the $\dot{N}_g$ background photons per second and square meter in the dark that have energy higher than the band gap is $J_0 = ef\dot{N}_g$, where $f \cong 2$ accounts for the fact that the waffle receives background photons from two sides. Therefore, the cell area exposed to background radiation is $fA_c \cong 2A_c$. As discussed in Landsberg (2003), a simpler empirical formula can be used for fast calculation of the dark saturation current density:

$$J_0 \left[\frac{A}{m^2} \right] = 1.5E9 \, \exp\left(-\frac{T_g}{T_c} \right) \tag{7.11}$$

The electrical circuit diagram of the PV-cell shown in Figure 7.10. can be solved using the Kirchhoff law, which requires that $J_{\text{ph}} = J_{\text{dark}} + J_{\text{sh}} + J_{\text{Load}}$. Therefore, one has the following current voltage characteristic for PV-cells:

$$J_{\text{ph}} = J_{\text{Load}} + J_0 \left[\exp\left(\frac{e(V_{\text{Load}} + J_{\text{Load}} R_s A_c)}{n_i k_B T} \right) - 1 \right] + \frac{V_{\text{Load}} + J_{\text{Load}} R_s A_c}{R_{\text{sh}}} \tag{7.12}$$

Here, the second term is the Shockley equation for diode current density (J_D) where n_i is the diode non-ideality factor (1 for ideal diode and 1—2.5 for non-ideal diode). For better accuracy the dark operation current is generated by two parallel diodes having the non-ideality factors n_1 and n_2, where n_1 is just superior to 1 and n_2 is around 2.

By setting $J_{Load}=0$ in Equation (7.12) the voltage over the load can be solved for; this will determine the open-circuit voltage, V_{oc}. Furthermore, the open-circuit voltage in dark conditions can be obtained when one sets $J_{Load}=0$ and $J_{ph}=0$. In practical PV devices the open-circuit voltage in dark conditions is negligible with respect to the open-circuit voltage of the illuminated cell. Therefore, for the estimation of the open-circuit voltage $R_{sh} \to \infty$ can be assumed. If this is done, Equation (7.12) gives the following approximate solution for the open-circuit voltage

$$V_{oc} \cong \frac{n_i k_B T_c}{e} \ln\left(1 + \frac{J_{ph}}{J_0}\right) \qquad (7.13)$$

The current–voltage diagram is a very important parameter for the PV-cell because it allows for maximization of energy transfer from the cell to the load for given operating conditions. The typical form of this curve is shown in Figure 7.11. It is made clear that in the two extreme situations (open and short circuit) the transmitted power to the load is zero. However, the power has a maximum at a certain load that corresponds to an optimal current density J_m and optimal cell voltage V_m. The maximum power is equal to the rectangular shaded area on the graph. The area of the rectangle $V_{oc} \times J_{sc}$ indicated with dashed line represents an ideal maximum power that would be transmitted to the load if the J–V characteristic of the cell is a straight line at $J=J_{sc}$. The fill factor is introduced as a useful tool to determine the maximum power output from the PV-array:

$$FF = (V_m J_m)/(V_{oc} J_{sc}) \qquad (7.14)$$

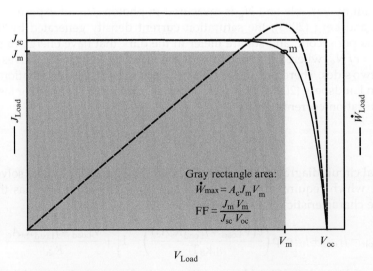

FIGURE 7.11   Typical J–V diagram for a PV-array and the $\dot{W}$–V curve showing the maximization of power output. The fill factor (FF) concept is introduced.

Here, the maximum power output from the PV-cell is $\dot{W}''_{max} = FF \, V_{oc}J_{sc}$. For fast calculations, the fill factor can be determined using some simple empirical correlations, as in Landsberg (2003) where the fill factor is correlated with the dimensionless open-circuit voltage $v_{oc} = (e V_{oc})/(k_B T_c)$ according to:

$$FF = \frac{v_{oc} - \ln(v_{oc} + 0.72)}{v_{oc} + 1} \left(1 - \frac{R_s J_{sc} A_c}{V_{oc}}\right) \qquad (7.15)$$

The PV-cell can be completely modeled under any irradiance and operating temperature provided that its external quantum efficiency, the diode non-ideality factor, and the shunt and series resistances are specified. For practical cells, the shunt resistance is given in the form of the product of ohmic resistance and the area of the cell. The order of magnitude of resistances values is 10^{-1}–$10^4 \, \Omega m^2$ for the shunt resistance and 0.1–2 $m\Omega m^2$ for the series resistance (see Nann and Emery, 1992, Table 2).

The external quantum efficiency varies with the type of the cell and depends on the manufacturer. Figure 7.12 shows spectral quantum efficiency from some manufacturers. There are many types of PV-cells, hence a comprehensive classification is difficult. In Figure 7.12 some of the most relevant PV-cell types are included. The amorphous silicon cell is one of the oldest ones. In general this type of cell does not work for all visible light spectrum. Upper frequencies, starting with blue color, are not useful for most of the PV-cells which are of amorphous silicon type. The figure shows the quantum efficiency of a recent amorphous silicon—nano-crystalline silicon cell (aSi/ncSi) developed by Kaneka (the data from the plot is approximated). One of the most promising technologies is based on crystalline silicone. The recently developed Sanyo HIT cell of cSi type shows an extraordinary wider spectrum with very high quantum efficiency. Other technology is based on GaAs semiconductors. Figure exemplifies the spectral performance of a GaAs cell from Alta, which can be used in the range of 500–900 nm.

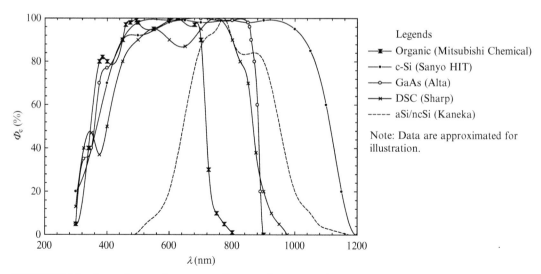

FIGURE 7.12 Spectral external quantum efficiency of some recently developed solar cells. *Data from Green et al. (2012).*

Dye-sensitized solar cells (DSC) were developed in recent years as a cheaper alternative to PV-cells. The principle of operation of DSC is photochemical rather than photovoltaic. These cells are thin film solar cells and use a metallo-organic anode as a photosensitizer that forms a semiconductor junction with an electrolyte. These cells have an excellent cost-performance ratio. An example is the DSC from Sharp, which has a high quantum efficiency of 450–800 nm. Another alternative to classical PV-cells is organic solar cells.

These cells are made with conductive organic polymers which are able to absorb light and dislocate electrons from HOMO (highest occupied molecular orbital) to LUMO (lowest unoccupied molecular orbital). Two dissimilar metallic conductors are applied to sandwich an organic selected polymer, such that an electrical field is created across the polymer. The LUMO electrons are attracted by the positive electrode, flow to the external circuit, and return to the negative electrode and back to the organic polymer where they occupy the HOMO level. These cells are cost effective and present high development potential, although their current efficiency and stability are rather low. Figure 7.12 illustrates an example of an organic solar cell from Mitsubishi chemical which has good quantic efficiency of between 500 and 700 nm.

The energy and exergy efficiencies are estimated for operation at maximum power. Hence, the energy efficiency is the maximum power divided to the solar energy input while the exergy efficiency is the maximum power divided by the solar exergy input. As the maximum power is determined with the fill factor, the open-circuit voltage, and the short-circuit current, the following mathematical formulas can be written for the efficiencies:

$$\begin{cases} \eta = \dfrac{FF \, V_{oc} J_{sc}}{I_N} \\[3mm] \psi = \dfrac{FF \, V_{oc} J_{sc}}{\dot{Ex}_N''} \end{cases} \tag{7.16}$$

where I_N and $\dot{Ex}_N''$ represent the irradiance and the exergy of light in a direction normal to the PV-cell surface, respectively.

As can be observed from Equation (7.16), the ratio between energy and exergy efficiency is the same as the ratio between exergy rate and irradiance: $\eta/\psi = \dot{Ex}_N''/I_N$. With Equation (7.7) the energy-to-exergy efficiency ratio becomes

$$\frac{\eta}{\psi} = 1 - \frac{T_0}{T_{rad}}$$

Table 7.3 gives the performance parameters of ten of the most efficient PV-cells. The efficiency of a PV-cell is determined in standard conditions for the AM 1.5 spectrum according to ASTM G-173-03 under global radiation at a tilt surface of 37°. Cell efficiency ranges from about 10% up to 36% (see the table). Nevertheless, the power generation efficiency of a PV-array consisting of several cells connected in parallel is slightly lower than the efficiency of the cell itself. Furthermore, the efficiency of a PV power plant (solar farm) is lower than the efficiency of a single PV-array. The decrease in efficiency is due to the additional losses caused by other components, such as diodes, wiring systems, inverter, and solar tracking mechanisms (when applicable).

TABLE 7.3 Manufacturer Performance Parameters of Selected PV-Cells

Manufacturer	Cell Technology	V_{oc} (V)	J_{sc} (A/cm^2)	FF	η
Sharp	Monolithic tandem GaInP/GaAs/GaInAs	3.006	14.05	87.5	36.9
AZUR	Monolithic 3-junction GaInP/GaInAs/Ge	2.691	14.70	86.0	34.1
Alta Devices	GaAs thin film	1.107	29.47	86.7	28.3
UNSW PERL	Crystalline silicon (c-Si)	0.706	42.7	82.8	25.0
Sun Power	n-type large crystalline silicon	0.721	40.5	82.9	24.2
Sanyo	n-type large crystalline silicon (Brand name: Sanyo HIT)	0.745	39.38	80.9	23.7
Kaneka	Thin film double junction a-Si/nc-Si	1.365	12.3	69.4	12.3
NIMS	Dye sensitized solar cell	0.743	21.34	72.2	11.3
Sharp	Dye sensitized solar cell	0.714	21.93	70.3	11.0
Mitsubishi Chemical	Organic thin film	0.899	16.75	66.1	10.0

Test conditions under AM 1.5 spectrum at 25 °C.
Data from Green et al. (2012).

The main components of a photovoltaic solar farm are shown in the electrical diagram in Figure 7.13. Each PV-array consists of several PV-modules, and each PV-module comprises a number of PV-cells. As the PV-modules of a PV-array can be illuminated differently due to shading or light blocking, diodes must be used to interconnect modules. Two types of diodes are used. The bypass diode allows for bypass current when the module generates a voltage lower than the pre-set voltage. The blocking diode does not allow current discharge from a module group that generates higher voltage to a module group that generates lower voltage. Note that each diode dissipates energy in the form of heat.

One or several electronic control blocks are required to adapt the load to the light input according to climacteric conditions. The control block assures that maximum power is extracted from the solar field and the DC voltage is converted at the level required for battery charging. The DC/AC inverter is used to connect the solar farm field to the grid for power supply. Of course, the net generated power is diminished due to the losses. The energy efficiency of the solar farm can be then calculated using the following equation

$$\eta_{farm} = \frac{\sum_i^N FF V_{oc} J_{sc} A_c}{I_N \sum_i^N A_c}$$

Table 7.4 gives the order of magnitude of efficiencies of the components involved in a photovoltaic installation. For electricity storage two cases can be considered: electricity storage in batteries or in reversible fuel cells. In the latter case the storage medium is hydrogen (and possibly oxygen). The efficiency of the storage is defined as the energy retrieved per energy stored. In a reversible fuel cell system one must consider around 56% (energy) and 52% (exergy) efficiency for hydrogen/oxygen generation from water electrolysis during the

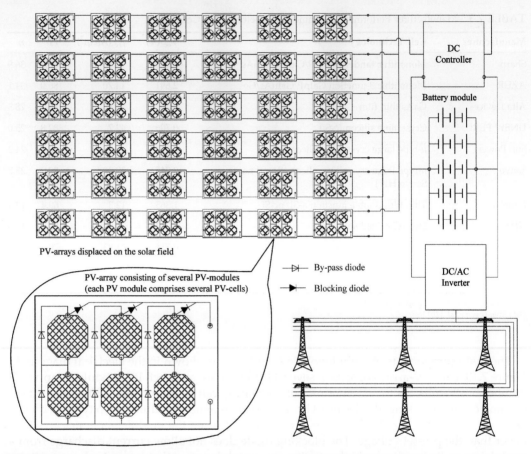

FIGURE 7.13 Electrical wiring for a photovoltaic solar farm (simplified diagram). Blocking and by-pass diodes must be used to interconnect PV modules.

TABLE 7.4 Average Energy and Exergy Efficiencies of the Components of a Photovoltaic Plant

Component	Energy Efficiency	Exergy Efficiency
PV-array	Monthly power generation as delivered to the grid per incident solar radiation per unit of area 11.2%–12.4%	Monthly power generation as delivered to the grid per incident solar exergy per unit of area 9.8%–11.5%
Charge regulators	DC power output over DC power input 85%–90%	DC power output over DC power input 85%–90%
Inverter	AC power delivered to the grid per DC power input 85%–90%	AC power delivered to the grid per DC power input 85%–90%
Electricity storage (battery, reversible fuel cell)	Energy retrieved per energy stored (Note: in reversible fuel cells hydrogen is the storage medium) 25%	Exergy retrieved per exergy stored. 23%

Source: Based on data from Joshi et al. (2009a).

storage phase and around 60% efficiency of electricity retrieval during fuel-cell operation of the storage system. Thus the storage efficiencies are roughly 25% (energy) and 23% (exergy). If batteries are used for storage about the same efficiency is expected.

EXAMPLE 7.2

In this example the calculation of PV-cell efficiency is demonstrated for actual radiation conditions that differ from the standard ASTM G-173-03. The following data are given:

- Outside temperature $T_0 = 287.5$ K.
- Cell temperature $T_c = 353.15$ K.
- Band gap wavelength $\lambda_g = 1200$ nm.
- Spectral external quantum efficiency $\Phi_{e,\lambda}$, according to Figure 7.8 for Sanyo HIT cell.
- Diode non-ideality factor is $n = 1.5$.
- Cell series resistance is given as $R_s A_c = 10^{-5} \Omega m^2$.
- Global normal radiation for AM 2.103 spectrum for Toronto area with sun position at noon, zenith angle 61.7°, azimuth 183°, standard atmosphere model, ground reflectance from lawn grass, atmospheric transmittance 0.6, $I_{sc} = 1348.92$ W/m^2.
- Cell orientation: normal to the beam; global normal radiation 951.86 W/m^2.

  - First, the spectral irradiance data is calculated with SMARTS software (note, the software is free to download from NREL at http://www.nrel.gov/rredc/smarts/)
  - Then global normal radiation is calculated by integration of the spectral global irradiation and normal direction; it is obtained $I_N = 951.86$ W/m^2
  - Then the entropy rate of the incident radiation is calculated (see Example 7.1)

$$\dot{S}'' = \frac{S_\lambda}{hc} \int_0^\infty I_{N,\lambda} d\lambda = 0.1542 \frac{W}{K m^2}$$

  - The temperature of the incident radiation is calculated with $\dot{S}'' T_{rad} = I_N \rightarrow T_{rad} = 6171$ K
  - With Equation (7.7) the exergy of the radiation is $\rightarrow \dot{E}x'_N = I_N(1 - T_0/T_{rad}) = 915.3$ W/m^2
  - The temperature corresponding to the band gap results from $hc = \lambda_g k_B T_g \rightarrow T_g = 11,930$ K
  - The saturation current density from Equation (7.11) is $J_0 = 1.798 \mu A/m^2$
  - The open-circuit voltage results from Equation (7.13) $\rightarrow V_{oc} = 0.8776$ V
  - The dimensionless open-circuit voltage is $v_{oc} = 28.84$
  - The fill factor calculated with Equation (7.15) becomes $FF = 0.8491$
  - The short-circuit current density calculated with Equation (7.10) becomes $J_{sc} = 402.1$ A/m^2
  - The maximum power per unit of cell surface is $\dot{W}'' = FF V_{oc} J_{sc} = 299.6$ W/m^2
  - The efficiencies from Equation (7.16) are $\eta = 31.21\%$ and $\psi = 32.73\%$

7.2.3.2 Concentrated Photovoltaic Systems

The major problem with current photovoltaic technology is the high investment cost when compared to conventional power generation systems. The cost is high due to the requirement for advanced manufacturing specific to the semiconductor industry. Hence a method to reduce the hardware cost is required to make photovoltaic technology more economically competitive with respect to conventional systems. One method to reduce the cost is by using solar concentration.

The photonic current increases if the solar radiation is concentrated over a PV-cell. It is a demonstrated fact that the cell operates more efficiently when exposed to concentrated radiation. At the same time more power is generated from a more intense flux of photons.

As a smaller area of PV-cell is required the overall system cost is smaller despite the fact that an optical concentrator is required for each cell. The benefits of C-PV (concentrated photovoltaic) systems are:

- A high-end solar cell can be used (multi-junction, tandem cells), which has 30% or more efficiency under standard conditions (these cells are too costly when used without concentration).
- The cell efficiency increases under concentrated radiation because the cell temperature is higher.
- Less cell area is required due to light concentration.
- Cogeneration (or PV/T—photovoltaic thermal) can be applied to cogenerate useful heat and make the system more economically viable;for a concentration ratio of 1.75 the cell temperature becomes around 100 °C.

A preferred method for solar concentration with PV-cells is the use of lenses or nonimaging concentrators. Therefore, a C-PV system can use both direct and diffuse radiation and better harvest the solar light, compared to systems that use imaging concentrators (which are able to concentrate only the direct radiation). Solar tracking systems are not required when the concentration is low (up to 10 suns). The preferred solar concentration system is with compound parabolic concentrators (CPC).

Table 7.5 gives the energy efficiency and power density for commercial PV systems dedicated to concentrated photovoltaic application. As seen, the concentration factor goes from 10 suns to about 400 suns and the efficiency increases to over 40% for high-concentration systems. Due to a better efficiency, the generated power increases from around 200 W/m^2 for a system with low concentration up to more than 400 W/m^2 for high-concentration systems. Note that the most common PV-arrays with no concentration generate around 100 W/m^2 under standard AM 1.5 conditions.

The light concentration leads to a substantial increase of the light energy flux incident on the cell. For example, the incoming energy flux is ∼180 kW/m^2 for the triple-cell "Solar Junction" system listed in Table 7.5. As the cell has an efficiency of about 43% it results that 57% of the incoming radiation is converted into heat; this is ∼240 kW/m^2. If one assumes a heat loss coefficient of 1000 W/m^{2}K for free convection and radiation between the hot cell back side and the environmental air at 25 °C, then the temperature of the cell must be about 250 °C. This is quite hot and will destroy the cell. Normally the cell temperature should be kept below 90 °C with the current technology. Cooling the cell with thermal fluids in forced convection must be applied. If a cell cooling system recovers 40% of the heat rejected by the cell,100 kW/m^2 of heat is generated. With an LMTD of 30 °C the system can generate hot water at 60 °C and maintain the cell at 85 °C, provided that the overall heat transfer coefficient is 3300 W/m^{2}K. This is feasible with advanced heat transfer technology. As a consequence, a

TABLE 7.5 Some PV-Cells, PV-Submodules and PV-Modules for Cconcentrated Photovoltaics

Manufacturer	Cell Technology	Concentration Factor	Efficiency	Power Density (W/m²)	
				Cell	Aperture
ENTECH	GaIn/GaAs/Gesubmodule	10 suns	27.0%	2700	270
UNSW	Laser-grooved large arc c-Si module	11 suns	21.7%	2387	217
DuPont	Split-spectrum tandem GaInP/GaAs with GaInAsP/GaInAssubmodule	20 suns	38.5%	7700	385
ENTECH	12-cell c-Si module	79 suns	20.5%	16,195	205
Amonix	Back-contact c-Si cell	92 suns	27.6%	25,392	254
Fraunhofer ISE	GaAs single cell	117 suns	29.1%	34,047	257
Spectrolab	Lattice-matched two-terminal triple-junction GaInP/GaInAs/Ge cell	364 suns	41.6%	151,424	416
Solar Junction	Triple-cell two-terminal GaInP/GaAs/ GaInNAs cell	418 suns	43.5%	181,830	435

Test conditions under AM 1.5 spectrum at cell temperature maintained at 25 °C.
Data from Green et al. (2012).

high-end concentrated PV system tilt installed on a roof can generate roughly 25 kWh per day for a 10 m² effective surface; in addition at least 25 kWh of hot water at 60 °C can be generated, which may meet the heating needs for space and sanitary hot water of a household.

One of the key aspects of concentrated photovoltaics is the development of cost effective compound parabolic concentrators which funnel the radiation without focusing it. Compound parabolic concentrators (CPC) consist of two compounded parabolas and can be constructed with either spherical or cylindrical geometry. Figure 7.14 shows an assembly comprising a CPC and PV-cell that operates under concentrated light. In cross-section the CPC appears as two parabolas with the focus of the left parabola being on the surface of the right parabola and viceversa. In the figure, θ_c is the half-angle of acceptance, β is the tilt angle, and θ is the angle of incidence between the direct radiation and the aperture surface.

The CPC can channel on the absorber three types of radiations derived from sunlight: direct, diffuse, and ground-reflected radiation. According to Duffie and Beckman (2013), the concentrated radiation intensity on the absorber (PV-cell) of the CPC is given as follows:

$$I_c = CR_{CPC}^{N_r}\left(I_{dc,CPC} + I_{diff,CPC} + I_{grd,CPC}\right) \tag{7.17}$$

where C is the concentration, $\mathcal{R}_{CPC}$ is the reflectance of the reflector surface, N_r is the average number of reflections, and the subscripts refer to concentrated radiation (c), direct and circumsolar on CPC (dc,CPC), diffuse radiation on CPC (diff,CPC), and ground-reflected radiation (grd,CPC). The diffuse and circumsolar term is given by the direct normal radiation

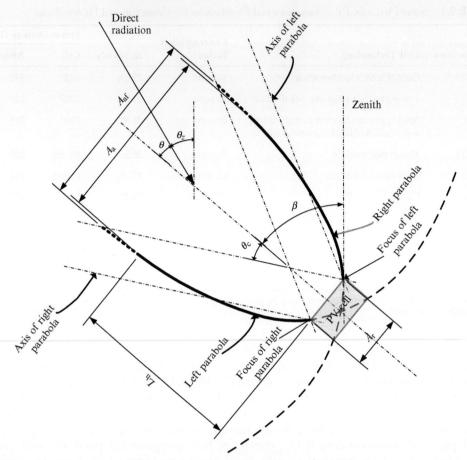

FIGURE 7.14 Concentrated-PV arrangement consisting of a CPC and PV-cell.

$I_{dc,n}$ and the cosine of the angle of incidence and a control factor f_{dc}, which is zero when the direct radiation is not present on the aperture and 1 when it is present, that is when the following inequality is true: $\beta - \theta_c \le \arctan(\tan(\theta_z)\cos(\gamma_s) \le \beta + \theta_c$ where θ_z is the zenith angle and γ_s is the azimuth angle. The $I_{dc,CPC}$ is given by

$$I_{dc,CPC} = f_{dc} I_{dc,n} \cos\theta \tag{7.18}$$

The diffuse radiation term depends on the tilt angle, the concentration factor, and an orientation factor f_{diff} that accounts for the tilt angle and the acceptance angle. The orientation factor for the diffuse radiation is 1 if the concentrator does not see the ground (i.e., $\beta + \theta_c < 90°$). If the concentrator sees the ground there is additional diffuse radiation accepted on the concentrator's aperture. Thus, the orientation factor is $f_{diff} = 0.5(1 + C\cos\beta)$ if the concentrator does see the ground, $\beta + \theta_c \ge 90°$. According to Duffie and Beckman (2013) $I_{diff,CPC}$ is

$$I_{diff,CPC} = f_{diff}\frac{I_{diff,h}}{C} \tag{7.19}$$

Furthermore, if the concentrator sees the ground there is accepted radiation that is due to the reflection of direct radiation. If one denotes $\mathcal{R}_g$ the ground reflectance then the ground-reflected radiation is, according to Duffie and Beckman (2013), equal to:

$$I_{grd,CPC} = 0.5 f_{grd} \mathcal{R}_g I_{dc,n} \left(\frac{1}{C} - \cos \beta \right) \tag{7.20}$$

The intensity of direct radiation depends also on the incidence angle, which is the angle between the aperture surface and the direction of sun. If the aperture is due south or due north then the incidence angle results from $\cos \theta = \cos(\phi - \beta) \cos \delta \cos \omega + \sin(\phi - \beta) \sin \delta$. In this equation δ is the solar declination angle. According to Duffie and Beckman (2013), the declination angle can be calculated as a function of the number of day of the year n counted from the 1st of January according to the equation: $\delta = 23.45 \sin[0.986(284 + n)]$. Also the incidence angle depends on the hour angle ω, which represents the angular displacement of the sun east or west of the local meridian due to rotation of the earth on its axis at 15° per hour, morning negative, afternoon positive. In addition, the incidence angle depends on the geographical latitude—that is, the angular location north or south of the equator, the north being positive and the south negative: $-90° < \phi < 90°$.

The concentration is defined as the area of the aperture divided by the area of the absorber; $C = A_{ap}/A_{abs}$. The ideal maximum concentration C_{max} depends on the acceptance angle which is $2\theta_c = \theta_s = 47$ mrad for seeing only the solar disc. For the disk focus ($C_{max,d}$) and strip focus ($C_{max,s}$) the maximum concentration ratio is (see Rabl, 1976):

$$C_{max,d} = \frac{n^2}{\sin^2 \theta_s} = 45{,}032 \text{ and } C_{max,s} = \frac{n}{\sin \theta_s} = 212 \tag{7.21}$$

where n is the refraction index of the medium.

One can observe from Equation (7.21) that for a refraction index of one (atmospheric air) the maximum concentration is 212 for line focus and 45,032 for point focus. If the medium has a refraction index higher than 1 then the ideal limit of concentration increases proportionally with n for line focus concentrators and proportionally with n^2 for point focus concentrators. The concentration of actual CPCs is smaller than the ideal due to the optical errors which are on the order of magnitude $\theta_{err} \cong 1°$. The actual concentrations for the disk concentrator C_d and strip concentrator C_s are, respectively, defined as follows:

$$C_d = f_t \frac{n^2}{\sin^2(\theta_c + \theta_{err})} \text{ and } C_s = f_t \frac{n}{\sin(\theta_c + \theta_{err})} \tag{7.22}$$

Here, the acceptance angle is superior to the view angle of the solar disk, namely $\theta_c > \theta_s$. Also the f_t is the truncation factor which depends on the height truncation fraction. If the CPC height is truncated with 20% then $f_t \cong 97–99\%$ for concentration ratios up to 20 (see Rabl, 1976). Truncation is technically advantageous up to 10–20% because it increases the acceptance angle with a relatively small decrease of concentration ratio, but with important reduction of reflective surface. For example, for a concentration of 14.5, the acceptance angle of the CPC is ~4° with a reflector surface per aperture ratio of ~16.

When the height is truncated by 20% then the reflector-to-aperture ratio reduces to 12 but the concentration ratio maintains to 14.2. Also the average number of reflections reduces when the concentrator surface is truncated. The minimum number of average reflections is

$N_{r,min} = 1 - C^{-1}$ while the maximum varies inversely proportional to the acceptance angle and the concentration. For concentrations from 2 to 20, $N_{r,max}$ is in the range of $\sim$0.5–1.5. From the geometry in Figure 7.14 it results that $H_t \tan \theta_{c,t} = H \tan \theta_c$, where H, H_t, θ_c, $\theta_{c,t}$ are the non-truncated and truncated heights and acceptance angles, respectively.

EXAMPLE 7.3

Concentrated photovoltaics are installed on a PV-array comprising $N = 20$ cells of 1 cm^2. Each PV-cell has attached a truncated CPC with the disc concentration having

- Acceptance angle $2\theta_c = 24\,°$, truncation factor $f_t = 99\%$, optical error $\theta_{err} = 1\,°$.
- Reflectance $\mathcal{R} = 0.88$, average reflection number $N_r = 1.5$, tilt angle $\beta = 45\,°$.
- Zenith angle $\theta_z = 36.85\,°$, azimuth angle $\gamma_s = -52.49\,°$, air mass $AM = 1.7$.
- Cloudiness/haziness factor $f_{ch} = 3$, atmospheric depletion factor $f_{at} = 0.3$.
- Number of sunshine hours for the day $n_{sh} = 5$.
- Cosine of the incidence angle $\cos \theta = 0.824$.
- Temperature of diffuse radiation $T_{diff} = 9000$ K.
- Temperature of direct beam and circumsolar radiation $T_{dc} = 6000$ K.
- Cell temperature $T_c = 80\,°C$, band gap temperature $T_g = 11{,}000$ K.
- Average external quantum efficiency $\bar{\Phi}_e = 0.85$, diode non-ideality factor $n_i = 1.5$.
- PV-cell series resistance in area $R_s A_c = 10^{-5}\,\Omega m^2$.

  - In isotropic atmosphere the refraction index is approximated to 1. Therefore, the concentration ratio of the CPC will result from Equation (7.22), $C_d = 0.99 \sin^{-2}(12\,° + 1\,°) = 19.56$.
  - The product of concentration ratio and average reflection is $C\mathcal{R}^{N_r} = 19.56 \times 0.88^{1.5} = 16.15$.
  - The control factor f_{dc} for direct radiation is determined based on $\beta - \theta_c = 33\,°$, $\beta + \theta_c = 57\,°$ and $\arctan(\tan(\theta_z)\cos(\gamma_s)) = 24.53\,°$; therefore, $f_{dc} = 1$ [see the comments on Equation (7.18)].
  - The ratio between global and direct radiation is calculated based on the number of sunshine hours [see Equation (7.2)], $f_{gd} = 10.5 - 1.12 \times 6 = 3.78$.
  - Assume that the extraterrestrial radiation is equal to the yearly average $I_{ext} = I_{sc} = 1367$ W/m^2. The direct normal radiation results from Equation (7.2) which must be divided with $f_{gd} \cos \theta_z$ to convert the radiation on a horizontal plane to a normal plane $\rightarrow I_{dc,n} = 1367 \times \exp[-2 \times (4.5 \times 10^{-4} 2^2 - 9.67 \times 2 + 0.108) \times 3 + 0.3] = 588.5$ W/m^2.
  - With Equation (7.18) $\rightarrow I_{dc,CPC} = 485$ W/m^2.
  - For 5 hours of sunshine the direct radiation can be assumed to be 75% of global radiation [see the comments from Equation (7.2)]; thus the diffuse radiation on a horizontal plane is $I_{diff,h} = I_{dc,n} \cos \theta_z (1/0.75 - 1) = 117.7$ W/m^2.
  - Test if the concentrator sees the ground: $\beta + \theta_c = 57\,° < 90\,° \rightarrow$ it does not see the ground, thus one has $f_{diff} = 1 \rightarrow$ from Equation (7.19) $I_{diff,CPC} = 8$ W/m^2.
  - The total concentrated radiation on a PV-cell according to Equation (7.18) is $I_c = 7962$ W/m^2.
  - From Equations (7.6) and (7.10) the following relationship can be derived for constant $\bar{\Phi}_e$

$$J_{sc} = C\mathcal{R}^{N_r} \bar{\Phi}_e \frac{e}{hc} \int_0^\infty \lambda I_\lambda \, d\lambda = C\mathcal{R}^{N_r} \frac{e}{S_\lambda} \left(\frac{I_{dc,CPC}}{T_{dc}} + \frac{I_{diff,CPC}}{T_{diff}} \right) = 4823 \frac{A}{m^2}$$

  - The saturation current results from Equation (7.11) $J_0 = 1.5E9 \exp(-11{,}000/353.15) = 44.5\,\mu A/m^2$.
  - The open-circuit voltage results from Equation (7.13) $V_{oc} = 0.8445$ V.

EXAMPLE 7.3   *(cont'd)*

- The filling factor results from Equation (7.15) with $v_{oc} = 27.75 \rightarrow FF = 0.80$.
- The cell energy efficiency according to Equation (7.16) is $\eta = FF\, V_{oc}J_{sc}/I_c = 40.9\%$.
- The power generated per unit of aperture is $\dot{W}'' = \eta(I_{dc,CPC} + I_{diff,CPC}) = 201.8$ W/m^2.
- The PV-array area $A = 20$ cells $\times 1$ cm$^2 \times C_d = 391$ cm^2.
- The PV-array output power $\dot{W} = 201.8 \times 0.0391 = 7.9$ W.
- If the cell operates with no concentration then one calculates for the same insolation conditions: $J_{sc} = 298.6$ A/m^2, $V_{oc} = 0.717$ V, $FF = 0.826$, $\eta = 0.359$ and $\dot{W}'' = 177$ W/m^2.
- In conclusion, the concentrated PV gains for the studied case has 5% efficiency and 201.8/177 $= 14\%$ in power generated per unit of PV-array area.

7.2.3.3 *Photovoltaic-Thermal Systems*

PV-arrays become more economically attractive if they are used to cogenerate power and heat. These systems are known as PV/T (photovoltaic-thermal). In many applications PV/T systems are used without light concentration. In this case, PV-arrays are cooled with air; the air is warmed up to temperature in the range of $\sim$27–55 °C and can be used for various applications, including space heating, water heating, crop drying, etc.

Photovoltaic thermal is a win–win application because due to cooling the efficiency of the PV-cell increases, and at the same time, useful heating is obtained. The construction of PV/T system can be achieved in various ways. The simplest possibility is to attach a duct at the sun-exposed surface of the PV-array. The duct is made of two transparent glasses and in between the glasses flows air or water which is heated. The contact thermal resistance between the glass and the PV-array can be as high as 10 kW/m^{2}K; hence most of the heat transfer resistance occurs at the fluid side.

Another construction option for non-concentrated PV/T is the unglazed integrated photovoltaic and thermal system (IPVTS). Such systems can reach 50% thermal efficiency and $\sim$8% electrical efficiency (as reviewed in Joshi et al., 2009b). Higher thermal and electrical output is obtained with concentrated-PV-thermal (C-PV/T) technique. In this instance, compound parabolic concentrators can be used to focalize light onto the PV-cell.

Any arrangement for PV/T implies grouping the PV-cells in modules. The efficiency of the PV-module will have slightly lower efficiency than the PV-cell due to ohmic losses. The energy and exergy efficiencies of the PV/T system can be calculated by

$$\begin{cases} \eta = \dfrac{FF\, V_{oc}J_{sc} + (UA)_c \text{LMTD}}{I_N} \\[3mm] \psi = \dfrac{FF\, V_{oc}J_{sc} + (1 - T_0/T_c)(UA)_c \text{LMTD}}{\dot{Ex}''_N} \end{cases} \tag{7.23}$$

where $(UA)_c$ represents the product between the superficial heat transfer coefficient between the PV-array and the heat transfer fluid (U) and the heat transfer area associated with the PV-array (A). The LMTD is the log-mean temperature difference between the PV-array surface and the heat transfer fluid (or space to be heated).

Water, glycol-water, or air are the typical heat transfer fluids used in PV/T systems. Hot water can be used for space heating, sanitary water heating, greenhouse heating, solar drying,

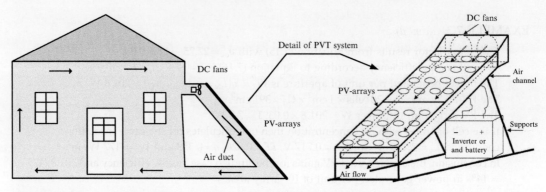

FIGURE 7.15   PV/T system for air heating and power. *Modified from Joshi et al. (2009b).*

solar stills, or other purposes. A typical PV/T arrangement that uses air as a heat transfer fluid is illustrated in Figure 7.15. The warmed air produced by such a system is at sufficient temperature for heating of a living space, as in the arrangement.

If water is used as a cooling medium then an arrangement can be made to combine PV technology with a flat plate solar thermal collector. Thus, water at its lower temperature can flow underneath a PV panel such that through heat transfer it cools the panel and improves its efficiency. Water is preheated in this way and then passed through flat plate solar thermal collectors for further heating. The diagram of such a system, used for space and water heating in a residence, is suggested in Figure 7.16.

As mentioned above, several constructive arrangements can be selected for PV/T systems. Figure 7.17 shows the glass-to-Tedlar® system, in which solar radiation is absorbed by a solar cell and then conducted to the base of the Tedlar® for thermal heating of air flowing below the Tedlar®. Note that Tedlar® is the brand name of a polyvinyl fluoride polymer film that is used as a back sheet for the PV modules. Figure 7.18 shows the configuration of a glass-to-glass PV/T system, a case in which solar radiation is absorbed by a solar cell and the black surface of the insulating base and the flowing air is heated by convective heat from the black surface

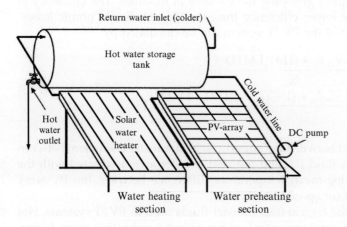

FIGURE 7.16   Hot water and power production with PV/T—solar thermal collector combination. *Modified from Joshi et al. (2009b).*

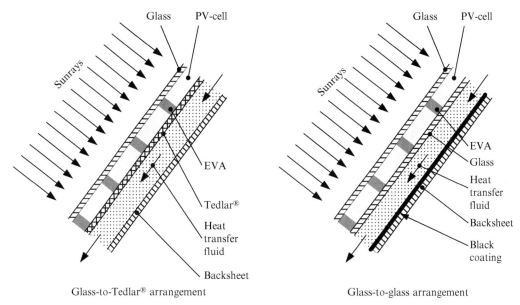

FIGURE 7.17 Two possible arrangements for PV/T systems with heat transfer fluid flowing at the back of PV-array.

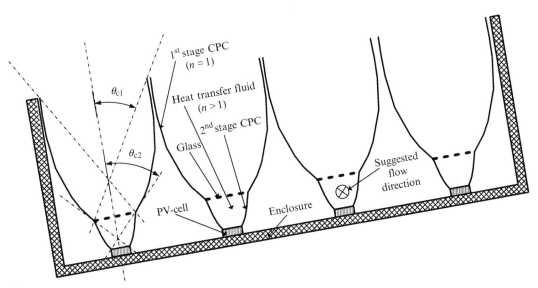

FIGURE 7.18 Concentrated PV/T system with heat transfer fluid above the PV-cell and double-stage CPC concentrator.

as well as heat conducted from the solar cell through a glass cover below the solar cell. In both systems shown the heat transfer fluid is circulated at the back of the PV-array, although arrangements may exist to circulate the fluid above the panel. Note that each individual cell of the PV-module is encapsulated in an ethylene-vinyl acetate polymer which confers mechanical support, electrical insolation, protection from environmental exposure, and optical coupling.

Concentrated PV/T systems may be even more attractive than PV/T without concentration due to the fact that they can operate more efficiently and produce heating at higher temperatures. Concentrated PV systems with heat transfer fluid above the PV-cell and a double-stage CPC concentrator where suggested in Rabl (1976). In this case the first stage of the CPC concentrator operates in air and has an acceptance angle θ_c with the maximum ideal concentration given by $1/\sin(\theta_{c1})$ for two-dimensional concentrators. At the aiming spot of the first stage CPC there is a transparent glass that seals the heat transfer fluid. Due to the geometrical property of the parabolic surfaces no rays will be reflected out by the glass.

Once passed through the glass (with small absorption), the light rays cross the heat transfer fluid with a refractive index $n > 1$. As a consequence of the reflection on the surface of the second-stage concentrator, the concentrated light reaches the PV-cell. For the second stage, it is suggested that the concentration limit drawing the acceptance angle is slightly larger than for the first stage. This means that all rays passing through the glass will be concentrated on PV-cells. Hence, the concentration limit or the second stage is $n/\sin(\theta_{cc})$ and the total ideal concentration becomes $n/[\sin(\theta_{c1})\sin(\theta_{c2})]$. The heat transfer fluid is heated by the PV-cell which, due to its dark color, absorbs heat in proportion to $1 - \eta$, where η is the power generation efficiency of the cell. The heat transfer to the coolant fluid is much more intense than the heat losses from the back of the cell. According to Rabl (1976), for CPC concentrators the heat losses at the back of the cell are on the order of 1–5 W/m²K. However, the heat transfer coefficient in forced convection may be on the order of hundreds of W/m²K.

The benefit of C-PV/T is even better when a government encourages the technology. In Ontario, the Ontario Power Generation Authority (OPGA) has put in place the "Standard Offer Program (SOP)," through which those who generate electricity from renewable energy can sell it to the grid to a guaranteed price, protected for inflation, of 11–42 ¢/kWh (as compared to 5¢/kWh for conventional electricity) under a contract with OPGA for twenty years. Since its start in 2006, this program, the first of its kind in North America, has boldly encouraged Ontarians to invest in local power generation.

A typical variation of cell temperature correlated to insolation and PV/T outlet air temperature is shown in Figure 7.19. The energy and exergy efficiencies obtained with the PV/T system are presented in Figure 7.20. To facilitate data interpretation, superimposed on the same figure is the variation of global solar radiation. In the morning, when solar radiation starts to grow in intensity and the panel surface temperature is reduced, the electric efficiency of the system is high (10.5%). At noon, when the solar radiation reaches its highest value, as does the panel's temperature, the electrical efficiency of the system is the lowest.

In conclusion, the degradation of the electric output due to the increase in cell temperature is more important than the increase in power generation due to higher solar radiation intensity at noon. The beneficial effect of the continuous cooling of the panel with air can be observed in the early afternoon, when the radiation level is still high but the temperature of the panel stabilizes, as does the electric efficiency.

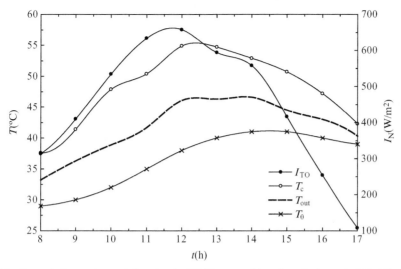

FIGURE 7.19 Recorded experimental data for the PV/T system [data from Joshi et al. (2009b)].

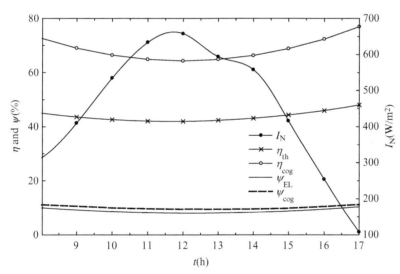

FIGURE 7.20 Energy and exergy efficiencies for a PV/T system for air heating and power. *Data from Joshi et al. (2009b).*

As observed in Figure 7.20, the system reaches maximum cogeneration energy efficiency during the evening (75%) and maximum cogeneration exergy efficiency during the mornings (11%). The average thermal efficiency is 44% and the average electric energy efficiency 9.5%.

The economic effectiveness of the PV/T option can be judged with respect to the PV-only application. Assuming N_{pbk} being is the number of years for payback, the PV-only system will generate $N_{pbk}\eta_e\bar{I}_{yr}$ electrical energy for each square meter of aperture, where $\bar{I}_{yr}$ is the annual averaged light intensity (in kWh/m^2 year). The capital cost of the square meter of PV-array depends on the packing factor (PF), which is the cell density per unit of PV-array surface. As during the payback period the capital cost (CC) must paid, one has

$$CC = N_cC_c = A_{PV}\ PFC_c = N_{pbk}\eta_e\bar{I}_{yr}\ LEC$$

where A_{PV} is the area of the PV-array, A_c is the cost of the cell and LEC is the levelized electricity cost (in \$/kWh). The capital cost (CC_1) for the concentrated PV/T system that generates power with efficiency $\eta_{e,1}$ and heat with efficiency η_t, and has a concentrator cost that is a fraction f_c of the capital cost is

$$CC_1 = N_{c,1}\left(C_c + f_cC_c\right) = \frac{A_{PV}}{C}PF\,C_c\left(1 + f_c\right) = \left(\eta_{e,1}LEC_1 + \eta_t LHC\right)N_{pbk}\bar{I}_{yr}$$

where C is the concentration ratio and LHC is the levelized heating cost. Dividing the two expressions the following relationship is obtained with simple rearrangement

$$\frac{LEC_1}{LEC} = \frac{\eta_e}{\eta_{e,1}}\frac{1+f_c}{C}\frac{1}{1 + \dfrac{\eta_t}{\eta_{e,1}}\dfrac{LHC}{LEC_1}}$$

If one assumes that simple non-tracking CPC concentrators are used with a concentration ratio on the order of 20, the cost of a fraction $f_c \sim 20\%$ of the capital cost in the PV-cell (assumed C_c), a thermal efficiency η_t of the same order as the electrical efficiency $\eta_{e,1}$, and a levelized heating cost of 20% of the LEC_1, then from the above expression it results that $LEC_1 \cong LEC/C$. Of course, if the concentration is higher, then the cost fraction f_c must be more important because the optics should be more accurate and tracking may be needed. However, this simple calculation reveals the clear benefit of both solar concentration and cogeneration.

EXAMPLE 7.4

Assume that a house has installed CPC-concentrated PV/T collectors for a surface of $A_{coll} = 10\ m^2$ of roof on the south-facing side. The CPC are of double-stage design, as shown in Figure 7.18. The cells are cooled with water flowing in a laminar regime. Compare the annual output, payback period, and exergy efficiency in the case of an east-west sun-tracking system and non-tracking system. Make reasonable assumptions. The following numerical data are given:

- Annual average direct radiation normal $= 2\ MWh/m^2$.
- Annual average number of hours of sunshine $= 8\ h$.
- Annual average temperature of direct radiation $= 6000\ K$.
- Annual average temperature of diffuse radiation $= 8000\ K$.
- Latitude $= 45°$, roof angle $= 50°$.
- Mirror reflectance $= 0.9$, glass transmittance $\mathcal{T}_g = 0.99$, optical error $= 1°$.
- Average number of reflections $= 1.5$ for first stage, 1.2 for second stage

EXAMPLE 7.4   *(cont'd)*

- Half-acceptance angle $=20°$ and $21°$ for first and second stages, respectively.
- CPC height truncated by 20%.
- Average quantum efficiency of PV-cell $=0.85$ with cut-off wavelength 1200 nm.
- Heat transfer to water $=500\ W/m^2K$, cell temperature $=80\ °C$
- LMTD between cell and water $=20\ K$.
- Cell series resistance $=0.02\ m\Omega\ m^2$.

  - First the calculations for the system with sun tracking are made.
  - For 6 hours sunlight per day on average we calculate f_{gd} for Equation (7.2) as follows $0.25+0.5$ $(9-6)/(9-5)=1.45$.
  - The incidence angle is calculated with $\cos\theta=\cos(\phi-\beta)\cos\delta\cos\omega+\sin(\phi-\beta)\sin\delta$; consider the representative day for the year the Spring equinox with zero declination (equal to the annual average), $\cos\omega=0$ as sun is tracked, $\beta=50°$ (equal to roof angle) and $\phi=45°$ (latitude); thus the incidence angle is $\theta=5°$.
  - The annual average zenith is $\theta_z=\phi=45°$; thus the annual direct radiation on a horizontal surface is $I_{dc,h}=2\cos(\theta_z)=1.414\ MWh/m^2$ year $\rightarrow I_{g,h}=f_{gd}I_{dc,h}=2.051\ MWh/m^2$ year (global horizontal) and $I_{diff,h}=I_{g,h}-I_{dc,h}=0.6364\ MWh/m^2$ year.
  - As CPC tracks the sun $f_{dc}=1$ in Equation (7.18) and $I_{dc,CPC}=2\cos5°=1.992\ [MWhm^{-2}year]$.
  - For height truncation of 20% the factor f_t for Equation (7.22) is ~0.97, whereas the half-acceptance angle increases to $\theta_{c1,t}=\arctan(1.25\tan20°)=24.46°$.
  - The refraction index of water is $n=1.33$, thus the total concentration with Equation (7.22) is

$$C_s=f_t\frac{n}{\sin(\theta_{c1}+\theta_{err})\sin(\theta_{c2}+\theta_{err})}=0.97\frac{1.33}{\sin21°\sin22°}=9.61.$$

  - The orientation factor f_{diff} for Equation (7.19) depends on the tilt angle and acceptance angle of the truncated concentrator $\beta+\theta_{c1,t}=74.46°<90°\rightarrow f_{diff}=1$; the CPC does not see the ground.
  - With Equation (7.19) one has $I_{diff,CPC}=1\times0.6364/9.61=0.06622\ MWh/m^2$ year.
  - The total annual radiation on the sun-tracking CPC is calculated with Equation (7.18) which becomes $I_{CPC}=C_s\mathcal{R}_{CPC}^{N_{r,1}+N_{r,2}}\mathcal{T}_g(I_{dc,CPC}+I_{diff,CPC})=14.74\ MWh/m^2$ year.
  - Similar to the Example 7.3, the annual averaged short-circuit current density is calculated with

$$J_{sc}=C_s\mathcal{R}_{CPC}^{N_{r,1}+N_{r,2}}\mathcal{T}_g\frac{e}{S_\lambda}\left(\frac{I_{dc,CPC}}{T_{dc}}+\frac{I_{diff,CPC}}{T_{diff}}\right)=8.879\frac{A}{m^2}$$

  - The temperature corresponding to band gap energy is $T_g=hc/k_B\lambda_g=11{,}990\ K$.
  - The saturation current density from Equation (7.11) $J_0=1.5E9\exp(-11{,}990/353.15)=2.696\ \mu A/m^2$ open-circuit voltage results from Equation (7.13), with assumed $n_i=1.1\rightarrow V_{oc}=0.5024\ V$.
  - The filling factor results from Equation (7.15) with $v_{oc}=16.51\rightarrow FF=0.78$.
  - The cell energy efficiency according to Equation (7.16) is $\eta=FF\,V_{oc}J_{sc}/I_c=0.2361$.
  - The annual power generated $W=10\,m^2\times0.2361\times14.74\,MWh/m^2year=34.79\ MW/year$.
  - The heat rejected by the cell $Q_{rej}=(1-\eta)I_{CPC}A_{coll}=112.6\ MWh/year$.
  - The total area of the cell is approximated from aperture area and total concentration ratio.

Continued

EXAMPLE 7.4 *(cont'd)*

- Thus $A_{c,tot} = A_{coll}/C_s = 10/9.61 = 1.041$ m^2.
- The heat transferred to the water for one year is $Q = UA_{c,tot}\text{LMTD}\,t_{year} = 91.16$ MWh/year (here $t_{year} = 365 \times 24 \times 3600$).
- The heat lost in air $Q_{lost} = 21.41$ MWh/year; heat loss coefficient is 0.343 W/m^{2}K.
- The average power and heat generated daily are $W = 95.32$ kWh/day, $Q = 249.7$ kWh/day.
- The amount of power and heat generated by the system is sufficient to satisfy the heat and power demand of 8–10 apartments
- The average electricity cost (incl. tax) can be taken as 10¢/kWh, whereas for heating it is estimated to 3¢/kWh (here one takes into account that the average power generation efficiency for combustion sources is at least 30%; thus heating is three times cheaper than power).
- The annual savings are of 34.79×$100+91.16×$30 = $6214.
- The annual average rate of power generation is W/tyear = 3.972 kW.
- For 10 m^2 of PV area and a concentration ratio of 9.61 the installed (rated) capacity under the standard 1 kW/m^2 is 10,000/9.61 = 1.061 kW.
- For standard-rated conditions the current cell cost is around 75¢/W, that is the cost of cells is roughly $800.
- This becomes probably ten times larger when concentrators, supports, a water tank, piping, ducts, and other hardware are added.
- Accounting also for taxes is appears that the payback period may be as little as 1.5–3 years.

7.2.4 Concentrated Photothermal Systems

7.2.4.1 *Central Receiver Power Stations*

Solar tower systems are large-scale power generators consisting of a field with heliostats and a solar tower with installed optics and a solar receiver. Eventually the high-temperature heat generated by the concentrated radiation at the focal point is converted into power using a conventional or advanced thermodynamic cycle. Mainly a steam Rankine cycle is used, although air Brayton or carbon dioxide cycles can be used. Thermal energy storage methods are integrated with the system to allow for continuous operation despite the intermittence of sunlight.

Romero et al. (2002) reviewed this technology and showed that it is now at the verge of commercialization, as it is proven feasible for sites with annual insolation of around 2 MWh/m^2 year. At least fourteen remarkable prototype plants covering a range of thermal power from 2 to 30 MW have been successfully demonstrated in the last 30 years. A system lifetime of 45 years, comparable to nuclear and conventional fossil fuel power plants, is achievable. The minimum capacity factor is 20%, with a maximum over 70%.

The tower is sometimes called a central receiver and consists of a tower pillar with either a reflector installed at the top or a solar receiver that converts the radiation into heat. When the tower has a reflector at the top, the solar receiver is located at the base and receives the reflected concentrated radiation from the top mirror system.

TABLE 7.6 Characteristics of Developed Solar Tower Projects

System Name	Heat Transfer Fluid	Thermal Storage	Thermodynamic Cycle	Installed Power (MW$_e$)
Gemasolar (Spain)	Nitrate salt	Nitrate salt	Steam Rankine	20
PS10 (Spain)	Pressurized air	Ceramic beds	Air Brayton	10
Solar Two (USA)	Nitrate salt	Nitrate salt	Steam Rankine	10
Solar One (USA)	Steam	Oil on rock	Steam Rankine	10
THEMIS (France)	Hitec salt	Hitec salt	Steam Rankine	2.5
Sunshine (Japan)	Steam	Nitrate salt/water	Steam Rankine	1
EURELIOS (Italy)	Steam	Nitrate salt/water	Steam Rankine	1
CESA-1 (Spain)	Steam	Nitrate salt/water	Steam Rankine	1
TSA (Spain)	Pressurized air	Ceramic beds	Air Brayton	1
MSEE/Cat B (USA)	Nitrate salt	Nitrate salt	Steam Rankine	1
SSPS (Spain)	Liquid sodium	Sodium	Steam Rankine	0.5
Solgate (Spain)	Pressurized air	N/A (fossil hybrid)	Brayton/Rankine*	0.3

*Combined gas turbine—steam Rankine with solar preheating of air.
Data from Romero et al. (2002).

The countries that have developed experimental solar tower systems are Spain, the USA, Japan, Israel, Russia, and Italy. Table 7.6 gives the features of the tested systems. The major cost item of concentrated solar tower systems is represented by heliostat field which, according to Vogel and Kalb (2010), takes ~40% of the investment. In order to run efficiently thermal storage is required. The thermal storage must be designed such that the power plant receives heat steadily and the turbine operates at its design point—that is, with maximum efficiency. The most used storage medium consists of 40% potassium salt and 60% sodium nitrate. The molten salt is heated by concentrated radiation from ~300 to 550 °C. A hot and a cold tank must be used and a sufficient quantity of hot salt must be stored. Due to the thermocline storage system the capacity factor of the plant can reach values in the range of 20–70%, or even more.

A classification of central tower power stations is shown in Figure 7.21. Two types of heliostat fields are used, which are correlated to the selected type of receiver. For the external receiver, the solar field is placed all around the tower while the receiver surface extends for 360° at the top of the tower. When cavity or volumetric receivers are used the heliostat field extends at the equatorial side of the solar tower. Figure 7.22 illustrates the configuration of the two types of solar fields. The optical concentration of the heliostat field can reach up to 1000 suns, with an optical efficiency of up to 0.75.

The energy of incident radiation on the central receiver is smaller than the radiation energy incident on the solar field due to a multitude of types of optical losses that vary depending on

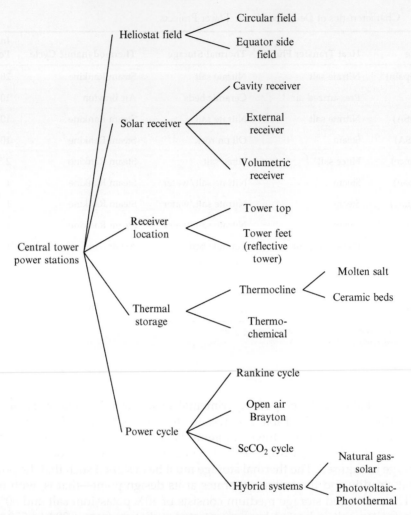

FIGURE 7.21   Classification of central tower solar power stations.

the heliostat position in the field and on the sun angle. The optical losses of the heliostat field can be incorporated into a compounded optical field loss factor defined by

$$F_{\text{field}} = f_\theta f_s f_b f_{\text{at}} f_\gamma f_\alpha$$

where the factors on the right refer to losses due to a momentarily incident angle of the sun with respect to reflecting surfaces (f_θ), shading of surfaces due to the position of other helio-stats in the field (f_s), partial blocking or reflected light (f_b), atmospheric attenuation of the light (f_{at}), optical aberration and light interception by the spectral splitter (f_γ), and the inci-dent angle of concentrated radiation on the receiver surface (f_α), respectively.

The factor F_{field} is an averaged one for all heliostats and can be determined by detailed solar field modeling based on known heliostat positions, optical characteristics, and known

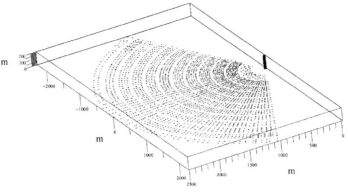

Equator-side heliostat field for cavity receiver

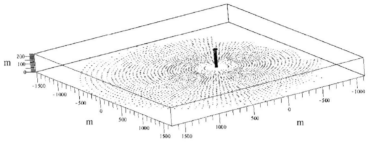

Circular heliostat field for external receiver

FIGURE 7.22 Equator side and circular configurations of heliostat fields for central tower power stations with concentrated radiation.

geographical location. Computer codes have been developed for performing optical system design of heliostat fields. An example is code DELSOL3 (Kistler, 1986) of Sandia National Laboratory, which is included in the System Advisor Modeling (SAM) program package provided by the National Renewable Energy Laboratory of the US Department of Energy.

Several constructive versions of heliostats have been proposed in the past. A heliostat is a large reflective surface able to concentrate the image of the solar disc onto the central reflector surface. Hence, the heliostat tracks the sun with the help of a two-axis system that follows both the zenith and azimuth angle.

With the advance of technology and mass production opportunities it is estimated that the cost of heliostats will drop well below \$90/m^2, which makes the central receiver system much more economically attractive. The energy and exergy fluxes reflected by the heliostat field depend on the optical properties of the mirrors. The heliostat reflects the majority of incident radiation according to its spectral reflectance, $R_{\lambda,h}$. Conventional heliostats use "second surface float glass wet silvered mirrors" which are very common in many harsh environment applications. They consist of a glass sheet back coated with a 700 Å silver layer, a 300 Å copper layer, and 25 μm black paint. The heliostat is made of many of these flat silver-on-glass mirrors, fixed on a metallic structure for a quasi-paraboloid surface with a focal point at the solar tower. The use of silver is justified by its high reflectance in visible range, at ~95%.

A more cost-effective option is that of using aluminum-based stretched metallic mirrors consisting of aluminum foil stretched over a circular metallic frame made of a rectangular steel profile (see Kolb et al., 2007). The aluminum foil is coated with a protective layer such as SiO_2 or TiO_2 (see Palik, 1997) which confers an extended lifetime without degrading the spectral reflectance. Two constructive options for metallic mirrors heliostats are shown in Figure 7.23. The spectral reflectance is calculated based on the extinction coefficient of the metal $k(\lambda)$ and the refraction index $n(\lambda)$ according to (Palik, 1997):

$$R_\lambda = \frac{(n-1)^2 + k^2}{(n+1)^2 + k^2} \qquad (7.24)$$

The reflectance of silver mirrors drops sharply in the UV region, whereas the reflectance of aluminum remains fairly high for all spectrum except for a small drop in the region of 800–900 nm. Table 7.7 gives the values of extinction and refraction coefficients and reflectance for aluminum and silver which are considered possible choices for our proposed system.

In most cases the solar receiver is placed atop the tower. The external receiver option generally consists of a number of vertical tubes assembled in registers. Through the tubes a molten salt is commonly circulated as the heat transfer fluid. Of course, this constructive solution is simple but will not allow high concentration due to the very large surface exposed to radiation.

Another solution is the use of a cavity receiver. In this case, concentrated light comes only from the equatorial direction as reflected by the field. There is a southward field in the

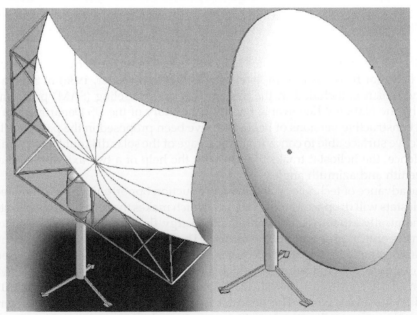

Petal sheets option   Stretched membrane option

FIGURE 7.23   Constructive options for heliostats made of reflective metallic mirrors.

TABLE 7.7 Extinction Coefficient and Refraction Index of Aluminum and Silver

Metal		Wavelength λ (µm)											
		0.2	0.3	0.4	0.5	0.6	0.7	0.8	0.9	1.0	2.0	4.0	10.0
Aluminum	k	2.3	3.61	4.86	6.08	7.26	8.31	8.45	8.30	9.58	20.7	39.8	89.8
	n	0.12	0.28	0.49	0.77	1.2	1.83	2.80	2.06	1.35	2.15	6.43	25.3
	R_λ	0.93	0.92	0.92	0.92	0.92	0.91	0.87	0.89	0.94	0.98	0.98	0.99
Silver	k	1.24	0.96	1.95	2.92	3.73	4.52	5.29	6.04	6.76	12.2	24.3	54
	n	1.07	1.51	0.17	0.13	0.12	0.14	0.16	0.18	0.21	0.65	2.3	13.3
	R_λ	0.26	0.16	0.87	0.95	0.97	0.97	0.98	0.98	0.98	0.98	0.98	0.98

Source: Palik (1997).

northern hemisphere, whereas in southern hemisphere it is a northward field. The aperture of the cavity receiver is relatively small, such that the radiative heat transfer is more intense than for the external receiver. Inside the cavity, the receiver is located. This can be designed in the form of a register of tubes with its surface exposed to the concentrated radiation.

One problem with the superficial light absorbers is the material strength limitation of the heat flux per unit of surface. Another issues regards the use of common molten salts, which must be pumped to the top of the high tower. There is, however, an alternative to the superficial receivers, namely volumetric receivers.

Volumetric receivers generally operate with heat transfer fluids in gaseous phase. An array of receivers is installed, each of them having a CPC for secondary concentration. A concave transparent glass is used to tightly confine the heat transfer fluid inside the receiver. Light is transmitted to the glass and interacts with a porous structure material inside through which the heat transfer flows. This arrangement allows for a volumetric light absorption rather than superficial light absorption; consequently, it can take much higher light fluxes. Note that with the superficial receiver the heat flux limitation is 200 kW/m^2 while with volumetric receivers the limit is on the order of 1 MW/m^2.

The energy and exergy balance equations for a heliostat field subsystem are $\dot{E}_1 = \dot{E}_2 + \dot{E}_{loss}$; $\dot{Ex}_1 = \dot{Ex}_2 + \dot{Ex}_d$, where $\dot{E}_1$ is the rate of energy incident on the total reflective surface of the field and $\dot{E}_2$ is the energy rate of the concentrated light on the receiver surface (with analog meaning for the exergies). The determination of energy losses and exergy destruction for a solar field allows for subsequent calculation of efficiencies according to the equations

$$\eta_{field} = \frac{\dot{E}_2}{\dot{E}_1} = 1 - \frac{\dot{E}_{loss}}{\dot{E}_1}, \ \psi_{field} = \frac{\dot{Ex}_2}{\dot{Ex}_1} = 1 - \frac{\dot{Ex}_d}{\dot{Ex}_1}$$

Note that not all the concentrated radiation is transferred to the heat transfer fluid; rather, a part is lost by radiation and convection in the ambient air. This defines the energy and exergy efficiency of the receiver, namely:

$$\eta_{rec} = \frac{\dot{Q}}{\dot{E}_2}, \ \psi_{rec} = \frac{\dot{Q}\left(1 - \dfrac{T_0}{T_{htf}}\right)}{\dot{E}_2}$$

where $\dot{Q}$ is the heat transfer rate to the fluid and T_{rec} is the average temperature of the heat transfer fluid.

As already mentioned, due to solar radiation intermittency and fluctuation there is a mismatch between power demand and power generation. The role of the energy storage unit is to compensate for the mismatch. The system is designed such that it allows for a continuous operation of the Rankine plant. There are two main known options for thermal energy storage in solar tower systems: storage in a thermocline (e.g., molten salt) or as hot air compressed in porous beds (see Vogel and Kalb, 2010). The molten salt poses difficulties of flow hydraulics and raises some stringent materials issues, as ionic species can dissociate and be corrosive. Air, although intrinsically safe to materials, requires large flow rates to maintain reduced temperature differences in conditions when the heat flux density at the solar receiver can reach 1 MW/m^2.

The use of ammonia for concentrated solar energy storage with solar dish systems has been proposed by Lovegrove (2004), based on the fact that the reaction is reversible and there are no side products. The forward reaction $NH_3 \rightarrow 1.5H_2 + 0.5N_2$ receives thermal energy when sun is available and the reverse reaction $1.5H_2 + 0.5N_2 \rightarrow NH_3$ delivers heat on demand. This process is sufficiently well understood and its thermodynamics are very favorable up to pressure of 200 bar. The receiver temperature is maintained high as the reaction evolves at a constant temperature.

Figure 7.24 provides a description of the ammonia-based chemical energy storage subsystem integrated into the overall system. Ammonia in liquid form (a) from the bottom of a storage tank is drawn and directed to the solar receiver. The forward reaction can be conducted at 20 bar for the more compact volumetric receiver applicable at that pressure. The reaction temperature is around 773–873 K. For better kinetics the conversion is maintained at 75% of equilibrium. Thus the returning stream in (b) comprises H_2, N_2, and NH_3. The backward reaction proceeds with a stoichiometric amount (c) drawn from the gases atop the liquid enclosed in a pressurized tank for storage with possible pressures of 50 bar (empty) and 250 bar (full). The reaction is facilitated by high pressures and will be conducted to 300 bar and 573–673 K.

Regarding power cycles, the most used is the steam Rankine type with regeneration. Commonly, two thermal reservoirs of molten salt are used, one hot and one cold, to facilitate the continuous operation of the turbine as shown in Figure 7.25 that illustrates the conceptual diagram of a molten salt/steam Rankine central receiver solar plant. The solar field is sized based on a solar multiple of two to four which means that the solar heat generation is two to

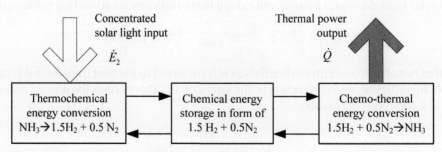

FIGURE 7.24　Thermochemical energy storage in ammonia.

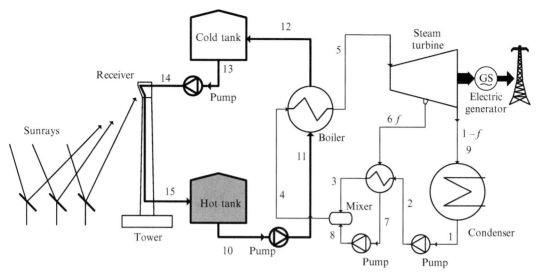

FIGURE 7.25 Conceptual diagram of a molten salt/steam Rankine central receiver solar plant.

four times higher than heat consumption by the Rankine cycle. Therefore, the accumulated heat allows not only for the steady operation of the cycle but also for preheating of the solar absorber just before the sunshine (Figure 7.25).

As stated above, molten salt is heavy, and its pumping atop the tower is difficult. Figure 7.26 shows a possible alternative. The heat transfer is air and the thermal storage is formed by solid materials such as ceramic and alumina. As known, alumina is very porous and will allow for a large heat transfer area with air. A special steam generator is devised to produce high-pressure steam with hot air. One main problem with this setup is the very large volume of thermal storage which may lead to an expensive construction.

Another alternative allows for the use of thermal storage in molten salt without the need for pumping it to the top of the tower. This is derived from the paper of Segal et al. (2004).

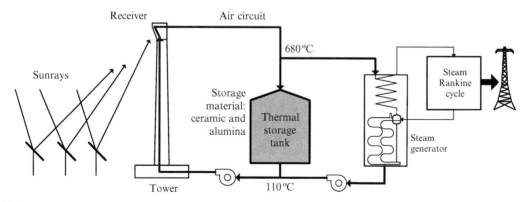

FIGURE 7.26 Central receiver plant with air as heat transfer medium and storage in solid material (ceramic bricks and alumina). *Modified from Romero et al.(2002)*

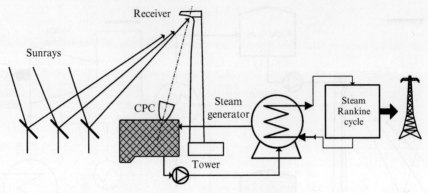

FIGURE 7.27   Reflective tower design. *Modified from Segal et al. (2004).*

In this case, a hyperboloid-shaped mirror placed atop of the tower is used to slightly diverge the concentrated radiation and direct it at the tower foot, where a secondary concentrator of CPC type is located. One possible option is to focus the concentrated light directly inside the molten salt reservoir, which acts as a cavity receiver itself, as shown in Figure 7.27. Phase-change materials can be placed inside an enclosure so that the thermal energy storage is more compact. Molten salt is then circulated through an external heat exchanger to transfer heat to a Rankine cycle. When natural gas can be made available onsite, hybridized systems can be used. In this case a combined gas–turbine and steam Rankine cycle is selected. Air is used as the heat transfer fluid and a volumetric receiver is placed atop the tower. When sunlight is not available the system works only with natural gas based on its combined gas–turbine and Rankine cycle, which uses a low-pressure heat recovery steam generator (see also Figure 5.67). If sunlight is available, air is preheated in the solar receiver atop the tower and then diverted toward the combustion chamber. As the air is at high temperature, the fuel requirement at the gas turbine combustion chamber will be minimal. During the daytime natural gas is only used to compensate for the fluctuations in solar radiation. Therefore, the combined cycle power plant will operate steadily without any essential requirement for thermal energy storage.

A hybrid central receiver solar power station is suggested in Figure 7.28. Eliminating the thermal storage requirement from the design will essentially reduce the system costs. Although it uses natural gas, the fuel consumption is much lower than for the same power generated to a conventional system, as are the emissions.

Better effectiveness can be obtained, if instead of heating air, a natural gas reforming process is conducted in the solar receiver. The thermochemical process of natural gas reformation requires catalysts. Hence, the reformer is heavier and it is best not to place it atop the tower, but rather at the tower foot. During the daytime, when concentrated solar radiation is available, a solar-driven methane reforming process is conducted. The resulting syngas is combusted in a combined cycle power plant. When solar radiation is not present, no methane conversion occurs and only natural gas is combusted. These kinds of systems present excellent economic potential because of the following advantages:

- No thermal storage is required.
- Power is produced continuously even when the solar radiation is low or absent.

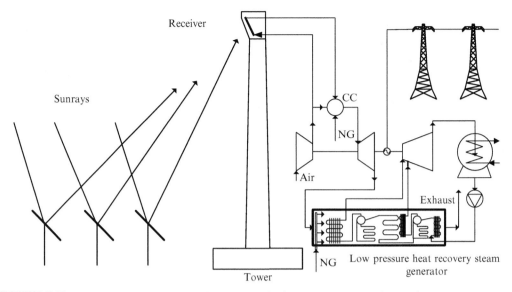

FIGURE 7.28   Hybrid, natural-gas-assisted concentrated solar power station with central receiver.

- There is not much need for receiver warm-up in the morning because hot un-combusted methane can be diverted from the combustion chamber to the receiver.
- The receiver operational parameters can be adjusted to maximize the output in accordance with the operational parameters.

A solar tower system configuration with natural gas reforming is illustrated in Figure 7.29. The system does not require thermal storage, except to minimize the start-up period in the

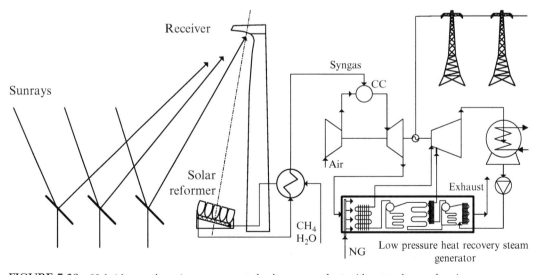

FIGURE 7.29   Hybrid central receiver concentrated solar power plant with natural gas reforming.

morning when the solar reformer must be preheated before sunrise. Wagar et al. (2011) show several systems for solar reforming of fuel with a view to upgrading their heating value for steady harvesting of fluctuating solar energy.

Other methods of hybridization of central receiver plants include the use of photovoltaic and heat engines with the help of a spectral splitter. Segal et al. (2004) suggest a method for this approach. In this case a reflective tower is used. At the tower foot a selective mirror is used that will reflect the wavelengths shorter than 600 nm on PV-arrays, preferably with mono-crystalline silicon cells illuminated under 500 suns. The cost of PV-cells would be "almost negligible" under these illumination conditions, as demonstrated in Segal et al. (2004). The longer shortwaves are concentrated with a CPC placed on the ground to provide heat to an advance cycle heat engine which generates additional power.

Figure 7.30 illustrates the solar splitter concept for central receiver CSP. Sunrays from the heliostat field are diverted toward the tower foot. The diverted rays are intercepted by a selective mirror (a cold mirror) which has a high refractive index for the upper spectrum so that wavelengths shorter than ~600 nm are diverted toward a large PV-array. The remaining spectrum is concentrated with CPC optics on a thermal receiver and used with a heat engine of advanced cycle to generate power.

Regardless of the type of central receiver system, an advanced—highly efficient—power cycle is required. Supercritical (or transcritical) CO_2 cycles have been developed in recent years and identified as one of the best options for central receiver CSP systems. The US Department of Energy has recently launched research programs in this direction, including cycle design and CO_2 turbine development. The advantage of the supercritical CO_2 cycle relies on the following.

- Carbon dioxide is very stable at high temperature and inert with respect to interaction with materials (e.g., with the receiver's components).
- The back work ratio of the cycle is excellent, with values being between those obtained with conventional steam Rankine cycles and those using a gas turbine (Brayton) cycle.

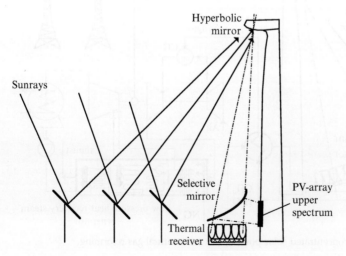

FIGURE 7.30   Solar splitter concept to combine PV with thermal receiver for a central tower plant.

FIGURE 7.31 Regenerative configuration of a transcritical carbon dioxide cycle.

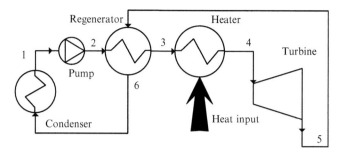

A transcritical carbondioxide plant configuration is presented in Figure 7.31. This is in fact a regenerative Rankine cycle where the working fluid is carbon dioxide. At the pump intake, carbon dioxide is a pressurized liquid (subcritical state). At the pump discharge, the pressure is supercritical, and at the turbine inlet, both the temperature and pressure are supercritical. Thus, at turbine intake, a supercritical fluid (carbon dioxide) is fed. After expansion, the pressure becomes subcritical while the temperature stays supercritical. The expanded flow regeneratively transfers heat to the preheating flow. Finally, the low-pressure carbon dioxide vapors reach saturation (state 6 on the figure) and are condensed. The cycle is called transcritical because the working fluid passes from a subcritical to a supercritical state and back.

Figure 7.32 illustrates the transcritical carbon dioxide cycle for some meaningful assumptions. Namely, it is assumed that: isentropic efficiencies are 0.9 and 0.8 for the pump and turbine, respectively, the pressure ratio is 4, the maximum temperature in the cycle is 1100 K, and the condensation is at 282.15 K. In these conditions the energy efficiency of the cycle is 46%.

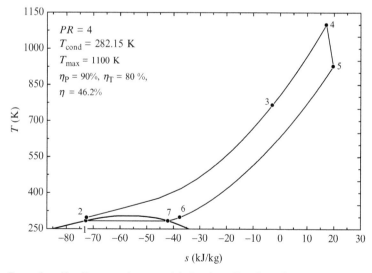

FIGURE 7.32 Exemplary T–s diagram of a transcritical carbon dioxide cycle.

The carbon dioxide cycle is very advantageous because it confers very high efficiency for a cycle with essentially five major components (therefore a reasonably cheap construction), but there are some challenges. The most difficult is the development of a turbine with a pressure ratio of about 4 and pressure difference of about 140 bar.

Regardless of the type of the power cycle (Brayton, Rankine, or other), the power plant must run relatively steadily for better effectiveness, despite sunlight fluctuation and intermittence. Here, thermal storage is crucially important. One can renounce thermal storage only if there is natural gas backup or a synthetic fuel such as hydrogen can be produced from sunlight. Thermochemical storage in reversible reactions may be a good option for reducing the investment in thermal storage (see Figure 7.24 above for reversible ammonia synthesis/cracking reaction).

EXAMPLE 7.5

Consider the system in Figure 7.28 in which air is preheated during daytime prior to the combustion chamber. Assume that for the average year the light intensity on the receiver is given by $I_{in} = 2.5$ [MW]$C \sin[0.1 (t-8)]$; daylight starts at 8 am, and after 6 pm there is no solar radiation. Air is heated up to 1000 K. The efficiency of the cascaded Rankine cycle is constant to 30% and the installed power generation is 250 MW. The low-pressure heat recovery steam generator recovers 50% of the heat rejected by the Brayton cycle, and the installed capacity of the Brayton cycle is 150 MW. Calculate the daily average natural gas consumption for generated MW of power and compare it to the nighttime specific consumption. Make reasonable assumptions.

Assumptions

- The ambient air is at $T_0 = 298.15$ K. Also the field factor is 0.6 and the concentration factor is 500 suns. The compression ratio for the Brayton cycle is 20; we will neglect the irreversibilities in the cycle. The maximum light input density is 1 MW/m^2 for a volumetric receiver with an emissivity of 0.5.
- Taking specific heat ratio 1.4 and a compression ratio of 20 the efficiency of the standard air Brayton cycle is $\eta_B = 1 - 20^{-0.4/1.4} = 57\%$.
- As the Brayton cycle output is 250 MW, the input heat must be $250/0.57 = 343$ MW.
- The recovered heat during nighttime is $343(1 - 0.57)/2 = 92$ MW.
- The required input by the Rankine cycle is $\dot{Q}_{in, Rankine} = 150/0.3 = 500$ MW.
- The natural gas amount fed to the LPHRSG (low-pressure heat recovery steam generator) is equivalent to $500 - 92 = 408$ MW.
- The total natural gas supplied at nighttime is $408 + 343 = 842$ MW.
- The number of hours of no sunshine, from 6 pm to 8 am, is 14.→The required MWh from natural gas is therefore $Q_{ng, night} = 14 \times 842 = 11,788$ MWh.
- During the daytime the Rankine cycle will require basically the same amount of natural gas as in the nighttime as the Brayton cycle will reject a similar amount of heat for the nighttime.
- The maximum irradiation on the receiver aperture is $I_{max} = 2.5 \, C \, F_{field} = 750$ MW; thus the surface of the absorber aperture must be $750/1 = 750$ m^2.
- The heat losses from the receiver are estimated with $\dot{Q}_{loss} = \varepsilon \sigma T^4 = 10$ MW.
- At any hour t during daytime the natural gas consumption in terms of energy rate results from $\dot{Q}_{ng} = 250/0.57 - 2.5 \, C \, F_{field} \sin[0.1 (t-8)] + \dot{Q}_{loss} + 408 = 856 - 750 \sin[0.1 (t-8)]$.
- For the 8 h of daylight the amount of natural gas consumption is $Q_{ng, day} = 1170$ MWh.

EXAMPLE 7.5   (*cont'd*)

- The total consumption of natural gas for the whole day is $Q_{ng} = 1170 + 11{,}788 = 12{,}958$ MWh.
- The rate of natural gas consumption is $12{,}958/24 = 540$ MW, which has to be compared to the rate of consumption of the natural-gas-only system, which uses 840 MW to generate the same effect of $\dot{W}_{net} = 150 + 250 = 400$ MW.
- The nighttime efficiency is $400/840 = 47.6\%$.
- The daytime energy input is of 540 MW from natural gas and an averaged 413 MW from sunlight; therefore, the efficiency is $400/(413 + 540) = 41.9\%$; the overall efficiency is 45.2%.
- In conclusion, the system has high efficiency, with >30% savings compared to natural gas only.

7.2.4.2 Through-Type Concentrated Solar Power Systems

Significant progress has been recorded at nine power generating stations totaling 354 MW of power generating capacity built in California's Mojave Desert by the 1980s. These systems use parabolic trough imaging concentrators to collect heat and generate steam for a Rankine cycle. On average, large-scale solar through steam power plants obtain 120–175 W of electric power per square meter of solar collector at 350 °C steam temperature. Through-type concentrators can be constructed in three main variants:

- As imaging single-axis tracking line focus concentrators that have an evacuated tube placed at the focal point.
- As non-imaging CPC concentrators with minimal azimuthal tracking and seasonal zenith angle tracking.
- Using Fresnel mirrors or V-type reflectors with line focus.

For temperatures higher than about 200 °C, non-imaging concentrators show major heat losses, and thus the preferred concentrators are of imaging type. Imaging concentrators of through type can concentrate the radiation on a line. The imaging concentrators must track the sun in order to be able to focus the sun on the receiver. The main characteristics and the performance of some parabolic through systems constructed and tested in the 1980s are indicated in Table 7.8.

Two methods of tracking the sun are used with line concentrators. In the first method, sensors provide optical feedback of the sun's position to allow for variable tracking. In the second method, the system is preprogrammed to follow the sun based on the local longitude, latitude, and time. Line focus concentrators in general track the sun in one direction, namely along the polar axis. The second axis alignment is adjusted daily or weekly. The azimuth tacking is applied to point-focus concentrators because it allows for constant two-axis tacking.

Most of the large-scale solar plants are hybridized with natural gas or coal combustion. Figure 7.33 presents a significant example of hybridized power plant. The net capacity of such systems, known as integrated solar combined cycle systems, is at least 10%, for a capital cost of three to five million dollars per installed MW.

Small-scale systems that use through-type concentrators are very attractive for small-scale applications. Specifically, through-type CPC concentrators are effective because they can be roof-mounted and require less tracking. Linked to ORC the CPC concentrator can provide

TABLE 7.8   Characteristics of Some Solar Through Concentrators

Collector	Acurex 3001	M.A.N. M480	Luz LS-1	Luz LS-2	Luz LS-2	Luz LS-3
Year	1981	1984	1984	1985	1988	1989
Area (m^2)	34	80	128	235	235	545
Aperture (m)	1.8	2.4	2.5	5	5	5.7
Length (m)	20	38	50	48	48	99
Receiver Diameter (mm)	51	58	42	7	7	7
Concentration Ratio	36:1	41:1	61:1	71:1	71:1	82:1
Optical Efficiency	0.77	0.77	0.734	0.737	0.764	0.8
Receiver Absorpivity	0.96	0.96	0.94	0.94	0.99	0.96
Mirror Reflectivity	0.93	0.93	0.94	0.94	0.94	0.94
Receiver Emittance	0.27	0.17	0.3	0.24	0.19	0.19
Operating Temp. (°C)	295	307	307	349	390	390

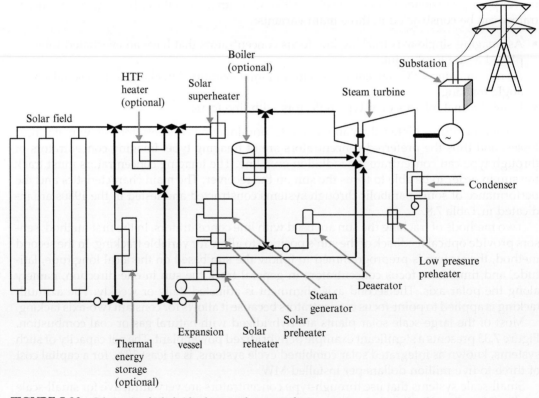

FIGURE 7.33   Schematic of a hybrid solar trough power plant.

enough thermal energy to generate power and cogenerate some heating to satisfy residential needs. A small-scale system for power and heating and hydrogen storage for residential applications is shown in Figures 7.34 and 7.35, respectively.

As it can be observed from Figure 7.35, a regenerative ORC generates power in (9) and heating (21) during the daytime only. The system is set such that during the daytime at most one-fourth of the generated power can be consumed, with the rest diverted toward the electrolyser (10) to generate hydrogen (14) which is stored in metal hydrides.

At nighttime hydrogen is consumed (16) and converted into power (17) by the PEM fuel cell. The system includes a water treatment unit for makeup water (12) and a water recycling subsystem that recovers water (18) from the PEM fuel cell while warming up (19) the metal hydride tank for hydrogen release. It is suggested that the system is cooled with water circulated through a ground loop that acts as a heat sink.

If one denotes I_T the incident global irradiation (annual average) on tilt surface, then the incident light is $\dot{E}_1 = I_T A$, where A is the total surface of solar collectors. Table 7.9 gives the energy balance equations for each of the subsystems. Note that the global radiation of 1800 MWh/m^2.year is representative of many locations where single house residences with sufficient roof surface (100 m^2) may exist; the roof must be southward oriented.

TABLE 7.9 Energy Balance Equations, Efficiency Definitions, and Descriptions

Component	Energy Balance Equation and Description	Efficiency
CPC	$\dot{E}_1 = \dot{E}_2 + \dot{E}_3$ where $\dot{E}_2 = \frac{\theta_{0.5}}{2\pi}\dot{E}_1$, only a fraction of the incident light is harvested depending on the acceptance angle.	$\eta_{CPC} = \dfrac{\dot{E}_2}{\dot{E}_1}$
SR (solar receiver)	$\dot{E}_2 = \dot{E}_4 + \dot{E}_5$ where $\dot{E}_5 = U\frac{A}{C}(T_{SR} - T_0)$, with U being the heat loss coefficient, T_{SR} the receiver temperature, C the concentration	$\eta_{SR} = \dfrac{\dot{E}_4}{\dot{E}_2}$
TMS (thermo-mechanical energy storage)	$\dot{E}_4 = \dot{E}_6 + \dot{E}_7$ where $\dot{E}_6 = \dot{m}_{TMS} x \Delta h_{lv}$ with $\dot{m}_{TMS}$ the mass flow rate of working fluid through thermosiphon loop, x the vapor quality at receiver exit, Δh_{lv} the latent heat of boiling. Heat loss $\dot{E}_7 = (UA)_{loss}(T_6 - T_0)$.	$\eta_{TMS} = \dfrac{\dot{E}_6}{\dot{E}_4}$
ORC	$\dot{E}_6 = \dot{E}_8 + \dot{E}_9$t The ORC is described separately in Figure 7.34. The pump and turbine isentropic efficiency is assumed at 0.8.	$\eta_{ORC} = \dfrac{\dot{E}_8}{\dot{E}_6}$
PIC (power inverter and control)	$\dot{E}_8 + \dot{E}_{17} = \dot{E}_{10} + \dot{E}_{20}$t The inverter produces AC current in 20 to supply the local grid, whereas it processes DC current at 8, 10, 17. When demand is high, fuel call and ORC can work together during the day. ORC never operates at nighttime.	$\eta_{PIC} = \dfrac{\dot{E}_{10}}{\dot{E}_8} = \dfrac{\dot{E}_{20}}{\dot{E}_8 + \dot{E}_{17}}$
PEME (electrolyser)	$\dot{E}_{10} = \dot{E}_{14} + Q_{loss}$Heat loss is released into the environment	$\eta_{PEME} = \dfrac{\dot{E}_{14}}{\dot{E}_{10}}$
FC (fuel cell)	$\dot{E}_{16} = \dot{E}_{17} + \dot{E}_{loss}$ where $\dot{E}_{loss} = Q_{19}h$ Heat released through the water is recovered and used to heat the hydride tank	$\eta_{FC} = \dfrac{\dot{E}_{17}}{\dot{E}_{16}}$
HR (heat recovery)	$\dot{E}_9 = \dot{E}_{21} + \dot{E}_{22}$a A part of the heat rejected by the ORC is recovered and used for water (or space) heating.	$\eta_{HR} = \dfrac{\dot{E}_{21}}{\dot{E}_9}$
SYS (overall)	$\dot{E}_1 = \dot{E}_{20} + \dot{E}_{21} + \dot{E}_{loss}$	$\eta = \dfrac{\dot{E}_{20} + \dot{E}_{21}}{\dot{E}_1}$

CASE STUDY 7.1

We consider the residential system as presented in Figures 7.34 and 7.35. A modeling case study is presented using the simplifying assumption that annual averaged values can be considered for the solar radiation. Other assumed modeling parameters are

- Half-acceptance angle of CPC: $_c = 60°$.
- Number of average reflections: $n_r = 1.1$.
- Concentration ratio CPC: $C = 5$.
- Total aperture area: $A = 100$ m^2.
- Annual irradiation: $I_T = 1800$ MWh/m^{2}year.
- Water heating temperature: $T_8 = 60°$C (see Figure 7.35).
- Collector temperature: $T_1 = T_2 = T_3 = 120°$C.
- Condensation temperature: $T_6 = T_9 = 20°$C (ground look cooled condenser).

The cycle is modeled based on balance equations written for steady operation. The thermodynamic cycle is illustrated in Figure 7.36. The cycle efficiency is $\eta_{ORC} = 28\%$ if no heat is cogenerated. The ORC efficiency depends on the amount of cogenerated heat (the heat demand). Note that the ORC self-regulates according to the heat demand. When there is no heat demand there is no cooling in branch 7–8 of the system (Figure 7.35). Vapor accumulates and the extracted fraction must reduce because the pressure on the branch tends to increase. When there is heat demand, there is also good condensation and the extraction fraction tends to increase.

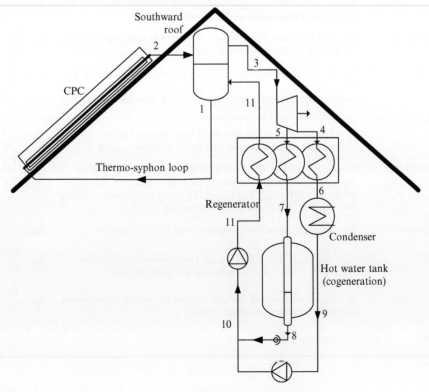

FIGURE 7.34   ORC system and thermosiphon configuration of solar collector.

Continued

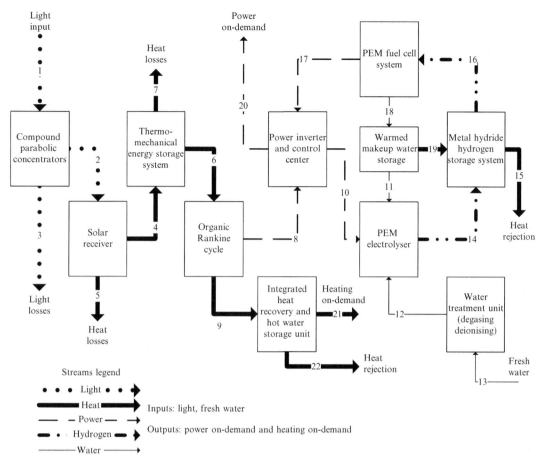

FIGURE 7.35   Detailed block diagram representation of the integrated ORC/electrolyser/fuel cell system with hydrogen generation.

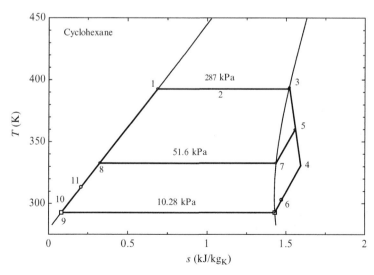

FIGURE 7.36   Thermodynamic cycle of cyclohexane-based ORC with cogeneration.

Continued

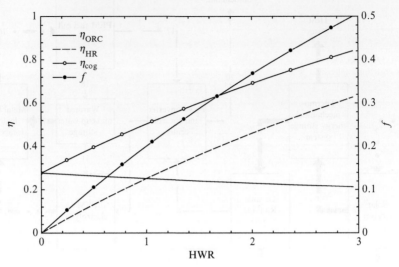

FIGURE 7.37   Energy efficiencies of ORC, heat recovery, cogeneration and vapor extraction fraction vs. heat to work ratio.

One introduces the parameter *HWR*—heat to work ratio—which represents the ratio between heat cogenerated and work generated. When this parameter is varied from 0 to 3 the ORC power efficiency degrades but more heat is recovered and delivered as useful product. Therefore, the cogeneration efficiency increases. Figure 7.37 shows the results of this parametric study. In addition the variation of extraction fraction *f* is indicated. When there is no extraction, then there is no cogeneration. If the extraction fraction is ∼0.5 then the cogeneration efficiency becomes more than 80%, even though the ORC efficiency for power generation degrades to 22%.

For estimation of overall system efficiency, the required balance equations are solved for each system component. In Table 7.10 the efficiency results are presented. For the CPC the required half-acceptance angle has been taken as 60°. Hence, one-third of the diffuse radiation is accepted by the CPC. As global radiation is assumed to be 1800 MWh/m^2 year and for AM 1.5 the direct beam radiation is 90% of the global radiation, it results that the CPC will accept 180 MWh/m^{2}year diffuse radiation. The CPC orientation is toward the equator in tilt position. As the average daylight is from 6 am to 6 pm, it means that the CPC sees the sun from 8 am to 4 pm, therefore accepting two-thirds of the direct beam radiation. Therefore, 1200 MWh/m^{2}year are received from direct beam, and 1380 MWh/m^{2}year is the total accepted radiation. Hence the CPC efficiency is 1200/1800=66.6%.

On the basis of 1800 MWh/m^{2}year the generated power can be calculated if one assumes one third of it is produced during daylight with an efficiency of $\eta_{PG}=14.9\%$ while two-thirds is generated with an efficiency of $\eta_{ODPG}=3.7\%$. Therefore, one calculates for the 10 m^2 collector area $0.149 \times 0.667 \times 1800 \times 10 = 1790$ MWh power generation. For heat generation one obtains 5652 MWh per year. Therefore, efficiency for the system that uses both direct power production during daylight and fuel cell generation at nighttime is 41.3%, whereas the power-only efficiency of the system calculated in a similar way is 9.94%.

Continued

TABLE 7.10 Energy Efficiencies of System and Components

Component	Assumptions and Remarks	Efficiency
CPC	AM 1.5 spectrum at 120° acceptance angle	$\eta_{CPC} = 66.7\%$
SR	Heat loss coefficient $U = 3$ W/m^{2}K	$\eta_{SR} = 95\%$
TMS	Assume 1% heat loss	$\eta_{TMS} = 99\%$
ORC	Determined according to Figure 7.37	$\eta_{ORC} = 25\%$
PIC	Assume electrical losses of 5% as shown in Yilanci et al. (2008)	$\eta_{PIC} = 95\%$
PEME	Assume practical value at optimal current density as in Yilanci et al. (2008)	$\eta_{PEME} = 50\%$
FC	Assume practical value at optimal current density as in Naterer et al. (2013)	$\eta_{FC} = 50$
HR	Determined according to Figure 5	$\eta_{HR} = 50\%$
SYS	The following system efficiencies are calculated:	$\eta_{PG} = 14.9\%$
	• Power generation efficiency $\eta_{PG} = \eta_{CPC}\eta_{SR}\eta_{TMS}\eta_{ORC}\eta_{PIC}$	$\eta_{H_2} = 7.5\%$
	• Hydrogen generation efficiency $\eta_{H_2} = \eta_{PG}\eta_{PEME}$	$\eta_{ODPG} = 3.7\%$
	• On-demand power generation efficiency $\eta_{ODPG} = \eta_{H_2}\eta_{FC}$	$\eta_{HG} = 31.4\%$
	• Heat generating efficiency $\eta_{HG} = \eta_{CPC}\eta_{SR}\eta_{TMS}\eta_{HR}$	$\eta_{SYS} = 35\%$
	• Cogeneration efficiency $\eta_{SYS} = \eta_{ODPG} + \eta_{HG}$	

7.2.4.3 Parabolic Dish Units

Although point concentrators are available, there is a cost and technological gap which needs to be closed in order to exploit their high efficiencies. A parabolic dish is a point-focus concentrator that consists of a parabolic reflector that tracks the sun and creates an image of the sun at the focal point. The performance factors of a dish system can be greatly degraded with changes in geometry; therefore, accuracy and rigidity are important in their design. The optical efficiency of the solar concentrator, concentration ratio, and the intercept factor, which will be defined rigorously in the subsequent paragraphs, are important factors in defining system performance.

On average, for an optical efficiency of 0.90–0.93, concentration ratio of 2000–5000, intercept factor of 0.98, and a lifetime of 30 years, a low-cost price estimate is $200–350/m^2. Another cost estimate is based on the Acurex concentrator with an optical efficiency of 0.86 and concentration ratio of 1900 costing $330/m^2 for a production scale of 100,000 units per year. Back-silvered glass is the standard design for the mirrors, which have 94% reflectivity. The reflector can be a single layer which is more efficient and more expensive, or it can be broken into components, which is cheaper but less efficient. Another option is the stretched membrane mirror manufacture of which involves a vacuum process. Singular-element stretched membrane mirrors have demonstrated optical efficiencies of 0.915, which is lower than that of the best glass-metal mirrors, but at a lower cost.

Regarding point concentrators, there have been a number of small-scale designs tried in the last decades. The OMNIUM-G concentrator has a 6-m-diameter paneled dish which provides

7–12 kW thermal under $1000 W/m^2$ insolation. "Test Bed" concentrator has an 11-m paneled dish and provides 76 kW thermal. A Lajet concentrator comprises 24 1.5-m-diameter dishes and delivers 33 kW thermal power. The ADVANCO concentrator provides 74 kW thermal using a 10.6-m-diameter paneled dish. General Electric's "Parabolic Dish Concentrator 1" uses a 12-m paneled dish to provide 72.5 kW heat. Power Kinetics has a 9-m square-shaped paneled concentrator that delivers 28 kW to a boiler under $0.88–0.94 kW/m^2$ insolation. The Acurex "Parabolic Dish Concentrator 2" uses an 11-m paneled dish providing 800 W concentrated solar heat per square meter of aperture. Boeing created a solar point concentrator equipped with 0.6×0.7 m mirror panels with optical efficiency of 0.8 up, providing a concentration ratio of 3000. ENTCH "Fresnel Concentrator Lens" uses panels of 0.67×1.2 m and can only provide an optical efficiency of 0.68 at a concentration ratio of 1500. Currently, the largest single-dish power system is Australia's "Big Dish," with an aperture of 20 m, producing 200 kW thermal feeding a 500 °C boiler.

In the design of the concentrating solar collector there is a trade-off between optics and heat losses. The optics of a paraboloid dish solar concentrator, including the definition of main parameters such as the rim angle, aperture, parabola equation, and focal distance are described with the help of Figure 7.38, which illustrates the profile of a paraboloid mirror targeting the sun. The image of the solar disc is incident on every point of the mirror surface under an angle of about 32 min ($\sim 0.54°$) and is specularly reflected toward the focal point under the same angle. In the focal plane, the solar disc image will be deformed, depending on the location of any particular point on the mirror.

The solar collector efficiency represents the ratio between the heat absorbed by the solar receiver $\dot{Q}_r$ and the incident solar radiation I_{T0}, normal on the collector's aperture of area A_a. Thus the thermal efficiency of the solar collector is $\eta_{coll} = \dot{Q}_r/(I_{T0}A_a)$. The solar collector can be divided in two subsystems. The first is the optical subsystem, having the role of concentrating the solar radiation on the small spot receiver. The efficiency of the optical subsystem—called the optical efficiency—can be defined as the intensity of the concentrated light over the intensity of the incident light. Because of optical losses due to reflectivity, transmittance, optical error, and shading one has $I_r A_r < I_{T0}A_a$; thus the ratio between the two is the

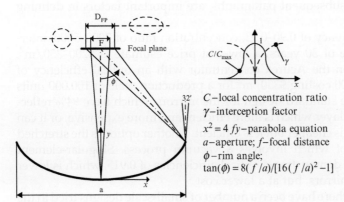

FIGURE 7.38 Optical model for paraboloid dish solar concentrator.

C – local concentration ratio
γ – interception factor
$x^2 = 4\,fy$ – parabola equation
a – aperture; f – focal distance
ϕ – rim angle;
$\tan(\phi) = 8(f/a)/[16(f/a)^2 - 1]$

optical efficiency $\eta_{opt} = (I_r A_r)/(I_{T0} A_a)$. The second subsystem is the thermal one; this has the role of converting the incident concentrated light radiation into heat.

Several thermal radiation and heat convection losses are characteristic of this subsystem. The receiver's absorbance, emittance, and the heat transfer coefficient by convection influence thermal efficiency the most; this is defined as the ratio of the absorbed heat and the concentrated solar radiation $(I_r A_r)$, written as $\eta_{th} = \dot{Q}_r/(I_r A_r)$. Thus, one can say that the solar collector efficiency is given by $\eta_{coll} = \eta_{opt} \times \eta_{th}$. The energy balance on the solar receiver, reads

$$I_r A_r = \dot{Q}_r + U A_r (T_r - T_0)$$

where U is the linearized heat loss coefficient of the receiver, assumed at temperature T_r, higher than that of the environment T_0. If one divides Equation (9.22) with $(I_{T0} A_a)$ the collector efficiency is obtained as (note that $C = A_a/A_r$ is the concentration ratio):

$$\eta_{coll} = \eta_{opt} \times \left[1 - \frac{U(T_r - T_0)}{C I_{T0} \eta_{opt}} \right]$$

which identifies

$$\eta_{th} = 1 - \frac{U(T_r - T_0)}{C I_{T0} \eta_{opt}}$$

For a point concentrator, the optical efficiency is large when most of the concentrated radiation is captured by the solar receiver surface placed in the vicinity of the focal point; that is when the hot spot exposed to the radiation has a large area. However, large hot spot area means large heat losses through radiation, or possibly trough convection.

The points from the rim circle produce the most deformed images; this maximum deformation is indicated with D_{FP} in Figure 7.38. The minimum deformation corresponds to the central point of the mirror (note that this point is generally shaded). Mirror reflectivity, shading, glazing transmissivity, absorbability, and other parameters influence the quality of the optical image formed on the focal plane. The relevant parameters for optical modeling of the solar dish are defined in Table 7.11. Based on the definition of the optical efficiency (above) and the definitions presented in Table 7.11 it results that

$$\eta_{opt} = \zeta \rho \gamma \tau \alpha$$

For design calculation the values of shading, reflectivity, transmissivity and absorbability are assumed at average values. The statistical dispersion of their values however, is taken into account through several kinds of optical errors. Thus, we will express the optical efficiency by two factors. The first factor is the optical factor introduced here as follows

$$\Omega = \zeta \rho \tau \alpha$$

The optical factor is calculated with average values for the reflectivity, transmissivity, and absorbability and it has values in the range of 0.75–0.85 depending on the quality of the optics.

TABLE 7.11　Relevant Parameters Characterizing the Optics of a Solar Dish

Name	Definition	Remarks
Shading factor	$\zeta = \dfrac{A_u}{A_a}$	The ratio between un-shaded mirror area and aperture area.
Reflectivity	$\rho = \dfrac{I_\rho}{I_{To}}$	Reflected solar radiation intensity over incident radiation intensity. Reflectivity accounts for imperfect specular reflection.
Intercept factor	$\gamma = \dfrac{\displaystyle\int_0^{A_{ab}} I \, dA}{I_\rho A_u}$	Not all concentrated radiation falls on the absorber surface. There is a practice of reducing the absorber surface in order to minimize the heat transfer. Thus, one accepts some loss of concentrated radiation. The intercept factor represents the ratio between the radiation power falling on the absorber and the power associated with the radiation reflected toward the focal point.
Transmissivity	$\tau = \dfrac{I_\tau A_{ab}}{\gamma I_\rho A_u}$	In a case where the absorber is covered by glazing (to reduce the heat loss via convection), a part of the incident (concentrated) radiation is lost due to transmission trough glazing. The transmissivity represents the ratio between transmitted power trough glazing and the radiation power incident on glazing.
Absorbability	$\alpha = \dfrac{I_c}{I_\tau}$	The ratio between the absorbed heat flux and the power associated with the radiation incident on the absorber.

The second factor is the intercept γ, which accounts for statistical dispersion of the reflectivity and other optical error. One remarks that

$$\eta_{opt} = \Omega\gamma$$

Figure 7.39 presents qualitatively the variation of the intercept factor in the focal plane, in the vicinity of the focal point (see the upper right corner). The intercept factor is zero at the focal point because no surface area can be associated with one single point; therefore, no radiation is "intercepted." The parameter x in the plot $\gamma(x)$ is the radial coordinate originating at the focal point. As x increases the amount of intercepted light increases over a disc centered at the focal point; thus the intercept factor increases and eventually reaches the unity value. The derivative of $\gamma(x)$ represents the local concentration ratio that has a maximum right at the focal point. The geometric concentration ratio is related to the optical system geometry, optical errors, and the intercept factor because γ is a function of A_r.

There are several kinds of optical errors that affect the quality of the solar image projected on the receiver surface. All of the optical errors listed in the following represent statistical means relative to the entire area of the solar concentrator. A first, unavoidable error is caused by the nonuniform angular distribution of solar radiation beam, δ_{sun}. Even though this is a small contribution to the overall optical error, it cannot be neglected; it has a typical value of one fourth of the solar angle, namely 2.3 mrad. Other errors are the slope error of the concentrator δ_{slope} (typically 2–3 mrad), the specularity δ_ω (typically 0.5–1 mrad), and the pointing error δ_p, meaning the receiver is not placed exactly at the focal point. These individual angular errors produce an overall optical error accounted for by $\delta^2 = \delta_{sun}^2 + \delta_{slope}^2 + \delta_\omega^2 + \delta_p^2$.

Using the above derivations, it can be shown that the collector efficiency can be analytically expressed as $\eta_{coll}/\Omega = \gamma(1 - F_{coll}/C\gamma)$, where $F_{coll} = U(T_r - T_0)/(I_{T0}\zeta\rho\tau\alpha)$ is a collector factor

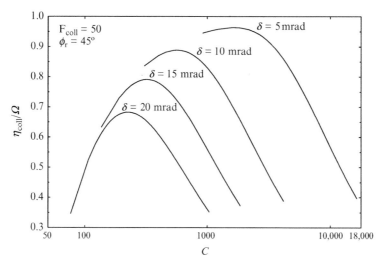

FIGURE 7.39 Point focus collector efficiency as a function of concentration ratio for several optical errors δ and rim angle $\phi_r = 45°$.

depending on the solar absorber temperature (assumed at the average), the insolation, and the optical properties.

Figure 7.39 illustrates the variation of collector efficiency (expressed as η_{coll}/Ω) with the concentration ratio for several optical errors δ. It is clearly observed that efficiency increases with the quality of surface (smaller δ); also it results that there is each time an optimal concentration ratio for which the collector efficiency is maximum. The existence of this maximum can be explained intuitively:

- If the concentration ratio is too low, then the concentrated radiation is low, and thus the heat absorbed by the receiver is low.
- If the concentration ratio is too high, then the receiver temperature is higher, and thus the heat losses to the ambient are higher; thus the heat absorbed by the receiver is also low.

In general the optical factor is high (0.8–0.95), which indicates that the point focus solar concentrators, can reach over 70% efficiency.

7.3 WIND ENERGY SYSTEMS

Wind energy is a significant renewable energy resource which has shown rapid growth in recent years. The cost of wind energy has achieved reasonable values such that wind energy farms are now constructed in many locations around the world. Germany, Denmark, and Spain are the lead countries in installing wind energy capacities. Wind energy can be regarded as a meteorological variable signifying the energy content of the wind.

The important parameter for meteorological modeling of wind is the wind velocity. In meteorology an atmospheric boundary is considered for predicting the local or regional winds by simulation. Other meteorological variables such as temperature, pressure, and humidity of the atmosphere are also important in wind occurrence. Such information together with measured velocity data are necessary for determining locations to install wind energy farms, estimating their capacity and efficiency.

Wind is in fact a form of mechanical energy. It has to be harvested by appropriately engineered devices and converted into other useful energies like electric energy. The major technical problem of a wind energy conversion system is represented by the fluctuating and intermittent nature of the wind. The harvesting system (typically a wind turbine) must react fast to the presence of wind or change in its direction and intensity. It is important to be able to store the mechanical energy harvested from wind. Several systems of wind energy storage are possible, where the mechanical energy resulting from wind is stored in various forms: kinetic (flywheels), electrochemical (batteries), chemical (hydrogen), thermomechanical (compressed air), etc. Nevertheless, the most used systems are electrical generators actuated by wind energy, having local storage capacities in batteries, and possibly being equipped with grid connecting systems. Wind turbine efficiency reaches around 35–50% with the current technology.

The harvested wind energy is normally converted into shaft rotation by a wind turbine. The mechanical energy can then be used directly. For smooth generation, the mechanical energy can be stored in some devices able to retrieve it in mechanical form, such as flywheels, hydro-storage, or compressed air. A typical direct use of wind energy is water pumping; others include grain milling, wood cutting, etc. The shaft rotating mechanical energy can be converted by appropriate electric generators into electrical energy. Common types of wind turbines are illustrated in Figure 7.40.

The most important wind turbine for power generation is the HAWT (horizontal axis wind turbine). The electrical energy generated by wind can be stored locally or transmitted to the grid or further used for hydrogen generation through water photosynthesis. Furthermore, hydrogen can be converted to ammonia though the well-matured Haber–Bosh process; this process requires additional consumption of electricity.

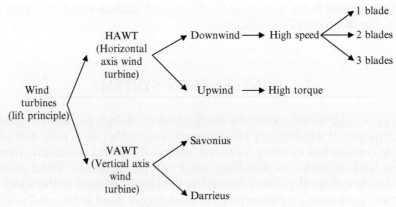

FIGURE 7.40   Conversion paths of wind energy.

FIGURE 7.41 Thermodynamic model for wind energy conversion.

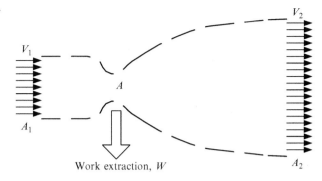

For thermodynamic analysis a fictitious boundary can be drawn around the turbine to delimit a control volume, as indicated in Figure 7.41. Outside this boundary, there is no significant modification of the velocity of air. On the contrary, inside the control volume the air feels the presence of the turbine; thus it accelerates or decelerates according to the laws of energy conservation and applicable constraints. Basically, the upstream area of the control volume (A_1) is much smaller than the downstream area (A_2). Thus the velocity upstream (V_1) is much higher than the velocity downstream (V_2).

The geometry of the control volume is similar to that of a nozzle which extracts the work W from the wind. The average velocity of the wind is denoted with $\overline{V} = (V_1 + V_2)/2$, and the mass flow rate of air is approximated with $\dot{m} = \rho A \overline{V}$. Thus the rate of variation of wind momentum is $\dot{m}(V_2 - V_1) = \rho A \overline{V}(V_2 - V_1)$. According to the second law of dynamics, the force exerted on the rotor is $F = -\rho A \overline{V}(V_2 - V_1)$. Based on the force and the average velocity it results that the work extracted by the rotor is $W = \rho A \overline{V}(V_2 - V_1) \times \overline{V}$. This work can be also calculated based on the kinetic energy variation of the air, that is $W = 0.5\rho A \overline{V}(V_1^2 - V_2^2)$. The equality of the two expressions for W results in $\overline{V} = (V_1 + V_2)/2$ which in fact justifies the definition of the average velocity as the arithmetic mean. Using the notation $\alpha = V_2/V_1$, one obtains simply that the work extraction is

$$W = \rho \frac{AV_1^3}{4}\left[(1+\alpha)(1-\alpha^2)\right]$$

The quantities ρ, A and V_1 should be assumed constant for the analysis; thus for a given wind velocity, outside temperature, and pressure (which fixes the air density) and a given wind turbine area, the generated work is a function of velocities ratio only, $W = W(\alpha)$. It is simple to show that the maximum of the work is obtained at $\alpha = 1/3$, which corresponds to

$$W_{\max} = \frac{8}{27}\rho\frac{AV_1^3}{4}$$

The efficiency of mechanical work production from wind energy must be equal to the useful shaft work generated divided to the wind energy, both given per unit of turbine surface area. The wind energy is the kinetic energy ($0.5\rho A V_1^3$). Thus the theoretical maximum energy efficiency of the wind energy conversion is

$$\eta_{\max} = \frac{8}{27}\rho\frac{AV_1^3}{4} \Big/ \frac{1}{2}\rho A V_1^3 = \frac{16}{27} = 59.3\% \tag{7.25}$$

Looking back to the equation giving the work extracted by the rotor, W, the quantity between square brackets defines the power coefficient of the rotor efficiency given by

$$C_P = \left[(1+\alpha)\left(1-\alpha^2\right)\right]/2. \tag{7.26}$$

The turbine efficiency becomes

$$\eta = C_P$$

As given above, the maximum value of $C_{P,max}$ is 0.593; however, the efficiency found in practice is in the range 35%–45%. With Equation (7.26) the work extraction reduces to

$$W = \rho C_P \frac{AV_1^3}{2}. \tag{7.27}$$

Note also that during a period of time (e.g., day, month, season) the wind velocity (V_1) varies. Another aspect regarding wind energy conversion relates to the variation of the wind speed at a certain location, during a year. The annual distribution probability of the wind velocity influences the total energy generated at a specific location during a year.

Weibull probability distribution can be used to model the occurrence of wind velocity based on two parameters k and c, according to $h(V) = k/c(V/c)^{k-1} \exp[-(V/c)^k]$; for parameter $k = 2$ this probability distribution function is called the Rayleigh distribution

$$h(V) = \frac{2V}{c^2}\exp\left[-\left(\frac{V}{c}\right)^2\right]. \tag{7.28}$$

The Rayleigh distribution for annual variation of wind velocity is exemplified in the plot in Figure 7.42 for the case $c = 7\,\text{m/s}$; in the plot, the annual probability of occurrence of wind velocity is correlated with the cube of velocity. Two vertical lines are shown at cut-in speed

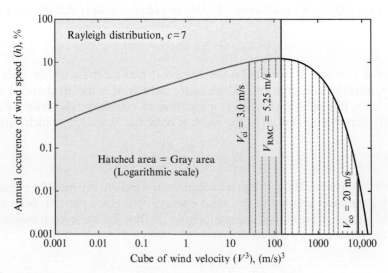

FIGURE 7.42 Definition of the root mean cube wind velocity.

(V_{ci}) assumed at 3 m/s and cut-out speed (V_{co}) assumed at 20 m/s. Outside the region delimited by V_{ci} and V_{co} the wind turbine does not operate. The hatched area on the plot, which corresponds with the wind turbine operation domain, can be calculated with the integral

$$\int_{V_{ci}}^{V_{co}} h(V)d(V^3) = 100\% \times V_{RMC}^3$$

where V_{RMC} one denotes the root mean cube wind velocity.

The gray area defines the root mean cube wind velocity as an average velocity that has 100% occurrence and generates the same amount of wind energy as the probable velocity profile described by the adopted probability distribution function. Thus the generated mechanical energy by the turbine operating at a constant rate throughout the year can be written as

$$W = \frac{1}{2}\rho C_p A V_{RMC}^3 \qquad (7.29)$$

where

$$V_{RMC}^3 = \int_{V_{ci}}^{V_{co}} h(V)d(V^3) \qquad (7.30)$$

When a wind turbine location is selected, it is important to know the average wind velocity in that location, because the turbine energy efficiency is specified by the manufacturer at a rated value which corresponds to a rated wind velocity (V_r). If the wind velocity is different than the turbine rating value, then a capacity factor can be used (as indicated by the manufacturer) to determine the generated power under the actual conditions. The capacity factor is defined as the percentage of the nominal power that the wind turbine generates in actual wind conditions. In this case the energy efficiency of the wind turbine becomes

$$\eta(V) = C_f(V) \qquad (7.31)$$

where $C_f(V)$ is the capacity factor expressed as a function of the actual wind velocity. The capacity factor is small at low and high velocities and typically has a maximum value at the design point.

Figure 7.43 exemplifies a typical variation of the wind capacity factor; on the same plot is superimposed an assumed annual occurrence of wind speed. It is possible to calculate the average wind velocity based on velocity occurrence; in this case, the average velocity is defined as the velocity which occurs at 100%; in other words, when the area between the $h(V) \times V$ curve is equal with the grey area shown in the figure. Therefore:

$$\overline{V} = \int_{V_{ci}}^{V_{co}} h(V)VdV \qquad (7.32)$$

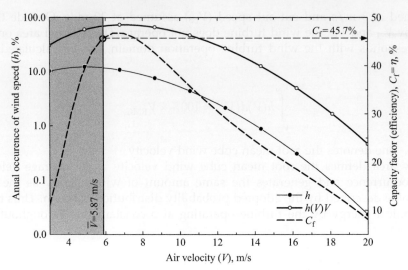

FIGURE 7.43   Case study showing the capacity factor and annual wind speed occurrence probability as a function of air velocity.

For the numerical example shown in the figure, the average wind velocity result is $\overline{V} = 5.87\,\text{m/s}$ which gives an average capacity factor of $C_f = 45.7\%$. In general one can write the average turbine efficiency as given by

$$\overline{\eta} = C_f(\overline{V}) \tag{7.33}$$

The average power generation of a wind turbine placed in certain location becomes

$$\overline{W} = \frac{1}{2}\rho A C_f(\overline{V}) V_{RMC}^3 \tag{7.34}$$

Another effect, the wind chill, can influence in some measure the efficiency of wind turbine generators. Faster cold wind makes the air feel colder than when wind is not present because it removes heat from our bodies faster due to intensified heat transfer by convection. Wind chill is a measure of this effect and is defined as the hypothetical air temperature in calm conditions (air speed, $V = 0$) that would cause the same heat flux from the skin as occurs at the actual air speed and temperature. Wind chill temperature can be estimated with

$$T_{wch} = 21.68 + 1.12T_0 - 38.58V^{0.16} + (0.83T_0 + 14.76)V^{0.16} \tag{7.35}$$

where T is temperature in °C, indices "wch" and "a" denote wind chill and surroundings air, and V is the wind speed in km/h.

Thus, a rigorous application of Equation (7.34) implies the estimation of the air density at wind chill temperature. The air density depends on the atmospheric pressure and temperature (a simple calculation of the air density can be based on the ideal gas law, $\rho = \rho_{ref} \times T_{ref}/T_0 \times P_0/P_{ref}$ where the index 0 represents the actual environment, ref is reference environment, and T is temperature in K and P is pressure). Due to expansion, air suffers additional cooling between upstream (1) and downstream states (2). Thus the average air density is

$\bar{\rho} = (\rho_1 + \rho_2)/2$, where $\rho_{1,2}$ are calculated with $(T_{1,2}, P_{1,2})$ and $T_{1,2}$ are wind chill temperatures calculated with Equation (7.35). Thus a more accurate energy efficiency expression, accounting for wind chill temperature, is

$$\bar{\eta} = C_f(\overline{V})\bar{\rho}/\rho_0 \tag{7.36}$$

where the index 0 indicates the temperature of the surroundings. The energy conversion of wind energy can be up to 2% higher than the turbine capacity factor.

The exergy efficiency of a wind turbine is defined as the work produced per exergy associated with the wind. The work generated by the turbine can be calculated, as discussed above, with $W = 0.5C_f(\overline{V})\bar{\rho}AV_{RMC}^3$. The exergy associated with the wind surrounding the turbine must comprise the kinetic energy of the wind (which is a form of exergy) and the thermomechanical exergy of the wind. The thermomechanical exergy of the wind must be calculated with respect to wind chill temperature, because this is the temperature of the medium in the vicinity of the turbine. The wind has a higher exergy upstream (Ex_1) than downstream (Ex_2). Thus the exergy that enters the system is the difference $(Ex = Ex_1 - Ex_2)$. Using appropriate expressions for thermomechanical and kinetic energy it results that

$$Ex = W + \bar{\rho}A\overline{V} \times \left\{ \overline{C}_p(T_1 - T_2) - T_0\left[\overline{C}_p \ln\left(\frac{T_2}{T_1}\right) - R \ln\left(\frac{P_2}{P_1}\right) - \overline{C}_p\left(1 - \frac{\overline{T}}{T_0}\right)\right]\right\} \tag{7.37}$$

where $\bar{\rho}A\overline{V}$ is the mass flow rate of air, $T_{1,2}$ are the wind chill temperatures upstream and downstream of the turbine, T_0 is the temperature of the surroundings, $\overline{T}$ is the arithmetic mean of temperatures $T_{1,2}$, pressures $P_{1,2}$ are upstream and downstream, respectively, $\overline{C}_p$ is the specific heat, and R is the ideal gas constant. The first term in braces represents the enthalpy difference, the second term is the entropy difference, and the irreversibility due to heat transfer between the stream of air and the surroundings; basically, the expression in braces is the specific thermomechanical exergy consumed by the system. The exergy efficiency becomes

$$\psi = \frac{W}{Ex} \tag{7.38}$$

Two basic categories of wind turbine design do exist, namely the horizontal axis wind turbine (HAWT) and the vertical axis wind turbines (VAWT). The horizontal axis wind turbines have emerged as the dominant technology. They have the rotor axis horizontal and in line with the wind direction. The common designs have the generator and the blade fixed on a rotating structure in the top of a tower. The mechanical system also comprises a gear box that multiplies the rotation of the blades. Compared to the VAWT, the HAWT achieves higher efficiency. The VAWT operates well in fluctuating wind amplitude provided that its direction remains quasi-constant.

One important feature of the VAWT is that it does not need to be pointed in the direction of the wind. Thus VAWT installation is appropriate where wind changes its direction very frequently. Another feature of the VAWT is that it allows placement of the generator and gear box in the lower side, closing the support system of the turbine.

The HAWT construction can differ in number of blades (generally two or three blades are used). The HAWT use airfoils to generate lift under the wind action and rotate the propeller.

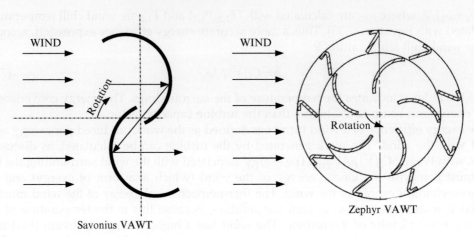

FIGURE 7.44   Operating principles of Savonius and Zephyr wind turbines.

The operation of some VAWT designs is also based on the aerodynamic lift principle, but there are designs that use form drag and aerodynamic friction for operation. Drag-type turbines are designed such that the form drag generates a torque; a typical drag turbine is the Savonius, much used in rural areas. The operating principle of the Savonius wind turbine is illustrated schematically in Figure 7.44. The generated torque is due to the difference in pressure between the concave and convex surfaces of the blades and reaction forces of the deflected wind coming from behind the convex surface. The efficiency of the Savonius turbine is over 30%.

A more elaborate design of the VAWT, namely the Zephyr, is also presented in Figure 7.33. It is designed to perform well in low wind speed and high turbulence conditions. For this reason, several Zephyr wind turbines can be installed close to each other to form a compact wind farm. The power coefficient of the Zephyr wind turbine is rather low (0.11) but the turbine has a good utilization factor, low cut-in speed, and high cut-off speed which make it competitive in urban areas.

Common models of lifting VAWT turbines are the triangular (delta) Darrieus turbine, the Giromill turbine, and the Darrieus–Troposkien turbine. The triangular Darrieus turbine has straight-blade geometry. The Giromill turbine is a low-load system with straight, vertical blades and an adjustable angle of attack to fit the best working conditions. The Darrieus–Troposkien turbine is the turbine most used for electric power generation. The blades of this turbines are airfoils assuming a Troposkien shape (which is a name derived from Greek meaning "the shape of a spinning rope"). Basically, the airfoil is a flexible blade connected to a vertical axis at two points (lower and upper); while rotating it approximates by revolution the shape of an ellipsoid. The blades can be made from light materials like aluminum, fiberglass, steel, or wood.

Apart from the wind turbine itself, wind power plants comprise several other important subsystems, such as yaw control, a speed multiplier (gear box), voltage–frequency and rotation controllers, voltage-raising and -lowering transformers (with the role of matching the generation, transmission, and distribution voltages at normal consumption levels), protection

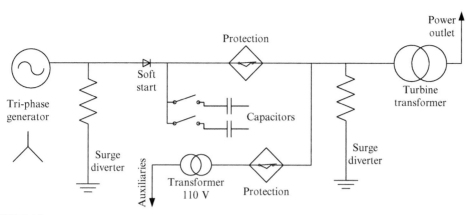

FIGURE 7.45 Simplified electric diagram of a wind power generation unit.

systems for over-current, over-speed, overvoltage, and atmospheric outbreaks, and other anomalous forms of operation. All these subsystems are sustained by a robust mechanical structure.

Figure 7.45 illustrates the simplified electric diagram of a wind power generation unit, including the main components and their functions. The system is equipped with general protection circuits, breakers, and over-current and reverse current protection. When a wind power generator is connected to the grid it is possible that—during low wind periods—the generated voltage is lower than the grid voltage. If this situation occurs the protection systems (see the figure) temporarily disconnect the system from the grid. Also if there is an over-current caused by a load too large, the thermal protection will disconnect the generator. The capacitors are used to correct the power factor of the generator and allow a soft start. The capacitors work together with a thyristor unit to soften the starting process of the generator. A small transformer is used to supply the auxiliary equipment. The system has a main transformer that allows connection of the unit to the grid. Local energy storage can be applied in batteries, in which case the system will include an inverter. Moreover, it is possible to couple several turbines to the main transformer, when it applies, rather than equipping each turbine with its own grid connection transformer.

The electrical system and electrical infrastructure of a wind power plant may account for 13–15% of the total installation costs. A breakdown of various costs for installing a 10-MW wind power unit is given in Figure 7.46. The electrical infrastructure is costly because it involves, among other things, massive work for grounding and lighting protection. The damage produced by lighting to wind farms and individual units is considerable. In Germany statistics show that lighting produces damage annually to 8% of installed wind turbines, while in Sweden this is 6% and in Denmark 4%.

The noise generated by wind power plants is also of major concern. The blades, the gearbox, the generator, the generator hub, the tower, and the auxiliary system all generate a significant amount of noise during operation. The noise of a wind power plant operating at 350 m far can be 10% higher than typical rural night background noise.

Considerable efforts have been deployed in recent years to develop techniques for reducing noise from wind turbines by proper design and installation. With current technology, the

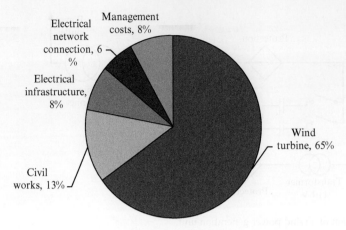

FIGURE 7.46 Breakdown of costs for a typical 10-MW power plant. *Source: Burton et al. (2001).*

noise level of the main power plant components ranges from 40 to 98 dB(A). Nevertheless, protection against noise increases the capital cost of the power plant significantly. By the year 2004, the investment cost was around 50¢ per installed watt and 6¢ per generated kWh.

By expanding wind power generation capacities in recent years, the cost of installed kW and the cost of generated kWh has reduced significantly. The investment cost is presently about 30% higher than that for natural gas power plants, while the generation cost is about 3.5¢/kWh.

7.4 GEOTHERMAL POWER GENERATION SYSTEMS

Geothermal energy is a form of thermal energy that is available on some regions of the earth's surface at temperature levels in the range of about 35–500 °C, even though most geothermal locations provide temperature levels up to 250 °C. Geothermal heat is used either for process heating or is converted into electricity through appropriate heat engines. A large palette of heat-consuming processes can be supplied by geothermal energy, including space and water heating applications, industrial processes, various supply procedures in agriculture and food industry, etc. Figure 7.47 illustrates the global utilization of geothermal heat.

It can be observed from the graph that most of the geothermal applications are ground source heat pump systems. In recent years there has been substantial development in geothermal energy, with major interest in the ground source heat pump, hydrogen production from geothermal energy, and installing electrical plants supplied by geothermal heat. Some historical facts regarding geothermal development in Canada are indicated in Table 7.12.

As a source of heat, the limit of geothermal energy conversion is governed by the Carnot factor. Thus, it is important to assess the range of the Carnot factor for geothermal reservoirs. Additionally, it is useful to analyze relevant irreversibilities specific to geothermal energy conversion. Thus one can obtain a general picture of the thermodynamic limits of geothermal energy conversion into work.

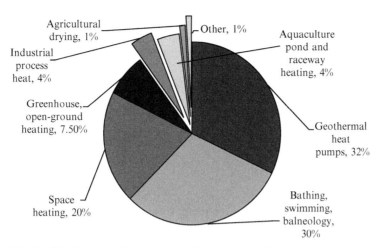

FIGURE 7.47 Global utilization of geothermal energy for heating applications.

TABLE 7.12 Geothermal Developments in Canada

Date	Facts
1886	In Banff (Alberta) hot springs were piped to hotels and spas.
1975	Drilling to assess high-temperature geothermal resources for electricity generation in British Columbia.
1976–1986	Ten-year federal research program assesses geothermal energy resources, technologies, and opportunities for Canada.
1990	Hydro Ontario funds a program to install geothermal heat pumps in 6749 residences.
1990s	Government take a greater interest in using renewable energy—including geothermal—as a way to decrease greenhouse gases and other emissions.
2004	Western GeoPower Corp. applies for government approvals to build a $340-million, 100-megawatt geothermal power plant at Meager Creek, northwest of Whistler (British Columbia). Manitoba government announces program to provide loans of up to $15,000 for installation of geothermal heat pump systems.

We will analyze, as a first step, the temperature level of geothermal sources and the nature of geothermal fluid. The average increase in temperature with depth—called the geothermal gradient—is about $0.03\,°C/m$, or $30\,°C/km$ for a few kilometers near the earth's surface. Values as low as about $10\,°C/km$ are found in ancient continental crust and very high values ($>100\,°C/km$) are found in areas of active volcanism.

Heat from the earth's depths is transported to the surface in three possible ways, characterized by the type of geothermal field: hot water, wet steam, and dry steam. Hot water fields contain reservoirs of water with temperatures between 60 and $100\,°C$, and are most suitable for space heating and agricultural applications. For hot water fields to be commercially viable, they must contain a large amount of water with a temperature of at least $60\,°C$ and lie within $2000\,m$ of the surface. Wet steam fields contain water under pressure and usually measure $100\,°C$.

These are the most common commercially exploitable fields. When the water is brought to the surface, some of the water flashes into steam, and the steam may drive turbines that can produce electrical power. Dry steam fields are geologically similar to wet steam fields, except that superheated steam is extracted from the aquifer. Dry steam fields are relatively uncommon. Because superheated water explosively transforms into steam when exposed to the atmosphere, it is much safer and generally more economical to use geothermal energy to generate electricity, which is much more easily transported.

$$ex = h(T,P) - h_0 - T_0[s(T,P) - s_0]$$

where T, P represent the temperature and the pressure at which the geothermal fluid is made available to the energy conversion system (power generator). The enthalpy part shown in the expression above represents the energy content of the stream, $e = h(T,P) - h_0$. With the index 0 one denotes the reference state. Thus, the energy and exergy efficiencies of the geothermal generator can be expressed by

$$\left. \begin{array}{l} \eta = \dfrac{\dot{W}}{\dot{m} \times e} \\[2mm] \psi = \dfrac{\dot{W}}{\dot{m} \times ex} \end{array} \right\}$$

where $\dot{W}$ represents the generated work rate and $\dot{m}$ is the mass flow rate of the geothermal fluid.

In order to obtain an estimate of conversion limit one has to consider an ideal thermodynamic cycle to which the geothermal energy is transferred to generate work. As the geothermal fluid is generally brine it appears logical to assume that during the heat transfer process the brine exchanges sensible heat. Thus the temperature of the brine decreases to T_0.

Figure 7.48 suggests a thermodynamic cycle where the geothermal fluid delivers sensible heat in a cooling process through which it reaches equilibrium with the environment at (T_0, P_0). We assume the process as a straight line in the aT–s diagram. The thermodynamic cycle that generates maximum work is in this case triangle-shaped; the maximum generated work is indicated with the triangular gray area. The efficiency is $\eta = (T - T_0)/(T + T_0)$.

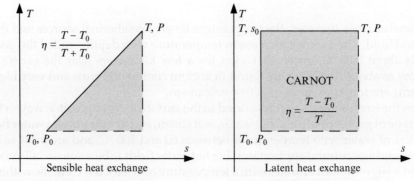

FIGURE 7.48   Thermodynamic cycles for maximum work extraction from geothermal energy.

TABLE 7.13 Thermodynamic Limits of Geothermal Energy Conversion

Geothermal Source	Temperature	Sensible Heat Exchange	Latent Heat Exchange
Low temperature	150 °C	17%	29%
Medium temperature	220 °C	24%	39%
High temperature	500 °C	44%	61%

Table 7.13 indicates figures for maximum conversion efficiency from low-, medium-, and high-temperature geothermal sources when sensible heat is extracted from the geothermal fluid. Note that in some cases the geothermal fluid is steam. One can assume in such cases that steam is condensed through an isothermal process after which the entropy reaches s_0. This process is followed by an isentropic expansion to (T_0, P_0). This is an idealization; such a process cannot occur in an actual system, but it is informative to know the conversion efficiency for this case (which is the upper limit). As this is a Carnot cycle the efficiency is $\eta = (T - T_0)/T$; this efficiency is also reported in the table.

This analysis shows that the efficiency of the geothermal energy conversion system must be much lower than 17–44% due to irreversibilities. Note that the energy efficiency of geothermal steam plants ranges from 10% to 17%, respectively. The energy efficiency of binary cycle plants ranges from 2.8% to 5.5%. These percentages are lower than in the case of steam power plants as binary plants are typically used for lower-temperature geothermal resources.

Many thermodynamic cycles have been developed for geothermal power generation. The selection of the cycle depends on the kind of geothermal fluid, its flow rate and the level of temperature. The geothermal fluid can be dry steam, low-pressure brine, or high-pressure brine. The geothermal reservoir may or may not require reinjection of the fluid after its use in the power plant. The power plant design must be such that it extracts as much exergy as possible from the geothermal fluid. Figure 7.49 presents a classification of geothermal power plants.

When the geothermal heat is available in the form of dry steam, its conversion into work can achieved with steam turbines. There are two cases: steam after expansion is condensed and

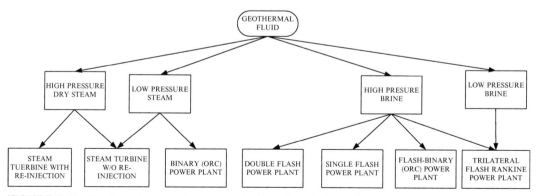

FIGURE 7.49 Geothermal power plants categorized based on the kind of geothermal fluid.

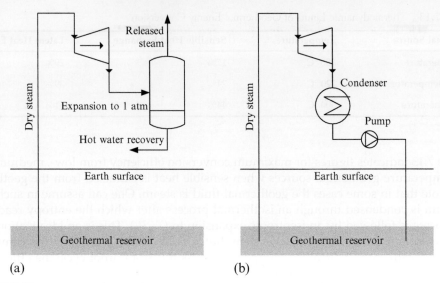

FIGURE 7.50   Dry power plants with direct expansion.

reinjected into the geothermal well, or it is simply released into the atmosphere. Figure 7.50 presents power plants diagrams with dry steam expansion with or without reinjection.

Some geothermal reservoirs reject steam at a high or lower pressure which can be expanded in a turbine to generate power. Depending on the type of geothermal well, reinjection of the geothermal fluid is or is not needed.

For example, if a geothermal site produces dry steam a turbine can be used to generate power as indicated in Figure 7.50. In the case that the steam pressure is considered low, or if there is no need to recycle the geothermal fluid, then, after expansion down to atmospheric pressure, the steam can be released to the atmosphere (Figure 7.50a). However, if the steam pressure and temperature are high enough, the amount of power generated allows for driving a recirculation pump and still generating satisfactory yield from the turbine. In this case—Figure 7.50b—the steam can be expanded in a vacuum, condensed, and the produced water pressurized in a pumping station and reinjected. Note that the reinjection pressure must be very high—a couple hundred bar. The reinjection is needed in many instances to keep the geothermal resource at steady production. If the geothermal reservoir is a hot rocky layer, then the water injected through one well is boiled and extracted as steam through another well.

The geothermal sources generating low-pressure steam that needs to be re-injected can be coupled to ORC generators as indicated in Figure 7.51. These cycles are also known as binary cycles as sometimes they operate with binary mixtures of refrigerants (e.g., ammonia-water). The dry steam extracted from the geothermal well at 9 is condensed (9–10) and subcooled (10–11) before water reinjection at 12. The heat released by the geothermal brine is then transmitted to a working fluid (toluene in the example). The working fluid is boiled and expanded as saturated toluene vapor at 1. After expansion it follows internal heat recovery in heat exchanger 2–3/5–6, then condensation (3–4), pressurization of the saturated liquid (4–5), preheating (5–6 and 6–7) and boiling (7–1).

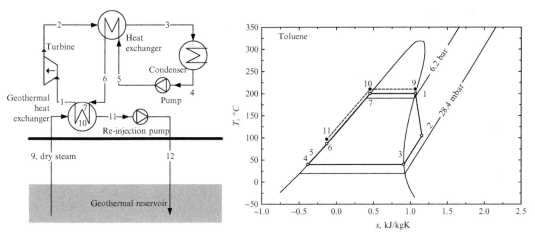

FIGURE 7.51   Geothermal binary cycle power plant driven by dry steam.

Calculations for the cycle are presented in Table 7.14, which includes all equations (energy and mass balance) needed to solve the problem. The calculated state point parameters are given in Table 7.15. The energy and exergy efficiencies are calculated by the following equations:

$$\left.\begin{aligned} \eta &= \frac{(h_1 - h_2) - (h_5 - h_4) - (h_{12} - h_{11})}{h_1 - h_6} \\ \psi &= \frac{(h_1 - h_2) - (h_5 - h_4) - (h_{12} - h_{11})}{h_9 - h_{11} - T_0(s_9 - s_{11})} \end{aligned}\right\}$$

The energy efficiency of this cycle is 24%, the exergy efficiency is 15%, and the Carnot efficiency calculated with the maximum temperature of geothermal brine is 38%. If the pressure of the geothermal brine is high enough, then steam-flash power plant cycles can be an effective solution for power generation.

Figure 7.52 presents two types of flash power plants, namely single-flash and double-flash kinds. In these systems, high-pressure geothermal brine is flashed at an intermediate pressure through an isenthalpic process (1–2). Thus saturated vapor flashes from the liquid and is directed toward the turbine. There it is expanded to low pressure and then condensed using heat sink as the ambient. After condensation, the resulting water is reinjected into the well. Note also that the steam resulting from flashing can be used for other processes than power generation. For instance it can be used to produce refrigeration, using ejectors in a refrigeration cycle. The double-flash cycle can be applied if the moisture content of the expanded steam is too high; this is done by expanding in two stages.

Another effective option for geothermal fields that generate high-pressure brine is the flash-binary power plant. This kind of power plant couples a flash cycle which operates with the geothermal brine as working fluid with a bottoming ORC in the way illustrated in Figure 7.53. After the flashing process, the steam is expanded in a turbine, while the resulting liquid is passed through a heat exchanger which preheats and boils the working fluid of the bottoming cycle. The system has two independent condensers. It is important to match the temperature profiles in the heat exchanger such that the available exergy is exploited at maximum benefit for improved system performance.

TABLE 7.14 Equations for Modeling the System in Figure 7.51

State	Equations	State	Equations
1	$T_1 = 200\,°C; x_1 = 1$ $s_1 = s(Toluene, T_1, x_1)$ $P_1 = P(Toluene, T_1, x_1)$ $h_1 = h(Toluene, T_1, x_1)$	6	$h_2 - h_3 = h_6 - h_5$ $P_6 = P_1$ $T_6 = T(Toluene, P_6, h_6)$ $s_6 = s(Toluene, P_6, h_6)$
2	$s_{2s} = s_1; P_2 = P_3; \eta_T = 0.8$ $h_{2s} = h(Toluene, P_2, s_{2s})$ $\eta_T (h_1 - h_{2s}) = h_1 - h_2$ $T_2 = T(Toluene, P_2, h_2)$ $s_2 = s(Toluene, P_2, h_2)$	7	$h_7 = h(Toluene, P_7, x_7)$ $P_7 = P_1; x_7 = 0; T_7 = T_1$ $s_7 = s(Toluene, T_7, x_7)$
3	$T_3 = 40\,°C; x_3 = 1$ $P_3 = P(Toluene, T_3, x_3)$ $h_3 = h(Toluene, T_3, x_3)$ $s_3 = s(Toluene, T_3, x_3)$	9	$T_9 = T_1 + 10; x_9 = 1$ $P_9 = P(Steam, T_9, x_9)$ $h_9 = h(Steam, T_9, x_9)$ $s_9 = s(Steam, T_9, x_9)$
4	$T_4 = T_3; x_4 = 0; P_4 = P_3$ $h_4 = h(Toluene, T_4, x_4)$ $s_4 = s(Toluene, T_4, x_4)$	10	$T_{10} = T_9; x_{10} = 0$ $P_{10} = P(Steam, T_{10}, x_{10})$ $h_{10} = h(Steam, T_{10}, x_{10})$ $s_{10} = s(Steam, T_{10}, x_{10})$
5	$P_5 = P_1; s_{5s} = s_4; \eta_P = 0.9$ $h_{5s} = h(Toluene, P = P_5, s = s_{5s})$ $\eta_P (h_5 - h_4) = h_{5s} - h_4$ $T_5 = T(Toluene, P_5, h_5)$ $s_5 = s(Toluene, P_5, h_5)$	11	$T_{11} = T_6 + 10; P_{11} = P_{10}$ $h_{11} = h(Steam, T_{11}, P_{11})$ $s_{11} = s(Steam, T_{11}, P_{11})$ $m_w (h_9 - h_{11}) = h_1 - h_6$

h, *enthalpy*; P, *pressure*; s, *entropy*; T, *temperature*; x, *vapor quality*; $\eta_{T,P}$, *efficiency (turbine, pump)*; m_w, *water mass flow rate*

TABLE 7.15 The Results Calculated for the System in Figure 7.51

State	s, kJ/kgK	T, °C	P, bar	h, kJ/kg
1	1.068	200	7.475	485.7
2	1.156	104.6	0.07907	356.7
3	0.9104	40	0.07907	271.8
4	−0.3799	40	0.07907	−132.3
5	−0.3796	40.23	7.475	−131.4
6	−0.1271	86.65	7.475	−46.42
7	0.4545	200	7.475	195.5
9	6.357	210	19.06	2798
10	2.425	210	19.06	897.7
11	1.267	96.65	19.06	406.3

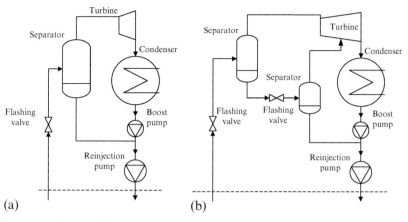

FIGURE 7.52 Types of steam-flash power plants (a) single flash, (b) double flash.

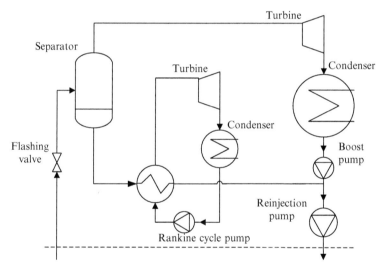

FIGURE 7.53 Flash-binary power plant.

7.5 BIOMASS ENERGY SYSTEMS

Any kind of fossilized living species is a form of biomass. It is one of the oldest energy sources on the earth and may become one of the most significant large-scale energy sources in the future. Biomass originates from the photosynthesis aspect of solar energy distribution and includes all plant life (terrestrial and marine), all subsequent species in the food chain, and eventually all organic waste. Biomass resources come in a large variety of wood forms, crop forms, and waste forms. The basic characteristic of biomass is its chemical composition, in such forms as sugar, starch, cellulose, hemicellulose, lignin, resins, and tannins.

Bioenergy (or biomass energy) may be defined as the energy extracted from biomass for conversion into a useful form for commercial heat, electricity, and transportation fuel applications. The simplest route to biomass energy conversion is by combustion to generate heat. The high-temperature heat can be further converted into work through heat engines. Other biomass conversion routes are illustrated in Figure 7.54. This combustion is a thermochemical process where it is called as gasification. Gasification converts the biomass into a gaseous fuel. Liquefaction produces a liquid fuel. Another route of biomass energy conversion is biochemical. In this case, the fermentation process (anaerobic or aerobic digestion) can lead to biogas, alcohol, and hydrogen generation. Also, the photosynthesis process conducted by some phototrophic organisms can eventually produce hydrogen.

Because energy crop fuel contains almost no sulfur and has significantly less nitrogen than fossil fuels (reducing pollutants causing acid rain [SO_2] and smog [NO_x]), its use will improve our air quality. An additional environmental benefit is in water quality, as energy crop fuel contains less (to none) mercury than coal. Also, energy crop farms using environmentally proactive designs will create water quality filtration zones, taking up and sequestering pollutants (e.g., phosphorus from soils that leach into water bodies).

Biomass generates about the same amount of CO_2 as do fossil fuels (when burned), but from a chemical balance point of view, every time a new plant grows CO_2 is actually removed from the atmosphere. The net emission of CO_2 will be zero as long as plants continue to be replenished for biomass energy purposes. If the biomass is converted through gasification or pyrolysis, the net balance can even result in removal of CO_2. Energy crops such as fast-growing trees and grasses are called biomass feedstocks. The use of biomass feedstocks can help increase profits for the agricultural industry.

The general model of biomass energy conversion into work consists of an adiabatic combustor and a reversible heat engine. Figure 7.55 shows a temperature versus air–fuel ratio (AFR) diagram for thermochemical conversion of biomass. Depending on temperature and the AFR the process used can be combustion, gasification, or pyrolysis. The biomass combusts in an adiabatic combustor which operates at the highest possible temperature (T_{ad}). The heat generated in this process is the net calorific value of the biomass (NCV).

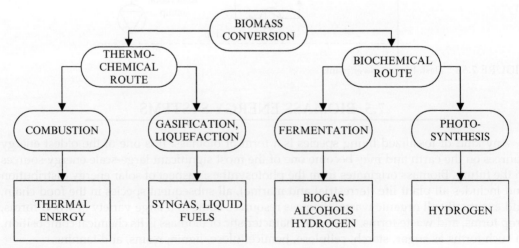

FIGURE 7.54   Routes to biomass energy conversion.

FIGURE 7.55 The temperature vs. air–fuel ratio (AFR) diagram for thermochemical conversion of biomass (data are approximated).

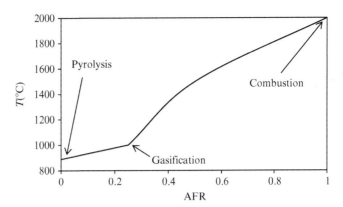

The work generated by this system is thus $\dot{W} = NCV(1 - T_0/T_{ad})$, and the efficiency of the biomass conversion process is given by the adiabatic flame temperature only, $\eta = 1 - T_0/T_{ad}$. Therefore, determining the thermodynamic limit of biomass energy conversion is equivalent to determining the adiabatic flame temperature for the biomass. We will analyze the biomass composition and the way in which this affects the net calorific value of biomass and the corresponding adiabatic flame temperature.

The biomass contains various biochemicals (amino acids, fiber, cellulose, sugars, glucose, and many others). Some living microorganisms and enzymes may be found in biomass. The most abundant chemical elements in biomass are carbon, hydrogen, oxygen, nitrogen, and sulfur. Other elements are also present, including metal atoms. When biomass is combusted, the metal atoms and other elements form ash.

The general chemical model of biomass is written as $C_{X_C}H_{X_H}O_{X_O}N_{X_N}S_{X_S}ash_{X_{ash}}(H_2O)_{X_w}$ where X_i is the number of constituents of species "i." If the moisture (water) is eliminated by a drying process, then the chemical representation of the dry biomass becomes $C_{X_C}H_{X_H}O_{X_O}N_{X_N}S_{X_S}ash_{X_{ash}}$. The molecular mass of dry biomass can be calculated with

$$M = 12 \times X_C + X_H + 16 \times X_O + 14 \times X_N + 32 \times X_S + M_{ash}X_{ash}$$

On a dry basis, the mass concentration of each major chemical element constituent is

$$w_C = 12 \times \frac{X_C}{M}; w_H = \frac{X_H}{M}; w_O = 16 \times \frac{X_O}{M}; w_N = 14 \times \frac{X_N}{M}; w_S = 32 \times \frac{X_S}{M}; w_{ash} = M_{ash} \times \frac{X_{ash}}{M}, \quad \text{where}$$

$w_{ash} = 0.5 - 12\%$ dry basis. The moisture content of biomass typically has a moisture concentration in the range $w_w = 0 - 50\%$ and can be expressed on wet basis as

$$M_{wet} = M + X_w \times 18$$

Thus,

$$w_w = \frac{18X_w}{M + X_w \times 18}$$

If the mass concentration of the moisture is known, then the molar concentration can be determined with

$$X_w = \frac{w_w M}{18(1 - w_w)}$$

The general chemical equation of biomass combustion with enriched oxygen and flue gas recirculation is $C_{X_C}H_{X_H}O_{X_O}N_{X_N}S_{X_S}$ ash$_{X_{ash}}(H_2O)_{X_w} + \lambda X_{O_2}[\zeta(O_2 + 3.76N_2) + (\zeta - 1)O_2] \rightarrow$ Prod $+ X_{ash}$ ash, where $\lambda \geq 1$ is the excess oxygen, 3.76 is the concentration of nitrogen in fresh air, $\zeta = 0 - 1$ is the fraction of fresh air, and "Prod" represents the reaction products in the form of flue gas. Assuming complete combustion one has: Prod $= X_C CO_2 + X_{H_2O}H_2O + X_S SO_2 + X_{N_2}N_2 + (\lambda - 1)X_{O_2}O_2$, where the stoichiometric coefficient for nitrogen is $X_{N_2} = X_N/2 + 3.76\lambda\zeta X_{O_2}$.

The plot in Figure 7.56 illustrates the variation of adiabatic flame temperature and conversion efficiency of a typical biomass with the moisture content. The energy and exergy efficiencies are calculated based on useful work generated and the energy and exergy input.

The maximum conversion efficiencies are obtained with dry biomass (moisture content 0%) and excess oxidant (air) as low as stoichiometric 1 ($\lambda = 1$). In these conditions the adiabatic flame temperature as well as the maximum biomass energy conversion efficiencies reach values around $T_{ad} = 1280 - 1300°C$, $\eta \cong 80\%$, and $\psi \cong 72\%$. Table 9.13 gives the characterization of some biomass resources.

Biomass can be converted to a gaseous fuel by two routes, namely the thermochemical and biochemical conversion routes. Through the thermochemical conversion route bio-syngas is obtained, while by biochemical conversion one can generate either biogas or hydrogen.

Several alternatives are available for generation of electric power from biomass. The simplest way to do this is by direct combustion of biomass, firing a steam generator which drives a turbine. Biomass combustion facilities can be also coupled with ORCs which are characterized by good turbine efficiency at low installed capacities.

When biomass is converted to biogas, then electric power generators based on internal combustion engines can be used for power production. It is also possible to couple biogas

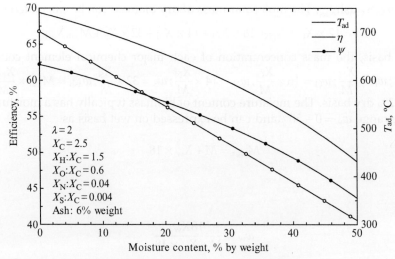

FIGURE 7.56 Variation of adiabatic flame temperature and energy and exergy efficiencies of biomass conversion with the moisture content.

production facilities with micro-power plants comprising gas turbines cascaded with a reciprocating internal combustion engine. For micro-power plants, alcohol and gasoline motors can be operated with methane without affecting their operational integrity. This adaptation is made by installing a cylinder of biogas in place of conventional fuel. For gas flow regulation, a reducer is placed close to the motor.

The combined heat and power generation (via biomass gasification techniques connected to gas-fired engines or gas turbines) can achieve significantly higher electrical efficiencies, between 22% and 37%, than those of biomass combustion technologies with steam generation and a steam turbine, 15–18%. If the gas produced is used in fuel cells for power generation, an even higher overall electrical efficiency can be attained, in the range of 25–50%, even in small-scale biomass gasification plants and under partial load operation.

Due to the improved electrical efficiency of energy conversion via gasification, the potential reduction in CO_2 is greater than with combustion. The formation of NO_x compounds can also be largely prevented, and the removal of pollutants is easier for various substances. The NO_x advantage, however, may be partly lost if the gas is subsequently used in gas-fired engines or gas turbines. Significantly lower emissions of NO_x, CO, and hydrocarbons can be expected when the gas produced is used in fuel cells rather than gas-fired engines or gas turbines.

Figure 7.57 shows a biomass gasifier where a gas production module converts biomass to a clean gas for power and steam generation with a combined Brayton cycle, steam expansion

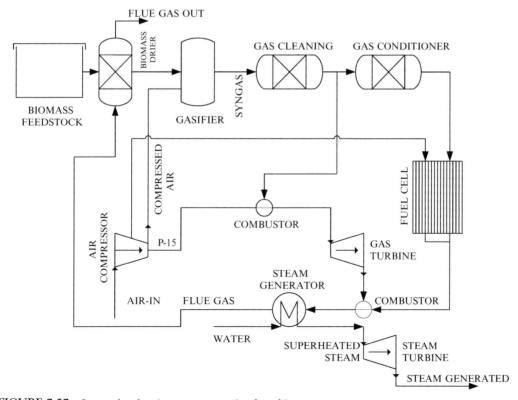

FIGURE 7.57   System for electric power generation from biomass.

turbine, and fuel cell. In this system, biomass is gasified first and then used in a gas turbine and a fuel cell system that operate in parallel. The un-combusted gases are directed toward a low-pressure combustor where additional combustion is applied at low pressure. The flue gases are used to generate high-pressure superheated steam that is expanded in a steam turbine. The resulting low-pressure steam can be used for some heating applications. Another part of the high-pressure steam is used for the gasification process.

7.6 OCEAN ENERGY SYSTEMS

The ability to directly extract energy from the world's oceans comes in four forms: tides, waves, ocean currents, and thermal energy. When tides come into the shore, they can be trapped in reservoirs behind dams. Then when the tide drops, the water behind the dam can be released just as in a regular hydroelectric power plant. Tidal energy shows good potential for electric energy generation. A classification of ocean energy conversion options for power generation is illustrated in Figure 7.58.

The kinetic energy resulting from the moon's (in addition to the sun's smaller) gravitational pull on the oceans during the earth's rotation, producing a diurnal tidal effect, creates the potential to utilize the kinetic energy of waves hitting continental shores for power generation. Wave energy can be collected with floating bodies that execute elliptic movement under the action of waves.

The heat cycle of tropical solar energy affects the oceans during the earth's rotation and generates kinetic energy that can be used directly to turn submerged turbine generators. The temperature gradient of the ocean with depth results in a temperature difference just large enough over reasonable depths to extract thermal energy at low efficiency. This is called ocean thermal energy conversion. Ammonia-water Rankine and Kalina cycles have been proposed for OTEC. Using depth seawater temperature as a sink at 4 °C, and the ambient temperature at 25–35 °C as a heat source, energy efficiency of about 5% can be obtained.

An OTEC system configuration is illustrated in Figure 7.59. The system is basically a Rankine cycle operating with ammonia as the working fluid. The system can be installed on a floating platform or on a ship. It uses the surface water as the source. Normally the water at ocean surface is at a higher temperature than in deep ocean. The surface water is circulated with pumps through a heat exchanger which acts as a boiler for ammonia. Water from deep ocean at 4 °C is pumped to the surface and used as heat sink in an ammonia condenser.

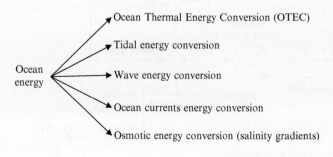

FIGURE 7.58 Classification of ocean energy conversion options for power generation.

FIGURE 7.59 Ocean thermal energy conversion system working with ammonia.

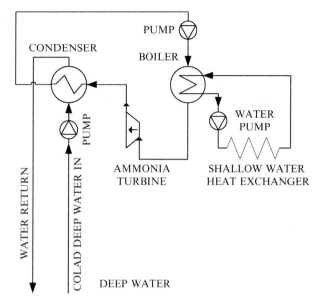

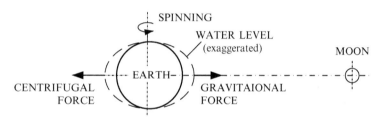

FIGURE 7.60 Formation of tides.

Apart from being an immensely large thermal energy storage system (which is an indirect form of solar energy storage), oceans are at the same time huge reservoirs of mechanical energy. This mechanical energy manifests in the form of tides, ocean currents, and ocean waves. The kinetic energy resulting from the moon's (in addition to the sun's smaller) gravitational pull on the oceans during the earth's rotation produces a diurnal tidal effect. The formation of tides is explained in Figure 7.60. Basically two bulges of water are formed at the equatorial belt due to the combined action of gravitational and centrifugal forces. Tidal energy shows good potential for electric energy generation. The associated energy can be converted into electricity by two methods:

- Generation of a difference in water level through impoundment. When tides come into the shore, they can be trapped in reservoirs behind dams. Then when the tide drops, the water behind the dam can be let out as in a regular hydroelectric power plant. As the tidal water level difference is rather small, Kaplan turbines are mostly used to generate power.
- Momentum transfer between the currents generated by tides and a conversion device like a water turbine. Ocean current harvesting systems are used to rotate propellers that are coupled to electric generators.

A tidal impoundment (barrage) system can have three methods of operation, depending on the phase at which it generates power: ebb generation, flood generation, and two-way generation. When ebb generation is applied the basin is filled during the flood tide. During the night additional water can be pumped into the basin, as a means of energy storage during off-peak hours. When the tide ebbs low enough, water is discharged over turbine systems which generate power. In flood generation, the dam gates are closed such that the water level increases on the ocean side until it reaches the maximum. Then water is allowed to flow through turbine systems and fill the basin while generating power. In two-way generation, electricity is generated in both the flood and ebb phases of the tide. In Table 7.16 some of the main tidal power generation systems that have been built and their principal characteristics are listed. The world largest tidal power generation site is in France at "La Rance," with an installed capacity of 240 MW, operating with 24 reversible turbines and a hydrostatic head of 5 m.

Ocean water currents are generated by tides, the earth spinning, the heat cycle of tropical solar energy,and superficial winds. The water current energy can be extracted through current turbines submerged in water. Basically, water current turbines are similar to wind turbines, with the difference that the density and the viscosity of water are 100 and 1000 times higher than that of air, respectively. Therefore, there are some differences between the operation of water current turbines and wind turbines. As for wind turbines, water current turbines are made with horizontal axis or vertical axis construction. The thrust generated by water current turbines is much higher than that specific to wind turbines. Therefore, their construction material must be more massive and resistant.

Surface winds, tides, and ocean currents contribute to the formation of ocean waves. The energy of waves can be collected with floating bodies that execute elliptic movement under the action of gravity and wave motion. The wave energy can be measured in terms of wave power per meter of wave front. The highest wave power on the globe appears to be in southern Argentina across the Strait of Magellan, where wave power density can reach 97 kW/m. Also, along the southwestern coast of South America, the wave power density is over 50 kW/m and often over 70 kW/m. On southwestern Australian coasts, the wave power reaches 78 kW/m. In the northern hemisphere, the highest wave intensity if found along the coast of western Ireland and the UK, with magnitudes around 70 kW/m. The southern coast of Alaska records wave power density up to 65–67 kW/m. The wave power density in other coastal regions varies from about 10–50 kW/m. This impressive amount of energy

TABLE 7.16   Some of the Major Tidal Impoundment Sites

Location	Head, m	Mean Power, MW	Production, GWh/year
Minas-Cobequid, North America	10.7	19,900	175,000
White Sea, Russia	5.65	14,400	126,000
Mont Saint-Michel, France	8.4	9700	85,100
San Jose, Argentina	5.9	5970	51,500
Shepody, North America	9.8	520	22,100
Severn, UK	9.8	1680	15,000

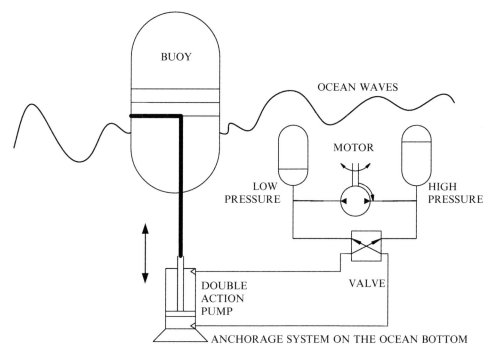

FIGURE 7.61 Principle of operation of buoy-type ocean wave energy conversion system.

can be converted to electricity with relatively simple mechanical systems that can be classified as one of two kinds, buoy type or turbine type.

The principle schematic of a buoy-type wave energy converter is suggested in Figure 7.61 and operates based on hydraulic-pneumatic systems. The buoy oscillates according to the wave movement at the ocean surface. It transmits the reciprocating movement to a double-effect hydraulic pump which is anchored rigidly on the ocean bottom. The pump generates a pressure difference between two pneumatic-hydraulic cylinders. A hydraulic motor generates shaft work by discharging the high-pressure liquid into a low-pressure reservoir. The shaft work turns an electric generator which produces electricity.

The energy of waves is correlated with the energy of surface winds. Two aspects are important in determining the interrelation between the wave height and the wind characteristics, namely the wind-water fetch (that is, the length over which the superficial wind contacts the water) and the duration of the wind. For example, if the wind blows constantly for 30 h at 30 km/h and is in contact with water over a length of 1200 km, the wave height can reach 20 m (see Da Rosa, 2009). The power density of the wave in deep ocean water can be approximated with

$$\dot{W} = 0.5 v \rho g h^2$$

while in shallow waters it is about half of this. In the equation, v is the eave speed, ρ is the water density, $g = 9.81 \, \text{m/s}^2$ is the gravity acceleration, and h is the height of the wave. Thus a 10-m wave propagating at 1 m/s carries a power density of 49 kW/m.

If the waves are high, an arrangement can be made in such a way that a difference of level is created constantly through impoundment. Thus the top of the wave carries waters over the impoundment, constantly filling a small basin at higher water level. The difference in water levels is turned into shaft work by a Kaplan turbine system.

Osmotic energy technology uses the energy available from the different salt concentrations in seawater and freshwater. Such resources are found in large river estuaries and fjords. The system uses a semi-permeable membrane that allows the salt concentrations to equalize, thus increasing pressure in the seawater compartment. The technology is still in early research and development stages (Achilli et al., 2009). Statkraft(2013) is one of the few industrial players in this sector, having set up the world's first prototype osmotic power plant in Norway. The key to further development lies in optimizing membrane characteristics. Today, the membranes generate only a few watts per square meter. The small number of players working on this technology and the need to improve membrane performance and reduce costs point to development prospects in the longer term. Research and development on osmotic power is also being carried out at the Tokyo Institute of Technology to develop new efficient membranes, and by RedStack in the Netherlands.

An osmotic power plant extracts power from salinity gradients by guiding freshwater and salt water into separate chambers, divided by a membrane. The salt molecules pull the freshwater through the membrane, creating a pressure on the salt-water side that can drive a turbine. Salinity gradients can also supply power through reversed electrodialysis (RED).

7.7 CONCLUDING REMARKS

In this chapter, we introduce renewable-energy-based power generation systems and discuss them for various options of power generation with numerous applications. We address the performance-related issues and explain how to assess the performance of such systems and applications through energy and exergy efficiencies. The available technologies for converting renewable energies into power are analyzed for each case. Numerous illustrative examples and cases studies are presented to highlight the importance of power system design, analysis, and assessment.

Study Problems

7.1 Describe the conversion paths of solar energy.

7.2 What is the thermodynamic limit of solar energy conversion with a blackbody receiver?

7.3 Define the fill factor.

7.4 Redo Example 7.1 for another set of data of your choice.

7.5 Rework Example 7.2 for an AM2 spectrum.

7.6 Define the concentration ratio.

7.7 Rework Example 7.4 for another set of data of your choice and perform a meaningful parametric study.

7.8 Define the cogeneration efficiency of PV/T systems.

7.9 Define field factor.

7.10 Calculate the optimum receiver temperature for concentrated solar power under reasonable assumptions.

7.11 Explain thermal and optical efficiency of solar concentrators and the difference between them.

7.12 Explain the wind chill effect on the energy and exergy efficiencies of wind turbines.

7.13 Describe the global utilization of geothermal energy.

7.14 Explain the thermodynamic limits of geothermal energy conversion.

7.15 Calculate a trilateral flash Rankine cycle with ammonia-water in EES or using ammonia enthalpy diagrams.

7.16 Calculate the adiabatic flame temperature for a biomass having X_C = 2, hydrogen per carbon of 2, oxygen per carbon of 0.1, nitrogen per carbon of 0.4, sulfur per carbon of 0.1, and 20% moisture by weight.

7.17 For the above case determine the maximum energy and exergy efficiencies of combustion.

Nomenclature

C	concentration ratio
C_p	power coefficient (−), specific heat (J/kgK)
COP	coefficient of performance
$\dot{Ex}$	energy rate, W
e	elementary electric charge, C
FF	filling factor
GCV	gross calorific value, MJ/kg
h	specific enthalpy, J/kg or J/mol
I	irradiation (W/m^2), current intensity (A)
k_B	Boltzmann constant
LEC	levelized electricity cost
M	molecular mass, kg/kmol
NCV	net calorific value, MJ/kg
Q	heat, J
R	radius (m), resistance (Ω)
T	temperature, K
U	heat transfer coefficient, W/m^{2}K
V	voltage (V), velocity (m/s)
W	work, J
w	moisture content

Greek letters

Ω	solid angle
α	absorptivity
γ	intercept factor
ε	emissivity
ζ	shading factor
η	energy efficiency
θ	dimensionless temperature
μ	chemical potential
ρ	reflectivity (−); density (kg/m^3)
σ	Stefan–Boltzmann constant, W/m^{2}K^4

τ transmissivity
φ subunitary factor
χ factor
ψ exergy efficiency
ω angle

Subscripts

0	reference state
cog	cogeneration
coll	collector
db	direct beam
dc	direct and circumsolar
diff	diffuse
diss	dissipation
e	emitted, or electric
ext	extraterrestrial
gt	global on tilt surface
max	maximum
oc	open circuit
opt	optical
PV	photovoltaic
r	receiver
rev	reversible
S	sun
SC	solar constant
sc	short circuit
T0	tilted surface
th	thermal

Superscripts

0	rate (per unit of time)
()″	per unit of surface

References

Achilli, A., Cath, T.Y., Childress, A.E., 2009. Power generation with pressure retarded osmosis: an experimental and theoretical investigation. J. Membr. Sci. 343, 42–52.

ASTM, 2008. G173-03 Reference Spectra Derived from SMARTS v. 2.9.2. Accessed from http://rredc.nrel.gov/solar/spectra/am1.5/ASTMG173/ASTMG173.html.

Burton, T., Sharpe, D., Jenkins, N., Bossanyi, E., 2001. Wind Energy Handbook. John Wiley and Sons, Ltd., West Sussex

Chen, Z.S., Mo, S.P., Hu, P., 2008. Recent progress in thermodynamics of radiation—exergy of radiation, effective temperature of photon and entropy constant of photon. Sci. China Ser. E: Technol. Sci. 51 (8), 1096–1109.

Da Rosa, A.V., 2009. Fundamentals of Renewable Energy Processes, second ed. Elsevier, Burlington, MA.

Duffie, J.A., Beckman, W.A., 2013. Solar Engineering of Thermal Processes, fourth ed. John Wiley and Sons, Hoboken, NJ.

Green, M.A., Emery, K., Hishikawa, Y., Warta, W., Dunlop, E.D., 2012. Solar cell efficiency tables (version 39). Prog. Photovolt. Res. Appl. 20 (1), 12–20. http://dx.doi.org/10.1002/pip.2163.

Gueymard, C., 1995. SMARTS2, simple model of the atmospheric radiative transfer of sunshine: algorithms and performance assessment. Report FSEC-PF-270-95, Florida Solar Energy Center, Cocoa, FL.

Gueymard, C.A., Myers, D., Emery, K., 2002. Proposed reference irradiance spectra for solar energy systems testing. Sol. Energy 73, 443–467.

Joshi, A.S., Dincer, I., Reddy, B.V., 2009a. Performance analysis of photovoltaic systems: a review. Renew. Sust. Energ. Rev. 13, 1884–1897.

Joshi, A.S., Tiwari, A., Tiwari, G.N., Dincer, I., Reddy, B.V., 2009b. Performance evaluation of a hybrid photovoltaic thermal (PV/T) (glass-to-glass) system. Int. J. Therm. Sci. 48, 154–164.

Kistler, B.L., 1986. A user's manual for delsols: a computer code for calculating the optical performance and optimal system design for solar thermal central receiver plants. Sandia National Laboratory, Livermore, Report SAND86-8018.

Klein, S.A., 2013. Engineering Equation Solver. F-Chart Software, Middleton, WI.

Kolb, G.J, Jones, S.A., Donnley, M.W., Gorman, D., Thomas, R., Davenport, R., Lumia, R., 2007. Heliostat cost reduction study. Sandia National Laboratory, Report 2393.

Landsberg, P.T., 2003. Ideal efficiencies. In: Markvart, T., Cartañer, L. (Eds.), Practical Handbook of Photovoltaics: Fundamentals and Applications. Elsevier, Oxford.

Lovegrove, K., 2004. Developing ammonia based thermochemical energy storage for dish power plants. Sol. Energy 76, 331–337.

Nann, S., Emery, K., 1992. Spectral effects on PV-device rating. Sol. Energy Mater. Sol. Cells 27, 189–216.

Naterer, G.F., Dincer, I., Zamfirescu, C., 2013. Hydrogen Production from Nuclear Energy. Springer, New York.

Palik, E.D. (Ed.), 1997. Handbook of Optical Constants in Solids. Elsevier Inc., Amsterdam

Rabl, A., 1976. Comparison of solar concentrators. Sol. Energy 18, 93–111.

Romero, M., Buck, R., Pacheco, J.E., 2002. An update on solar central receiver systems, projects, and technologies. J. Sol. Energy Eng. 124, 98–108.

Segal, A., Epstein, M., Yogev, A., 2004. Hybrid concentrated photovoltaic and thermal power conversion at different spectral bands. Sol. Energy 76, 591–601.

Singh, H.N., Tiwari, G.N., 2005. Evaluation of cloudiness/haziness factor for composite climate. Energy. 20, 1589–1601.

Statkraft, 2013. Osmotic power. Accessed from http://www.statkraft.com/energy-sources/osmotic-power/.

Vogel, W., Kalb, H., 2010. Large-Scale Solar Thermal Power. Technologies, Costs and Development. Wiley-VCH, Weinheim.

Wagar, W.R., Zamfirescu, C., Dincer, I., 2011. Thermodynamic analysis of solar energy use for reforming fuels to hydrogen. Int. J. Hydrog. Energy. 36, 7002–7011.

Yilanci, A., Dincer, I., Ozturk, H.K., 2008. Performance analysis of a PEM fuel cell unit in a solar-hydrogen system. Int. J. Hydrogen Energ. 33, 7538–7552.

Integrated Power Generating Systems

8.1 INTRODUCTION

Development of power generation systems which are more sustainable and effective with respect to economics and environment requires new design approaches. Technology has evolved very fast in the last decades, with discovery of new methods of power conversion: fuel cells, transcritical carbon dioxide cycles, concentrated solar radiation with spectral splitting, concentrated photovoltaics, power generation from ocean's salinity gradients, etc. Nevertheless, new concepts and developments regarding system integration for improved efficiency, reduced cost and environmental impact, and increased effectiveness are required. Cogeneration is also very important and has to be considered for power generation systems because it provides a better use of the energy resource, and this leads to better system economics and less environmental impact.

More benefits can be obtained from system integration, which entails combining several subsystems into a larger system in which the subsystems work together for better effectiveness or efficiency. Hence, the purpose of integration is to obtain better energy utilization factors, and better effectiveness with respect to separate systems. The importance of system integration for sustainable energy applications is stressed in Dincer and Zamfirescu (2011).

A non-exhaustive classification attempt of integration options for power generation systems is shown in Figure 8.1. Here, four main options of system integration are included: combining, cascading, multistaging, and hybridizing. Here are some examples of system integration in power generation, namely:

- Cascading gas turbine (GT) cycles with bottoming Rankine cycles for improved efficiency of power generation.
- Hybridizing solid oxide fuel cell (SOFC) with GT cycles for better fuel utilization and enhanced power production efficiency.
- Combining coal gasification with GT generators, again for better power generation efficiency.
- Multistaging of flash-expansion process for a Rankine cycle devoted to low-grade heat recovery applications (e.g., geothermal).

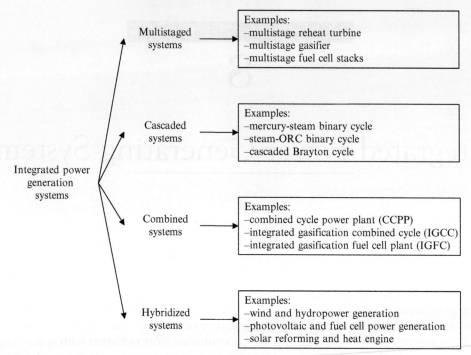

FIGURE 8.1   Main options of integration for advanced power generation systems.

One particular integration option is of a heat recovery subsystem into a power cycle to cogenerate heat as an additional output. An example can be given as follows. Let us consider a high-temperature thermal energy source. A power cycle operating between the heat source and environmental temperature can convert the thermal energy from the heat source into power, in accordance to thermodynamics laws. The Carnot factor limits the power conversion efficiency. If the heat rejected by the power cycle is recovered and used (e.g., for space heating or water heating or an industrial process), then due to its ability to generate power and heating the energy utilization factor becomes (in theory) 100%. Many fossil fuel and some nuclear power stations around the world are cascaded with heat recovery systems to deliver district heating.

There is potentially more benefit if the system is engineered to produce power from multiple kinds of energy sources. In this case completely different technologies must be combined to extract useful power in a synergistic manner. We discuss in this case about a *hybridized* system. Some examples of hybridized systems are: integrated wind and photovoltaic power generators; integrated systems that generate power and heat from geothermal and solar sources or from biomass combustion and concentrated solar radiation; concentrated solar collectors integrated with advanced Rankine power generators and with natural gas steam generators for backup power. Also, hybridized systems for power and hydrogen generation allow nuclear power stations have higher capacity factors and contribute better to grid load leveling. Some integrated systems for power generation that link nuclear reactors to hydrogen production systems and fuel cells are discussed elsewhere (Naterer et al., 2013).

In this chapter, integrated power generation systems are discussed. The problem is complex, as the power generation issues are specific to the types of energy sources and systems considered. Therefore, the advanced systems presented herein are based on case studies. However, regardless the specificity of each case, the methods for integration are essentially those shown in Figure 8.1. As a logical consequence, the chapter follows the same structure as in the figure and concludes with a comprehensive section of case studies.

8.2 MULTISTAGED SYSTEMS

In past decades extensive efforts have been devoted to enhancing the efficiency of a power cycle, closer to the limit imposed by the second law of thermodynamics. Of course, power cycles integrate many functional components, of which the most important is the prime mover. It is a demonstrated fact that for many power cycles the integration of multistage processes leads to an increase in power generation and reduction of irreversibilities.

Multistaging can be applied in many ways. One very well-known option is the use of multiple expansion-reheat stages (in general up to two or three; see Chapter 5). Figure 8.2 illustrates this case. As known, an optimum pressure ratio can be found for the expansion stage. If the number of stages is n, then the optimum pressure ratio is given by $\sqrt[n]{P_1 P_2}$, where $P_{1,2}$ is the low or high pressure, respectively.

Furthermore, increase of the expansion stages to more than three does not bring much benefit because the enhancement of generated power is too small to justify the additional investment. Analogous to multistage expansion there is multistage compression, which is a method of reducing the back work ratio in Brayton cycles. Multistage compression implies the integration of intercoolers within the cycle. Again, for economic reasons, the number of compression stages should be limited to two or three. Note that increase in the number of stages brings additional irreversibility within the power cycle. However, the overall irreversibility associated with a single-stage process is in most cases higher than for multistage processes.

Other opportunities for multistaging within Rankine cycles are with the processes of preheating and steam generation. A preheating process in stages is discussed in Chapter 5; it uses multiple feedwater heaters (open or closed), multiple pumps, and multiple points of steam extraction from the turbine. Figure 8.3 shows a multistage preheating loop that integrates three closed feedwater heaters with one open feedwater heater. Multistaging within a Rankine cycle can lead to energy efficiency increase of a few percentages, whereas the increase in exergy efficiency (or the reduction of irreversibilities) is much more pronounced.

Multistaging is also applied to steam (or vapor) generators. As known, due to the boiling region the LMTD for steam generators is high. If the steam generation is done in a single stage, this can lead to large irreversibility due to heat transfer at a large temperature difference. Figure 8.4 shows an example of the temperature profile along the heat transfer surface of a single-stage fired tube vapor generator. On the left side the case of steam generation is represented. The heating agent is combustion gas. The example is calculated for combustion gas with a composition of 65% N_2, 5% O_2, 15% H_2O, and 15% CO_2 and an inlet temperature of 288 °C. The steam pressure is 35 bar, with a saturation pressure of 242 °C; therefore, it is not possible to have large superheating in these types of vapor generators. A pinch point

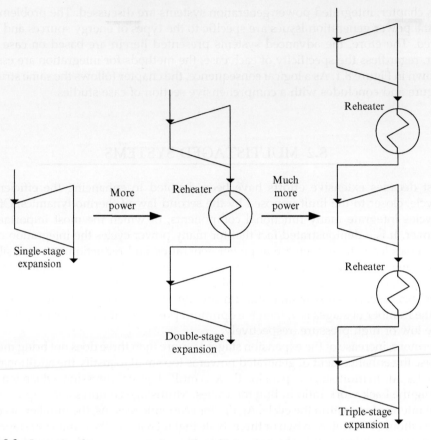

FIGURE 8.2   Increasing the power generation with multistage expansion.

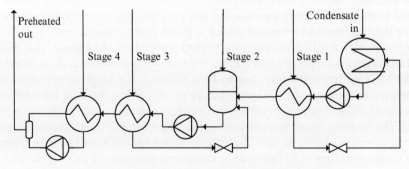

FIGURE 8.3   Multistage preheating commonly practiced in steam Rankine cycles.

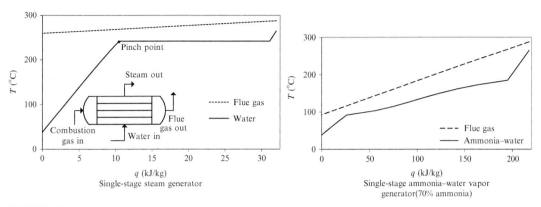

FIGURE 8.4 Temperature profile mismatch in two types of single-stage vapor generators.

temperature difference of 25 °C is assumed and the feedwater is 38 °C. In these conditions, the calculated temperature at fluid gas exit is 260 °C and for 1 kg/s of flue gas the transferred thermal power is 32 kW.

If instead of steam an ammonia–water vapor generator is considered with the same conditions for flue gas at the inlet, the same pinch point temperature difference, and the same operating pressure of 35 bar for the working fluid, then better performance is obtained, with reduced irreversibility. In this case, the flue gas temperature at the outlet is 93 °C and the transferred thermal power for 1 kg/s flue gas is 217 kW.

Although the irreversibilities due to heat transfer at the vapor generator for ammonia–water combinations are smaller than those for the pure stream, there is still significant exergy destruction. Better configurations are possible if multistaging is applied. In the case of pure steam, a multistage system will include preheating stages, calandria drum boiling stages, and superheating stages arranged optimally. For ammonia–water (or another zeotropic mixture), more benefit can be obtained with multistage distillation to reduce ammonia concentration prior to the turbine.

The temperature vs. heat transfer surface diagram of a multistage steam generator is shown in Figure 8.5. For the represented case, superheated steam is generated at two pressures, high and intermediate. The vapor generator comprises six distinct stages, and for each stage specific heat exchanger configurations are used. When the temperature difference across the heat exchanger becomes low, an extended heat surface may be required. This can be the situation in the vicinity of pinch point. If boiling occurs the preferred arrangement is calandria because it maintains lower vapor quality in the tubes and, therefore, a better heat transfer coefficient, corresponding to a nucleate boiling scenario. Of course, for each stage the optimized geometry of the heat exchangers that maximizes the heat transfer rate and minimizes the irreversibility can be found. Of special interest for steam generators is the optimal arrangement of the calandria boiler. In Kim et al. (2011), it is shown that the optimal spacing of the calandria tubes depends mainly on the horizontal flow length.

Gasification for power generation is another area where multistaging is applied with success because better syngas yield is obtained, with reduced tar. For example, in a two-stage gasification process the pyrolysis and gasification zones are clearly separated and the

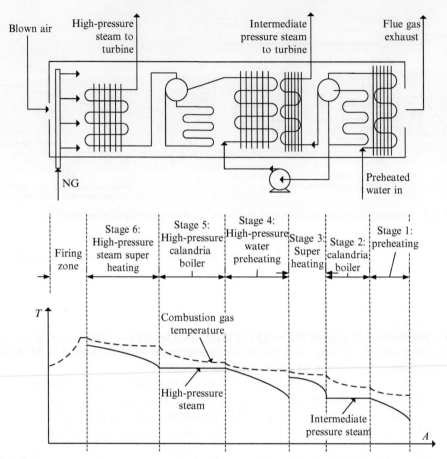

FIGURE 8.5 Principle of a multistage (dual-pressure) steam generator and its temperature vs. heat transfer surface area diagram.

possibility of supplying secondary air between the zones exists. Hence it results in higher temperature in the gasification zone and more favorable kinetics for tar decomposition and syngas generation. Figure 8.6 illustrates a power generation process applicable to heterogeneous solid fuel (coal, biomass, or a coal-biomass combination). The system—which is described in Henriksen et al. (2006)—uses a two-stage gasification process which is intrinsically integrated with a heat engine unit. The exhaust gases of the engine are used to provide heating to the first stage of gasification, the pyrolysis.

Multistaging in fuel cells is one strategy used to obtain better fuel utilization ratios. This is similar in purpose to staging turbines in Brayton and Rankine cycles with interstate reheating. It is well known that in fuel cells the fuel must be supplied in excess, otherwise the cell efficiency will be low. However, too much fuel excess leads to a low fuel utilization ratio (U_f), and hence, it is a source of loss. In some cases, the necessity of having a relatively modest U_f for better fuel cell efficiency is compensated for by cascading the stack with a GT where the leftover fuel is further combusted. Thus, an alternative solution to this is the staging of fuel cell stacks.

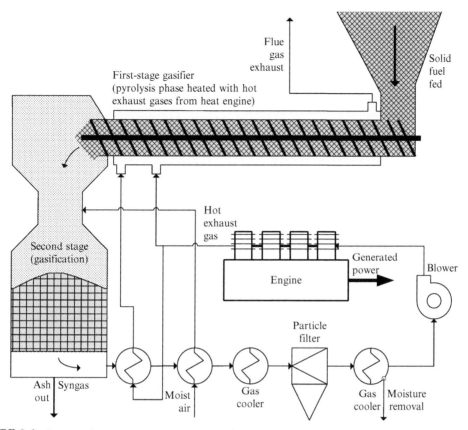

FIGURE 8.6 Integrated power generation system with two-stage gasifier. *Modified from Henriksen et al. (2006).*

Figure 8.7 shows two possibilities of staging of fuel cell stacks. In the two-stage configuration two stacks are used instead of one. The system can be operated such that it gives the same power output as the unstaged system. However, with the staged system there is the possibility of adjusting the load independently for each of the stages. In a first option of staging, both the anodes and cathodes are connected in series. This means that the second stage is supplied with a depleted fuel and depleted oxidant from the first stage. In a fuel cell, the fuel utilization ratio (the utilized fuel divided by the supplied fuel) is proportional to the current density: $J = U_f J_{in}$, where J_{in} is the current density when all fuel is consumed. Hence, the maximum power is $W_{max} = (EU_f)_{opt} J_{in}$ whereas the supplied energy rate density is $E_{th} J_{in}$, with $\Delta H = z F E_{th}$ being the reaction enthalpy, z is the number of charges per reaction, and F is the Faraday constant. It results that the efficiency of a fuel cell stack is

$$\eta = \frac{(EU_f)_{opt}}{E_{th}} \tag{8.1}$$

where the voltage over the load is denoted with E. The above relation really shows that the efficiency is higher when a higher fuel utilization ratio can be obtained at the optimum operation point. Using the fuel utilization value, the voltage vs. current density of a fuel cell

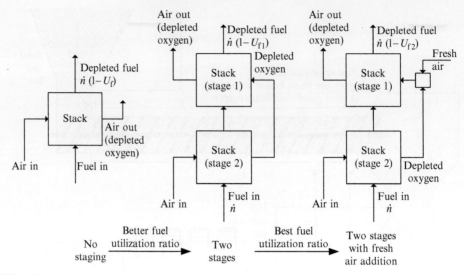

FIGURE 8.7   Staging of fuel cell stacks for better fuel utilization.

can be transposed into the form of voltage vs. fuel utilization ratio; these two diagrams have the same shape. Moreover, using a linearized form of the Nernst equation it is possible to predict the benefit of staging.

Following Standaert et al. (1998), we denote with $\mathcal{R}_{fc}(\Omega/m^2)$ any apparent ohmic resistance of the fuel cell (owing to compounded effect of activation, ohmic, and concentration overpotentials) and with E_{oc} the open circuit voltage given by the Nernst equation. Hence, the voltage over the load for an infinitesimal fuel cell area dA is $E\,dA = E_{oc}(U_f)\,dA - \mathcal{R}_{fc}J_{in}\,dU_f$. For most of operating range the Nernst potential is linear with the fuel utilization ratio (the only nonlinearity occurs close to the open- and short-circuit conditions). The linearization of the Nernst potential can be determined according to $E_{oc}(U_f) = E_{oc}(0) - LF\,U_f$, where LF is the linearizing factor and is measured in volts. The value of LF can be determined on a case-by-case basis. Some examples of LF values are taken from Standaert et al. (1998):

- 180mV for a molten carbonate fuel cell (MCFC) at 650 °C with 70% air/30% CO_2 cathode feed and 56% H_2/28% H_2O/8% CO_2/8% CO anode feed
- 181 mV for an oxygen-ion-conducting solid oxide fuel cell (SOFC−) at 1000 °C with 79.2% H_2/8% H_2O/8% CO_2
- 38 mV for a proton-conducting solid oxide fuel cell (SOFC+) at 100 °C with 80% H_2/20% H_2O
- 10 mV for a proton exchange membrane fuel cell (PEMFC) at 80 °C with 42.4% H_2/10.6% CO_2/47% H_2O
- 14 mV for a phosphoric acid fuel cell (PAFC) at 200 °C with 80% H_2/20% H_2O.

From the above considerations, it results that $\mathcal{R}_{fc}J_{in}\,dU_f = [E_{oc}(0) - LF\,U_f - E]\,dA$. Further derivations are detailed in Standaert et al. (1998), which lead to the following equations for the load voltage per stage and the power enhancement factor (PEF) of the multistaged systems:

$$\begin{cases} E = E_{oc}(0) - LF\left[1 - \exp\left(-\dfrac{LF}{\mathcal{R}_{fc}\,J_{in}}\right)\right]^{-1} U_f \\[2mm] \mathrm{PEF} = \dfrac{\dot{W}_N''}{\dot{W}''} \simeq 1 + \dfrac{(LF\,U_f)^2}{12\,\mathcal{R}_{fc}\,\dot{W}''}\left(1 - \dfrac{1}{N^2}\right) \end{cases} \tag{8.2}$$

where $\dot{W}''$ is the power density [A/m²] of the single-stage stack operating with fuel utilization ratio U_f, N is the number of stages of the multistaged stack arrangement of similar stacks which have the same fuel utilization ratio as the single-stage stack.

From Equation (8.2), the improvement factor is influenced by the number of stages according to the following increment factors: 0.75 for two stacks, 0.889 for three stacks, 0.937 for four stacks, 0.96 for five stacks, etc. Given that the improvement factor depends on the square of the linearization factor for the Nernst potential—see Equation (8.2)—it results that the staging is favorable when the LF is high. From the above-listed values, the linearization factors of SOFC− and MCFC are ten times larger in magnitude than the linearization factors of SOFC+, PEMFC, PAFC, etc.; hence, as expected, staging is more beneficial with SOFC− and MCFC than with other fuel cell types. This observation is confirmed by experimentation and detailed flow-sheet calculations.

Based on a flow-sheet calculation procedure presented by Au et al. (2003) multistage fuel cell stacks have the anodes in series and the cathodes parallel. When both the anode and cathode are in series, then the efficiency increase is of 0.6% points (case A), whereas 0.9% points of increase (case B) are obtained when only the anodes are in series. The reference case (no staging) generated 306.19 kW power (single stack). For case A staging, 160.16 kW was generated in the stage one stack and 148.94 kW was generated in the stage two stack, totaling 309.10 kW. For case B, 160.78 kW was generated in the first stage whereas the second stage produced 149.6 kW, with an overall 310.38 kW. The overall fuel utilization for all cases was set to 70%.

Multiflash systems for geothermal power generation are another example of system integration by means of multistaging. Figure 8.8 shows comparatively the concepts of

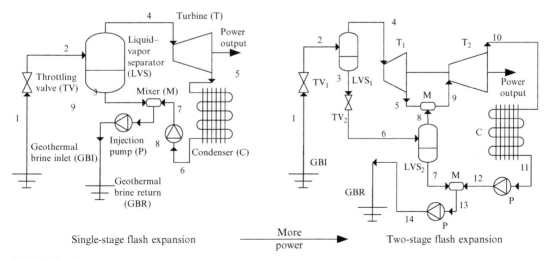

FIGURE 8.8 Single-stage vs. two-stage flash-expansion systems for geothermal power generation.

single-stage and double-stage expansion for power generation from geothermal brines. The hot brine in liquid phase is extracted at the surface from the geothermal well. In the single-stage power generation system, the brine is throttled and expands to an intermediate pressure in the two-phase domain. Hence, the brine lowers its temperature but generates hot saturated vapor. The vapor is expanded back to the pressure of the geothermal well in a turbine. Obviously, the expansion is in the two-phase domain and some tolerable amount of moisture is found at the turbine outlet.

For the two-stage system (see the figure), the expansion is done in two steps. The first step produces a vapor fraction (x_1) with respect to the inlet. The saturated liquid from the first-stage expansion is flashed one more time to a lower pressure and a fraction (x_2) of its mass flow rate is converted to vapor. The system has two turbines. The high-pressure turbine will expand a fraction of vapor (x_1) from the inlet mass flow rate of the geothermal brine. The second turbine expands more vapor, namely a fraction of $x_1 + (1 - x_1)x_2$ from the inlet brine flow rate. Consequently, the staging leads to additional power generation. In some cases the condensation heat can be used for district heating or other heating purposes. In other cases, the condenser is not necessary and humid vapor can be reinjected into the well.

The net power output from a flash-expansion system and the associated exergy destruction, regardless of the number of stages, results from the overall energy balance for the cycle. These can be written mathematically according to the following equations:

$$
\begin{cases}
EBE : \dot{m}_1 h_1 + \dot{W}_P = \sum_{stages} \dot{W}_T + \dot{Q}_C + \dot{m}_1 h_{inj} \\
ExBE : \dot{m}_1 (h_1 - T_0 s_1) + \dot{W}_P = \sum_{stages} \dot{W}_T + \dot{Ex}_{Q_C} + \dot{m}_1 \left(h_{inj} - T_0 s_{inj} \right) + \dot{Ex}_d
\end{cases}
\tag{8.3}
$$

where the equality of the brine inlet mass flow rate $\dot{m}_1$ and the reinjected brine mass flow rate are assumed. In addition, the brine does not change its chemical composition or physical constituency in the process. The net power output is $\dot{W}_{net} = \sum_{stages} \dot{W}_T - \dot{W}_P$. The energy input into the cycle is $\dot{E}_{in} = \dot{m}_1 (h_1 - h_{inj})$, whereas the maximum energy input is $\dot{E}_{in,max} = \dot{m}_1 (h_1 - h_0)$, where h_0 is the enthalpy of the brine in thermodynamic equilibrium with the reference environment. Following Ratlamwala and Dincer (2013), two types of energy efficiencies for multistage flash systems can be defined as follows:

$$
\begin{cases}
\eta_1 = \dfrac{\dot{W}_{net}}{\dot{m}_1 \left(h_1 - h_{inj} \right)}; \ h_{inj} > h_0 \\
\eta_2 = \dfrac{\dot{W}_{net}}{\dot{m}_1 \left(h_1 - h_0 \right)}; \ h_0 = h(T_0, P_0)
\end{cases}
\tag{8.4}
$$

One can observe the usefulness of the η_2 definition which allows for the comparison of the multistage flash system that uses the same geothermal source ($\dot{m}_1$ and h_1, fixed). It is easy to demonstrate that an optimum of flashing pressure can be determined which maximizes the work output in single-stage flash system. When multiple stages are used, then the intermediate pressures can be optimized for each stage. A simple case study is illustrated here, assuming that the turbine and the pump isentropic efficiency are 80% and that the brine inlet pressure is in the range of 2–3 MPa while the temperature is 520 K. These conditions are similar to those in the study of Ratlamwala and Dincer (2013). The optimization process is shown in Figure 8.9.

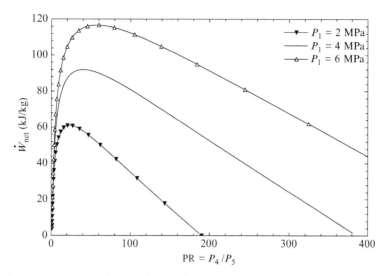

FIGURE 8.9 Optimization process for a single-stage flash steam expansion cycle for geothermal power generation (single-variable optimization: *PR*).

As seen in the figure, there is an optimum of pressure ratio across the turbine for any specified inlet brine pressure P_1 which maximizes the net generated work. The optimum pressure ratio displaces toward higher values when the brine inlet pressure increases. Figure 8.10 shows the *T–s* diagram of an optimized single-stage flash cycle with steam expansion designed according to the schematics from Figure 8.8. The cycle has an optimum pressure ratio over the turbine of 31 when the inlet brine temperature is assumed at 503 K.

Note that the energy efficiency of the optimized single-stage flash cycle illustrated in Figure 8.11 as calculated with Equation (8.4) is $\eta_2 = 8.8\%$, whereas the energy efficiency of

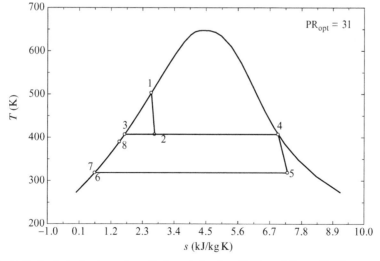

FIGURE 8.10 Optimized single-stage steam flash cycle for $P_1 = 3$ MPa and $T_{cont} = 318$ K.

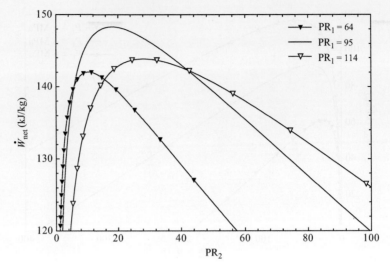

FIGURE 8.11 Optimization process of a two-stage steam flash cycle for geothermal power generation (two-variable optimization: PR_1, PR_2).

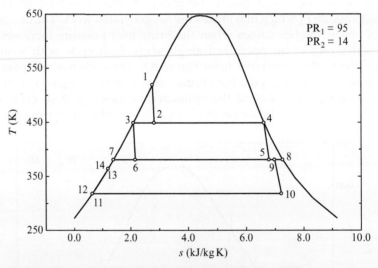

FIGURE 8.12 Optimized two-stage steam flash cycle for $P_1 = 4$ MPa, $T_1 = 520$ K.

the optimized two-stage cycle exemplified in Figure 8.12 is $\eta_2 = 12.2\%$. The exergy efficiencies are calculated respectively with

$$
\begin{cases}
\psi_1 = \dfrac{\dot{W}_{net}}{\dot{m}_1 \left(h_1 - h_{inj} - T_0 \left(s_1 - s_{inj}\right)\right)}; \; h_{inj} > h_0 \\[4mm]
\psi_2 = \dfrac{\dot{W}_{net}}{\dot{m}_1 \left(h_1 - h_0 - T_0(s_1 - s_0)\right)}; \; h_0 = h(T_0, P_0); \; s_0 = s(T_0, P_0)
\end{cases}
\tag{8.5}
$$

The exergy efficiencies for the exemplified cases are $\psi_2 = 35.9\%$ and 47.1%. Higher efficiency can be obtained if instead of two-stage flash a multistage arrangement is set. In Ratlamwala and Dincer (2013) multistage flash cycles up to quintuple systems are designed, and the incremental increase of efficiency is demonstrated. The schematic of a quintuple cycle is shown in Figure 8.13. Depending on the actual operating conditions and economic factors a trade-off problem can be used to determine the best cycle design and the number of stages. In addition, cogeneration can be applied if a heat recovery heat exchanger is installed at the last flash cycle. The condensate pump is not required in this case. Figure 8.14 shows a two-stage flash cycle with cogeneration based on the above-explained principle. Alternatives to flash Rankine cycles can be constructed with organic fluids which use flash expansion in a closed loop.

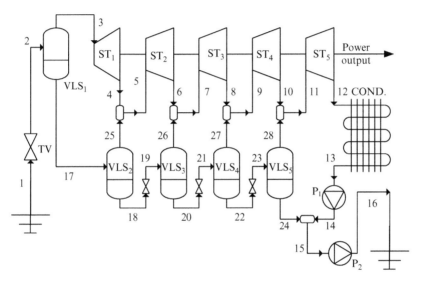

FIGURE 8.13   Quintuple flash cycle for geothermal power generation.

FIGURE 8.14   Cogeneration system for geothermal power and heat generation with a two-stage steam flash cycle.

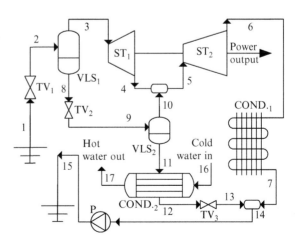

EXAMPLE 8.1

Here, the calculation of the power enhancement factor of a two-stage SOFC — system is exemplified. Assume the following data: the linearization factor for the Nernst potential is 400 mW, and the equilibrium potential for the linearized Nernst potential is $E_{oc}(0) = 0.8$ V. The current density at the inlet section is 2845 A/m^2, and the apparent resistance is 10^{-4} Ωm^2.

- The power generated for the single-stage system is given by $\dot{W}'' = E U_f J_{in}$. Using Equation (8.2) with the values for the known parameters,

$$\dot{W}'' = 2845 \left\{ 0.8 - 0.4 \left[1 - \exp\left(-\frac{0.4}{0.0001 \times 2845} \right) \right]^{-1} U_f \right\} U_f$$

- The above expression is a second-degree polynomial and has a maximum value of 795.3 W/m^2 at $U_{f,opt} = 0.8$.
- If N-staged fuel cells are used, the expected power enhancement factor is given by Equation (8.2); replacing the known parameters (including the utilization factor, which is the same as for the single-stage system), this equation becomes

$$PEF = 1 + \frac{(0.4 \times 0.8)^2}{12 \times 0.0001 \times 795.3} \left(1 - \frac{1}{N^2} \right)$$

- For the two-staged system, PEF = 1.08, which means that power generation increases by 8% with respect to the single-stage system, for the same fuel utilization factor.

8.3 CASCADED SYSTEMS

A common system integration scheme for power generation system is cascading. As discussed in the introduction, cascading refers to situation when one power generation subsystem receives (or recovers) energy from the other one to produce additional power. A simple representation of cascading is given in Figure 8.15 with two power generation cycles, considered in general of the same kind (e.g., two cascaded Rankine cycles).

The heat source for the bottom subsystem is a heat sink for the topping subsystem. Both subsystems generate power. Of course, one can cascade two or more subsystems. Binary cycles are common examples of cascaded systems in which the condenser of the topping cycle is the boiler of the bottoming cycle.

If the heat rejected by the topping cycle is recovered and used in a bottoming power cycle that operates between an intermediate temperature level and the ambient heat sink then more power can be generated with respect to one single cycle. Cascading necessitates adapting the working fluid to the external conditions of the power cycle. This aspect can be understood from a binary cycle and is well explained in general thermodynamics textbooks, such as Cengel and Boles (2010). Better match of the external conditions at source and sink can be obtained when two Rankine cycles are cascaded. Commonly the bottoming cycle runs with steam; for the topping cycle multiple options exist:

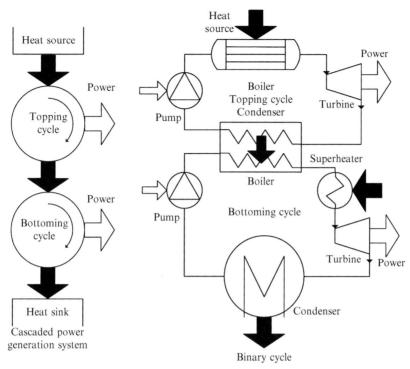

FIGURE 8.15 Cascaded power generation system and a binary cycle configuration.

- Mercury with critical temperature of 1750 K, critical pressure 172 MPa, normal boiling point 630 K.
- Sodium with critical temperature of 2573 K, critical pressure 35 MPa, normal boiling point 1156 K.
- Potassium with critical temperature of 2223 K, critical pressure 16 MPa, normal boiling point 1032 K.
- Cesium with critical temperature of 2043 K, critical pressure 13 MPa, normal boiling point 944 K.

The binary power plant can reach very high efficiency because higher temperatures are possible at the topping cycle when the vapor pressure in the boiler is not high. The most attractive of the binary cycles appears to be the steam-mercury power plant which has been proposed in the past for both stationary power generation and astronautic applications. The equation of state of mercury is given in Sugawara et al. (1962), where a set of mathematical relations are given to calculate the enthalpy, entropy, specific heat, and vapor pressure.

Table 8.1 gives the mercury equations necessary for cascaded cycle calculations. These equations give accurate predictions for temperatures up to 1073 K, which is a reasonable level for topping cycle. Figure 8.16 gives a qualitative representation of a cascaded cycle of binary type. One can observe the excellent match of temperature between the condensing and boiling fluids.

TABLE 8.1   Equations of State for Mercury (Enthalpy, Entropy, Specific Volume, and Vapor Pressure)

Parameter	Equation
Vapor pressure	$P\,[\text{kPa}] = 98.1 \times 10^{A(T_r)}; T_r = 0.01\,T\,[\text{K}]; A(T) = a - bT_r + cT_r^2 - dT_r^{-1} + eT_r^{-2} - fT_r^{-3}$ $a = 6.396;\ b = 0.12075;\ c = 3.915\text{E}{-3};\ d = 39.921;\ e = 29.055;\ f = 42.46$
Saturated liquid enthalpy	$h'[\text{kJ}\,\text{kg}^{-1}] = aT - bT^2 + cT^3 - d,\ \text{with}\ T\ \text{in}\ ^\circ\text{C}$ $a = 3.6187\text{E}{-3};\ b = 6.812\text{E}{-6};\ c = 4.033\text{E}{-9};\ d = 9.458$
Saturated liquid entropy	$s'[\text{kJ}\,\text{kg}^{-1}\text{K}^{-1}] = a\log_{10}T - bT + cT^2 - d,\ \text{with}\ T\ \text{in}\ ^\circ\text{C}$ $a = 8.3327\text{E}{-2};\ b = 1.3624\text{E}{-5};\ c = 6.050\text{E}{-9};\ d = 0.199739$
Vapor enthalpy	$h[\text{kJ}\,\text{kg}^{-1}] = aT_r - bp(cT_r^{-1} - dT_r^{-2} - e) + f\ \text{with}\ p = 0.010194\,P[\text{kPa}], T_r = 0.01\,T[\text{K}]$ $a = 10.368;\ b = 0.0981;\ c = 3.568;\ d = 7.53;\ e = 0.1071;\ f = 279.13$
Vapor entropy	$s[\text{kJ}\,\text{kg}^{-1}\text{K}^{-1}] = a\log_{10}T - b\log_{10}p - cp(dT_r^{-2} - eT_r^{-3}) - f,\ \text{with}\ p = 0.010194\,P[\text{kPa}],$ $T_r = 0.01\,T[\text{K}], a = 0.23874;\ b = 0.0955;\ c = 0.00981;\ d = 1.793;\ e = 5.02;\ f = 0.0831$
Vapor volume	$v[\text{m}^3\,\text{kg}^{-1}] = a(RT/P) - bT_r^{-1} + cT_r^{-2} + d;\ \text{with}\ T_r = 0.01\,T\,[\text{K}], P\ \text{in}\ \text{kPa}$ $a = 98.1;\ b = 0.01793;\ c = 0.0251;\ d = 0.001071;\ R = 0.041752\,\text{kJ/kgK}$

Source: Sugawara et al. (1962)

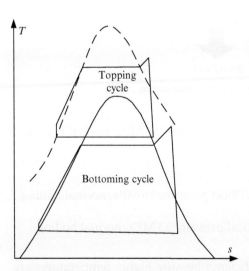

FIGURE 8.16   Qualitative representation of a binary Rankine cycle (an example of system integration by cascading).

According to Angelino and Invernizzi (2008), binary Rankine cycles with topping alkali metals or mercury can obtain superior efficiency to combined cycles because of their "full condensing nature." The efficiency increases to 50% when the turbine inlet is at the 873 K level and beyond 56% for 1273 K. In an example from Angelino and Invernizzi (2008), a steam Rankine cycle with superheat at 873 K has 43% energy efficiency when the boiling temperature is 625 K. When a topping potassium cycle is added, having condensation temperature at about 675 K, the overall efficiency increases to 52%, whereas the efficiency of the potassium cycle itself is 18%.

Another purpose of the binary cycle is to help recover part of the rejected heat from steam Rankine cycles when coupled in a cascaded fashion. In this respect, a bottoming cycle is cascaded to the condenser of the main steam Rankine cycle. This system is denoted as a low-temperature binary cycle to distinguish it from the high-temperature binary cycle discussed above, where a topping cycle is added to reach higher source temperatures. For low-temperature binary cycles, the bottoming cycle will operate with an organic fluid, typically from the class of refrigerants. One example of this kind of cascading is shown in Figure 8.17 from Al-Sulaiman (2013), where a steam cycle (topping) is integrated with an organic Rankine cycle (bottoming). One major advantage of integrating a bottoming ORC with a steam Rankine cycle comes from the fact that steam condensation can be conducted at much closer to atmospheric pressure. This will simplify the condenser construction but also will reduce the size of the turbine and the condenser because the steam volumetric flow rate decreases. Apparently the best efficiency has been obtained with R134a as the working fluid (over 30%) for the bottoming ORC. Other investigated fluids were R600, R600a, R152a, R290, and R407c.

Another type of cascaded cycle is the binary Brayton cycle. This cascading is analogous to the binary Rankine cycle discussed above, with the difference that the cascaded cycles are of Brayton type. El-Maksoud (2013) analysed this type of binary Brayton cycle.

Figure 8.17 shows an example of integrated power generation with a cascaded Brayton cycle. The topping cycle uses a fuel for combustion with the role of heat source. Note that the cycle features an isothermal combustion chamber and an isothermal regenerator. The isothermal combustion chamber may be the final segment of the combustor. It is suggested in El-Maksoud (2013) that isothermal combustion may be achieved by placing a number of fuel injectors along the gas flow by the last part of the combustion chamber. The temperature is maintained constant, and enthalpy is added to the flow. So, the gas pressure decreases during isothermal combustion. Hence, the gas enters the topping turbine at slightly lower pressure than in the combustion chamber, but with a higher enthalpy. After the expansion, the still hot combustion gases are cooled at constant pressure while they transfer heat to the bottoming (cascaded) cycle. Figure 8.18 is a qualitative representation of the cascaded cycle system using temperature vs. entropy coordinates.

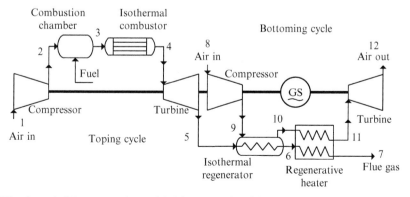

FIGURE 8.17   Cascaded Brayton cycle. *Modified from El-Maksoud (2013).*

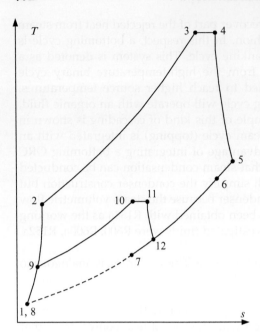

FIGURE 8.18   Qualitative representation of the cascaded Brayton cycle in *T–s* diagram. *Modified from El-Maksoud (2013).*

The bottoming cycle compresses air from the surroundings as represented by process 8–9 in the figure. Further, air is heated at constant pressure in a typical regenerator, according to the process 9 - 10. Next, air is heated isothermally in the nonconventional regenerative heat exchanger according to the process 10–11. As shown in El-Maksoud (2013), the binary Brayton cycle has an efficiency of over 50% when the temperature at turbine inlet (TIT) becomes higher than 1300 K; for 1200 K < TIT < 1300 K the energy efficiency is >45%.

EXAMPLE 8.2

We attempt to calculate the energy and exergy efficiency of a cascaded mercury-steam cycle (high-temperature binary cycle) when the heat source is provided at a constant temperature of 1000 K. Reasonable assumptions will be made.

- Assume boiling of steam at 575 K and condensation at 300 K and a Rankine cycle configuration as in Figure 5.13 from Chapter 5. This cycle has been solved in Example 5.4 for condensation at 303 K and boiling at 473 K.

- Example 5.4 is reworked assuming superheating to 875 K; the steam extraction fraction is optimized 56% for a maximum energy efficiency of 40%. The enthalpy of preheated water at the steam generator inlet is 700.1 kJ/kg, the saturated vapor enthalpy is 2800 kJ/kg, and the enthalpy at the turbine inlet is 3680 kJ/kg.

- Assume a 25-K temperature difference for the cascading heat exchanger (condenser/boiler); hence, the condensation of mercury should be at 575 + 25 = 600 K.

- Vapor pressure of mercury in the condenser results from Table 8.1 (first equation); it is obtained that $P_{cond} = 57$ kPa.

EXAMPLE 8.2 (*cont'd*)

- The enthalpy of saturated liquid and vapor for mercury at condensation pressure are calculated. The equations from Table 8.1 are used: $h' = -8.862$ kJ/kg, $h'' = 341.3$ kJ/kg.
- The entropy of saturated liquid and vapor: $s' = 0.02576$ kJ/kg K, $s'' = 0.602$ kJ/kg K.
- The specific volume of saturated liquid at the mercury condenser is found in Sugawara et al. (1962), namely $v' = 0.00007795$ m^3/kg.
- The pump mass-specific work is given by $w_P = v'\Delta P = 0.000077(6724 - 57) = 0.52$kJ/kg.
- The specific enthalpy at the pump outlet results from $w_P = h_P - h' \rightarrow h_P = -8.342$ kJ/kg.
- At 1000 K the vapor pressure of mercury is 6724 kPa (67.2 bar).
- The enthalpy of saturated liquid and vapor for mercury at boiling point are: $h' = -8.878$ kJ/kg, $h'' = 381.6$ kJ/kg and the saturated vapor entropy is $s'' = 0.4491$ kJ/kg K.
- The vapor quality at isentropic discharge is given by $0.4491 = [(1 - x_s)s' + x_s s'']_{\text{cond}}$. It results that the vapor quality $x_s = 0.7346$.
- The enthalpy at the isentropic turbine outlet is $h_s = (1 - x_s)h' + x_s h'' = 248.4$ kJ/kg.
- The specific work of the turbine is $w_T = 381.6 - 288.4 = 93.27$ kJ/kg; the net work output of the mercury cycle is $w_{\text{net,Hg}} = w_T - w_P = 93.27 - 0.52 = 92.75$ kJ/kg.
- The vapor quality at the turbine exit results from $h_T = (1 - x)h' + xh'' \rightarrow x = 0.85$.
- The mass flow rate of mercury with respect to the mass flow rate of steam results from the energy balance on the condenser-boiler: $\dot{m}_{\text{Hg}}\left(h_T - h'\right) = \dot{m}_{\text{steam}}(2800 - 700.1)$. Hence, one has $\dot{m}_{\text{Hg}} = 7.065\dot{m}_{\text{steam}}$.
- Heat input to the topping cycle vapor generator is
  $$\dot{Q}_{\text{in,Hg}} = \dot{m}_{\text{Hg}}\left(h''_{\text{boiler}} - h_P\right) = \dot{m}_{\text{Hg}}(381.6 + 8.342) = 2755\dot{m}_{\text{steam}}.$$
- Heat input to superheat steam $\dot{Q}_{\text{in,steam}} = \dot{m}_{\text{steam}}(3680 - 2800) = 880\dot{m}_{\text{steam}}$.
- Total heat input: $\dot{Q}_{\text{in}} = 3635\dot{m}_{\text{steam}}$.
- Power generated by topping cycle $\dot{W}_{\text{Hg}} = \dot{m}_{\text{Hg}}w_{\text{net, Hg}} = 655.3\dot{m}_{\text{steam}}$.
- Heat input to bottoming cycle: $\dot{Q}_{\text{bot}} = \dot{m}_{\text{steam}}(3680 - 700.1) = 2980\dot{m}_{\text{steam}}$.
- Net power output generated by bottoming cycle: $\dot{W}_{\text{steam}} = \eta\dot{Q}_{\text{bot}} = 1192\dot{m}_{\text{steam}}$.
- Total power generated: $\dot{W}_{\text{gen}} = \dot{W}_{\text{Hg}} + \dot{W}_{\text{steam}} = 1847\dot{m}_{\text{steam}}$.
- Energy efficiency: $\eta = \dot{W}_{\text{gen}}/\dot{Q}_{\text{in}} = 51\%$.
- Carnot factor at heat source (assume $T_{\text{so}} = 1100$ K, $T_0 = 288$ K): $\eta_C = 0.9745$.
- Exergy input: $\dot{\text{Ex}}_{\text{in}} = \eta_C\dot{Q}_{\text{in}} = 3543\dot{m}_{\text{steam}}$.
- Exergy efficiency: $\psi = \dot{W}_{\text{gen}}/\dot{\text{Ex}}_{\text{in}} = 52.1\%$.

8.4 COMBINED SYSTEMS

CCPPs are already established technologies. They combine GTs with Rankine cycle generators. These are common examples of integration of power generation systems by combining two or more thermodynamic cycles or technologies. In this section we will study combined systems that generate power or power and heating.

Table 8.2 gives a combination matrix for CCPPs. In the table six power generation systems are indexed in an order that corresponds with their typical operating temperature (the highest temperature within the system). Accordingly, topping internal combustion engines (Diesel,

TABLE 8.2   Combination Matrix for Combined Cycle Power Generation

Process	T_{max} (K)	Toping / Bottoming ID	1	2	3	4	5	6
Internal combustion engine	1,800	1	■	√	✗	✗	√	√
Brayton cycle (gas turbine)	1,200	2	√	■	✗	✗	√	√
SOFC	1,000	3	✗	√	■	√	√	√
MCFC	850	4	✗	√	✗	■	√	√
Steam Rankine cycle	800	5	✗	✗	✗	✗	■	√
Organic Rankine cycle	550	6	✗	✗	✗	✗	√	■

Otto) can be combined with bottoming GT or Rankine cycles. GTs can be combined with Rankine cycles. SOFCs can be combined with MCFCs which operate at lower temperature and can combust the unburned gases from the topping SOFC. Both MCFC and SOFC can be combined with bottoming GTs to combust the unburned gases and produce more power. In addition, high-temperature fuel cells can be combined with Rankine cycles. A steam Rankine cycle can be combined or cascaded with a bottoming ORC or vice versa.

Besides the potential combined systems mentioned in Table 8.2, many other possibilities exist. For example, one can combine SOFC, a GT, and a Rankine cycle for three-stage power generation, where the SOFC is the topping cycle, the GT the intermediate cycle, and the Rankine cycle the bottoming cycle. As many vapor cycles exist, there is a large pool for selection of a suitable Rankine cycle.

Here we will discuss some conventional and some advanced combined systems. We start with the CCPP. These systems were introduced in Section 5.4 of Chapter 5 and consists of a bottoming GT combined with a steam Rankine cycle. The component that integrates the two power cycles is the heat recovery steam generator. There are many versions of this combined cycle which differ in the number of stages for the Brayton and Rankine cycles and the number of steam pressure stages in the heat recovery steam generator.

One of the features of the CCPP is that the high exergy destructions in the combustion chamber (which are about 80% or more of total exergy destructions) are compensated by an efficient heat recovery system for steam generation integrated with a Rankine cycle. Figure 8.19 presents a very basic CCPP. We will demonstrate the thermodynamic analysis of the CCPP based on the simplified diagram in the figure. The analysis starts with the GT and follows the pathway of the air. A detailed case study analysis of this cycle is presented at the end of this chapter based on Ahmadi et al. (2011).

Table 8.3 shows the main equations for thermodynamic modeling. When a general hydrocarbon fuel (C_mH_n) is considered, the combustion equation for a given fuel-air ratio (FAR$=n_f/n_{air}$) is: $FAR\,C_mH_n + pO_2 + qN_2 + rH_2O + sCO_2 + tAr \rightarrow p'O_2 + q'N_2 + r'H_2O + s'CO_2 + uNO + vCO + tAr$.

Based on the combustion equation, the following stoichiometric coefficients in the product gas can be determined: $p' = p - FAR\,m - 0.25\,FAR\,n - v - u$, $q' = q - u$, $r' = r + 0.5\,FAR\,n$,

FIGURE 8.19 Simplified diagram of a combined cycle power plant (CCPP).

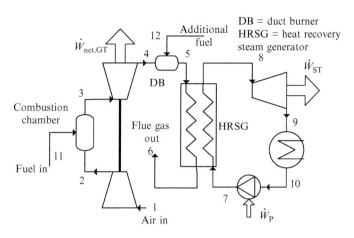

DB = duct burner
HRSG = heat recovery steam generator

TABLE 8.3 Modeling Equations for CCPP

Equipment	Assumptions and equations
Compressor	$MBE: \dot{m}_1 = \dot{m}_2$ $EBE: \dot{m}_1 h_1 = \dot{m}_2 h_2$ $ExBE: \dot{m}_1 ex_1 = \dot{m}_2 ex_2 + \dot{Ex}_d$ $\dfrac{T_2}{T_1} = 1 + \dfrac{r^\kappa - 1}{\eta_C}$, with $\kappa = \dfrac{\gamma - 1}{\gamma}$, $r = \dfrac{P_2}{P_1}$; $\dot{W}_C = \dot{m}_1 C_{pa}(T_2 - T_1)$ Assumed: $(P, T)_1 = (P, T)_0$, γ is assumed constant and $C_{pa} = f(T)$
Combustion chamber	$MBE: \dot{m}_3 = \dot{m}_2 + \dot{m}_{11}$ $EBE: \dot{m}_2 h_2 + \dot{m}_{11} LHV = \dot{m}_3 h_3 = m_{cg} h_{cg} + (1 - \varepsilon_{cc}) \dot{m}_{11} LHV$ $ExBE: \dot{m}_2 ex_2 + \dot{m}_{11} ex_{11} = \dot{m}_3 ex_3 + \dot{Ex}_d$ Assumed: adiabatic, non-isobaric combustion chamber, $\varepsilon_{cc} \cong 0.94$
Gas turbine	$MBE: \dot{m}_3 = \dot{m}_4$ $EBE: \dot{m}_3 h_3 = \dot{m}_4 h_4 + \dot{W}_{GT}$ $ExBE: \dot{m}_3 ex_3 = \dot{m}_4 ex_4 + \dot{W}_{GT} + \dot{Ex}_d$ $\dfrac{T_4}{T_3} = 1 - \eta_{GT}(1 - r^\kappa)$, with $r = \dfrac{P_3}{P_4}$; $\dot{W}_{GT} = \dot{m}_3 C_{pcg}(T_3 - T_4)$ Assumed: γ is assumed constant and $C_{pcg} = f(T)$
Duct burner	$MBE: \dot{m}_4 + \dot{m}_{12} = \dot{m}_5$ $EBE: \dot{m}_4 h_4 + \dot{m}_{12} LHV = \dot{m}_5 h_5 = \dot{m}_{cg} h_{cg} + (1 - \varepsilon_{cc}) \dot{m}_{12} LHV$ $ExBE: \dot{m}_4 ex_4 + \dot{m}_{12} ex_{12} = \dot{m}_5 ex_5 + \dot{Ex}_d$ Assumed: adiabatic, isobaric combustion chamber, $\varepsilon_{cc} \cong 0.94$
Heat recovery steam generator	$MBE: \dot{m}_5 + \dot{m}_7 = \dot{m}_6 + \dot{m}_8$; $\dot{m}_5 = \dot{m}_6$; $\dot{m}_7 = \dot{m}_8 = \dot{m}_w$ $EBE: \dot{m}_5 h_5 + \dot{m}_w h_7 = \dot{m}_5 h_6 + \dot{m}_w h_8$ $ExBE: \dot{m}_5 ex_5 + \dot{m}_w ex_7 = \dot{m}_5 ex_6 + \dot{m}_w ex_8 + \dot{Ex}_d$ Assumed: adiabatic HRSG
Steam turbine	$MBE: \dot{m}_8 = \dot{m}_9 = \dot{m}_w$ $EBE: \dot{m}_8 h_8 = \dot{m}_9 h_9 + \dot{W}_{ST}$ $ExBE: \dot{m}_8 ex_8 = \dot{m}_9 ex_9 + \dot{W}_{ST} + \dot{Ex}_d$ $\eta_{ST} = \dfrac{h_8 - h_9}{h_8 - h_{9s}}$; $h_{9s} = h(P_9, s_8)$; $\dot{W}_{ST} = \dot{m}_w (h_8 - h_9)$ Assumed: adiabatic, non-isentropic process; $\eta_{ST} \cong 0.8 - 0.9$

Continued

TABLE 8.3   Modeling Equations for CCPP—Cont'd

Equipment	Assumptions and equations
Condenser	MBE : $\dot{m}_9 = \dot{m}_{10} = \dot{m}_w$
	EBE : $\dot{m}_9 h_9 = \dot{Q}_{CD} + \dot{m}_{10} h_{10}$; $\dot{Q}_{CD} = \dot{m}_{cw} C_{pcw}(T_{2w} - T_{1w})$
	ExBE : $\dot{m}_9 ex_9 = \dot{Ex}_Q + \dot{m}_{10} ex_{10} + \dot{Ex}_d$
	Assumed: isobaric process; $\dot{Q}_{CD} = UALMTD; LMTD \cong 15\,^{\circ}C$
Pump	MBE : $\dot{m}_{10} = \dot{m}_7 = \dot{m}_w$
	EBE : $\dot{m}_w h_{10} + \dot{W}_P = \dot{m}_w h_7$
	ExBE : $\dot{m}_w ex_{10} + \dot{W}_P = \dot{m}_w ex_7 + \dot{Ex}_d$
	$\eta_P = \dfrac{h_{7s} - h_{10}}{h_7 - h_{10}}$; $h_{7s} = h(P_7, s_{10})$; $\dot{W}_P = \dot{m}_w(h_7 - h_{10})$
	Assumed: adiabatic, non-isentropic process; $\eta_P \cong 0.8 - 0.9$

$s' = FAR\,p + s - v$. Subsequent fuel injection in the duct burner will lead to combustion which can be modeled in a similar fashion as for the main combustion chamber, taking in account for the uncombusted fraction of fuel in combustion gases.

Figure 8.20 shows a qualitative representation of the temperature vs. entropy diagram for the simplest CCPP. The allowed heat source temperature with this cycle corresponds to the combustion chamber temperature; therefore, it surpasses 1000 K. Also, the cycle features excellent heat recovery from the exhaust gas. Due to this fact, the cycle has efficiency that surpasses 60%. Better efficiency is obtained when multistage heat recovery steam generators are used in conjunction with combined cycles. Figure 8.21 shows the layout of a dual-pressure CCPP. In this power plant, the GT unit is directly connected to the duct burner to minimize the flow length and heat losses. With the dual-pressure HRSG the exergy losses are much reduced and the heat recovery from flue gases is maximized. At the same time, the preheating zone of the Rankine cycle is simplified, with no need for regeneration.

The CCPP operates only with gaseous or liquid fuels because it includes a GT cycle. However, this cycle can be adapted to solid fuels (coal, biomass) if it is integrated with a

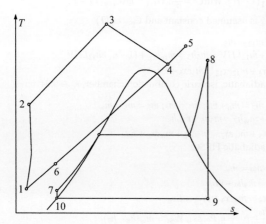

FIGURE 8.20   Temperature-entropy diagram for a simple combined cycle power plant (CCPP).

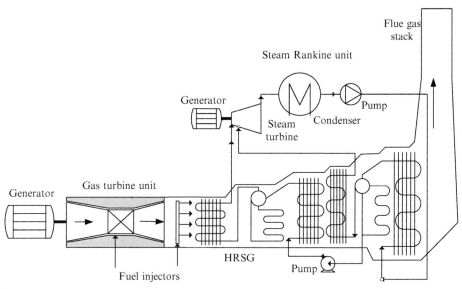

FIGURE 8.21 Layout of a dual-pressure combined cycle power plant.

gasifier. This system forms an IGCC. The system integration of these combined cycles is very beneficial because of various features:

- Coal/biomass gasification can be made advantageously under pressure using a solid fuel slurry with water, where water plays the role of gasification agent.
- An air separation unit can be used in the gasifier and oxy-gasification applied, which is very advantageous.
- Heat recovery from the gasifier can be used for the water preheating section in the steam generator.
- Carbon dioxide can be removed from product gas and further sequestrated; hence, the greenhouse gas emissions can be slightly reduced.
- The CCPP is part of the integrated system. Therefore, high power generation efficiency is obtained.
- Coal or biomass can be dried when this applies using low-grade heat recovered from flue gases.
- Solid fuels such as coal and biomass are economically more attractive and more widely available than natural gas, which confers better energy security.

Although the current capital cost of the IGCC is as high as 2000$/kW (see Christou et al., 2008), perspectives exist that this cost is reduced. Due to the high cost, direct combustion of coal or biomass appears to be the best economical choice for investment today. However, taking into account that the efficiency of the IGCC is highly superior (up to 10% efficiency points higher), long return of investment periods may be reduced by the intervention of governments to favor the application of the technology.

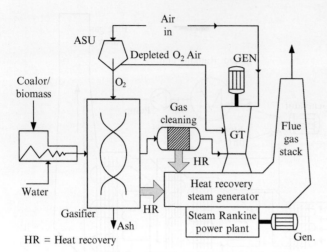

FIGURE 8.22 Simplified schematic of an integrated gasification combined cycle power plant.

A preferred integration scheme for the IGCC is shown in Figure 8.22. The heat recovery feature from a gasified and gas-cleaning system is clearly observed. This recovered heat together with the heat recovered from the hot combustion gases discharged by the GT subsystem is very beneficial for the heat recovery steam generator. Moreover, during the flow through the heat recovery steam generator, the combustion gases have a longer residence time, a fact that contributes to the completeness of combustion. The addition of nitrogen to the GT module is made at an intermediate pressure. The temperature of nitrogen is rather reduced after the air separation unit. Therefore, it helps cool the compression process between compression stages. The presence of nitrogen in the system lowers the combustion temperature and reduces the NOx.

Of course, various types of gasifiers can be used in an IGCC depending on what is most suitable for the actual case. In general a pressurized gasification process is required such that the syngas is injected into the combustion chamber at the required pressure. Moving bed, fluidized bed and entrained flow gasifiers have been used in the past with IGCC, of which the last choice is considered the best (see Christou et al., 2008) because the achieved temperature is the highest, >1500 K. At this temperature, syngas cannot be fed into the GT plant but must be cooled. This is achieved using heat recovery, where a part of the steam needed in the Rankine cycle is produced. Moreover, the colder syngas is cleaned to remove particulate matter, sulfur, and nitrous compounds such that the combustion products are less harmful to the environment compared to coal or biomass combustion.

Advanced concepts of IGCC were recently proposed by Guan et al. (2010). The practical problem to be solved to improve the efficiency of IGCC regards the partial combustion process conducted in the gasifier. In this process the exergy destructions are the highest. Due to the coal conversion step, the IGCC has lower efficiency than a CCPP which uses natural gas or a liquid fuel instead. The efficiency limit of IGGC is 50% as compared to the over 60% achievable by CCPP.

The reduction of irreversibility in a gasifier by recycling hot GT exhaust gases is suggested for the advanced IGCC cycle (A-IGCC) proposed in Guan et al. (2010). This arrangement pushes the power conversion efficiency to 60%, which is comparable to the CCPP. In an A-IGCC system the gasification agent is steam or a steam-oxygen mixture. This reduces the amount of oxygen required and the gasification temperature from approximately 1500 K (commonly used for the oxygen gasification process) to 1000 K, which is beneficial for low-grade coal (see Guan et al., 2010). Furthermore, the bottoming Rankine cycle is replaced with a steam turbine because the focus is to produce steam.

The use of a steam turbine instead of a complete Rankine plant greatly enhances the system efficiency because instead of converting low-grade heat into power as with the bottoming cycle of IGCC, the lower-grade heat is used mainly to generate steam, which in turn is converted into an upgraded gaseous fuel generated with reduced exergy destructions.

The advanced IGCC system of Guan et al. (2010) is shown in Figure 8.23. This system features a triple-stage gasifier. The first-stage gasification process is char combustion, which is performed in the riser of a circulating fluidized bed that uses sand as a heat transport/inert medium. Oxygen is blown at the riser bottom together with a part of the recirculated gases. The gases are separated in a cyclone above from the hot sand particles which separate and fall in a downer.

The downer is a pyrolysis reactor. Solid fuel (coal, biomass, or any coal-biomass combination) is forced into the downer where it undergoes pyrolysis under the presence of hot steam and hot sand particles. The result of the pyrolysis is the production of tar, char, and volatile gases. At the bottom of the downer a cyclone separator is placed to extract the tar and volatiles from the char. Further, the volatiles and char are directed to a thermo-catalytic steam reformer and converted into syngas. The char and sand fall in the gasifier section, which operates as a bubble bed with superheated steam injected from below. As a consequence of heat input from steam and sand, the char converts partially to syngas, which is extracted from the top side of the gasifier. The unreacted char falls into a downer together with sand particles and moves into the riser section.

A gas-cleaning module is used to clean the produced synfuel and product gas from the riser. Then the synfuel is fed into the GT generator. After exit from the GT, the uncombusted gases further combust in the heat recovery steam generator (HRSG). Note that the HRSG is designed such that it only supplies the required amount of steam to the gasifier, pyrolyser, and reformer. In doing so, some work is generated by a steam turbine. Therefore, the HRSG and steam turbine act as an open-loop steam Rankine engine. Within the HRSG pressurized water is preheated and fed into a drum boiler then superheated and expanded in the steam turbine. The expanded low-pressure steam is reheated and supplied to the process.

Based on the data from Guan et al. (2010), Figure 8.24 shows the energy and exergy balances for the A-IGCC. Note that the energy and exergy efficiencies are 59% and 62%, respectively. In addition, the integrated power plant operates regeneratively because it produces steam from heat recovery. The recovered energy becomes 35% of energy input while the recycled exergy is 15% of the exergy input.

Another cascading option uses high-temperature fuel cells (SOFC, MCFC or AFC). In Chapter 4, a number of integrated systems that combine fuel cells with GT cycles were presented. Two main schemes are possible for integration, as indicated in Figure 8.25. When high-temperature fuel cells are used (SOFC, MCFC), then the system combines with a GT and a fuel processor module (which partially reforms the fuel to hydrogen). When lower-temperature alkaline fuel cells (AFCs) are used then the heat recovered from exhaust gases

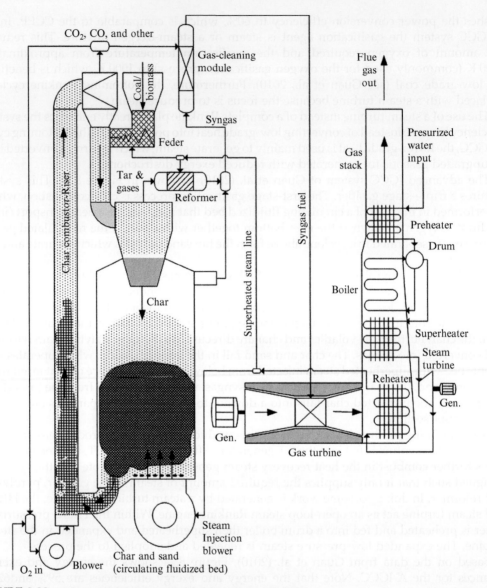

FIGURE 8.23 Advanced IGCC plant. *Modified from Guan et al. (2010).*

can be transferred to a bottoming organic Rankine cycle (ORC). As an option, a combustor can be inserted into the exhaust gases to combust any unreacted fuel, as in the system shown in Figure 4.45 (Chapter 4).

A simplified schematic of a combined SOFC—GT power plant is shown in Figure 8.26. In this system natural gas is fed into the SOFC anode (6) while at the cathode (3) a preheated air is fed as an oxidant. The resulting gases, with partially uncombusted hydrocarbon, are fed into a combustor where some additional fuel (5, natural gas) is injected. The produced hot

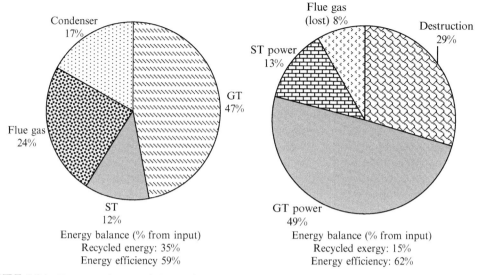

FIGURE 8.24 Energy and exergy balances for A-IGCC plant. *Data from Guan et al. (2010).*

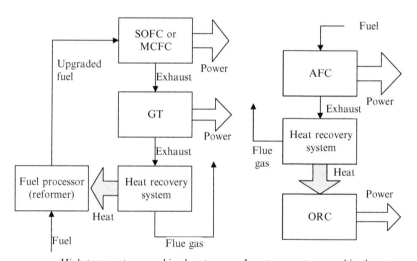

High-temperature combined systems   Low-temperature combined systems

FIGURE 8.25 General layout of fuel-cell-based combined systems.

FIGURE 8.26 Simplified schematic of a combined SOFC-GT power plant.

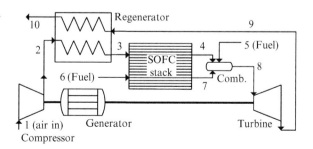

combustion gases (8) are expanded in a GT to generate work. The still hot exhaust gases (9) pass through a regenerator heat exchanger where they have the function of preheating the compressed air (process 2–3).

The thermodynamic modeling of the SOFC-GT system poses some difficulties in formulating the equations for the SOFC and the downstream combustor. Here, we give in Table 8.4 the relevant assumptions and the main equations for SOFC-GT modeling. Haseli et al. (2008) showed that there is an optimal efficiency of SOFC-GT combined systems which occurs at a compression ratio of around 4, whereas for conventional GT plants this maximum occurs at approximately 6 when the turbine inlet temperature is around 1250 K. Consequently, the SOFC-GT plants have much higher energy efficiency. This can be observed from the plot in Figure 8.27, which shows that the combined SOFC-GT system has a gain of at least 30% efficiency.

TABLE 8.4 Main Assumptions and Equations for Simple SOFC-GT Model

Item	Assumptions and equations
Electrochemical equations	Assumption: natural gas can be modeled as methane Overall reaction: $CH_4 + 2O_2 \rightarrow CO_2 + 2H_2O$ Cathode reaction: $0.5O_2 + 2e^- \rightarrow O^{2-}$ Anode reactions: $CH_4 + 4O^{2-} \rightarrow 2H_2O + 8e^-$; $CO + O^{2-} \rightarrow CO_2 + 2e^-$; $H_2 + O^{2-} \rightarrow H_2O + 2e^-$
Cell voltage under operating conditions	Assumption: $z = 8$ electrons are exchanged per reaction. See Chapter 4 for FC modeling. The concentration overpotential can be neglected. $$E = \frac{\Delta G}{8F\Delta H} + \frac{RT}{8F} \ln\left(\frac{P_{CH_4} P_{O_2}^2}{P_{CO_2} P_{H_2O}^2}\right) - V_{act}^0 \ln\frac{J}{J_0} - \mathcal{R}J$$
Mass balance equation SOFC	$\dot{m}_3 + \dot{m}_6 = \dot{m}_4 + \dot{m}_7 = (1 - U_f)\dot{m}_6 + \dot{m}_{cg}$ where U_f is the fuel utilization factor and $\dot{m}_{cg}$ is the mass flow rate of non-combustible species
Energy balance equation SOFC	$\dot{m}_3 h_3 + \dot{m}_6 h_6 = \dot{W}_{FC} + \dot{m}_4 h_4 + \dot{m}_7 h_7$ where $h_6 = U_f LHV + (1 - U_f)h_{f,in}$; $h_t = (1 - U_f)h_{f,out}$; h_f = thermal enthalpy of fuel
Exergy balance equation SOFC	$\dot{m}_3 ex_3 + \dot{m}_6 ex_6 = \dot{W}_{FC} + \dot{m}_4 ex_4 + \dot{m}_7 ex_7 + \dot{Ex}_d$ where ex_6 and ex_7 include chemical exergy component and $\dot{Ex}_d$ is the destroyed exergy rate
Mass balance equation Comb.	Assumption: 100% fuel utilization factor in the combustor $\dot{m}_4 + \dot{m}_5 + \dot{m}_7 = \dot{m}_8$ with $\dot{m}_7 = (1 - U_f)\dot{m}_6$
Energy balance equation Comb.	Assumption: non-adiabatic combustion chamber: $\dot{Q}_{loss} = (1 - \varepsilon_{cc})[(1 - U_f)\dot{m}_6 + \dot{m}_5]LHV$ $\dot{m}_4 h_4 + (1 - U_f)\dot{m}_6 h_{f,out} + [(1 - U_f)\dot{m}_6 + \dot{m}_5]LHV = \dot{Q}_{loss} + \dot{m}_8 h_8$
Exergy balance equation Comb.	Assumption: heat loss occurs at $T_8 \rightarrow \dot{Ex}_{Qloss} = (1 - T_0/T_8)\dot{Q}_{loss}$ $\dot{m}_4 ex_4 + (1 - U_f)\dot{m}_6 ex_6 + \dot{m}_5 ex_5 = \dot{Ex}_{Qloss} + \dot{m}_8 ex_{8 + \dot{Ex}_d}$; $ex_{5,6}$ include chemical exergy
Overall balance equations	$MBE: \dot{m}_1 + \dot{m}_5 + \dot{m}_6 = \dot{m}_{10}$ $EBE: \dot{m}_1 h_1 + (\dot{m}_5 + \dot{m}_6)LHV + \dot{W}_{comp} = \dot{m}_{10} h_{10} + \dot{W}_{GT} + \dot{W}_{SOFC}$ $ExBE: \dot{m}_1 ex_1 + (\dot{m}_5 + \dot{m}_6)ex_{fuel} + \dot{W}_{comp} = \dot{m}_{10} ex_{10} + \dot{W}_{GT} + \dot{W}_{SOFC} + \dot{Ex}_d$

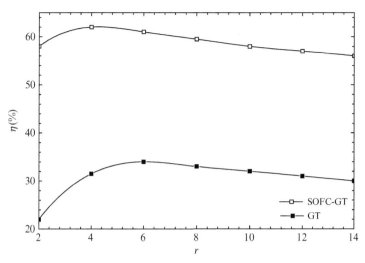

FIGURE 8.27   Energy efficiency comparison of conventional gas turbine (GT) power plants and combined SOFC-GT power plants. *Data from Hasseli et al. (2008).*

An integrated gasifier-fuel cell-GT-Rankine cycle (IGFC) is one of the alternative options to the IGCC for generating power from solid fuels. The energy efficiency of the IGFC is lower than that of the SOFC-GT because of the energy losses in the gasification process. According to El-Emam et al. (2012) this system has an efficiency of around 35–40%, that is with approximately 20% efficiency points fewer than the SOFC-GT. Furthermore, the exergy efficiency is in the range of 20–30%. The major importance of these systems is their ability to use coal and biomass as widely available fuels. Specifically, the use of low-rank coal is favored in these systems to achieve 40% energy conversion efficiency.

The simplified diagram of an IGFC power plant is shown in Figure 8.28. In a common design approach the heat recovery from the combustion gases is used to generate additional power with the bottoming steam Rankine cycle. There are other recent options which recycle some recovered heat to assist the gasification process. A cycle alternative studied in El-Emam et al. (2012) considers co-fueling of the cycle with coal and natural gas, where the natural gas is used as additional fuel for the GT module. The schematic of the IGFC system studied in El-Emam et al. (2012) is shown in Figure 8.29.

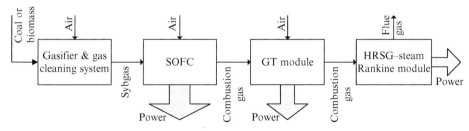

FIGURE 8.28   Simplified functional diagram of IGFC combined power plant.

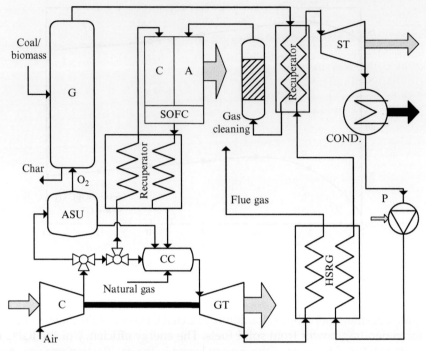

FIGURE 8.29   IGFC power plant. *Modified from El-Emam et al. (2012).*

In an IGFC such as in Figure 8.29, the fuel fed in the anode is mixed with recirculated fuel, which contains uncombusted methane. As a result, steam reformation and a water–gas shift reaction occur at the anode: $CH_4 + H_2O \rightarrow 3H_2 + C.$ and $CO + H_2O \rightarrow H_2 + CO_2$. These reactions are at equilibrium. The generated SOFC power output can be calculated with the procedure given in Table 8.5, summarized from Colpan et al. (2007).

The SOFC method of calculation given in Table 8.5 requires the following input parameters:

- RR = recirculation ratio (percent of recirculated flow from anode flow rate output).
- U_f = fuel utilization fraction in the SOFC.
- ζ_{CH_4} and ζ_{CO} = extent of methane and CO conversion, respectively.
- $y_{j,i}$ = species concentration at the anode inlet, $j = CH_4, CO_2, CO, H_2, H_2$.
- $\dot{n}_a$ = molar flow rate at anode inlet.

The typical exergy destructions of the components of IGFC power plants with respect to total exergy destruction are shown in Figure 8.30. The exergy efficiency of IGFC can be increased if heat is cogenerated. Odukoya et al. (2011) showed that the predicted energy efficiency of an IGFC unit is 82% while exergy efficiency is 72% when steam is cogenerated. An alternative way to increase the efficiency of an IGFC is indicated in Guan et al. (2010): high-temperature thermal energy is recovered from the SOFC and GT and recycled to the gasifier. As shown in Guan et al. (2010), the heat recovered from the SOFC to the gasifier represents 44% of energy input, and 14% of the energy input is recycled internally from the GT to the

TABLE 8.5 Calculation Procedure for Power Generated by Syngas-Fueled SOFC

Item	Assumptions and equations
Molar flow rate of utilized fuel	$\dot{n}_{f,a} = U_f \dfrac{y_{H_2O,i}\dot{n}_a + 3\zeta_{CH_4} + \zeta_{CO}}{1 - RR + RRU_f}$
Electric current generated by the cell	$I = 2F\dot{n}_{f,a}$, where F is Faraday number and I results in Amperes

Equilibrium constants for reforming and water-gas shift

$$K_j = \exp\left(\sum_{i=0}^{4} A_i T^{4-i}\right)$$
$$j = CH_4, CO$$

	Steam reforming		Water-gas shift	
	A_0	−2.6312E−11	A_0	5.47E-12
	A_1	1.2406E-7	A_1	−2.547E-8
	A_2	−2.2523E-4	A_2	4.6374E-5
	A_3	0.195027	A_3	−3.915E-2
	A_4	−66.139488	A_4	13.209723

Determine molar concentrations of species at anode exit

A nonlinear system of equations must be solved. The equations are

$n_{a,e} = n_{a,i} + 2\zeta_{CH_4}$,
$n_{CH_4,e} = n_{CH_4,i} - \zeta_{CH_4}, n_{CO,e} = n_{CO,i} + \zeta_{CH_4} - \zeta_{CO}, n_{CO_2,e} = n_{CO_2,i} + \zeta_{CO}$,
$n_{H_2,e} = n_{H_2,i} + 3\zeta_{CH_4} + \zeta_{CO} - n_{f,a}, n_{H_2O,e} = n_{H_2O,i} - \zeta_{CH_4} - \zeta_{CO} + n_{f,a}$, e=exit, i=inlet

Molar fractions: $y_{j,e} = n_{j,e}/n_{a,e}$, $n_{je} = n_a C_j + y_{j,e} RR\, n_{a,e}, j = CH_4, CO_2, CO, H_2, H_2O$

$$K_{CH_4} = \frac{y_{H_2,e}^3 y_{CO,e}}{y_{CH_4,e} y_{H_2O,e}}\left(\frac{P}{P_0}\right)^2, \quad K_{CO} = \frac{y_{H_2,e} y_{CO_2,e}}{y_{CO,e} y_{H_2O,e}}$$

Calculate power generated by SOFC

Assumption: the overpotentials are neglected

$$\dot{W} = EI \text{ where } E = 1.19 + \frac{RT}{2F}\ln\left(\frac{y_{H_2,e}\, y_{O_2,e}^{0.5}}{y_{H_2O,e}} P^{0.5}\right)$$

Note: Method based on Colpan et al. (2007).

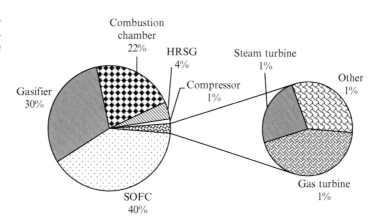

FIGURE 8.30 Typical exergy destruction fractions for the components of IGFC power plants. *Data from El-Emam et al. (2012).*

Combustion chamber 22%
HRSG 4%
Steam turbine 1%
Gasifier 30%
Compressor 1%
Other 1%
SOFC 40%
Gas turbine 1%

gasifier. The total internally recycled exergy is 43% of exergy input. The energy efficiency of the advanced IGFC (which includes internal regeneration) is predicted to be 66% while the exergy efficiency is 69%.

EXAMPLE 8.3

In this example we illustrate the efficiency calculation for an IGFC power plant under some assumptions. The overall efficiency will be determined. The IGFC comprises a steam gasifier that operates at 10 bar, an SOFC, a GT, and a heat recovery steam generator. There is no steam turbine in this plant; the HRSG only generates pressurized steam for the gasifier.

- Assume system is fed with coal-biomass fuel represented by formula: $C_{5.5}H_{7.4}O_2N_{0.05}S_{0.01}$.

- Assume produced syngas is given by $H_2 + 0.8\,CO + 0.6\,CH_4$.
- Steam gasification is applied with partial oxidation with λ the stoichiometric number for air.
- Assume gasification equation with air as oxidant:

$$(5.5C + 3.7H_2 + O_2 + 0.025N_2 + 0.01S) + 4.1H_2O + \lambda(O_2 + 3.76N_2) \rightarrow$$
$$3.73(H_2 + 0.8CO + 0.6CH_4) + 0.28CO_2 + 0.01SO_2 + (0.025 + 1.88\lambda)N_2 + (\lambda + 1.467)O_2$$

- Assume the temperature of gasification is 1100 K and coal/biomass is fed at T_0 while steam/air is fed at 1100 K. From thermodynamic tables or Engineering Equation Solver software we read enthalpies for each species and write the energy balance as follows

$$EBE : 0 - 4.5 \times 215.8 + 103.4\lambda =$$
$$-3.73 \times 197.3 - 0.28 \times 360.1 - 0.01 \times 262.4 + 21.4(0.025 + 1.88\lambda) + 22.7(\lambda + 1.476)$$

- Above equation is solved for λ and one gets: $\lambda = 14.6$ mol air per mole fuel.
- Assume negligible energy for syngas purification → syngas molar composition fed to the SOFC is 41.67%H_2, 33.33%CO, and 25% CH_4.
- Assume fuel utilization factor in the fuel cell of 60%; assume extent of reaction for steam reforming of 0.249 and extent of reaction for water–gas shift reaction of 0.45.
- For fuel cell operation at 1100 K one calculates the equilibrium constants for reforming and shift reactions according to the equations from Table 8.5 → $K_{CH_4} = 11.7$, $K_{CO} = 1.44$.
- Assume reasonable values for molar fractions of hydrogen, CO, and CH_4 at SOFC exit; the guessesare14%, 25%, 0.04% for H_2, CO, and CH_4, respectively. We will check if with these guesses the reaction extends for reforming and gas shift and if the fuel utilization factor is reasonable.
- We calculate the equilibrium molar fraction of steam from the definition equation for K_{CH_4} as given in Table 8.5: $K_{CH_4}y_{2CH_4}y_{2H_2O} = y_{2H_2}^3 y_{2CO}PR^2$.
- It results that $y_{H_2O} = 17.4\%$. Similarly, the molar fraction of CO_2 is calculated from K_{CO} and 24.8% is obtained. Next the conservation of carbon atoms at the anode (inlet/outlet) is imposed as follows: $y_{1CO} + y_{1CH_4} = n_2\left(y_{2CO} + y_{2CH_4} + y_{2CO_2}\right)$, where n_2 represents the number of moles of outlet flow at the anode per 1 mol of inlet stream. It results that $n_2 = 1.17$.
- The utilized hydrogen in fuel cells is calculated from $(1 - U_f)y_{1H_2} = n_2 y_{2H_2} \rightarrow U_f = 60\%$ (this is a reasonable value).

EXAMPLE 8.3 *(cont'd)*

- According to the procedure from Table 8.5 the extent of reforming reaction results from the equation: $n_2 y_{2CH_4} = y_{1CH_4} - \zeta_{CH_4} \rightarrow \zeta_{CH_4} = 0.249$; further $n_2 y_{2CO} = y_{1CO} + \zeta_{CH4} - \zeta_{CO} \rightarrow \zeta_{CO} = 0.256$; these values are reasonable; hence, the molar fraction guesses are validated.
- The cell voltage results from the equation given in Table 8.5 $\rightarrow E = 1.157$ V.
- The SOFC current is given by $I = 3.73 \times 2FU_f \dot{m}_1 = 429,053 \dot{m}_1$.
- The power generated by the SOFC is $\dot{W}_{FC}[W] = EI = 128,353 \, \dot{n}_f$ [kmol/s], where $\dot{n}_f$ is the molar rate of coal/biomass fuel fed into the gasifier.
- The efficiency of the SOFC results from its operational voltage $\eta_{SOFC} = 0.79$.
- The heat rejected by the SOFC is

$$\dot{Q}_{FC} = \frac{1 - \eta_{FC}}{\eta_{FC}} \dot{W}_{FC} = 33,955 \, \dot{n}_f$$

- The uncombusted H_2 and CO are further combusted in the combustion chamber according to the following equation, written for 1 mol of fuel input into the gasifier:

$$y_{2H_2} H_2 + y_{2CO} CO + 0.5 \left(y_{2H_2} + y_{2CO} \right) O_2 \rightarrow y_{2H_2} H_2 O + y_{2CO} CO_2$$

- The energy balance for the combustion chamber at 1100 K (the same as the SOFC) determines the heat released by the combustion process $\rightarrow \dot{Q}_{comb}[W] = 40,037 \, \dot{n}_f$ [kmol/s].
- This heat and the heat rejected by the fuel cell is used internally to heat the combustion gases; therefore, the amount transferred by the SOFC and combustion chamber to the combustion gases is

$$\dot{Q}_{cg} = \dot{Q}_{FC} + \dot{Q}_{comb} = 294,614 \, \dot{n}_f$$

- In addition to the fuel, the combustion chamber is fed with air from the fuel cell cathode, non-combustible gases from the anode and residual gases from the gasification process. Based on the chemical equations for gasification and the SOFC process, it results that the number of moles of gas that exit the combustion chamber is $\dot{n}_{comb} = 46 \dot{n}_f$.
- The energy balance requires $\dot{Q}_{cg} = \dot{n}_{comb} C_p \Delta T_{cg}$, where ΔT_{comb} is the temperature increase over the combustion chamber.
- The temperature of the combustion gases result from

$$T_{cg} = 1000 + \frac{49,147 \, \dot{n}_f}{46 \, \dot{n}_f C_p} = 1148 \text{K}$$

- The energy efficiency of the GT for a compression ratio of 6.4 and exergy efficiency of 0.7 is given by $\eta_{GT} = 0.7(1 - 6.4^{-\kappa})$ with $\kappa = 0.2857 \rightarrow \eta_{GT} = 29\%$.
- The work generated by the GT is $\dot{W}_{GT}[W] = \eta_{GT} \dot{Q}_{cg} = 21,319 \dot{n}_f$ [kmol/s].
- The temperature of the turbine exhaust gases is $T_{teg} = 6.4^{-\kappa} T_{comb} = 675$ K.
- The amount of steam required by the gasifier results from the assumed stoichiometry discussed above $\dot{n}_{steam} = 4.5 \, \dot{n}_f$.

Continued

EXAMPLE 8.3   *(cont'd)*

- The amount of heat required to boil and superheat a molar flow rate $\dot{n}_{steam}$ of saturated liquid water at 10 bar is $\dot{Q}_{bsh} = \dot{n}_{steam}\left(h_{@1100\ K} - h_{@10\ bar}\right) = 279{,}971\,\dot{n}_f$. It is also assumed that gasification is conducted at 10 bar.
- The heat recovered from combustion gases when cooled from T_{teg} to $T_{pinch} = T_{sat} + 10K$ is $\dot{Q}_{teg} = \dot{n}_{comb}C_p\left(T_{teg} - T_{pinch}\right) = 294{,}614\ \dot{n}_f$. This heat compensates for the amount required to boil and superheat the saturated liquid water $\dot{Q}_{bsh}$ and for a heat loss fraction of 5% of $\dot{Q}_{teg}$, which is reasonable.
- The heat required to preheat water from T_0 to $T_{sat} = 453.6$ K is $\dot{Q}_{ph} = 53{,}483\ \dot{n}_f$.
- The exit temperature of flue gases results from $\dot{Q}_{ph} = \dot{n}_{comb}C_p\left(T_{pinch} - T_{fg}\right)$; from here it results that the flue gas temperature is $T_{fg} = 424.2$ K, which is reasonable.
- The lower heating value of fuel is calculated from the energy balance for the stoichiometric fuel oxidation reaction:

$$(5.5C + 3.7H_2 + O_2 + 0.025N_2 + 0.01S) + 6.36O_2 \rightarrow 5.5CO_2 + 3.7H_2O + 0.025N_2 + 0.01SO_2$$

- In the above reaction all species are at T_0, and water is in vapor phase; hence,

$$EBE : 0 = LHV - 5.5 \times 393.4 - 3.7 \times 241.8 - 0.01 \times 296.8 \rightarrow LHV = 299.154\ kJ/kmol$$

- Net power generation (auxiliary power demand neglected) is:

$$\dot{W}_{net} = \dot{W}_{FC} + \dot{W}_{GT} = 149{,}672\ \dot{n}_f$$

- Power generation energy efficiency is:

$$\eta = \frac{\dot{W}_{net}}{LHV\,\dot{n}_f} = \frac{149{,}672\ \dot{n}_f}{299{,}154\ \dot{n}_f} = 50\%$$

8.5 HYBRID SYSTEMS

The most comprehensive manner of integrating systems is by hybridizing. In this case two or multiple dissimilar power generation systems work together synergistically while they convert energy from one source or a variety of sources. Hybridization can integrate very dissimilar energy sources, such as photovoltaic generators and wind turbines, hydro and wind, solar and natural gas, coal and concentrated solar power, solar photovoltaics and hydrogen fuel cells, biomass and coal, geothermal and solar, and ocean thermal and solar. The list is open.

Hybrid energy systems offer in general better energy security, as they relate diverse energy sources for better sustainability. Usually, hybrid systems are designed in conjunction with renewable energies, with the goal of compensating for the fluctuating and intermittent nature of those energies. Sometimes these hybrid systems are called "integrated renewable energy systems." In addition, hybrid systems with renewable energies are often designed to meet the peak demand when they run in conjunction with a conventional power generation system. There are two main possibilities in this respect:

- The hybrid system is primarily based on renewable energies and uses a conventional power generation subsystem (e.g., diesel generator) in the periods when there is a peak demand or not sufficient availability of the primary resource.
- The hybrid system is primarily based on a conventional power generator and uses some renewable energy subsystems to supplement the power generation or to save fuel, especially during the off-peak periods.

The following types of hybrid systems with wind energy can be proposed for applications:

- Wind turbines (farms) and solar-PV arrays.
- Wind farms and hydro storage.
- Wind farms and compressed air energy storage.
- Wind turbines and diesel generators.
- Wind turbines and hydrogen and fuel cell systems.

In a hybrid photovoltaic-wind power plant the two integrated system PV-generators and wind turbines share a common electrical energy storage, electric power conversion (inverter) facility, and, depending on the case, grid connection system.

The common practice is to use a battery bank for electrical energy storage. By making use of both solar and wind resources, the integrated system has a higher capacity factor than independent systems and a lower cost, because they share a common infrastructure for power conversion, storage, and distribution. Figure 8.31 shows a simplified schematic of a hybrid PV-wind turbine power plant. The system may include an engine-driven generator (e.g., diesel type) that operates in coordination with demand and wind and solar energy availability.

Wind and hydro hybrid systems use wind power to pump water at a higher level in an artificial lake. Hydraulic turbines and generators are installed to produce power on demand. These systems are connected to the regional grid. Basically, the role of wind-hydro systems is to smooth the flow of power generation by compensating for the fluctuating nature of wind. The energy efficiency of these systems is rather high as they involve only mechanical-electrical energy conversion.

Wind–diesel hybrid power generators have been implemented for grid independent systems of rather small size, at the community level. In these systems the two technologies share the same electrical infrastructure, consisting of inverters, controllers, systems for connection to a small local grid, and battery banks for power storage. The main aim of the wind–diesel hybrid is to reduce the consumption of diesel fuel; therefore, in many cases, the wind component has a larger share of total power generation.

Another wind-based hybrid energy system combines wind power generation with compressed air energy storage. Compressed air storage is a thermomechanical process in which energy is charged and retrieved, respectively, through mechanical work and heat. Air is compressed during wind availability hours and stored in large underground reservoirs, which may be naturally occurring caverns, salt domes, abandoned mine shafts, depleted gas and oil fields, or manmade caverns. There is good experience accumulated in designing large storage reservoirs in underground salt caverns. Such caverns can be mined by using aqueous solutions that dissolve salt and generate the desired shape of the reservoir. As a second option,

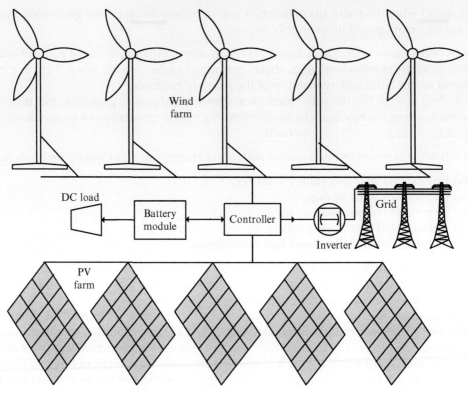

FIGURE 8.31   Hybrid PV-wind power plant.

large quantities of compressed air can be stored in aquifers, above the water level, if the geo-morphological configuration is such that above the aquifer there is a hard rock that impedes air leakages. Retrieval of compressed air energy—by converting back to electricity—is usually made with GT systems which have a fast start-up and because the availability of compressed air will have a substantially reduced back work ratio. The layout of a hybrid wind-compressed air storage system is shown in Figure 8.32.

One of the major problems with wind and photovoltaic power generators is the difficulty of storing large amounts of electrical energy in batteries. This is where hybridization with hydrogen and fuel cells provide a solution. Case study 7.1 in Chapter 7 illustrates the integration of PV-arrays with an electrolyser and hydrogen storage in metal hydride and fuel cells. Wind- and/or PV-hydrogen fuel cell systems represent one attractive option for wind power at community level settings. In this case the wind turbine generators are connected to an electrolyser unit to produce hydrogen when wind is available. Hydrogen is stored, commonly in metal hydride tanks. Then, hydrogen is used in a fuel cell system to generate power on demand. These systems can be further hybridized with diesel-engine-driven generators. Figure 8.33 shows an example of a hybridized power producing system that integrates wind, solar-PV, and diesel generators.

Another interesting hybridization option with solar energy and fuel cells is that which combines concentrated solar collectors with thermochemical reformers of fuels to generate

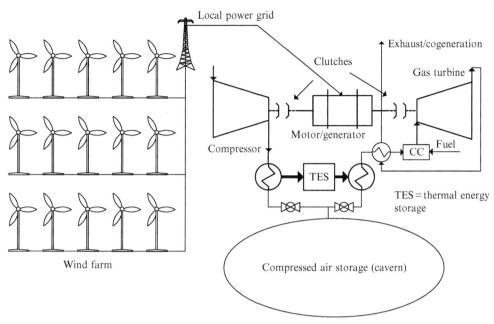

FIGURE 8.32 Layout of a hybrid wind-compressed air storage system.

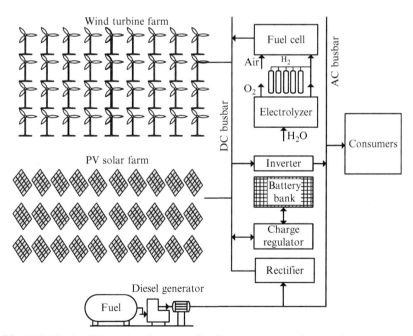

FIGURE 8.33 Hybrid solar-PV, wind turbine, and diesel power integrated generation system.

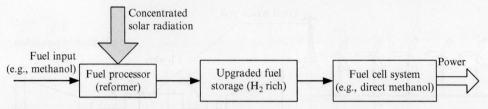

FIGURE 8.34   Hybrid solar–fossil fuel power generation with concentrated solar reforming and fuel cell system.

and store hydrogen and produce power on demand with fuel cells or thermal engines. A hybrid system that uses natural gas reforming with concentrated solar heat is described in Chapter 7, Figure 7.29. Another fuel that is easy to reform is methanol, which is one of the most promising candidates for hydrogen generation from reforming because the process of reformation is relatively straightforward. Figure 8.34 show a hybrid system with concentrated solar reforming and a fuel cell. The solar energy harvested is retrieved in the form of an upgraded fuel, with a higher calorific value than the supplied fuel. Candidate fuels for solar reforming with moderate temperature levels are methanol and ammonia, for which reforming temperatures of 200–400 °C are necessary. For higher temperatures of 700–1000 °C solar reforming of ethanol, propane, and natural gas becomes possible (see Wagar et al., 2011). Table 8.6 gives the potential fluids and reforming reactions.

The simplest system uses thermo-catalytic cracking of ammonia, a reaction that evolves at equilibrium and produces no side products. A catalytic membrane reactor can be used to enhance the reaction rate and the yield. The most attractive system appears to be that using methanol, as methanol is simple to store and direct methanol fuel cells are readily available. The required temperature is obtained using line focus concentrators with single-axis solar tracking. System control can be obtained by flow rate adjustment in such a way that the reformer temperature is maintained at the desired level in accordance with the intensity of solar radiation. At nighttime, the accumulated reformate is used; moreover, the fuel cell can be directly fed with methanol-water. As suggested in the figure, at nighttime the solar reformer is bypassed using a valve.

Some schemes to recover more thermomechanical enthalpy from reformate that is produced by solar reforming were investigated by Wagar et al. (2011). In Figure 8.35 a hybrid power generation system which uses a point focus two-axis sun tracking concentrator is shown. The concentrator focuses light on a solar reformer which increases both the thermomechanical enthalpy and the heating value of the fuel when sun is available. A part of the hot product gas is expanded in a GT to generate power. A regenerative system is designed which uses a network of heat exchangers to preheat the fuel prior to the reaction. The main power generator is a fuel cell or a combustion engine which is fed with reformatted fuel.

TABLE 8.6   Some Candidate Fuels for Solar Reforming/Fuel Cell Power Generation

Fuel	Reforming (cracking) reaction	Temperature	Pressure	LHV (kJ/g)
Ammonia	$NH_3 \rightarrow 1.5H_2 + 0.5N_2$	350–550 °C	1–20 bar	18.6
Methanol	$CH_3OH + H_2O \rightarrow 3H_2 + CO_2$	280–400 °C	4–20 bar	19.9
Propane	$C_3H_8 + 6H_2O \rightarrow 10H_2 + 3CO_2$	700–1.00 °C	3–25 bar	46.35
Methane	$CH_4 + 2H_2O \rightarrow 4H_2 + CO_2$	700–1.00 °C	3–25 bar	50.0

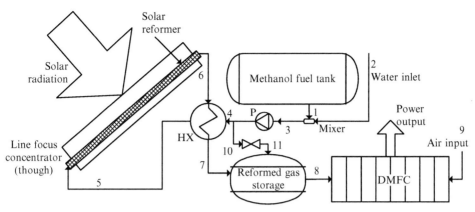

FIGURE 8.35 Simplified schematics of hybrid one-through solar reformer/fuel cell system.

Of course there are much many options for integrating power generation systems in a hybrid manner, depending on the applications. For example, in farm settings where there is manure and residual crop/biomass hybridized systems can be designed to accept biogas-biomass as input. The following main components may be integrated to form a hybrid power generation system for farm settings: wind turbine, PV-arrays, diesel or propane generator, micro-hydro, bio-digester (biogas producer), solar-assisted biomass drier-gasifier, concentrated solar power, etc.

EXAMPLE 8.4

We analyze the hybrid system from Figure 8.36 under reasonable assumptions for household application:

- Collector area 10 m²; daylight duration 12 h with sinusoidal profile.
- Maximum radiation intensity 800 W/m2; single-axis sun tracking with 20% optical losses.

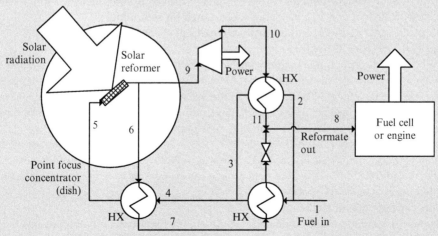

FIGURE 8.36 Schematic of a hybrid power generation system with solar fuel reforming and additional turbine. *Modified from Wagar et al. (2011).*

Continued

EXAMPLE 8.4   *(cont'd)*

- Heat leak coefficient of solar concentrator 1 W/m^2K; concentration factor 40.
- The fuel is methanol. Reformer tube diameter 15 mm; tube length 4 m.
- Average density 7000 kg/m^3, specific heat 450 J/kg K.
- Operating temperature of reformer 575 K; in the morning the tubes are at T_0.

  - Calculate the total solar receiver area $A_r = 20\,m^2/40$ [concentration ratio] $= 0.5\,m^2$.
  - Calculate the number of receivers $NDL = A_r \rightarrow N = 0.5/(0.015 \times 4) = 4$.
  - Corrected value of tube length $6 \times 0.02 \times L = 0.5 \rightarrow L = 4.167\,m$.
  - Calculate the total mass of receiver tube (incl. catalyst): $m = 0.25 N \rho \pi D^2 L = 21\,kg$.
  - Calculate the amount of heat required to warm up the receiver tubes

$$Q = mC_p(T_r - T_0) = 55 \times 450 \times (575 - 298) = 2684\,kJ$$

  - Calculate the time required to warm up the receiver tubes in the morning (energy balance based on t, daylight time)

$$3600 \times 0.8\,I_{max}\,A \sin\left(\frac{\pi}{12}t\right) = mC_p\frac{dT}{dt} + U\frac{A}{C}(T - T_0) \rightarrow \text{replacing the values it results in}$$

$$\frac{d(T - T_0)}{dt} = 1635 \sin\left(\frac{\pi}{12}t\right) - 2.6 \times 10^{-5}(T - T_0) \rightarrow \text{integrating from 0 to } t$$

results in the time required for warming up ~69 min (in the morning).
  - After the warm-up the mass flow rate of methanol is adjusted such that the receiver temperature remains constant regardless of the level of solar radiation intensity. The total solar energy harvested in the form of reformed fuel during the 10.9 h of operation (from after the warm-up period till sunset) is

$$E_{ref} = 3600\left\{\frac{12}{\pi}I_{max}\,A\left[\cos\left(\frac{1.14\pi}{12}\right) + 1\right] - 11 U\frac{A}{C}(T_r - T_0)\right\} = 145.2\,MJ$$

  - Methanol consumed for the reforming process results from the assumed equation for reforming $CH_3OH + H_2O \rightarrow 3H_2 + CO_2$. Hence, there are 3 mol H_2 per mole of methanol.
  - For LHV equivalent of 145.2 MJ barely 1.2 kg of hydrogen is used, that is 0.6 kmol; hence, 0.2 kmol methanol or 6.4 kg methanol is consumed for reforming.
  - Electricity generated with a direct methanol fuel cell of efficiency 50% is $E_{gen,1} = 0.5 \times 145.9 = 72.5$ MJ.
  - The household consumption is approximated for the following consumers: refrigerator (700 W, continuous run), water heater and furnace blowers (600 W, 6 h run/day), sewage pump (900 W, 1 h/day), electric range (3000 W, 2 h/day), other kitchen appliances (1000 W, 1 h/day), washing machine and drier (averaged 3000 W, 1 h/day), multimedia and computers (1500 W, 5 h/day), light bulbs (400 W, 5 h/day). The total daily consumption is of $E_{cons} = 148.7$ MJ.
  - Additional electricity is needed: $E_{add} = 148.7 - 72.5 = 76.1$ MJ, or a consumption of methanol given by $E_{add} = \eta\,m_{CH3OH}\text{LHV}$, or $76.1 = 0.5\,m_{CH3OH}\,19.9 \rightarrow m_{CH3OH} = 7.65$ kg; however, 6.4 kg methanol and its LHV energy are entered into the system for reforming; hence, the only difference is $m_{add} = 7.65 - 6.4 = 1.25$ kg.
  - Daily consumption of methanol: $m_{day} = 7.65$ kg.
  - Methanol consumption for non-integrated system (methanol fuel cell only) results from the energy balance $E_{cons} = \eta\,m\,\text{LHV} \rightarrow m = 14.94$ kg.
  - Methanol savings $m_{sav} = 14.94 - 7.65 = 7.3$ kg/day or 0.23 kmol/day.
  - Greenhouse gas reduction is 0.23 kmol/day or 10 kg/day.

8.6 CASE STUDIES

8.6.1 Integration Options of a Coal-Fired Power Plant

In this case study four methods of system integration for coal-fired power plant are introduced, analyzed thermodynamically, and compared to a reference conventional coal-fired power plant. Note that system integration requires a sound theoretical basis to quantitatively assess its benefit with respect to a reference case. This assessment is multi-criteria because it must consider economic and environmental impact factors in addition to the pure technical performance assessment criteria.

In this respect the thermodynamic analysis methods of entropy generation minimization or exergy efficiency maximization intrinsically account for the potential impact of a substance or energy stream on the environment and can, therefore, be applied for benefit assessment of system integration. Chapter 2 of this book introduces some criteria for sustainability assessment of energy systems in Section 2.4. There, environmental impact indicators (see Table 2.3 for definition of abiotic resource depletion, global warming potential, stratospheric ozone depletion, toxicity, photo-oxidant formation, acidification, and eutrophication) and indicators such as improvement potential, sustainability index, depletion number, greenization factor, and exergy efficiency are discussed in detail.

8.6.1.1 Reference System: Conventional Coal-Fired Power Plant

Nanticoke Power Generation Station (NPGS) in Ontario is a coal-fired power plant of conventional type located on the north shore of Lake Erie. After thirty years of operation, this plant is presently undergoing a retrofitting program aimed at converting its coal-fired steam generator to natural gas only and/or natural gas/biomass co-firing systems. This station was introduced in Chapter 5 of this book (Table 5.16 and Figure 5.29). For this case study a reference power generation station similar to NPGS is used as a reference. Here we show the diagram of the reference conventional coal-fired power plant in Figure 8.37. For this system the turbines and pumps are assumed to have 80% isentropic efficiency and 90% mechanical efficiency. The calorific value of coal is 33 MJ/kg and its chemical exergy is 34 MJ/kg (9.44 kWh/kg).

This case study is modified from Dincer and Zamfirescu (2012). The reference power plant is cooled with lake water available at 15 °C. The plant is analyzed thermodynamically based on mass and energy and exergy balance equations. The parameters at each state point are given in Table 8.7, assuming that the mass flow rate of coal is 1 kg/s. The maximum exergy destruction in the coal-fired power plant occurs in the steam generator. For example, the steam generator of NPGS is responsible for 75% of the total exergy destructions of the plant (see Table 5.18).

The steam generator is a complex, multistage system (see the discussion in the second section of this chapter). However, the steam generator can be modeled in a simplified manner according to the model shown in Figure 8.38. The main processes within the steam generator are: coal combustion (1-2-4), water preheating (15-25-26), water boiling (26-27), water superheating (27-5), water reheating (6-7) and regenerative heat recovery from flue gas (31-4).

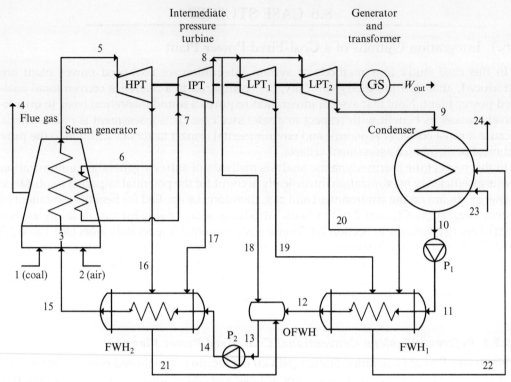

FIGURE 8.37   Diagram of the reference coal-fired power plant for the case study.

TABLE 8.7   Stream Parameters for Reference Power Plant

Stream	Content	T (°C)	P (MPa)	h (MJ/kg)	ex (MJ/kg)	$\dot{m}$ (kg/s)
1	Coal	15	0.101	32.764	34.181	1
2	Air	15	0.101	0.000	0.000	13.37
3	Hot flue gas	592.8	0.101	1.926	1.384	14.20
4	Stack gas	119.44	0.101	0.105	0.088	14.20
5	Steam	538	16.2	3.495	1.585	9.07
6	Steam	323.36	3.65	3.162	1.210	8.21
7	Steam	538	3.65	3.638	1.501	8.21
8	Steam	360.5	1.03	3.292	1.118	7.36
9	Steam	35.63	0.0045	2.502	0.175	6.19
10	Water	35.63	0.0045	0.099	0.003	7.36

Continued

TABLE 8.7 Stream Parameters for Reference Power Plant—Cont'd

Stream	Content	T (°C)	P (MPa)	h (MJ/kg)	ex (MJ/kg)	$\dot{m}$ (kg/s)
11	Water	35.73	1	0.101	0.005	7.36
12	Water	124.86	1	0.533	0.083	7.36
13	Water	165.86	1	0.738	0.147	9.07
14	Water	169.28	16.2	0.765	0.171	9.07
15	Water	228.24	16.2	1.073	0.291	9.07
16	Steam	323.36	3.65	3.162	1.209	0.86
17	Steam	423.23	1.72	3.413	1.253	0.32
18	Steam	360.5	1.03	3.293	1.118	0.54
19	Steam	223.0	0.241	3.029	0.817	0.69
20	Steam	86.0	0.0345	2.771	0.509	0.48
21	Water	188.33	1.21	0.855	0.189	1.18
22	Water	55.56	0.0133	0.195	0.013	1.16
23	Water	15	0.101	0.000	0.000	372.72
24	Water	23.3	0.101	0.040	0.001	372.72

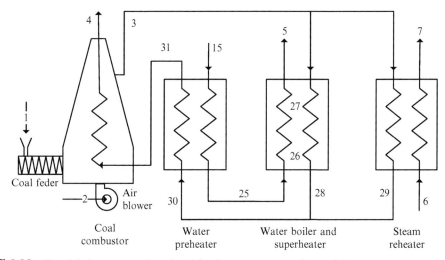

FIGURE 8.38 Simplified representation of coal-fired steam generator for modeling purpose.

The coal combustion process can be modeled based on a simplified chemical equation as follows:

$$C + \lambda \times (O_2 + 3.76N_2) \rightarrow CO_2 + (\lambda - 1)O_2 + 3.76 \times \lambda N_2$$

where $\lambda > 1$ is the excess air ratio. Using this chemical equation and the molar enthalpies of each relevant stream (including formation enthalpy), the following balance equation for the coal combustor can be written

$$\dot{n}_1 h_1 + \dot{n}_2 h_2 = \dot{Q}_{wph} + \dot{Q}_{sg} + \dot{Q}_{rh} + \dot{n}_4 h_4 \qquad (8.6)$$

where $\dot{Q}$ is the heat duty for each subsystem shown in the model in Figure 8.38 (wph: water preheater, sg: water boiler and steam superheater, rh: steam reheater).

Using Equation (8.6) and the state point parameters given in Table 8.7, the temperature profile of the steam generator is determined and represented in the temperature vs. heat duty diagram shown in Figure 8.39. The flue gas is at more than 1650 °C in state 3 and leaves the heat exchanger network at about 1350 °C in state 31. The temperature differences on the heat exchangers are higher than 800 °C. This explains the large exergy destructions in the steam generator of the reference plant. This badly degrades the exergy efficiency of the plant. Using the known formula for exergy efficiency (net power output vs. exergy input), one calculates the exergy efficiency of the plant at 36%.

The specific carbon dioxide emissions can be determined from the assumed chemical equation (above) which shows that for each mole of coal one mole of CO_2 is generated. If one denotes with $\dot{m}_C$ the mass flow rate of coal, then the net generated power results from chemical exergy and exergy efficiency according to $\dot{W}_{net} = \psi ex_C^{ch} \dot{m}_C$. Furthermore, the molar rate of coal results from $\dot{m}_C = \dot{n}_C M_C$, where $M_C = 12$ kg/kmol assimilates with the molecular mass of coal.

If coal is consumed at the rate $\dot{n}_C$ then carbon dioxide is expelled at the mass flow rate given by $\dot{m}_{GHG} = M_{CO_2} \dot{n}_C$. Therefore, one can approximate the environmental impact factor for the

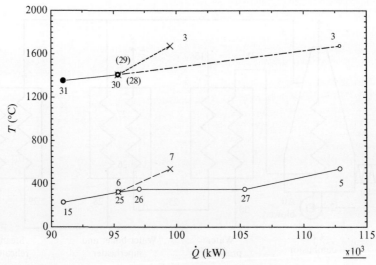

FIGURE 8.39   Temperature profile in the reference power generation plant.

power plant to be the ratio of the rate of carbon dioxide emission to the rate of net power generation

$$EI\left[\frac{kg_{CO_2}}{kWh}\right] \cong \frac{M_{CO_2}}{\psi \, ex_C^{ch} M_C} = \frac{0.389}{\psi} \tag{8.7}$$

where ex_C^{ch} can be expressed in kWh/kg. For the reference power plant, one calculates $EI_{ref} = 0.389/0.36 = 1.08 kg_{CO_2}/kWh$.

8.6.1.2 System Integration Option 1: Integration of an Advanced Coal Gasifier with a Rankine Plant

An incremental improvement of the reference system can be obtained if the coal-fired steam generator is replaced with an advanced gasifier. No other changes to the system are to be done: the temperature profile and flow rates for water boiling (15-25-26-27-5) and steam reheating (6-7) are fixed. Hence, the Rankine plant will have the same net power generation which, for 1 kg/s coal feed (assumed above), is $\dot{W}_{net} = 0.36 \times 9.44 = 3.39$ MW.

A membrane gasifier is used which generates pure hydrogen at low pressure; however, the gasification process occurs at higher pressure (approximately 20 bar), a fact that allows for better kinetics and faster flow rate diffusion through the hydrogen-selective membrane. The temperature in the gasifier varies in the range of 800–1800 °C. A combustion chamber for hydrogen with preheated air is integrated with the gasifier. The system is shown in Figure 8.40.

The auto-thermal coal gasifier is fed with coal, steam, and air. The gasifier comprises an integrated hydrogen-selective palladium membrane system with the role of enhancing the hydrogen conversion by shifting the chemical equilibrium forward and generating a pure stream of hydrogen. Hydrogen is further combusted with preheated air and the resulting combustion gases are supplied to the heat exchanger sections of the steam generator.

A small part of the combustion gases, comprising steam, oxygen, and nitrogen is diverted back to the gasifier, where steam and oxygen are used as reactants. The gases released out of the steam generator/reheating system are still hot and are, therefore, used to preheat the combustion air. The combustion gases are also used to preheat coal and water for the gasification process. Given that hydrogen combustion products are cleaner, it is possible to utilize their heat extensively until water condenses and can be separated and recovered. The chemical reaction for auto-thermal coal gasification is written as

$$C + 2\alpha H_2O + (1-\alpha)(O_2 + 3.76N_2) \rightarrow CO_2 + 2\alpha H_2 + 3.76(1-\alpha)N_2$$

where α is a stoichiometric factor calculated based on the energy balance for the auto-thermal gasifier (enthalpy of products is the same as the enthalpy of reactants).

Keeping the reaction temperature lower defavors formation of nitrogen oxides (NOx). Moreover, the GHG-containing flue gases at 1000 °C can be used to preheat the reactants up to 950 °C. Therefore, the reaction temperature is 1000 °C and the reactants are fed at an assumed 950 °C. The adiabatic flame temperature of the hydrogen combustor is set to 1200 °C. With these assumptions, the mass and energy balance equations for each of the subsystems of the integrated gasifier-combustor-steam generator system are solved to obtain the mass flow rates, compositions, and thermodynamic parameters at each state point. It is found

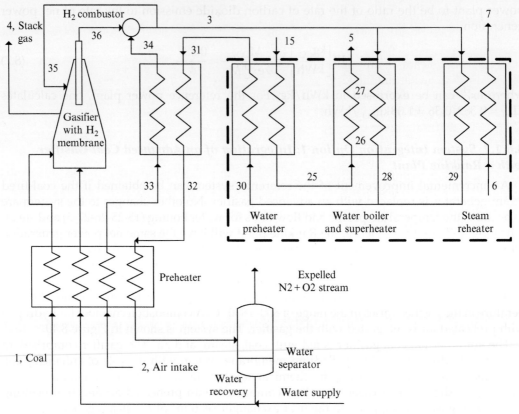

FIGURE 8.40   Integrated membrane-gasifier/combustor with steam generator of Rankine plant.

that the coal consumption rate is smaller than the one for the reference case due to better efficiency and the ability to apply internal heat regeneration, including the cooling of a part of the combustion products to below 100 °C. The temperature profile in the integrated steam generator is shown in Figure 8.41. The system operates under an essentially lower-temperature difference than in the reference case. Because of a decrease in the coal consumption rate, there is a corresponding decrease of exergy input into the system. It is found that the exergy efficiency of the system increases to 45%. The environmental impact factor calculated with Equation (8.7) is $EI = 0.86 \, \text{kg}_{CO_2}/\text{kWh}$. Furthermore, one calculates the exergy improvement factor and the system greenization factor with

$$\text{ExIF}_1 = \frac{\psi - \psi_{\text{ref}}}{\psi_{\text{ref}}} = 0.250 \quad \text{and} \quad \text{GF}_1 = \frac{\text{EI}_{\text{ref}} - \text{EI}}{\text{EI}_{\text{ref}}} = 0.203$$

8.6.1.3 System Integration Option 2: Coal/Biomass Gasification with Rankine Plant

More environmental benefit from integrated system 2 is obtained when, instead of coal, the fuel is a mixture of coal and biomass. Because biomass is virtually carbon neutral, the

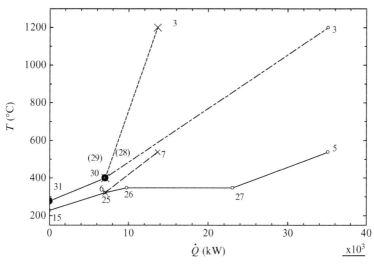

FIGURE 8.41   Temperature profile in the integrated gasifier-steam generator system.

greenhouse gas (GHG) emissions of the system decrease. The system remains the same as the one indicated in Figure 8.40 and the temperature at which hydrogen is generated is maintained at 1000 °C.

Cereal crops, wood, grass, straw, sawdust, and other biomasses can be used to obtain a net calorific value of 7–16 MJ/kg with an exergy content of 7.3–19.2 MJ/kg. The coal gasification process with the additional energy input from biomass can be modeled based on the following energy balance equation

$$h_C + 2\alpha h_{H_2O} + f_b NCV_b + (1-\alpha)(h_{O_2} + 3.76 h_{N_2}) = h_{CO_2} + 2\alpha h_{H_2} + 3.76(1-\alpha)h_{N_2} \qquad (8.8)$$

From the energy balance expressed by Equation (8.8) the stoichiometric factor α can be determined as a function of the parameter f_b which represents the stoichiometric fraction of biomass. If no biomass is fed, the situation treated above as integrated system option 1is found. If more biomass is fed, the allothermal gasification limit is reached for which $\alpha = 1$ and 2 mol of hydrogen is generated for the mole of coal.

The impact of factor f_b on the environmental impact factor is shown in Figure 8.42 together with the stoichiometric coefficient α from Equation (8.8) for an assumed NCV of 10 MJ/kg. This plot helps in deciding on the value of the biomass fraction f_b. For a biomass fraction higher than 0.7 allothermal gasification is approached with an environmental impact factor reduced below 0.25, but the consumption of biomass becomes too high. Rather, with a biomass fraction of 0.15 the consumption of biomass is reasonable and the environmental impact factor is decreased from 0.86 to approximately 0.73. In this case the stoichiometric factor α becomes 0.32 while the exergy efficiency of the system is about the same, 45%. Therefore,

$$ExIF_2 = 0.250 \text{ and } GF_1 = \frac{1.08 - 0.73}{1.08} = 0.321$$

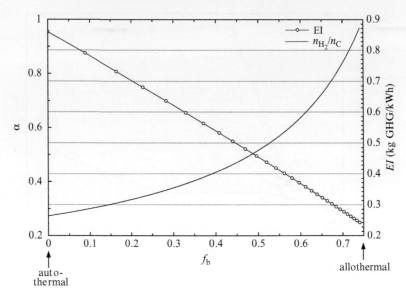

FIGURE 8.42   Environmental impact factor and steam gasification stoichiometric factor as a function of biomass vs. coal fraction for an integrated gasifier/steam generator.

8.6.1.4 System Integration Option 3: Gasifier + SOFC + GT + Rankine Plant

The power generation system can be incrementally improved by integrating an SOFC (solid oxide fuel cell) and a GT with the gasifier and Rankine cycle. The SOFC uses hydrogen from the gasifier while the expelled gases are combusted with additional air and directed to the GT. The GT turns an air compressor. The compressed air is fed into the fuel cell system and as well into the gasifier. The gases expelled by the SOFC comprise unreacted hydrogen, nitrogen, and oxygen. Therefore, there is no need for additional air for combustion. After the GT, the hot gas is diverted toward the steam generator and reheater. Again, it is assumed that the steam generator works at the same parameters as in the initial configuration (on the steam side). The process diagram for the integrated system is presented in Figure 8.43.

In this system, hydrogen produced by the gasifier is fed first to the SOFC. The uncombusted hydrogen is further combusted with oxygen from the cathode stream and the combustion products are expanded in a GT. Note that the compressor of the GT feeds air both to the gasifier and to the SOFC. The gases expelled by the GT are used to heat the steam generator of the Rankine cycle. The gasifier is similar to that describe above.

The following assumptions are made for the integrated system modeling:

- The isentropic efficiency of all turbines is 0.8 and of compressor is 0.7.
- The energy efficiency of the SOFC is 50%.
- The GT operates at 6 bar.
- The fuel is 15% biomass and 85% coal.
- Steam and fuel are preheated at 250 °C prior to gasifier entrance.
- Air is preheated at 538 °C prior to the gasifier inlet.

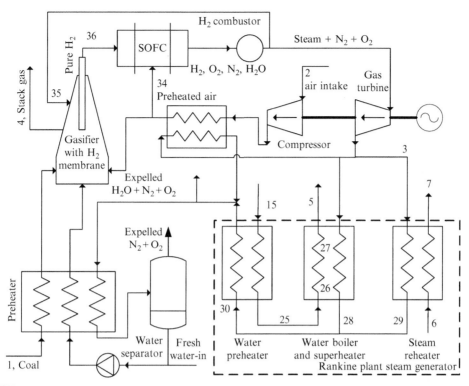

FIGURE 8.43 Integrated generation system with advanced gasifier, SOFC, GT, and Rankine cycle.

- The electrical power generated by the SOFC is given by $W_{el} = -\eta_{SOFC} \dot{n}_{H_2} \Delta G$.
- The following balance equation can be written for SOFC-combustor assembly.

$$h_{H_2} + 0.5\lambda(h_{O_2} + 3.76 \times h_{N_2}) = h_{H_2O} + 0.5(\lambda - 1) \times h_{O_2} + 1.88\lambda \times h_{N_2} - \eta_{SOFC}\Delta G$$

When the balance equation is solved for the excess air coefficient a value of $\lambda = 3.52$ is found and the combustion gas composition is determined to be 11% H_2O, 74% N_2, and 15% O_2. Furthermore, the back work ratio of the compressor is 0.37, while 7% of the mass flow rate of compressed air goes to the gasifier and the rest to the SOFC.

The system exergy efficiency increases dramatically due to a better match of the temperature profiles in the integrated gasifier-steam generator. Figure 8.44 shows the calculated temperature profile. Here, the performance of the integrated system is determined as follows:

- Exergy efficiency = 65%.
- Environmental impact factor = 0.41 kg_{CO_2}/kWh.
- Exergy improvement factor $ExIF_3 = 0.805$.
- Greenization factor $GF_3 = 0.62$.

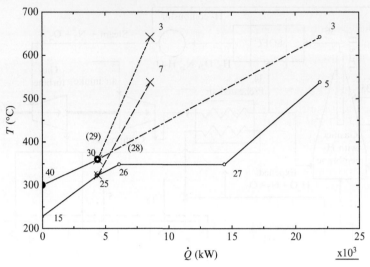

FIGURE 8.44   Temperature profile in gasifier-steam generation system of Option 3.

8.6.1.5 System Integration Option 4: G + SOFC + GT + Two Rankine Cycles + Cogeneration

The integrated system in Option 3 allows for more heat recovery of low-grade heat. In this context the application of an organic Rankine cycle or another cycle such as an ammonia–water cycle that works well with small temperature differences between the heat source and heat sink is opportune. Here a trilateral ammonia–water flash cycle is considered for heat recovery and heat cogeneration. This cycle, also described in Zamfirescu and Dincer (2008), uses, for this particular application, an ammonia–water mixture of 30% ammonia by mass as the working fluid. This system allows for a temperature glide from 40 to 80 °C at the heat sink side, thus being capable of generating hot service water at 65 °C. The integrated system is shown in Figure 8.45.

The figure shows that the second Rankine cycle is connected to the heat recovery system for lower-temperature combustion gases. In process 24–25, the exhaust gases are cooled. Their heat is transferred to coal for preheating prior to the gasifier but also to water for boiling and superheating prior to the gasifier. The rest of the heat is transferred to the heater of the ammonia–water Rankine cycle. The efficiency of this cycle is as low as 5% but its presence in the system allows for heat cogeneration, and thus the coal utilization factor increases from 67% (option 3) to 80% (option 4). The exergy efficiency of the system can be determined based on a balance equation: it is, as reported in Dincer and Zamfirescu (2012), 65%.

In terms of kWh generated, 85% of the total is in the form of power and 15% for heating. We will consider as a reference system for power and heating a combination of coal-fired power generation (for power) and a natural gas furnace (for heating). With an efficiency of 90% the natural gas furnace has a specific emission of 3.6 kg CO_2 per kWh. Therefore, the specific emissions for the cogeneration system are $EI_{ref,co} = 0.85 \times 1.08 + 0.15 \times 3.6 = 1.46 kg_{CO_2}/kWh$. The emissions for the integrated system are $0.41 kg_{CO_2}/kWh_e$ and $0.07 kg_{CO_2}/kWh_{th}$. Therefore, the total emissions of the integrated system are $EI = 0.85 \times 0.41 + 0.15 \times 0.07 = 0.36 kg_{CO_2}/kWh$.

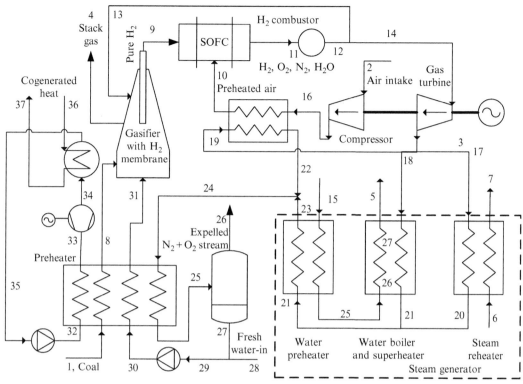

FIGURE 8.45 Integrated cogeneration plant with gasifier, SOFC, GT, and two Rankine cycles.

The exergy of cogenerated heat, based on a Carnot factor of 0.179 ($T_0=298\,\mathrm{K}$, $T=353$ K) and generation of 15% heating and 85% power, is 3% of exergy input. Therefore, the total generated exergy (power and heating) is 68% of exergy input, and this is the exergy efficiency. In conclusion, the exergy improvement factor is $\mathrm{ExIF_4}=(0.68-0.36)/0.36=0.89$. The greenization factor is $\mathrm{GF_4}=(1.46-0.36)/1.46=0.81$.

8.6.1.6 Comparison of the Integrated Systems

The analyzed systems are compared in Table 8.8 with respect to the exergy efficiency improvement factor and the greenization factor. In the table an additional system is added. This is in fact the same configuration as System 4 except that stack gas 4 passes through a carbon

TABLE 8.8 Comparison Table of Integrated Systems for the Case Study

| Parameter | Reference | System | | | | | Target |
		1	2	3	4	5	
ExIF	0	0.25	0.25	0.81	0.89	0.4	1.78
GF	0	0.20	0.32	0.62	0.81	1	1

capture system which extracts CO_2 and compresses it for storage. This system consumes a part of the generated power, which is an approximately 40% power generation penalty.

It is useful to represent the system performances on a chart with the axes being the exergy efficiency improvement factor and the greenization factor, respectively. If this is done, then the reference system is represented with a point at coordinates (0, 0). Furthermore, one identifies a target system at coordinates (1.78, 1) which represents an ideal system with no environmental impact and 100% exergy efficiency. Here the exergy efficiency improvement factor of 1.78 is obtained if—hypothetically—the exergy efficiency becomes 100%. Then, with respect to the reference system ExIF = $(1 - 0.36)/0.36 = 1.78$.

The chart representation in Figure 8.46 reveals the existence of Pareto frontiers, which represent the collection of optimized systems for any fixed degree of integration. It shows that both the exergy efficiency and greenization improve with the degree of system integration (or with system size and complexity). However, if the degree of integration is fixed, then ExIF and GF have opposite trends, as is in particular shown for Systems 4 and 5 (see Table 8.8).

The major gain in system performance (both ExIF and GF) is obtained when the SOFC+GT are integrated into the system. Points 1–5 on the chart represent the performance of the studied integrated systems. Although the Pareto frontiers in this study are drawn qualitatively, the trend presented here is general. This aspect is further discussed in the case study in Section 8.6.2, where a CCPP is analyzed.

8.6.2 Exergoeconomic and environmental optimization of a CCPP

Here we present a multiobjective optimization case study for a single-pressure CCPP which is based on work by Ahmadi et al. (2011). The system was described in Section 8.4 and to the schematic in Figure 8.19. The modeling equations are given in Table 8.4 and the thermodynamic cycle in Figure 8.20. Here the analysis is developed further by considering

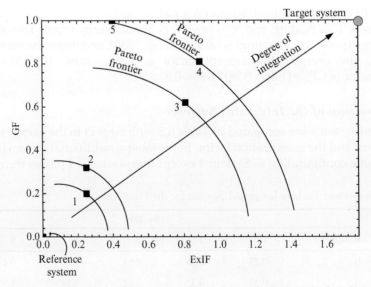

FIGURE 8.46   Impact of system integration on system's technical and environmental performance.

the costs and environmental impact indicators. In this case the specific cost of the system must be expressed as a function of thermodynamic design parameters.

The relevant economic and environmental impact parameters are introduced in Table 8.9. The economic equation allows for setting up a linear system that can be solved for cost rates at each state point within the power plant. As the power plant is the one represented in Figure 8.19, there are twelve material streams. In addition, there is an exergy cost associated with work generation by the turbine and work consumption of the compressor and pump. Moreover, the rate of NOx and CO pollution can be determined based on the equations given in the table. Furthermore, the carbon dioxide emission is correlated to the fuel consumption and the assumed chemical equation for combustion.

For the multiobjective optimization problem, three-objective functions are considered, the first being the exergy efficiency that must be maximized

$$f_{obj,1} = \psi = \frac{\dot{W}_{net}}{\dot{m}_f \, ex_f^{ch}}$$

The second objective function expresses the system cost rate and has four components: fuel cost rate ($\dot{C}_f$), total capital cost rate ($\sum \dot{Z}$), destruction cost rate ($\dot{C}_d$), and the cost rate associated with environmental impact ($\dot{C}_{EI}$). The cost of fuel is determined from the mass flow rate of consumed fuel and the specific cost of fuel $\dot{C}_f = C_f \dot{m}_f$. The destruction cost rate is proportional to exergy destruction according to $\dot{C}_d = C_f \dot{Ex}_d$ (see Ahmadi et al., 2011). The environmental impact cost has two components, the environmental cost penalty associated with NOx and CO emissions, respectively. This cost rate is: $\dot{C}_{EI} = C_{NOx} \dot{m}_{NOx} + C_{CO} \dot{m}_{CO}$. Here, as mentioned in Ahmadi et al. (2011), the following specific cost penalties are assumed for NOx and CO emissions: $C_{NOx} = 6.853$ \$/kg and $C_{CO} = 0.02086$ \$/kg. Hence, the second objective function to be minimized is

TABLE 8.9 Economic and Environmental Impact Parameters for Modeling

Parameter	Definition	
Capital cost	Z = the purchase cost, given in \$. It can be determined or assumed for each system component.	
Capital recovery factor	$CRF = \dfrac{i(1+i)^n}{(1+i)^n - 1}$	where i is the interest rate and n is the number of years of operation. It can be assumed that all system components have the same n.
Capital cost rate	$\dot{Z}\left[\dfrac{\$}{s}\right] = \dfrac{Z\,CRF\,MF}{3600N}$	where $MF > 1$ is the maintenance factor (typically $MF \cong 1.06$) and N is the annual number of hours of operation.
Cost balance equation	$C_f \dot{m}_f + C_{ex}\left(\sum_j \dot{Ex}_j\right)_{in} + \dot{Z} = C_{ex}\left(\sum_j \dot{Ex}_j\right)_{out}$	C_{ex} = specific cost of exergy [\$/MJ] C_f = specific cost of fuel [\$/kg] $\dot{C} = C_{ex}\dot{Ex}$ = cost rate [\$/s]
Specific NOx emissions	$m_{NOx}\left[\dfrac{g}{kg_{fuel}}\right] = 6.7 \times 10^{13}\dfrac{\exp(71{,}100/T_{ad})}{P_{cc}^{0.05}(\Delta P_{cc}/P_{cc})^{0.5}}$	T_{ad} = adiabatic flame temperature (K) P_{cc} = combustion chamber pressure (bar)
Specific CO emissions	$m_{CO}\left[\dfrac{g}{kg_{fuel}}\right] = 8.95 \times 10^{10}\dfrac{\exp(7800/T_{ad})}{P_{cc}^2(\Delta P_{cc}/P_{cc})^{0.5}}$	T_{ad} = adiabatic flame temperature (K) P_{cc} = combustion chamber pressure (bar)

$$f_{obj,2} = \dot{C}_{tot} = \dot{C}_f + \sum \dot{Z} + \dot{C}_d + \dot{C}_{EI}$$

The third objective function is the specific carbon dioxide emissions that must be minimized. This function is defined according to

$$f_{obj,3} = \dot{m}_{CO2} = \frac{\dot{m}_{CO_2}}{\dot{W}_{net}}$$

The objective functions depend on the following parameters that play the role of "decision variables" (see Figure 8.19): GT pressure ratio r, compressor isentropic efficiency $\eta_{s,c}$, GT isentropic efficiency $\eta_{s,gt}$, steam turbine isentropic efficiency $\eta_{s,st}$, pump isentropic efficiency $\eta_{s,p}$, GT inlet temperature T_{tit}, duct burner mass flow rate $\dot{m}_{db}$, main steam pressure P_s, main steam temperature T_s, pinch point temperature difference ΔT_{pp}, and condenser pressure P_c. Of course, the decision variables are constrained by reasonable ranges. The multiobjective optimization problem is formulated as follows:

$$\left\{ \max f_{obj,1}, \; \min f_{obj,2}, \; \min f_{obj,3} \, | \, r < 22, \; \eta_{s,c} < 0.9, \; \eta_{s,gt} < 0.9, \; \eta_{s,gt} < 0.9, \; T_{tit} < 1550 \text{ K}, \; \dot{m}_{db} \right.$$

$$\left. < 2 \text{kg/s}, \; P_s < 110 \text{ bar}, \; T_s < 700 \, K, \; 5\text{K} < \Delta T_{pp} < 20\text{K}, \; 5 \text{ kPa} < P_c < 15 \text{ kPa} \right\} \quad (8.9)$$

The optimization problem is solved with specific algorithms. Ahmadi et al. (2011) used a non-dominated sorted genetic algorithm and determined the Pareto frontiers expressed in two forms: (i) total cost rate as a function of exergy efficiency and (ii) total cost rate as a function of the specific carbon dioxide emission rate. These forms can be combined into a 3D representation of a Pareto frontier surface which plots the optimal solutions. For the considered optimization problem the Pareto frontier is shown in Figure 8.47, based on the data from Ahmadi et al. (2011).

The exergy efficiency increases together with the total cost rate. However, the cost escalates very much for exergy efficiency higher than 56%. The cost does not vary significantly for efficiency between 52% and 56%, but it increases sharply for efficiency above 56%. The conclusion is that the optimum system will have 56% efficiency based on maximum efficiency vs. minimum cost trade-off.

Regarding GHG emissions, these decrease steadily from 344 to 340 kg/MWh when the cost rate increases from 4880\$/h to 4920\$/h. However, the increase in cost rate is very abrupt if one wants to reduce the emissions rate below 340 kg/MWh. For example, one finds that for 339.75 kg/MWh the cost rate escalates from 4920\$/h to 5300\$/h. This reveal that a cost rate of 4900\$/h yields a good trade-off between cost rate and environmental impact. When all three-objective functions are considered together, a good trade-off solution can be found for a cost rate of 4900\$/h which yields 57.5% exergy efficiency and $340.3 \, \text{kg}_{CO_2}/\text{MWh}$.

8.6.3 Optimization of a Closed Multistage Flash ORC with Two-Phase Flow Expanders

Here we demonstrate the optimization of a multistage flash organic Rankine cycle which uses a two-phase flow expander instead of throttling valves. The cycle is comparatively assessed with respect to the commonly used cycle that uses throttling valves instead of expanders. The working fluid is R 245fa and the brine is available at 6 bar and a temperature of 150 °C.

FIGURE 8.47 Pareto 3D surface frontier of optimal solutions for three-objective optimization problem formulated in Equation (8.9). *Data from Ahmadi et al. (2011).*

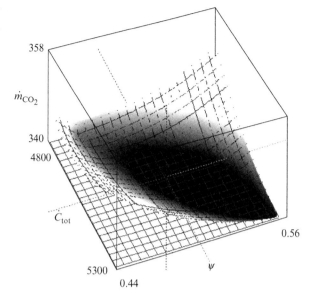

Figure 8.48 shows the system considered in this case study. The figure represents a quintuple closed-loop flash Rankine cycle. Two-phase expanders are used to flash saturated liquid and generate some power. This is an alternative and better option than using throttling valves to flash the saturated liquid, because throttling valve operation is isenthalpic and does not have ability to recover work from the expanding flow.

In this study it is assumed that the expanders have an isentropic efficiency of 0.7, and the turbines and the pump operate with 0.85 isentropic efficiency. The system is to be compared to the more common one that uses throttling valves instead of expanders. The number of stages is to be varied from 1 to 6 and the influence of the number of stages is to be analyzed. The system performance is assessed with the help of exergy efficiency, defined with:

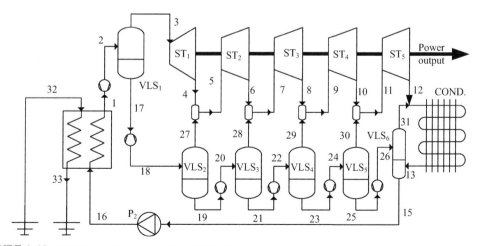

FIGURE 8.48 Multistage closed-loop flash ORC with two-phase expanders.

$$\psi = \frac{\dot{W}_{net}}{\dot{m}_b(ex_{32} - ex_{33})} \tag{8.10}$$

where $\dot{m}_b = \dot{m}_{32} = \dot{m}_{33}$ is the mass flow rate of the brine and ex_{29} is the specific exergy of the brine at the inlet. The brine is modeled as pure water. The reference state is taken at $T_0 = 298.15\,K$ and $P_0 = 101.325\,kPa$. The exergy of the brine at the heat exchanger inlet is

$$ex_{32} = h_{32} - h_0 - T_0(s_{32} - s_0)$$

We assume that the mass flow rate of brine is 1 kg/s, the pinch point temperature difference is 10 °C, the brine inlet temperature is $T_{29} = 150\,°C$, and $P_1 = 600\,kPa$. In addition, the temperature difference between the brine and working fluid at the pump inlet section is $\Delta T = T_{30} - T_{15} = 10\,°C$. Also, the condensation temperature is assumed to be $T_c = 30\,°C$.

Mass and energy balance equations must be written for each of the system components so that the thermodynamic state parameters and flow rates are determined for each stream. The optimization problem is formulated as follows:

$$\max\left\{\psi(PR_i, x_1) \,|\, PR_i = \frac{P_i}{P_{i-1}} > 1 \,(\text{per stage } i),\, i = 1,\dots,5,\, x_1 \geq 1\,(\text{vap. quality in 1})\right\} \tag{8.11}$$

The optimization is done for three cases as follows:

- Isentropic efficiency of the expanders is taken as zero; this case corresponds to a system that has throttling valves instead of expanders.
- Isentropic efficiency of all expanders is assumed to 0.7.
- Two cases are considered at state 1 (first expander inlet): (i) saturated liquid and (ii) a two-phase flow with certain quality.

Here we show the optimization results in Figure 8.49 in the form of curves $\psi = f(n)$, where ψ represents the maximum exergy efficiency optimized for each value of n (number of stages).

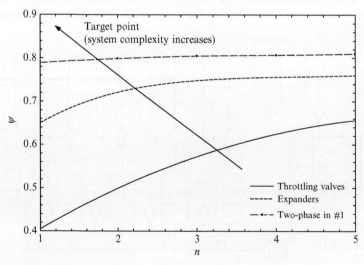

FIGURE 8.49 Optimization results for the multistage power plant; the curves represent the maxima of exergy efficiency as a function of the number of stages.

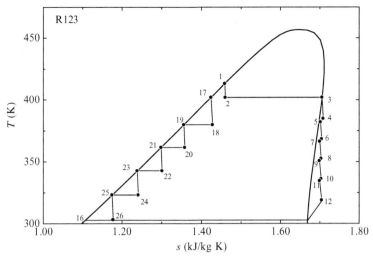

FIGURE 8.50   Optimized five-stage flash organic Rankine cycle

The first curve in the graph (continuous line) represents the variation of maxima of exergy efficiency as a function of the number of stages for a system that uses throttling valves. Here it is observed that the exergy efficiency increases when the number of stages is increased. This is a trade-off problem for selection of the best system because the increase in the number of stages means an increased investment cost. On the other hand the increase in exergy efficiency leads to higher productivity and, hence, higher revenues from the system. Therefore, the curve $\psi(n)$ represents a Pareto frontier. The target point for design selection is toward higher exergy efficiency and reduced number of stages.

The second curve in the plot (dashed line) represents the variation of exergy efficiency maxima with the number of stages for the system that uses the expander instead of throttling valves. Therefore, reduced exergy destruction is obtained and some additional power is produced. The efficiency is higher and the curve is less sensitive to the number of stages.

The last curve is drawn for the case when at the first expander intake (1) is a two-phase flow. This flow is obtained by allowing heating along process 16-1 (see Figure 8.48) to enter in a two-phase region (incipient boiling) until certain vapor quality is reached. It is observed in the optimization problem formulation—Equation (8.11)—that vapor quality at 1 is chosen as an optimization variable. The last system has the highest exergy efficiency; also, the exergy efficiency is less sensitive to the number of stages. Figure 8.50 shows the optimized cycle with five stages and saturated liquid at the intake of the expander in state 1; the exergy efficiency of this cycle is 76%, although its energy efficiency is only 15%.

8.7 CONCLUDING REMARKS

In this chapter, integrated power generation systems are introduced and discussed. The importance of integration to obtain higher efficiency and system sustainability is emphasized. Four major methods of system integration are identified and discussed in separate sections:

multistaging, cascading, combining, and hybridizing power generation systems. Of course, any combination of these basic methods can be used to design more complex and more robust systems capable of achieving higher performance. The chapter ends with three illustrative case studies that further demonstrate the benefits of system integration.

Study Problems

1. Design a transcriticalRankine cycle with carbon dioxide having condensation at 30 °C and a temperature at turbine inlet of 800 °C. Assess the benefit of turbine staging with interstage reheating to 800 °C. One can assess this comparatively, that is by comparing efficiencies (energy, exergy) of the single-stage system with a two-stage or three-stage expansion system.

2. Compare the exergy efficiency of a steam generator to that of an ammonia–water vapor generator starting from the diagrams in Figure 8.4.

3. Consider the system in Figure 8.5. Develop a design of a duct burner that produces steam at two pressures and try to maximize its exergy efficiency.

4. Starting from the work of Henriksen et al. (2006) do a design for the integrated system presented in Figure 8.6.

5. Consider the cogeneration system in Figure 8.14. Determine the thermodynamic properties at each state point under reasonable assumptions and calculate the system efficiencies.

6. Rework Example 8.1 for a three-stage fuel cell system.

7. Make a design for a cascaded mercury-steam cycle. Consider Example 8.2 as a starting point for the design. Implement routines for a mercury equation of state from literature sources, e.g., Sugawara et al. (1962).

8. Consider the cascaded Brayton cycle in Figure 8.17 and the cycle represented qualitatively in Figure 8.18. Using reasonable assumptions, determine the efficiency of this cycle and compare it with a regenerative Brayton cycle operating under similar conditions.

9. Consider a dual-pressure combined cycle power plant such as the one represented in Figure 8.21. Using balance equations (see Table 8.4) and reasonable assumptions determine the plant efficiency and assess its specific emission parameters.

10. Use literature sources (e.g., the work by Guan et al., 2010) to design an advanced IGPP based on the layout in Figure 8.23. Compare the plant performance (efficiency and specific emission) to that of the plant shown in Figure 8.22 under similar working conditions.

11. Consider a combined fuel cell and GT plant of the layout presented in Figure 8.26. The SOFC power is 10 MW, and it operates with natural gas at 1000 K. Determine the size of the compressor, the duty of the regenerator, and the efficiency of the overall combined plant. Compare the efficiency of the combined plant with that of the SOFC alone. Make reasonable assumptions.

12. Using literature data and reasonable assumptions make a design for the integrated system in Figure 8.29. Determine its efficiency and specific emissions and compare them with those of a conventional coal-fired power plant.

13. Rework Example 8.3 for an assumed sulfur-free fuel represented by the formula $C_{6.5}H_{7.4}O_3N_{0.07}$. The produced syngas has the molar composition of 48% H_2 and 36% CO, with the rest being CH_4.

14. For a hybrid solar and wind system similar to the one shown in Figure 8.33 determine the installed power of the rectifier, inverter, electrolyser, and fuel cell and the capacity of the battery bank and hydrogen storage canisters, provided that the wind energy input is 10 MW and solar energy input is 20 MW. Estimate the required size of land for wind and solar energy collection assuming a location of your choice in North America.

15. Determine the parameters of the system in Figure 8.35 when it is used for a residence in the Ontario region and the solar collector is 15 m^2. Try to select the actual fuel cell system from a manufacturer and determine the required capacity of the pump, the heat exchanger duty, and the size of the reformate storage tank.

16. Using literature sources, design the system in Figure 8.36 for two cases: (i) ammonia is reformed and (ii) methane is reformed. Compare the efficiencies and the costs.

17. Assume that for the integrated system in Figure 8.45 carbon dioxide is captured and compressed to supercritical state for sequestration. Determine the energy penalty and the values for the exergy improvement factor. Make reasonable assumptions.

18. Rework case study 8.6.3 for a geothermal source at 100 °C with R134a as working fluid.

Nomenclature

A	area, m^2
C	concentration ratio
$\mathcal{C}$	specific cost, \$/MJ
C_p	molar specific heat, kJ/mol K
CRF	capital recovery factor
E	electric potential, V
EI	environmental impact factor, kg_{CO_2}/kWh
ExIF	exergy efficiency improvement factor
$\dot{Ex}$	exergy rate, W
ex	specific exergy, kJ/kg or kJ/mol
F	Faraday constant, C/mol
FAR	fuel-to-air ratio
f_b	stoichiometric fraction of biomass
G	molar specific Gibbs free energy, J/mol
GF	greenization factor
H	molar specific enthalpy, J/mol
h	specific enthalpy, J/kg
i	interest rate
J	current density, A/m^2
K	equilibrium constant
LF	linearizing factor
LHV	lower heating value
LMTD	log mean temperature difference, K
M	molar mass, kg/kmol
m	specific emission factor, g/kg or g/kWh
$\dot{m}$	mass flow rate, kg/s
N	number of stages
N	annual number of hours of operation, h

n	number of moles, mol
$\dot{n}$	molar flow rate, mol/s
P	pressure, kPa
PEF	power enhancement factor
PR	pressure ratio
$\dot{Q}$	heat flux, W
R	universal gas constant, J/mol K
$\mathscr{R}$	specific ohmic resistance [Ω/cm^2]
r	compression ratio
s	specific entropy, J/kg K
T	temperature, K
t	time, s
U	overall heat transfer coefficient, W/m^{2}K
U_f	fuel utilization factor
v	specific volume, m^3/kg
$\dot{W}$	power, W
$\dot{W}''$	power density, W/m^2
x	vapor fraction or vapor quality
y	molar fraction
Z	purchase cost, $
$\dot{Z}$	capital cost rate, $/s

Greek Letters

α	stoichiometric factor
γ	specific heat ratio
ε	effectiveness
κ	adiabatic expansion exponent
λ	excess air coefficient
η	energy efficiency
ψ	exergy efficiency
ζ	extend of reaction

Subscripts

0	reference state
a	air
act	activation
ad	adiabatic
air	air
b	brine
C	compressor
cc	combustion chamber
cg	combustion gases
co	cogeneration
comp	compressor
d	destroyed
e	exit
ex	exergy
GT	gas turbine
i	inlet
in	inlet
inj	injection

f	fuel
fc	fuel cell
fuel	fuel
loss	losses
net	net generation
oc	open circuit
opt	optimal
out	outlet
P	pump
Q	heat
r	reduced
ref	reformed or refernece
rh	reheater
sg	steam generator
SOFC	solid oxide fuel cell
ST	steam turbine
T	turbine
th	thermal
w	water
wph	water preheater

Superscript

0	standard state
ch	chemical

References

Ahmadi, P., Dincer, I., Rosen, M.A., 2011. Exergy, exergoeconomic and environmental analyses and evolutionary algorithm based multi-objective optimization of combined cycle power plants. Energy 36, 5886–5898.

Al-Sulaiman, F.A., 2013. Energy and sizing analyses of parabolic trough solar collector integrated with steam and binary vapor cycles. Energy 58, 561–570.

Angelino, G., Invernizzi, C., 2008. Binary conversion cycles for concentrating solar power technology. Solar Energy 82, 637–647.

Au, S.F., Woudstra, N., Hemmes, K., 2003. Study of multistage oxidation by flowsheet calculations on a combined heat and power molten carbonate fuel cell plant. Journal of Power Sources 122, 28–36.

Cengel, Y.A., Boles, M.A., 2010. Thermodynamics: An Engineering Approach. McGraw-Hill, New York.

Christou, C., Hadjipaschalis, I., Poullikkas, A., 2008. Assessment of integrated gasification combined cycle technology competitiveness. Renewable and Sustainable Energy Reviews 12, 2459–2471.

Colpan, C.O., Dincer, I., Hamdullahpur, F., 2007. Thermodynamic modeling of direct internal reforming solid oxide fuel cells operating with syngas. International Journal of Hydrogen Energy 32, 787–795.

Dincer, I., Zamfirescu, C., 2011. Sustainable Energy Systems and Applications. Springer, New York.

Dincer, I., Zamfirescu, C., 2012. Potential options to greenize energy systems. Energy 46, 5–15.

El-Emam, R.S., Dincer, I., Naterer, G.F., 2012. Energy and exergy analyses of an integrated SOFC and coal gasification system. International Journal of Hydrogen Energy 37, 1689–1697.

El-Maskoud, R.M.A., 2013. Binary Brayton cycle with two isothermal processes. Energy Convers. Manage. 73, 303–308.

Guan, G., Fushimi, C., Tsutsumi, A., Ishizuka, M., Matsuda, S., Hatano, H., Suzuki, Y., 2010. High-density circulating fluidized bed gasifier for advanced IGCC/IGFC—Advantages and challenges. Particuology 8, 602–606.

Haseli, Y., Dincer, I., Naterer, G.F., 2008. Thermodynamic modeling of a gas turbine cycle combined with a solid oxide fuel cell. International Journal of Hydrogen Energy 33, 5811–5822.

Henriksen, U., Ahrenfeldt, J., Jensen, T.K., Gøbel, B., Bentzen, J.D., Hindsgaul, C., Sørensen, L.H., 2006. The design, construction and operation of a 75kW two-stage gasifier. Energy 31, 1542–1553.

Kim, Y., Lorente, S., Bejan, A., 2011. Steam generator structure: Continuous model and constructal design. International Journal of Energy Research 35, 336–345.

Naterer, G.F., Dincer, I., Zamfirescu, C., 2013. Hydrogen Production from Nuclear Energy. Springer, New York.

Odukoya, A., Dincer, I., Naterer, G.F., 2011. Exergy Analysis of a Gasification-Based Combined Cycle with Solid Oxide Fuel Cells for Cogeneration. International Journal of Green Energy 8, 834–856.

Ratlamwala, T.A.H., Dincer, I., 2013. Comparative efficiency assessment of a multi-flash integrated system based on three efficiency definitions. Int. J. Low Carbon Tech. 8, 238–244.

Standaert, F., Hemmes, K., Woudstra, N., 1998. Analytical fuel cell modelling; non-isothermal fuel cells. J. Power Sourc. 70, 181–199.

Sugawara, S., Sato, T., Minamiyama, T., 1962. On the equation of state of mercury. Jpn. Soc. Mech. Eng. 5, 711–718.

Wagar, W.R., Zamfirescu, C., Dincer, I., 2011. Thermodynamic analysis of solar energy use for reforming fuels to hydrogen. International Journal of Hydrogen Energy 36, 7002–7011.

Zamfirescu, C., Dincer, I., 2008. Thermodynamic analysis of a novel ammonia–water trilateral Rankine cycle. ThermochimicaActa 477, 7–15.

9

Multigeneration Systems

9.1 INTRODUCTION

Today, more than ever, the development of affordable, reliable, sustainable, and secure power generation systems is of great concern because the global energy demand increases steadily with the growth of population and increased living standards. Energy security poses a global problem as the conventional fossil fuel and nuclear resources are concentrated in a limited number of geopolitical regions around the globe, while many other countries rely mostly on fuel imports. It is obvious today that the problem of power generation requires a multidimensional solution that can address simultaneously the needs for higher efficiency, lower pollution, usage of a large diversity of energy supply (as opposed to only one type of energy) and, of course, increased economic attractiveness of alternative energy systems as an inducement to substantial investment.

Multipurpose power generation systems have emerged in recent years as a robust solution to these problems. These are integrated systems that combine key power cycles with key processes such that ultimately the power generation platform is able to (i) convert diverse energy sources, and (ii) produce power as the main output and in addition a multitude of other marketable products.

Multigeneration systems are an extension of the integrated systems treated in Chapter 8, where power-only and combined power and heat production (CHP) were discussed. CHP is the simplest multigeneration system, which besides power cogenerates heat as a product. Depending on the actual system, when cogeneration is applied the energy utilization factor increases from 15–30% (power only) to 50–80% when heat is cogenerated (although the ideal limit is 100%, as heat generation is not subject to the Carnot factor).

The key point is to seek alternatives to produce useful commodities in addition to power generation. For example, Hasan et al. (2002) proposed an ammonia-water cycle that simultaneously generates power and cooling. Ratlamwala et al. (2010) developed a power and cooling system that integrates an absorption refrigerator with a fuel cell. These systems have better energy conversion efficiency than those with two independent systems (a power cycle and a refrigerator).

Trigeneration systems add one more output to the cogeneration system: they produce power, heating, and cooling. In fact, trigeneration integrates a power cycle with a cooling cycle and a heat recovery system. Besides cooling and heating there are many types of outputs that can be considered for multipurpose power generation systems: desalination (freshwater and salt water), hydrogen and other synthetic fuels, other chemicals (chlorine, oxygen, sodium hydroxide), etc.

Some recent papers on trigeneration systems are given subsequently. Khaliq et al. (2009a) proposed a trigeneration system that works with heat recovery from industrial waste. The system is an extension of the cogeneration system with ammonia-water formerly proposed by Hasan et al. (2002). Further, Khaliq et al. (2009b) proposed a trigeneration system for heating, cooling, and power that includes an Li–Br absorption refrigerator, a heat recovery steam generator, and a steam Rankine cycle (SRC), and uses 425–525 °C recovered flue gas from industry. The exergy efficiency ranges between 35% and 52%, depending on the inlet gas temperature. It is shown that the exergy efficiency varies from 45% to 52% if the pinch point in the heat recovery steam generator is reduced from 50 to 10 °C. A trigeneration system that integrates a gas turbine cycle, a steam turbine cycle, and a single-effect absorption chiller was analyzed by Ahmadi et al. (2011) and showed improved efficiency and less environmental pollution when compared to independent energy conversion systems. Al-Sulaiman et al. (2010) showed a system integration of a solid oxide fuel cell with a Rankine cycle and an absorption refrigerator for trigeneration.

Cetinkaya (2013) studied two types of trigeneration systems. "System 1" uses concentrated solar radiation to produce power, hot water, and desalination (freshwater) and integrates two power cycles (organic Rankine cycle (ORC) and steam Rankine) and a key process (desalination). "System 2" produces power, hot water, and synthetic fuel using coal and biomass as energy input. This system integrates three power cycles (Brayton, steam Rankine, and ORC) with two key processes (gasification and Fischer–Tropsch (FT) synthesis). The study confirmed that multigeneration gives better results than single generation because the studied systems have almost double efficiency with respect to individual single generation units that produce the same outputs.

In addition, recent advances in the thermochemical water-splitting cycle suggest the possibility of their integration with a nuclear reactor for trigeneration of power, hydrogen, and oxygen. An example of a trigeneration system of this kind was presented in Zamfirescu et al. (2010).

Polygeneration systems or multigeneration systems produce more than three outputs. An integrated power generation system that produces power, space heating, hot water, cooling, and hydrogen was introduced in Ahmadi et al. (2013a). In another system Ahmadi et al. (2013b) used biomass as the primary energy source to generate power, hot water, cooling, and hydrogen. The system combines a biomass combustor with an ORC, an absorption chiller and a proton exchange membrane electrolyser (PEME) to produce hydrogen. In a recent paper, Ozturk and Dincer (2013a,b) presented a poly-generation system to produce six outputs, which were power, hot water, space heating, cooling, hydrogen, and oxygen. The system consists of a SRC, an ORC, absorption cooling and heating, high-temperature steam electrolysis (HTSE) and a proton exchange membrane fuel cell. The exergy efficiency of the optimized system is superior to 57%.

Dincer and Zamfirescu (2012) demonstrated the benefit of increasing the number of outputs from one (power-only) to six (power, hot water, space heating, cooling, hydrogen, and salt) when the energy input was in the form of intermediate-temperature heat around 200 °C.

It was shown that, compared to single generation, cogeneration increases the GHG (greenhouse gas) mitigation about 2–4 times and the payback period decreases 2.8 times.

Multigeneration systems must be designed in such a way that the additional investment for production of outputs in addition to power is justified by higher revenues, a shorter payback period, lower levelized electricity cost, or lower environmental impact. Therefore, assessment criteria must be established for multigeneration systems to assess their benefits from multiple points of view and compare them with competing systems. Furthermore, special multi-objective, multivariable optimization techniques must be applied to improve the design of power systems with multiproduct generation. These aspects are all treated in this chapter, which is based on ten illustrative case studies that demonstrate the design methods, comparative assessment, and optimization of multigeneration systems.

9.2 KEY PROCESSES AND SUBSYSTEMS FOR MULTIGENERATION

In Chapters 4–8 many types of power generation systems, conventional and advanced, have been introduced, and some key processes have been mentioned, such as fuel synthesis, drying, desalination, gasification, etc. Here the key processes and subsystems which can play a role in designing power systems with multiproduct generation are inventoried. In addition, performance criteria are developed for assessment of multigeneration systems.

A simplified representation of multigeneration systems is sufficient for modeling their main features. Figure 9.1a illustrates a single generation system that produces power only. The figure of merit for this system is energy efficiency or energy utilization factor, that is $\eta = \dot{W}/\dot{E}_{in}$. Similarly one has $\psi = \dot{W}/\dot{E}x_{in}$ as exergy efficiency, and other performance indicators such as specific cost $c = C/\dot{W}$ (measured in \$/kW) or specific emissions factor $SEF = \mathcal{M}/\dot{W}$ (where $\mathcal{M}$ is the emissions produced, measured depending on the case in kg CO_2 equivalent or as compounded emissions accounting for CO_2, CO, NO_x, SO_x, and VOCs; VOCs = volatile organic compounds).

FIGURE 9.1 Simplified representation of a (a) single generation and a (b) cogeneration system.

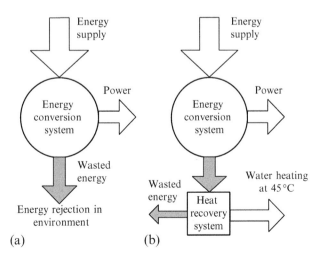

Assume that energy supply and power generation rates (or amounts) remain the same, but the waste energy is partially recovered and used for water heating. This is the system in Figure 9.1b. Therefore, in an incremental manner, this system will produce marketable heating (hot water) besides power and its figure of merit shows improvement because for the same $\dot{E}_{in}$ and $\dot{Ex}_{in}$: (i) the output is $\dot{W} + \dot{Q}_{wh}$ instead of $\dot{W}$ in terms of energy, and (ii) the output is $\dot{W} + \dot{Q}_{wh}(1 + T_0/T_{wh})$ instead of $\dot{W}$ in terms of exergy. Therefore, specific emissions are reduced, and if the additional investment is not high, and if the revenue from heating is reasonable, then the specific power cost is also reduced.

The systems shown in Figure 9.2 represent two incremental improvements of the cogeneration system in Figure 9.1b. The trigeneration system in Figure 9.2a recovers more heat and delivers heating at two levels: water heating at 45 °C and space heating at 25 °C. Furthermore, the quadruple generation system from Figure 9.2b uses a part of heat recovery to generate cooling with an absorption refrigerator. Of course, it is assumed that the main energy conversion system rejects a part of the heat at a temperature level of around 100 °C or a little more, which is necessary to drive the absorption refrigeration machine.

The systems shown in Figure 9.3 implement quintuple and sextuple generation as they produce hydrogen and hydrogen and desalination, respectively, in addition to power, water heating, space heating, and cooling. Here, cascaded systems are suggested for integration. Power generation is produced by the topping cycle. The bottoming systems use heat recovered from the topping system to drive absorption refrigeration and a heat engine (e.g., ORC) which generates additional power to run an electrolyser and to produce hydrogen. The heat

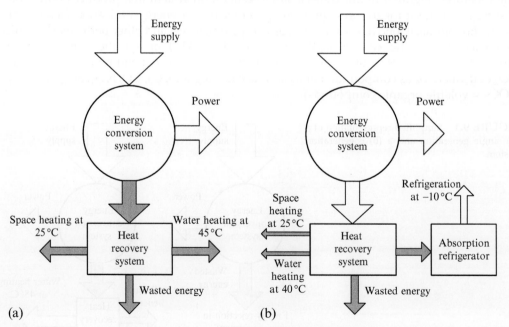

FIGURE 9.2 System improvement by incremental addition of generated products from trigeneration (a) to quadruple generation (b).

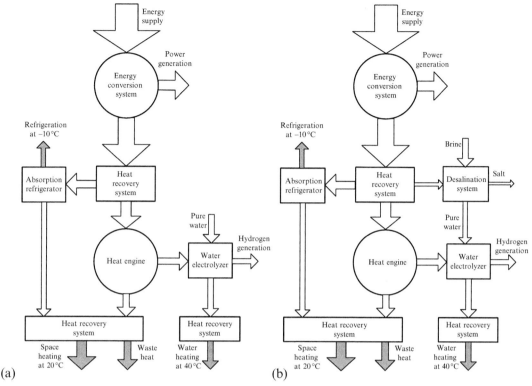

FIGURE 9.3 Integrated power generation systems with multiple products: (a) quintuple generation system and (b) sextuple generation system.

rejected by the bottoming energy conversion systems is used for low-grade applications: hot water production at 45 °C and space heating at 25 °C.

The systems in Figures 9.1–9.3 show how power systems with additional product generation can be conceptualized. The systems are designed such that the output incrementally increases as one proceeds from system 1a to system 3b. For example, the net generated output can be expressed in the form of the rate of energy embedded in the products. This is, for the sextuple power generation system, the following sum

$$\dot{E}_{gen} = \dot{W} + \dot{Q}_{wh} + \dot{Q}_{sh} + \dot{Q}_{c} + \dot{m}_{sf}HHV_{sf} + \dot{m}_{s}h_{s}$$

where the subscripts stands for water heating (wh), space heating (sh), cooling (c), synthetic fuel (sf), salt (s), $\dot{W}$ is power generation, $\dot{Q}$ is heating or cooling rate, $\dot{m}$ is mass flow rate, and h is specific enthalpy (including formation enthalpy term). Similarly, there is a net exergy generation as follows:

$$\dot{Ex}_{gen} = \dot{W} + \dot{Q}_{wh}\left(1 - \frac{T_0}{T_{wh}}\right) + \dot{Q}_{sh}\left(1 - \frac{T_0}{T_{sh}}\right) + \dot{Q}_{c}\left(\frac{T_0}{T_c} - 1\right) + \dot{m}_{sf}ex_{sf}^{ch} + \dot{m}_{s}ex_{s}^{ch}$$

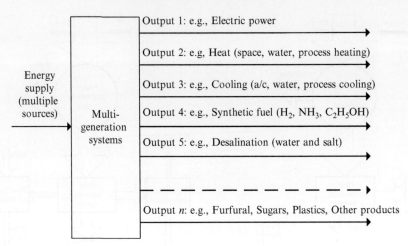

FIGURE 9.4   Layout of a power generation system with multiple outputs.

An abstract representation of a multigeneration system is shown in Figure 9.4, where it is suggested that from one or multiple energy supplies many dissimilar outputs are produced. The system configuration must be determined by the engineer who designs the system and optimized in such a way that the benefit of the multigeneration system is maximized. The system will include a number of power generation cycles which may be considered as key subsystems.

The benefit of system integration for multigeneration can be demonstrated for a simple case study with practical relevance: supply of the energy needs of a household in terms of power, heating, hot water, and air conditioning. A simple thermodynamic analysis leads to the results presented subsequently. Assume a conventional household that consumes 11 MWh electric power annually (including the power required for air conditioning in summer). The household heating demand is 18 MWh per year for space and water heating. The multigeneration system produces power, space and water heating in winter, and power, water heating, and air conditioning in summers. Four systems are considered for the analysis, among which three are conventional (systems 1, 2, 3) and one is based on solar energy (system 4). The systems are described as follows:

- System 1: Natural gas heated. Grid-connected household with natural gas water and space heating system and electrical air conditioning.
- System 2: Heating oil based. Grid-connected household with heating-oil water and space heating system and electrical AC.
- System 3: Electrical heat pump based. Grid-connected household with electrical HP/AC system for water and space heating system and air conditioning.
- System 4: Solar energy based (alternative energy system). Integrated, grid-connected trigeneration system that delivers some energy to the grid during daytime and fully uses the grid at night. Production: power, water, and space heating in winter; power, water heating, and AC in summer. It consists of a ∼5-m-diameter dish for solar concentration, a heat engine and heat recovery system, and a thermal storage system. It has 5 kW electric and 15 kW thermal power generation.

TABLE 9.1   Comparison of Conventional and Integrated Trigeneration Systems for Household

Item	System 1 (Natural Gas)	System 2 (Heating Oil)	System 3 (Electrical Heat Pump)	System 4 (Solar Trigeneration)
Investment	$13,000	$13,000	$6000	$40,000
Operation costs	$28,000	$41,000	$38,000	−$5000 (alternative energy credit)
Lifetime cost	$41,000	$54,000	$44,000	$35,000
Electricity cost	10¢/kWh	10¢/kWh	10¢/kWh	2¢/kWh
Thermal energy cost	$18/MWh	$47/MWh	$40/MWh	$30/MWh

For all systems a lifetime of 25 years is assumed, and for the alternative energy system #4 there is a tax credit of $5000 for the lifetime to compensate for GHG mitigation and renewable energy generation fed to the grid. The systems are compared in Table 9.1.

A non-exhaustive list of power generation cycles to be considered for integration into a multigeneration system is as follows:

- Steam Rankine and ORCs (see Chapter 5).
- Gas turbines, open or closed Brayton cycles (see Chapter 5).
- Supercritical and transcritical carbon dioxide cycles (see Chapters 5 and 10).
- Ammonia-water power cycles (see Chapter 10).
- Fuel cell systems (see Chapter 4).
- Water electrolysis systems (see Chapter 4).
- Renewable-energy-based power generation systems (see Chapter 7).

Within multigeneration systems one or several power cycles are integrated with selected processes. There are a number of key processes that may play a substantial role in the development of multigeneration systems. Here is a non-exhaustive list of processes involved in multigeneration:

- Combustion and gasification (of coal and of biomass).
- Direct coal liquefaction and plastics-to-fuel conversion.
- Heat recovery steam (or vapor) generation.
- Methanol synthesis and methanol reforming.
- Ammonia synthesis and ammonia cracking (reversible/equilibrium process).
- FT synthesis.
- Water desalination.
- Solar light processing (concentration, collimation, and spectral splitting).
- Biochemical processes (digestion, bio-photolysis).

Combustion characteristics of fossil fuels and biofuel/biomasses are given in Chapter 3. Combustion modeling is treated in Chapter 5, based on energy balance equations. Fluidized bed combustion systems are also a good choice for integration with multigeneration, especially when coal rank is low. Some advanced chemical looping combustion systems are described in Chapter 10.

Gasification is described in Chapter 4 as a method of hydrogen production, in Chapter 7 as one of the most effective methods of biomass energy conversion, and in Chapter 8 ina framework of integrated power generation systems. Co-gasification of biomass and coal is another example of integration which takes advantage of the high oxygen content of biomass that allows for a lower oxygen requirement in the pyrolysis phase.

Direct coal liquefaction is a process that directly converts coal to a liquid hydrocarbon fuel. This process is attractive for multigeneration because it occurs at a relatively low temperature (400–500 °C), although the pressure is high ($\sim$150 to $\sim$600 bar). The process occurs catalytically by coal hydrogenation: mC(s) + $(m+1)$H$_2$(g) $\rightarrow$ C$_m$H$_{2m+2}$(l).

Indirect liquefaction of coal and biomass can be conducted by cascading gasification with a FT process. This process uses syngas as input and converts it thermocatalytically to a liquid hydrocarbon fuel with a high cetane number. This can replace conventional diesel fuel. Another conversion process of interest which generates diesel fuel is plastics-to-fuel conversion, through which wasted plastic materials are converted to diesel by thermocatalytic cracking.

Methanol and ammonia synthesis are considered very established processes and excellent candidates for integration with multigeneration systems. Methanol can be formed from syngas under a thermocatalytic reaction which occurs at 75 bar and $\sim$250 °C. Ammonia can be synthesized from hydrogen and nitrogen at 200–300 bar and about 500 °C. Both ammonia and methanol can be used as fuels. Methanol can be reformed to hydrogen using steam reforming. Ammonia can be cracked thermocatalytically to generate hydrogen. These are excellent options to store renewable energy within integrated multigeneration systems.

A biochemical platform in a biorefinery represents a set of processes by which biomasses can be treated to generate a multitude of products and fuels and can be integrated with fuel combustion facilities and power generators. This includes lignin production, enzymatic hydrolysis of cellulose, and fermentative processes that ultimately lead to ethanol synthesis and other biochemical byproducts (see Figure 3.17).

Brine and sea water desalination is done mainly for freshwater production but in some cases for salt extraction. Desalination can be integrated with power generation systems in many ways because multiple methods of desalination exist: multistage-flash distillation, multi-effect distillation, vapor compression distillation, reverse osmosis, electrodialysis and nano-filtration. Desalination requires low-grade heat, and, therefore, it can be combined with a power cycle with partial recovery of the rejected heat.

9.3 ASSESSMENT AND OPTIMIZATION OF MULTIGENERATION SYSTEMS

The very general system representation in Figure 9.4 allows for writing mathematical expressions for performance indicators to use for assessment of multigeneration systems. These indicators are as follows:

- Energy efficiency.

$$\eta = \frac{\dot{W} + \sum \dot{Q}_H + \sum \dot{Q}_C + \sum \dot{m}_f HHV_f + \sum \dot{m}_{cp} h_{cp}}{\sum \dot{E}_{source}} \tag{9.1}$$

- Exergy efficiency.

$$\psi = \frac{\dot{W} + \sum \dot{Q}_H \left(1 - \frac{T_0}{T_H}\right) + \sum \dot{Q}_C \left(\frac{T_0}{T_c} - 1\right) + \sum \dot{m}_f ex_f + \sum \dot{m}_{cp} ex_{cp}}{\sum \dot{E}x_{source}} \tag{9.2}$$

- Specific emissions (per unit of power generation or, depending on the case, per unit of exergy content in products, or defined in a similar fashion according to the actual system).

$$SEF = \frac{\dot{M}}{\dot{W}} \text{ or } SEF_{ex} = \frac{\dot{M}}{\dot{W} + \sum \dot{Q}_H \left(1 - \frac{T_0}{T_H}\right) + \sum \dot{Q}_C \left(\frac{T_0}{T_c} - 1\right) + \sum \dot{m}_f ex_f + \sum \dot{m}_{cp} ex_{cp}} \tag{9.3}$$

- Payback period—defined as the number of years required to return the investment cost (IC) if the products are sold at market price; it depends on the annual operational costs (AOC) and annual rate of income from multiproduct regeneration:

$$PBP = \frac{IC}{\sum c_H \dot{Q}_H + \sum c_C \dot{Q}_C + \sum c_f \dot{m}_f + \sum c_{cp} \dot{m}_{cp} - AOC} \tag{9.4}$$

- Levelized electricity cost—defined as the selling price of a unit of power to balance the investment cost (IC) and the operational costs (OPC) for the lifetime (LT). The OPC are proportional to the energy consumption $OPC = LT \sum c_{source} \dot{E}_{source}$, where c_{source} is the specific cost of energy supply, per kind

$$LEC = \frac{IC + LT \left[\sum c_{source} \dot{E}_{source} - \sum c_H \dot{Q}_H - \sum c_C \dot{Q}_C - \sum c_f \dot{m}_f - \sum c_{cp} \dot{m}_{cp} \right]}{LT \ \dot{W}} \tag{9.5}$$

- total exergetic cost rate—defined based on exergy-economic balance equation (see Dincer and Rosen, 2013). It depends on fuel cost rate ($\dot{C}_f$), cost of exergy destruction ($\dot{C}_d$), cost of environmental impact ($\dot{C}_{ei}$), and capital cost rate ($\dot{Z}_{tot}$)

$$\dot{C}_{tot} = \dot{C}_f + \dot{C}_d + \dot{C}_{ei} + \dot{Z}_{tot} \tag{9.6}$$

Note that the factor $\dot{M}$ in Equation (9.3) represents a compounded emissions rate that accounts for the emissions specific to the analyzed system. Furthermore, in Equation (9.5) c_H, c_C, c_f, and c_{cp} represent the market price (for selling) of the additional products heating, cooling, fuels, and chemicals, respectively.

Design of multigeneration systems requires the application multi-objective, multivariable optimization methods. The engineering goal is to maximize some performance indicators, such as efficiencies, and to minimize other performance indicators, including payback period, levelized electricity cost, total cost rate (exergy-economic), and specific emissions factor. The performance indicators (or assessment criteria) constitute themselves in objective functions. In general it is impossible to optimize all objective functions because they vary in opposite directions. For example, efficiency increases lead to an expensive system. Hence, the best design selection must be made based on a trade-off analysis.

The multi-objective optimization requires the determination of the Pareto front which indicates the best solutions. Here one needs to recognize optimization variables, which are the parameters that influence the system performance. In common multi-objective optimization of engineering systems with fixed configuration, the parameters to be optimized are the design (e.g., the duty of the heat exchangers) and operational parameters (e.g., mass flow rates). Integrated and multigeneration systems bring an additional dimension to multi-objective optimization because intrinsically these systems aim to improve system configuration (architecture). The additional optimization variable brought to the table by multigeneration systems is the design configuration. In general, system configurations with more outputs generated tend to perform better.

Figure 8.46 illustrates how the degree of integration improves the system configuration so that the system performs better and the successive Pareto fronts approach the target point for optimization. Analogous Pareto fronts of multigeneration systems are shown in Figure 9.5a, with incrementally higher performance as the number of outputs increases. A three-dimensional representation of optima is more useful. This is shown in Figure 9.5b. Two

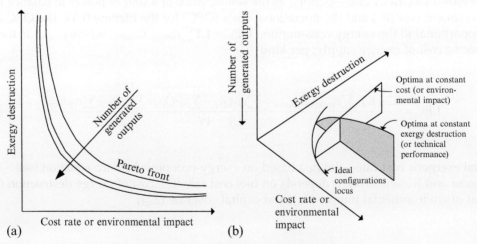

FIGURE 9.5 Multi-objective optimization of multigeneration systems [modified from Bejan and Lorente (2006)]: (a) two-dimensional Pareto fronts, (b) three-dimensional optimal representation.

surface cuts are shown. The first represents the optimized system with fixed economic or environmental impact indicators (the vertical line hatched surface). Only exergy efficiency is optimized along this cut. The curve suggests that exergy destructions reduce when the number of outputs increases. Another cut is at fixed exergy destruction (or fixed exergy efficiency). Again, the increase in number of outputs enhances the revenue and reduces the environmental impact. These two cuts describe a 3D Pareto surface; see Figure 8.47 as an example.

Multigeneration systems increase the profitability due to an increase in the number of generated outputs. This strategy of design improvement is comparable to "survival by increasing the flow territory" and with the increase in the "freedom to morph" mentioned in Bejan and Lorente (2006). Indeed, when the number of generation outputs is increased the system expands into a new space, because it can become a player in new markets as in addition to power it generates other diversified products. Implicitly, if the output number increases the system compactness must decrease or the system svelteness increase. Furthermore, the multigeneration system becomes more complex. Therefore, it has more opportunity to morph, that is the number of optimization variables increases. Because of more freedom to morph, the Pareto surface expands toward better systems.

There are several analytical and numerical methods for optimization of complex systems, as reviewed in Dincer and Rosen (2013). The classical method is applicable to functions that are continuous until the first and second derivatives. Because in most cases the optimization problem is a multivariable one, gradient methods should be applied. Multi-objective problems can be reduced to a single objective by formulating a compound objective function or by applying constraints. Assume for example that for a multigeneration system the objective function is the sustainability index $SI = 1 - \psi$, introduced in Dincer and Rosen (2013). The second objective function is the levelized cost of electricity LEC, Equation (9.5). Both these parameters depend on a number of optimization variables $x_i, i = 1 \ldots n$. Both these performance parameters must be minimized. Therefore, the multi-objective optimization problem can be formulated mathematically as follows:

$$\min \{SI(x_i), LEC(x_i), i = 1 \ldots n | N = \text{fixed, other constraints}\} \qquad (9.7)$$

where N represents the number of outputs of the multigeneration system.

The multi-objective optimization problem in Equation (9.7) can be transformed into m single-objective optimizations if the levelized electricity cost is constrained to m values across its constrained range of variation. For example, $LEC = LEC_j, j = 1 \ldots m$. These problems are formulated as follows:

$$\min \{SI(x_i), i = 1 \ldots n | LEC(x_i) = LEC_j, j = 1 \ldots m, N = \text{fixed, other constraints}\} \qquad (9.8)$$

Depending on the nature of the constraints, Lagrange multiplier or Khun–Tucker methods can be applied to solve the optimization problem in Equation (9.8). The results of the optimization will describe a Pareto front. Besides these methods, there are several numerical techniques that can solve Equation (9.8) or even Equation (9.7) directly. One of the common methods is based on evolutionary algorithms with a random but educated search. Among these evolutionary algorithms, genetic algorithms appear to be particularly useful for multi-objective optimization. With these algorithms the search progresses according to the theory of the natural selection of the fittest. These methods can be applied to discontinuous functions, which is a great advantage when complex systems are optimized. Nevertheless, the algorithm search is educated rather than completely random. For multigeneration systems, besides the performance parameters, three other parameters that describe the system configuration should be considered. These are the number of outputs, system compactness, and system scale. The search strategy take into consideration that better systems have less compactness, more generated outputs, larger scale, and a higher degree of design freedom (see Bejan and Lorente, 2006).

9.4 CASE STUDIES

9.4.1 Thermally Driven Multigeneration System

This is a case study based on work done by Dincer and Zamfirescu (2012) which illustrates the benefit of multigeneration systems with an incremental number of outputs. The energy supply is heat at intermediate-temperature level, derived from waste heat recovery. Therefore, renewable energy is used. The configurations of considered systems are according to Figures 9.1–9.3.

Equations (9.1) and (9.2) are used for energy and exergy efficiencies calculation for the considered system under some reasonable assumptions, described below. Because the system uses only renewable energy it is not meaningful to analyze the specific emissions factor defined by Equation (9.3). Rather, the system will have a positive environmental impact because carbon emissions are mitigated when a conventional power generation system is replaced with this one. Therefore, we introduce here a carbon mitigation factor, CMF, that quantifies the mitigated GHG emissions expressed in a kg CO_2 equivalent per kWh of electric power. For the Canadian grid the carbon emissions factor is $SEF_w = 0.2\,kg_{CO_2}/kWh$.

One of the objective functions to be minimized for the system design will thus be the carbon mitigation factor. According to the explanations given above it can be demonstrated that the carbon mitigation factor for the cogeneration system (power and heating) can be expressed with

$$CMF = SEF_w \left(\eta_w + \eta_H \, \tilde{M} \right) \tag{9.9}$$

where η_w represents the power generation efficiency, and η_H is defined here as the heating generation efficiency $\eta_H = \dot{Q}_H / \dot{E}_{in}$. Furthermore, the dimensionless factor $\tilde{M}$ is introduced to denote the ratio between GHG mitigation due to the avoidance of conventional GHG-emitting heating technologies and SEF_w. The factor $\tilde{M}$ is ~0.5 for methane combustion and 1.0 for electrical heating.

The energy conversion system is a heat engine connected to a heat source with a temperature in the range of 150–350 °C and a fixed heat sink temperature of 50 °C. The energy efficiency of the heat engine itself is correlated to the heat source temperature T_{so} and the temperature level for heat recovery (see Figure 9.1b), denoted T_H. This is calculated with

$$\eta_w = \eta_{1a} = (1.055 - 0.001\,T_{so}) \left(1 - \frac{T_H}{T_{so}} \right) \tag{9.10}$$

where the subscript 1a refers to the single generation system shown in Figure 9.1a.

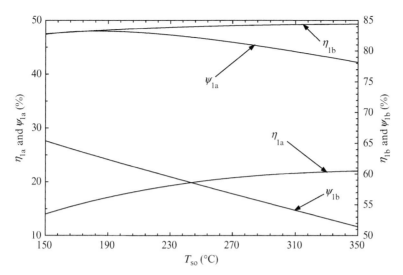

FIGURE 9.6 Energy and exergy efficiencies of single generation (Figure 9.1a) and cogeneration (Figure 9.1b) systems. *Data from Dincer and Zamfirescu (2012).*

If one denotes with f the recovered fraction of the heat rejected by the heat engine, then η_H from Equation (9.9) becomes $\eta_H = f\eta_w$. Furthermore, the energy efficiency of the cogeneration system—which represents itself one of the objective functions to be studied—is given by

$$\eta_{1b} = \eta_w + f(1 - \eta_w) \tag{9.11}$$

where the subscript 1b refers to the cogeneration system shown in Figures 9.1b and 9.6.

The first remark refers to the exergy efficiency of the single generation system. For the considered heat engine one observes that the exergy of the heat source increases faster than the energy efficiency of the heat engine; consequently, the exergy efficiency of the single generation system ψ_W has a maximum of 48% at a 195 °C source temperature.

The exergy efficiency of the system with cogeneration results from the Carnot factor of the heat source and from combining Equations (9.2) and (9.10)

$$\psi_{1b} = \psi_w + f(1 - \eta_w)\left(1 - \frac{T_0}{T_H}\right) \bigg/ \left(1 - \frac{T_0}{T_{so}}\right) \tag{9.12}$$

where

$$\psi_{1a} = \psi_w = (1.055 - 0.001\,T_{so})\left(1 - \frac{T_0}{T_H}\right) \bigg/ \left(1 - \frac{T_0}{T_{so}}\right) \tag{9.13}$$

The energy utilization efficiency of the cogeneration system is in the range of 80–85%, whereas the exergy efficiency is about double that of the exergy efficiency of the single generation system. The exergy efficiency of the cogeneration system is higher at a lower source temperature; for 150 °C, it reaches 65%.

Figure 9.7 shows the GHG mitigation results by applying renewable-energy-based single and cogeneration systems. Three cases are assumed: (i) no cogeneration applied;

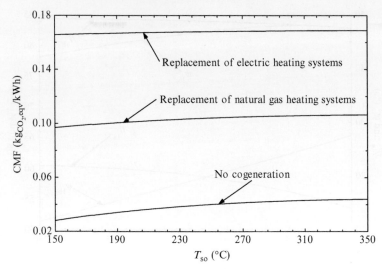

FIGURE 9.7   Carbon mitigation factor for the cogeneration system (Figure 9.1b). *Data from Dincer and Zamfirescu (2012).*

(ii) the renewable-energy-based system with cogeneration replaces natural gas heaters; and (iii) the system replaces electric heaters. It is observed that by applying cogeneration in the form of hot water the incremental increase of mitigation is on the order of 0.6 kg CO_2/kWh, which represents an increase in mitigation of 250%. If electric heaters are replaced with a cogeneration system, the obtained GHG mitigation is on average 0.17 kg CO_2/kWh, which represents a 425% increase with respect to the single generation system.

The plot shown in Figure 9.8 shows the Pareto frontiers for the single and cogeneration systems on the same plot, in terms of exergy efficiency versus carbon mitigation factor (CMF). For determination of the payback period some specific case assumptions are made in Dincer and Zamfirescu (2012). Sixteen percent heat engine efficiency is assumed. It is also assumed that the system is implemented in Ontario, where the natural gas price for heating is predicted to increase with a rate of 5.5% annually. The engine is driven by renewable heat from concentrated solar radiation. The number of clear sky days is assumed to be 276 per year, and the average direct beam radiation is 5.5 kWh/m^2 daily. The assumed heat recovery ratio is $f = 77\%$. The system sells back a portion of its electrical power to the grid at times of peak power, and the extremely high sell-back price of about $0.80/kWh from Ontario's Feed-in Tariff program may be greatly appreciated. The last factors considered in the financial calculation are the initial cost and salvage value of the system and the operation and maintenance costs of the system over its 30-year life.

The initial investment is $10k, with an assumed 50% paid through a government grant for renewable energy. The salvage value is taken as 15% and the system maintenance as 1%, both from the initial system costs. A tax credit of $30 per ton of CO_2 mitigated is assumed.

Under these assumptions the payback time is 7 years, and 63% of the payback period is due to heating generation using a renewable source instead of natural gas, 30% is due to electricity generation from renewable resources instead of using the grid, and 7% is due to recovered

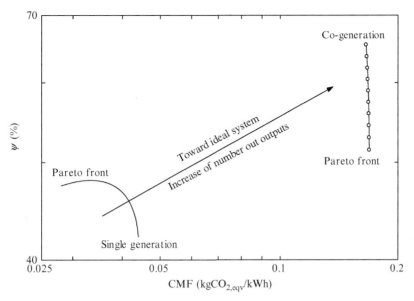

FIGURE 9.8 Pareto frontiers for single and cogeneration systems represented in terms of exergy efficiency versus carbon mitigation factor (CMF).

carbon tax, which is proportional to GHG mitigation by the system. Under the assumption of proportionality, it is approximated that:

- The payback time due to power generation effect is given by $PB_w = 1.096\ \eta_w$, years per installed capacity in terms of kW of renewable energy consumption.
- The payback time due to heating generation effect is given by $PB_H = 0.568\ \eta_H$, years per installed capacity in terms of kW of renewable energy consumption.
- The payback time due to GHG mitigation is $PB_{GHG} = 0.192\ \mathcal{M}_{GHG}\ \eta_H$, per installed capacity in terms of kW of renewable energy consumption.

The results indicating the economic benefit of the cogeneration system are shown in Figure 9.9. The payback period per unit of kW installed capacity ($\overline{PBP}$) is plotted against the source temperature. If no heat is generated, the payback period can be assimilated with the curve indicated with "Due to power sell-back"; the average of this curve is 0.2 year per installed capacity in terms of kW of renewable energy consumption. The curve corresponding to the cogeneration system is labeled "Cumulated." The average value in the cogeneration case is 0.57 year per installed capacity in terms of kW of renewable energy consumption. This is an increase of about 285% over the single generation case. Furthermore, one observes that the effect of cogenerated heat decreases slightly with the increase in source temperature, whereas the effects of GHG mitigation and of "green" power production both increase.

For the case of trigeneration of power, hot water, and space heating (system in Figure 9.2a), it is assumed that the system delivers heat at 35 °C to warm air at 25 °C and that the heat recovery factor is $f = 95\%$. For the quadruple generation system of Figure 9.2b the heat rejected by the condenser of the absorption refrigerator is recovered and used for air heating. Hence,

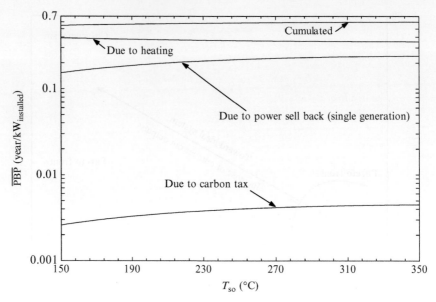

FIGURE 9.9   Payback period of the cogeneration system per kW of energy source. *Data from Dincer and Zamfirescu (2012).*

the energy efficiency system trigeneration and quadruple generation systems are set to remain the same. The effect of output increment in multigeneration systems is illustrated in Figure 9.10 by making a comparison between cogeneration and trigeneration systems.

For a 350 °C heat source temperature the calculated exergy efficiency of the quadruple generation system is 60% while that of the trigeneration system is only 52%. Note that it is assumed that 20% of the recovered heat is used for water heating and 70% is used to drive the absorption refrigerator. Thus 35% of the heat rejected by the heat engine is converted into cooling effect, while the rest is converted into space heating effect; therefore, 45% of the rejected heat is recovered and used for space heating.

In Figure 9.11 it is remarked that the increase in level of multigeneration (that is, increase in the number of products) is beneficial in two aspects: (i) it increases the exergy efficiency, and (ii) the required source temperature decreases slightly with the increase in degree of multigeneration. Indeed, it is made clear from the figure that the system with quadruple generation reaches a maximum of exergy efficiency of 78% with a 207 °C source temperature; on the contrary, the single generation system reaches a maximum 59% at 220 °C.

For the quintuple generation system considered in Figure 9.3a, a bottoming power cycle is included that operates between 100 and 30 °C. It is assumed that the heat rejected by the topping heat engine is 95% recovered. From this recovered heat, 10% is used to drive the absorption refrigerator which generates cooling and rejects the condensation heat at a temperature level of 30 °C. This heat is used for space heating at 20 °C. The remaining 90% of the recovered heat is used to drive the bottoming heat engine, which also rejects heat at 30 °C for space heating. The power generated by the bottoming heat engine is supplied to a water electrolysis system with electrical efficiency assumed at 85%. The heat rejected by the

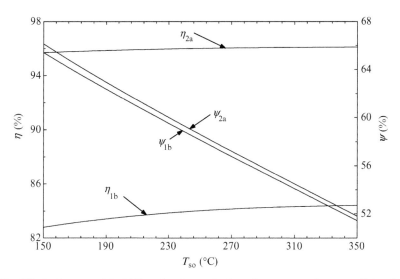

FIGURE 9.10 Effect of incremental addition of system outputs on the energy and exergy efficiencies as functions of source temperature. *Data from Dincer and Zamfirescu (2012).*

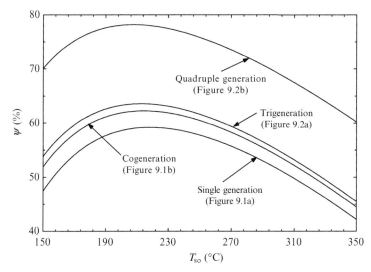

FIGURE 9.11 Comparison of single generation, cogeneration, trigeneration, and quadruple generation systems according to exergy efficiency.

electrolyser is recovered and used for water heating; note that this recovered heat is made available at 50 °C to heat water at 40 °C, thus keeping a 10 °C-temperature difference to drive the heat transfer process.

As reported in Dincer and Zamfirescu (2012), there is not much difference in exergy efficiency between the quadruple generation system and the quintuple generation system, as the exergy efficiency of quintuple generation reaches 79% and occurs at a 191 °C source temperature. An assumption is made that 95% of the heat rejected by the topping heat engine is recovered; from this 90% is used to generate electricity in the bottoming heat engine. For a system with sextuple generation (Figure 9.3b) the chemical exergy of 14.3 kJ/kg of salt and an average salt concentration of 3.5% in oceans are taken into account for evaluation of the exergy efficiency. The desalination process requires 0.040 kWh/kg of water in the form of power input. Based on the electrolyser efficiency and the power generated by the bottoming heat engine, the hydrogen generation rate is determined. From this information, the required mass flow rate of water is calculated and subsequently the production rate of salt is determined.

Figure 9.12 summarizes the optimum source temperature findings together with the maximum exergy efficiency for each system. The incremental increase in exergy efficiency with the number of generated products is noted. The additional benefit of reducing the source temperature (even slightly) is also emphasized. The augmented exergy efficiency is more accentuated when the number of outputs increases from one to four, after which it stabilizes. This shows that the increase in number of outputs over four must be considered on a case-by-case basis, taking in account, besides the exergy efficiency, the overall system cost and the environmental impact.

It is important to compare the integrated multigeneration systems with a separate system that generates exactly the same outputs. In terms of exergy efficiency the assembly of independent systems is determined according to the following equation (written here for the sixtuple generation case):

$$\psi_{sep} = \frac{\dot{W} + \dot{Ex}_{wh} + \dot{Ex}_{sh} + \dot{Ex}_C + \dot{Ex}_{H_2} + \dot{Ex}_S}{\sum_1^6 \dot{Ex}_{in}} \tag{9.14}$$

Here it is reasonable to assume that all energy inputs are delivered by the same renewable source (this means that the source has the same temperature for all cases). As observed in Figure 9.13, the exergy efficiency of the multigeneration systems is higher than that of the

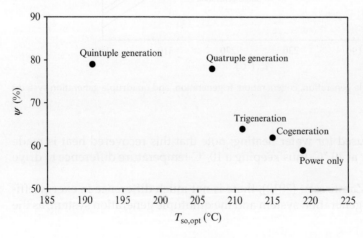

FIGURE 9.12 Maximum exergy efficiency and optimum source temperature of power generation system with incremental number of additional outputs. *Data from Dincer and Zamfirescu (2012).*

FIGURE 9.13 Exergy efficiency of integrated multigeneration systems (ψ) compared with exergy efficiency of separate systems (ψ_{sep}).

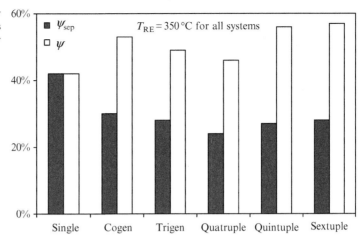

single generation case. Also, the difference between the exergy efficiency of separate technologies and multigeneration systems increases with the number of generated outputs.

This case study aims to demonstrate that integrated multigeneration systems offer advantages such as better efficiency, better cost effectiveness, a better environment, and better sustainability. System design using integration for generation of multiple products is more beneficial than generation with separate technologies.

9.4.2 Micro-Gas-Turbine-Based Multigeneration System

In this case study the multi-objective optimization of a multigeneration system for power, space heating, hot water, cooling, and hydrogen is presented based on the work by Ahmadi et al. (2013b). The system is based on a micro gas turbine as the main unit of power generation. In addition, the system integrates a dual-pressure heat recovery steam generator, an absorption chiller, an ejector refrigeration cycle, a domestic water heater, and a PEME.

Figure 9.14 shows the studied multigeneration system. The micro gas turbine comprises the compressor, the combustion chamber, and the gas turbine and is responsible for main power generation by the integrated system. The fuel used is natural gas. A double-pressure heat recovery steam generator is cascaded with the gas turbine. The superheated high-pressure steam is used to produce power in a steam turbine. The superheated low-pressure steam is diverted to a single-effect absorption chiller that generates cooling as a product. Further, steam is diverted to a domestic water heater to generate hot water. A secondary heat recovery vapor generator (HRVG) is used to generate saturated vapor in an ORC operating with R123. The ORC is integrated with an ejector refrigerator which generates a cooling effect for space cooling as a product. Furthermore, the power generated by the ORC is used internally in full by a PEME for hydrogen production.

9.4.2.1 System Modeling

The integrated multigeneration system is divided into six subsystems: Brayton cycle, SRC, ORC, absorption chiller, domestic water heater, and PEME. The system is modeled in steady state. The main modeling equations for these subsystems are given in Table 9.2.

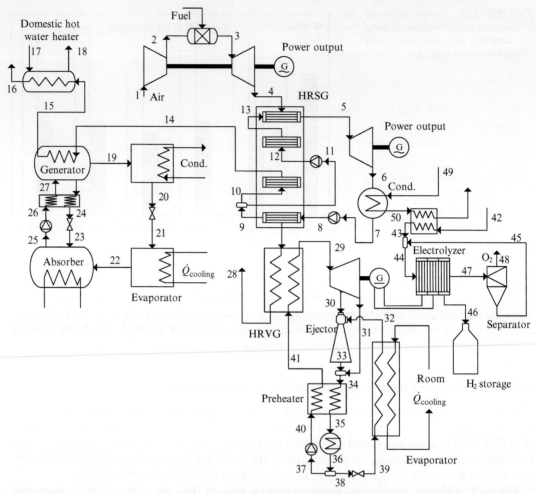

FIGURE 9.14 Micro-gas-turbine-based multigeneration system. *Modified from Ahmadi et al. (2013a).*

The dual-pressure HRSG with two economizers (LP and HP) and two evaporators (LP and HP) is used in the multigeneration cycle to provide both low- and high-pressure steam. As a dual-pressure type, the HRSG has two pinch points (one at low and another at high pressure) which are assumed at reasonable values. There are four heat transfer segments for which separate energy balance equations are written.

A lithium–bromide absorption cycle is integrated into this system. The modeling of the absorption cycle involves mass balance equations for the working fluid and for the lithium–bromide. Basically, these equations are $\sum \dot{m}_{in} = \sum \dot{m}_{out}$ and $\sum (\dot{m}x)_{in} = \sum (\dot{m}x)_{out}$, where x is the mass concentration of LiBr. In addition, energy balance equations are written for each component of the system. The cooling effect results from $\dot{Q}_{cooling} = \dot{m} (h_{22} - h_{21})$.

The domestic water heater is a very basic subsystem consisting of a simple component. It is modeled based on an energy balance equation which reads: $\dot{m}_{lp}(h_{15} - h_{16}) = \dot{m}_{w}(h_{18} - h_{17})$.

TABLE 9.2 Modeling Equations for Brayton Cycle

Component	Equations and Remarks
Air compressor	$\dfrac{T_2}{T_1} = 1 + \dfrac{1}{\eta_{ac}}(r^\kappa - 1), \kappa = \dfrac{\gamma - 1}{\gamma}, \gamma = 1.4, \dot{W}_{ac} = \dot{m}_a C_p(\overline{T})(T_2 - T_1)$
Combustion chamber	$\dot{m}_a h_2 + \dot{m}_f \mathrm{LHV} = \dot{m}_g h_3 + (1 - \varepsilon_{cc})\dot{m}_f \mathrm{LHV}, P_3 < P_2, n_f = \lambda n_a$ Species assumed in combustion gases: CO_2, H_2O, CO, NO, O_2, N_2, Ar
Gas turbine	$\dfrac{T_4}{T_3} = 1 - \eta_{gt}(1 - r^{-\kappa}), \kappa = \dfrac{\gamma_g - 1}{\gamma_g}, \dot{W}_{gt} = \dot{m}_g C_{p,g}(\overline{T})(T_3 - T_4), \dot{m}_g = \dot{m}_a + \dot{m}_f$

TABLE 9.3 Modeling Equations for Steam Rankine Cycle

Component	Equations and Remarks
HSRG	MBE : $\dot{m}_w = \dot{m}_{lp} + \dot{m}_{hp}$, EBEs : $\dot{m}(h_9 - h_8) = (\dot{m}_g C_p \Delta T)_d, \dot{m}_{lp}(h_{11} - h_{10})$ $= (\dot{m}_g C_p \Delta T)_c$ EBEs : $\dot{m}_{hp}(h_{13} - h_{12}) = (\dot{m}_g C_p \Delta T)_b, \dot{m}(h_5 - h_{13}) = (\dot{m}_g C_p \Delta T)_a$
Steam turbine	EBE : $\dot{m}_w h_5 = \dot{W}_{st} + \dot{m}_w h_6, \dot{m}_{st} = \eta_{st}\dot{m}_s$ η_{st} assumed in a practical range
Condenser	EBE : $\dot{m}_6 h_6 = \dot{Q}_{cond} + \dot{m}_7 h_7$
Water pump	EBE : $\dot{m}_w h_7 + \dot{W}_{wp} = \dot{m}_w h_8, \dot{W}_{wp} = \eta_{wp}\dot{W}_s$ η_{wp} assumed in a practical range

For the ORC all modeling equations are very basic and will not be written here. However, modeling the ejector requires some special considerations. The ejector uses the Venturi effect of a converging-diverging nozzle to convert the mechanical energy (pressure) of a motive fluid to kinetic energy (velocity), creating a low-pressure zone that draws in and entrains a suction fluid (Table 9.3).

The ejector operates adiabatically with no work transfer and negligible gas velocity at the entrance and exit sections. According to Figure 9.14, the primary motive flow is extracted vapor at #30 and the suction flow is cold vapor at #32. The entrainment ratio expresses the relationship between the main flow rate and suction flow rate: $\dot{m}_{32} = \omega \dot{m}_{30}$. The nozzle efficiency of the ejector is defined as follows:

$$\eta_{nz} = \frac{V_{nz}^2}{2(h_{30} - h_{nz})} \tag{9.15}$$

The velocity of primary flow at the nozzle exit results from conservation of momentum, which reduces to $V_{nz} = (1 + \omega)V_{mix}/\sqrt{\eta_{mix}}$, where V_{mix} is the velocity of the mixed flow and η_{mix} is the mixing chamber efficiency. After the mixing chamber, the flow is compressed in the diffuser and ideally reaches enthalpy h_s when compression is isentropic. The less than ideal result is accounted for by the efficiency of the diffuser η_{diff} which is defined by

$$\eta_{diff} = \frac{h_s - h_{mix}}{h_{33} - h_{mix}} \tag{9.16}$$

The electrolyser is modeled using a basic electrochemical equation. No water preheating occurs in the electrolyser; rather, water is preheated before the inlet with the help of a heat recovery heat exchanger (see Figure 9.14). The Nernst equation gives the reversible potential as a function of the operating temperature according to:

$$V_0 = 1.229 - 0.00085(T_{PEME} - 298) \tag{9.17}$$

The activation and ohmic overpotentials can be calculated using equations specific to PEME as follows:

- Membrane conductivity $\sigma = (0.51\Lambda - 0.326)\exp\left[1268\left(\frac{1}{303} - \frac{1}{T}\right)\right]$.
- Moisture content across membrane $\Lambda(x) = \Lambda_c + \frac{x}{D}(\Lambda_a - \Lambda_c)$.
- Ohmic resistance $R_\Omega = \int_0^D \frac{dx}{\sigma(\Lambda(x))}$.
- Activation potential $V_{act} = \frac{RT}{F}\sinh^{-1}\left(\frac{J}{2J_0}\right)$.
- Exchange current density $J_0 = \mathcal{A}\exp\left(-\frac{E_{act}}{RT}\right)$.

Another important modeling aspect regards the mixing and chemical exergy. When two or more streams mix, then the specific exergy of the mixed stream is calculated based on the chemical exergy of the mixing streams and the mixing exergy as follows:

$$ex_{mix}^{ch} = \sum y_i ex_i^{ch} + RT_0 \sum y_i \ln y_i \tag{9.18}$$

The multigeneration system is assessed according to two criteria: exergy efficiency and total cost rate. The exergy efficiency of the multigeneration system equals the ratio of all exergy content in products to the exergy input. Mathematically, the exergy efficiency is expressed as

$$\psi = \frac{\dot{W}_{net} + \dot{Ex}_H + \dot{Ex}_{C,A} + \dot{Ex}_{C,ORC} + \dot{Ex}_{H_2} + \dot{Ex}_{18}}{\dot{m}_f ex_f} \tag{9.19}$$

where $\dot{W}_{net}$ is the net generated power. $\dot{Ex}_H$ is the exergy rate in the heating product of the ORC condenser as $\dot{Ex}_H = \dot{Q}_{cond}\left(1 - \frac{T_0}{T_{cond}}\right)$. $\dot{Ex}_{C,A}$ and $\dot{Ex}_{C,ORC}$ are exergy rates of cooling by the absorption chiller and ORC evaporator as $\dot{Ex}_C = \dot{Q}_{cooling}\left(\frac{T_0}{T_{evap}} - 1\right)$. $\dot{Ex}_{H_2}$ is the exergy rate in the generated hydrogen $\dot{Ex}_{H_2} = \dot{m}_{H_2}ex_{H_2}^{ch}$. $\dot{Ex}_{18}$ is the exergy rate in generated hot water as $\dot{Ex}_{18} = \dot{m}_{18}[h_{18} - h_0 - T_0(s_{18} - s_0)]$.

The total cost rate is a parameter constructed in such a way that it assesses the economic performance and the environmental impact of the system together. Similar to the case study in Chapter 8, Section 8.6.2, the total cost rate defined here has three terms: capital cost rate ($\dot{Z}$), fuel cost rate ($\mathcal{C}_f$), and environmental cost rate ($\dot{C}_{env}$). This is defined with:

$$\dot{C}_{tot} = \dot{Z} + \dot{C}_f + \dot{C}_{env} \tag{9.20}$$

where $\dot{Z} = \sum \dot{Z}_k$ is the sum of the capital cost of each of the system components as defined in Table 8.8 from Chapter 8. $\dot{C}_f = \dot{m}_f \mathcal{C}_f$ is the cost rate of fuel and $\mathcal{C}_f$ is the mass specific fuel cost. $\dot{C}_{env} = \dot{m}_{CO_2}\mathcal{C}_{CO_2} + \dot{m}_{CO}\mathcal{C}_{CO} + \dot{m}_{NO_x}\mathcal{C}_{NO_x}$ is the environmental cost rate.

9.4.2.2 System Optimization

The multigeneration system is optimized for maximization of the exergy efficiency and minimization of the cost rate. The configuration of the system is fixed during the optimization. The variables to be optimized and their constraints are given in Table 9.4. Hence, the multivariable optimization problem is formulated as follows:

$$\{\max\psi(x_i),\min\dot{C}_{tot}(x_i),i=1\ldots15|x_i \text{ and constraints as in Table9.4}\} \qquad (9.21)$$

The result of the multi-objective optimization is expressed in the form of the Pareto frontier that is given as the total cost as a function of exergy efficiency or *vice versa*. The problem is solved in Ahmadi et al. (2013a) using an evolutionary multi-objective optimization algorithm. The regression curve $\dot{C}_{tot}=f(\psi)$ is:

$$\dot{C}_{tot}\left[\frac{\$}{h}\right]=\frac{692.4\psi^3-2294\psi^2+1492\psi-129.7}{\psi^5+51.48\psi^4-136.1\psi^3+130.7\psi^2-55.98\psi+9.27} \qquad (9.22)$$

Figure 9.15 shows the Pareto front of the optimized solutions in terms of exergy efficiency and total cost rate of generated products. Four particular points, A, B, C, and D, are indicated on the Pareto front. These are particular optimization solutions, as follows:

- Point A corresponds to a minimum cost rate; this is the cheapest system among the best solutions.
- Points B and C represent a compromise between cost and performance (here exergy efficiency).
- Point D shows the system with maximum exergy efficiency but also with the higher cost.

TABLE 9.4 Optimization Variables and Their Constraints

Item	Variable Name	Symbol	Constraints
1	Compressor pressure ratio	r	$r<22$
2	Compressor isentropic efficiency	η_{ac}	$\eta_{ac}<0.9$
3	Gas turbine isentropic efficiency	η_{gt}	$\eta_{gt}<0.9$
4	Gas turbine inlet temperature	TIT	$TIT<1550\,K$
5	High-pressure pinch point temperature difference	ΔT_{hp}	$10\,K<\Delta T_{hp}<22\,K$
6	Low-pressure pinch point temperature difference	ΔT_{lp}	$12\,K<\Delta T_{hp}<22\,K$
7	Pressure of high-pressure steam	P_{hp}	$P_{hp}<40\,bar$
8	Pressure of low-pressure steam	P_{lp}	$P_{hp}<5.5\,bar$
9	Steam turbine isentropic efficiency	η_{st}	$\eta_{st}<0.9$
10	Pump isentropic efficiency	η_{p}	$\eta_{p}<0.9$
11	Steam condenser pressure	P_{cond}	$8\,kPa<P_{cond}<10\,kPa$
12	Absorption chiller evaporator temperature	T_{ev}	$2\,^{\circ}C<T_{ev}<6\,^{\circ}C$
13	ORC turbine inlet pressure	P_{ORC}	$500\,kPa<P_{ORC}<750\,kPa$
14	ORC turbine extraction pressure	P_{ex}	$180\,kPa<P_{ex}<250\,kPa$
15	ORC evaporator pressure	P_{ev}	$20\,kPa<P_{ev}<35\,kPa$

Source: Ahmadi et al. (2013a).

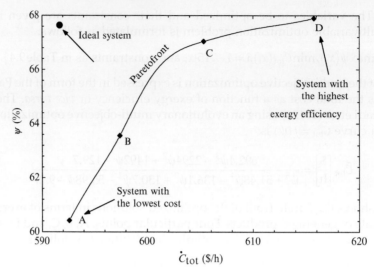

FIGURE 9.15   Pareto front of the optimization solutions for the multigeneration system. *Data from Ahmadi et al. (2013a).*

Table 9.5 gives numerical details of optimized solutions for the Pareto front. In the table the specific carbon emissions (SCE) are given in kilograms of carbon dioxide equivalent per kWh. Many details can be studied for optimal solutions, which can help determine the best system. One relevant detail is shown graphically in Figure 9.16. Here, the SCE of the optimized systems are plotted against the cost rate. It is clearly observed that there is a cost rate for which the emissions are maximized. In addition, the power generated around the point with minimum carbon emissions is relatively constant. This observation may recommend the best system design selection to be between points C and D at a cost rate of 612 $/h and an efficiency of larger than 67%.

9.4.3 Integrated Biomass-Fueled Multigeneration System

In this case study a biomass-fueled multigeneration system that produces four commodities besides power is presented. The additional products are water heating, space heating, cooling, and hydrogen. The case study is taken from a previous work reported in Ahmadi et al. (2013b). The system comprises a biomass combustor, an ORC, an absorption chiller, a PEME to produce hydrogen, and a domestic water heater for hot water production. Figure 9.17 shows the system.

Pine sawdust is used as biomass fuel, and the combustion heat is transferred to an ORC with n-octane and to an LiBr absorption chiller. Low-grade heat is recovered from the combustion gases and transferred to the PEME with the purpose of preheating the freshwater. The heat rejected by the ORC is used for heat generation as a byproduct. The power generated by the ORC is used to satisfy residential power demand and to generate hydrogen through a PEME, and hydrogen is stored for later usage. A domestic water heater is also integrated, which uses heat recovered from flue gases.

TABLE 9.5 Optimized Variables and Performance Parameters for Solutions A, B, C, and D of Multi-Objective Multivariable Optimization of the Multigeneration System (Figure 9.15)

Optimized Variable	Optimal Solution				Performance Parameter	Orptimized Parameters			
	A	B	C	D		A	B	C	D
r	14.9	14.9	14.9	14.97	ψ (%)	60.42	63.56	67.04	67.90
η_{ac}	0.88	0.88	0.88	0.87	$\dot{C}_{tot}$ ($/h)	592.61	597.44	605.31	615.75
η_{gt}	0.9	0.89	0.9	0.9	$\dot{Ex}_d$ (MW)	14.9	14.4	13.9	13.8
TIT (K)	1498	1495	1499	1496	SCE (kg$_{GHG}$/kWh)	136.9	130.1	116.5	114.8
ΔT_{hp} (K)	14.98	14.9	14.9	4.46	$\dot{W}$ (MW)	10.3	10.8	11.4	11.4
ΔT_{lp} (K)	14.92	14.9	14.8	14.95	$\dot{Q}_{heating}$ (MW)	4.8	5.2	6.6	6.8
P_{hp} (bar)	12.29	23.4	29.9	29.9	$\dot{Q}_{cooling}$ (kW)	930	915	904	930
P_{lp} (bar)	2.01	2	4.9	4.9	$\dot{m}_{dwh}$ (kg/s)	0.83	0.82	0.83	0.85
η_{st}	0.77	0.78	0.87	0.88	$\dot{m}_{H_2}$ (kg/day)	17.04	17.04	30	31
η_p	0.84	0.84	0.75	0.87					
P_{cond} (kPa)	10	9.86	8.04	8.1					
T_{ev} (°C)	5	1.1	2.31	2.1					
P_{ORC} (kPa)	718	689	503	506					
P_{ex} (kPa)	249	248	246	249					
P_{ev} (kPa)	24.84	34.9	21.2	27.6					

Source: Ahmadi et al. (2013a).

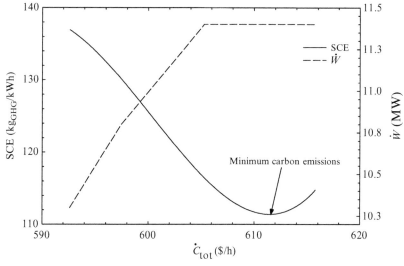

FIGURE 9.16 Specific carbon emissions and net generated power of the optimized multigeneration system (Figure 9.14) plotted against the total cost rate. *Data from Ahmadi et al. (2013a).*

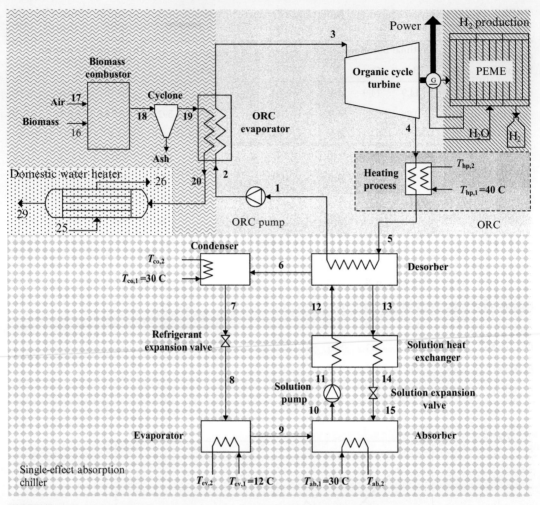

FIGURE 9.17   Integrated biomass-based multigeneration system for power, hot water, space heating, cooling, and hydrogen. *Modified from Ahmadi et al. (2013b).*

The following steps are performed in this analysis: thermodynamic modeling of the system, system assessment through exergy and environmental impact, and assessment of the benefit of multigeneration with respect to power-only generation. Biomass is represented by the following simplified chemical equation: $C_xH_yO_z(H_2O)_\alpha$. The following chemical equation for combustion is assumed:

$$C_xH_yO_z(H_2O)_\alpha + \lambda(O_2 + 3.76\,N_2) \rightarrow aCO_2 + bH_2O + cN_2$$

The biomass has 10% moisture by weight and the following dry-basis elemental composition: 50.54% C, 7.08% H, 41.11% O, 0.57% S. The enthalpy of the biomass with the dry-basis lower heating value, denoted LHV, stream #16 in Figure 9.17, is:

$$h_{16} = xh_{CO_2} + 0.5yh_{H_2O} + \text{LHV}$$

The energy balance equation for the adiabatic combustor can be solved to determine the adiabatic flame temperature:

$$h_{16} + \alpha h_{H_2O} + \lambda O_2 + 3.76\lambda N_2 = a h_{CO_2} + b h_{H_2O} + c h_{N_2}$$

The modeling of all other system components assumes steady state operation and is based on balance equations. The rate of exergy input into the system is calculated based on the wet-basis LHV and the quality factor according to the equation by Szargut et al. (1988):

$$\dot{E}x_{in} = \dot{m}_{16}LHV \frac{1.0414 + 0.0177\left(\dfrac{H}{C}\right) - 0.3328\left(\dfrac{O}{C}\right)\left[1 + 0.0537\left(\dfrac{H}{C}\right)\right]}{1 - 0.4021\left(\dfrac{O}{C}\right)} \tag{9.23}$$

The multigeneration system is assessed based on two criteria: exergy efficiency and specific GHG emissions (SCE) per unit of generated energy (this quantifies the environmental impact). These performance parameters are defined mathematically by:

$$\psi = \frac{\dot{E}x_{gen}}{\dot{E}x_{in}} \text{ and SCE}\left[\frac{kg\,CO_{2,eqv.}}{kW}\right] = \frac{\dot{m}_{GHG}}{\dot{E}_{in}} \tag{9.24}$$

where the generated rates of energy and exergy by the multigeneration system are:

$$\dot{E}_{gen} = \dot{W} + \dot{E}x_{heat} + \dot{E}x_{cool} + \dot{E}_{26} + E_{H_2} \text{ and } \dot{E}x_{gen} = \dot{W} + \dot{Q}_{heat} + \dot{Q}_{cool} + \dot{E}x_{26} + Ex_{H_2}$$

For system modeling, a reference case is assumed for which the isentropic efficiencies of the ORC turbine and pump are both 85%, the inlet turbine temperature is 400 °C, and 5% of net ORC power is used to operate the PEME. Reference case parameters are given in Table 9.6.

The biomass feed rate is set to 0.3 kg/s, and the ORC fluid mass flow rate is determined to be 4.28 kg/s. In this condition the following performance parameters are determined for the system: heat duty 2617 kW, cooling duty 611 kW (absorption chiller COP is 0.43), output power 671 kW, hydrogen production rate 3.14 kg/day, generated hot water flow rate 0.56 kg/s, exergy efficiency 22.2%, and specific carbon dioxide emissions 518.8 kg_{GHG}/MWh.

The system performance is studied with respect to the following parameters: pinch point temperature difference, turbine inlet pressure, and pump inlet temperature. In addition, the multigeneration system exergy is compared with the cogeneration and single generation efficiencies. For single generation exergy efficiency the net exergy output is $\dot{E}x_{gen} = \dot{W}$ whereas for cogeneration $\dot{E}x_{gen} = \dot{W} + \dot{E}x_{heat}$. The results of this comparison are shown in Figure 9.18 and demonstrate that the exergy efficiency of the multigeneration system is almost double that of power-only generation.

Figure 9.19 shows parametric curves of SCE plotted against exergy efficiency when only one parameter is varied, namely the pressure at turbine inlet, P_3. These curves show that the increase of P_3 is beneficial in terms of both exergy and the environment: it increases efficiency and decreases the carbon emissions. Furthermore, the multigeneration system is clearly the most beneficial, while cogeneration is better than single generation, both technically and environmentally.

TABLE 9.6 State Parameters for Multigeneration System in Reference Case (as presented in Figure 9.17)

State	$\dot{m}$ (kg/s)	P (kPa)	T (°C)	h (kJ/kg)	s (kJ/kgK)	ex (kJ/kg)
1	4.29	28.09	85	140.5	0.429	12.65
2	4.29	2000	85	144.3	0.4312	15.82
3	4.29	2000	400	1241	2.57	474.9
4	4.29	28.09	335.7	1080	2.617	300.4
5	4.29	28.09	85	469.2	1.346	67.96
6	0.262	7.424	80	2649	8.481	126.6
7	0.262	7.424	40.11	168	0.5737	1.487
8	0.262	0.6812	1.5	168	0.6116	−9.794
9	0.262	0.6812	1.5	2503	9.114	−208.3
10	28.73	0.6812	34.6	93.07	0.1977	0.9138
11	28.73	7.424	34.6	97.19	0.1977	5.038
12	28.73	7.424	67.6	159	0.3989	6.883
13	28.47	7.424	80	185.6	0.4665	10.25
14	28.47	7.424	45.62	123.2	0.264	8.236
15	28.47	0.6812	35.62	123.2	0.202	26.69
16	0.3	100	20	−9442	1.38	18,765
17	1.63	100	20	298.6	6.864	0
18	1.93	100	1800	2832	7.847	1898
19	1.93	100	1800	2837	7.847	1898
20	1.93	100	125	400.4	5.884	46.75
21	35.63	100	20	104.8	0.3669	0
22	35.63	100	35	146.7	0.5049	0.7069
23	31.08	100	20	104.8	0.3669	0
24	31.08	100	30	125.8	0.4365	0.1839
25	0.5596	200	20	104.8	0.3669	0
26	0.5596	200	50	209.5	0.7036	4.303
27	1.93	100	90	364	5.898	6.203

Source: Ahmadi et al. (2013b).

The effect of pump inlet temperature and pinch point temperature difference on system performance is shown in Figures 9.20 and 9.21, respectively. The increase of pump inlet temperature reduces the heat duty and increases the cooling duty. Furthermore, increase of the pinch point temperature difference decreases both the heating and cooling duties.

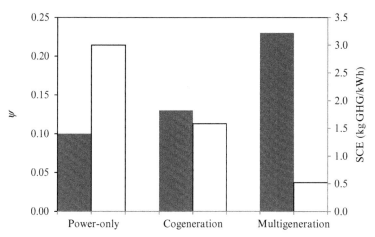

FIGURE 9.18 Comparison of multigeneration system with single and cogeneration systems in terms of exergy efficiency (ψ) and specific carbon emissions (SCE). *Data from Ahmadi et al. (2013b).*

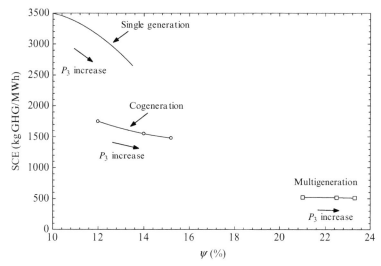

FIGURE 9.19 Benefit of multigeneration in terms of exergy efficiency and specific carbon emissions for varied ORC turbine inlet pressure P_3. *Data from Ahmadi et al. (2013b).*

9.4.4 Solar-Based Coal Gasification System with Multigeneration

This case study presents a concentrated solar-based coal gasification system with multigeneration of power, space heating, hot water, cooling, hydrogen, and oxygen, as, taken from Ozturk and Dincer (2013a). The system integrates the following main subsystems: solar power tower for concentrated solar radiation and high-temperature heat generation, coal gasification and hydrogen syngas storage, gas turbine power generator, SRC with heat recovery steam generator, ORC, hydrogen PEME, hydrogen storage system and hydrogen

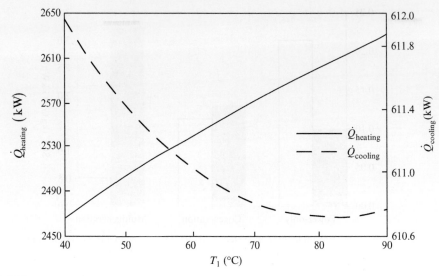

FIGURE 9.20 Effect of pump inlet temperature on heating and cooling load. *Data from Ahmadi et al. (2013b).*

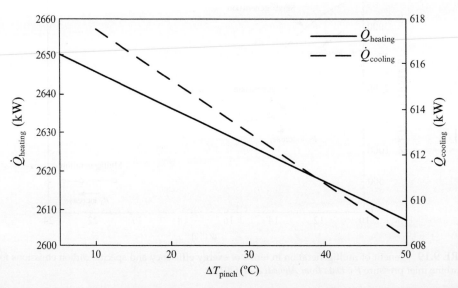

FIGURE 9.21 Effect of pinch point temperature difference on heating and cooling loads. *Data from Ahmadi et al. (2013b).*

fuel cell, oxygen separation unit, hot water tank generator, and absorption cooling and heating system.

The overall system is shown in Figure 9.22. The solar power tower has two functional subsystems, namely the heliostat field and the central receiver. The solar receiver generates high-temperature molten salt which passes from HEX-I to HEX-II to HEX-III (see the figure) and

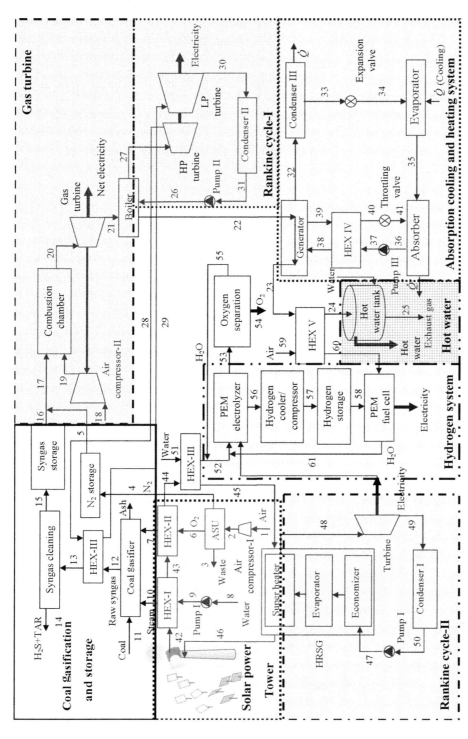

FIGURE 9.22 Solar-based coal gasification system with multigeneration. *Modified from Ozturk and Dincer (2013a).*

transfers heat to the following processes: steam generation for the gasifier, oxygen preheating for the gasifier, and the heat recovery steam generator for which the level of temperature at the hot side is 650 °C. Hydrogen generated by the PEME at 5 bar and 85 °C is compressed with a four-stage compressor with intercooling at 40 °C. The byproduct oxygen has relatively negligible energy and exergy content and can be stored for medical requirements or other purposes, such as oxy-fuel combustion.

Hot steam generated by the solar tower system is used for the coal gasification process with the addition of pure oxygen from the air separation unit. Of course, when sun is not available or the intensity of solar radiation is low, stored syngas from coal gasification is combusted in the gas turbine, with the addition of nitrogen (formerly obtained from the air separation unit). The gas turbine is coupled with a HRVG integrated to an ORC heated with hot combustion gases at 690 °C. In a subsequent heat recovery stage the combustion gases heat the vapor generator of a LiBr-water absorption cycle which generates cooling and heating effects. Furthermore, after the absorption machine vapor generator, the combustion gases are diverted to a hot water tank where they generate hot water.

Thermodynamic analysis based on balance equations must be performed first to assess the system performance of a reference case design. Then a parametric study is performed to determine the best operating conditions and to assess the benefit of multigeneration.

Two particular modeling elements of this integrated system regard the solar tower and the coal gasifier. The reflected solar radiation by the heliostat field is $\dot{Q}_r = nA_h I_n \rho \alpha$, where n is the number of heliostats, A_h is the area of one heliostat, I_n is the intensity of the normal radiation, ρ is the reflectance, and α is the absorbance of the optical surface of the heliostat. The energy balance for the solar receiver reads:

$$\rho_r \alpha_r F_r \dot{Q}_r = \dot{Q}_u + A_r \left[U_L (T_r - T_0) + \varepsilon \sigma (T_r^4 - T_0^4) \right]$$

where $\dot{Q}_u$ is the heat rate transferred to the heat transfer fluid, ρ_r, α_r are the reflectance and absorbance of the receiver, F_r is the intercept factor, T_r, A_r the receiver temperature and surface area, and U_L is the convective heat leak coefficient.

The ultimate analysis of coal considered for this case study and the gasifier parameters and syngas composition are given in Table 9.7; the lower heating value is 22.74 MJ/kg, and the higher heating value is 24.069 MJ/kg. For the chemical exergy of coal, the following equations are derived based on Szargut et al. (1988):

$$\begin{cases} ex_{coal}^{ch} = (LHV + h_w)\beta + y_S \left(ex_S^{ch} - h_S \right) + y_{ash} ex_{ash}^{ch} + y_w ex_w \\ \beta LHV = (LHV + 2442 y_w)\varphi + 941 y_S \\ \varphi = 1.037 + 0.1882 \dfrac{y_{H_2}}{y_C} + 0.061 \dfrac{y_{O_2}}{y_C} + 0.0404 \dfrac{y_{N_2}}{y_C}. \end{cases}$$

In the gasifier the overall chemical reaction of coal, pressurized steam, and oxygen occurs as follows:

$$y_C C + y_{H_2} H_2 + y_{O_2} O_2 + y_{N_2} N_2 + y_S S + z_{O_2} O_2 + z_{H_2O} H_2O \rightarrow$$

$$z_{pg} \left(y_{H_2} H_2 + y_{CO} CO + y_{CH_4} CH_4 + y_{H_2S} H_2S + y_{CO_2} CO_2 + y_{H_2O} H_2O + y_{N_2} N_2 \right) + y_{tar} Tar$$

TABLE 9.7 Gasifier Parameters for Case Study 9.7

Coal Analysis		Gasifier Parameters		Product Gas Composition (Rough Syngas)	
Element	Molar Fraction	Parameter	Value	Component	Molar Fraction
Carbon	0.671	Steam/coal mass ratio	0.35	H_2	0.2742
Hydrogen	0.2269	Air/coal mass ratio	2.45	CO	0.4331
Oxygen	0.02	Coal feeding rate	52.26 kg/s	CH_4	0.0216
Nitrogen	0.006	Operating pressure	5.5 MPa	H_2S	0.0071
Sulfur	0.002	Gasification temperature	1150 °C	CO_2	0.0371
Ash	0.028	Syngas LHV	10.1 MJ/kg	H_2O	0.0505
Moisture	0.0461	–	–	N_2	0.1761

Source: Ozturk and Dincer (2013a).

For the gas turbine cycle, a compression ratio of 36 is assumed, with an isentropic efficiency of 0.82, and the isentropic efficiency of the turbine is assumed to be 0.88. In addition, the turbine inlet temperature is set to 1247 °C and combustion gases at turbine exit are at 690 °C. The effectiveness of the combustion chamber is 0.95, with a fuel consumption of 9.6 kg/s.

The exergy efficiency definitions for the multigeneration system are given in Table 9.8. The reference environment temperature is taken to be 25 °C and the pressure is taken as 101.3 kPa. The results show that the exergy destruction of the coal gasifier, combustion chamber, and boiler are higher than in other system components. The reasons for exergy destruction in the coal gasifier and combustion chamber are generally gasification and combustion, which means chemical reaction and heat transfer through large temperature differences. Exergy destruction in the boiler subsystem is due to the large temperature difference between the exhaust gas and working fluid inside the boiler walls.

According to the exergy destruction rates and exergy loss ratios, development applications would be chosen for these system components. Minimization of exergy destruction rate is essential to enhance the performance of the multigeneration energy system, with a return by decrease in the cost and emissions. This helps reduce the associated environmental impacts. The integrated solar power tower and coal gasification system for multigeneration purposes offer heat recovery capabilities from waste heats and exhaust gas.

The impact of coal gasifier efficiency on the overall efficiency of the system can be observed in Figure 9.23. The energy and exergy efficiencies for this process increase linearly as the gasification temperature increases. The exergy contained in the coal sample is transferred by the gasification process into the chemical and the physical exergy of the synthesis gas. Thus, exergy content in the syngas is increased with increase in the gasifier temperature.

Rankine cycle efficiency also has an important impact on the efficiency of the overall system. Here, the efficiency of a Rankine-I cycle (see Figure 9.22) and the overall system efficiencies are plotted against the turbine inlet temperature, T_{12}. This shows that increasing the turbine inlet temperature increases the energy and exergy efficiencies of the system

TABLE 9.8   Exergy Efficiency Definitions for Multigeneration System and Its Subunits

Process	Exergy Efficiency Definition Equation
Solar power tower	$\psi_{SPT} = \dfrac{\dot{Ex}_{HEX-I}^{Q} + \dot{Ex}_{HEX-II}^{Q} + \dot{Ex}_{HEX-III}^{Q} + \dot{Ex}_{HSRG}^{Q}}{\dot{Ex}_{solar}^{Q}}$
Gasifier	$\psi_{G} = \dfrac{\dot{m}_{pg}ex_{pg}^{ch}}{\dot{Ex}_{HEX-I}^{Q} + \dot{Ex}_{HEX-II}^{Q} + \dot{m}_{coal}ex_{coal}^{ch}}$
Rankine cycle I	$\psi_{R-I} = \dfrac{\dot{W}_{R-I}}{\dot{Ex}_{HSRG}^{Q}}$
Gas turbine	$\psi_{GT} = \dfrac{\dot{W}_{GT}}{\dot{m}_{f}ex_{f}^{ch}}$
Rankine cycle II	$\psi_{R-II} = \dfrac{\dot{W}_{R-II}}{\dot{Ex}_{bolier}^{Q} + \dot{Ex}_{HEX-III}^{Q}}$
Absorption cooling and heating	$\psi_{abs} = \dfrac{\dot{Ex}_{cooling}^{Q} + \dot{Ex}_{heating}^{Q}}{\dot{Ex}_{generator}^{Q} + \dot{W}_{pump-II}}$
Hydrogen production	$\psi_{H_2} = \dfrac{\dot{m}_{H_2}ex_{H_2}^{ch}}{\dot{Ex}_{HEX-VI}^{Q} + \dot{W}_{PEME}}$
Multigeneration system	$\psi = \dfrac{\dot{W}_{R-I} + \dot{W}_{GT} + \dot{W}_{GT} + \dot{Ex}_{cooling}^{Q} + \dot{Ex}_{heating}^{Q} + \dot{m}_{H_2}ex_{H_2}^{ch} + \dot{Ex}_{hot-water}^{Q}}{\dot{Ex}_{solar}^{Q} + \dot{m}_{coal}ex_{coal}^{ch}}$

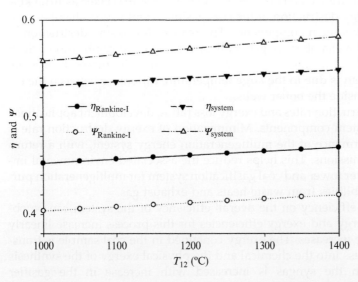

FIGURE 9.23   Impact of gasification temperature on efficiency of gasifier and overall system. *Data from Ozturk and Dincer (2013a).*

components and the whole system. The increase of the HRSG outlet temperature increases the quality of the working fluid which operates the turbine; therefore, the hydrogen production rate during PEM electrolysis increases depending on the electricity coming from the turbine (Figure 9.24).

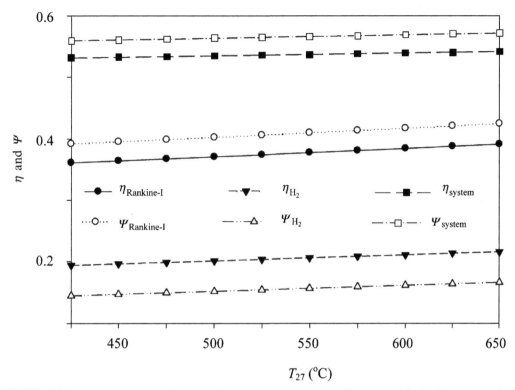

FIGURE 9.24 Impact of Rankine-II cycle turbine inlet temperature on efficiencies of the Rankine cycle and the overall system. *Data from Oztürk and Dincer (2013a).*

The variation of the energy and exergy efficiencies for the gas turbine, the Rankine system II, the absorption refrigeration system and the whole system with varying gas turbine inlet temperature is shown in Figure 9.25. Energy efficiency of the subsystems and the whole system remain constant, whereas exergy efficiency increases with increase in the gas turbine inlet temperature. The thermal energy transfer rate increases by absorbing greater thermal energy from the exhaust gas in the boiler and generator.

Note that increasing the fuel use in the combustion chamber and increasing the gas turbine inlet temperature lead to a higher gas turbine exhaust gas temperature. Thus, the temperature of the working fluids of the Rankine system II and absorption cycle would increase. Therefore, exergy efficiency of the subsystem and the whole system increases with an increase of the gas turbine inlet temperature. Increasing the temperature in the gas turbine inlet and increasing the fuel use in the combustion chamber increase the power output of the multigeneration system rapidly. Higher inlet temperature for the gas turbine at a fixed pressure gives a higher temperature for the gas turbine exhaust gas, and consequently more steam is generated in the boiler.

The benefit of multigeneration can be contemplated from the bar chart in Figure 9.26. There, it can been seen that the energy and exergy efficiencies of each of the generating subunits are smaller than that of the integrated system. Indeed, when the subunits work

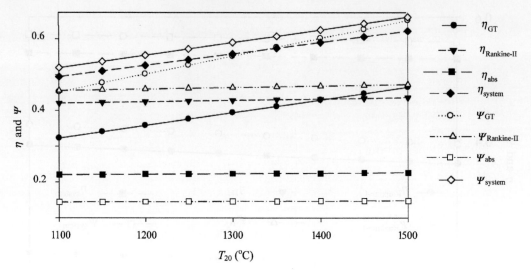

FIGURE 9.25   Impact of gas turbine inlet temperature (T_{20}) on energy and exergy efficiencies of Rankine-II cycle, Brayton cycle, absorption cycle, and overall system. *Data from Ozturk and Dincer (2013a).*

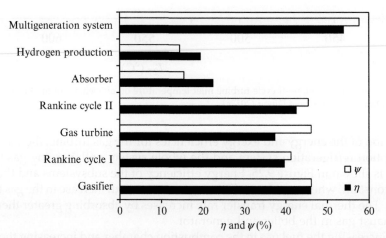

FIGURE 9.26   Exergy and energy efficiency of the multigeneration system in Figure 9.22 and that of its subsystems. *Data from Ozturk and Dincer (2013a).*

synergistically within the multigeneration system much of the energy can be regenerated internally through heat recovery. This opportunity leads to higher efficiency for the integrated system.

9.4.5 Solar-Based Multigeneration System with Hydrogen Production

In this case study a multigeneration system that integrates a SRC, an ORC (with isobutane), absorption cooling and heating, and electrolytic hydrogen production and implementation is demonstrated. The subunits are integrated with a concentrated solar radiation system. Six outputs are generated, namely power, hot water, space heating, refrigeration, hydrogen, and oxygen. This case study is derived from the work by Ozturk and Dincer (2013b).

The multigeneration system is shown in Figure 9.27. The thermal energy of solar radiation is first collected and concentrated using a parabolic dish collector for use in the boiler and reheater of the Rankine cycle. A two-stage turbine (low and high pressure) is considered in this Rankine cycle. A part of power output is transferred to a HTSE system for hydrogen production. Steam produced by the parabolic dish collector II (PDC-II) has a temperature of 320 °C when it leaves the reheater (shown at point 11 in Figure 9.27). Therefore, energy content of this steam is used for the ORC in order to produce electricity. A single-effect absorption refrigeration cycle via LiBr-water provides both cooling and heating loads.

The heat requirement for the HTSE is supplied by solar energy via a PDC-II series. A membrane separator is introduced to separate hydrogen from steam, while steam is recycled to the HTSE system. The produced hydrogen stream is cooled to 40 °C with the help of the water which in turn is heated to 80 °C and then utilized in the hot water tank as a generated product. Another system output is pure oxygen resulting from the HTSE, which is stored under pressure in a separate tank. Heat is recovered from cooling the oxygen stream from 85 °C (HTSE exit) to 45 °C (prior to feeding it to storage).

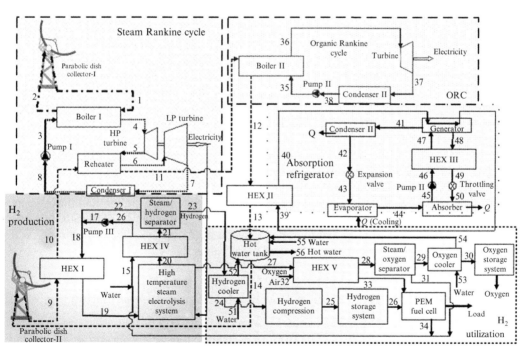

FIGURE 9.27 Solar-based multigeneration system with hydrogen production. *Modified from Ozturk and Dincer (2013b).*

The cooling and heating subsystem is a single-effect LiBr-water absorption refrigerator cascaded with the hot water tank. Heat input of the absorption system is provided by the energy content of the steam, shown at point 12. The absorption cycle consists of six components: generator, condenser, absorber, solution heat pump, expansion valve, and evaporator.

The working fluid at 130 °C (point 13) enters the hot water tank to produce hot water for domestic needs.

The system modeling is performed at steady state for a reference case based on balance equations written for each of the system components. The exergy input into the system is the sum of the thermal exergy provided by the PDC-I and PDC-II (see Figure 9.27). Therefore, the exergy efficiency of the multigeneration system is written as

$$\psi = \frac{\dot{W}_{SRC} + \dot{W}_{ORC} + \dot{W}_{PEMFC} + \dot{m}_{H_2}ex_{H_2}^{ch} + \dot{m}_{O_2}ex_{O_2}^{ch} + \dot{E}x_{hot,H_2O}^{Q} + \dot{E}x_{sp.heat.}^{Q} + \dot{E}x_{cool}^{Q}}{\dot{E}x_{PDC-I}^{Q} + \dot{E}x_{PDC-II}^{Q}} \qquad (9.25)$$

where $\dot{W}_{PEMFC}$ represents the power generated by the PEMFC. The system is controlled in such a way that when power demand is high and solar energy is not sufficient or absent, hydrogen is consumed from the storage tank to satisfy the power need. However, any hydrogen surplus of the system can be valorized or used in other ways. The $\dot{m}_{H_2}$ represents the rate of hydrogen generated as product (excluding the hydrogen which is used internally by the system for power generation).

Table 9.9 gives process flow data for the multigeneration system, and in Table 9.10 the results of the analysis are given, namely the exergy destruction rates for each of the system components as well as the exergy destruction fractions (ExDF), exergy efficiencies of the subsystems and work or heat transfer rates (depending on the component type).

Even though exergy efficiencies of the two parabolic collectors are 55.2% and 59.95%, their contribution to overall exergy destruction is the highest with respect to other components. The HTSE system and hydrogen compression system also show high ExDF. On the other hand, the analysis shows that the absorption cooling and heating subsystem does not generate considerable exergy destructions. This is due to the fact that the absorption cycle utilizes waste steam recovery.

Figure 9.28 shows the main results of the analysis: the exergy efficiency of the multigeneration system and its subsystems as a function of the fluctuating solar radiation intensity (I_n is the normal direct beam radiation intensity). This result confirms that the energy and exergy efficiencies of the integrated multigeneration system are higher than the efficiency of each of its subsystems. For example, the energy and exergy efficiencies of the SRC are ~44% and 41%, respectively. This is compared to the energy and exergy efficiencies of the multigeneration system, which are ~52% and 56%, respectively.

9.4.6 Solar-Based Trigeneration of Power, Heating, and Desalination

In this case study a system that generates power, heating, and desalination (freshwater) is presented based on the work by Cetinkaya (2013). The system uses concentrated solar radiation with heliostat-tower receiver technology for its energy input. Figure 9.29 shows the system schematics. It consists of five energy conversion subunits. First it is the heliostat field and the solar tower receiver which generate high-temperature heat that is transferred to a molten

TABLE 9.9   Process Flow Data for Multigeneration System (Figure 9.27)

State	P (kPa)	T (°C)	$\dot{m}$ (kg/s)	$\dot{E}$ (kW)	$\dot{Ex}$ (kW)	State	P (kPa)	T (°C)	$\dot{m}$ (kg/s)	$\dot{E}$ (kW)	$\dot{Ex}$ (kW)
1	4500	750	7.784	30,480	13,505	29	4000	108.4	5.195	370.4	1523
2	4500	188	7.784	5413	1131	30	4000	45	5.195	50.56	1477
3	15,200	38	7.357	498.6	119.4	31	101.3	108.4	0.9672	2503	475
4	15,200	600	7.357	25,565	11,741	32	101.3	25	2.624	0.003156	0.0034
5	1000	227.9	7.357	20,508	6338	33	101.3	90	2.624	714	68.14
6	1000	450	7.357	24,027	8122	34	101.3	80	2.624	603.8	49.69
7	7	39.01	7.357	17,381	769.8	35	3250	13.76	34.24	934.8	135.5
8	7	38	7.357	399.5	7.79	36	3250	146.8	34.24	18,965	3456
9	4500	750	7.784	30,479	13,505	37	410	79.5	34.24	16,529	436.7
10	4500	502.9	7.784	26,010	10,526	38	410	11.7	34.24	700.6	32.9
11	4500	318.1	7.784	22,491	8557	39	101.3	60	0.3936	57.61	3.138
12	4500	160	7.784	4461	814	40	101.3	160	0.3936	1059	204.1
13	4500	130	7.784	3459	526.9	41	5.616	85	0.321	819.9	28.43
14	4500	50	7.784	843.4	66.56	42	5.616	35	0.321	13.42	1.882
15	101.3	90.19	2.76	753.3	72.08	43	1.069	6	0.321	11.16	1.682
16	5000	261	2.76	2855	777.2	44	1.069	6	0.321	772.7	53.52
17	5000	266	2.76	7445	2819	45	1.069	35	2.89	109.7	84.29
18	5000	285	5.989	16,583	6312	46	5.616	38	2.89	126.9	84.89
19	5000	580	5.989	21,050	8841	47	5.616	63	2.89	272.5	96.07
20	5000	557	0.6559	7683	7930	48	5.616	85	2.568	435.4	219.2
21	5000	299	3.253	9160	3501	49	5.616	50	2.568	271.1	199
22	5000	299	3.229	9091	3474	50	1.069	45	2.568	248	197.5
23	5000	299	0.6559	5206	6518	51	101.3	26	11.23	46.99	0.0932
24	5000	40	0.6559	2736	5761	52	101.3	80	11.23	2584	212.7
25	10,300	42	0.6559	2773	6371	53	101.3	26	1.61	6.737	0.0134
26	506.5	85	0.2624	1259	1576	54	101.3	80	1.61	370.5	30.49
27	4000	557	5.195	2733	2672	55	101.3	26	15.31	64.05	0.1271
28	4000	108.4	5.195	2376	3529	56	101.3	80	15.31	3523	289.9

Data from Ozturk and Dincer (2013b).

TABLE 9.10 Exergy Destruction Rate, Exergy Destruction Fraction (ExDF), Exergy Efficiency, and Work or Heat Transfer Rates for Components of the Multigeneration System (Figure 9.27)

Subsystem	$\dot{E}x_d$(kW)	ExDF(%)	ψ(%)	$\dot{W}$or $\dot{Q}$(kW)	Subsystem	$\dot{E}x_d$(kW)	ExDF(%)	ψ(%)	$\dot{W}$or $\dot{Q}$(kW)
PDC-I	2523	15.8	55.2	25,064	Evaporator	75.08	0.47	40.84	761.5
Boiler-I	752.2	4.7	93.92	25,064	Absorber	131	0.82	21.42	928.3
HPT	345.8	2.1	93.6	5056	Pump-III	10.99	0.069	34.41	17.23
LPT	705.3	4.4	90.41	6646	Throttling valve	1.551	0.01	96.58	23.13
Condenser-I	108.9	0.68	85.71	16,979	HEX-III	8.961	0.05	55.53	145.5
Pump-I	60.14	0.38	56.03	99.16	HTSE system	1434	9.01	49.57	7536
PDC-II	2832	17.8	59.95	29,632	SteamH$_2$ sep.	424	2.66	65.06	1838
Reheater	185.2	1.16	90.59	3519	HEX-IV	709.6	4.46	52.7	2102
HEX-I	450.2	2.8	84.89	4468	Pump-IV	254	1.59	44.49	458
Boiler-II	244	1.5	42.89	1802	HEX-V	328.4	2.06	53.13	714
ORC turbine	583.2	3.66	80.68	2436	Steam-O$_2$ sep.	772	4.8	42.55	1253
Condenser-II	397.4	2.5	22.86	15,826	H$_2$ comp.	814.8	5.1	12.26	1750.6
Pump-II	153	0.96	60.77	234	H$_2$ storage	66.81	0.42	24.74	216.4
HEX-II	86.19	0.54	69.98	1001	PEMFC	269.9	1.7	37.03	1042
Generator	49.38	0.31	75.42	1000	Hot water tank	413.7	2.6	41.19	3458
Condenser-III	26.55	0.167	21.94	806.5	H$_2$ cooler	570.3	3.58	27.15	2537
Expansion valve	0.2001	0.001	97.36	2.255	O$_2$ cooler	122.34	0.77	57.71	363.8

Data from Ozturk and Dincer (2013b).

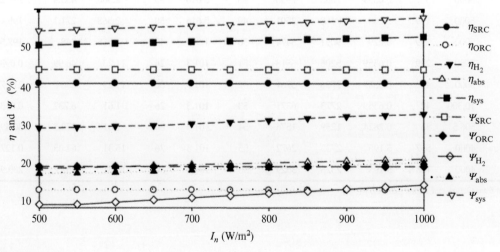

FIGURE 9.28 Product generation efficiency of the multigeneration system and its subsystems. *Data from Ozturk and Dincer (2013b).*

salt heat transfer fluid. Two thermal storage reservoirs are used, one for the cold heat transfer fluid and one for the hot heat transfer fluid. Of course, the temperature of the "cold" heat transfer fluid is rather elevated.

The main power generation unit is the reheat SRC, with its condenser cooled by stream A–B. The heat required by the SRC is provided by the heat transfer fluid at its hottest temperature. A secondary power generation unit is the ORC, which in fact uses lower grade heat from the colder heat transfer fluid stream. In addition, the heat transfer fluid loop includes a heat exchanger to transfer low-grade heat to the desalination unit, which is of reverse osmosis type.

For heat transfer, the molten salt used is 60% NaNO$_3$ 40% KNO$_3$ by weight. The highest temperature of the molten salt within the loop is assumed to 580 °C (state 11 in Figure 9.29).

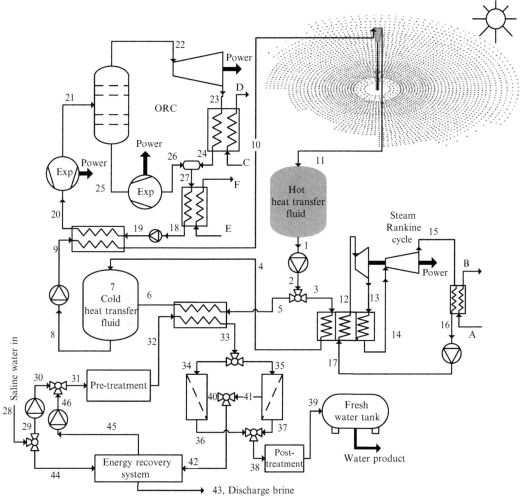

FIGURE 9.29 Solar-based multigeneration system for power, heating, and desalination. *Modified from Cetinkaya (2013).*

The system uses 60,000 heliostats, each of 2.3 m^2 surface and 10% energy efficiency. Also, reasonable isentropic efficiencies are assumed for the prime movers. Hence, the state parameters are determined for each state based on mass, energy, and exergy balance equations. Based on data from Cetinkaya (2013) the exergy destruction is the highest in solar concentration systems, with 23% of the total exergy destruction. However, the cumulated exergy destruction in all heat exchangers is 50% of the total. Figure 9.30 shows the distribution of exergy destructions among the main types of system subunits.

The energy and exergy efficiencies of the main subunits of the multigeneration system are shown in the bar chart in Figure 9.31. The improvement of system efficiency due to increase in the number of generated outputs is shown in Figure 9.32. As seen, when the number of outputs increases the efficiencies do the same. The power-only case presupposes that only Rankine cycles operate to produce power as the sole output. For the cogeneration case, power

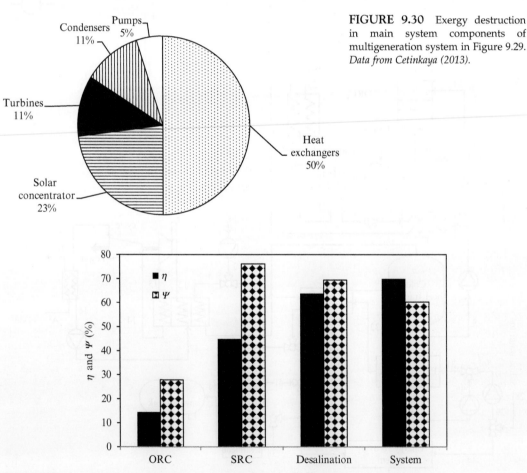

FIGURE 9.30 Exergy destruction in main system components of multigeneration system in Figure 9.29. *Data from Cetinkaya (2013).*

FIGURE 9.31 Energy and exergy efficiencies of the main subunits of the multigeneration system in Figure 9.29. *Data from Cetinkaya (2013).*

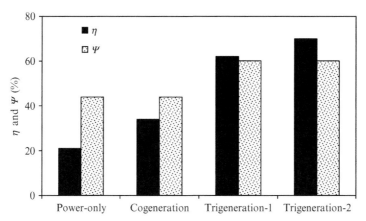

FIGURE 9.32   Improvement of system efficiency due to an increase in the number of generated outputs. *Data from Cetinkaya (2013).*

and desalination are produced. In the trigeneration-1 case, power, desalination, and heating are achieved by cogeneration. Trigeneration-2 produces some more power from internal energy recovery.

9.4.7 Multigeneration System for Power, Hot Water, and Fuel from Biomass and Coal

The system presented in this case study integrates a coal-biomass gasifier, a Brayton cycle, a SRC, an ORC, and a FT synthesis unit. The produced outputs are power, hot water, and synthetic fuel. The produced fuel is in fact a mixture of olefins and paraffin that form a precursory for liquefied petroleum gas, diesel fuel, and wax. This case study is based on work by Cetinkaya (2013). The multigeneration system is shown in Figure 9.33.

The fuel used as the energy source for the system is a blend of coal and biomass which is gasified using air with excess nitrogen (or oxygen-depleted air). After gasification, the resulting syngas is cleaned by flying ash separation and acid gas and sulfur compound removal. A part of the syngas is combusted in a gas turbine to generate power (stream #10 in the figure). The gas turbine cycle is cascaded with a SRC. The other part (stream #24) is directed to a shift reactor where steam is provided by a FT process that occurs thermocatalytically and exothermically on iron catalysts. A water loop is used to transfer heat from the FT reactor to an ORC such that the reaction temperature is maintained at the optimal value to maximize synthetic fuel production.

The composition of product gas (stream #5) includes hydrogen, carbon monoxide, carbon dioxide, nitrogen, methane, flying ash, and minor sulfur-based compounds. For the case study, the assumed ash composition is 62% SiO_2, 21% Al_2O_3, 11% Fe_2O_3, 6% MgO. The enthalpy of ash with this composition is taken into account for the energy balance of the gasifier.

All the balance equations are written by assuming a steady state operation of the system. For the combustion chamber an effectiveness of 90% and 20% excess air are assumed with respect to stoichiometric combustion. The isentropic efficiencies of the compressors and turbines are taken as 80% whereas the intercooler and the reheater (see the Figure 9.33) have an

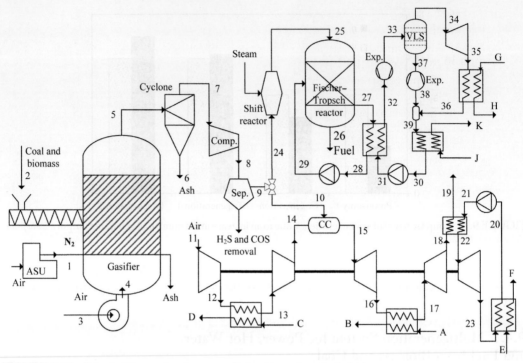

FIGURE 9.33   Trigeneration system for power, hot water, and synthetic fuel.

assumed effectiveness of 85%. Approximately half of the generated synthetic gas is diverted to the FT synthesis unit. However, the hydrogen concentration in the syngas must be increased to reach optimal operating conditions for the fuel synthesis reaction.

A water-gas shift reactor is used to conduct the reaction $CO + H_2O \leftrightarrow H_2 + CO_2$, which occurs at equilibrium. This reaction increases the molar ratio of hydrogen to carbon monoxide to 2 prior to the feed into the FT reactor. Heat is transferred from by the FT unit to the ORC with the help of heat exchangers 27-28-31-32. The ORC is of flash type, comprising two expanders and a turbine, all of assumed isentropic 0.8.

The assumed fuel consisting of biomass (straw) and coal is represented with the chemical formula CH_mO_n, which is sufficiently accurate for determination of the mass and energy balance of the system. Nitrogen reactions are neglected when the calculation focuses on energy balance. As nitrogen reactions can be neglected, the chemical equation for the gasification process is:

$$CH_mO_n + \alpha O_2 + \beta H_2O \rightarrow \zeta_1 H_2 + \zeta_2 CO + \zeta_3 CO_2 + \zeta_4 CH_4 + \zeta_5 C_2H_4 + \zeta_6 C_2H_6 + \zeta_7 CH_pO_q \quad (9.26)$$

where from stoichiometric balance one obtains

$$1 = \zeta_2 + \zeta_3 + 2\zeta_5 + 2\zeta_6 + \zeta_7$$

$$m + 2\beta = 2\zeta_1 + 4\zeta_4 + 6\zeta_6 + p\zeta_7$$

$$n + 2\alpha + \beta - \zeta_2 + 3\zeta_3 + q\zeta_7$$

For the case considered in Cetinkaya (2013) the reaction heat for gasification is determined from stoichiometric energy balance. However, in order to assure that correct results are obtained, experiments were conducted with straw and various types of coal to determine the composition ranges of the product gas. The average of the reaction enthalpy is -40 kJ/mol.

Based on thermodynamic modeling the energy and exergy efficiencies of each system component are determined. Figure 9.34 reports the efficiencies and Figure 9.35 shows the ExDF. The FT process has the highest efficiency among the system components because it produces fuels with rather high chemical exergy contents and in addition allows for heat recovery at ~575 K (Carnot factor 0.48). The heat recovered from the FT unit is ~1.7 kJ/kg of fuel product.

The benefit of the increase in the number of outputs is shown in the bar chart of Figure 9.36. The first bar in the chart is energy efficiency of the single generation unit, which only

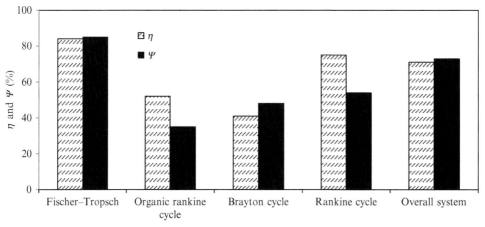

FIGURE 9.34 Energy and exergy efficiencies of the main system components and the overall system in Figure 9.33. *Data from Cetinkaya (2013).*

FIGURE 9.35 Exergy destruction fractions by components of the multi-generation system in Figure 9.33. *Data from Cetinkaya (2013).*

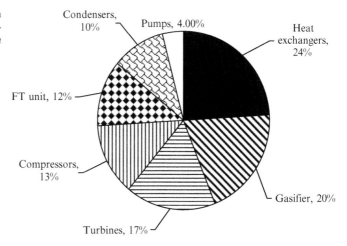

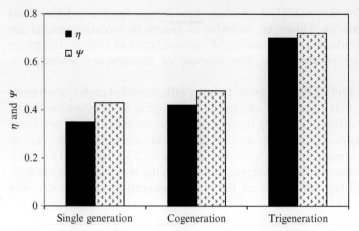

FIGURE 9.36 The benefit of integrated multigeneration in energy and exergy efficiencies. *Data from Cetinkaya (2013).*

produced power as useful product; the other generated products by the system in Figure 9.33 are not considered in this case. The second bar set in the chart represents the efficiencies of the system with cogeneration, when power and hot water are the products. The efficiency improves slightly. For the system with trigeneration the additional output is the FT fuel. The trigeneration system has much higher efficiency then the cogeneration system (~70%).

9.4.8 SOFC-ORC-Absorption Trigeneration System Fueled with Natural Gas and Biomass

In this case study a trigeneration system for power, heating, and cooling is presented based on the work of Al-Sulaiman et al. (2010). The system integrates a solid oxide fuel cell coupled with an ORC and an absorption chiller and is co-fueled with natural gas and biomass, as shown in Figure 9.37. The waste heat from the SOFC is recovered and used to transfer heat to the ORC. Furthermore, the waste heat rejected by the ORC is recovered and used for heating purposes and to drive the absorption refrigerator. The ORC works with n-octane, which has a critical temperature of 569 K.

The main assumptions for modeling of the cycle include: steady state operation, isentropic efficiency of the ORC turbine and pump 80% (both), effectiveness of the ORC boiler 80%, effectiveness of the solution heat exchanger of the absorption refrigerator 70%, electric generator efficiency 95%, DC/AC inverter efficiency of the SOFC 95%, fuel utilization factor in the SOFC 85%, inlet stream temperature of the SOFC 1000 K, and conversion process within the fuel cell at equilibrium.

As the system is co-fueled, the exergy efficiency definition is written as follows:

$$\psi = \frac{\dot{W}_{net} + \left(T_0/\overline{T}_{cooling} - 1\right)\dot{Q}_{cooling} + \left(1 - T_0/\overline{T}_{heating}\right)\dot{Q}_{heating}}{\dot{m}_{CH_4}ex^{ch}_{CH_4} + \dot{m}_{wood}ex^{ch}_{biomass}} \tag{9.27}$$

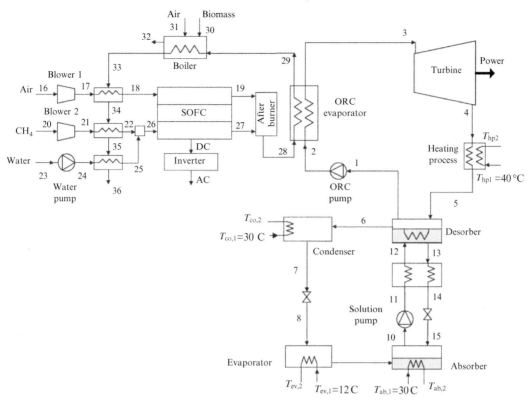

FIGURE 9.37   Natural gas and biomass co-fueled trigeneration system with SOFC, ORC, and absorption chiller. *Source: Al-Sulaiman et al. (2010).*

where $\dot{W}_{net} = \eta_{inv}\dot{W}_{SOFC} + \eta_{gen}\dot{W}_{T} - \left(\dot{W}_{blow,1} + \dot{W}_{blow,2} + \dot{W}_{pmp,w} + \dot{W}_{pmp,ORC}\right)/\eta_{mot}$ is the net generated, η_{inv} is the inverter efficiency, η_{gen} the efficiency of electrical generator, and η_{mot} is the average value of the efficiency of the electrical motors that drive the blowers and the pumps.

The parameters influencing the overall exergy efficiency the most are the current density and the fuel cell inlet temperature. The correlation between exergy efficiency, the fuel cell temperature, and current density can be seen in the plot shown in Figure 9.38.

The exergy efficiency of the trigeneration system is compared with exergy efficiency of the single and cogeneration systems as shown in Figure 9.39. If the cooling and heating effects are disregarded, the exergy efficiency of the system is 46%. If cooling and power are the products, then the exergy efficiency increases to 57%. Further, with power and heating (but no cooling) the exergy efficiency is 71%, whereas for the case of cooling, heating, and power production, the exergy efficiency becomes 74%. As reported in Al-Sulaiman et al. (2010), the benefit of this trigeneration system results from the fact that the gain in the exergy efficiency when trigeneration is used instead of the power-only cycle is from 3% to 25%, depending on the operating conditions.

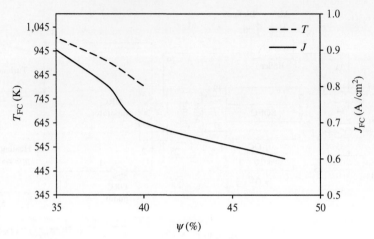

FIGURE 9.38　The influence of fuel cell inlet temperature and of current density on the overall trigeneration exergy. *Adapted from Al-Sulaiman et al. (2010).*

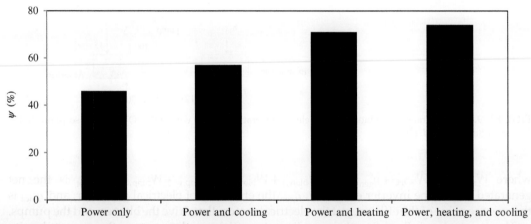

FIGURE 9.39　The benefit of multigeneration with the integrated system represented in Figure 9.37. *Data from Al-Sulaiman et al. (2010).*

9.4.9 Hybrid Solar/Syngas Trigeneration with ORC, SOFC, and Absorption

Here a trigeneration system is presented which produces power, hot water at 55–60 °C and cooling (at 4 °C), and integrates an internal reforming tubular solid oxide fuel cell, a combustor and HRVG, a reheat-type ORC with R245fa, a parabolic solar through-collector with a heat transfer fluid loop, and an LiBr-water absorption chiller. The system diagram is shown in Figure 9.40.

This case study is based on the work by Al-Sulaiman et al. (2010). The energy sources used to run the system are syngas and concentrated solar radiation produced via a solar field with parabolic through-concentrators. The syngas source is assumed to be biomass

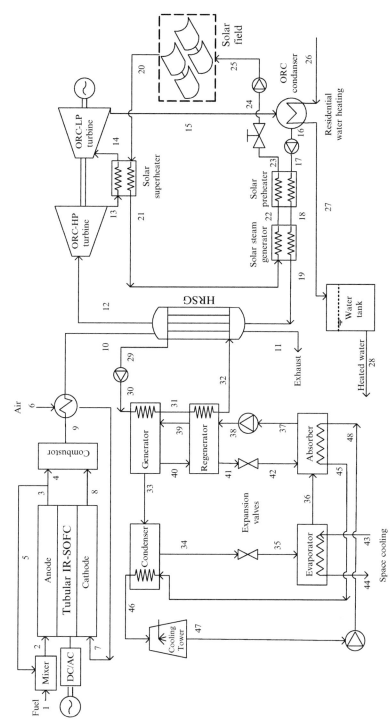

FIGURE 9.40 Hybrid solar/natural gas trigeneration system with SOFC, ORC, and absorption chiller. *Modified from Ozcan and Dincer (2013).*

gasification and has the following composition: 21% CH_4, 40% H_2, 20% CO, 18% CO_2, 1% N_2. The SOFC operates at ambient pressure. Therefore, air compression is not required. This saves important energy. The system is modelled in steady state based on mass, energy, and exergy balance equations. The exergy efficiency of the trigeneration system is given as follows:

$$\psi = \frac{\dot{W}_{SOFC} + \dot{W}_{ORC} + \dot{Ex}_{heating} + \dot{Ex}_{cooling}}{\dot{Ex}_{fuel} + \dot{Ex}_{solar}} \tag{9.28}$$

For a reference case calculation of the exergy efficiencies of the system and subsystems are shown in Figure 9.41. The use of exhaust gases for heat recovery leads to a 15.7% increase in the system's exergy efficiency. From total exergy generated by the system in the form of useful product 58% is due to heating, 29% to SOFC power, 8% to ORC power, and 5% is the exergy associated with the cooling effect. The overall energy efficiency of the system is 85.1% and exergy efficiency is 32.6%. The impact of using solar radiation on system efficiency is important. With solar radiation assistance the system efficiency increases by 12–16%, and, hence, significantly more power is generated.

9.4.10 Ammonia-Water-Based Trigeneration System for Waste Heat Recovery

This case study is based on the work reported by Khaliq et al. (2009a). It presents an innovative trigeneration system based on a modified ammonia-water cooling and power cycle previously proposed by Hasan et al. (2002). The modified version of the system integrates a heat recovery unit. Hence, it produces three useful outputs, as power, heating, and cooling. The cycle is illustrated in Figure 9.42. It is assumed that the heat source is recovered from an industrial process, or from combustion gases.

Combustion gases at a temperature level in the range of 375–475 °C are directed to heat exchanger 1–2, which transfers some thermal energy for process heating (e.g., steam generation). Further, combustion gases are directed to the vapor superheater of the combined absorption cooling and power cycle 2–3 and then to the vapor generator 3–4. Superheated ammonia vapor in #10 is expanded in a turbine for work production. During the expansion process the stream enters into the thermodynamic two-phase region and reduces its temperature below the environment. Thus, cold stream #11 enters the evaporator where the liquid phase boils off and the vapor is superheated at low pressure and temperature such that a cooling effect occurs (11–12). The vapor #12 is then absorbed into the weak solution (#16), and strong ammonia-water solution #5 is formed. Ammonia is carried to high pressure by the strong solution where, in the vapor generator, ammonia desorbs and weak solution #14 returns. The electrical-to-thermal energy ratio of the system varies between 10% and 20%. The exergy efficiency of the system is 34% and the energy efficiency is 42%.

9.4.11 Gas Turbine Trigeneration System with Absorption Cooling, Heating, and Power

This case study is based on the work by Khaliq et al. (2009b). A trigeneration system comprising a special Brayton cycle, and an $LiBr/H_2O$ absorption refrigeration cycle is presented.

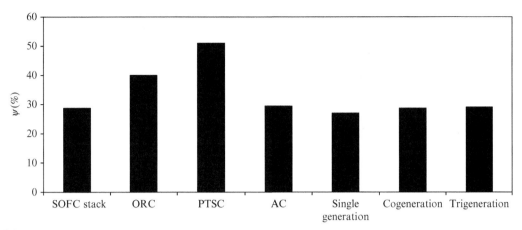

FIGURE 9.41 Exergy efficiency of main system components and of system for single generation, cogeneration, and trigeneration. *Data from Ozcan and Dincer (2013).*

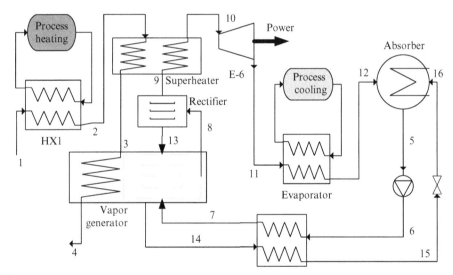

FIGURE 9.42 Ammonia-water-based trigeneration system for waste heat recovery.

The system schematic is shown in Figure 9.43. The Brayton cycle implements a gas turbine which is installed on the same shaft as an air expansion turbine.

The compressed air is split into two parts. One part, steam (#5), runs an air refrigeration cycle: it is cooled to close to ambient temperature (process 5–23) and expanded to ambient pressure while it cools below the ambient temperature (stream #24). Air is then mixed with intake air in a mixing chamber MC (see Figure 9.43). Hence, the intake air in the compressor (#2) is colder than the surroundings; this reduces the compression work and improves the cycle efficiency.

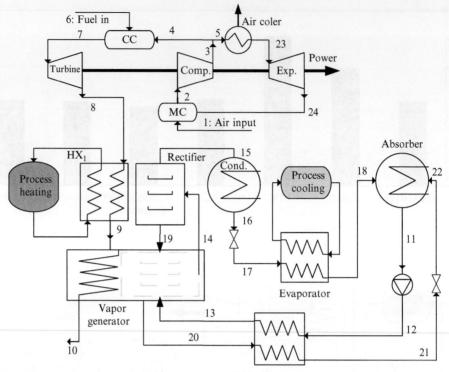

FIGURE 9.43   Trigeneration cycle with gas turbine. *Modified from Khaliq et al. (2009b).*

The combustion gases leaving the gas turbine are used for process heating, and subsequently they are diverted to the vapor generator of an Li–Br-water absorption refrigerator. Hence, the trigeneration system produces power, heating, and cooling effects.

For thermodynamic modeling, mass, energy, and entropy balances are written for each component of the system. In the mixing chamber it is assumed that a 2% pressure drop takes place. The compression process is assumed to be polytropic. An electrical to thermal energy ratio can be defined by the expression:

$$R_{ET} = \frac{\dot{W}_{net}}{\dot{Q}_{hp} + \dot{Q}_C} \tag{9.29}$$

The assumptions made are: 310 K with 60% relative humidity, gas turbine pressure ratio of 10 with 1473 K inlet temperature, fixed specific heat ratio of combustion gases $\gamma = 1.33$ and of air compressor process $\gamma = 1.4$, temperature difference in heat exchanger 5 K, and pinch point 25 K.

The exergy efficiency can reach 51%, while the fuel utilization efficiency approaches 90%. The thermodynamic analysis reported in Khaliq et al. (2009b) demonstrates that the utilization and exergy efficiencies increase while the electrical to thermal energy ratio decreases with the extracted mass rate of air (inlet air cooling). Moreover, the electrical to thermal

FIGURE 9.44 Exergy destruction fractions for the main components of the trigeneration system in Figure 9.43. *Data from Khaliq et al. (2009b).*

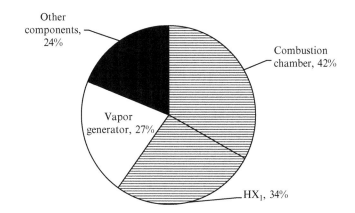

energy ratio and exergy efficiency are sensitive to process heat pressure. The process heat pressure should be high for better performance, based on the first and second laws. The exergy destruction results in Figure 9.44 demonstrate that 42% irreversibility occurs in the combustion chamber, 34% in HX_1, 27% in the generator of the absorption refrigeration system, and 24% in components.

9.5 CONCLUDING REMARKS

In this chapter, multigeneration systems are introduced and discussed from the efficiency, cost, and environmental impact viewpoints. A number of case studies are given to highlight the importance of multigeneration. Several criteria are also introduced for assessment of the multigeneration systems for various outputs. Exergy efficiency is one of the most important parameters for assessment and comparing one multigeneration system with another, especially with conventional systems. The benefit of the multigeneration system is emphasized more when the system is compared with independent systems that generate the same outputs. Due to their ability to regenerate energy internally, the integrated multigeneration systems are always better than independent single generation systems.

A number of key processes that can be integrated into multigeneration systems are discussed, namely combustion and gasification, direct coal liquefaction, heat recovery vapor generation, FT synthesis, water desalination, and bio-photolysis. In addition to energy and exergy efficiencies, the levelized electricity cost can be determined for the system lifetime. If more products can be valorized, the levelized electricity cost decreases with application of multigeneration systems. Another key criterion is the specific emissions factor, which quantifies the emissions (or pollution) per unit of generated exergy. The increase in the number of outputs leads to better system economics and less environmental impact provided that the integrated system is designed in such a way that more internal regeneration can be applied.

Multi-objective, multivariable optimization of multigeneration systems for power and diverse product generation is presented based on the case studies considered to exemplify the concepts and their effects.

Study Problems

9.1. Develop a multigeneration system similar to that in Figure 9.3b but with seven outputs: power, water heating, space heating, cooling, hydrogen, salt, and drinking water. Extend the analysis from Section 9.2 to determine the emissions factor, exergy efficiency, and levelized electricity cost. Select an actual case and optimize the system based on multi-objective optimization technique.

9.2. Design a solar-energy-based multigeneration system. Start with a power-only system and increase the number of outputs up to six. Optimize each system based on two criteria. Determine Pareto fronts, one for each number of generations. A curve similar to that in Figure 9.5a can be obtained.

9.3. Extend the results presented in Figure 9.8 up to quadruple generation. In this respect propose novel ways to increase the number of outputs of the system.

9.4. Consider the system in Figure 9.14. Obtain the Pareto fronts for summer and winter conditions for a location of your choice where the average temperature in summer differs substantially from that in winter.

9.5. For the system in Figure 9.17 change the absorption refrigerator from LiBr-water to ammonia-water. Redo the system optimization for exergy efficiency and SCE. Study the effect of increase in the number of outputs.

9.6. Optimize the multigeneration system in Figure 9.22 with respect to cost and specific emissions and obtain Pareto fronts.

9.7. Consider the system in Figure 9.27. Change the absorption cycle to ammonia-water and select another working fluid for the ORC. Do a multi-objective optimization for maximum exergy efficiency, maximum generated revenue, and minimum environmental impact.

9.8. For the system in Figure 9.29 assume that the working fluid of the ORC is isobutane. Optimize the ORC for maximum exergy efficiency under imposed fixed boundary conditions for operation within the integrated multigeneration system. Report the influence of ORC exergy efficiency on the overall efficiency of the system.

9.9. Expand problem (9.8) by optimizing the system for minimum environmental impact and maximum exergy efficiency for a fixed capital cost.

9.10. Optimize the system in Figure 9.33 for maximum exergy efficiency and minimum environmental impact.

9.11. Using the multigeneration system presented in Figure 9.33 as a model, develop six power generation systems with incremental number of outputs. Consider in the analysis the environmental cost due to emissions, the operating and the capital cost, and determine the revenue rate. The lifetime of the system 30 years. Determine Pareto fronts for optimization of the systems with respect with exergy efficiency and revenue rate, using as a parameter the number of generated outputs.

9.12. Modify the system in Figure 9.37 by changing the SOFC with MCFC and the absorption cycle with an ammonia-water cycle, and optimize the system for maximum exergy efficiency.

9.13. Modify the system in Figure 9.40 by replacing the parabolic through concentrators with a solar tower-heliostat field. Optimize the system such that the revenue is maximized for a 40-year lifetime. Include the cost penalty due to environmental impact in the analysis.

9.14. Perform a comprehensive thermodynamic analysis based on energy and exergy methods for the trigeneration system in Figure 9.42 and optimize the operating parameters such that the exergy efficiency is maximized.

9.15. Consider that the absorption chiller integrated in the system in Figure 9.43 is of LiBr-water type. Perform a comprehensive thermodynamic analysis with energy and exergy methods. Optimize the system such that the power generation efficiency is maximized.

Nomenclature

$\mathcal{A}$	pre-exponential factor
A	area, m^2
AOC	annual operational costs, \$/year
$\dot{c}$	cost rate, \$/s
C_p	specific heat, kJ kgK
c	specific cost, \$/unit
$\mathcal{D}$	diffusivity, m^2/s
$\dot{E}$	energy rate, kW
$\dot{Ex}$	exergy rate, kW
$\dot{ex}$	specific exergy, kJ/kg
F	Faraday constant, C/mol
F_r	intercept factor of the receiver
h	specific enthalpy, kJ/kg
HHV	higher heating value, MJ/kg
IC	investment cost, \$
J	current density, A/m^2
LEC	levelized electricity cost, \$/kWh
LHV	lower heating value
LT	life time, years
$\dot{M}$	mass flow rate of environmental emissions, kg/s
$\dot{m}$	mass flow rate, kg/s
N	number of generated outputs
OPC	operational cost, \$
P	pressure, kPa
PBP	payback period, years
$\dot{Q}$	heat rate, kW
R	universal gas constant, J/molK
R_Ω	ohmic resistance, Ω
s	specific entropy, kJ/kgK
SCE	specific carbon emissions, kg/kW
SEF	specific emissions factor, kg/kW
SI	sustainability index
T	temperature, K
TIT	turbine inlet temperature, K
U	overall heat transfer coefficient, W/m^2K
V	velocity, m/s
V	voltage, V
$\dot{W}$	power, kW
x	distance, m
y	molar fraction
$\dot{Z}$	capital cost rate, \$/s

Greek letters

α　absorbance
β　stoichiometric factor
γ　specific heat ratio
ε　effectiveness or emissivity
ζ　stoichiometric factor
κ　parameter defined in Table 9.1
Λ　membrane moisture fraction
η　energy efficiency
ψ　exergy efficiency
ρ　reflectance
σ　electrical conductivity of membrane, m/Ω
σ　Stefan Boltzmann constant, W/m^2K^4
ω　ejector mass entrainment ratio

Superscript

0　　standard state
a　　anode
ac　　air compressor
act　　activation
c　　critical or cathode
C　　cooling
cc　　combustion chamber
ch　　chemical
cp　　cooling process
cond　condenser
d　　destruction
ei　　environmental impact
f　　fuel
g　　gas
gen　generation
gt　　gas turbine
H　　heat source
hp　　high pressure
lp　　low pressure
mix　mixer
net　　net generation
nz　　nozzle
ORC　organic Rankine cycle
PEME　proton exchange membrane electrolyser
s　　salt
sep　separate technologies
sf　　synthetic fuel
sh　　space heating
SRC　steam Rankine cycle
st　　steam turbine
w　　power
wh　　water heating
wp　　water pump

Superscript

ch　chemical

References

Ahmadi, P., Rosen, M.A., Dincer, I., 2011. Greenhouse gas emission and exergo-environmental analyses of a trigeneration energy system. International Journal of Greenhouse Gas Control. 5, 1540–1549.

Ahmadi, P., Dincer, I., Rosen, M.A., 2013a. Thermodynamic modeling and multi-objective evolutionary-based optimization of a new multigeneration energy system. Energy Convers. Manag. 76, 282–300.

Ahmadi, P., Dincer, I., Rosen, M.A., 2013b. Development and assessment of an integrated biomass-based multigeneration energy system. Energy. 56, 155–166.

Al-Sulaiman, F.A., Dincer, I., Hamdullahpur, F., 2010. Energy analysis of a trigeneration plant based on solid oxide fuel cell and organic Rankine cycle. Int. J. Hydrog. Energy. 35, 5104–5113.

Bejan, A., Lorente, S., 2006. Constructal theory of generation of configuration in nature and engineering. J. Appl. Phys.. 100, 041301.

Cetinkaya, E., 2013. Experimental Investigation and Modeling of Integrated Tri-generation Systems. PhD Thesis, University of Ontario Institute of Technology, Oshawa, Ontario, pp. 214.

Dincer, I., Rosen, M.A., 2013. Exergy. Energy, Environment and Sustainable Development. Elsevier, New York.

Dincer, I., Zamfirescu, C., 2012. Renewable-energy-based multigeneration systems. Int. J. Energy Res. 36, 1403–1415.

Hasan, A.A., Goswami, D.Y., Vijayaraghavan, S., 2002. First and second law analysis of a new power and refrigeration thermodynamic cycle using a solar heat source. Sol. Energy. 73, 385–393.

Khaliq, A., Kumar, R., Dincer, I., 2009a. Performance analysis of an industrial waste heat-based trigeneration system. Int. J. Energy Res.. 33, 737–744.

Khaliq, A., Choudhary, K., Dincer, I., 2009b. Exergy analysis of a gas turbine trigeneration system using the Brayton refrigeration cycle for inlet air cooling. Proceedings of IMechE, Part A: Journal of Power and Energy. 224, 449–461.

Ozcan, H., Dincer, I., 2013. Thermodynamic analysis of an integrated SOFC, solar ORC and absorption chiller for trigeneration applications. Fuel Cells. 13, 781–793.

Ozturk, M., Dincer, I., 2013a. Thermodynamic assessment of an integrated solar power tower and coal gasification system for multi-generation purposes. Energy Convers. Manag.. 76, 1061–1072.

Ozturk, M., Dincer, I., 2013b. Thermodynamic analysis of a solar-based multi-generation system with hydrogen production. Appl. Therm. Eng.. 51, 1235–1244.

Ratlamwala, T.A.H., Gadalla, M.A., Dincer, I., 2010. Evaluation and analysis of an integrated PEM fuel cell with absorption cooling. In: Proceedings of the 10th International Conference Enhanced Buildings Operations, Kuwait, October 26–28, paper # ESL-IC-10-10-80.

Szargut, J., Morris, D.R., Steward, F.R., 1988. Exergy Analysis of Thermal, Chemical, and Metallurgical Processes. Hemisphere Publishing Corporation, New York.

Zamfirescu, C., Naterer, G.F., Dincer, I., 2010. Upgrading of waste heat for combined power and hydrogen production with nuclear reactors. J. Eng. Gas Turbines Power. 132 (102911), 1–9.

Novel Power Generating Systems

10.1 INTRODUCTION

The perpetual growth of the global economy and populations and the increasing demand for commodities to maintain contemporary living standards (electricity, hot water, fuel, materials, temperature maintenance, transportation vehicles, etc.) pose an unprecedented strain on traditional methods of power production. In the past decades extensive efforts have been devoted to addressing these issues through the development of novel power generation systems able to convert energy from various sources in a more efficient and sustainable manner than the conventional systems.

Nowadays, there is a global *ad hoc* initiative to provide clean, efficient alternatives to existing commodity production methods, primarily driven by the highly pollutant combustion of fossil fuels. Exploring the potential of sustainable energy resources such as biofuels, biomass, municipal waste, solar, geothermal, wind, ocean heat, and industrial waste heat represents a key development strategy.

One of the major challenges to attaining high performance is the reduction of the irreversibilities of energy conversion in key processes such as combustion, heat transfer at source-side heat exchangers of power plants, and electrochemical energy conversion. This requires the development of novel methods able to match energy source characteristics with power cycle characteristics in a manner that minimizes exergy destruction. A common example is the problem of pinch point temperature difference in conventional steam power plants, which illustrates the mismatch between the temperature profile of combustion gases and the boiling water temperature profile. This mismatch leads to high exergy destruction.

Internal combustion engines are commercially available in a wide range of capacities; however, they are limited to liquid or gas fuels. New developments in this area are clearly identified, especially with respect to the use of nonconventional fuels, including cryogenic liquefied natural gas, methanol, hydrogen, and ammonia. Advanced split cycles are being developed, as are rectilinear free-piston engine generators. An integrated system that uses ammonia as fuel and a free-piston generator is elucidated in Dincer and Zamfirescu (2012). A free-piston engine developed at Sandia National Laboratory is described in Van Blarigan (2000).

Another research focus is on externally supplied heat engines able to be coupled with a variety of sources, including any combustible material and noncombustible heat sources. In the past, power cycles that used heat externally were denoted as "external combustion heat engines." The typical example is the steam power cycle. Heat engines operating with an external heat supply have a great deal of potential as a method of generating power utilizing the mentioned resources. In such heat engines, heat is transferred from the external source to the working fluid using a heat exchanger. This type of heat engine is versatile in terms of the heat source and generating capacity and can be used for cogeneration.

Novel power cycles with external heat supply, such as ammonia–water cycles, organic Rankine cycles, supercritical water cycles, and supercritical and transcritical carbon dioxide cycles, are expected to play a major role in the future. An advanced power cycle with an external heat supply is presented in Zamfirescu and Dincer (2008a). The system shows reduced exergy destruction and can be applied to low-grade heat sources including geothermal, low concentration solar collectors, and ocean thermal energy conversion. Smouse et al. (2013) developed a novel supercritical carbon dioxide power cycle with cryogenic compression, which is coupled to an external oxy-combustion system of natural gas.

A direct conversion method of thermal energy sources is the thermoelectric effect (also known as the Seebeck effect). Although this system is not widespread, it shows excellent potential, particularly with concentrated solar radiation. Some recent developments in this area are reported by Sano et al. (2003). These systems can be applied to regenerative combustion processes to gain efficiency, as shown in the study by Weinberg et al. (2002).

Cleaner combustion technology plays an increased role. Several novel developments in this area can be remarked. Definitely, one of the important directions is the development of chemical looping combustion-based power generation. These systems allow for intrinsic separation of carbon dioxide from gaseous fossil fuels (either natural gas or syngas) and can also be applied to hydrogen combustion in advanced gas turbine cycles. A novel application of chemical looping combustion is the sorbent energy transfer system recently reported by Copeland et al. (2001).

Other developments include advances in coal and natural gas combustion technologies with integrated gasification which may involve hybrid fuelling systems for power plants. Han et al. (2011) showed an integrated system based on the partial fossil fuel conversion principle: the use of partial coal gasification in conjunction with partial natural gas integrated with combined gas and steam power cycles. There are many types of integrated systems for clean combustion of fossil fuels, such as the integration of direct carbon fuel cells with hydrogen systems for power generation as outlined in Muradov et al. (2010). Research advances in magneto-hydrodynamic and thermoacoustic power generators are also remarked.

In this chapter, an overview is given of some selected systems for power generation that have emerged in recent years and have good promise for future development. The systems included here are based on a literature survey. Some of the presented systems come from the research work of the authors.

10.2 NOVEL AMMONIA–WATER POWER CYCLES

The development of a power cycle able to use thermal energy as input is of major importance, as there is a large palette of these kinds of sources, ranging from concentrated solar power, geothermal energy, combustion gases, to recovery of process heat. What is important in such applications is to match the temperature profiles at sink and source such that the available exergy can be exploited to maximum benefit for improved system performance.

Using a mixture like ammonia–water is attractive in this context because of the opportunity to adjust the temperature variation during liquid–vapor phase change by regulating the ammonia concentration in the proper way. It is possible to match the temperature profiles of the fluids that exchange heat at sink or source level. This is a way to minimize the destroyed exergy in the heat exchangers.

In the past the Kalina cycle has been proposed. This cycle uses a turbine similar to a steam turbine. The expanded vapors are condensed in a rather complicated distillation and condensation subsystem. Recent results from Zamfirescu and Dincer (2008b) show that ammonia–water can be used in the flash Rankine cycle with a simpler configuration provided that a suitable working machine (expander) is developed. The cycle configuration may be adapted to actual application: if cogeneration is required, the four-component Rankine configuration may be suitable; if power-only generation is required, then a regenerative five-component cycle can be configured.

An interesting system configuration is the "trilateral flash cycle" proposed by Smith et al. (1996). This cycle runs on the organic working fluid R134a, which is expanded in a screw machine right after preheating. Here, the saturated liquid is flashed into two phases right in the expander, which generates power. The resulting vapor–liquid mixture is then fully condensed and the liquid is pumped to high pressures and heated to the saturation temperature. This cycle is attractive for two reasons: it matches the temperature profiles at source side in a "perfect" manner because it does not include boiling, and it operates at reasonable pressures such that its implementation is economically feasible for low-power applications.

Here we present a novel version of trilateral flash cycle which uses ammonia–water mixture as working fluid instead of an organic fluid and investigate the opportunity to match both the temperatures at sink and source levels with the aim of overall system performance maximization. An experimental system has been constructed by the authors for testing performance of this cycle as presented subsequently.

The diagram of the ammonia–water trilateral flash cycle is shown in Figure 10.1. There is no boiling process of the working fluid. Instead, it is heated until the liquid reaches the

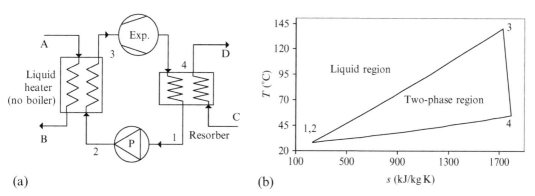

(a)   (b)

FIGURE 10.1   Ammonia–water trilateral flash power cycle schematic and T–s diagram: (a) System schematic and (b) Thermodynamic cycle for a specific case.

saturation. Saturated liquid #1 is pressurized with a pump (P) and reaches state #2 at approximately the same temperature. A liquid heater is used to heat the liquid to saturation point #3. It is assumed that the heat is provided by a fluid that exchanges sensible heat (combustion gases or a fluid stream from any process). The exergy efficiency of the liquid heater is very high because there is no boiling process and the two streams match temperature profiles and keep the LMTD to a minimum (LMTD: logarithmic mean temperature difference).

The saturated liquid is expanded in a volumetric expander—typically of scroll or screw type—that produces power at its shaft and a liquid–vapor mixture at its outlet in state 4. The process is similar to a flash Rankine cycle but uses a two-phase expander instead of a throttling valve. In the resorber cooling is applied to the two-phase mixture using the heat sink stream. As a consequence, a combined process of condensation and absorption occurs that eventually results in the release of a saturated liquid in state #1.

A double-stage configuration of the ammonia–water trilateral flash cycle for cogeneration purposes has been recently studied by Cetinkaya (2013). This system is shown in Figure 10.2. After the flashing into the first expander, process #3–4, the two-phase ammonia–water mixture is separated in the vapor–liquid separator (VLS) into liquid (at the bottom) and vapor (at the top). The saturated vapors can be superheated (optionally). The vapors are further expanded using a turbine, #6–7. The saturated liquid at intermediate pressure is flashed in the secondary expander, which generates work. The expanded mixture is cooled by the turbine according to process #7–8, during which ammonia vapors resorb in liquid phase. At the same time, heat is cogenerated and transferred to stream C–D. Further heat rejection occurs in the heat exchanger at #11–1.

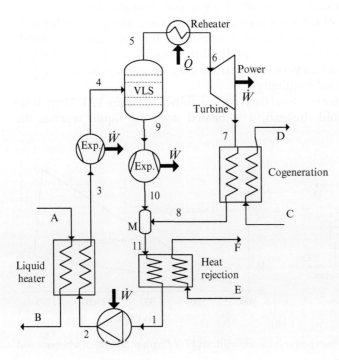

FIGURE 10.2 Double-stage configuration of the ammonia–water flash Rankine cycle. *Modified from Cetinkaya (2013).*

The key element for the implementation of this cycle is the expander that performs process #3–4 (Figure 10.1) and processes #3–4 and 9–10 (Figure 10.2). Several attempts have demonstrated that it is possible and beneficial to flash liquid in refrigeration systems using screw, scroll, or centrifugal machines, and isentropic efficiencies on the order of 0.7 have been achieved. Scroll machines are considered excellent devices, capable of expanding to two phases. The authors and coworkers developed several prototypes of a scroll expander converted from refrigeration scroll compressors that were successfully tested by expanding two phase flows (Hogerwaard et al., 2013; Oralli et al., 2011; Oralli et al., 2014; Tarique et al., 2014).

The modeling of the ammonia–water trilateral flash cycle is presented in Zamfirescu and Dincer (2008b). The modeling involves particular elements related to the behavior of the ammonia–water mixture. The energy balance for each cycle component is

$$\left.\begin{array}{l} \dot{W}_P = \dot{m}(h_2 - h_1) \\ \dot{m}_{so}(h_{1,so} - h_{2,so}) = \dot{m}(h_3 - h_2) \\ \dot{W}_E = \dot{m}(h_3 - h_4) \\ \dot{m}_{si}(h_{2,si} - h_{1,si}) = \dot{m}(h_4 - h_1) \end{array}\right\} \tag{10.1}$$

The isentropic efficiency of the pump is the parameter that relates the thermodynamic states at pump intake and discharge according to the following equation:

$$\eta_{sP} = \frac{h_2 - h(T_1, \zeta, x = 0)}{h(P_2, s_1, \zeta) - h(T_1, \zeta, x = 0)}$$

where ζ is the ammonia concentration by weight in the working fluid and x is the vapor quality.

The expander is modeled assuming a reasonable value for its isentropic efficiency. The expanded flow is in two phases. The ammonia concentration in vapor ζ'' is higher than the concentration in liquid phase ζ'. For an isentropic discharge of the expander #4s, there are the following equations:

$$\left.\begin{array}{l} s_{4s} = s_3 \\ x_{4s} = \dfrac{\xi - \xi'(T_{4s}, P_4)}{\xi''(T_{4s}, P_4) - \xi'(T_{4s}, P_4)} \\ s_{4s} = (1 - x_{4s})h'(T_{4s}, P_4) + x_{4s}h''(T_{4s}, P_4) \end{array}\right\} \tag{10.2}$$

Equation (10.2) can be solved for mixture temperature, vapor quality, and specific entropy of the mixture. Furthermore, the enthalpy of the mixture is calculated: $h_{4s} = (1 - x)h'(T_4, P_4) + xh''(T_4, P_4)$. Now, using the isentropic efficiency of the expander, the enthalpy at the expander discharge results from:

$$h_4 = h_3 + (h_{4s} - h_3)\eta_E \tag{10.3}$$

If enthalpy at the expander discharge is determined, then the mixture temperature and vapor quality result from the following system of two nonlinear equations:

$$\left.\begin{array}{l} h_4 = (1 - x_4)h'(T_4, P_4) + x_4 h''(T_4, P_4) \\ \xi = (1 - x_4)\xi'(T_4, P_4) + x_4 \xi''(T_4, P_4) \end{array}\right\} \tag{10.4}$$

The cycle can be assessed based on energy and exergy efficiencies. These are usually defined as the net power generated divided by the heat input and exergy input, respectively. Since there is no pinch point in the heat source heat exchanger it is logical to assume that the thermomechanical exergy of the hot stream is close to zero in state B, when it leaves the heat exchanger (see Figures 10.1 and 10.2). When cogeneration is considered, the output exergy is assumed to be the sum of net power generated and the exergy rate in the heated fluid for cogeneration—state D in the figures.

Here, we compare the ammonia–water trilateral flash cycle (TFC) with four organic Rankine cycles and a Kalina cycle operating under the same conditions. Organic Rankine cycles with the following working fluids were considered for comparison: R141b (1,1-dichloro-1-fluoroethane), R123 (2,2-dichloro-1,1,1-trifluoroethane), R245ca (1,1,2,2,3-pentafluoropropane), and R21 (dichlorofluoromethane). Table 10.1 gives the comparative assessment of the cycles.

The results from the table show that the cycle has the best energy efficiency under the imposed conditions when the working fluid is R141b (10%), whereas for ammonia–water the energy efficiency is 8%. However, looking at exergy efficiency, the ammonia–water TFC has the highest value, about double that of the other considered cycles. When heat is cogenerated, the TFC reaches 71% exergy efficiency with ammonia–water. As seen in the table, the ammonia–water TFC develops the maximum amount of work with respect to the other cycles. This shows that in fact the cycle has better ability to use the thermal energy carried by the hot stream of the heat source.

The ammonia–water TFC, optimized for a 150 °C source inlet temperature (by assuming that the heat source fluid is a geothermal brine modeled as hot pressurized water), is represented in exergy rate versus heat rate in the diagram shown in Figure 10.3. In order to form this diagram, the flow enthalpy in state 1 has been set to zero, that is point $(\dot{Q}, \dot{Ex})_1 = (0, \dot{Ex}_1)$ in the diagram. The optimized cycle performance parameters are those given in Table 10.1. The optimal ammonia concentration is 0.47.

In Figure 10.4, two TFC cycles are compared, namely steam (a) and ammonia–water with 50% ammonia concentration (b). Both cycles operate within the same boundary conditions and have the same energy efficiency of 8%. However, the exergy efficiency of the steam cycle is 23%, compared to 30% for the ammonia–water. The increase in exergy efficiency of 7% is

TABLE 10.1 Performance Comparison of Various Power Operating Cycles

| Parameter | Organic Rankine Cycles | | | | NH$_3$–H$_2$O cycles | |
	R141b	R123	R245ca	R21	TFC	Kalina
η (%)	10	9	9	9	8	3
ψ (%)	13	16	16	13	30	13
ψ_{cog} (%)	27	36	40	51	71	56
$\dot{Q}_{\text{so}}$ (kW)	132	179	189	198	477	373
$\dot{W}$ (kW)	13	17	18	18	38	13

Note: Mass flow rate of heat source fluid (pressurized hot water) is $\dot{m} = 1\,\text{kg/s}$ kg/s; inlet temperature 150 °C.
Data from Zamfirescu and Dincer (2008h)

FIGURE 10.3 Exergy rate versus heat rate diagram for the optimized ammonia–water TFC for hot source fluid of 150 °C and 1 kg/s [data from Zamfirescu and Dincer (2008b)]. Optimized ammonia concentration is 0.47.

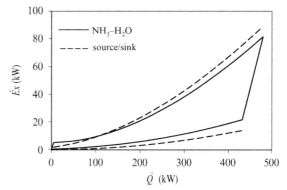

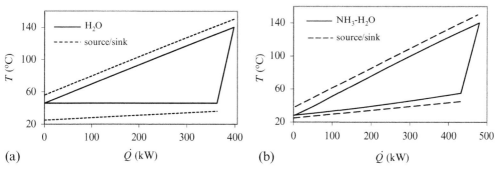

(a)   (b)

FIGURE 10.4 Temperature versus heat rate diagram for TFC with steam and with ammonia–water: (a) TFC with steam as working fluid and (b) TFC with ammonia–water (50%) [data from Zamfirescu and Dincer (2008b)].

significant and is due to a reduced LMTD at the sink-side heat exchanger used for the ammonia–water cycle; this can be easily observed when (a) and (b) are compared.

The double-stage TFC has better efficiency than the simple-stage TFC. Cetinkaya (2013) performed a parametric study which indicated that for geothermal applications with 150 °C brine temperature the double-stage cycle enhances exergy efficiency by a factor of 1.5 with no reheat. Some slight increase in efficiency (2–3%) can be obtained if the saturated vapors are reheated.

A long-term experimental research program is underway at the University of Ontario Institute of Technology (UOIT) to develop the trilateral flash cycle with ammonia–water and an ammonia–water scroll expander. Three prototypes were built and tested and more tests are underway for the current prototype heat engine. Some preliminary tests were performed with R134a and confirmed the cycle feasibility. The research is currently underway and further results are expected to confirm the efficiency of the ammonia–water heat engine.

The last heat engine prototype was designed to use low-grade sustainable heat sources working with ammonia–water as the working fluid. A scroll expander was built by modifying the design of a refrigeration compressor to work as a prime mover. Figure 10.5 shows the process diagram of the heat engine prototype. A photograph of the overall system is shown in Figure 10.6.

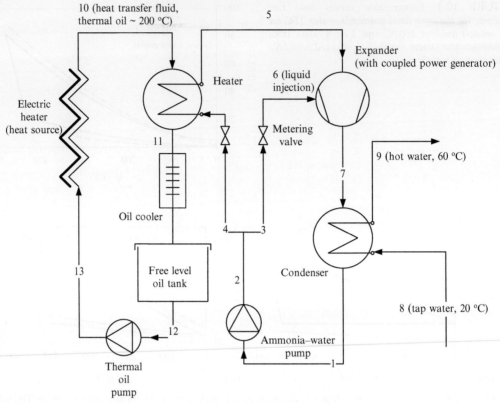

FIGURE 10.5   Schematics of the test prototype ammonia–water heat engine developed in our Sustainable Energy Systems Laboratory at University of Ontario Institute of Technology.

The heat engine includes an ammonia–water gear pump operated with a variable speed motor, where the non-adiabatic pressurization process #1–2 occurs. The sub-cooled liquid in #2 is divided into a small stream in #3 and a main stream in #4. The stream in #3 is used for liquid injection in the expander, which plays several roles: it helps lubricate the compliant scrolls, it helps seal the leakage paths for gas, it cools the scrolls, and it reinforces the pressure of the expanding scroll cavity by partial flashing and vapor generation. While injection is required for the operation of scroll, short-time operation without injection is possible under test conditions. A metering valve is placed in the line at port #6 to adjust the injection flow. The main working fluid stream in #4 is preheated, boiled, and superheated in a single heat exchanger, performing the process #4–5. To further the expansion, the low-pressure ammonia–water stream in #7 is cooled using tap water flow as heat sink process #8–9. The working fluid is fully condensed in #1.

The heat source consists of a hot thermal oil heater that transfers heat at the boiler according to process #10–11. After exiting the boiler the oil is cooled to below 80 °C in an extended-surface oil cooler prior to feed into the oil pump in #12. The oil circuit has a free vessel for thermal expansion purpose. After the oil pump, the oil is fed into an electrically heated oil heater that plays the role of heat source. The heat flux of the heater can be adjusted

FIGURE 10.6 Ammonia–water heat engine prototype at University of Ontario Institute of Technology.

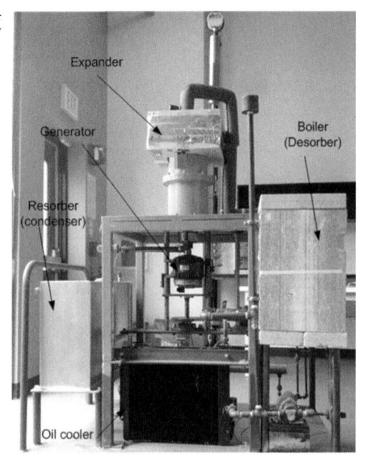

by controlling the electrical power input to maintain an oil temperature up to around 200 °C for the experiment. The overall system is designed to expect an expander shaft power of 1 kW, the total capacity of the heat exchangers at heat source is 20–30 kW, and the condenser capacity is in the heat exchanger range.

The scroll expander selected for this project was originally a hermetically sealed refrigeration scroll compressor, Bristol H20R483DBE. The compressor was modified to work as an expander in the proposed cycle. To convert it to an expander, the working fluid had to flow in the reverse direction. The check valve in the discharge line of the compressor was removed to allow the process fluid to enter the intake side of the scroll unit. The scroll wrap of the scroll expander unit is shown in the Figure 10.7 photograph.

There were two pressure relief valves which were made of brass. These valves were removed and the holes were plugged with steel blinds to preclude corrosion damage by ammonia–water, which was the process fluid for the proposed heat engine. The shaft of the scroll machine previously attached to the motor armature was detached. Due to its eccentric design features, the shaft was balanced by a load made of brass. This load was replaced with steel load of the same weight to preclude the corrosive effect of the process fluid. The shaft was then repositioned by

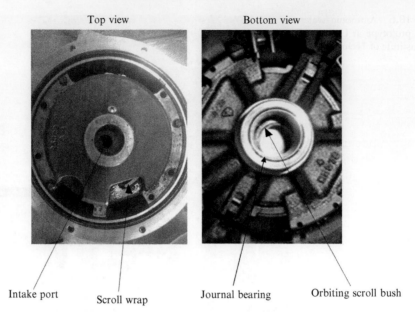

FIGURE 10.7   Scroll wrap of UOIT scroll expander prototype modified from refrigeration compressor unit Bristol H20R483DBE.

fixing one end to the orbiting wrap of the scroll unit, and the other end was supported on a base plate with a thrust bearing. The orbiting wrap changed orbiting motion to rotational motion through this shaft. The shaft therefore had a tendency to experience transverse movement in addition to the longitudinal movement. A needle bearing was therefore positioned on the upper side of the shaft to stabilize the rotation of the shaft.

The whole unit with the shaft was enclosed in a cylindrical casing that was rebuilt into a two split-type cylindrical body. The upper portion was covered with a cap that was welded to a flange. This feature was designed to facilitate the assembly and disassembly of the expander unit. The upper part was the cap of the expander casing. Two inlet ports were constructed in the cap to accommodate the two process fluid supply channels. A side port was welded to the shell to attach the main flow piping and a top port for insertion of a thin tube for injecting working fluid in liquid phase into the intake port. This design modification was performed as a substitute method for oil lubrication of the rotating parts. The flow of the injected liquid was controlled by a metering valve.

The shaft end was connected to the male part of a magnetic coupling. The magnetic coupling was used to transmit the shaft power to a generator without physical contact. This arrangement was used to confine the process fluid within the leak-proof closed loop. Figure 10.8 shows the magnetic coupling and the expander clutch arrangement that connected the expander shaft to the electric generator. The female part of the magnetic coupling was attached to the shaft of the generator and the generator unit was assembled with a clutch mechanism to move the generator unit up or down to disengage the magnetic coupling. This provision was made to start the engine without load, with reduced starting torque.

FIGURE 10.8 Magnetic coupling and generator arrangement.

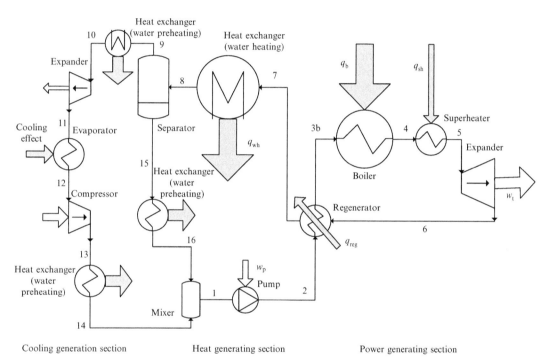

FIGURE 10.9 Ammonia–water-based trigeneration system.

Recent theoretical development of the seam research group at UOIT led to a novel ammonia–water trigeneration system concept based on a modified Rankine cycle with two-phase expansion. The cycle is shown in Figure 10.9 and generates power, sanitary hot water, and cooling. Note that, in terms of exergy, cooling is relatively more valuable than heating. For example, if $T_0 = 298\,K$ and cooling is provided at 273 K, the associated exergy of a 1 kJ heat transfer is given by $1 - T_0/T = 1 - 298/273 = -1.09\,kJ$, whereas if heating is provided at 323 K then the exergy is $1 - T_0/T = 1 - 298/323 = 0.92\,kJ$. Ammonia–water is an attractive working fluid for a trigeneration system because it is realized as an excellent working fluid for the Rankine power cycle and is also a brilliant refrigerant for the vapor compression cycle.

Ideally, a trigeneration system must operate with weak ammonia–water solution during power generation and very strong ammonia–water solution during cooling. In the studied system, the power generation section has an ammonia concentration of around 0.25, whereas during cooling the ammonia concentration is around 0.75.

In this cycle the temperature and enthalpy of the expanded stream at #6 is high enough to provide the thermal energy needed for preheating and part of the boiling/desorption process. Hence, in the regenerative heat exchanger, the heat #6–7 is transferred to process #2–3–3b. Further, in state #7, the working fluid is cooled and condensed/resorbed until a two-phase mixture in state #8 consisting of saturated vapor #9 comes into equilibrium with saturated liquid at #15. Since vapor is much more concentrated with ammonia than liquid, they are extracted in a vapor–liquid separator and cooled by heat transfer using the thermal sink to a temperature of $T_{10}=T_1$.

The two-phase mixture from state #10 is expanded in a scroll expander (able to operate in two-phase) until it reaches a low temperature and pressure in #12. It follows the evaporation process #12–13, where the heat for cooling effect is absorbed. Further, there is a two-phase compression process with parameters selected such that $T_{13}=T_9$, and in state #13 it becomes saturated vapor. Cooling is then applied to the saturated vapor until it reaches state 14, with $T_{14}=T_1$. The weak ammonia–water solution separated at state #15 is cooled to $T_{16}=T_1$. When the streams at state #14 and #16 are adiabatically mixed, a saturated liquid is formed in state #1 at suction of the pump. The net generation of work by the system is given by the work of the turbine and refrigeration expander (process at #11–12) minus the work consumed by the refrigeration compressor and the pump.

There is heat rejected by the cycle, with temperature decreasing from T_8 to T_1. This heat can be recovered from heat exchangers #9–10, #13–14, and #15–16, which can work in parallel fashion; the recovered heat can be used for preheating water or other useful heating process, depending on the application. Mass, energy, and exergy balances allow for determination of the thermodynamic states at each of the cycle points. The state points parameters are given in Table 10.2.

The performance parameters of the trigeneration system are given in Table 10.3. The cooling effect of this system is 226 kJ per kg of working fluid, compared with a power output of 69 kJ/kg and heat generation of 1930 kJ/kg. The exergy destruction is 253 kJ/kg and the system exergy efficiency is 49%.

TABLE 10.2 State Point Parameters for Trigeneration System in Figure 10.9

State	T (K)	P (bar)	v (m³/kg)	Q	h (kJ/kg)	s (kJ/kg.K)	ex (kJ/kg)
1	298.2	0.6473	0.001106	0	−65.21	0.319	−0.04405
2	298.2	2.654	0.001106	−0.001	−64.88	0.3193	0.1813
3	336.4	2.654	0.001137	0	99.17	0.8366	9.999
3b	348.1	2.654	0.0514	0.08083	291.5	1.396	35.46
4	393	2.654	0.6744	1	2395	6.929	489.6
5	393.2	2.654	0.6747	1.001	2396	6.931	490.2

Continued

TABLE 10.2   State Point Parameters for Trigeneration System in Figure 10.9—Cont'd

State	T (K)	P (bar)	v (m³/kg)	Q	h (kJ/kg)	s (kJ/kg.K)	ex (kJ/kg)
6	353.1	0.6473	2.423	0.9577	2234	7.13	268.5
7	351.5	0.6473	2.03	0.8039	1878	6.119	213.6
8	333.2	0.6473	0.6741	0.2746	620.5	2.478	41.93
9	333.2	0.6473	2.453	1	1767	6.723	−77.51
10	298.2	0.6473	1.417	0.6362	861.8	3.896	−139.5
11	240.2	0.03236	21.86	0.6041	3.9	3.9	−998.6
12	273.3	0.03236	34.95	0.8571	1298	6.501	−479.4
13	333.1	0.6473	2.448	0.9981	1762	6.71	−77.98
14	298.2	0.6473	1.417	0.6362	861.8	3.896	−139.5
15	333.2	0.6473	0.001061	0	186.8	0.8718	87.14
16	298.2	0.6473	0.001042	−0.001	40.41	0.4075	79.17

TABLE 10.3   Performance Parameters of the Trigeneration System in Figure 10.9

Quantity	Definition equation	Value
Net generated work	$w_{net} = w_{e1} + w_{e2} - (w_p - w_c)$	69.1 kJ/kg
Generated heat	$q_h = q_{wh} + q_{wph}$	1930 kJ/kg
Generated cooling	$q_{cool} = q_8(h_{12} - h_{11})$	226 kJ/kg
Heat input	$q_{in} = q_b + q_{sh}$	2105 kJ/kg
Energy input	$ex_{in} = ex_b + ex_{sh}$	537.1 kJ/kg
Generated heat exergy	$ex_h = ex_{wh} + ex_{wph9-10} + ex_{wph13-14} + ex_{wph15-16}$	176 kJ/kg
Exergy of cooling effect	$ex_{cool} = q_{cool}\left(1 - T_0/\bar{T}_{15-16}\right)$	−19.13 kJ/kg
Exergy efficiency	$\psi = \dfrac{\dot{W}_{net} + \dot{E}x_{heating} + \left|\dot{E}x_{cooling}\right|}{\dot{E}x_{in}}$	49%

10.3 SOLAR THERMOELECTRICAL POWER GENERATION

The thermoelectric effect consists of generating electromotive force under the influence of a temperature gradient. A thermoelectric device is comprised of one p-doped and one n-doped semiconductor connected to form a p–n junction arranged as shown in Figure 10.10. Two electrodes are attached to the semiconductors. The junction is put in contact with a heat source while the electrodes are in thermal contact with a heat sink. In this setup the electromotive force allows electrons to be transported internally from the positive to the negative electrodes. The electrons can flow from the negative to the positive electrode through the external circuit.

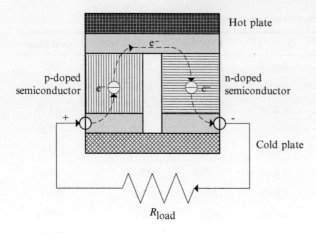

FIGURE 10.10 Seebeck (thermo-electric) effect.

The Seebeck coefficient C_S relates in a linear fasihon the gradient of temperature across the device with the generated electromotive force E_{emf}. Since the direction of the electromotive force opposes the temperature gradient, the Seebeck effect is described mathematically as $E_{emf} = C_S \operatorname{grad}(T)$. The Seebeck coefficient is on the order of 1 μV/K. The current density through the device depends on internal ohmic losses and the electromotive force as $J = -\sigma [\operatorname{grad}(E) + C_S \operatorname{grad}(T)]$, where σ is the electric conductivity.

The power generated by the thermoelectricl device with surface area A is given by the product of current density and electromotive force as follows:

$$\dot{W} = -\sigma A C_S \operatorname{grad}(T)\left[\operatorname{grad}(E) + C_S \operatorname{grad}(T)\right]$$

When the thermoelectric cell is put in contact with the heat source at temperature T_h and heat sink at temperature T_c then the ideal energy efficiency of the cell is given by:

$$\eta = \frac{\dot{W}}{\dot{Q}_{in}} = \left(1 - \frac{T_c}{T_h}\right)\frac{ZT - 1}{ZT + \dfrac{T_c}{T_h}} \tag{10.5}$$

which is given by Kraemer et al. (2011). The parameter ZT is a figure of merit defined by the following equation:

$$ZT = \sqrt{1 + 0.5\frac{C_S^2 \sigma(T_h + T_c)}{k}}$$

where σ is the electrical conductivity and k is thermal conductivity of the semiconductor.

From Equation (10.5), if the cold side of the cell is kept as the reference temperature the exergy efficiency of the ideal cell becomes

$$\psi = \frac{\dot{W}}{\dot{Q}_{in}\left(1 - \dfrac{T_0}{T_h}\right)} = \frac{ZT - 1}{ZT + \dfrac{T_c}{T_h}} \tag{10.6}$$

It appears from Equation (10.6) that the exergy efficiency of a cell heated to 200 °C with the cold side maintained at standard temperature is ~24%, which is an encouraging figure. There are many applications of thermoelectric devices for power generation, especially with renewable energy. Li et al. (2010) describe a system with concentrated solar radiation that generates electricity via the Seebeck effect. The thermoelectric device consists of a number of p–n junctions connected in series. The system aperture consists of a Fresnel lens surface which tracks the sun and concentrates solar radiation on planar mirrors as shown in Figure 10.11. The radiation is eventually reflected onto the hot surface of the thermoelectric generator. The heat sink is water coolant. Since water warms up, cogeneration is obtained from the system in form of hot water.

Bismuth–tellurium semiconductors are used for contracting the thermoelectric device. The power generation efficiency is rather small, on the order of 10%, but the system hardware is cheap. The required concentration ratio of sunlight is on the order of 200–400, for which hot surface temperatures of 650–900 K are obtained.

A promising design is demonstrated by Kraemer et al. (2011), which shows that flat panels can be constructed with thermoelectric devices for an efficiency of about 4% under peak AM1.5G (1000 W/m^2) solar radiation intensity with hardware that appears to be very cost effective. The voltage–current diagram of the cells is approximately linear, with open circuit voltage on the order of 100 mV and generated power on the order of 1–2 kW/m^2 under concentrated radiation.

Weinberg et al. (2002) propose a very interesting power generation system which uses a thermoelectric device inside a combustion chamber. In this case the temperature of the hot

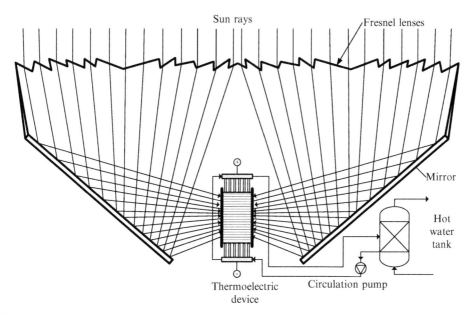

FIGURE 10.11 Thermoelectric power generator with concentrated solar radiation and hot water cogeneration [*modified from Li et al. (2010)*].

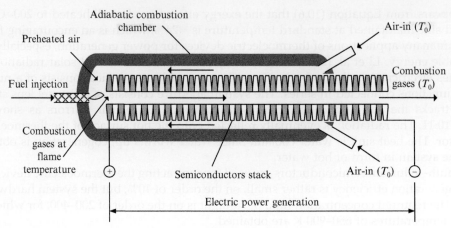

FIGURE 10.12 Thermoelectric power generator integrated with a combustion chamber *[modified from Weinberg et al. (2002)].*

surface of the device can reach 1200 K, in which case the demonstrated power generation efficiency is above 20%. The system is configured in the form of a regenerative heat exchanger integrated with a combustion chamber and a thermoelectric device.

Figure 10.12 shows the thermoelectric power generator integrated with a regenerative combustion chamber. As seen in the figure, air enters the combustion chamber at environmental temperature T_0. The air then flows into the annular space between the thermally insulated chamber mantel and the coaxial thermoelectric device that forms a separation wall between the annulus and the core. Being in thermal contact with the hotter semiconductor parts of the cold side of the thermoelectric generator, the air flow heats up. The preheated air reaches the combustion zone, where combustion gases are generated at a high temperature, on the order of 1500 K. The combustion gases flow though the core, thus heating the hot side of the thermoelectric device.

10.4 CHEMICAL LOOPING COMBUSTION FOR POWER GENERATION

Chemical Looping Combustion (CLC) emerged in recent years as one of the most promising technologies for clean power generation. This process consists of two successive chemical reactions that are cyclically repeated: oxidation of a selected metal to produce a metal-oxide (M_xO_y), and metal-oxide reduction with hydrogen or a hydrocarbon.

The oxidation of a metal is a highly exothermic process. For example, the oxidation of nickel ($2Ni + O_2 \rightarrow 2NiO$) produces 467 kJ per mol of oxygen. If natural gas is used to reduce the nickel oxide ($4NiO + CH_4 \rightarrow CO_2 + 2H_2O + 4Ni$), then an endothermic reaction occurs because cracking the molecular bonds of the fuel and of the metal oxide requires more energy than that released by the water formation reaction. However, if hydrogen is used to reduce the metal oxide ($NiO + H_2 \rightarrow H_2O + Ni$) the chemical reaction is slightly exothermic, generating heat of about 20.7 kJ/mol H_2.

In general, the reduction reaction enthalpy represents only a small fraction of the heat produced by the oxidation reaction, making the overall power generation process very attractive. Because the only products of the reduction reaction are CO_2 and H_2O, the CLC of hydrocarbons is a clean energy generation process that allows for cheap separation of CO_2 via steam condensation.

The implementation of a power cycle able to conduct and extract power from the CLC process first requires selection and design of chemical reactors. Coupled fluidized bed reactors are used for oxidation and reduction, respectively. A bed of particles made from selected metals deposited on ceramic supports is circulated between the two reactors. Important research is presently devoted to the development of long-service metallic carriers of oxygen, using Cu, Co, Ni, Fe, and Mn. Various solid sorbents can also be used to conduct redox reactions.

In general, air is compressed up to about 20 bar and directed to the oxidation reactor where the oxygen reacts with the metallic particles. On the other side, the fuel is delivered to the reduction reactor. By using hydrogen fuel Ishida and Jin (2000) obtained an impressive efficiency of 63.5% with their CLC engine. This is a 12% higher efficiency than that of a gas turbine with H_2/O_2 combustion. Moreover, the CLC engine is environmentally superior because the process is NO_x-free.

There are many integration schemes in the literature that couple CLC with gas turbines for power generation. Here we show in Figure 10.13 a system proposed by Copeland et al. (2001). This system consists of two reactors (reduction/oxidation) for the CLC integrated with a gas turbine and a two-stage steam turbine. Air is compressed #1–2 and steam is added to it prior to injection into the oxidation reactor ("Oxy." in the figure).

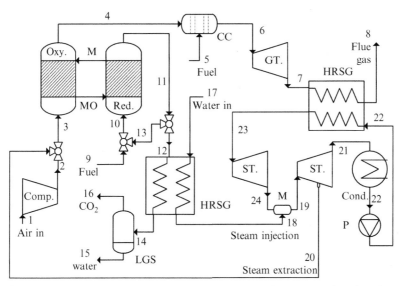

FIGURE 10.13 Chemical-looping-based power generation system with intrinsic carbon dioxide separation [*modified from Copeland et al. (2001)*]

The product gases of the oxidation reaction are still rich in oxygen; more fuel is added in a combustion chamber prior to expansion and power production in the gas turbine #6–7. Hot exhaust gases at high pressure are passed through a heat recovery steam generator to run a steam Rankine cycle.

Gaseous fuel (natural gas or syngas) is added to the oxidation reactor and combined with recycled product gases in order to achieve better fuel economy. The hot product gases (stream #12) pass through a heat recovery steam generator which produces intermediate-pressure steam (stream #18) which is injected between stages for expansion in steam turbines. In the steam Rankine cycle the condensate is pressurized and converted to high-pressure super-heated steam #23 which expands in the high-pressure turbine #23–24. Low-pressure steam is extracted from the low-pressure turbine to humidify the compressed air in #20.

According to Copeland et al. (2001) the system achieves a power generation efficiency of 57% when it is fuelled with natural gas. The highest process temperature is 1175 K and the reduction temperature is around 675 K. The greatest advantage of the system is the ability to separate carbon dioxide intrinsically in the liquid gas separator (LGS) which collects condensed water, leaving above a gas of almost pure carbon dioxide.

The chemical looping system used in the system in Figure 10.13 is primarily a coupled fluidized bed reaction system. However, other systems can be used, such as packed bed reactors. In Zamfirescu (2007) a rotating packed bed chemical looping combustion reactor is mentioned. The system uses a Ljungstrom-type wheel containing a selected metal (e.g., Ni, Fe, Co, or Mn), deposited on a porous structure. Reduction and oxidation reactions take place successively as the wheel turns while the injection ports for air and fuel are fixed.

Under the action of centrifugal forces air dissipates through the porous wheel carrying a reduced metal and eventually a depleted air (N_2) exits through the fixed peripheral port, since reaction occurs in the well structure. The fuel port is displaced with respect to the air injection port at a location where reduction occurs and eventually steam and carbon dioxide are formed. In this way the metal oxide deposited on the porous matrix is reduced. As a process, the oxidation is faster than the reduction. Therefore, the position of the fixed ports and the wheel rotation speed must be calculated such that the overall process is cyclical: complete oxidation is followed by complete reduction. For safety reasons, no direct contact between fuel and oxidant is allowed.

An option for designing the porous structure of the wheel consists of depositing the metallic carriers on ceramic particles that are placed inside the wheel; the wheel's walls are permeable to gas streams. The size of the particles is variable. In contact with the walls, large particles are used to avoid clogging and to allow gas distribution/collection, while in the core, fine particles are used to enhance the mass transfer surface.

10.5 LINEAR ENGINE POWER GENERATORS

In this section we introduce some other emerging power generation systems which appear to have good potential for future development. To begin with, let us consider free-piston linear generators. These machines consist of a free-piston engine which includes a permanent magnet attached to the double-effect piston enclosed in a cylinder. An electrical wire is coiled over the cylinder which is manufactured in a material that facilitates passage of the magnetic field. Due to the reciprocal movement of the magnet, electric voltage is induced in the coil and power is generated over an external circuit.

One of the advantages of the permanent magnet linear engine generator with a free piston is the fact that the system is able to operate in reverse, that is as a motor. Therefore, the engine is self-starting provided that electric power is applied in the starting-up phase. Van Blarigan (2000) demonstrates a free-piston linear generator operating with various fuels. The unit has the following features:

- When hydrogen fuelled
  - Compression ratio is adjusted to 33
  - Energy efficiency is 57%
- When used with biogas (13.3% H_2, 3.7% CH_4, 21.5% CO, 13.4% CO_2, 48.1% N_2)
  - Compression ratio is adjusted to 24
  - Energy efficiency is 37%
- When used with ammonia directly
  - Compression ratio is adjusted to 48
  - Energy efficiency is 52%.

Dincer and Zamfirescu (2012) describe an integrated power generation system based on a free-piston linear engine generator fuelled with ammonia. This system generates power, heating, and cooling and is shown in Figure 10.14. The system includes a thermally insulated ammonia tank, two throttling valves (#1 and #7), the linear spark ignition engine generator, an ammonia pre-heater for NO_x reduction, a selective catalytic reactor, a heat recovery heat exchanger, an electronic block (E), a drive/generator (D/G), a mechanical transmission system (T), an electrical accumulator battery (B), an ammonia compressor (C), an air-cooled condenser ("Cond."), and an ammonia evaporator to generate cooling effect.

Prior to fueling the linear generator the ammonia is throttled so that it serves as a coolant of the combustion engine. The NO_x contained in the exhaust gas is reduced in the selective

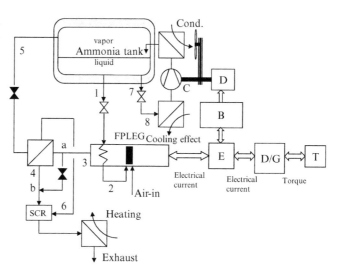

FIGURE 10.14 Trigeneration system with linear engine generator fuelled with ammonia [modified from Dincer and Zamfirescu (2012)].

catalytic reductor (SCR) using small quantities of ammonia taken from the tank (#5) and preheated in the heat exchanger (#5–6). The clean exhaust gases can be cooled down to a temperature close to ambient temperature (water can be condensed) and their heat recovered for heating purposes.

The electronic block (E) controls the system operation by relating the linear engine generator to the electrical drive/generator (D/G) and the electrical accumulator/battery (B). The mechanical transmission system (T) can produce or consume torque, depending on the case.

The system integrates a high-COP ammonia air-conditioning system that takes advantage of the existent ammonia-fuel. Its compressor (C) is driven by the electrical drive (D) (COP = coefficient of performance). The air-cooled condenser ("Cond.") discharges back into the fuel tank. From there ammonia liquid is extracted, throttled in #7, and delivered to the evaporator to generate cooling.

10.6 CONCLUDING REMARKS

In this chapter, several novel power generation systems are presented to discuss the potential for power generation in a more sustainable manner. Ammonia–water is attractive as a working fluid in power cycles because of the opportunity it offers to adjust the ammonia concentration such that the LMTD over heat source or heat sink heat exchangers is minimized. Three novel ammonia–water power cycles are also presented. Thermoelectrical power generators are introduced, and two systems are described. The first is powered by concentrated solar radiation. The second uses a novel combustor to directly generate power. A chemical looping combustion system integrated with a power cycle is discussed. Finally, some emphasis is placed on linear engine power generators for practical applications.

Study Problems

10.1 Consider the TFC system represented in Figure 10.1. The cycle is heated with Duratherm 450 thermal oil at 175 °C and cooled with air at 25 °C. Determine optimized cycles for the following working fluids: ammonia–water with 10%, 20%, 30%, 40% ammonia concentration, R141b, R123, R21, cyclohexane. Compare the exergy efficiency of these cycles.

10.2 For the system in Figure 10.2 assume that the working fluid is ammonia–water with 25% ammonia concentration and the heat source is combustion gas at 125 °C. Optimize the cycle and determine exergy efficiency.

10.3 Repeat problem 10.2 with cyclohexane as the working fluid.

10.4 Do a thermodynamic analysis of the trigeneration system in Figure 10.9 using meaningful assumptions.

10.5 Consider a thermoelectric system driven by concentrated solar radiation, similar to that shown in Figure 10.11, which generates power and hot water. Use Equation (10.5) and meaningful assumptions for thermodynamic modeling. Assess the system exergy, accounting for the two outputs (power and hot water). The aperture is set to 25 m^2.

10.6 A thermodynamic design is required for the integrated thermoelectric combustion system described in Figure 10.12. The system is fuelled with syngas. Determine the design parameters for a 5 kW power generator and extend the system by including cogeneration of space heating and hot water.

10.7 For the system with chemical looping combustion shown in Figure 10.13 do a thermodynamic analysis based on energy and exergy methods. Assume that the net generated power is 100 MW.

10.8 Using meaningful assumptions determine the exergy efficiency of the trigeneration system presented in Figure 10.14.

Nomenclature

A	area, m^2
Cs	Seebeck constant
E	electric potential, V
$\dot{Ex}$	exergy rate, kW
ex	specific exergy, kJ/kg
h	specific enthalpy, kJ/kg
k	thermal conductivity, W/m K
$\dot{m}$	mass flow rate, kg/s
s	specific entropy, kJ/kg.K
T	temperature, K
$\dot{W}$	power, W
x	vapor quality
ZT	figure of merit in Equation (10.5)

Greek letters

η	energy efficiency
ζ	ammonia concentration
ψ	exergy efficiency
σ	electrical conductivity, Ω^{-1}

Subscripts

0	reference state
b	boiling
c	compressor
cog	cogeneration
cool	cooling
E	expander
h	heat
in	input
P	pump
si	heat sink
so	heat source
sh	superheating
wph	water preheater

Superscript

$'$	liquid
$''$	vapor
ch	chemical

References

Cetinkaya E. 2013. Experimental Investigation and Modeling of Integrated Trigeneration Systems. PhD Thesis, University of Ontario Institute of Technology, pp. 214.

Copeland, R.J., Alptekin, G., Cesario, M., Gershanovich, Y., 2001. A Novel CO_2 separation system. In: First National Conference on Carbon Sequestration, May 14–17. National Energy Technology Laboratory, San Ramon, California, pp. 1–23.

Dincer I., Zamfirescu C. 2012. Methods and apparatus for using ammonia as sustainable fuel, refrigerant and NOx reduction agent. US Patent 61/064,133.

Han, W., Jin, H., Lin, R., 2011. A novel power generation system based on moderate conversion of chemical energy of coal and natural gas. Fuel. 90, 263–271.

Hogerwaard, J., Dincer, I., Zamfirescu, C., 2013. Analysis and assessment of a new organic Rankine cycle system with/without cogeneration. Energy, the International Journal. 62, 300–310.

Ishida, M., Jin, H., 2000. A novel gas turbine cycle with hydrogen-fueled chemical-looping combustion. International Journal of Hydrogen Energy. 25, 1209–1215.

Kraemer, D., Poudel, B., Feng, H.-P., Caylor, J.C., Yu, B., Yan, X., Ma, Y., Wang, X., Wang, D., Muto, A., McEnaney, K., Chiesa, M., Ren, Z., Chen, G., 2011. High-performance flat-panel solar thermoelectric generators with high thermal concentration. Nature materials. 10, 532–538.

Li, P., Cai, L., Zhai, P., Tang, X., Zhang, Q., Niino, M., 2010. Design of a concentration solar thermoelectric generator. Journal of Electronic Materials. 39, 1522–1530.

Muradov, N., Smith, F., Choi, P., Bokerman, G., 2010. A novel power generation system based on combination of hydrogen and direct carbon fuel cells for decentralized applications. In: World Hydrogen Energy Conference.

Oralli, E., Tarique, M.A., Zamfirescu, C., Dincer, I., 2011. Analysis and modeling of conversion of scroll compressor into expander for low-power Rankine cycle. International Journal of Low-Carbon Technologies. 6, 200–206.

Oralli, E., Dincer, I., Zamfirescu, C., 2014. Modeling and analysis of scroll compressor conversion into expander for Rankine cycles. International Journal of Green Energy. 11, 1–22.

Sano, S., Mizukami, H., Kaibe, H., 2003. Development of high-efficiency thermoelectric power generation system. Komatsu Technical Report. 49, 1–7.

Smith, I.K., Stosic, N., Aldis, C.A., 1996. Development of the trilateral flash cycle system Part 3: The design of high efficiency two-phase screw expanders. Proc Instn Mech Engrs, Part A. 210 (A2), 75–93.

Smouse, S., Lani, B., Brun, K., 2013. A Novel Supercritical Carbon Dioxide Power Cycle Utilizing Pressurized Oxy-combustion in Conjunction with Cryogenic Compression. National Energy technology Laboratory, Pittsburgh, PA US DoE Project DE-FE0009395.

Tarique, M.A., Dincer, I., Zamfirescu, C., 2014. Experimental investigation of a scroll expander for an organic Rankine cycle. International Journal of Energy Research. http://dx.doi.org/10.1002/er.3189.

Van Blarigan, P., 2000. Advanced internal combustion engine research. In: Proceedings of the U.S. DOE Hydrogen Program Review. National Renewable Energy Laboratory, San Ramon, California, pp. 639–657.

Weinberg, F.J., Rowe, D.M., Min, G., 2002. Novel high performance small-scale thermoelectric power generation employing regenerative combustion systems. Journal of Physics D: Applied Physics. 35, L61–L63.

Zamfirescu, C., 2007. Preliminary design of a new-concept chemical looping combustion reactor based on Ljungstrom principle. In: Third International Energy, Exergy and Environment Symposium – IEEES-3, Évora, Portugal July 1–5.

Zamfirescu, C., Dincer, I., 2008a. Using ammonia as a sustainable fuel. Journal of Power Sources. 185, 459–465.

Zamfirescu, C., Dincer, I., 2008b. Thermodynamic analysis of a novel ammonia–water trilateral Rankine cycle. Thermochimica Acta. 477, 7–15.

APPENDIX

A

Conversion Factors

Quantity	SI to English	English to SI
Length	1 m $=$ 39.37 in 1 m $=$ 3.28 ft	1 in $=$ 0.0254 m 1 ft $=$ 0.3048 m
Area	1 m^2 $=$ 10.764 ft 1 m^2 $=$ 1550.0 in^2	1 ft^2 $=$ 0.00929 m^2 1 in^2 $=$ 6.452 $\times$ 10^{-4} m^2
Volume	1 m^3 $=$ 264.172 gal 1 m^3 $=$ 35.315 ft^3	1 gal $=$ 0.00378 m^3 1 ft^3 $=$ 0.0283 m^3
Mass	1 kg $=$ 2.204 lbm 1 kg $=$ 0.0685 slug	1 lbm $=$ 0.454 kg 1 slug $=$ 14.594 kg
Force	1 N $=$ 0.225 lbf 1 N $=$ 10^5 dyne	1 lbf $=$ 4.448 N 1 dyne $=$ 10^{-5} N
Dynamic viscosity	1 Ns/m^2 $=$ 1000 cP	1 cP $=$ 0.001 Ns/m^2
Kinematic viscosity	1 m^2/s $=$ 10,000 St	1 St $=$ 1 E $-$ 4 m^2/s
Pressure	1 Pa $=$ 1.4504 $\times$ 10^{-4} lb$_f$/in^2 1 Pa $=$ 0.020886 lb$_f$/ft^2 1 Pa $=$ 4.015 $\times$ 10^{-3} in water 1 Pa $=$ 2.953 $\times$ 10^{-4} in Hg	1 lb$_f$/in^2 $=$ 6894.6 Pa lb$_f$/ft^2 $=$ 47.88 Pa 1 in water $=$ 249.07 Pa 1 in Hg $=$ 3386.39 Pa
Energy	1 J $=$ 2.778 kWh 1 J $=$ 9.478E $-$ 4 BTU	1 kWh $=$ 3.6E6 J 1 BTU $=$ 1055.056 J
Power	1 W $=$ 3.4123 Btu/h 1 W $=$ 0.737 562 Ib$_f$ $\cdot$ ft/s 1 W $=$ 0.00136 hp (metric) 1 W $=$ 0.000284 ton refrigeration	1 Btu/h $=$ 0.2931 W 1 lb$_f$ $\cdot$ ft/s $=$ 1.3558 W 1 hp (metric) $=$ 0.735499 kW 1 ton refrigeration $=$ 3.516 85 kW
Heat flux	1 W/m^2 $=$ 0.3171 Btu/h $\cdot$ ft^2 1 W/m^2 $=$ 0.8598 kcal/h $\cdot$ m^2	1 Btu/h $\cdot$ ft^2 $=$ 3.1525 W/m^2 1 kcal/h $\cdot$ m^2 $=$ 1.163 W/m^2
Temperature	$T(K) = T(°C) + 273.15$ $T(K) = T(°R)/1.8$ $T(K) = [T(°F) + 459.67]/1.8$ $T(°C) = [T(°F) -32]/1.8$	$T(°C) = T(K) − 273.15$ $T(°R) = 1.8\, T(K)$ $T(°F) = 1.8[T(K) − 273.15] + 32$ $T(°F) = 1.8\, T(°C) + 32$
Temperature difference	1 ΔK $=$ 1 Δ°C $=$ 1.8 Δ°F $=$ 1.8 Δ°R	1 Δ°F $=$ 1 Δ°R $=$ 0.555 Δ°C $=$ 0.555 Δ K
Specific heat	1 J/kg $\cdot$ K $=$ 2.3886 $\times$ 10^{-4} Btu/lb$_m$ $\cdot$ °F	1 Btu/lb$_m$ $\cdot$ °F $=$ 4187 J/kg $\cdot$ K
Heat transfer coefficient	1 W/m^2 $\cdot$ K $=$ 0.1761 Btu/h $\cdot$ ft^2 $\cdot$ °F 1 W/m^2 $\cdot$ K $=$ 0.8598 kcal/h $\cdot$ m^2 $\cdot$ °C	1 Btu/h $\cdot$ ft^2 $\cdot$ °F $=$ 5.678/m^2 $\cdot$ K 1 kcal/h $\cdot$ m^2 $\cdot$ °C $=$ 1.163 W/m^2 $\cdot$ K

Thermophysical Properties

TABLE B.1 Thermodynamic Properties of Selected Substances

Substance	R	TPT	TPP	ΔH_{fus}	NBP	ΔH_{boil}	P_c	T_c	v_c	Z_c	ω	a	b
Hydrogen	4124	13.96	73.59	454.5	20.37	448.7	12.96	33.15	0.0336	3.135	−0.2187	26,785	0.0165
Para-hydrogen	4124	13.8	70.41	432.5	20.27	446.6	12.86	32.94	0.0327	3.228	−0.2186	26,670	0.0165
Oxygen	259.8	54.36	1.463	243	90.19	213.2	50.43	154.6	0.00217	3.661	0.0222	149,769	0.0198
Nitrogen	296.8	63.15	125.2	215.5	77.36	199.2	33.96	126.2	0.00280	3.936	0.0372	148,225	0.0240
Helium	2077	1	1	1	4.222	20.7	2.275	5.195	0.01196	3.967	−0.385	3751	0.0147
Neon	412	24.56	432	88.16	27.11	85.77	26.8	44.49	0.00188	3.62	−0.037	23,345	0.0107
Argon	208.1	83.81	688.9	172.2	87.3	169.4	48.63	150.7	0.00161	4.002	−0.002	147,587	0.0200
Xenon	63.32	161.4	827.6	96.72	164.9	95.87	58.4	289.7	0.001	3.142	0.0036	454,364	0.0320
Fluorine	218.8	53.48	2.413	196.3	85.03	175.2	52.4	144.4	0.00163	3.691	0.0504	125,813	0.0178
Ammonia	488.2	195.5	61.02	1486	239.8	1369	113.3	405.4	0.00516	3.381	0.2558	458,377	0.0231
Hydrogen sulfide	244	187.7	231.9	574.7	212.9	546.6	89.63	373.4	0.00298	3.401	0.0961	491,620	0.0269
Krypton	99.21	115.8	732	108.8	119.8	107.5	55.1	209.4	0.00121	3.101	−0.002	251,616	0.0245
Methane	518.3	90.69	117	544.2	111.7	510.8	45.99	190.6	0.00734	2.923	0.0114	249,574	0.0268
Methanol	259.5	175.6	0.0018	1314	338.2	1103	81.04	513.4	0.00581	2.828	0.5646	1.028×10^6	0.0409
Ethane	276.5	90.37	0.0114	595.7	184.6	489.4	48.72	305.3	0.00451	3.838	0.0995	604,771	0.0405
Ethanol	180.5	159.1	0.00002	1017	351.4	848.8	61.48	513.9	0.00611	2.466	0.6442	1.358×10^6	0.0540
n-Octane	72.79	216.4	0.01399	411.5	398.6	302.4	24.97	569.3	0.00390	4.249	0.3928	4.103×10^6	0.1475
Isobutane	143	113.6	0.00021	482.9	261.5	365.5	36.4	407.8	0.0049	3.271	0.1853	1.444×10^6	0.0724
Propane	188.6	85.48	N/A	1	231.1	425.3	42.47	369.8	0.00505	3.245	0.1524	1.017×10^6	0.0563
Isopentane	115.2	112.7	N/A	N/A	301	342.5	33.7	460.4	0.00446	3.524	0.2266	1.988×10^6	0.0883

TABLE B.1 Thermodynamic Properties of Selected Substances—Cont'd

Substance	R	TPT	TPP	ΔH_{fus}	NBP	ΔH_{boil}	P_c	T_c	v_c	Z_c	ω	a	b
n-Butane	143	134.9	0.00693	494	272.6	384.7	37.96	425.1	0.00373	4.29	0.2	1.505×10^6	0.0724
n-Decane	58.44	243.5	0.01452	392.6	447.4	276.3	21.03	617.7	0.00379	4.519	0.4903	5.735×10^6	0.19
n-Dodecane	48.81	263.6	0.00631	381.5	489.5	256.3	18.17	658.1	0.00514	3.436	0.574	7.534×10^6	0.2343
Isobutane	143	113.6	0.00021	482.9	261.5	365.5	36.4	407.8	0.0049	3.271	0.185	1.444×10^6	0.0724
Isopropanol	138.4	0	N/A	N/A	355.6	663.1	47.62	508.3	0.00363	4.062	0.668	1.715×10^6	0.0690
Ethylene	296.4	104	1.226	568.2	169.3	482.2	50.4	282.3	0.00484	3.43	0.086	499,944	0.0362
Cyclohexane	98.79	279.5	52.49	402.4	353.9	356	40.75	553.6	0.00375	3.571	0.2091	2.378×10^6	0.0878
n-Heptane	82.98	182.6	0.00176	441.1	371.5	316.9	27.27	540.1	0.00556	2.956	0.3479	3.382×10^6	0.1281
n-Hexane	96.49	177.8	0.0035	618.8	342.4	329.9	30.58	507.9	0.00400	3.998	0.3117	2.666×10^6	0.1074
n-Nonane	64.83	219.7	0.0046	410.3	424.2	288.4	22.81	594.6	0.00482	3.5	0.4468	4.898×10^6	0.1686
n-Pentane	115.2	143.5	0.00012	466.1	309	358	33.64	469.7	0.00540	2.977	0.2499	2.073×10^6	0.0903
Propylene	197.6	87.95	N/A	N/A	225.4	439.2	46.65	365.6	0.00458	3.38	0.1407	905,579	0.0506
R22	96.15	115.7	0.004715	301.9	232.3	232.5	49.89	369.3	0.00203	3.503	0.2208	864,055	0.0478
R134a	81.49	168.9	3.545	263.7	247.1	217	40.59	374.2	0.00170	4.408	0.3269	1.09×10^6	0.0596
R290	188.6	85.48	1	1	231.1	425.3	42.47	369.8	0.00505	3.245	0.1524	1.0×10^6	0.0563
R404a	85.19	−555	1	1	226.6	199.7	37.35	345.3	0.00193	4.065	0.275	1.0×10^6	0.0598
R218	44.22	125.5	0.02119	141.1	236.5	105.1	26.4	345	0.00169	3.417	0.3201	1.4×10^6	0.0845
R245fa	62.03	171.1	0.1303	258.5	288.3	196	36.51	427.2	0.00203	3.561	0.379	1.5×10^6	0.0756
R507a	84.1	−555	1	1	226.1	196.3	37.14	343.9	0.00199	3.903	0.2793	1×10^6	0.0599
R508b	87.16	−555	1	1	184.9	163	39.25	287.1	0.0017	3.735	0.2272	664,053	0.047

R=, gas constant (J/kg K); TPT, triple point temperature (K); TPP, triple point pressure (bar); $\Delta H_{fus}=$, fusion heat at triple point (kJ/kg); NBP, normal boiling point (K); $\Delta H_{boil}=$, boiling enthalpy at normal boiling point; P, pressure (bar); T, temperature (K); v, specific volume (m³/kg); Z, compressibility factor; subscript "c", critical point; ω, accentric factor; a, b, constants in Peng-Robinson equation of state.

TABLE B.2   Properties of Air at Atmospheric Pressure

T (°C)	ρ, (g/dm³)	μ, (μPa s)	c_p, (J/kg K)	k, (mW/m K)	Pr	β, (K⁻¹)	c, (m/s)	γ
−200	47.61	5.201	1002	6.667	0.7821	0.01367	171.6	1.401
−170	33.77	7.311	1002	9.668	0.7581	0.009695	203.7	1.401
−140	26.16	9.296	1002	12.53	0.7435	0.00751	231.4	1.401
−110	21.35	11.16	1003	15.26	0.7332	0.006129	256.2	1.401
−80	18.03	12.93	1003	17.87	0.7251	0.005177	278.8	1.401
−50	15.61	14.6	1003	20.37	0.7186	0.004481	299.6	1.401
−20	13.76	16.19	1003	22.77	0.7131	0.00395	319.1	1.401
10	12.3	17.7	1004	25.09	0.7086	0.003532	337.4	1.4
40	11.12	19.15	1006	27.32	0.7049	0.003193	354.7	1.4
70	10.15	20.54	1008	29.49	0.702	0.002914	371.2	1.398
100	9.334	21.88	1011	31.6	0.6999	0.00268	386.8	1.397
130	8.639	23.18	1014	33.64	0.6985	0.00248	401.8	1.395
160	8.041	24.43	1018	35.64	0.6978	0.002309	416.2	1.393
190	7.52	25.64	1023	37.59	0.6977	0.002159	429.9	1.39
220	7.063	26.82	1028	39.5	0.6982	0.002028	443.2	1.387
250	6.658	27.96	1034	41.37	0.6991	0.001912	456	1.384
280	6.296	29.07	1041	43.2	0.7004	0.001808	468.3	1.381
310	5.973	30.16	1047	45.0	0.7019	0.001715	480.3	1.378
340	5.68	31.22	1054	46.77	0.7038	0.001631	491.9	1.374
370	5.415	32.26	1061	48.51	0.7058	0.001555	503.1	1.371
400	5.174	33.28	1069	50.23	0.7079	0.001486	514.1	1.367
430	4.953	34.27	1076	51.92	0.7101	0.001422	524.7	1.364
760	3.371	44.21	1147	69.34	0.7314	0.0009679	629	1.334
790	3.276	45.04	1153	70.84	0.7328	0.0009406	637.5	1.332
820	3.186	45.86	1158	72.33	0.7341	0.0009148	646	1.33
850	3.101	46.67	1163	73.82	0.7353	0.0008904	654.3	1.328
880	3.02	47.48	1168	75.29	0.7364	0.0008672	662.6	1.326
910	2.944	48.28	1172	76.75	0.7374	0.0008452	670.7	1.324
940	2.871	49.07	1177	78.21	0.7383	0.0008243	678.8	1.323
970	2.802	49.85	1181	79.65	0.7391	0.0008044	686.7	1.321
1000	2.736	50.63	1185	81.09	0.7398	0.0007855	694.5	1.32

T, temperature; ρ, density; μ, dynamic viscosity; c_p, specific heat; k, thermal conductivity; Pr, Prandtl number; β, volume expansion coefficient; c, sound speed, γ, adiabatic expansion coefficient; $\omega = -0.009278$, accentric factor. Critical parameters: $T_c = 132.531\,K$, $P_c = 37.86\,bar$, $v_c = 2.917\,dm^3/kg$.

TABLE B.3 Thermophysical Properties of Pure Water at Atmospheric Pressure

T (°C)	ρ, (kg/m³)	μ, (Pa s)	c_p, (J/kg K)	k, (W/mK)	Pr	β,(K⁻¹)	c, (m/s)	σ, (N/m)
5	1000	0.001519	4200	0.5576	11.44	0.00001135	1426	0.07494
10	999.7	0.001307	4188	0.5674	9.642	0.00008743	1448	0.07422
15	999.1	0.001138	4184	0.5769	8.253	0.0001523	1467	0.07348
20	998.2	0.001002	4183	0.5861	7.152	0.000209	1483	0.07273
25	997.1	0.0008905	4183	0.5948	6.263	0.0002594	1497	0.07197
30	995.7	0.0007977	4183	0.603	5.534	0.0003051	1509	0.07119
35	994	0.0007196	4183	0.6107	4.929	0.000347	1520	0.0704
40	992.2	0.0006533	4182	0.6178	4.422	0.0003859	1528	0.06959
45	990.2	0.0005963	4182	0.6244	3.994	0.0004225	1534	0.06877
50	988	0.0005471	4181	0.6305	3.628	0.0004572	1537	0.06794
55	985.7	0.0005042	4182	0.636	3.315	0.0004903	1538	0.0671
60	983.2	0.0004666	4183	0.641	3.045	0.0005221	1537	0.06624
65	980.6	0.0004334	4184	0.6455	2.81	0.0005528	1534	0.06536
70	977.8	0.000404	4187	0.6495	2.605	0.0005827	1529	0.06448
75	974.9	0.0003779	4190	0.653	2.425	0.0006118	1523	0.06358
80	971.8	0.0003545	4194	0.6562	2.266	0.0006402	1514	0.06267
85	968.6	0.0003335	4199	0.6589	2.125	0.0006682	1504	0.06175
90	965.3	0.0003145	4204	0.6613	2	0.0006958	1491	0.06081
95	961.9	0.0002974	4210	0.6634	1.888	0.000723	1475	0.05987
100	0.5896	0.00001227	2042	0.02506	0.9996	0.002881	472.8	0.05891

T, temperature; ρ, density; μ,dynamic viscosity; c$_p$, specific heat; k, thermal conductivity; Pr, Prandtl number; β, volume expansion coefficient; c, sound speed; σ, superficial tension; ω=0.3443, accentric factor. Critical parameters: T$_c$=373.984°C, P$_c$=220.64 bar, v$_c$=3.106 dm³/kg. Triple point parameters: T$_t$=0.01°C, P$_t$=611.732 Pa.

TABLE B.4   Thermophysical Properties of Water at Saturation

T	P	σ	ρ_L	c_{pL}	k_L	μ_L	Pr_L	β_L	v_V	c_{pV}	k_V	μ_V	Pr_V	β_V	γ
0.01	0.00612	75.64	1000	4.23	547.5	1792	13.84	N/A	206.0	1.87	17.07	9.216	1.008	3.672	1.33
10	0.0123	74.22	999.7	4.19	567.4	1307	9.645	0.0872	106.3	1.87	17.62	9.461	1.006	3.548	1.33
20	0.0234	72.73	998.2	4.18	586	1002	7.154	0.209	57.78	1.88	18.22	9.727	1.004	3.435	1.33
30	0.0425	71.19	995.6	4.18	602.9	797.7	5.535	0.305	32.9	1.89	18.88	10.01	1.003	3.332	1.33
40	0.0738	69.59	992.2	4.18	617.8	653.3	4.423	0.386	19.53	1.90	19.59	10.31	1.002	3.239	1.33
50	0.1234	67.94	988	4.18	630.4	547.1	3.629	0.457	12.04	1.92	20.36	10.62	1.001	3.156	1.33
60	0.1993	66.24	983.2	4.18	640.9	466.6	3.045	0.522	7.674	1.94	21.18	10.93	1	3.082	1.33
70	0.3118	64.48	977.7	4.19	649.5	404	2.605	0.583	5.045	1.96	22.06	11.26	0.9995	3.017	1.33
80	0.4737	62.67	971.8	4.19	656.2	354.5	2.266	0.64	3.409	1.98	23	11.59	0.9993	2.962	1.33
90	0.7012	60.81	965.3	4.20	661.3	314.5	2	0.696	2.362	2.01	24	11.93	0.9994	2.917	1.34
100	1.013	58.91	958.4	4.22	665.1	281.9	1.787	0.75	1.674	2.04	25.08	12.27	1	2.882	1.34
110	1.432	56.96	951	4.23	667.6	254.8	1.615	0.804	1.211	2.08	26.22	12.61	1.001	2.857	1.34
120	1.985	54.96	943.2	4.25	669.1	232.1	1.474	0.858	0.892	2.12	27.44	12.96	1.004	2.843	1.35
130	2.7	52.93	934.9	4.27	669.7	213	1.357	0.912	0.669	2.17	28.73	13.3	1.007	2.84	1.36
140	3.612	50.85	926.2	4.29	669.4	196.6	1.259	0.968	0.509	2.23	30.09	13.65	1.012	2.849	1.36
150	4.757	48.74	917.1	4.31	668.3	182.5	1.178	1.026	0.393	2.30	31.54	13.99	1.02	2.872	1.37
160	6.177	46.59	907.5	4.34	666.4	170.3	1.109	1.087	0.3071	2.37	33.06	14.34	1.029	2.908	1.39
170	7.915	44.4	897.5	4.37	663.7	159.6	1.051	1.152	0.243	2.46	34.66	14.68	1.042	2.959	1.40
180	10.02	42.19	887.1	4.40	660.2	150.2	1.002	1.221	0.194	2.56	36.34	15.03	1.057	3.027	1.42
190	12.54	39.94	876.1	4.44	655.9	141.8	0.9607	1.296	0.156	2.67	38.09	15.37	1.077	3.114	1.44
200	15.54	37.67	864.7	4.49	650.7	134.4	0.9269	1.377	0.127	2.80	39.93	15.71	1.1	3.221	1.46
210	19.06	35.38	852.8	4.54	644.7	127.7	0.8994	1.467	0.104	2.94	41.85	16.06	1.129	3.351	1.48
220	23.18	33.06	840.3	4.60	637.6	121.6	0.8777	1.567	0.086	3.11	43.86	16.41	1.162	3.508	1.52

TABLE B.4 Thermophysical Properties of Water at Saturation—Cont'd

T	P	σ	ρ_L	c_pL	k_L	μ_L	Pr_L	β_L	v_V	c_pV	k_V	μ_V	Pr_V	β_V	γ
230	27.95	30.73	827.3	4.67	629.5	116	0.8615	1.679	0.071	3.30	45.96	16.76	1.203	3.697	1.55
240	33.45	28.39	813.5	4.76	620.3	110.9	0.8508	1.807	0.06	3.51	48.15	17.12	1.25	3.923	1.59
250	39.74	26.04	799.1	4.86	609.8	106.2	0.8455	1.954	0.050	3.77	50.46	17.49	1.306	4.195	1.64
260	46.89	23.68	783.8	4.97	598	101.7	0.846	2.126	0.042	4.06	52.9	17.88	1.373	4.523	1.70
270	55	21.33	767.7	5.11	584.5	97.55	0.853	2.33	0.036	4.41	55.49	18.28	1.453	4.923	1.77
280	64.13	18.99	750.5	5.28	569.4	93.56	0.8673	2.576	0.031	4.83	58.28	18.7	1.549	5.415	1.86
290	74.38	16.66	732.2	5.48	552.3	89.71	0.8908	2.88	0.026	5.33	61.31	19.15	1.666	6.032	1.97
300	85.84	14.35	712.4	5.74	533.1	85.95	0.9262	3.266	0.022	5.97	64.67	19.65	1.813	6.822	2.11
310	98.61	12.08	691	6.08	511.3	82.22	0.9782	3.773	0.018	6.78	68.45	20.21	2.001	7.861	2.29
320	112.8	9.858	667.4	6.54	486.8	78.46	1.054	4.47	0.015	7.87	72.84	20.84	2.251	9.278	2.54
330	128.5	7.697	641	7.2	458.8	74.57	1.17	5.489	0.013	9.41	78.1	21.6	2.603	11.31	2.91
340	145.9	5.62	610.8	8.24	426.9	70.45	1.359	7.116	0.011	11.8	84.68	22.55	3.137	14.43	3.47
350	165.2	3.66	574.7	10.1	389.7	65.88	1.711	10.1	0.0088	15.9	93.48	23.81	4.058	19.77	4.47
360	186.6	1.872	528.1	14.7	344.4	60.39	2.574	17.11	0.007	25.2	106.6	25.71	6.085	30.79	6.74
370	210.3	0.3846	453.1	41.7	280.7	52.25	7.765	45.88	0.0050	70.4	132.6	29.57	15.7	64.53	18.0
373.9	220.4	0.0004	349.4	N/A	210.8	41.95	277.1	87.51	0.0034	340.1	180	37	69.91	104.5	84.5

T, *temperature* (°C); P, *pressure* (bar); σ=, *surface tension* (mN/m); ρ, *density* (kg/m³); c_pL, *specific heat* (kJ/kg K); c_p, *specific heat* (kJ/kg K); k, *thermal conductivity* (mW/m K); μ, *dynamic viscosity* (μPa s); Pr, *Prandtl number*; β, *volumetric expansion coefficient* (mK⁻¹); v, *specific volume* (m³/kg); γ, *isentropic expansion coefficient of saturated vapor*; indices: L, *liquid*; V, *vapor*.

TABLE B.5 Thermophysical Properties of Hydrogen at Saturation

T	P	σ	ρ_L	c_{pL}	k_L	μ_L	Pr_L	β_L	v_V	c_{pV}	k_V	μ_V	Pr_V	β_V	γ
14	0.08	2.977	77.0	6.992	74.36	24.30	2.29	10.58	7.6332	10.56	14.51	1.24	0.90	74.32	1.69
15	0.13	2.819	76.1	7.335	81.62	21.68	1.95	11.26	4.7044	10.69	15.21	1.27	0.89	70.67	1.70
16	0.21	2.656	75.3	7.717	87.55	19.39	1.71	12.01	3.0742	10.85	15.94	1.31	0.89	67.82	1.72
17	0.32	2.490	74.3	8.132	92.26	17.40	1.53	12.87	2.1066	11.03	16.69	1.34	0.89	65.69	1.74
18	0.47	2.321	73.4	8.574	95.85	15.69	1.40	13.85	1.4996	11.25	17.49	1.37	0.88	64.23	1.77
19	0.13	2.819	76.1	7.335	81.62	21.68	1.95	11.26	4.7044	10.69	15.21	1.27	0.89	70.67	1.70
20	0.91	1.977	71.3	9.563	100.1	12.97	1.24	16.18	0.8288	11.85	19.21	1.44	0.89	63.20	1.84
21	1.22	1.804	70.1	10.13	101	11.92	1.20	17.60	0.6370	12.25	20.15	1.48	0.90	63.63	1.89
22	1.59	1.630	68.9	10.76	101.1	11.02	1.17	19.26	0.4981	12.75	21.16	1.51	0.91	64.74	1.95
23	2.04	1.458	67.6	11.48	100.7	10.26	1.17	21.23	0.3951	13.36	22.25	1.54	0.93	66.61	2.03
24	2.58	1.286	66.2	12.31	99.8	9.61	1.18	23.60	0.3171	14.12	23.43	1.68	1.01	69.39	2.12
25	3.21	1.117	64.7	13.28	98.51	9.03	1.22	26.55	0.2569	15.07	24.72	1.73	1.05	73.32	2.23
26	3.94	0.952	63.1	14.46	96.94	8.51	1.27	30.30	0.2098	16.29	26.14	1.78	1.11	78.77	2.37
27	4.78	0.790	61.3	15.95	95.2	8.00	1.34	35.28	0.1722	17.90	27.73	1.85	1.19	86.38	2.56
28	5.74	0.634	59.3	17.91	93.4	7.49	1.44	42.19	0.1418	20.09	29.53	1.92	1.31	97.27	2.82
29	6.82	0.484	57.1	20.69	91.64	6.95	1.57	52.50	0.1168	23.24	31.62	2.01	1.47	113.61	3.19
30	8.04	0.343	54.5	25.02	90.02	6.35	1.76	69.50	0.0957	28.15	34.11	2.11	1.74	140.12	3.76
31	9.42	0.213	51.4	32.99	88.65	5.65	2.10	102.69	0.0775	36.81	37.28	2.26	2.23	189.42	4.75
32	10.96	0.097	47.1	53.14	87.64	4.84	2.93	193.05	0.0606	55.68	42.11	2.50	3.31	308.86	6.89
33.1	12.88	0.0017	35.3	200	1000	N/A	N/A	981.45	0.0367	131.74	59.93	3.41	7.50	1036.71	15.44

T, temperature (K); P, pressure (bar); σ, surface tension (mN/m); ρ, density (kg/m³); c_p, specific heat (kJ/kg K); k, thermal conductivity (mW/m K); μ, dynamic viscosity (μPa s); Pr, Prandtl number; β, volumetric expansion coefficient (mK⁻¹); v, specific volume (m³/kg); γ, isentropic expansion coefficient of saturated vapor; indices: L, liquid; V, vapor.

TABLE B.6 Thermophysical Properties of Oxygen at Saturation

T	P	σ	ρ_L	c_pL	k_L	μ_L	Pr_L	β_L	v_V	c_pV	k_V	μ_V	Pr_V	β_V	γ
55	N/A	22.30	1303.5	1.455	201	747.53	5.41	N/A	80.029	0.93	4.48	4.15	0.86	18.28	1.39
60	0.01	20.98	1283.1	126.4	194.3	582.46	379.02	3.49	21.463	0.95	4.98	4.55	0.86	16.87	1.39
65	0.02	19.66	1260.7	1.65	187.2	460.77	4.06	3.58	7.2177	0.96	5.49	4.96	0.87	15.70	1.38
70	0.06	18.348	1237.9	1.657	180.1	373.97	3.44	3.70	2.8933	0.97	5.99	5.36	0.87	14.70	1.38
75	0.15	17.039	1214.9	9.855	172.9	310.37	17.69	3.83	1.3306	0.98	6.50	5.75	0.86	13.84	1.39
80	0.30	15.739	1191.4	1.66	165.8	262.53	2.63	N/A	0.6819	0.97	7.03	6.15	0.85	13.13	1.40
85	0.57	14.450	1167.5	1.669	158.6	225.61	2.37	4.15	0.3809	0.97	7.56	6.54	0.84	12.58	1.41
90	0.99	13.176	1142.9	1.683	151.3	196.38	2.18	4.36	0.2281	0.97	8.12	6.94	0.83	12.22	1.43
95	1.63	11.919	1117.7	1.704	144	172.66	2.04	4.59	0.1445	0.98	8.70	7.33	0.83	12.03	1.46
100	2.54	10.6829	1091.5	1.731	136.7	152.96	1.94	4.88	0.0959	1.00	9.31	7.73	0.83	12.04	1.49
105	3.79	9.4707	1064.4	1.765	129.3	136.23	1.86	5.21	0.0661	1.04	9.96	8.13	0.85	12.26	1.52
110	5.43	8.2861	1035.9	1.808	121.9	121.74	1.81	5.62	0.0470	1.09	10.66	8.55	0.88	12.71	1.57
115	7.56	7.1328	1006.0	1.861	114.4	108.97	1.77	6.13	0.0342	1.16	11.42	8.98	0.92	13.42	1.62
120	10.22	6.0154	974.2	1.927	106.9	97.53	1.76	6.76	0.0254	1.26	12.26	9.43	0.97	14.49	1.70
125	13.51	4.9387	939.7	2.013	99.43	87.09	1.76	N/A	0.0192	1.38	13.19	9.91	1.04	16.05	1.80
130	17.49	3.9088	902.6	2.152	92.02	77.59	1.81	9.52	0.0146	1.56	14.26	10.44	1.14	18.36	1.95
135	22.25	2.9333	861.0	2.35	84.54	68.69	1.91	11.69	0.0112	1.81	15.50	11.06	1.29	21.96	2.17
140	27.88	2.0224	813.1	2.682	76.97	60.20	2.10	15.40	0.0086	2.22	17.04	11.82	1.54	28.07	2.52
145	34.48	1.192	755.0	3.337	69.11	51.85	2.50	23.04	0.0065	3.00	19.09	12.88	2.02	40.28	3.17
150	42.19	0.472	675.5	5.225	60.16	42.90	3.73	46.63	0.0047	4.95	22.39	14.71	3.26	74.61	4.73
154	49.31	0.036	547.4	15.42	48.04	32.13	10.31	N/A	0.0031	10.63	29.63	19.14	6.86	218.2	8.44

T, temperature (K); P, pressure (bar); σ, surface tension (mN/m); ρ, density (kg/m³); c_p, specific heat (kJ/kg K); k, thermal conductivity (mW/m K); μ, dynamic viscosity (μPa s); Pr, Prandtl number; β, volumetric expansion coefficient (mK⁻¹); v, specific volume (m³/kg); γ, isentropic expansion coefficient of saturated vapor; indices: L, liquid; V, vapor.

TABLE B.7   Thermophysical Properties of Nitrogen at Saturation

T	P	σ	ρ_L	c_{pL}	k_L	μ_L	Pr_L	β_L	v_V	c_{pV}	k_V	μ_V	Pr_V	β_V	γ
64	0.15	11.9581	863.7	2.002	171.5	297.50	3.47	4.78	1.2869	1.06	5.71	4.44	0.82	16.13	1.41
65	0.17	11.7169	859.6	1.994	169.5	282.08	3.32	4.82	1.0950	1.06	5.82	4.51	0.82	15.95	1.41
70	0.39	10.5262	838.5	2.005	159.5	220.26	2.77	5.11	0.5274	1.08	6.35	4.88	0.83	15.24	1.43
75	0.76	9.3624	816.7	27.05	149.5	176.75	31.99	N/A	0.2825	1.11	6.91	5.26	0.84	14.82	1.44
80	1.37	8.2276	793.9	2.054	139.5	145.11	2.14	N/A	0.1642	1.14	7.49	5.65	0.86	14.69	1.47
85	2.29	7.1240	770.1	2.094	129.6	121.31	1.96	N/A	0.1018	1.19	8.11	6.06	0.89	14.89	1.51
90	3.60	6.0544	745.0	2.147	119.6	102.83	1.85	6.94	0.0663	1.26	8.79	6.48	0.93	15.46	1.56
95	5.41	5.0222	718.3	7.073	109.7	88.02	5.68	N/A	0.0449	1.35	9.55	6.94	0.98	16.54	1.64
100	7.78	4.0318	689.4	2.301	99.71	75.77	1.75	9.15	0.0313	1.55	10.44	7.43	1.10	18.14	1.67
105	10.83	3.0892	659.0	2.432	90.15	65.72	1.77	10.21	0.0222	1.44	11.49	7.98	1.00	18.19	1.80
110	14.66	2.2025	625.6	2.736	80.72	56.96	1.93	12.90	0.0160	1.16	12.80	8.63	0.78	20.83	3.96
115	19.37	1.3844	583.8	3.211	70.45	48.23	2.20	18.78	0.0115	2.02	14.51	9.44	1.32	31.69	3.57
120	25.11	0.6578	527.5	4.351	59.19	39.02	2.87	33.00	0.0080	3.74	16.98	10.62	2.34	56.88	4.00
125	32.07	0.0829	430.2	11.79	44.96	27.35	7.17	136.45	0.0049	9.65	22.52	13.33	5.71	187.71	N/A
126	33.65	0.0083	378.6	20.16	38.93	22.77	11.79	266.83	0.0039	13.27	26.63	15.37	7.66	308.90	8.52

T, temperature (K); P, pressure (bar); σ, surface tension (mN/m); ρ, density (kg/m^3); c_p, specific heat (kJ/kg K); k, thermal conductivity (mW/m K); μ, dynamic viscosity (μPa s); Pr, Prandtl number; β, volumetric expansion coefficient (mK^{-1}); v, specific volume (m^3/kg); γ, isentropic expansion coefficient of saturated vapor; indices: L, liquid; V, vapor.

TABLE B.8 Thermophysical Properties of Carbon Dioxide at Saturation

T	P	σ	ρ_L	c_{pL}	k_L	μ_L	Pr_L	β_L	v_V	c_{pV}	k_V	μ_V	Pr_V	β_V	γ
−56.6	5.18	15.7767	1178.5	1.953	180.4	256.7	2.78	N/A	0.0727	0.90	10.97	10.95	0.90	6.11	1.44
−50	6.82	14.3248	1154.6	14.29	171.8	229.3	19.07	3.27	0.0558	0.95	11.51	11.31	0.93	6.30	1.46
−45	8.32	13.2367	1135.9	1.989	165.3	210.7	2.54	3.36	0.0460	0.98	11.94	11.58	0.95	6.50	1.48
−40	10.05	12.1654	1116.5	2.011	158.9	193.8	2.45	N/A	0.0383	1.03	12.41	11.87	0.98	6.74	1.51
−35	12.02	11.1118	1096.4	4.99	152.5	178.5	5.83	N/A	0.0320	1.07	12.90	12.16	1.01	7.04	1.54
−30	14.28	10.0769	1075.7	2.377	146.3	164.2	2.67	3.93	0.0270	1.13	13.42	12.46	1.05	7.41	1.57
−25	16.83	9.0619	1054.2	2.114	140	151.3	2.28	4.38	0.0228	1.19	13.99	12.78	1.09	7.87	1.62
−20	19.70	8.0682	1031.7	2.165	133.8	139.3	2.25	4.74	0.0193	1.27	14.60	13.12	1.14	8.44	1.67
−15	22.91	7.0974	1008.0	2.227	127.6	128.3	2.24	5.18	0.0165	1.36	15.27	13.47	1.20	9.15	1.73
−10	26.49	6.1514	982.9	2.306	121.4	118	2.24	5.73	0.0140	1.47	16.01	13.86	1.27	10.06	1.81
−5	30.46	5.2326	956.2	2.406	115.1	108.4	2.27	6.44	0.0120	1.61	16.84	14.30	1.37	11.23	1.92
0	34.85	4.3440	927.4	2.539	108.7	99.39	2.32	7.37	0.0102	1.79	17.77	14.79	1.49	12.79	2.05
5	39.69	3.4893	896.0	2.72	102.2	90.81	2.42	8.67	0.0087	2.03	18.86	15.36	1.65	14.95	2.23
10	45.02	2.6737	861.1	2.985	95.46	82.55	2.58	10.59	0.0074	2.37	20.14	16.06	1.89	18.12	2.49
15	50.87	1.9043	821.2	3.405	88.32	74.43	2.87	13.71	0.0062	2.89	21.74	16.95	2.25	23.17	2.88
20	57.29	1.1924	773.4	4.159	80.48	66.15	3.42	19.52	0.0051	3.77	23.83	18.19	2.87	32.32	3.51
25	64.34	0.5584	711.8	5.903	71.43	57.23	4.73	33.72	0.0041	5.54	26.94	20.16	4.14	53.05	4.69
30	72.14	0.0582	602.4	13.83	57.69	44.68	10.71	102.99	0.0029	10.16	33.86	25.03	7.51	128.22	6.80
30.9	73.64	0.0025	521.3	18.05	49.07	37.25	13.71	141.42	0.0024	11.77	39.21	29.16	8.75	163.45	5.80

T, temperature (°C); P, pressure (bar); σ, surface tension (mN/m); ρ, density (kg/m³); c_p, specific heat (kJ/kg K); k, thermal conductivity (mW/m K); μ, dynamic viscosity (μPa s); Pr, Prandtl number; β, volumetric expansion coefficient (mK⁻¹); v, specific volume (m³/kg); γ, isentropic expansion coefficient of saturated vapor; indices: L, liquid; V, vapor.

TABLE B.9 Thermophysical Properties of Ammonia at Saturation

T	P	σ	ρ_L	c_{pL}	k_L	μ_L	Pr_L	β_L	v_V	c_{pV}	k_V	μ_V	Pr_V	β_V	γ
−77.6	0.061	44.78	733.6	3.95	821.2	564.1	2.713	1.514	15.68	2.062	19.64	6.84	0.7182	5.23	1.325
−65	0.1563	41.51	719.3	4.162	774.8	430.6	2.312	1.586	6.478	2.102	19.82	7.162	0.7597	4.99	1.328
−55	0.3013	38.95	707.8	4.332	739.3	357.5	2.095	1.63	3.498	2.147	20.07	7.433	0.7951	4.86	1.331
−45	0.5447	36.42	696	4.387	704.6	303	1.887	1.717	2.008	2.205	20.43	7.715	0.8328	4.77	1.336
−35	0.9307	33.93	683.9	4.439	670.8	261.2	1.728	1.804	1.215	2.277	20.88	8.004	0.8729	4.72	1.343
−25	1.515	31.46	671.4	4.489	638	228.3	1.606	1.896	0.7695	2.365	21.44	8.3	0.9154	4.716	1.352
−15	2.362	29.04	658.6	4.538	605.9	201.7	1.511	1.995	0.5068	2.47	22.12	8.599	0.9603	4.757	1.364
−5	3.549	26.65	645.4	4.589	574.7	179.8	1.436	2.079	0.3452	2.595	22.92	8.902	1.008	4.846	1.379
5	5.16	24.29	631.8	4.643	544.3	161.4	1.377	2.199	0.2421	2.74	23.85	9.209	1.058	4.989	1.399
15	7.288	21.98	617.6	4.705	514.7	145.6	1.331	2.338	0.1741	2.908	24.92	9.519	1.111	5.193	1.424
25	10.03	19.71	602.9	4.781	485.7	131.8	1.297	2.499	0.1279	3.104	26.16	9.834	1.167	5.466	1.456
35	13.51	17.49	587.5	4.873	457.4	119.6	1.275	2.688	0.0956	3.333	27.58	10.16	1.228	5.826	1.497
45	17.82	15.32	571.3	4.993	429.6	108.8	1.264	3.006	0.0726	3.604	29.21	10.5	1.295	6.295	1.549
55	23.1	13.2	554.1	5.143	402.3	98.99	1.265	3.335	0.0557	3.93	31.11	10.86	1.372	6.912	1.615
65	29.48	11.14	535.8	5.34	375.4	90.06	1.281	3.768	0.0432	4.334	33.33	11.25	1.463	7.738	1.702
75	37.09	9.154	516.1	5.608	348.6	81.83	1.316	4.359	0.0336	4.851	35.97	11.7	1.578	8.876	1.819
85	46.09	7.24	494.5	5.987	322	74.18	1.379	5.21	0.0262	5.55	39.21	12.23	1.732	10.52	1.983
95	56.65	5.414	470.2	6.56	294.6	66.84	1.489	6.525	0.0203	6.56	43.31	12.9	1.954	13.05	2.227
105	68.95	3.695	442.1	7.52	266.1	59.7	1.687	8.801	0.0156	8.168	48.83	13.82	2.312	17.35	2.622
115	83.22	2.113	407.4	9.492	235	52.35	2.114	13.67	0.0116	11.14	56.96	15.22	2.977	26.03	3.358
125	99.7	0.7327	357.8	16.29	196.6	43.8	3.629	31.43	0.0082	18.25	71.25	17.84	4.57	50.81	5.121
127.5	104.2	0.437	338.7	21.71	183.4	40.94	4.848	46.11	0.0074	21.75	76.35	18.82	5.363	65.59	6.022
130	108.9	0.1754	310.5	34.41	165.1	37.05	7.72	81.46	0.0065	26.74	84.63	20.44	6.457	90.37	7.258
132.2	113.1	0.0017	248.5	64.71	129.5	29.55	14.77	168.7	0.0055	32.93	97.33	22.96	7.77	127.8	8.808

T, temperature (°C); P, pressure (bar); σ, surface tension (mN/m); ρ, density (kg/m^3); c_p, specific heat ($kJ/kg\ K$); k, thermal conductivity ($mW/m\ K$); μ, dynamic viscosity ($\mu Pa\ s$); Pr, Prandtl number; β, volumetric expansion coefficient (mK^{-1}); v, specific volume (m^3/kg); γ, isentropic expansion coefficient of saturated vapor; indices: L, liquid; V, vapor.

TABLE B.10 Standard Chemical Exergy of Some Elements

Element	B	C	Ca	Cl_2	Cu	F_2	Fe	H_2	I_2	K	Mg
ex^{ch} (kJ/mol)	628.1	410.27	729.1	123.7	132.6	505.8	374.3	236.12	175.7	336.7	626.9
ELEMENT	Mo	N_2	Na	Ni	O_2	P	Pb	Pt	Pu	Si	Ti
ex^{ch} (kJ/mol)	731.3	0.67	336.7	242.6	3.92	861.3	249.2	141.2	1100	855.0	907.2

Source: *Rivero R., Garfias M., Standard chemical exergy of elements updated. Energy 31:3310-3326.*

TABLE B.11 Formation Enthalpy, Entropy, Gibbs Energy, and Chemical Exergy of Some Substances

Substance name	Chemical formula	M	$\Delta_f H^0$	$\Delta_f S^0$	$\Delta_f G^0$	ex^{ch}
Acetone	C_3H_6O	59	−217.14	295.6	−152.69	1788
Acetylene	C_2H_2	26	228.19	200.9	210.67	1267
Ammonia	NH_3	17	−46.14	99.5	−16.5	337.9
n-Butane	C_4H_{10}	58	−125.78	309.90	−16.529	2805
Calcium carbonate	$CaCO_3$	100.1	−1207	91.7	−1128	17.52
Carbon monoxide	CO	28	−110.53	197.6	−137.17	275.06
Carbon dioxide	CO_2	44	−393.49	213.8	−394.35	198.38
Copper oxychloride	Cu_2OCl_2	214	−384.65	154.35	−369.7	21.08
Cupric oxide	CuO	79	−156.06	42.59	−128.29	6.27
Copper(II) chloride	$CuCl_2$	134	−218.0	108.07	−173.83	82.47
Cuprous chloride	CuCl	100	−136.82	87.45	−119.44	75.00
Ethanol	C_2H_5OH	47	−234.94	280.60	−167.71	1363
Iron(III) oxide	Fe_2O_3	159.7	−824.20	205.1	−742.22	12.26
Lithium bromide	LiBr	86.85	−351.14	74.01	−341.83	101.37
Magnetite	Fe_3O_4	231.5	−1118	146.1	−1015	115.61
Methane	CH_4	16	−74.60	186.40	−50.53	831.98
Nitrogen dioxide	NO_2	46	34.19	240.20	52.31	56.56
Nitrous oxide	N_2O	44	81.59	220.00	103.71	106.34
n-Octane	C_8H_{18}	131	−208.74	5.734	16.25	7549
Propane	C_3H_2	44	−104.67	130.7	−24.29	2151
Sodium chloride	NaCl	58.44	−411.24	223.1	−384.20	14.353
Sulphur dioxide	SO_2	64	−296.79	248.2	−300.0	313.10
Uranium hexafluoride	UF_6	352	−2198	227.8	−2069	644.96
Water	H_2O	18	−285.81	69.94	−237.12	0.750

M, molecular mass (kg/kmol); $\Delta_f H^0$, formation enthalpy (kJ/mol); formation entropy, $\Delta_f S^0 = s^0 - \sum_{elements} v s^0_{f,element}$, J/mol K, where v is the number of moles of the element in chemical equation of formation of the substance; Gibbs free energy of formation, $\Delta_f G^0 = \Delta_f H^0 - T_0 \Delta_f S^0$, (kJ/mol); chemical exergy, $ex^{ch} = \Delta_f G^0 + \sum_{elements} v \times ex^{ch}_{element}$, (kJ/mol); reference temperature, $T_0 = 298.15 K$; reference pressure, $P_0 = 1 bar$; superscript 0 in above notations refers to standard pressure.

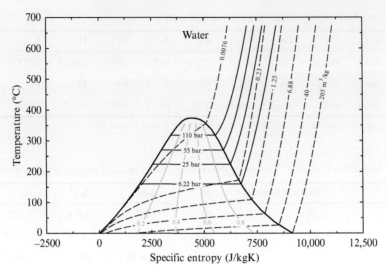

DIAGRAM B.1   Temperature-entropy diagram of water.

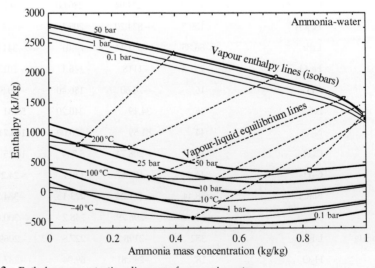

DIAGRAM B.2   Enthalpy-concentration diagram of ammonia-water.

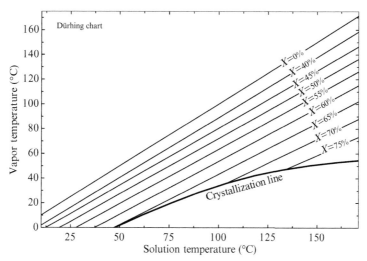

DIAGRAM B.3 Dürhing chart for Lithium bromide/water solution.

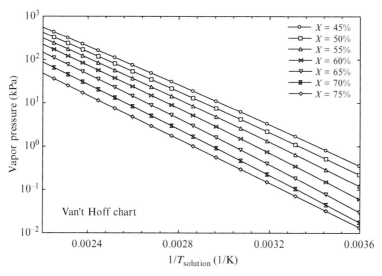

DIAGRAM B.4 Van't Hoff chart of vapor pressure of LiBr/water solution.

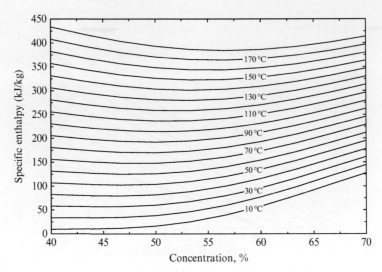

DIAGRAM B.5   Enthalpy-concentration diagram for LiBr-water.

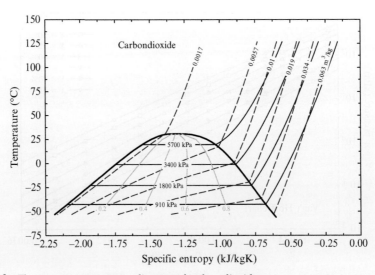

DIAGRAM B.6   The temperature entropy diagram of carbon dioxide.

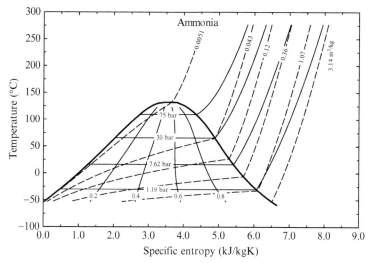

DIAGRAM B.7 The temperature entropy diagram of ammonia.

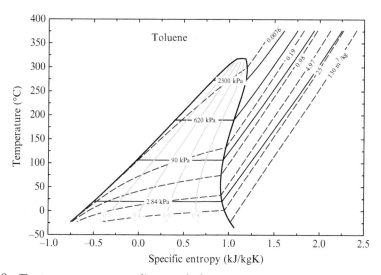

DIAGRAM B.8 The temperature entropy diagram of toluene.

Index

Printed and bound by CPI Group (UK) Ltd, Croydon, CR0 4YY

08/05/2025

01864884-0003